THE HISTORY OF
NORTH AMERICAN SMALL
GAS TURBINE AIRCRAFT ENGINES

Richard A. Leyes II
William A. Fleming

With contributions by A. Stuart Atkinson
Foreword by Hans von Ohain

  Smithsonian Institution

A copublication of the National Air and Space Museum
and the American Institute of Aeronautics and Astronautics

American Institute of Aeronautics and Astronautics, Inc., Reston, Virginia

1   2   3   4   5

Library of Congress Cataloging-in-Publication Data

Leyes, Richard A., 1948–
    The history of North American small gas turbine aircraft engines / Richard A. Leyes II,
William A. Fleming; with contributions by A. Stuart Atkinson; foreword by Hans von
Ohain.
        p. cm.
    Includes bibliographical references and index.
    ISBN 1-56347-332-1 (alk. paper)
    1. Airplanes—United States—Turbojet engines—History. 2. Aircraft gas-turbines—
History. I. Fleming, William A., 1921– II. Title.
TL709.3.T83 L48   1999   629.134'353'097—dc21   99-045475

Cover design by Sara Bluestone

CONTENTS

PREFACE

The publication of *The History of North American Small Gas Turbine Aircraft Engines* represents an important milestone for the National Air and Space Museum (NASM) and the AIAA. For the first time, there will be an authoritative—even definitive—study of small gas turbine engines, arguably one of the most significant spheres of aeronautical technology in the second half of the twentieth century. It is noteworthy that this important reference work, written by Rick Leyes and William Fleming, found its genesis at NASM. At the Museum, there is not only an extraordinary aero propulsion collection, but—as mirrored in this study—an active program of scholarly research. The AIAA became the logical publisher of this superlative history, given the AIAA's long and impressive program of publications in technical literature.

When first proposed more than a decade ago, a history of small gas turbine engines fit well with several institutional goals at NASM. For Rick Leyes, the Curator for Aero Propulsion, the project grew out of his own personal research in collaboration with William Fleming, who originally suggested focusing on small gas turbines. All agreed such a book would fill an obvious gap in the historical literature. The newly inaugurated Smithsonian History of Aviation Book Series at NASM offered a logical source of support and encouragement. The book series editors took a keen interest in *The History of North American Small Gas Turbine Aircraft Engines,* offering editorial and financial support for the project. Ultimately, the project earned the sustained support of three NASM directors, James Tyler, Martin Harwit, and Donald Engen, who properly viewed the project as a unique vehicle for the Museum to contribute to our understanding of aeronautical technology and its global impact.

We learn that the small gas turbine engine technology, as a distinct sector within the larger aviation industry, emerged slowly in the 1940s and 1950s. Small gas turbine engines made an impact on both civil and military aviation, being a vital component in the development of business jet aircraft, helicopters, missiles, remotely piloted vehicles, and a variety of other small aircraft. No less important, this same technology played a critical role in shaping the nature of air warfare during the Vietnam War.

The authors also explain the corporate background for the evolution of the small gas turbine technology. Seven corporations, as it turned out, played a major role in this history. How these corporations overcame various engineering challenges and responded to uncertain markets is a compelling story. The motive force behind the rapidly evolving technology became the quest to design small, lightweight, powerful and highly efficient power plants. Leyes and Fleming reconstruct this story for the first time, even as they provide authoritative reference data on the technology.

One is impressed with the research behind *The History of North American Small Gas Turbine Aircraft Engines*. The research was exhaustive, the product of a carefully designed plan: the authors explored the primary corporate records, conducted oral history interviews of key project managers and engineers, and analyzed many independent historical sources, human and documentary, which shed light on the technology. For these reasons, the book is and will remain a benchmark reference volume.

Rick Leyes has contributed in manifold ways to the preservation and study of the National Air and Space Museum's unique ensemble of aero engines, following in the path of Robert Meyer who pioneered the establishment of the Museum's aero propulsion collection. Rick's exemplary research and management of this project moves the NASM aero propulsion collection to a new level. His book is timely and welcome, a major contribution to historical literature. Rick Leyes is also a private pilot and the holder an FAA Airframe and Power Plant license. William A. Fleming, an aeronautical engineer, is well known for his professional work at the Lewis Flight Propulsion Laboratory and the NASA lunar landing program, among other contributions to aerospace engineering. Together, Leyes and Fleming have labored with a small team to bring this book to fruition. An important team member and contributor to the book project was Stuart Atkinson, a Navy engine engineer.

As the editor of the NASM history of aviation book series, I have been associated with this project from its inception. I know of no research project or historical work that surpasses *The History of North American Small Gas Turbine Aircraft Engines* as an optimal museum effort—the blend of collections based research and historical writing.

This book, I feel, constitutes a major contribution to aerospace studies. It will have a long shelf life.

Von Hardesty
Editor
Smithsonian History of Aviation Book Series

FOREWORD

From the beginning of their work, Mr. Leyes and Mr. Fleming kept me informed about the progress of their research. I enjoyed reading the various chapters of the book in draft form. They recognized the significance of the small gas turbine as an important sector of aerospace technology.

I fully agreed with the authors' view that the small gas turbine could not be considered simply as a scaled-down version of a large gas turbine. They correctly pointed out some essential differences between large and small gas turbine types which are based upon special structural, manufacturing, and aerodynamic characteristics. These points became even clearer during the progress of their research work.

At the beginning of their project on the small gas turbine history, the co-authors searched for previous work in this field. They found that little or none existed. So, with complete autonomy and a great deal of originality, they established an excellent research program including a process for the evaluation of their findings.

This program was based on two objectives that the authors established for the book. One objective was to show how the small gas turbine aircraft engine industry emerged and evolved. The other was to identify the major contributions that this industry, its corporations, and their products have made.

It was for me a great pleasure to observe how thoroughly the entire program was executed. All seven major North American companies producing small gas turbine aircraft engines [Teledyne CAE, Lycoming, General Electric Small Air-

Dr. Hans von Ohain with a reproduction of his Heinkel HeS 3B turbojet engine which is in the National Air and Space Museum collection. National Air and Space Museum photo by Dale Hrabak, Smithsonian Institution (SI Neg. No. 84-12203-8).

craft Engines, Williams International, Pratt & Whitney Canada, Allison, and Garrett (AlliedSignal Engines)] had been visited. The evolution of these companies and their products was carefully investigated through interviews, corporate reports and many other technical and historical sources. This was a systematic research endeavor of great magnitude.

I am convinced that the book is an outstanding achievement, from a historical as well as technical viewpoint, and that the objectives of the authors were fulfilled. Further, it can be considered as the standard reference work for small gas turbines.

Hans von Ohain
August 27, 1995

Dr. Hans von Ohain was one of the two independent inventors of the first jet aircraft engines. On August 27, 1939, his HeS 3B engine powered the Heinkel He 178 on the world's first flight of a turbojet-powered aircraft.

Dedicated to the Memories of

Vice Admiral Donald D. Engen
William A. Fleming
Paul E. Garber
Harvey H. Lippincott
Robert B. Meyer, Jr.
Dr. Hans von Ohain
Sir Frank Whittle

INTRODUCTION | 1

WHY WAS THIS BOOK WRITTEN?

For many decades, aeronautics has been the domain of popular historical literature. Recently, scholarly books have appeared covering many important themes. As we approach the centennial of aviation, it is noteworthy that there are very few studies on aeronautical technology, books that examine important technological breakthroughs and the impact of such breakthroughs on world social and military history. The invention and development of the gas turbine aircraft engine has been widely recognized as one of the important technological breakthroughs in the history of aviation. The authors have extensively surveyed the published literature on aero propulsion and have identified the major books relevant to the invention and development of gas turbine aircraft engines and growth of the industry.

The following books contain historical surveys of aircraft turbine engine development or are basic references. The history of the invention of the jet engine by the two, independent inventors, Sir Frank Whittle in England, and Dr. Hans von Ohain in Germany is well documented. The U.S. Air Force publication, *An Encounter Between the Jet Engine Inventors, Sir Frank Whittle and Dr. Hans von Ohain, 3-4 May, 1978*, is an excellent review of their early engine development work. A collection of writings of several of the principal pioneers in the field of jet aviation was presented by Boyne and Lopez in *The Jet Age*. In *Jet Propulsion Progress*, Neville and Silsbee provided a brief outline of the international,

1

historical, and technical development of jet propulsion. Sawyer's work, *The Modern Gas Turbine*, contained a short history of early gas turbine engines. In *Development of Aircraft Engines, Development of Aviation Fuels—Two Studies of Relations between Government and Business*, Schlaifer and Heron reviewed the development of the aircraft gas turbine through the 1940s. A broad historical survey of aircraft gas turbines was presented in Smith's text, *Gas Turbines and Jet Propulsion*. In *The Origins of The Turbojet Revolution*, Constant addressed the factors contributing to the development of the aircraft gas turbine. Gunston's book, *The Development of Jet and Turbine Aero Engines*, also covered the invention and early development of the gas turbine aircraft engine. Jane's *All the World's Aircraft* series and Paul H. Wilkinson's *Aircraft Engines of the World* series annually tabulated invaluable reference information relating to the world's aircraft engines. Otis and Vosbury's *Encyclopedia of Jet Aircraft and Engines* followed in the tradition of Wilkinson as an international gas turbine engine reference book and dictionary.

Several company histories have been prepared, including William R. Travers's two histories of General Electric (GE) engines, neither of which have been published. The first, *The General Electric Aircraft Engine Story*, was the predecessor of GE's *Seven Decades of Progress: A Heritage of Aircraft Turbine Technology*, which presented the history of the company and its development of turbine engines through the 1970s. The second was *Engine Men*. GE updated its engine history in *Eight Decades of Progress: A Heritage of Aircraft Turbine Technology*. Five of the North American engine corporations have also written or sponsored similar corporate histories. The Avco Corporation (a parent corporation of Lycoming) published the book, *Avco Corporation—The First Fifty Years*. In his book, *Continental! Its Motors and Its People*, Wagner covered the history of the company beginning in 1902 with its piston engine development and production up to its recent small gas turbine activities. *Out of Thin Air: Garrett's First 50 Years* by Schoneberger and Scholl portrayed all facets of Garrett's history up to the mid-1980s, including its small gas turbine aircraft engine work. *Power: The Pratt & Whitney Canada Story* by Sullivan and Milberry and *Allison Power of Excellence* by Sonnenburg and Schoneberger described the histories of their respective companies from their beginnings until each history was published.

There are also several relevant biographical and autobiographical works. Whittle's book, *Jet: The Story of a Pioneer*, traced the author's role in developing the first British turbine aircraft engine. Heiman's *Jet Pioneers* highlighted the pioneering roles of several early turbine engine designers. Hooker's book *Not Much of an Engineer: An Autobiography* reviewed the author's considerable technical achievements on Rolls-Royce and Bristol turbine engines. Neumann's *Herman the German: Enemy Alien U.S. Army Master Sergeant #10500000* explained the author's contribution to the Collier Trophy winning GE J79 engine. Anselm Franz's *From Jets to Tanks—My Contributions to the Turbine Age* covered his career at both Junkers and Lycoming.

All of these books make major contributions to the history of the gas turbine engine, but each is limited in scope. The historical surveys focus primarily on large gas turbine engines. Corporate histories briefly describe their particular products and highlight corporate contributions, but none exclusively address small engines as a class or deal with any of the engines in depth. Similarly, the biographical and autobiographical works deal with only those engines with which the authors were associated and generally emphasize large gas turbine engines. Each book in Jane's series and Wilkinson's series is limited to those engines in development or production for a particular year. Only two works, the *Development of Aircraft Engines, Development of Aviation Fuels* and the *Origins of the Turbojet Revolution*, address larger thematic issues relevant to the development of gas turbine aircraft engines. The first explores the relations between government and the aircraft engine industry. The second addresses the origin of the turbojet as a significant example of the historical process of technological change.

There remains a significant void in the history of gas turbine aircraft engines. Although small gas turbine aircraft engines have been recognized as a separate category distinct from large turbine aircraft engines, and although some of the corporations in this industry are among the largest in the country, no book has been devoted solely to the history of this industry and its engines as a distinct category.

The small gas turbine aircraft engine emerged in the 1940s as a distinct class of turbine engines. These engines have been defined by the authors in terms of the various airframe applications in which they were used. The applications, which are discussed in more detail in Chapter 2, are:

- Remotely Piloted Vehicles (RPVs), Unmanned Aerial Vehicles (UAVs), and Target Drones
- Decoy, Tactical, and Strategic Missiles
- Military Trainer Aircraft
- Special Purpose Aircraft
- Selected Small Military Aircraft
- Helicopters
- General Aviation Aircraft
- Regional Transport Aircraft

Following the late 1940s, the powerful, lightweight, and reliable small gas turbine engine emerged as a major technology development leading to a new generation of aircraft, rotorcraft, and missiles. These new air vehicles have had a profound national and international impact. It was the small turbine engine that provided the foundation for the development and expansion of the regional air transportation industry. The utility of the helicopter in both its military and commercial roles was vastly improved and its use was greatly expanded by the development of the small turbine engine. These small engines also powered the majority of executive aircraft and have significantly increased the effectiveness of such aircraft as business tools. Jet trainers and small fighters, helicopters, remotely piloted reconnaissance and target drone aircraft, and missiles have had a vital impact on military operations and national defense. Thus, the evolution of the small turbine engine and the companies that produce them are an important part of aviation history.

THE OBJECTIVES OF THIS BOOK

The objectives of this book are: (1) to show how the small gas turbine aircraft engine industry, its corporations, and their engines emerged and evolved in the context of the influences that shaped them; (2) to present the history of the design and development of each company's engines and to identify their principal airframe applications; (3) to identify the major contributions that the industry, its companies, and their engines have made.

THE APPROACH

The background research for this book generated a great deal of information, including many events and detailed technical developments that had occurred over a period of more than 50 years. An approach needed to be selected which presented this comprehensive history in a manner that was both understandable and useful to a spectrum of readers, each of whom would have a different level of interest.

As pointed out by Alfred D. Chandler, Jr., historians have the basic responsibility for "setting the record straight" by providing an accurate record of historical events.[1] He also reminds us that a valid historical analysis must be compara-

1. Alfred D. Chandler, Jr., *The Visible Hand—The Managerial Revolution in American Business* (Cambridge, Massachusetts and London, England: The Belknap Press of Harvard University Press, 1977), 6.

tive and that comparisons must be made of enterprises in the same industries.[2] This is the primary, structural approach that was selected for this book. The histories of the major corporations are established in individual chapters. The concluding chapter then provides a broader overview of the entire industry. Within the context of the historical development of the industry, a comparative, critical analysis of individual corporations and their engines is possible.

While Chandler's approach provides the basic framework, this book is actually a blend of different approaches to history. It incorporates elements of traditional business histories, histories of technology, and aircraft engine histories within a broad context of the development of a major industry, important technologies, influential historical events, and the impact of these engines on society.

This book traces the history of the development of each corporation's commercial and military engine products, the significant technologies incorporated in these engines, and identifies key people involved in the development of the engines. Included are discussions on the technological challenges that were and are peculiar to the design and manufacture of small gas turbine engines and how these challenges were overcome. The changing marketshare of corporations over time is analyzed. Each corporate history also addresses the contributions that important engines made to the aerospace industry, the military and commercial sector, the economy, and to society.

THE SCOPE OF THIS BOOK

As pointed out in the AIAA paper, "Rationale for Determining Information Associated with Aircraft Gas Turbine Engines That is of Historical Importance," the gas turbine aircraft engine industry has grown into a vast international complex involving a very large group of people with an interest in many different aspects of the industry.[3] Because of the magnitude of the industry, any practical history about it must be limited in scope. The reference paper also emphasized that the selection of information to be included in any gas turbine aircraft engine history is not a simple or trivial matter, and it outlined the events and data

2. Alfred D. Chandler, Jr., *Scale and Scope—The Dynamics of Industrial Capitalism* (Cambridge, Massachusetts and London, England: The Belknap Press of Harvard University Press, 1990), 10.

3. A. Stuart Atkinson, "Rationale for Determining Information Associated with Aircraft Gas Turbine Engines That is of Historical Importance," (AIAA Paper No. 90-2754 presented at the AIAA/SAE/ASME/ASEE 26th Joint Propulsion Conference, Orlando, Florida, 16-18 July 1990), 1.

that the writer considered to be of historical importance. It is significant that, although this history and the reference paper were developed completely independently of each other, the events and the data selected for inclusion in this history and those identified as important in the paper are remarkably consistent.

This history is confined to small gas turbine aircraft engines as defined in Chapter 2. It includes generally known companies that are or were at one time in the small gas turbine aircraft engine business. It includes pioneer companies, some of which entered the industry in the 1940s and left later. However, it concentrates on the seven major corporations that began developing small gas turbine engines for aircraft during the 1950s, and covers the histories of these corporations through the early to mid-1990s. The work of other companies that entered the industry later and made small contributions is also included. This book is limited to North American companies, but does address their major international partnerships and license agreements.

METHODOLOGY

The authors took particular care in devising and executing a research methodology that would ensure that the history was comprehensive, accurate, and objective. Their first step was to use the considerable library, archive, and collection resources of the National Air and Space Museum as well as available company information to conduct a preliminary review of each corporation and its products.

On the basis of this background research, questionnaires were developed for each corporation outlining desired additional information. These were sent in advance to the corporations to enable them to select and prepare the people whom the authors should interview. The authors also requested interviews with specific corporate officials and active or retired individuals of historic importance. For those companies no longer in existence or active in the small gas turbine business, the authors contacted, when possible, former company officials or key engineers for their review and comment on material that had been prepared on their company.

For the most part, among the seven major corporations, interviews were scheduled with presidents and CEOs, principal program managers, engineers, scientists, marketing persons, and retirees who played key roles during the formative years of small gas turbine engine development. The persons selected for interviews typically had very long associations with the corporations as well as considerable responsibility for engine design and program management. Each major engine program was represented by one or more individuals. Every effort was made to locate people who were pioneers and who made especially important early contributions.

With appreciation, it is noted that all corporations cooperated fully with the authors in this project. Scores of in-depth interviews were conducted and extensive correspondence was carried out with many people who were not interviewed. The credits provided in the Acknowledgements list these people. The authors consider themselves to be particularly fortunate in that they were able to obtain much important historical information directly from the pioneers who created the history and to record it before it was forever lost. One or both of the authors made visits to the corporations, and the interviews were conducted on site and in person. In some cases, interviews were conducted at the homes of retired engineers or over the telephone. These interviews provided the essential story line of the engine program narratives, but specific details were derived from many other sources. With the consent of the interviewees and corporations, all interviews were tape recorded and transcribed. The tapes and transcriptions were made part of NASM's archives.

Non-proprietary and publically-available engineering papers, published articles, corporate reports, engine brochures, statistics and specifications, photographs, etc. were solicited and provided by the corporations as further sources of information for the book. These articles were also made part of NASM's archives. In some cases, historically significant aircraft engines were donated to NASM as a direct result of visits to corporations, further enriching the National Collection.

To ensure a balanced approach, the authors conducted additional independent research for information other than that provided by the corporations. In the process of writing this book, the authors sought to blend oral histories, corporate literature, articles published in aerospace periodicals, engineering papers published by professional engineering organizations, military histories, correspondence with historians and engineers, corporate histories, and statistics and specifications in standard aerospace reference books.

Each author prepared the initial draft of specific sections of the book. Each author's drafts were then reviewed by his counterpart. Thus, the final draft of each section represented a joint authorship point of view. Draft copies of the chapters were sent to the corporations who circulated the draft to all interviewees for review in order to ensure factual and technical accuracy. In most cases, a referee, who was typically a managerial engineer with a direct and long association with the small turbine engine programs of that company, was jointly selected by the authors and the company to integrate all comments into a single document. This set of comments was returned to the authors in order to correct and improve the manuscript as required. However, even with these reviews, there were some cases where conflicting information existed in company records and other information sources. In such cases, the authors attempted to resolve the conflict by selecting what they believed to be the most reliable source. These decisions and all other final editorial decisions with respect to the draft chapters were made by the authors.

This major research project was funded almost entirely by the Smithsonian Institution. In particular, the authors are grateful to the Smithsonian Institution Office of Fellowships and Grants for receiving a Scholarly Studies Award, the National Air and Space Museum, the Hallion-Gorn Fund of the Smithsonian History of Aviation Book Series, and the Admiral Dewitt C. Ramsey Fund for underwriting this effort. In acknowledging the contribution of the Ramsey Fund, it is also appropriate to recognize the historic contribution to jet engine development made by then Captain Ramsey, Assistant Chief of the Bureau of Aeronautics, who, in November 1941, asked Westinghouse for a proposal calling for a jet engine design study, which ultimately became the first all-American-designed turbojet engine, the 19A Yankee. A contribution was also made by the Naval Air Propulsion Association which we sincerely appreciate. All of this funding contributed importantly to ensuring that the research was independent of the corporations being studied.

AUTHOR AND PROJECT MANAGER

Richard A. Leyes II is the Curator for Aero Propulsion at the National Air and Space Museum, Smithsonian Institution. He has a B.A. in Economics from the University of Wisconsin and an M.S. in Industrial and Labor Relations from Cornell University. For the Smithsonian Institution, Mr. Leyes has done aircraft engine research and writing, collecting, and exhibitions. He also holds a private pilot's license and a Federal Aviation Administration Airframe and Power Plant license.

AUTHOR

William A. Fleming is a graduate of Purdue University with a B.S. in Aeronautical Engineering. He was a pioneer and leader in jet engine research during the 1940s and 1950s at the National Advisory Committee for Aeronautics (NACA) Lewis Flight Propulsion Laboratory. He later directed development of the plan for the National Aeronautics and Space Administration's (NASA) Apollo manned lunar landing program and served in senior management positions at NASA Headquarters. Mr. Fleming's publications include *Future Aeronautics and Space Opportunities—Volume 1 Space* and approximately 30 NACA research reports. Following retirement from NASA, he spent twelve years as a management consultant.

CONTRIBUTOR

An important contributor to the book is A. Stuart Atkinson, who holds a B.S. in Mechanical Engineering from the University of Arkansas and a Master of Business Administration from George Washington University. For over 37 years, Mr. Atkinson held engineering positions in the Department of the Navy associated with aeronautical research and development, including senior managerial positions with the Naval Air Systems Command. Most of his career was in the field of aero propulsion, during which time he developed the technical and program requirements for Navy engines and contributed to the development of many of the small turbine engines sponsored by the Navy.

THE SMALL GAS TURBINE AIRCRAFT ENGINE 2

What is a Small Engine?

From the beginning of their evolution, large and small gas turbine aircraft engines have been recognized throughout the aircraft engine industry as separate categories of engines. Not only did small engines have unique problems and distinct advantages related to their size, but they were used for a category of applications quite different from those of large engines.[1]

Following the introduction of gas turbines for aircraft in the 1940s, in little more than a decade, the large turbine engine quickly took over the traditional reciprocating engine role in military fighters and bombers and began making inroads on the civil transport aircraft markets for high-power engines. Following that time, the large turbine engine grew in size and capabilities far beyond the possibilities open to the reciprocating engine.[2] For example, by the 1990s, the thrust of large turbine engines used in military combat aircraft had grown to the 30,000 lb thrust class, very large turbofans used in civil transports were in the 100,000 lb thrust class, and turboprops had grown to more than 8,000 shp.

But, at the low-power end of the aircraft engine market, the piston engine was considerably more difficult to oust. As a result, the evolution of the small engine came more slowly.[3] Nevertheless, over time, the small gas turbine engine

---

1. "Small Gas Turbine Progress," *Aviation Age*, Vol. 23, No. 2 (February 1955): 26–59.

2. "Small Turbines Growing Market," *Flying Review*, Vol. 24, No. 3 (November 1968): 1.

3. "Small Turbines Growing Market," 1.

became the power plant of choice for: Remotely Piloted Vehicles (RPVs), Unmanned Aerial Vehicles (UAVs), and target drones; decoy, tactical, and strategic missiles; military trainer aircraft; special purpose aircraft; selected small military aircraft; helicopters; general aviation aircraft; and regional transport aircraft. The small turbine engine provided these aircraft with greater operational capabilities in terms of speed, payload, altitude, and reliability than the piston engine. Small turbojet and turbofan engines provided missile systems with range and endurance capabilities not achievable with rocket propulsion.[4]

Because small gas turbine aircraft engines have been traditionally associated with distinct categories of aircraft applications, the authors followed this convention and defined small turbine engines by their airframe applications. One exception to this classification included small turbine engines that were used as auxiliary or boost engines on some piston engine fighters, larger aircraft, and high-speed rotorcraft.[5]

REMOTELY PILOTED VEHICLES, UNMANNED AERIAL VEHICLES, AND TARGET DRONES

RPVs and UAVs were aircraft of the type that the pilot did not fly in, but controlled from another aircraft or from a station on the ground or aboard a ship. Reconnaissance RPVs and UAVs were typically used for aerial photography, surveillance, and intelligence gathering. Target drones, often RPVs, were pilotless target aircraft. Drones were utilized for target practice by ground or ship gun and missile crews and by pilots for aerial combat practice. RPVs, UAVs, and drones were variously powered by turbojet, turbofan, or turboshaft engines.

DECOY, TACTICAL, AND STRATEGIC MISSILES

Operational missiles were typically powered by small turbojet or turbofan engines. Included were turbofan-powered low- and high-altitude, long-range, strategic cruise missiles, a variety of turbojet-powered low-altitude tactical mis-

---

4. "General Electric Presents: Small Turbines for Varied Markets," *Interavia*, Vol. XVI (July 1961): 966-969.

5. An auxiliary engine, also known as a boost or booster engine, was not a prime propulsion engine. In the early 1940s, some small U.S. turbojet engines were designed originally as booster engines for some piston engine fighters to provide them with extra power for short periods during combat. Later, some larger aircraft powered by piston engines were equipped with small turbojet auxiliary engines to assist in takeoff at highly-loaded conditions. Some rotorcraft designed for high-speed flight were also fitted with small turbojet engines to assist in high-speed, forward flight.

siles, decoy missiles to protect aircraft, and other missiles with highly specific missions. Tactical missiles ranged from tiny, low-cost, fiber optic guided missiles that were being developed in the early 1990s for use against helicopters and ground targets to larger and more expensive standoff missiles used against aircraft, ground, and sea targets. Cruise missiles, which could perform a variety of missions, were typically armed with either conventional or nuclear explosives.[6]

MILITARY TRAINER AIRCRAFT

Trainer aircraft were used for training military pilots. These included subsonic and supersonic jet aircraft and turboprop trainer aircraft. Jet aircraft were powered by either small turbojet or turbofan engines. Supersonic aircraft engines incorporated an afterburner.

SPECIAL PURPOSE AIRCRAFT

Typically, these were military Vertical Takeoff and Landing (VTOL) aircraft, including turbine-powered platforms and fans, rotorcraft, propeller-driven convertiplanes, liftjets and tiltjets, and vectored and deflected jets. They were powered by a variety of turbine engine types.

SELECTED SMALL MILITARY AIRCRAFT

Among these were a large variety of small fixed-wing aircraft, including small lightweight fighters and close air support, anti-submarine warfare, observation, utility, and small personnel transport aircraft. They were powered by a variety of turbine engine types.

HELICOPTERS

This application encompassed all types of helicopters, both civil and military, that were normally powered by turboshaft engines. Categories included light, intermediate, medium, and heavy helicopters. Military helicopter missions included close air support and attack, general utility, troop and cargo transport, observation, rescue, anti-submarine warfare, and assault. Examples of civilian helicopter roles included executive and commercial transport, law enforcement

6. Exceptions to this definition of missiles were Army, Navy, and Air Force missiles deployed in the 1950s that were powered by modifications of man-rated engines. The Allison J33 and J71, Pratt & Whitney J52, and GE J79, classified as "large" engines at the time, were used in such missiles.

and border patrol, offshore oil and natural gas rig support and service, remote resources exploration and extraction, logging, construction, rescue, aerial application, air ambulance work, fire fighting, sight-seeing, and general utility.

GENERAL AVIATION AIRCRAFT

This category included all turbine-powered civil aircraft except those used for scheduled air transportation. Among the many varieties of general aviation aircraft were business aircraft, agricultural aircraft, sport aircraft, utility aircraft, etc. Civil helicopters were also in the general aviation category, and, to a certain extent, this category overlaps the helicopter category. These aircraft and rotorcraft, depending on the design, were variously powered by turbojet, turbofan, turboprop, or turboshaft engines.

REGIONAL TRANSPORT AND COMMUTER AIRCRAFT

Powered by turboprop and turbofan engines, regional aircraft were used in commercial operations for relatively short-haul flights, including those feeding into commercial airline hub operations.

CHANGES IN SMALL ENGINES OVER TIME

The engines used for "small engine class" applications were physically small when initially introduced in the 1940s and 1950s. Likewise, they produced relatively low thrusts, typically from about 150 to 1,000 lb, and power outputs, from about 150 to 900 shp. However, as airframe and engine technology advanced and the list of engine applications lengthened, there was both a progressive increase and decrease in the size and output of small turbine engines. By the early 1990s, thrust outputs ranged from less than 100 lb for miniature turbojets for small tactical missiles to more than 7,000 lb for turbofans powering executive and regional transport aircraft. Similarly, the power output of "small" turboshaft and turboprop engines extended from about 250 to 6,000 shp. In fact, by that time, the maximum thrust and power outputs of engines classified as "small" exceeded the outputs of the so-called "large engines" that existed during the 1940s and early 1950s.

Small Turbine Engine Design and Manufacturing Challenges

With large jet engines demonstrating high performance in military aircraft during the late 1940s and 1950s, U.S. military services recognized the need for smaller engines to power smaller aircraft. Thus, the military services turned to engine manufacturers to develop small turbine engines for applications such as missiles, drones, trainer aircraft, and helicopters. Many of these types of aircraft would not have achieved the forms they ultimately assumed without the small gas turbine engine.[7] In fact, both the military services and airframe designers placed a great deal of pressure on the engine industry to produce these small power packages for them.[8]

The manufacturers' response was fueled by the opportunity to acquire military contracts that would fund the development of their small engines and the potential of a profitable market for them once they were developed. By the late 1950s, with a number of small turbine engines successfully operating in military aircraft, a potentially profitable market for the engines in civil aircraft began to emerge.

There were other incentives for companies to enter the business. In some cases, manufacturers of small piston engines saw a need to get into the small gas turbine business to maintain their aircraft engine market position and profitability level. Other companies, already established in the manufacture of large engines, seized the opportunity to further expand their business base by applying their technical expertise to the development of small turbine engines. One company even began developing small turbojet and turbofan engines using its own funds with the philosophy that once it had successfully developed an engine, there would be a market for it. That company, Williams International, captured the U.S. long-range cruise missile engine market. These incentives and potentials led an array of engine companies to enter the small turbine business during the 1950s.

The small turbine engine filled what had been viewed as a gap in the power spectrum. It offered a number of potential technical payoffs for small fixed-wing aircraft and for helicopters. It provided both types of aircraft the potential of flying faster, at higher altitudes, and with greater payloads than their piston-powered counterparts. In addition, the small turbine engine had the potential of providing greater reliability, far longer Time Between Overhaul (TBO), and smoothness of operation.[9] For missiles, it could provide unmatched range and payload capabilities.

---

7. "General Electric Presents: Small Turbines for Varied Markets," 966.

8. A. T. Gregory, "The Future of Small Turbojet Engines," (Paper based on a presentation made at the Air Force-Aircraft Industry Conference on Turbojet Propulsion Systems, Santa Barbara, Cal., 13-15 November 1956), 2.

9. "General Electric Presents: Small Turbines for Varied Markets," 966-969.

Although the design and development of large gas turbine aircraft engines presented some of the most complex and challenging problems in the industry, developers of small turbine engines were confronted by a set of design challenges unique to small engines. Small turbine engines were quite different mechanically from their large engine counterparts. Nevertheless, during the infancy of the small engine industry, engineers working on large turbine engines generally viewed small engines as scaled-down versions of large engines. They believed that, in the scaling-down process, component efficiencies would drop, fuel consumption would rise, and power-to-weight ratio would be low because rotor blade tip clearances, blade profiles, and smoothness necessary for good aerodynamic efficiencies would approach or exceed cost effective limits of manufacturing processes. Many large engine designers considered the likelihood of building an efficient and cost effective small engine to be slim. Another entrenched view was the philosophy that "bigger is better." Thus, the large engine proponents continued pursuing the development of still larger engines, leaving the small engine designers to wrestle with their unique design and manufacturing challenges.

In contrast, other engineers believed that the size effects on small engine performance were exaggerated. They recognized that the scaling-down approach was not viable and that small engines required a quite different design approach. In addition, they held a strong belief that efficient small engines were technically feasible and economically viable. These views were augmented by the business incentives for developing small engines. With this faith, these engineers began to put together the engineering and manufacturing techniques and technology base needed for the development and manufacture of small turbine engines.

An overriding factor in the design of small engines was cost. According to Dr. Anselm Franz, former Vice President of Avco Lycoming Stratford Division, significant differences between the applications of large and small engines existed that influenced the relative importance of cost considerations in their design.[10] For example, the large turbojet had no competition from the reciprocating engine in terms of its ability to satisfy requirements for high speed flight in early military applications, thus its cost, compared to that of piston engines, was not an important design factor. However, the small turbine engine, specifically turboprop and turboshaft engines, competed directly with the reciprocating engine in terms of both cost and performance. As a result, cost became a driving factor in the design of selected small engines.

---

10. Dr. Anselm Franz, "Some Experiences in the Development and Application of Lycoming's T53 Gas Turbine Engine," *American Helicopter*, Vol. 46, No. 4 (March 1957): 6-7.

Two basic aspects of engine cost were important in the design and manufacturing of small engines: the initial cost of the engine and its direct operating costs. Two major elements of an engine's initial price were the cost of its development and production tooling, both relatively high for turbine engines. To offset those costs, it was important to take advantage of the potential of an engine in the low power range to be used for a variety of applications. In this way, development and tooling costs could be spread over a number of aircraft applications, increasing production volume and reducing the unit production price of an engine. Major factors contributing to direct operating cost were engine and component lifetime, maintenance requirements, and fuel consumption. Thus, in the design of the small turbine engine it was important that emphasis be placed on reliability, ease of maintenance, and cycle efficiency.[11]

Two disadvantages that the small turbine engine faced in displacing the piston engine in fixed-wing aircraft and helicopters were that its manufacturing cost was greater and its fuel economy lower than those of its piston engine counterpart. Those disadvantages still existed by the 1990s. Consequently, to penetrate the piston engine market, it was necessary to convince airframers that the turbine engine's higher cost and lower fuel economy would be offset by the benefits of its ability to provide greater airspeed, altitude, and payload capability and its potential for higher reliability and longer TBO.

The use of turbine engines in the form of turboshafts to power helicopters was a new use of the turbine engine that emerged in the early 1950s. Up to that time, aircraft turbine engines were either turbojets or turboprops that powered fixed-wing aircraft. In addition, they were typically single-shaft engines. Use of the turboshaft to power helicopters introduced a new set of problems and design challenges. One relatively new concept was the free turbine to provide better low speed torque and operating characteristics than single-shaft engines; this resulted in the need for dual-shaft designs. The gearbox design for the turboshaft was different than that for turboprops, and the nature of the interaction between the turboshaft engine and the helicopter rotors required different and more complex control systems.

Other challenges faced by the early designers of small engines were shaped by factors inherent to the physical size of small turbine engines. A number of these factors were recognized to have an adverse affect on engine performance. The principal factors that limited the performance of small engines, prevented direct scaling of their design and performance from large engines, required new design

---

11. Franz, "Some Experiences in the Development and Application of Lycoming's T53 Gas Turbine Engine," 6-7.

approaches, and dictated the requirement for special design and manufacturing approaches were:

- Size effects on aerodynamic performance
- Manufacturing limitations
- Combustor design constraints
- Turbine blade cooling and thermal stresses in turbine disks
- Mechanical design problems

The smaller the size of the engine, the greater was the impact of these factors.

The effect of Reynolds Number on compressor and turbine efficiency was considered to be a problem for small turbine engines, mitigated to some extent by the cube-square law. In accordance with that law, the weight of an engine varied as the cube of its diameter and the thrust or power as the square of its diameter. Thus, in so far as the cube-square (3/2) law held, a smaller size engine would be lighter per pound of thrust than its corresponding large engine counterpart, tending to improve its thrust-to-weight ratio. However, because the accessories in the early small engines were not proportionately smaller and structural materials were not proportionately thinner, some felt that the 3/2 law might more accurately be characterized as a 5/4 law.[12]

SIZE EFFECTS ON AERODYNAMIC PERFORMANCE

The thermodynamic cycle on which the performance of a gas turbine engine is based is not directly affected by engine size itself. For a given cycle pressure ratio and turbine temperature, the theoretical fuel efficiency is unaffected by size, if the efficiencies of the engine components remain constant. However, three problems were encountered in trying to achieve high compressor and turbine efficiencies. One problem was the relatively adverse boundary layer thickness and friction coefficients due to low Reynolds Numbers within the engine.[13] Another was the difficulty in minimizing blade tip leakage with mechanically

---

12. A. T. Gregory, "Small Turbojet Engines- A Big Factor in Aviation," (SAE Paper No. 499 presented at the SAE Golden Anniversary Aeronautic Meeting, New York, 18-21 April 1955), 5.

13. The Reynolds Number of fluid flow over an airfoil is proportional to the density of the fluid, its velocity relative to the airfoil, and a significant dimension, which is airfoil chord. Reynolds Number is inversely proportional to fluid viscosity. Early aerodynamic experiments demonstrated that when the Reynolds Number of the flow over an airfoil was reduced sufficiently, the drag coefficient rose. In the late 1940s, it was experimentally

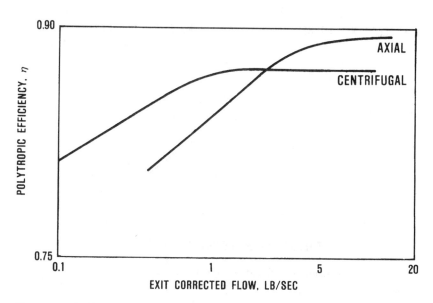

Compressor Performance is Corrected Flow Sensitive. Photograph courtesy of Garrett.

practical clearances. The third problem was the need to obtain the desired surface smoothness of compressor and turbine blades without encountering unreasonable manufacturing costs.[14]

The combined effect of these three factors on the efficiency of axial- and centrifugal-flow compressors of varying sizes is illustrated in the figure titled, "Compressor Performance is Corrected Flow Sensitive." The Exit Corrected Flow is proportional to the Reynolds Numbers within the compressors and thus to engine size. To provide a reference on the relation between Exit Corrected Flow and engine size, the point at which the curves for the axial and centrifugal compressors intersect corresponds to an airflow in the range of many small engine designs. The figure illustrates two points. One point is the degree to which compressor efficiency can be expected to decrease as engine size is reduced. The other point is that very small centrifugal compressors can be ex-

---

demonstrated that turbojet engine axial compressor and turbine efficiencies decreased as altitude increased due to the reduction of Reynolds Number within the engine. See: Lewis E. Wallner and William A. Fleming, "Reynolds Number Effects on Axial-Flow Compressor Performance," (NACA Lewis Flight Propulsion Laboratory, Cleveland, Ohio, 1949), 1-24.

14. "General Electric Presents: Small Turbines for Varied Markets," 967.

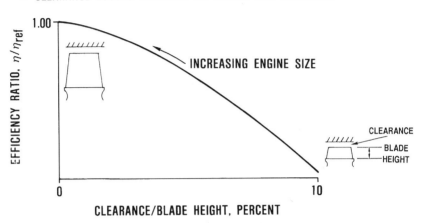

Component Performance is Affected by Clearance Effect. Photograph courtesy of Garrett.

pected to be more efficient than correspondingly small axial compressors. This is because the efficiency of centrifugal compressors is less sensitive to blade tip clearance than that of axial compressors.[15]

The effect of tip clearance alone on component efficiency is illustrated in the figure titled "Component Performance is Affected by Clearance Effect," which illustrates how efficiency can be expected to decrease as tip clearance increases.[16] Some quantitative measurements of the tip clearance effect were made by Sunstrand on its 4-inch diameter turbojet, the TJ20 Gemjet developed in the late 1980s. That engine employed a back-to-back single-stage centrifugal compressor and radial turbine in a single casting known as a monorotor. Desiring to facilitate high volume production of the monorotor at low cost, Sunstrand investigated the effect of increasing the tip clearance by 0.010 inches as a means of relaxing machining tolerance requirements. That increase in tip clearance reduced compressor efficiency by 1.8 percent and turbine efficiency by 1.7 percent.[17]

Reductions in Reynolds Number within the compressor also reduce the

15. Donald Comey, interview by Rick Leyes and William Fleming, 28 March 1990, Allison Interviews, transcript, National Air and Space Museum Archives, Washington, D.C.

16. Comey to Leyes and Fleming, 28 March 1990.

17. A. Jones, H. Weber, E. Fort, "GEMJET—A Small, Low Cost, Expendable Turbojet," (AIAA Paper No. AIAA-87-2140 presented at the AIAA/SAE/ASME/ASEE 23rd Joint Propulsion Conference, 29 June - 2 July 1987, San Diego, Cal.), 5.

compressor stall margin. One example of this effect was experienced in the Teledyne CAE J69 turbojet used to power the Cessna T-37 military trainer. The J69 performed well at low altitudes. However, the engine did have a tight surge margin, and the compressor was subjected to inlet flow distortion from the S-shaped T-37 inlet duct design. Those two factors, when combined with the reduced Reynolds Number within the compressor at high altitude, resulted in compressor surge during high-speed maneuvers at altitudes above 20,000 feet. That problem ultimately led to a complete redesign of the J69 compressor.[18]

MANUFACTURING LIMITATIONS

Manufacturing small engine parts while holding close tolerances was found to be a difficult task. To determine the effects of manufacturing on performance, a study was conducted at Garrett. The results are shown in the figure titled "Manufacturing Effects Impact Component Performance." The "Baseline" column in the figure indicates the tolerances used for the baseline turbine. The "Small Scale" column indicates the corresponding tolerances in the small turbine, which was scaled down from the baseline turbine by a factor of 0.3. The "Manufacturing Required Tolerance" column indicates the minimum practical tolerance that could be obtained in manufacturing the part without encountering undue costs. The efficiency penalties resulting from the deviation in manufacturing tolerances from the desired values resulted in a more than a 6 percent reduction in turbine efficiency. As might be expected, experience with the compressor was similar.[19]

COMBUSTOR DESIGN CONSTRAINTS

There were two main problems encountered in the design of combustors for small engines. The first related to the reduced availability of air for use in controlling the turbine inlet temperature profile, as illustrated in the figure titled, "Combustor Design Highly Dependent on Size." In a combustor, one portion of the air is introduced into the primary combustion zone at the forward end of the combustor and burned with the fuel. A second portion of the air is injected along the walls of the combustor liner to cool it. The remaining portion of the air is injected into the dilution zone to cool the combustion gases and control the radial temperature profile approaching the turbine. As combustor size is re-

---

18. Eli Benstein, interview by Rick Leyes, 9 August 1988, transcript, Teledyne CAE Interviews, National Air and Space Museum Archives, Washington, D.C.

19. Comey to Leyes and Fleming, 28 March 1990.

TURBINE GEOMETRY PARAMETER	BASELINE	SMALL SCALE	MANUFACTURING REQUIRED TOLERANCE	EFFICIENCY PENALTY (POINTS)
SCALE	1	0.3	0.3	
TRAILING EDGE THICKNESS, IN.	0.042	0.014	0.030	– 1.0
AIRFOIL TOLERANCE, IN.	± 0.003	± 0.001	± 0.003	– 0.7
FLOW-PATH TOLERANCE, IN.	± 0.005	± 0.0015	± 0.005	– 0.5
FILLET RADIUS, IN.	0.06	0.02	0.045	– 0.44
TIP CLEARANCE, IN.	0.013	0.0043	0.013	– 3.64
TOTAL				– 6.28

Manufacturing Effects Impact Component Performance. Photograph courtesy of Garrett.

duced, the amount of its surface area that needs to be cooled increases relative to the engine airflow. As a result, there remains less dilution air for use in controlling the radial temperature profile. Combustor designers thus found it more difficult to achieve the desired temperature profile in small engines.

The second problem related to the fuel nozzles. To obtain a uniform fuel spray, which promotes efficient burning and good ignition characteristics, it is desirable to have a relatively large number of nozzles placed sufficiently close together around the combustor to provide a good fuel spray pattern. However, because of the low fuel flow rates in small engines, particularly at high altitudes, use of a large number of nozzles results in a very low fuel flow through each of them. As a result, relatively small orifice sizes are required in the nozzles to promote a good spray pattern. Small fuel orifices lead to carbon deposit problems and plugging of fuel nozzles. The combustor designer was faced with making tradeoffs between the number of fuel nozzles used and the fuel nozzle orifice size in achieving good combustor performance.[20]

Teledyne CAE avoided the fuel nozzle problem with its slinger injection system. That system used the centrifugal force and tangential velocity of the hollow rotor shaft, into which fuel was injected at low pressure, to inject highly atomized fuel into the combustor through holes in the shaft. As a result, the degree of fuel atomization varied only with engine speed and was unaffected by the reduction in fuel flow rate with altitude.[21] Williams International also employed a slinger injection type of system in its engines.

20. Comey to Leyes and Fleming, 28 March 1990.

21. H. C. Maskey and F. X. Marsh, "The Annular Combustion Chamber With Centrifugal Fuel Injection," (SAE Paper No. 444C presented at the SAE Automotive Engineering Congress, Detroit, Michigan, 8-12 January 1962), 2-4.

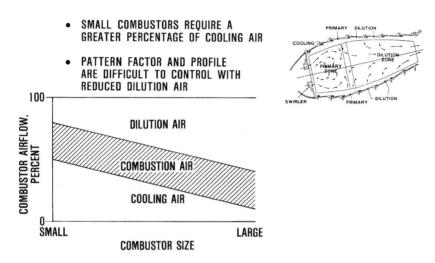

Combustor Design Highly Dependent on Size. Photograph courtesy of Garrett.

TURBINE COOLING AND THERMAL STRESSES

The designers of turbines for small engines encountered two mechanical design challenges. One was related to cooling of the turbine blades and the other was concerned with thermal stresses encountered in the turbine disk. The cooling of the tiny turbine blades, often as small as one to one and one-half inches in length with blade chords of less than an inch, posed a major design problem. As a result, it was not until the 1960s that designers of small engines introduced turbine blade cooling. Even then, the amount of cooling that could be provided was significantly less than in large engines because it was not physically possible to use the complex blade cooling schemes employed in large engines. The small turbines were restricted to lower operating temperatures than large engine turbines, hence lower cycle pressure ratios and thermal efficiencies.

A blade cooling design that is characteristic of the type used initially in cooling small turbine blades is compared to one used in large engines in the figure titled, "Small Engines Have Distinct Technology Challenges." The large blade illustrated in the figure was for a 7,000-lb-thrust engine, the small blade for a 1,200-lb-thrust engine. A five-cent coin, approximately three-fourths of an inch in diameter, is included in the figure to illustrate the size of the blades. The large blade contained a three-pass serpentine flow passage, turbulators within the flow passages to cause mixing of the hot air near the wall and cooler air in the center of the passages to promote wall cooling, showerhead cooling from

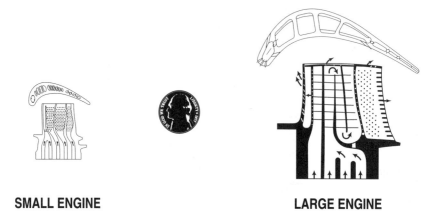

SMALL ENGINE   **LARGE ENGINE**

Small Engines Have Distinct Technology Challenges. Photograph courtesy of Garrett.

orifices in the blade leading edge, and a cooling passage near the trailing edge with trailing edge discharge orifices. In contrast, the small blade had only single pass cooling with discharge of the air at the blade tip.[22] However, as blade casting technologies evolved and fabrication methods improved, increasingly intricate cooling systems were designed into small turbine blades. For example, by the late 1980s, the core turbine rotor and stator blades in General Electric's T700 turboshaft engine had a serpentine blade cooling system similar to that illustrated in the figure for the large engine.

The occurrence of high thermal stresses in the turbine disk during engine start-up is a problem inherent in small engines. In both large and small engines the temperature of the turbine rim, which holds the blades in place, heats up quickly following start-up, and the disk begins to grow in diameter and the rim in circumference. Meanwhile the bore temperature remains relatively cool, resulting in little or no growth. Because the distance between the rim and the bore of the small turbine disk is much smaller than that of the large turbine, the thermal strain per inch is much greater during startup for the small engine. As a result, to obtain the same low-cycle thermal fatigue failure rate as in large engines, the designers of small turbines must design for much higher thermal stresses in the turbine disk than are encountered in large engines.[23]

A number of small engines were troubled with low-cycle fatigue turbine fail-

22. Comey to Leyes and Fleming, 28 March 1990.

23. Mike Hudson, interview by Rick Leyes and William Fleming, 6 June 1991, transcript, National Air and Space Museum Archives, Washington, D.C.

ures. One of the earliest instances occurred in 1950 when Garrett's first gas turbine auxiliary power unit, the GTC43/44, experienced cracks in the turbine
rim. This cracking was traced to tension and compression failures in the rim
during startup caused by thermal fatigue. Garrett's solution at that time was to
cut deep gashes in the rim to relieve the stresses.[24]

MECHANICAL DESIGN PROBLEMS

A factor obvious to the small engine designer was that the thickness of compressor, combustor, and turbine casings and flanges could not be scaled down
from those used in large engines. Internal pressures were approximately the
same in small and large engines, thus, it was necessary that shell casings of small
engines be approximately as thick as those of large engines to withstand the internal pressures. Likewise, flanges in small engines carried the same loads as in
large engines. As a result, the small engine paid an inherent structural weight
penalty.[25]

The design of bearings and seals also posed scaling problems for the small engine designer. Bearings could not be scaled down directly for small engines,
otherwise they would be so small that they could not carry the loads imposed
on them. Thus, at the bearing positions, the rotor shaft had to be sufficiently
large in diameter to accept the size bearing needed. Scaling down labyrinth seals,
the type normally used in large engines, resulted in excessive leakages. For example, in the Lycoming T53 turboshaft engine, leakage was so severe that the
designers resorted to a positive seal in which a carbon disk held in place by a
bellows was allowed to ride lightly on the rotating shaft.[26]

A problem that confronted designers of small two-spool engines having concentric shafts was vibration of the internal shaft. Some considered this as the
most difficult single problem encountered in the development of small engines.
There were two factors that, when compounded, could induce excessive vibration. One resulted from the need to keep the internal shaft small enough in diameter to fit within the outer shaft. As shaft diameter was reduced, its natural
frequency of vibration also decreased, often to within the rotating speed range

24. H. J. Wood, Sherman Oaks, California, to Neil Cleere, letter, 21 February 1990, National
    Air and Space Museum Archives, Washington, D.C., 4.

25. Comey to Leyes and Fleming, 28 March 1990.

26. Heinz Moellmann, interview by Rick Leyes and William Fleming, 12 October 1989,
    Textron Lycoming Interviews, transcript, National Air and Space Museum Archives,
    Washington, D.C.

of the shaft. In addition, the centrifugal shaking force due to any unbalance in the shaft varied as the square of the shaft's rotating speed. The smaller the engine, the higher was its rotating speed, thereby producing substantial increases in the unbalanced shaking force. Early engine designers approached this problem in a variety of ways, the end result being to raise the natural frequency of the shaft above its rotating speed range while minimizing shaft unbalance.

Another mechanical problem that challenged the small engine designer was the development of smaller and lighter engine controls that had the same degree of reliability as those of larger engines. As complex as were the controls of large engines, small engine controls faced a more critical accuracy problem because of the lower rates of fuel flow through them.[27] In some cases, especially in turboshaft engines for helicopters, fuel control development was far more complicated than that for turbojet or turboprop engines. Small turbine engine companies including Lycoming, Allison, and General Electric pioneered their fuel control development by working in conjunction with established fuel control companies. With continued effort over the years, engine controls became smaller, lighter, and more complex with reliability equal to that of large engine controls.

OVERCOMING ITS PROBLEMS

The inherent problems of scale in the small engine were gradually overcome by an aggressive technology development effort by the industry. That effort was supported both by the use of company funds and by technology demonstration programs that received funding from the military. The advances produced by these efforts allowed the small engine to overcome the problems related to its size and to attain outstanding performance.

---

27. "General Electric Presents: Small Turbines for Varied Markets," 966–967.

GENESIS OF THE INDUSTRY 3

The 1940s Environment

The small gas turbine aircraft engine industry in North America grew out of the embryonic aircraft gas turbine industry that emerged in the U.S. at the beginning of the 1940s. That industry adopted policies and procedures in use by the military to govern the development, procurement, and testing of piston engines for their aircraft.

POLICIES AND PROCEDURES

In the early 1940s, the overwhelming external influence on the fledgling aircraft gas turbine industry in the U.S. was the threat of war and World War II itself. The sole customers for these new types of engines were the Army and the Navy. Within these organizations, aircraft engine acquisition was under the direction of two small groups of managers. The Army's group was within the Power Plant Laboratory at Wright Field, Ohio. For the Navy, it was the Power Plant Branch (later the Power Plant Division) of the Bureau of Aeronautics in Washington, D.C. Both of these organizations were experienced, had been operating since the early 1920s, and were almost fully occupied with managing the development and production of the reciprocating aircraft engines essential for the war effort. The relationships that existed between these organizations and the aircraft engine industry are well documented.[1] The National Advisory Committee for Aeronautics (NACA) supported these two organizations and the engine companies with which they had contracted by testing engine components and complete engines in its test facilities.[2]

The Power Plant Laboratory at Wright Field and the Power Plant Division within the Navy's Bureau of Aeronautics functioned with little policy or management guidance from higher authority within their services and with no formal technical policy direction. Most decisions involving the development of aircraft engines were made by the personnel within these organizations. The "line" managers in both the Army and Navy organizations were officers who were transferred to other assignments within their services every few years. However, each service had a staff of professional civil servants that retained the "corporate knowledge" of the organization. The coordination between the two services and the exchange of information was, for the most part, completely informal.

Despite the absence of formal policy guidance and management control from high levels within the services, standard contracting procedures and specifications had been established for the development and procurement of reciprocating engines. These included the General Army-Navy Specifications for Engines. After agreement between the services on these specifications, they were circulated to industry for comments and recommendations. The final general specifications approved by both services were the basis for the preparation of all Detailed Model Specifications for the development and production of each engine configuration.

The qualification tests for piston engines were endurance tests that were run for the purpose of assuring that the durability and reliability of the particular engine model were acceptable. The Preliminary Flight Rating Test (PFRT), a 50-hour test, assured that the engine was acceptable for flight. The Model Qualification Test (MQT), a 150-hour test, assured that the engine was acceptable for production. As the endurance characteristics of piston engines, with their superchargers, were basically insensitive to altitude, both the PFRT and the MQT were run at sea level. The initial specifications for man-rated gas turbine engines were modeled directly after the specifications for piston engines. For example, the PFRT and MQT specified for early gas turbine engines were sea-level endurance tests of 50 hours and 150 hours, respectively.

1. Robert Schlaifer and S. D. Heron, *Development of Aircraft Engines, Development of Aviation Fuels*, (Boston: Graduate School of Business Administration, Harvard University, 1950), 7-14.

2. Virginia P. Dawson, *Engines and Innovation*, (Washington, D.C.: National Aeronautics and Space Administration, 1991), 19-58.

Standard piston engine procurement practices were also used in the acquisition of early turbine engines. One significant practice of the services was that of contracting with an airplane company for the development and production of a specific aircraft, and contracting separately with equipment contractors, including those for engines. Engines and equipment were then supplied, via the government, to the airplane company as Government Furnished Equipment (GFE). The airplane company was then required to integrate the equipment supplied by the government into the complete "weapon system" that it was responsible for delivering to the government.

Two principal characteristics of the military's piston engine design and development practices carried over into the early years of the gas turbine engine. One was the ceaseless effort to produce engines with more power and higher altitude capability without sacrificing specific weight or specific fuel consumption.[3] There was very limited freedom to optimize engines for particular missions, and the same basic engines were used in all types of military aircraft (fighters, bombers, and patrol planes). Thus, with engines being provided by the government to the aircraft builder as GFE, piston engine design was relatively insensitive to the specific aircraft in which they would be used. Rather, the engines were developed for a general size or type of aircraft. In general, the development of early turbojet and turboprop engines followed this practice.

The other main piston engine design and development characteristic that carried over to early gas turbine engines was the focus on achieving acceptable performance and durability as validated by a successful qualification test. Reliability and maintainability were generally addressed after the fact as problems arose during service operation and were then corrected by design changes of the offending parts and/or by revised maintenance and operating procedures. Basically, engine production cost was a minor factor in procurement decisions. Environmental factors such as noise and pollution received no attention.

In the context of these policies and procedures, there were no cooperative efforts between engine companies in the U.S. In fact, such efforts were not only discouraged, but they were prohibited by anti-trust laws except under very limited and strictly controlled circumstances. Competition was very keen, and companies chose not to share proprietary information.

3. Bureau of Aeronautics—Navy Department, *Naval Aircraft, Equipment, and Support Facilities, Vol. IV, Aircraft Power Plants: Ten Year History and Program of Future Research and Development* (10 September 1945), 7-20.

The aircraft turbine engine industry in the U.S. emerged during an era in which there were ample funds for research and development. With the U.S. absorbed in supporting the efforts of World War II, money for military engine development was plentiful through 1945. For example, funds for the Navy's research and development was over $26 million in 1945.[4] Translated into 1994 dollars that funding level corresponded to between $400 and $500 million per year.[5] The practice for piston engines, and for early gas turbines, was to apply these research and development (R&D) funds to the development of an engine through proof of feasibility, which normally meant through the first run of the original engine model. Subsequent engine development efforts were supported by production funds.

Because aircraft turbine engines were on the cutting edge of technology during the early to mid-1940s, major emphasis was placed on development of these new engines and their technology. As a result, turbine engine research and development had a high priority and received greater emphasis and funding than did other areas of military aviation at that time, such as avionics, aerodynamics, and aircraft structures. Consequently, there was ample R&D money to fund engine development, which meant that there were sufficient funds for the military to develop a number of different turbine engines during the early to mid-1940s.

In addition to the sea-level test facilities required to demonstrate the endurance characteristics of piston engines, altitude test facilities were also required to demonstrate their performance characteristics. Sea-level test facilities for piston engines were plentiful and relatively inexpensive. Similarly, sea-level facilities were readily constructed for the early turbine engines. However, the fact that the air consumption of a turbine engine was several times greater than that of a piston engine having the same power output presented a national problem with regard to altitude test facilities. Industry had no facilities for testing piston engines at altitude and the only facility able to test piston and gas turbines at simulated altitude was the Altitude Wind Tunnel at the NACA Aircraft Engine Research Laboratory in Cleveland that became operational in early 1944. The Navy's Aeronautical Engine Laboratory in Philadelphia and the Army's Power Plant Laboratory at Wright Field had facilities used for testing reciprocating engines at altitude conditions. However, they only had the capability to test small gas turbines over a limited range of simulated flight conditions.

4. Bureau of Aeronautics, *Aircraft Power Plants: Ten Year History and Program of Future Research and Development*, 152.

5. The translation of 1945 dollars to 1994 dollars was based on estimated inflation factors ranging from 15 to 19.

The Development of the Gas Turbine Aircraft Engine in the U.S.

The history of the development of the gas turbine aircraft engine has been well documented.[6] However, in order to put the history of small gas turbine aircraft engines into perspective, a few highlights are germane.

Hans von Ohain and Frank Whittle, working independently, but in parallel, invented the first jet aircraft engines. In 1937, von Ohain in Germany and Whittle in Britain successfully demonstrated their engines. Derivatives of those engines powered the first jet aircraft. German technology was, of course, kept from the U.S., and Whittle's engine was developed under the rules of military secrecy in Great Britain.

During the 1930s, technical studies on aircraft turbine engines were done in the U.S. Both turboprops and turbojets were evaluated, but only up to maximum airspeeds common for propeller driven aircraft during that period. For this and other technical reasons, the turbine engine was not considered to be a viable aircraft power plant. However, the studies generally failed to evaluate the weight-saving potential of the turbine engine or to realize the high-speed flight possibilities of jet-powered aircraft. Studies by the engine community focused on incremental performance improvements in relative isolation from potential improvements in aircraft. Had potential future engines and aircraft been analyzed as a whole system, the advantages of a turbojet aircraft might have been realized earlier.

About 1939, an experimental two-stage turbosupercharger built by Turbo Engineering Corporation (TEC) was bench tested by DeLaval Steam Turbine Company; part of the TEC turbo test consisted of rigging it as an independent turbine engine which was capable of running under its own power. In 1940, Lockheed undertook a private initiative on its L-1000 turbojet, and NACA began construction of a Campini-type ducted fan engine. The U.S. power plant community in 1940, both the industry and the military, was still generally unaware of the current turbojet developments in Germany and Great Britain, cloaked in wartime security restrictions.[7] However, by 1941, the military's interest in aircraft gas turbines was sufficient for them to initiate design studies which were undertaken by General Electric (GE), Westinghouse, Allis-Chalmers, TEC, and Northrop. In 1941, Pratt & Whitney also began preliminary layout of a free-piston reciprocating Diesel compressor gas turbine turboprop (designated

---

6. Schlaifer and Heron, *Development of Aircraft Engines, Development of Aviation Fuels*, 321–508.

7. In 1937, during a visit to Great Britain, A. R. Smith of GE had seen the initial Whittle engine. This information was apparently not widely shared.

PT-1), and Wright Aeronautical unsuccessfully attempted to license for manufacture the Whittle engine.

A key event event in the development of the gas turbine was initiated by Chief of the Army Air Corps General Henry H. Arnold. On February 25, 1941, General Arnold wrote a letter to Vannevar Bush, Chairman of both the National Defense Research Committee (NDRC) and National Advisory Committee for Aeronautics, in which Arnold reported that the Germans were making considerable progress in jet propulsion and emphasized the importance and urgency of studying that subject.

Bush responded to General Arnold's concerns a month later by establishing the NACA Special Committee on Jet Propulsion under W. F. Durand. Then 82 years old and dean of the American engineering community, Durand was Professor Emeritus of Mechanical Engineering at Stanford University and a member of the NACA (indeed, he had been its founding director in 1915). Membership of the Special Committee included representatives from the three American companies with experience in the manufacture of industrial power turbines, Westinghouse, Allis-Chalmers, and the Schenectady, N.Y. Steam Turbine Division of GE.

Following the Special Committee's initial considerations of the subject, in July 1941, it was decided that each of the three turbine companies should proceed with a detailed study of an engine of the type that it preferred. After the preliminary studies were made, it would be decided whether one or more should be developed. Then, in September 1941, the Special Committee recommended that the military services award contracts for the development of all three engines.

Westinghouse had proposed a simple turbojet (which later became the 19A) and also a ram jet. Allis-Chalmers proposed a turbine-driven ducted fan, but its engine never went beyond the design stage. GE proposed a turboprop, which became the TG-100 (military designation T-31). All three companies proposed a straight-through design using axial-flow compressors and turbines on a single shaft. At the time they submitted their proposals, none of the three companies, including GE's Steam Turbine Division, were aware that the Whittle engine and its design data had been sent to GE's turbosupercharger group in Lynn, Mass., where development of an American prototype turbojet was to take place.

In the spring of 1941, General Arnold had visited England where he learned about the Whittle engine and witnessed taxiing tests and short flights of the Gloster aircraft that it powered. Arnold immediately set into motion the process for acquiring and developing this technology in the U.S. In September 1941, the Army contracted with GE to build an American version of the Whittle-designed

Power Jets W.1.X. turbojet engine. The derivative GE engine, known as Type I, successfully ran on April 18, 1942; it was the first jet engine to run in the U.S.[8]

Immediately after the Japanese attack on Pearl Harbor on December 7, 1941, Westinghouse began development work on a turbojet engine for the Navy. It became the 19A Yankee, the first all-American-designed turbojet engine, and it ran for the first time on March 19, 1943.

GE and Westinghouse subsequently became the vanguard of the gas turbine aircraft engine industry in the U.S. To put this into achievement into perspective, it is important to point out that between 1942 and 1945, a number of new and unorthodox engines were developed to one extent or another by a variety of companies. The military services had under contract some 15 different engine designs, including seven turboprops, five turbojets, and three pulse jets. In addition, there were several design studies of ram jets. A Navy propulsion plan evaluating the situation in 1945 stated:

> In considering the turbine program as a whole, it is obvious that this type engine is so new that the final configuration cannot be predicted. There are a wide variety of fields of endeavor which have possibilities of offering eventual improvement. All of these fields must be and will be investigated. The basic job in the next three to five years will be to prove the simplest form of gas turbine as a usable engine.[9]

Sorting out if and where these new types of engines should be used came into focus soon after the end of World War II. Although the Navy's original and immediate post-War interest in turbines emphasized the development of turboprops, the simplicity and high speed capability provided by the turbojet led to its near monopolization of the short-term development efforts of both the Army and the Navy.

Pioneer Small Gas Turbine Aircraft Engine Companies in the 1940s

The generally known North American companies that developed and built small gas turbine aircraft engines–as opposed to large ones—between the early 1940s and the 1990s are illustrated in the accompanying chart. Those companies that entered the business during the 1940s are identified as pioneer companies.

8. The Type I first ran on March 18, 1942, but it stalled well below full speed.

9. Bureau of Aeronautics, *Aircraft Power Plants: Ten Year History and Program of Future Research and Development*, 154-160.

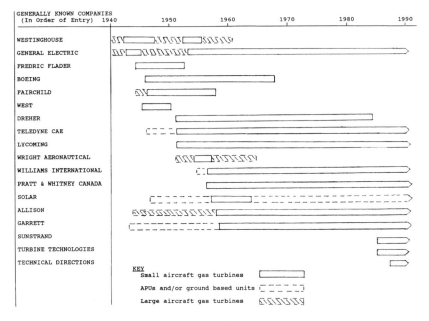

Years Spent in the Small Gas Turbine Aircraft Engine Business.

In late 1942, Westinghouse began developing a 9.5-inch diameter turbojet engine based on earlier development work begun in December 1941 on the 19A turbojet engine. A 9.5 engine successfully powered a Navy TD2N-1 drone on August 17, 1945. In late 1943, GE ran a Type B turbosupercharger in boot-strap (i.e. it was analogous to a self-sustaining turbine engine); following that in the spring of 1944, GE built a small turbojet engine known as the B-1, derived from a Type B-31 turbosupercharger. Two B-1 engines powered a Northrop JB-1A flying bomb during an unsuccessful flight on December 7, 1944.[10] After these early efforts, Westinghouse phased out of small gas turbine engines in the 1950s, and GE did not resume its small gas turbine efforts until 1953.

In the 1940s, other U.S. companies also became active in studying, develop-ing, and manufacturing small gas turbine engines for aircraft propulsion. Much of their work was sponsored by the military, and engines were developed for both piloted and pilotless aircraft. These companies included Fredric Flader, Boeing, Fairchild, and West Engineering. Each of these companies eventually phased out of the small gas turbine aircraft engine business.

10. D. M. Carpenter and P. V. D'Alessandro, *JB-1A Thunderbug: WWII Flying Wing Bomb* (Jet Pioneers of America, ISBN No. 0-9633387-8-1), 12-13.

Allison got its start in the large gas turbine aircraft engine business in June 1944 when it was designated as a second production source for GE's J33 turbojet engine. It did not start work on small gas turbine aircraft engines until the late 1950s.

Another relevant activity underway during the 1940s was small gas turbine component and non-aircraft research and development. In early 1943, Garrett, then known as AiResearch, began work on Project A, a two-stage compressor for aircraft cabin pressurization which later led to turbine environmental control systems, jet engine starters, and APUs. Soon after World War II, Continental Aviation and Engineering (CAE) began development of its first gas turbine starter, the TR125, although it was unsuccessful. In 1946, Solar began development of a turbine-driven APU.

Westinghouse

By 1941, when Westinghouse entered the aircraft gas turbine engine business, the company had over 40 years experience in the development and manufacture of industrial steam turbines. Engineers in development engineering had studied the possibilities of developing gas turbines for land and industrial applications in 1939 and 1940. That work was followed by a study early in 1941 on a gas turbine for aircraft propulsion and the preparation of a proposal in the late fall of 1941 for a marine gas turbine propulsion unit.[11]

Early in 1941, Dr. L. W. Chubb, Director of the Westinghouse Research Laboratories, was appointed a member of the NACA Special Committee to investigate the possibilities of using the gas turbine as a jet propulsion engine for use in aircraft. Dr. Stewart Way of the Westinghouse Research Laboratories made his thrust calculations for a turbojet engine in notes dated July 19, 1941.[12] Way made reports to the Committee on a ram jet engine (Research Memo SM-101 dated July 21, 1941) and a turbojet engine (Research Memo SM-106 dated August 14, 1941). Subsequently, Westinghouse was assigned to "investigate jet propulsion in which the full propulsion effect is obtained from hot turbine gases."[13]

On November 5, 1941, the Captain D. C. Ramsey, Assistant Chief of the Bu-

11. "The History of Westinghouse in the War Aviation Gas Turbine Division, Engineering Department," undated document (circa 1945), National Air and Space Museum Archives, Washington, D.C., 1-2.

12. Dr. Stewart Way, Whitehall, Montana, to Rick Leyes, letter with enclosures, 21 November 1992, National Air and Space Museum Archives, Washington, D.C.

13. "The History of Westinghouse in the War Aviation Gas Turbine Division, Engineering Department," 4-5.

reau of Aeronautics, asked Westinghouse for a proposal calling for a design study of internal combustion turbines utilizing only jet energy for the propulsion of aircraft. The proposal was to be based on a unit of approximately 600 thrust horsepower at 25,000' altitude and 500 mph. The design of the complete jet engine was undertaken by Westinghouse Steam Division engineers under R. P. Kroon, Manager of Development Engineering, and the combustion research was assigned to the Research Laboratories under the direction of Dr. Way. On December 8, 1941, the day after Japan attacked Pearl Harbor, Westinghouse took the initiative and immediately started development of its engine without an official letter of intent or contract. On January 7, 1942, the Navy issued Westinghouse a letter of intent to proceed with a design study for its turbojet.[14]

19A TURBOJET

Within a week after December 7, 1941, Westinghouse had established a project team of 14 engineers and 9 skilled experimental mechanics. This team was under the direction of Reinout P. Kroon, Manager of Development Engineering in the Steam Division of Westinghouse located at the South Philadelphia works. The entire Development Engineering Department was immediately involved in development of the 19A engine.[15] This group included: Oliver E. Rodgers (mechanical design details), who was assisted by A. S. Thompson; W. R. New (fundamental flow studies and test house and equipment design); A. H. Redding (compressor); Charles A. Meyer (turbine and engine thermodynamics); Dr. C. C. Davenport (bearing and lubrication system design); C. D. Flagle (accessories design); Norman Mochel (materials development); Joe F. Chalupa (drafting, layout, manufacturing, and production); H. B. Sadin (test house and equipment design); B. V. Anoschenko (in charge of the experimental laboratory); John Rivell (test engineer); Mark Benedict (engine installation); P. G. deHuff (metallurgical and material studies); H. J. Clyman.[16] Dr. Way made the first design

14. Bureau of Aeronautics, *Aircraft Powerplants: Ten Year History and Program of Future Research and Development*, 28. The document, "The History of Westinghouse in the War Aviation Gas Turbine Division, Engineering Department," 6-7, indicates that the Navy letter of intent was received on December 9, 1941. The reason for the discrepancy regarding the letter of intent date of issuance is not apparent.

15. Robert L. Wells, Kennett Square, Pennsylvania, to Rick Leyes, letter, 5 December 1993, National Air and Space Museum Archives, Washington, D.C.

16. Stewart Way, "Early Development of the American Turbojet Engine," unpublished paper, undated (circa 1993), National Air and Space Museum Archives, Washington, D.C. Also, "The History of Westinghouse in the War Aviation Gas Turbine Division, Engineering Department," 7-9.

studies, and in cooperation with Dr. A. E. Hershey, was responsible for combustion research. Jet engine work at Westinghouse remained under the Development Engineering until February 1945 when the Aviation Gas Turbine Division was formed and Kroon was named Manager of Engineering in this new division.[17]

On August 10, 1942, the design of the 19A (military designation J30) was begun. The 19A had a six-stage axial compressor, 24-can annular combustor, single-stage axial-flow turbine; it developed approximately 1,100 lb thrust, weighed 825 lb, and was 19 inches in diameter (hence its name 19A). On October 22, 1942, by amendment to its design study contract, Westinghouse was authorized to construct two 19A units. The engine ran for the first time on March 19, 1943. In March 1943, the number of 19A engines on order was increased to a total of six. On July 5, 1943, the first 19A completed a 100-hour endurance test. The No. 2 19A unit was flown on a Chance-Vought FG-1 Corsair on January 21, 1944 in its original function as a booster engine.

BOOSTER TURBOJET ENGINES AND THE NAVY'S TRANSITION TO JET PROPULSION

In its early consideration of gas turbine aircraft engines in the 1940s, the Navy was primarily interested in the development of a turboprop engine for aircraft operation from carriers. The Navy Bureau of Aeronautics generally recognized the potential advantages of the mechanical simplicity, high thrust at what was then considered to be high flight speeds (400–450 mph), and the lower weight of the jet engine as compared to the piston engine. However, the Navy believed that a turbojet-powered aircraft was not desirable due to poor take-off thrust, relatively poor climb capability, and restricted range due to high fuel consumption. It did appear that the turbojet might provide conventional engine fighters with extra power (i.e. as a booster jet) for short periods during combat, which would minimize the disadvantages of its high fuel consumption. Accordingly, the Navy contracted with Westinghouse and the Turbo Engineering Corporation for turbojet engines designed primarily as boost units. The contract with Westinghouse resulted in the 19A.[18]

In June 1941, the Navy had issued a contract to TEC for a comprehensive design study on using centrifugal compressors and radial flow turbines for gas tur-

17. Wells to Leyes, 5 December 1993.

18. Bureau of Aeronautics, *Aircraft Power Plants: Ten Year History and Program of Future Research and Development*, 25–42.

bines. The results of this study were important in guiding the Navy's future policy in pointing out that, for aircraft applications, the simplest gas turbines were the most attractive. On January 9, 1942, the Navy issued a letter of intent to TEC which led to an October 19, 1942 contract for the design and construction of two (booster) turbojet engines with an approximate thrust of 1,000 lb and weight of 400 lb. The engine design consisted of a single-stage centrifugal, mixed-flow compressor and single-stage radial flow turbine with internal blade cooling. However, between 1942 and 1944, nearly all of TEC's efforts were focused on developing its turbosupercharger that the Navy had contracted for production. Thus, TEC's jet engine project lagged behind, and the Navy canceled it in September 1944.

By 1943, it was evident that despite the fact that jet-powered fighters would be inferior in both takeoff and range, there was no use having good takeoff and range if the fighter having them would be out-performed in combat by enemy jet-powered fighters.[19] Furthermore, because the design of jet aircraft depended upon the availability of jet engines with adequate thrust, it was only when first generation jet engines, such as the Westinghouse 19 series as well as those made by GE, became available that jet aircraft could be built in the U.S. Consequently, the Navy began development of jet-powered aircraft. During 1943, the McDonnell XFD-1, Interstate XBDR-1, Northrop XP-79B, Douglas XBTD-2, and Ryan XFR-1 were either designed or considered for the installation of the Westinghouse 19B.

On January 7, 1943, the Navy issued a letter of intent to McDonnell for engineering, development, and tooling for two aircraft to be powered by turbojet engines. The prototype aircraft, known as the McDonnell XFD-1, later became the FH-1 Phantom. The sponsorship of this project was the first genuine Navy commitment for the use of jet engines as primary propulsion. The McDonnell FH-1 Phantom became the Navy and Marine Corps' first operational pure jet fighter.

19B/XB TURBOJET

On March 8, 1943, the Navy made a commitment followed on May 26, 1943 by a contract signed with Westinghouse to initiate the design of an improved version of the Westinghouse 19A, known as the 19B. The 1,365-lb-thrust 19B was intended for application to either a pure jet aircraft or as a booster unit. Whereas the 19A had no airplane accessories as it was intended as a booster unit, the 19B

19. Schlaifer and Heron, *Development of Aircraft Engines, Development of Aviation Fuels*, 471.

Cutaway of Westinghouse 19XB-3B turbojet engine. National Air and Space Museum, Smithsonian Institution (SI Neg. No. 95-1220-2).

incorporated both engine and airplane accessories. The combustion chamber was simplified into a single annular chamber, and the design of the compressor and turbine were improved considerably. The first 19B was tested on March 18, 1944 and was delivered on July 13, 1944. The 19B was flown on a Martin JM-1 testbed aircraft on September 28, 1944. The McDonnell XFD-1 Phantom, powered by two 19Bs, made its first flight on January 26, 1945. The Northrop XP-79B was also powered by two 19Bs, but crashed on its first flight on September 12, 1945. A total of 28 19B engines were ordered.

On February 28, 1944, Westinghouse submitted to the Navy a proposal for improving the performance and decreasing the weight of the the 19B with a reconfigured engine designated 19XB. The 1,600 lb thrust of the 19XB was achieved primarily by adding four additional axial-flow compressor stages. A Navy letter of intent was issued on April 7, 1944, and the first 19XB went to test on October 11, 1944. Both the 19B and 19XB were tested at NACA's altitude wind tunnel in Cleveland. The 19XBs were originally contemplated for production as a replacement for the GE I-16 in the Ryan FR-1 airplane. The 19XB was used in the Grumman XTB3F-1, Douglas XB-42A (as auxiliary engines), and Northrop X-4. The primary application for the 19XB engine was the McDonnell FH-1 Phantom. A total of 261 19XB engines were made.

Because the Westinghouse 19A was originally designed as a booster engine and the 19B as either booster or a prime propulsion unit, a brief history of their development has been included. The 9.5, a derivative of the 19A and designed from the start as a very small turbojet engine, is the primary focus of this Westinghouse history.

9.5 TURBOJET

With work on the 19A proceeding, on November 27, 1942, Kroon visited the Bureau of Aeronautics in Washington. There he was informed of a jet-powered light fighter plane that was being designed by the Bureau. To obtain the required performance, the airplane was to have extremely "clean" aerodynamic design with 10 to 12 jet engines housed entirely within the wings. That called for engines about one-half the size of the 19A. An airframer (McDonnell) had been found who claimed that this all-jet aircraft could be built in nine months. Westinghouse was asked to investigate how long it would take to design and build the engines for one plane and how long it would take to deliver 300 to 400 units a month.[20]

Though not yet officially authorized, Westinghouse began work immediately on a turbojet half the diameter of the 19A. The engine was designated 9.5A. Design work on the 9.5A was done using the marginal time of the same group of engineers that designed the 19A engine. Oliver E. Rodgers was responsible for the original mechanical design of both engines.[21] Some time after the work had started on the 9.5A, representatives of McDonnell went to the Navy and partially convinced them that larger jet engines were more practical. In February 1943, the Navy indicated that it did not want any 9.5A engines, but said that since part of the work was done that it should be finished.[22]

By May 1943, the Bureau of Aeronautics had designed an "Aerial Ram" (missile) in two sizes for use in combat. The smaller one was to use the 9.5A engine and the larger one the 19.[23] On July 19, 1943, the Bureau of Aeronautics

20. R. P. Kroon, "Trip Report, Bureau of Aeronautics, Washington D.C.," 30 November 1942, National Air and Space Museum Archives, Washington, D.C. While not specifically stated in the trip report, the airframer is evidently McDonnell. The first project designs of the prototype McDonnell XFD jet aircraft incorporated three 9.5 engines in each wing.

21. Oliver E. Rodgers, Kennett Square, Pennsylvania, to Rick Leyes, letter, 7 December 1993, National Air and Space Museum Archives, Washington, D.C.

22. "The History of Westinghouse in the War Aviation Gas Turbine Division, Engineering Department," 20-21.

23. Radm. Delmar S. Fahrney, "The Genesis of the Cruise Missile," *Astronautics and Aeronautics*, Vol. 20, No. 1 (January 1982): 34-53.

Westinghouse 9.5A turbojet engine was selected to power the Gorgon II-B and III-B air-to-air missiles. The 9.5A was a half-size version of the Westinghouse 19A, the first all-American-designed turbojet engine. This engine is in the National Air and Space Museum collection. National Air and Space Museum, Smithsonian Institution (SI Neg. No. 94-1149-18A).

established a project at the Naval Aircraft Factory in Philadelphia to design, develop, and test a missile powered by the 9.5A, known as the Gorgon. The 9.5A was selected to power the Gorgon II-B and III-B air-to-air missiles; the 9.5B powered the Martin TD2N-1 high-speed target drone.

Work on the 9.5A had been resumed following a Navy letter of intent with Westinghouse (signed on June 26, 1943) covering the development of the X9.5A and the delivery of six engines.[24] Based on the demand for 16 of the 9.5 engines for flight tests in the Gorgon airframes, on January 8, 1944, Westinghouse proposed building an additional 20 engines; on March 4, 1944, the Navy issued a letter of intent for these engines. The first engine was ultimately delivered on July 31, 1944. The X9.5A, given the military designation XJ32-WE-2, had a six-stage axial-flow compressor, annular combustor, and single-stage tur-

24. "Westinghouse Jet Engine Type 9.5A (J-32)," unpublished document, undated (circa 1945), National Air and Space Museum Archives, Washington, D.C.

bine. It had a diameter of 9.5 inches (hence its name 9.5), a length of 52 inches, weighed 145 lb, produced about 260 lb of thrust with a 1.6 sfc, and had a specification design life of three hours.[25]

Several problems were encountered during development and acceptance testing of the engine. There were problems with the thrust bearing requiring modification of its design. Also, the governor hunted causing erratic operation.[26] Although several design modifications were tested to correct these problems, no fully satisfactory solutions were found at that time. The engine also had a dead band between 17,000 and 23,000 rpm. Rated speed was 36,000 rpm. Within the dead band, a rapid opening of the throttle resulted in a large increase in fuel flow, no substantial increase in engine speed, and afterburning through the turbine and out the exhaust nozzle.

Westinghouse delivered its first X9.5A engine for testing in July 1944. The unit passed a three-hour Navy acceptance test on July 7, 1944. In January 1945, the X9.5A passed its 150-hour type test. Two 9.5A units were delivered to the Naval Aircraft Factory before the middle of January 1945, but their high-altitude performance was considered unsatisfactory.

The first flight of a 9.5-powered TD2N-1 drone took place on June 27, 1945.[27] While the engine operated, the radio-control system did not, and the drone crashed. The first successful flight of the TD2N-1 was on August 17, 1945. Subsequent experimental flights successfully demonstrated the capability of the TD2N-1. However, due to the engine's high cost and continuing engine development delays, the TD2N-1 program was canceled in March 1946.[28] The Gorgon II-B and III-B missiles were not developed because the engine was not considered satisfactory. The Navy decided that the Gorgons should be powered either by liquid rockets or by pulsejets. Later, ramjet propulsion was introduced into the Gorgon program.

In September 1944, the Army Air Forces had taken its initial steps in procuring four X9.5A engines for testing models of the Northrop XP-79 and several

---

25. The 9.5 was actually designed with a much longer life in mind in order to be able to conduct development tests.

26. "Westinghouse Jet Propulsion Motor Navy Development Acceptance Test—Model 9.5A," (Acceptance test report submitted to United States Navy, Bureau of Aeronautics by Aircraft Apparatus Section, Development Engineering, Westinghouse Electric and Mfg. Co., 25 September 1944).

27. William F. Trimble, *Wings for the Navy: A History of the Naval Aircraft Factory* (Annapolis, Md: United States Naval Institute, 1990), 282-283.

28. Fredrick I. Ordway, III and Ronald C. Wakeford, *International Missile and Spacecraft Guide* (New York: McGraw-Hill Book Company, Inc., 1960), 180-182.

high-speed, jet-propelled medium bombers in the 20-foot-diameter wind tunnel at Wright Field. However, a problem in use of the engine for such tests was the relatively short life (less than five hours) of the combustion chamber liner and excessive creep in the turbine disk around the blade roots. An increase in engine life permitting the engine to complete its 150-hour type test in January 1945 and making it suitable for testing the wind tunnel model was attributed largely to redesign of the thrust bearing.

By April 1946, the X9.5A had been superseded by the X9.5B model. The principal difference between the "A" and "B" model was a larger combustion chamber on the "B" (the diameter was increased from 9.5 inches to 11.75 inches). The larger combustor resulted in an increase in maximum operating altitude from 15,000 feet for the "A" to 35,000 feet for the "B." The 9.5B produced between 275 to 300 lb thrust. The Army Air Forces ordered 10 9.5B engines.[29]

With its improved engine life and high-altitude operating capability, Westinghouse offered the X9.5B engine for use as a booster unit for existing aircraft powered by reciprocating engines. However, finding no market for the engine as a booster unit, it was discontinued about 1948, thereby terminating work at Westinghouse on small engines of its own design. By that time, the company had turned its full attention to the development of its new and more powerful engines for military combat aircraft.

A total of 24 9.5A and 20 9.5B engines were built. Although the 9.5A and 9.5B engines were never used widely, they comprised the first family of small turbojet engines successfully developed and produced in the United States. Because of proprietary restrictions on the engine companies, information on the technical and design characteristics of the engines were not available to others developing small engines during the 1940s. Nevertheless, the 9.5A and B engines demonstrated that small, reliable turbojet engines having a reasonable level of efficiency could be built.

XJ81 TURBOJET

About 1953, Westinghouse acquired a license from Rolls-Royce to manufacture and sell that company's 1,740-lb-thrust Soar turbojet engine in the U.S. It was also in 1953 that Northrop began design of the XQ-4 supersonic target drone. The XQ-4 required a turbojet with approximately 1,800 lb thrust. There were no U.S. turbojets in that thrust class at the time.

---

29. "Case History of Turbojet Engines XJ30-WE, XJ32-WE, XJ34-WE Series (X19B, X19XB, X9.5 and X24C Series)," (Historical Study 99, United States Air Force, Headquarters Air Force Logistics Command, January 1947), 4-6.

The sequence and interrelationship of the Westinghouse acquisiton of its license for the Rolls-Royce Soar and the establishment of power plant requirements for the XQ-4 is uncertain. Nevertheless, in the absence of a suitable U.S. engine, the Rolls-Royce RB93/2 (R.Sr.-2), with an Air Force designation of XJ81-WE-3, was offered by Westinghouse and was selected by the Air Force as the interim power plant for the XQ-4. Although the XJ81 was approved as the interim power plant for the XQ-4, that decision visualized the earliest possible availability of an American engine as the power plant for follow-on Q-4 production.

The XQ-4 target drone was designed to operate in the low-supersonic speed range with a maximum speed of Mach 1.55. A total of 24 of these prototype drones were manufactured by Northrop. Flight test programs on them began in 1958 at Holloman Air Force Base in New Mexico.[30]

The XJ81-WE-3 engine had been designed by Rolls-Royce as a lightweight turbojet with a thrust of 1,740 lb and a weight of 304 lb. The engine had a seven-stage axial-flow compressor, an annular combustor, and a single-stage axial turbine. All 24 of the engines supplied for the XQ-4 drone were manufactured by Rolls-Royce and procured through Westinghouse. The XQ-4 prototypes were the only application of the XJ81-WE-3 engine in the U.S., and no engines of this type were ever built by Westinghouse.

LEAVING THE JET ENGINE BUSINESS

Westinghouse continued its development of large engines for Navy aircraft until 1960 when the company decided to go out of the aircraft gas turbine business. The first of these engines following the 19B and 19XB was the highly successful and reliable 24C (military designation J34) used by the Navy in its fighter aircraft. The J34 was also used in the Navy's Rockwell T-2 Buckeye trainer aircraft. That engine was followed by the larger afterburning J40 and J46 engines. However, both of those engines encountered problems, and therefore, their delivery was late. The J40 problem was one that had a major effect on the company's image. The McDonnell F3H Navy fighter, in which the engine was to be used, grew by more than 30% in weight. This required a corresponding increase in engine thrust. However, the engine modifications made to increase thrust led to critical problems and schedule delays; the result was that the J40 never reached production. Westinghouse was criticized for not anticipating enough growth potential in the initial design of the engine.

30. Richard A. Botzum, *50 Years of Target Drone Aircraft* (Newbury Park, Cal.: Publications Group of Northrop Corporation, Ventura Division, 1985), 82-83.

A number of reasons have been expressed for the Westinghouse decision to leave the jet engine business. Among them were the company's tarnished image resulting from the J40 and J46 problems, the company's failure to anticipate the need for thrust growth in its design of new engines being developed, failure to anticipate the need for adequate testing and manpower resources for advanced technology programs, the inability to adjust its long-standing steam turbine manufacturing practices to conform to the aircraft engine industry, and the company's inability to compete with the industry leaders, Pratt & Whitney (P&W) and GE. But, perhaps the reasons for Westinghouse leaving the jet engine business were best expressed by Reinout Kroon in a letter written to Lewis Smith in 1990. He wrote:

> The J40 story is sort of sad. We were able to make the required thrust, as I recall, well. But the airplane which was to use the engine became much heavier than anticipated, and that required more thrust. This we saw no way of providing. We were criticized for not anticipating engine growth in our design.
>
> This may indeed have been one of the reasons why Westinghouse got out of the business. But I think the more basic reason was that after the war, finances began to dry up. The Services decided that, in the long run, they could only afford two suppliers of jet engines, Pratt & Whitney and GE. It was probably a wise decision. Westinghouse management had much difficulty adapting to the style of aircraft engine manufacturing, which was so different from the turbine activity they had been used to.[31]

In closing the Westinghouse story, tribute should be paid to Reinout Kroon who was considered the father of Navy aircraft gas turbine development. He was also one of the pioneers of jet engines in America, having led the development of the first American designed turbojet. It is of historic interest to note that prior to the first run of that engine, the 19A, he wrote the following words on the flyleaf of the engine's log book:

> This book is to record the starting of the first jet propulsion motor of American design. May this motor prove as successful as we, who dedicated ourselves to its development, so fervently hope. May our jet motors do their part in the fight for freedom and a better world.
>
> R.P.K.
>
> 3/17/43[32]

31. Lewis F. Smith, "History of Westinghouse Aviation Gas Turbines," unpublished paper, undated (circa 1990), National Air and Space Museum Archives, Washington, D.C., 7-8.

32. Robert L. Wells, "Tribute to Reinout P. Kroon," (Unnumbered Paper presented at the ASME/ICTI Meeting, Cincinnati, Ohio, May 1993), 10.

Fredric Flader

Before founding his own company in 1944, Fredric Flader, an aeronautical engineer, had a 24 year career in the design and development of a variety of aircraft. He began his professional career in 1920 as an aircraft designer for the Army at McCook Field in Dayton, Ohio. He moved to the Buhl-Verville Aircraft Corp. in Detroit in 1925, and later to the Eberhart Airplane & Motor Company, the Consolidated Aircraft Corp., and the Curtiss-Wright Corporation's Airplane Division in Buffalo. During the years from 1932 to 1944, which he spent with Curtiss-Wright, he rose to the position of Chief Engineer and was responsible for the designs of a number of aircraft.

After leaving Curtiss-Wright, Flader's next step was the founding of Fredric Flader, Inc., in 1944, to develop small gas turbine engines. At that time, he received a contract from the Army Air Forces to develop a 5,900 shp turboprop engine, designated the XT33-FF-1. Soon afterward, Fredric Flader dedicated a new plant in Tonawanda, N.Y. for use in developing his small gas turbine engine and work proceeded on its design. However, while the XT33 engine was still in the design stage and before any hardware was fabricated, the Air Forces canceled the contract.

About the time work had begun on the XT33, Flader also received a contract from the Navy to develop an 8-inch diameter gas turbine engine to power an alternator. This engine was a simple single-shaft engine with an axial-flow compressor, through-flow combustor, and single-stage turbine. Unfortunately for Flader, the Navy withdrew its support before the engine had reached the prototype stage, and further development was discontinued.[33]

In late 1946, the Army Air Forces began developing the requirements and performance characteristics for three unmanned aircraft that were to be used by the Army Air Forces for a variety of purposes. One of these was a small high performance target aircraft that was to be jet powered and capable of 600 miles per hour at 15,000 feet altitude with a service ceiling of 40,000 feet. This jet-powered target aircraft was designated the XQ-2. In early 1947, a request for quotation was released by the Army Air Forces for design and development of the XQ-2. Following an intensive competition, the Ryan Aeronautical Company was awarded a contract in August 1948 to develop the XQ-2, subsequently the Ryan Q-2 Firebee.[34]

---

33. Fred T. Sarginson, telephone interview by Rick Leyes and William Fleming, 21 September 1988, transcript, National Air and Space Museum Archives, Washington,D.C.

34. William Wagner, *Lightning Bugs* (Fallbrook, Cal: Armed Forces Journal International in cooperation with Aero Publishers, Inc., 1982), 86-87.

Fredric Flader XJ55-FF-1 turbojet engine having a single-stage supersonic compressor. National Air and Space Museum, Smithsonian Institution (SI Neg. No. 93-16136).

Meanwhile, in keeping with its procurement policy to contract separately for airframe and engines, on February 7, 1947, the Power Plant Laboratory at the Wright Air Development Center issued a Purchase Request to initiate design and development of an engine for the XQ-2. On June 23, 1947, Fredric Flader, Inc., was awarded a competitive contract to develop this engine, given the military designation of XJ55-FF-1.[35]

The XJ55-FF-1 engine had a single-stage supersonic compressor of the shock-in-rotor type, an annular combustor, and a single-stage turbine. It was rated at 700 lb thrust with a 1.65 cruise sfc. The unique feature of this engine was its single-stage supersonic compressor that was designed to provide a pressure ratio of approximately 2.75.[36]

The characteristic of a supersonic compressor was that, at and near design operating speed, the flow entering the rotor stage was supersonic over the entire blade span. The aerodynamic design was such that a shock wave occurred in the rotor stage, thereby producing a characteristic abrupt pressure rise across the shock wave as the velocity of the air became subsonic. It was the occurrence of this shock wave in the rotor stage that provided the supersonic compressor its high pressure ratio per stage.

35. William D. Downs, "Development of the XJ55-FF-1 Engine," (Technical Note No. WADC-55-169, Wright Air Development Center, Wright-Patterson Air Force Base, Ohio, December 1953), 1.

36. The National Air and Space Museum has an XJ55 engine in its collection.

The supersonic compressor was revolutionary at that time and was as yet untried. From 1946 to 1948, research engineers at the NACA's Lewis Flight Propulsion Research Laboratory in Cleveland, Oh. and Langley Aeronautical Laboratory at Langley Field, Va. had designed and tested supersonic compressors. These early investigations of the supersonic compressor showed encouraging results. Further, the XJ55-FF-1 compressor rotor blade design was based on methods developed at NACA's Langley laboratory.[37]

In justifying the use of this new and revolutionary technology in its engine, the Fredric Flader company made the following statement in a letter to the Army Air Forces dated April 26, 1947, two months before contract award:

> Prior to submitting a primary bid on the engine with a supersonic compressor, we have checked the development status of this type of compressor thoroughly, and we are convinced that this development has progressed to a point where its incorporation in this design proposal is warranted.[38]

The advantage of the supersonic compressor was that the pressure ratio per stage that could be achieved was several times that obtainable with a conventional subsonic axial-flow compressor stage. However, to achieve this high pressure ratio, it was necessary that supersonic flow be established over the entire span of the rotor stage. It had been found by the NACA that the ability to achieve this supersonic flow condition in the rotor with corresponding high performance was extremely sensitive to small variations in area ratios and passage contours of the rotor stage.[39] Thus, small variations in rotor stage dimensions from the aerodynamic optimum had a significant impact on engine thrust and fuel economy.

By the time the prototype model of the XJ55-FF-1 was ready for testing, Flader had taken over the Packard test laboratory located in Toledo, Oh. The laboratory had been used earlier by Packard to perform development tests of its XJ49 (PT-205) large ducted fan engine that was later discontinued.[40]

37. Melvin J. Hartmann and Robert C. Graham, "Performance of Axial-Flow Supersonic Compressor of XJ55-FF-1 Turbojet Engine. I—Preliminary Performance of Compressor," (Research Memorandum Number SE9A31, Lewis Flight Propulsion Laboratory, National Advisory Committee for Aeronautics, Cleveland, Ohio, 1949), 1-2.

38. Downs, "Development of the XJ55-FF-1 Engine," 65.

39. Melvin J. Hartmann and Edward R. Tysl, "Performance of Axial-Flow Supersonic Compressor of XJ55-FF-1 Turbojet Engine. III—Over-all Performance of Compressor," (Research Memorandum Number SE9G19, Lewis Flight Propulsion Laboratory, National Advisory Committee for Aeronautics, Cleveland, Ohio, 1949), 7.

40. A Packard XJ49 engine is in the National Air and Space Museum collection.

Testing of the XJ55 began in June 1948. During that period of time, the Flader engineers were working closely with research engineers at NACA's Flight Propulsion Research Laboratory, and an early version of the XJ55-FF-1 compressor was tested at the laboratory. It was discovered that the compressor performance near rated speed was significantly below the design value. In addition, the efficiency decreased as rotor speed was increased from approximately 50% to 100% of rated speed, which was contrary to NACA's earlier hypothesized and experimentally substantiated trends. Also, after about 35 hours of testing, the thin leading edge of the rotor blades was found to be curled over, apparently due to high internal aerodynamic forces.[41] NACA propulsion laboratory engineers concluded that the poor high-speed performance of the compressor was due to the fact that it did not attain the condition of supersonic operation with supersonic flow over the entire rotor blade span. This performance deficiency was attributed to the fact that the contraction ratio of the rotor (the ratio of the flow area between the rotor blades at the stage entry to the minimum area of the flow passage between the rotor blades within the stage) was too high, caused in part by the presence of a thick subsonic boundary layer that constricted the flow.[42]

In addition to this work on the compressor, problems were also encountered in the combustor and turbine. Poor combustor operation resulted in afterburning and unstable combustion. Turbine development problems centered on finding turbine materials that had suitable high temperature life and durability. In mid-1949, Flader delivered two derated engines to the Air Force that produced only 450 lb thrust. Although accepted by the Air Force, these two engines were unacceptable for use in the XQ-2 target vehicle.[43]

After further development work, two tests were made with an improved model of the engine in January 1952 to demonstrate to the Air Force the engine's ability to achieve the contractually specified thrust of 700 lb. The first test was performed on January 24, 1952, during which the specified thrust was achieved. The second and final test of the engine was performed a few days later on January 31, and was observed by W. J. Brown, T. Perry, and W. D. Downs of

41. Hartmann and Tysl, "Performance of Axial-Flow Supersonic Compressor of XJ55-FF-1 Turbojet Engine. III," 1-6.

42. Robert C. Graham and Melvin J. Hartmann, "Performance of Axial-Flow Supersonic Compressor of XJ55-FF-1 Turbojet Engine. IV—Analysis of Compressor Operation Over a Range of Equivalent Tip Speeds from 801 to 1614 Feet per Second," (Research Memorandum Number 0SE9J14, Lewis Flight Propulsion Laboratory, National Advisory Committee for Aeronautics, Cleveland, Ohio, 1949), 7-10.

43. Downs, "Development of the XJ55-FF-1 Engine," 64.

the Air Force.[44] During this final test run, the contractually specified thrust of 700 lb was achieved for a period of one minute, as certified by the Air Force observers. Unfortunately, it had been necessary to exceed the design turbine temperature to achieve this level of thrust, and after one minute of operation at this condition, the turbine failed and the engine literally blew up.[45]

Although the engine had met its specified thrust, a number of problems existed that made it unacceptable in its then current form. The compressor performance was far from acceptable. Care was required when accelerating the engine to avoid compressor surge. The compressor's low efficiency at design speed resulted in an sfc of approximately 2.0, more than 20% above the design specification. In addition, the engine exceeded its weight specification, had no self-contained controls, fuel pump or lubrication system, and it exceeded the size of its competitor engines. Fredric Flader estimated that further work to meet the engine performance, weight, and size specifications would require an additional three years.[46]

At that time, the Fairchild YJ44 and the Marboré II (newly acquired by CAE from Turboméca of France) were competitor engines available for use in the Air Force XQ-2 target vehicle. However, modification of the XQ-2 was required to accommodate either of these engines in an aerodynamically optimum configuration. Past experience had demonstrated that airframe development time was less than development time for engines. Thus, due to the urgency of proceeding with the XQ-2, the Air Force concluded that it would be more advantageous to modify the XQ-2 to accommodate either the XJ44 or Marboré II than to pursue further development work on the XJ55-FF-1. In February 1952, the Air Force Wright Air Development Command recommended that further work on the engine be brought to a conclusion as soon as possible. Accordingly, the contract with Flader was terminated soon afterward.[47] Cancellation of the XJ55-FF-1 engine terminated Flader's development of aircraft gas turbine engines.

Fredric Flader, Inc., also developed two industrial gas turbine engines during the early 1950s. One of these engines was developed to drive a compressor that would pump natural gas in a pipeline. The engine developed 5,500 shp, had a 10-stage axial-flow compressor, and a two-stage turbine. The engine was satisfactorily demonstrated, but was never used.

44. Downs, "Development of the XJ55-FF-1 Engine," 29.

45. Sarginson to Leyes and Fleming, 21 September 1988.

46. Downs, "Development of the XJ55-FF-1 Engine," 59.

47. Downs, "Development of the XJ55-FF-1 Engine," 59-60.

The other engine was a two-stage axial-flow gas turbine developed for the Linde Air Products Company to drive an air compressor. One such engine was installed and performed satisfactorily in a Linde plant. However, it was seldom used because the operators were fearful of a machine operating at such a high rotating speed. It finally destroyed itself during an electrical power outage when the electrically driven oil pump in the lubrication system shut down with the gas turbine still running.[48]

Following discontinuation of the XJ55-FF-1, Flader also performed a variety of engineering consulting tasks for the Wright Aeronautical and Allison companies. In 1955, Flader and its plant in Tonawanda, N.Y. were taken over by the Eaton Manufacturing Company. Eaton wished to use Flader as an R&D and prototype manufacturing plant to produce small batches of high alloy turbine blades. In 1957, Eaton decided to discontinue use of Flader's facilities. Fredric Flader, Inc. was terminated on Labor Day in 1957.

Fredric Flader's contribution to small gas turbine aircraft engines was his effort to develop the first engine using a supersonic compressor. Unfortunately, physical realities of the technology led to failure of his endeavor. Nevertheless, the development effort on the XJ55, which benefitted from supersonic compressor design data provided by engineers at NACA's Lewis Flight Propulsion Laboratory in Cleveland and testing of the compressor at the laboratory, led to the conclusion that the supersonic compressor was not the path to follow. In fact, during the tests at Cleveland, NACA engineers observed that efficiency of the compressor improved when operating at transonic flow conditions. Through later research by NACA and others, transonic compressors were perfected, and by 1990, were used in nearly every small axial-flow gas turbine aircraft engine. It is interesting to note that, during the early 1970s, Lycoming successfully demonstrated a high pressure ratio compressor with two axial stages and one centrifugal stage in which the second axial stage was supersonic with the flow fully sonic from the hub to the tip of the rotor blades.

Boeing

As early as 1943, when the existence of the turbojet was finally unveiled in the U.S., the Army Air Forces became interested in the possibilities of using gas turbine engines in large aircraft, a type of plane in which Boeing had specialized. Further, if large jet-powered military aircraft proved successful, it was recog-

48. Sarginson to Leyes and Fleming, 21 September 1988.

nized that the airlines would want them. As a result, Boeing began considering the development of large jet-propelled aircraft.

To make plans for the future, it was essential to understand jet engines and to have answers to the problems concerning them. Thus, Boeing sent its engineers into the field to search for information about this new type of engine. What they found was conflicting information about gas turbines and a lack of agreement among manufacturers and laboratories on such things as fuel economy, life expectancy, safe operating temperatures, and maximum thrust or power attainable from gas turbines. Dissatisfied with the information they were able to find, Boeing set out to develop answers for itself by initiating its own jet propulsion research.

In June 1943, Boeing inaugurated its study of jet propulsion. A decision was made to build and test various gas turbine engine components. To support that effort, development of a jet propulsion laboratory was begun at Boeing's Plant 1 in Seattle. The laboratory contained engine and component test cells, machine shops, and an assembly area. Initially, small axial- and centrifugal-flow compressors and other engine components were built and tested. Eventually, this research was expanded to embrace the assembly and testing of complete experimental power units, culminating in the design and construction of two small gas turbines, the Model 500 turbojet and the Model 502 turboprop. The first complete engine, the Model 500, was assembled in 1947.[49]

Although this work began as a project in pure research to acquaint Boeing engineers with the problems of gas turbine engines, it became apparent that these engines were significant in themselves. Thus, development of these two engines proceeded through the mid- and late-1940s. The first engine to run was the Model 500, a 150-lb-thrust turbojet weighing 85 lb. That engine was closely followed by the Model 502, a 160-shp free turbine turboprop engine weighing 140 lb and having a 1.8 sfc. In 1947, Boeing publicly announced these two engines, offering them for a variety of uses. The Model 500 turbojet was considered suitable as a starter engine for larger jet engines and also as a propulsion system for missiles, glide bombs, and small aircraft. The Model 502 turboprop was offered for use in light aircraft, although it was anticipated that early use would be as an auxiliary power unit in large aircraft and as a ground power unit for stationary and vehicular applications. The concept of using the engine as a helicopter power plant had not as yet materialized.[50]

The development of these engines benefitted from the close cooperation received from Boeing's several engineering units. These units included the acous-

---

49. Gordon S. Williams, "Propulsion," *Boeing Magazine*, Vol. XVII, No. 7 (July 1947): 9-11.

50. "Boeing Develops Baby Turbojet," *Aviation Week*, Vol. 47, No. 4 (July 28, 1947): 31.

tics, vibration, electronics, process, and metallurgy groups, each with its own laboratory and specialists. These groups, together with Boeing's jet propulsion laboratory, resulted in a research organization considered by Boeing to be unparalleled at that time by any other potential U.S. aircraft turbine manufacturer.[51]

MODEL 500 TURBOJET

The Model 500 engine was a single-shaft turbojet with a single-entry centrifugal compressor stage, two through-flow can-type combustors, and a single-stage axial turbine. The Model 500 engine, which was first run in 1947, was rated at 150 lb thrust, had a 1.3 sfc, and weighed 85 lb.[52] By 1954, the thrust had been uprated to 195 lb, and engine weight had increased to 120 lb. Performance improvements came principally as a result of modifications in compressor and turbine design.[53] Although development and marketing of this engine continued until the mid-1950s, no applications materialized for it. Consequently, Boeing's small engine work became focused on its free turbine turboprop and turboshaft engines, derivatives of the Model 500.

MODEL 502 TURBOPROP AND TURBOSHAFT

The Model 502 was developed initially as a turboprop and later as a turboshaft engine. Of these two, the turboshaft version became most widely used. This engine used the Model 500 power section with the addition of an opposed single-stage free power turbine and a gearbox. Simplicity, ruggedness, and low cost guided the design of Boeing's engines, hence the Model 502's single-stage power section compressor and turbine stages on one shaft with the opposed single-stage free power turbine and gearbox on another.

Initially run in 1947, the first version of the Model 502 weighed 140 lb and produced 120 shp with a 1.8 sfc. By 1949, at which time Boeing delivered six production engines to the Navy, the power output had been increased to 160 shp, and the sfc was reduced to 1.5. A number of refinements were found necessary to make the engine useful, principally a complete aerodynamic redesign and thickening of the turbine blades, structural and aerodynamic modification of the compressor, complete redesign of the accessory section and gear box, and

51. Edward. G. Wells, "Packaged Power," *Boeing Magazine*, (December 1949): 3-5, 14.

52. *Jane's All the World's Aircraft—1948* (New York: The MacMillan Company, 1948), 67d-68d.

53. *Jane's All the World's Aircraft—1954-55* (London: Jane's All the World's Aircraft Publishing Co. Ltd., 1955), 336-337.

Boeing Model 500 turbojet engine. National Air and Space Museum, Smithsonian Institution (SI Neg. No. 93-16137).

the addition of a third engine bearing. In addition, a major share of attention was given to the fuel control system to provide a reliable unit capable of satisfying the control requirements.

After 10 years of development beginning with the first laboratory test engine, the power output had doubled as represented by the Model 502-10C turboshaft version of the engine. Power output had increased from 120 to 240

Boeing Model 502-11B turboshaft, which was an early free power turbine engine. National Air and Space Museum, Smithsonian Institution (SI Neg. No. 71-2657).

shp, sfc had decreased from 1.8 to 1.02, and weight had grown from 220 to 320 lb. Overhaul life and production cost was approaching those for piston engines.[54] Further improvements that were embodied in the Model 502-10VC turboshaft by the mid-1960s had further increased the output to 300 shp and reduced the sfc to 0.98.

Although Boeing had financed the design and initial development of the Model 502 engine, the Navy Bureau of Ships began sponsoring the engine's further development, beginning in 1947, with a view to its potential use as a power source on naval vessels.[55] At that time, it was given the military designation of T50. Stimulated by the Navy's interest in its engine, Boeing gave increased attention to the further development and marketing of its small gas turbine. That growing attention was signaled by the appointment of Jay Morrison, Staff Assistant to William M. Allen, Boeing President, as the Manager of the gas turbine project, giving the program the status of an independent division.[56]

The first production contract for the Model 502 was received by Boeing in 1950 for engines to be used on Navy minesweepers. On these ships, two Model

54. Wallace E. Skidmore, "Evolution of a Gas Turbine," (SAE Paper No. 31 presented at the SAE Annual Meeting, Detroit, Michigan, 14-18 January 1957), 1-8.

55. S. D. Hage and Donald W. Finlay, "Gas Turbines for Lightplanes," *Flying*, Vol. 47, No. 2 (August 1950): 54-55.

56. "Boeing Readies Turbine for Market," *Aviation Week*, Vol. 52, No. 15 (April 10, 1950): 45.

The world's first gas turbine-powered helicopter, the Kaman K-225 powered by a Boeing Model 502-2E turboshaft engine. This helicopter is in the National Air and Space Museum collection. National Air and Space Museum, Smithsonian Institution (SI Neg. No. 93-16138).

502-6 engines fed power to a combining gearbox that drove a power generator. The Model 502 also had several other non-aircraft propulsion uses, mainly in turboshaft-powered air supply carts for starting aircraft turbine engines, as the power source for fire truck water pumps, and as the power plant for a 60,000-lb truck and a locomotive.

Soon after receiving the production contract from the Navy for the ship-borne version of the engine, a turboshaft version of the Model 502 was used to power a helicopter. On December 10, 1951, the Kaman K-225 helicopter powered by the Model 502-2E (YT50) turboshaft made its initial flight, becoming the world's first gas turbine-powered helicopter.

The stimulus for development of the K-225 came from a discussion between Charles Kaman and a group of his Navy friends in Washington that took place around 1950. Kaman extolled the benefits that the helicopter industry would accrue once the gas turbine became available to power helicopters. One of the group members challenged him with, "Well, why don't you do something about it?"[57]

That challenge led Kaman's engineers to the discovery of the Model 502 engine as a suitable power plant for a light helicopter. The engine was installed

57. Charles H. Kaman, *Kaman: Our Early Years* (Indianapolis, Indiana: The Curtis Publishing Company, 1985), 85-90.

in the K-225, which was identical to the Kaman K-190 except that the piston engine was replaced by the Model 502 turboshaft. Kaman submitted a proposal to the Navy's Bureau of Aeronautics for construction of the K-225, an experimental helicopter. According to Kaman's history of his company's early years, "That proposal pointed out that, short of any production potential with these particular hardware elements, the opportunity to demonstrate the unquestionable contribution of turbine power to helicopters was at hand."[58] Three months after receipt of a contract from the Bureau of Aeronautics, the K-225 flew for the first time. As a result of its use in the successful demonstration flights during 1951, Boeing's turboshaft engine played a key role in introducing the era of turbine-powered helicopters. The original K-225 helicopter is part of the National Air and Space Museum's collection in Washington, D.C.

Later, the Kaman HTK-1, a twin-engine helicopter, was experimentally powered by two Model 502-2 engines. First flown with the Model 502 engine in March 1954, the HTK-1 became the first twin turbine-powered helicopter to fly.[59]

In 1952, Kaman began flight tests in a pilotless helicopter program sponsored by the Navy. The purpose of those tests was to demonstrate the feasibility of using gas turbine-powered drone helicopters for Anti-Submarine Warfare (ASW) surveillance. That work led to the subsequent issuance of a Navy requirement for such a helicopter and a procurement competition between Kaman and Gyrodyne of Long Island. Gyrodyne won the competition and developed the Model 502 (military designation T50) powered DSN-3 for the Navy's Drone Anti-Submarine Helicopter (DASH) program. The three key reasons for selection of the T50 for the DSN-3, as expressed by Boeing, were reliability, low cost, and availability.[60]

The DSN-3 drone helicopter powered by the 270-shp Model 502-10V (military designation T50-BO-4) was first flown on April 6, 1961. Later, the DSN-3 was redesignated the QH-50C, which used the uprated 300 shp Model 502-10VC (military designation T50-BO-8A) engine. The Model 502-10VC was the first U.S. turbine engine in the 200- to 350-shp range to receive an approved 150-hour qualification rating.[61]

58. Kaman, *Kaman: Our Early Years*, 87.

59. Kaman, *Kaman: Our Early Years*, 85-87, 90-91.

60. Richard L. Sweeney, "Boeing's Turbine Family," *Flying*, Vol. 70, No. 2 (February 1962): 34-35, 111-114.

61. *Jane's All the World's Aircraft—1963-64* (London: Sampson Low, Marston & Company, Ltd., 1963), 325-326, 476.

Installation of a Boeing Model 502-8 engine in the Cessna XL-19B Bird Dog, the world's first turboprop-powered light aircraft. National Air and Space Museum, Smithsonian Institution (SI Neg. No. 93-16141).

The turboprop version of the Model 502 was installed initially in a Cessna XL-19B Bird Dog, which became the world's first turboprop-powered light aircraft. The standard piston engine-powered version of the Bird Dog, the L-19, had distinguished itself as a liaison and observation plane in the Korean conflict. Flight tests of the XL-19B using the 210-hp Model 502-8 engine began in November 1952. The XL-19B set a world's light plane altitude record of 37,063 feet on July 16, 1953. Development of that experimental turbine-powered light airplane was a joint development project of the Cessna and Boeing companies and the Army and Air Force. By 1960, the 300-shp Model 502-10F turboprop was being used to power the Army's rocket-launched Radioplane RP-77D, a radio-controlled aerial target for gunnery and missile personnel training.[62]

By the time production of the Model 502 was completed in 1965, a total of 1,500 engines of all types had been produced, including turboshaft, turboprop, and ground power versions of the engine.[63]

MODEL 520 TURBOSHAFT

In early 1954, work began on the design of the Model 520 series engines, developed using both Navy and Boeing funds. The initial version for helicopter applications was the 430-shp Model 520-2 (military designation T60-BO-2). The engine had a double-entry centrifugal compressor, two reverse-flow combustion chambers, a single radial gas generator turbine, and a single axial free turbine driving the power takeoff. The new double-entry compressor provided an increase in airflow and pressure ratio of about 25% over those for the Model 502 engine.

Three turboshaft and two turboprop versions of the Model 520 were developed with rated power outputs ranging from 430 to 550 shp with sfcs from 0.72 to 0.65 respectively.[64] Although the Model 520 series engines had been successfully developed and the Model 520-2 had passed its 50-hour PFRT by 1960, the Navy's planned applications for the engines did not materialize. Consequently, the engines never reached production and further development was discontinued in the early 1960s.

62. S. Daniel Hage, "Boeing Aircraft Gas Turbines," *Newsletter*, American Helicopter Society, Inc., Vol. 6, No. 11 (November 1960): 3-4.

63. *Jane's All the World's Aircraft—1966-67* (London: Sampson Low, Marston & Company, Ltd., 1967), 521.

64. Hage, "Boeing Aircraft Gas Turbines," 5-6, 9.

MODEL 550 TURBOSHAFT

The 365-shp Model 550 (military designation T50) engine, introduced in the early 1960s, was a direct derivative of the Model 502. Developed initially as a much improved and lighter version of the Model 502, the engine was re-designated the Model 550. The engine embodied several design changes: a single axial stage was added ahead of the single-sided centrifugal stage, the two reverse-flow can-type combustors were redesigned, and a radial-type turbine was used to drive the compressor. Comparing the Model 550 with the Model 502-10VC, the design changes raised the rated power output by 65 shp, decreased engine weight by 84 lb, and reduced the sfc from 0.98 to 0.80.

The Model 550 was put into production in the mid-1960s, and the military version of the engine, the Model 550-1 (military designation T50-BO-12), was used to power the Gyrodyne QH-50D drone, which had been uprated to provide increased range and payload over those of the QH-50C. The QH-50D was the final production version of the Gyrodyne helicopter. Production of the Model 550 continued until April 1968 to supply engines for the QH-50D, the engine's sole application.[65]

SALE OF THE TURBINE DIVISION

By the end of 1965, the Turbine Division staff had grown to 1,800 people. Boeing had invested heavily in the development of a gas turbine manufacturing and test facility with the capability for fabrication, assembly, and testing of 100 gas turbine engines per month. In addition to a diversified array of manufacturing equipment, the facility included 14 engine test cells, component test rigs, and a turbine spin pit.

However, Boeing's Turbine Division had experienced an up and down history accompanied by a lackluster track record. By 1965, it had recorded 17 consecutive unprofitable years and had gained the reputation as a graveyard for a procession of managers. At that time, Bill Allen, Boeing's President, offered Malcom Stamper, an outstanding engineer, the job of running the division and turning it around.

Stamper had come to Boeing in 1962 after spending 24 years at General Motors. In 1965, the Turbine Division was one of several jobs offered to him, in addition to several benefits, to counter an offer Stamper had received from another company. Stamper selected the Turbine Division, viewing it as a challenge. Al-

65. *Jane's All the World's Aircraft—1968-69* (New York: McGraw-Hill Book Company, 1968), 455, 639.

though it had a terrible track record, it was doing $25 million worth of national and international business annually with both military and commercial customers, and he felt it could become profitable.

When Stamper accepted the job, Allen told him, "Make it the best in the world or let's get rid of it."[66] During its years of operation, Boeing had never made a full commitment to the division, but Allen gave Stamper that commitment when he accepted the job. However, when Stamper told him a year later that the division needed a $74 million transfusion for a complete turnaround, Allen pronounced its death sentence, "We don't have the money. We're going to build the largest airplane in the world (the 747), and we need all our resources for that program. Sell it."[67]

On January 25, 1966, Boeing announced plans to phase out its gas turbine engine operations over a period of several years. All production commitments were to be met and arrangements were to be made for support of equipment in service as long as required. At that time, Boeing had about 1,200 turbine engines in service for a wide variety of applications. In addition, members of the staff were assured of continued employment with opportunities for reassignment to other divisions of the company as the phaseout progressed.

A phase-out group was established with orders to assess all turbine contracts to determine whether they should be canceled or completed. Each turbine customer was visited, given an explanation of why Boeing was closing down the division, and promised support of the products that they had bought already.[68]

On December 7, 1966, an agreement was reached between Boeing and the Caterpillar Tractor Co. for the continued development, manufacture, sales, and service of Boeing's Model 551/553 engines.[69] These engines represented a series of advanced vehicular turbines that Boeing had planned for production in the 1970s. However, design and development work on the engines was terminated following the announcement of plans to phase out of the turbine business.

Assembly of Boeing's last production turbine engine, a Model 550-1 (military designation T50-BO-12), was completed in April 1968. On March 1, 1969,

66. Robert J. Serling, *Legend and Legacy—The Story of Boeing and Its People* (New York: St. Martin's Press, 1992), 243-244.

67. Serling, *Legend and Legacy—The Story of Boeing and Its People*, 244.

68. Serling, *Legend and Legacy—The Story of Boeing and Its People*, 243-244.

69. "Basic Agreement Between The Boeing Company and Caterpillar Tractor Co. Relating to Model 551/553 Engines," 7 December 1966, National Air and Space Museum Archives, Washington, D.C., 1-10.

Boeing signed an agreement with Steward-Davis, Inc., transferring all rights to develop, manufacture, sell, and service Model 502 and T50-BO-10/12 engines to Steward-Davis, Inc. Steward-Davis was a major turbine overhaul and repair firm located in Long Beach, Cal.[70]

After 25 years of gas turbine research, development, and production and having become a recognized manufacturer of small turbine engines, Boeing discontinued its gas turbine business. In addition to the early and continued use of their engines in helicopters and aircraft, Boeing's gas turbines provided auxiliary power for ground-based and ship-board applications. For example, the cart-mounted version of the Model 502 engine had become widely used by airlines to start their aircraft engines pneumatically.

Boeing was a true pioneer in the field of aircraft gas turbines. In the early 1950s, Boeing engines led the vanguard of turbine-powered helicopters and light aircraft as well as the application of the gas turbine to a variety of marine and land based uses. The Model 502 turboshaft was used in the world's first turbine-powered helicopter, and the turboprop version of the engine powered the world's first turboprop-powered light aircraft.

Fairchild

It was in the mid-1940s that Fairchild's engine division began design studies of a small, expendable turbojet engine. By that time, the Fairchild Engine and Airplane Corporation had a background of nearly 20 years experience in the design, development, and manufacture of aircraft engines.

On February 9, 1920, after several years of experimentation in the aerial camera field, Sherman M. Fairchild formed the Fairchild Aerial Camera Corporation. On September 26, 1925, the name of the company was changed to the Fairchild Aviation Corporation to reflect its status as an aviation company. At that time, Fairchild became interested in aircraft engines, having acquired the rights to develop and manufacture the Caminez cam drive engine. The engine was successfully developed, and two experimental models were flown in 1928 on a demonstration tour. However, in 1929, plans to market that engine were abandoned. With development of the cam drive engine, engineering and manufacturing facilities of its Engine Division were established in a small, modern factory at Farmingdale, N.Y., on Long Island.

70. "Agreement Between The Boeing Company and Steward-Davis, Inc. Relating to Model 502 and Model T50-BO-10/12 Engines," 1 March 1969, National Air and Space Museum Archives, Washington, D.C., 1-28.

Soon afterward, work began on the design and development of two Ranger engine models. The first was a 6-cylinder, inverted, in-line, air-cooled engine (Models 6-370 through 6-410) and the second, design of which began in 1932, was a 12-cylinder, inverted V (Model V-770), in-line, air-cooled engine. A limited number of these engines were manufactured during the 1930s, most used in Fairchild aircraft. In 1939, a quantity order was received from the Army Air Corps for the 175-hp 6-cylinder L-440 engine used in the Fairchild PT-19 primary trainer. That order was followed in 1940 by an order from the Navy for the 565-hp 12-cylinder V-770 engine used in the Curtiss SO3C-1 scout airplane. Production of these engines for the military continued until 1944 when the Army Air Forces canceled its training program, and the Navy canceled its scout plane. During World War II, the company also produced parts for the Rolls-Royce Merlin aircraft engine, manufactured a piston engine-powered auxiliary power plant used in U.S. bombers and military transports, and 5-inch-diameter rocket motors for the Navy.[71]

Soon after World War II, as an outgrowth of a Navy contract for development of the XH-1850, an experimental 24-cylinder, air-cooled engine, Fairchild performed a design study of various turboprop engine configurations. That was followed by the first phase of design and development of the XT46-R-2, a 9,850-shp turboprop engine with multiple power sections driving a single gearbox. It went no further than the initial design and development phase. At the same time, Fairchild recognized a need for a low-priced expendable turbojet engine for guided missiles, and, in 1946, began preliminary design studies of such an engine. These led to the award of a contract by the Navy, in June 1947, for the design and development of an expendable turbojet engine for an air-to-underwater torpedo-carrying missile.[72] That engine became the J44.

In addition to the Fairchild Engine Division's work on turbojets, Fairchild's Stratos Division began development of a gas turbine-powered auxiliary power unit in 1951 as a step to broaden its position as an equipment supplier. That unit was developed from a French gas turbine of 160 shp for which Stratos had obtained sole U.S. production and sales rights.

71. General History of Fairchild from 1920 to 1951, 1-37, Fairchild Public Relations Collection, Accession Number 1989-0060, Box 2, National Air and Space Museum Archives, Washington, D.C.

72. Ranger Aircraft Engines Division of Fairchild Engine and Airplane Corporation, "Proposal for Production Study of Expendable Turbojet Engine (in response to) United States Air Force, Air Materiel Command Request for Quotation, Purchase Request No. 96037," (11 May 1949), 1-2.

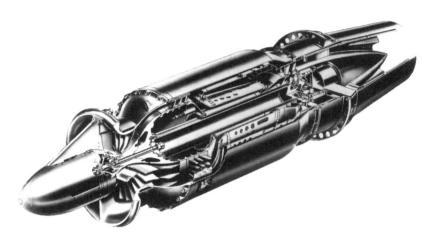

Cutaway of the Fairchild J44 turbojet engine used in the Ryan Firebee target drone. National Air and Space Museum, Smithsonian Institution.

J44 TURBOJET

The J44 engine was created under rigid design limitations of weight, size, and thrust in conformance with the requirements of the Navy missile that it was to power. It was the first U.S. turbojet engine designed from the start for expendability. The engine was 22 inches in diameter, weighed 280 lb, and had a sea-level-rated thrust of nearly 1,000 lb. It was required that the engine have a demonstrated life of 10 hours.

Basic factors influencing the engine's design were the requirements for simplification of manufacturing to achieve low cost, and minimum use of strategic materials. To simplify the production process, manufacturing process engineers worked with the design engineers from the initiation of the design. Steps were taken to design the engine with comparatively few assemblies and to eliminate machining operations wherever possible. For example, the single-stage mixed-flow compressor used in the engine was manufactured from an aluminum casting requiring no profiling of blade contours. The diffuser section following the compressor was a single-piece casting of magnesium with integrally-cast curved vanes that required no machining. The combustor used a single annular basket fabricated from sheet metal with no machining required. Commercially available oil burner-type fuel nozzles were used in the combustor. The turbine nozzle diaphragm was a one-piece steel casting with integrally cast blades. The single-stage axial turbine rotor consisted of a forged steel disk to which the turbine blades were welded. The rotor shaft connecting the compressor and turbine was a piece of stainless steel tubing with flanges electrically flash welded to each end.[73] Because of its simplicity to manufacture, only three months were required to set up tooling for the J44's first production line.

One of the military requirements was that the engine be enclosed in a shell having a smooth outer surface, eliminating the need for additional cowling in missile applications. Fairchild engineers reasoned that they might as well put this enclosure to work. The result was a structural monocoque aluminum alloy shell, which formed a pressure chamber around the engine and eliminated the need for an internal main frame: the shell constituted both the cowling of the engine and its main supporting structure.[74]

Design of the engine began in June 1947, and the first complete engine was run in August 1948. A production order was received from the Navy in January 1948, and delivery began in early 1949. During 1949, the J44 underwent service tests, and it made its first flight in 1950 powering the Petrel air-to-underwater, torpedo-carrying missile developed by Fairchild. The missile was propelled by the J44 as a winged aircraft until it reached the point at which the torpedo was to be launched into the water. At that time, the wings and engine were blown off, and the torpedo entered the water.

By 1950, the J44 had also been selected to power the Ryan Firebee high-speed target drone being developed jointly by the Army, Navy, and Air Force. The first flight of the Firebee powered by the J44 was in 1951. During 1951 and 1952, hundreds of flight tests were conducted using the J44 in the Navy's missile and in the Firebee drone. The J44-powered Firebee target drone was operational by the mid-1950s, and the Petrel missile reached operational status in 1956.[75]

The intense simplification of the engine design resulted in a number of problems related to manufacturing practices. Because engine test runs and flights of missile and drone engines were relatively short, test time on the engine built up slowly. Thus, many of the problems encountered did not show up until a quantity of engines had been produced.

The principal problems encountered involved the turbine nozzle box, compressor outlet diffuser, engine rotor shaft, and lubrication system. Although the

73. Ranger Aircraft Engine Division, "Proposal For Production Study of Expendable Turbojet Engine (in response to) United States Air Force, Air Materiel Command Request for Quotation," 2-4.

74. Irving Stone, "Fairchild Stresses Simplicity in J44 Engine," *Aviation Week*, Vol. 59, No. 22 (November 30, 1953): 30-39.

75. "Fairchild J44 Turbojet Engines," *Thrust* (A Quarterly Periodical Reporting on Jet Power Developments at Fairchild Engine Division, Fairchild Engine and Airplane Corporation, Deer Park, Long Island, N.Y.), Vol. 1, No. 1 (Summer 1957): 11.

Ryan Firebee target drone powered by Fairchild J44 turbojet engine. National Air and Space Museum, Smithsonian Institution.

initial cast turbine nozzle box was satisfactory during early engine tests, as soon as production was initiated, difficulty was experienced due to variations in nozzle flow area that caused a mismatching between the turbine and compressor. This problem was caused by the fact that the engine could not tolerate the variations permitted in production casting tolerances. The solution was to machine the surfaces of the guide vanes, bend them in an arbor press, and check the flow areas in a flow rig.

A similar problem was encountered with the cast compressor outlet diffuser. After producing a quantity of engines, it was found that the small variations in the flow area between the vanes of the diffuser often resulted in compressor surging and mismatching between the compressor and turbine. It became necessary to machine the passages between the diffuser vanes to obtain adequate dimensional control of the diffuser flow area.

The seamless steel tube with its welded flanges used for the engine rotor presented a serious balancing problem due to wall thickness variations, as much as

.022 inches. To solve this problem, heavy wall tubing was procured with flanges formed on each end. The flanges were then machined and the tubing was machined both inside and out. Surprisingly, the cost of this shaft in production was only one-half that of the original finished part.

The oil system was a throw-away type system carrying sufficient oil for a 10-hour flight. Oil was fed from a tank pressurized with air bled from the compressor discharge. The oil flowed through tubing that had an orifice at each of the two engine bearings. Although this system worked satisfactorily at low altitudes, when flying at altitudes of 40,000 feet and above, the greatly reduced pressure in the oil tank made it necessary to select orifice sizes for each flight condition to obtain proper oil flow to the bearings. The solution was to use the compressor discharge air to produce a fine oil mist within the oil tank that was conveyed through tubes to the two bearings. This system was relatively insensitive to outside pressure and temperature.[76]

By 1955, Fairchild had accumulated over 2,200 hours of running time on J44 engines. Much of this operating time had been accumulated in the company's test facilities at Deer Park. In addition, qualification and altitude calibration tests were run at the Aeronautical Engine Laboratory of the U.S. Naval Air Material Center located in Philadelphia. An official 12.5-hour Qualification Test of the XJ44-R-10 engine model used in the Petrel and Firebee was completed on December 9, 1952. An improved version of the engine, the J44-R-20, passed its official 12.5-hour Qualification Test on October 11, 1954.

With a view to applying its J44 experience to the development of a thrust assist or main propulsion unit for manned aircraft, the development of such an engine, the J44-R-3 (commercial designation FT-101E), was begun in 1954 as a corporation sponsored project. The objective of this project was to develop from the J44-R-20 an engine having improved durability and performance. The aerodynamic design features of the J44-R-20 were incorporated into the J44-R-3 engine. However, the J44-R-3 was altered structurally to increase its life. These alterations included a machined, forged inducer, an Inconel combustion chamber, Vitallium turbine nozzle blades, carburized accessory gears, a heavy turbine blade footing, improved bearings, and an Inconel tailcone. The J44-R-3 had a sea level maximum thrust of 1,000 lb with a 1.55 sfc.[77]

76. A. T. Gregory, "Small Turbojet Engines—A Big Factor in Aviation," (SAE Paper No. 499 presented at the SAE Golden Anniversary Aeronautic Meeting, New York, 18-21 April 1955), 1-4.

77. "Fairchild J44-R-3," (Paper presented at the J44-R-3 Engine Conference, Fairchild Engine Division, Fairchild Engine and Airplane Corporation, Farmingdale, Long Island, N.Y., 14 June 1955), 1-15.

Qualified for use in manned aircraft in 1955, the J44-R-3 became the first jet engine in its power and weight class to receive an Approved Type Certificate from the Civil Aeronautics Administration. Between 1954 and 1955, the J44 powered the Bell Model 65 VTOL Air Test Vehicle in a series of experimental flights. That vehicle is in the National Air and Space Museum collection. The Model 65 was powered by two J44 engines that were rotated to provide vertical thrust for takeoff and horizontal thrust for forward flight. In 1954, a J44-R-3 was mounted atop the fuselage of a Fairchild C-82 cargo aircraft to provide added thrust for takeoff and a safety margin in the event of a main engine failure during takeoff. In April 1957, a C-82 with a J44-R-3 mounted on top of the fuselage was commissioned by TWA as a flying maintenance base for use in Europe. The success of the C-82 installation led to the mounting of a J44-R-3 on each wing tip of the Fairchild C-123B cargo aircraft in 1955 to improve its take-off and climb performance.[78] Although the engine performed satisfactorily in these three applications, its use in them was limited, and there were no subsequent demands for the engine.

A major use of the J44 during the 1950s was in the Firebee target drone used by the three military services. In addition to the J44-R-20 engine, the Teledyne CAE J69 turbojet was also used in the Firebee. However, by the late 1950s, use of the J44 in the Firebee was phased out because its specific fuel consumption could not compete with that of the J69 engine. Production of the Petrel missile was also phased out by that time. About 1958, Fairchild received its final order for J44s, and the last batch of J44 engines and spare parts were manufactured in 1959.

XJ83 TURBOJET

In December 1952, the Air Force began work on the concept of decoy missiles that could simulate the radar image of its large bomber aircraft. One concept was for a ground-launched decoy missile with intercontinental range. The other was for an air-launched decoy with a range approaching 500 miles. Contracts for development of these two decoys were awarded following separate competitions for each. In December 1955, a contract was signed with Fairchild for development of the XSM-73 (WS-123A) Bull Goose long-range surface-launched decoy missile. In February 1956, McDonnell was selected to develop the GAM-72 (ADM-20A) Quail air-launched decoy missile.[79]

78. "Fairchild J44 Turbojet Engines," 6-12.

79. Kenneth P. Werrell, *The Evolution of the Cruise Missile* (Maxwell Air Force Base, Alabama: Air University Press, September 1985), 124-125.

Meanwhile, attention was being given to providing a power plant for these decoy missiles. To minimize development risk, the Air Force elected to award two competitive contracts. Following a procurement competition, in November 1954, a contract was awarded to GE for development of the J85, an engine in the 2,500-lb-thrust class with a thrust-to-weight-ratio goal of 10:1. Fairchild was also awarded a contract to develop a competing engine, the J83.[80] Fairchild had proposed a lightweight engine of relatively conventional design. The proposed GE engine had a more advanced design, involving more risk, but producing a higher thrust-to-weight ratio.

The Goose was powered by the J83. Although the Goose was required to also be able to use the J85, all of its test flights were made using the J83. The Quail was designed only for use of the smaller and lighter J85.[81]

The J83 had a seven-stage transonic axial-flow compressor, an annular combustor, and a two-stage turbine. Its rated sea level thrust was 2,450 lb with a 0.94 sfc, and it weighed 363 lb (YJ83-R-1). The engine was running by early 1957, and in June 1957, it was first flown in the Goose from the Atlantic Missile Range. The J83 made 15 flights in the Goose.

The engine had also been selected to power the Canadair CL-41 trainer aircraft. However, a decision was made in 1958 to power the two prototype CL-41s instead with the Pratt & Whitney JT12 turbojet. Later, production versions of the CL-41 were powered by the Orenda-built GE J85. Thus, by late 1958, the only application for the J83 was the Bull Goose.

The fate of the J83 was sealed in November 1958 when Fairchild's President, Richard Boutelle, was notified by the Air Force Deputy Assistant Secretary for R&D, William Weitzen, that the J83 program was to be terminated immediately. Air Force reasons were that there were neither funds nor program requirements to support the continued development of two engines. Although both the J83 and J85 had problems at that time, it was the opinion of the Air Force that the J85 could reach its performance goals faster and with greater assurance than could the J83. Further, the J85 could satisfy both the Goose and Quail requirements, while the J83 could fit only into the Goose. Finally, in view of those factors plus the fact that the Northrop T-38 trainer powered by J85 engines was programmed to become a prime aircraft system in the Air Force inventory, the J85 was selected to be retained over the J83.[82] An even further nail in the J83's coffin was that, the following month, the Department of Defense

80. Nick Constantine, "The J85 Story," unpublished, undated paper, National Air and Space Museum Archives, Washington, D.C.

81. Werrell, *The Evolution of the Cruise Missile*, 124–125.

82. Constantine, "The J85 Story."

canceled the Goose missile because of budgetary pressures and the fact that the Goose could not simulate the B-52 on enemy radar.[83]

TERMINATION

With cancellation of the Air Force contract and with no other applications for the engine, Fairchild discontinued further work on the J83. Having no forthcoming revenue sources, in January 1959, the Fairchild Engine Division laid off approximately 2,000 of its 2,600 employees. The remaining personnel were retained to complete manufacture of the final batch of its J44 engines and its V-32, a reciprocating engine-powered auxiliary power plant. The company's works, which had been moved to Deer Park, Long Island in the mid-1950s, suspended all operations in the summer of 1959, and the facility was disposed of at that time.[84]

Fairchild was in the small jet engine business for nearly 15 years and was one of the pioneers that contributed to the early demonstration of the small turbojet's viability for both unmanned and manned aircraft. A part of this pioneering work was the early development of casting technology used in the fabrication of J44 parts to minimize the cost of the engine.

West Engineering

Among the early small gas turbine aircraft engines developed in North America was one developed by Edward West, Jr. About 1946, he obtained 6,000 unused war-surplus Type B aircraft engine turbosuperchargers manufactured by GE. He formed the West Engineering Co., Inc. and was awarded a Navy contract to develop a turbojet engine by modifications to the turbosupercharger design. The primary modifications included the attachment of a single reverse-flow combustion chamber that received the air from the turbosupercharger's single-entry, centrifugal-flow compressor and discharged the heated, high-pressure air to drive the single-stage turbine of the turbosupercharger. The modifications also included a fuel injection system, intake and exhaust ducts, and a redesigned turbine to improve efficiency.[85]

83. Werrell, *The Evolution of the Cruise Missile*, 124-125.

84. *Jane's All the World's Aircraft—1959-60* (London: Sampson Low, Marston & Company, Ltd., 1959), 526.

85. Scholer Bangs, "From Supercharger to Turbojet," *Aviation Week*, Vol. 48, No. 15 (April 12, 1948): 23-24.

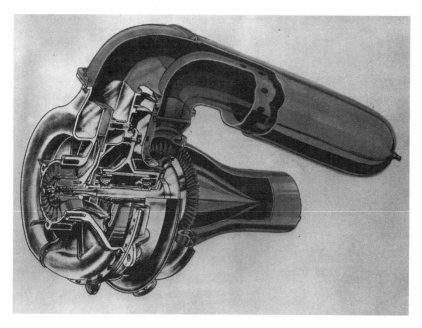

To fulfill a need for a cheap, expendable engine to be used in target drones, the Navy contracted the development of a small, lightweight, experimental jet engine. Utilizing surplus GE Type B turbosuperchargers and adding a combustion chamber, West Engineering Company developed this jet engine. National Air and Space Museum, Smithsonian Institution (15099).

The 1946 Navy contract called for developing the engine through a modified qualification test and delivery of three engines. The engine, designated XJ38-WS-2, was to provide 214 lb thrust at 26,000 rpm. The engine weight was 180 lb and the sfc was 1.7.[86] The Navy was interested in this engine as a cheap, short-life, expendable engine for powering target drones. There was also private interest in using this engine in a light twin-jet racing plane. However, the engine was never flown on any aircraft. Further, the engine was not flight qualified due to the lack of funding from the Navy for qualification testing.

With the objective of increasing the engine's thrust to 400 lb for use in a small twin-engine aircraft that he had conceived, West continued develop-

---

86. E. M. Powers, "Report to the Commanding General, Army Air Forces: American and British Gas Turbine Development," (Former Confidential Report), 1 July 1947, National Air and Space Museum Archives, Washington, D.C., 7 (Tables).

ment efforts until 1950 when he terminated his work due to the lack of finances.[87]

Subsequent Growth of the Industry: Companies that Entered the Small Gas Turbine Aircraft Engine Industry in the 1950s

In the 1950s, a number of other North American companies began dedicated efforts that focused on the development of small gas turbine aircraft engines. Seven emerged to become important components of the industry. They were CAE, Lycoming, GE, Williams International, Pratt & Whitney Canada, Allison, and Garrett. These companies and their engines are addressed in subsequent chapters.

Several other companies began limited development of small gas turbine aircraft engines in the 1950s. They were Dreher, Wright Aeronautical, and Solar.

Dreher

About 1951, Max Dreher, an aeronautical engineer, established the Dreher Engineering Company in Santa Monica, Cal. and began developing miniature turbojet engines. A series of these engines was developed, and by 1969, the latest in this series was the TJD-76A Baby Mamba rated at 55 lb thrust with a 1.5 sfc. The Baby Mamba had a single-stage centrifugal-flow compressor, a straight-through-flow annular combustor, and a single-stage axial-flow turbine. The engine was only six inches in diameter and weighed 17 lb. Test flights of the engine were made on the Prue 215A sailplane, which was used as a flying testbed.[88]

During the early 1970s, the TJD-76A was followed by the -76B/C/D/E models, all with thrust ratings of about 55 lb. The -76B, incorporating several mechanical and aerodynamic improvements, was used for research and development testing. The -76C/D/E models were developed to meet the low-cost and short-life requirements of small drones and other expendable vehicles. The -76D/E models, with performance similar to that of the -76B/C, weighed only about 9.9 lb.[89]

87. Edward West, Jr., Cambria, Cal., to Rick Leyes, returned annotated letter from Leyes to West, 5 November 1996, National Air and Space Museum Archives, Washington, D.C.

88. *Jane's All the World's Aircraft 1969-70* (London: Sampson Low Marston & Co. Ltd., 1969), 733.

89. *Jane's All the World's Aircraft 1975-76* (London: Macdonald & Co., Ltd., 1975), 739-740.

About 1970, Dreher began developing a higher thrust derivative of the Baby Mamba. The TJD-79A Baby Python was rated at 120 lb thrust with a 1.3 sfc.[90] Similar in configuration to the Baby Mamba, it was 11 inches in diameter and weighed 28 lb. The larger Baby Python was considered to be more useful in sailplane applications than the Baby Mamba. Component testing of the TJD-79A continued through the mid-1970s, but further work on the engine was discontinued soon afterward.

Max Dreher worked on the development of his miniature turbojet engines over a period of about 35 years. However, by the mid-1980s, he had not been successful in finding an application for them.[91] No record was found of further work by Dreher on his engines beyond that time.

Wright Aeronautical

Despite the historical heritage of Wright Aeronautical, which goes back to the Wright brothers, and the many outstanding aircraft engines that the company developed and produced from its beginning through World War II, the corporation had a relatively small role within gas turbine aircraft engine industry. The corporation's efforts to enter the small gas turbine aircraft engine industry were unsuccessful. Nevertheless, in the interest of completeness, a brief history of the company's known efforts to market small turbine engines is included.

The Wright Company was organized on November 26, 1909, and its founders included Orville and Wilbur Wright. In the succeeding years, the company grew, and during World War I, it produced approximately 10,000 Hispano-Suiza engines for Allied aircraft. In 1923, the Wright Company merged with the Lawrance Company which had been successful in developing radial air-cooled engines for both the Army and Navy. An important product that emerged from this merger was the Wright Whirlwind Model J-5 engine, which powered the *Spirit of St. Louis* on its history-making flight across the Atlantic. Following a merger in 1929 with the Curtiss Airplane Company, the Curtiss-Wright Corporation was formed. The engine division of Curtiss-Wright became the Wright Aeronautical Corporation, which, in 1952, was renamed the Wright Aeronautical Division.[92] Wright excelled during the 1930s and 1940s in the development and production of radial air-cooled engines. Toward the end of

90. *Jane's All the World's Aircraft 1969-70*, 733.

91. *Jane's All the World's Aircraft 1984-85* (London: Jane's Publishing Co. Ltd., 1984), 866.

92. For the sake of simplicity, the corporation is hereafter referred to as Wright.

World War II, the company had produced over 72,000 R-1820 engines for the Boeing B-17 bomber and 31,000 R-3350 engines for the Boeing B-29.[93]

There was no serious effort by Wright to enter the aircraft gas turbine business until 1950 when President Roy T. Hurley traveled to Britain and obtained licenses from Bristol for its Olympus turbojet engine and Armstrong Siddeley for its Sapphire turbojet engine to manufacture and sell their engines in the U.S.[94] Licenses for the Python, Mamba, and Double Mamba turboprop engines were also obtained from Armstrong Siddeley. Wright did development work on the Olympus (military designation YJ67) for the Air Force, but it was not military qualified. Other Wright military turbine engine projects that were paper designs or which were partially tested, but did not go into production, included the XJ51, XJ59, XJ61, XT35, XT43, XT44, YT47, and XT49. The only turbojet on which Wright did any significant development work and was produced in quantity was the Armstrong Siddeley Sapphire, which was military qualified as the J65.

Company efforts in the small gas turbine aircraft engine business were very limited. In the mid-1950s, Wright obtained a license to market the 1,900-2,460-lb-thrust Armstrong-Siddeley Viper turbojet in the U.S. The engine had a seven-stage axial-flow compressor, an annular combustor, and a single-stage axial turbine. The Viper was given the Wright designation TJ34. Wright offered the Viper for the Ryan Firebee (Q2) target drone, the Fairchild Goose decoy missile, and as a booster engine for the Martin P5M aircraft. Although the Viper had the advantage of being a production and lower-cost engine, it was heavier and had higher fuel consumption than the competing GE J85, Fairchild J83, and P&WA JT12 engines.[95] As a result, Wright was unsuccessful in marketing the Viper. Wright also entered the Army's 1958 competition for a 250-shp turboprop/turboshaft engine, which was won by the Allison design that became the T63. Wright made an unsuccessful proposal offering a turboprop/turboshaft engine designated the SE103.

In the late 1950s, Wright was briefly successful in marketing the 4,850-lb-thrust Bristol Orpheus turbojet. The Orpheus had a seven-stage axial-flow

93. H. L. Linsley, *The Birth of Flight*, unpublished paper reproduced by the Wright Aeronautical Corporation, Service Department Training School, Paterson, New Jersey, 6 July 1944, National Air and Space Museum Library, Washington, D.C. The company history through 1944 is summarized from this report.

94. Sir Stanley Hooker, *Not Much of an Engineer* (Shrewsbury, England: Airlife Publishing Ltd., 1984), 141-144.

95. Wright Aeronautical Division, *Sales Department Organization & Recruitment Program Interim Report*, 30 November 1956, Aviation Hall of Fame of New Jersey Archives, Teterboro, New Jersey.

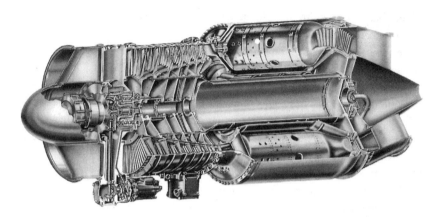

Curtiss-Wright TJ37 turbojet engine. National Air and Space Museum, Smithsonian Institution.

compressor, an annular combustor, and a single-stage turbine. Two Orpheus engines were used to power an early Air Force Utility Transport Experimental (UCX) prototype built by Lockheed and designated CL-329. The CL-329 was designed to be powered by engines located on both sides of the aft fuselage. However, in the mid-1950s, there existed no U.S.-made turbojet engines in the required thrust class. Because Wright was planning to build under license the Bristol Orpheus engine as the TJ37, and pending the availability of domestically-produced power plants, two prototype versions of the CL-329 were each fitted with two British-built Orpheus 1/5 turbojets. The first of these prototypes flew on September 4, 1957.[96]

Unfortunately, reductions in the military budget during 1957 prevented the Air Force from proceeding with its planned acquisition of the UCX aircraft, and the 1958 economic recession in the U.S. delayed development of the business jet market, another market for which a civilian version of the UCX was planned. Wright's TJ37 engines were not military qualified and were competing against at least two engines, the J83 and J85, which were in military sponsored development programs. The original specifications for the UCX were for a four-engine aircraft, and a four-engine aircraft was perceived by some in the Air Force as having greater reliability. Despite these disadvantages, Wright main-

96. The authors recognize that the Orpheus might be considered a large turbojet engine for its time. However, it is mentioned because it was used to power an aircraft that became a business jet, and is therefore within the category of small engines as defined by the authors.

tained that a two-engine aircraft had a lower first cost and lower direct operating cost than a four-engine aircraft.[97]

Because the Air Force was reluctant to acquire an aircraft powered by foreign-built turbojets and because the licensed manufacturing of Orpheus engines by Wright did not materialize, Lockheed began studying four-engine versions of its aircraft using U.S. engines in the 2,500 to 3,000-lb-thrust class that were becoming available. Pratt & Whitney Aircraft JT12 turbojet (military designation J60) engines were selected to test the four-engine configuration, which first flew in January 1960. Successful results led to Lockheed's decision to fit JT12s to production aircraft.[98] This aircraft became the Lockheed JetStar, a successful business jet aircraft; the military version was designated C-140. For Wright, the opportunity to produce an engine for at least one small jet aircraft had been lost.

In 1956, it was announced that the Curtiss-Wright Corporation had organized a new Turbomotor Division for the development of engines of up to 7,500 lb thrust for aircraft, helicopters, missiles, and drone applications.[99] The organization of this division followed from the acquisition of Turbomotor Associates in Hempstead, N.Y. When in full operation, the plan was to split gas turbine operations with Wright concentrating on larger power plants and the new group working on smaller engines. No small turbine aircraft engines are known to have come from this acquisition.

THE DEMISE OF WRIGHT

With a lineage dating from the Wright brothers, Wright had become a leading aviation company by the end of World War II manufacturing aircraft, aircraft engines, and propellers. However, the company did not make a successful post-war transition into the gas turbine era, and it progressively discontinued manufacturing its various product lines. Among the reasons given for the corporation's demise in the aviation business were the failure of its management to recognize the importance of the jet engine market in time to gain a foothold in

97. Wright Aeronautical Division, *TJ37A1*, 28 February 1958, Aviation Hall of Fame of New Jersey Archives, Teterboro, New Jersey.

98. Rene J. Francillon, *Lockheed Aircraft Since 1913* (Annapolis, Md: Naval Institute Press, 1957), 394-396.

99. "Curtiss Organizes Small Turbine Unit," *Aviation Week*, Vol. 64, No. 5 (January 30, 1956): 67, 69.

it and the control of top management by Wall Street bankers and lawyers who were devoted more to profit than to aerospace systems.[100]

In 1965, T. Roland Berner, CEO and President of Curtiss-Wright, decided that Wright would no longer pursue research and development work on gas turbine aircraft engines. According to then Director of Marketing, John C. Schettino:

> Berner was not convinced that the military services would support another major engine company (other than the two major competitors at the time, GE and P&WA) in the business. He was convinced that the advanced research and development that Wright had underway would be picked up by other companies and would not be lost to the industry. He also decided that Wright would continue the maintenance and support of all Wright engines in service operation for as long as the military services desired, but that all research and development effort would be discontinued.[101]

Solar

Solar was involved only briefly with small gas turbine aircraft engines during the late 1950s. However, it is appropriate that the company be recognized as a part of the small gas turbine industry in view of its contributions over more than four decades in the development and manufacture of turbine units for a variety of industrial purposes.

Solar Turbines Inc. evolved from the Prudden-San Diego Airplane Company, founded in 1927 by George Prudden with a syndicate of seven San Diego businessmen. As a result of differences in management philosophy between Prudden and the other investors, he left the company in November 1928, and in March 1929, the shareholders changed the company name to Solar Aircraft Company. The company's initial product was a small all-metal tri-motor transport aircraft, which, because of the great depression, was never marketed successfully. During the years following 1929, the company underwent further name changes accompanied by changes in its products, markets, and ownership.

100. Robert W. Fausel, *Whatever Happened to Curtiss-Wright?* (Manhattan, Kansas: Sunflower University Press, 1990), 77-80.

101. John C. Schettino, Silver Spring, Maryland, to Stuart Atkinson, notes from phone conversation, 30 October 1995, National Air and Space Museum Archives, Washington, D.C.

Unable to market its airplane, Solar turned to other products. That decision was to set a new course for the company that ultimately led to Solar's entry into the turbomachinery business. In 1930, the company received its first order for stainless steel aircraft engine manifolds from the Navy. During the 1930s, Solar fabricated various hot parts for aircraft engines. By mid-1939, with a work force of 226 people, Solar was fabricating engine cowlings, flame dampers, tailpipes, and shrouds for a variety of military and transport aircraft. With the military's demand for manifolds and other engine parts manufactured by Solar during World War II, its employment grew to 5,000 people. However, with the cancellation of military contracts at war's end, the employment level abruptly dropped to 850.

At that time, Solar's management regrouped and developed a plan for peacetime operation. With the implementation of that plan, Solar craftsmen began working on an unusual mixture of products. Among them were stainless steel race car bodies, meat and dairy processing equipment, and other stainless steel products including caskets. Solar salesmen at the time reportedly carried business cards selling caskets with the motto "Go to hell in Inconel." Work also continued on the fabrication of manifolds, and by April 1947, Solar had produced its 350,000th manifold.

In view of its experience in fabricating stainless steel hot parts for engines, in 1944, Solar was selected to produce high-temperature components of the GE I-40 turbojet engine. That work was followed by a Navy contract to develop an afterburner for the Westinghouse J34 turbojet engine. As a result of that contract, Solar became the first U.S. company to produce a practical afterburner. Solar-developed and -built afterburners were also used on the Allison J33 and J35, the Canadian Orenda, and the Bristol Olympus engines. This work on afterburners led to Solar's entry into the turbomachinery business.[102]

In the mid-1940s, Edmund Price, Solar's President, came to the conclusion that the gas turbine was going to be the prime mover of the future. Price selected Paul Pitt, who had joined the company as a field engineer in 1942, to spearhead the development of a small gas turbine engine. Pitt went on to lead the gas turbine development at Solar, becoming Chief Engineer in 1955, Vice President of Engineering and Research in 1960, and Senior Vice President of the company in 1977. Morris Sievert, who was Solar's President at that time, hailed Pitt's contributions, "He is the father of the gas turbine at Solar. In his years here, he has been responsible for bringing together the most talented tech-

---

102. *Growing Through Turbulent Times* (San Diego, Cal.: Solar Turbines Incorporated, 1990), 2-49.

nical team, and organizing the most effective technical design program in the industry."[103]

The team that Pitt had assembled in 1946 began work at that time on development of an 80-hp turbine-driven auxiliary power unit for the Army Air Force's B-36 bomber. Intended to produce 50 kW at altitudes up to 40,000 feet, it was an axial machine with two gas turbines arranged in parallel driving a common gearbox. Only three of the units were built and ultimately the contract was canceled.[104] That work was followed by the development of a 250 kW axial-flow gas turbine for the Navy to drive a shipboard emergency generator. With design beginning in mid-1947, the engine was first run in 1949. That engine, designated the T-400, became the first American gas turbine engine to be installed in a Navy ship.[105] Other versions of this engine were used to supply power for minesweepers and to drive various landing craft.

In 1947, Solar engineer Leon Wosika joined with a consultant, Dr. Eric Balje, to design a unique centrifugal-flow compressor and radial turbine assembly. The unit they designed was called the MPM-45 radial turbine, later named Mars, which was capable of delivering 50 shp. In their design, the compressor and turbine rotors were assembled back-to-back, making the assembly unusually compact. This same basic design was carried over to a number of kindred machines such as the 55-shp Titan, 350-shp Spartan, and 10-kw (13.5-shp) Gemini. During the decade following the initial design of the Mars, Solar also developed several larger axial-flow gas turbine units with power outputs up to 10,000 shp.[106]

TITAN T62-S-2 TURBOSHAFT

Design of Solar's first and only fully developed gas turbine for aircraft propulsion was begun in 1957. In 1956, the Navy had developed the concept of small one-man and drone helicopters that could operate from both land and shipboard. This concept called for craft that were lightweight, compact, and easy to operate. It was visualized that the power plant would be a 55-shp gas turbine mounted in a vertical position to drive the rotor and that the engine would be started with a hand crank.

103. *Growing Through Turbulent Times*, 26, 53.

104. *Growing Through Turbulent Times*, 26.

105. R. R. Peterson and P. G. Carlson, "Design Features of a 250 kW Gas Turbine Engine for Driving Shipboard Emergency Generator," (ASME Paper No. 51-A-105 presented at the ASME Annual Meeting, Atlantic City, N.J., 25-30 November 1951), 1-3.

106. *Growing Through Turbulent Times*, 26-35.

Following a competition held by the Navy's Bureau of Aeronautics, Solar was selected to develop the engine. Selection of Solar came primarily from its extensive experience in the design, development and production of a line of small gas turbines in the range of 45 to 500 shp and the merit of its proposed design, which incorporated a back-to-back mounted centrifugal compressor and radial turbine resulting in a lightweight, compact arrangement with only two main shaft bearings. Also, Solar was the only competitor that had experience with hand starting, a feature incorporated in its 45-shp T45 gas turbine unit used by the Navy to power a portable fire pump.

Principal members of the design team were: Paul A. Pitt, Solar's Chief Engineer; Paul G. Carlson, Chief Project Engineer; Pich Pichel, Project Engineer in charge of managing the project; Leon R. Wosika, Chief of Preliminary Design and responsible for the concept and basic design of the engine; Colin Rodgers, Aerodynamicist, who was responsible for aero-thermo design of the compressor; Aubry Stone, Aerodynamicist, who was responsible for the design of the turbine components; and Paul Shumate, Test Engineer, who had responsibility for engine development testing.

The engine, designated the T62-S-2 Titan, used a back-to-back centrifugal compressor and radial turbine of the type initially designed in 1947 for the Mars unit. This arrangement was selected because of its inherent ruggedness, low cost, and good aerodynamic characteristics for starting. The back-to-back compressor and turbine rotors were overhung from a forward antifriction two-bearing system. A reverse-flow annular combustor was wrapped around the turbine section. The engine's design arrangement made it both compact and light at only 52 lb.

To minimize the possibility of fuel injector orifice clogging at the engine's relatively low fuel flows, air atomizing injectors were used. Lubricating oil was supplied from an external tank and metered to the bearings as a mist, then vented overboard into the exhaust gas stream. Engine rpm was controlled by a simple flyball governor with only two settings, idle speed and full speed. Thus, the pilot needed only to manipulate the collective pitch of the helicopter rotor through a simple lever to demand power from the constant-rpm engine.

Development problems of the T62 were relatively routine except for those special features associated with its vertical mounting, its high rotating speed of 56,700 rpm, and its small annular combustor. Vertical mounting of the engine led to much experimentation to find bearings that would provide adequate life. These bearing problems led to adaptation of the air-oil mist system for bearing lubrication and cooling. The problem of avoiding fuel injector clogging at the engine's low fuel flow rates led to the development of special air atomizing, low pressure, vaporizing injectors.

Experimental Gyrodyne YRON-1 Rotorcycle one-man helicopter powered by Solar YT62-S-2 turboshaft engine. Photograph courtesy of Solar

The YT62-S-2 engine was first run in March 1958 and first flew in August 1959 in the Navy's one-man Gyrodyne YRON-1 Rotorcycle helicopter. The engine also passed its 40-hour PFRT in 1959. Although the Rotorcycle helicopter and its YT62 power plant were considered successful, the Navy elected to abandon its requirement for the helicopter while it was still in development testing. Consequently, only a small number of development and flight test engines were manufactured by Solar.[107]

Following discontinuation of that program, Solar redirected work on the T62 Titan to aircraft on-board APU applications. Production of the engine as an APU began in 1962, and during the years that followed, it was used in a number of military, commercial, and private aircraft. The T62 engine was progressively uprated and evolved into a family of T62 models, some with power ratings exceeding 200 shp.[108] Production of T62 engines by Solar continued until the en-

107. Paul A. Pitt, San Diego, Cal., to Rick Leyes, letter with attachments, 31 July 1992, National Air and Space Museum Archives, Washington, D.C., 1.

108. Pitt to Leyes, 31 July 1992.

gine family was transferred as part of the Turbomach Division sale to the Sunstrand Corporation in July 1985.

A significant contribution of the T62 was its early demonstration that a gas turbine could be designed to operate in a vertically-mounted position. From a technical viewpoint, the T62 established a benchmark for the design of small, lightweight turbine engines using back-to-back radial compressor and turbine components. The success of that design was illustrated by the fact that it was still being used in the T62 family of engines 25 years after it was originally conceived.[109]

TITAN YT66-S-2 TURBOSHAFT

Soon after the contract for the YT62-S-2 was awarded by the Navy, it was amended to include the development of a free power turbine version of the engine that was to be designated the YT66-S-2. The contract modification was made because the Navy believed that a two-shaft arrangement would better suit the power plant needs of flying platform types of craft then being considered as tactical weapons. The engine was to drive a fixed-pitch, flying platform fan via a main reduction gearbox. The two-shaft arrangement was considered to have part load characteristics that would give the pilot better maneuverability control than would a single-shaft engine.[110]

To provide as much commonality with the YT62 single-shaft engine as possible, it was determined that its back-to-back centrifugal compressor and radial-inflow turbine assembly and combustor should be used for the gas generator portion of the YT66. However, various alternatives existed for design of the power turbine. One alternative was the conventional two-shaft radial power turbine arrangement with an interstage duct, a stationary axial nozzle ring, and either concentric-shaft or hot-end power output drives. To reduce the complexity offered by the conventional power turbine arrangement, a simplified power turbine design was sought.

The arrangement chosen separated the radial-inflow turbine, which drove the compressor, from an exducer type turbine, which drove the power output shaft. The absence of interstage ducting between the inflow turbine and the exducer (or second) turbine nozzle with this arrangement kept length, weight, and cost to a minimum and maximized the number of engine parts common with the YT62. In 1964, a patent for this exducer turbine arrangement was awarded to Solar's Leon Wosika, who was co-designer of the concept with Aubry Stone, who per-

109. Pitt to Leyes, 31 July 1992.

110. Pitt to Leyes, 31 July 1992.

formed the aero-thermodynamic analysis. The exducer type turbine was a new and unproven concept. However, it appeared theoretically sound, and in view of its apparent advantages over conventional power turbines, it was selected for use in the YT66. The power turbine drove a forward-mounted reduction gearbox through a concentric shaft.[111]

Initial tests of the engine showed that the output power of the YT66 did not meet the 55-shp design target. In addition, aerodynamic interference between the gas generator and power turbines set up destructive vibration of the power turbine. Modifications were made allowing full-speed testing and power turbine calibrations to demonstrate that the engine design was sound. However, with only two development engines having been built, the program was canceled by the Navy before the problems were fully resolved.[112]

TRANSITIONS AT SOLAR

From the early 1960s to the mid-1980s, a number of transitional actions occurred at Solar. In early 1960, Solar had been purchased by the International Harvester Company; it became an operating division of the company in 1963. In 1965, design was begun on a 3,000-hp gas turbine for gas transmission pipeline and gas compression applications, and design of a 10,000-hp unit was begun in 1973. In addition to its gas turbines, Solar also continued to manufacture a number of other aerospace components. Then, in March 1973, the company took a major step in restructuring itself and its markets. At that time, Solar's management decided to phase out all aerospace component and subcontracting business. A major factor leading to this decision was that aerospace work had become so competitive that Solar was unable to capture a profitable share of the subcontracting business. On the other hand, there were growing opportunities for use of gas turbines in oil, gas, and related industries being uncovered by Manager of Sales, O. Morris Sievert. It was decided that the company would henceforth concentrate its engineering expertise and gas turbine inventory on the rapidly expanding turbomachinery markets. Its objective was to establish itself as a leading manufacturer of industrial equipment.

In the spring of 1975, a Radial Engine Group was established to focus on design and manufacture of radial engines. That group was renamed the Turbo-

111. Colin Rodgers, "Performance and Applications of the Exducer Power Turbine," (SAE Paper No. 750208 presented at the SAE Automotive Engineering Congress and Exposition, Detroit, Michigan, 24-28 February 1975), 1-4.

112. Pitt to Leyes, 31 July 1992.

mach Division in 1980. Another major transition for Solar came in 1981 when International Harvester sold the Solar Division to the Caterpillar Tractor Company. Following this acquisition, Caterpillar transferred responsibility for its Model 5600 gas turbine engine family to Solar. That program was an outgrowth of Boeing's Model 551/553 engine family acquired by Caterpillar in 1966 after Boeing decided to discontinue its Gas Turbine Division. Then in July 1985, following a long-time relationship between Turbomach and Sunstrand, Caterpillar accepted an offer from the Sunstrand Corporation to buy Solar's Turbomach Division. The sale included all of Solar's active radial flow engines, the Titan, Gemini, and original small Mars unit.[113]

By 1990, Solar had been in the gas turbine business for over 40 years, was briefly a part of the aircraft gas turbine industry, and had restructured itself to become recognized as a world leader as a supplier of gas turbines in the 1,000- to 14,000-hp range for a variety of industrial applications.

The Seven Major Small Gas Turbine Aircraft Engine Corporations

By the early 1990s, seven major North American companies in the small gas turbine aircraft engine industry had evolved and matured, Teledyne CAE, Lycoming, GE, Williams International, Pratt & Whitney Canada, Allison, and Garrett (AlliedSignal Engines). They reached positions of major importance and exerted a significant influence on the aviation industry. Chapters 4 through 10 address the work of these seven companies in the order in which they began their continuing work on small gas turbine engines as prime propulsion components for aircraft.

Miniature and Model Turbine Engines

In the late 1980s, six companies began developing miniature engines for small, tactical missiles under the auspices of the U.S. Army Missile Command. These included Allison, Teledyne CAE, Williams International, Sunstrand Power Systems, Technical Directions, Inc., and Turbine Technologies, Inc. About the same time, full working turbojet engines small enough to propel model aircraft became available. This history is summarized in Chapter 11.

---

113. *Growing Through Turbulent Times*, 37-61.

TELEDYNE CAE | 4

Company Origin and Growth

After graduation from the Armour Institute of Technology in Chicago in 1902, 20-year-old Ross W. Judson decided to build a career around his interest in designing and constructing gasoline engines for the newly developing automobile industry. (At that time, automobile companies built chassis but had no engine production capability of their own.) In September 1902, Judson entered into a partnership with his brother-in-law, Arthur W. Tobin. As the engine business grew, in May 1904, the partners incorporated to form the Autocar Equipment Company.

Prior to his start in the gasoline engine business, Judson had visited Europe, where he had been captivated by the designs of European cars and engines. He applied many European features to his engines, and his advertising promoted his "Continental Motors." To reflect the "continental" features of its engines, in February 1905, the company was renamed the Continental Motor Manufacturing Company. The company grew rapidly, and in relatively few years was supplying engines for nearly half of the cars built in the U.S. In 1916, the company's name was shortened to Continental Motors Company.[1] Carl Bachle, an engineer who joined Continental about 1927, and who knew Judson at that time, regarded the company's founder as all important to the company in that

1. William Wagner, *Continental! Its Motors and Its People* (Fallbrook, Cal: Armed Forces Journal International in cooperation with Aero Publishers, Inc., 1983), 5-19.

Early employees of Teledyne CAE from left to right: Henry Maskey, Chief Engineer; Carl Bachle, V.P. of Engineering; Dorothy Bush, Secretary for Art Wild; Carl Schiller, Purchasing; Toby Herringshaw, Purchasing (seated); Pete Orlando, Designer; Eli Benstein, Performance (seated). Most of these employees were directly or indirectly involved in the early Teledyne CAE J69 turbojet engine program. Not pictured, but key individuals directly involved in the original J69 program included: Art Wild, President; Frank Marsh, engineering development; Jim Shields, design; Don Todd, aerodynamics. Photograph courtesy of Walt Dindoffer.

era. Bachle attributed Continental's success in selling its engines to the automobile companies to Judson's high-living, high-entertaining, super salesmanship.[2]

In the early 1920s, Continental's management observed a transition from steam power to gasoline power in portable construction machinery. As the automobile companies began to build their own engines, Continental added industrial engines for use in tractors, combines, road-building equipment, and construction machines to its product line. As technology evolved and new markets appeared, product diversification became Continental's hallmark

By the mid-1920s, Judson and his Executive Vice-President, W. R. Angell, assessed the newly emerging market for aircraft engines and decided to pursue it as part of Continental's diversification strategy. In a 1977 letter, Carl Bachle wrote, "In 1927, we all dreamed of an Aircraft Industry but didn't visualize the way it has gone. At that time, it was 'an aircraft in every garage' like the Ford Model T."[3]

2. Carl Bachle, interview by Rick Leyes, 9 August 1988, Teledyne CAE Interviews, transcript, National Air and Space Museum Archives, Washington, D.C.

3. Carl Bachle, Toledo, Ohio, to Les Waters, Teledyne Continental Motors, letter, 2 December 1977, National Air and Space Museum Archives, Washington, D.C.

In August 1929, the Continental Aircraft Engine Company was formed as a subsidiary of Continental Motors to develop and produce the company's aircraft engines. Just before this, Continental had announced its new 165-hp seven-cylinder radial air-cooled engine, the Model A-70. This engine was accepted quickly, and by 1931, more than 100 engines had been produced. In 1930, development began on the horizontally-opposed Model A-40 engine. The horizontally-opposed design became extremely popular, and during the 1930s, Continental developed a number of improved horizontally-opposed engine models. The most outstanding of these engines was the A-65, which saw wide use in aircraft built by Piper, Aeronca, Luscombe, Taylorcraft, Stinson, and others.

Having survived the worst years of the Great Depression, the company faced new bankruptcy in the late 1930s. The recession of 1938-39 made a powerful impact. Fortunately, a few years earlier, Continental had converted its W-670 air-cooled radial aircraft engine for use in tanks. In September 1939, war erupted in Europe, and two months later, the Arms Embargo Act was repealed, allowing U.S. manufacturers to sell war materiel to Britain and France. Unprecedented orders for military equipment soon pushed Continental to the limit of its capability to build engines. The demand for tank engines, as well as for other ground vehicle and aircraft engines, brought Continental back to its feet.

To eliminate delays encountered in the development of a new liquid-cooled aircraft engine for the military, in 1940, Continental formed the Continental Aviation and Engineering (CAE) Corporation with controlling interest held by Continental Motors. This new entity was to perform all of Continental Motors' development and production work on aircraft-type engines of over 500 hp at its Detroit plant.[4]

As it became evident during the late 1940s that there was a potential future market for small gas turbine engines, CAE undertook the development of two small turboshaft engines. Although these early efforts were not fruitful, the 1950s marked CAE's entry into a growing market for small gas turbine aircraft engines. This success followed the company's acquisition of a license from Turboméca of France to manufacture and sell, within the U.S., a family of Turboméca's small turbojet and turboshaft engines.

Following the company's initial development work on small gas turbine engines during the late 1940s, CAE began construction of a new research facility at its Detroit plant. That facility was completed and occupied in 1954. However, with a growing demand by the military for deliveries of its J69 turbojet and its

4. Wagner, *Continental! Its Motors and Its People,* 23-103.

MA-1 and -1A air compressor units used to start jet engines, CAE found that there was insufficient manufacturing space at the Detroit plant.

During a search for an alternate facility, CAE was offered the use of a government-owned production plant at Toledo, Ohio. Air Force Plant 27 had been built during World War II for the Lycoming Division of Avco. From 1943 to 1949, it had been occupied by the Packard Motor Co. for production of aircraft engines and later for a short period by Frederic Flader Co. for development work on a jet engine. Subsequently, A. O. Smith Corp. had used the plant for several years to produce landing gears for Air Force B-47 and B-52 bombers. The plant had been idle for a year when CAE took over use of it in the spring of 1955. All manufacturing activities were moved to the Toledo plant while research and development remained at Detroit.[5]

A 375,000-square-foot facility on a 79-acre tract, the Toledo plant contained a number of test cells that had been used for its earlier aircraft engine activities. As the backlog of orders for J69s rose, the Air Force funded additions to the plant, including test cells and additional machine tools and equipment. In 1957, facilities at Toledo were further improved by the construction of a well-equipped component test laboratory. With those additions, the plant included 30 on-site test cells, including 14 component test rigs and two altitude test chambers capable of testing full-scale engines from sea level to 90,000 feet at Mach numbers up to 2.0.[6] In 1958, CAE phased out of Detroit and located all activities in the Toledo plant.

Gas Turbine Engines

By the end of World War II, Continental's success as a piston engine company had spanned over four decades. However, at the end of the war, Continental's—and CAE's—sales and earnings fell sharply. CAE began stressing its research and development work, nearly all of which was funded by the federal government, and began evaluating new products and promising new markets. It was in this environment that the company took its first step into the field of gas turbine engines.

Soon after the end of World War II, CAE began development of its first gas turbine, the TR125, a 125-shp unit for the Air Force, intended to drive an electric generator for starting large jet engines. The TR125 was built and tested, but

5. Wagner, *Continental! Its Motors and Its People*, 155-156.

6. Deloras D. Eaton, "This Is Teledyne," unpublished paper, National Air and Space Museum Archives, Washington, D.C., 1-4.

work on it was soon discontinued because it had been designed with too low a pressure ratio. Resulting poor fuel economy made it uncompetitive with gas turbine units being offered by Garrett. About 1950, a 50-hp gas turbine to drive portable fire pumps was developed for the Navy. This unit was also discontinued because of poor performance.[7] Carl Bachle, a senior CAE engineer at that time, described the technology on which the engineers based the designs of these engines as coming strictly from the literature and textbooks.

By 1951, Continental's interest in the small gas turbine engine market had grown to the point that the company was determined to expand its product line in this direction. It elected to acquire a license to manufacture and market already developed and proven small gas turbine engines for aircraft and ground power units.

By the early 1950s, Carl Bachle, who kept abreast of technical developments in Europe, had become familiar with the reputation of Turboméca in France as a developer of small gas turbine engines for aircraft and ground power units. During a trip to England in 1951, Bachle visited Air Commodore Francis Rodwell Banks, who had been director of engine research and development for the British Air Ministry during World War II. Bachle asked Banks, who was also familiar with Turboméca, whether he believed the company was as technically capable as had been reported. Banks responded affirmatively and suggested that Continental contact Turboméca in France. Continental followed his suggestion; the technical infusion from Turboméca allowed the company to enter the small gas turbine engine field very quickly.[8]

Société Turboméca S. A., of Bordes, France, had been founded by Josef Szydlowski, a young Polish engineer, whose interest in gas turbines started in 1927. The Turboméca company was formed in 1938 to produce turbo compressors; later, gas turbines were added. The company was forced to cease development and production during the World War II German occupation of France. At the close of the war, it resumed operations, and, by 1951, had a line of several small gas turbine engines with extremely attractive performance characteristics. These included excellent fuel economy, long service life, ability to use a wide range of fuels, high power-to-weight ratio, small size in relation to power, and interchangeability of parts among various models.[9]

In mid-1951, after months of negotiations and testing in the U.S. and France, Continental Motors developed a licensing agreement with Turboméca for the technical rights and for U.S. production and marketing of a family of eight small

7. Bachle to Leyes, 9 August 1988.

8. Bachle to Leyes, 9 August 1988.

9. Wagner, *Continental! Its Motors and Its People*, 138-139.

gas turbine engines. In October of that year, Continental sub-licensed CAE to produce eight Turboméca engines. The following is a list of these engines with their performance ratings at that time and the CAE and military designations subsequently assigned to them:[10]

Turboméca Name	Engine Type	Performance Rating	CAE Model	Military Designation
Marboré II	turbojet	880 lb	352	XJ69-T-9
Palas	turbojet	330 lb	320	–
Arbizon	turbojet	550 lb	324	–
Aspin II	ducted fan	790 lb	420-1	–
Artouste I	turboshaft	280 shp	210-1	XT51-T-1
Artouste II	turboshaft	425 shp	220-2	XT51-T-3
Palouste II	air compressor	185 air hp	140	MA-1A
Autan	air compressor	325 air hp	160	–

Having acquired exclusive U.S. development and production rights for eight Turboméca engines in 1951, CAE immediately set forth to penetrate the military marketplace. With these engines as its product line, CAE's objective was to find a market for as many of them as possible. Since its establishment in 1940, CAE had focused its sales efforts on the military marketplace. Thus, it logically turned to the military to market its gas turbines.

Bachle had initially considered the Artouste II, later the XT51-T-3, as the most important of the Turboméca engines.[11] He believed that it was a logical replacement for the Continental 0-470 horizontally-opposed reciprocating engine. He found that although the Artouste II could do everything that the 0-470 could, it cost five times as much and used twice as much fuel.

The first Turboméca engine that CAE produced was the Palouste II, an engine that Bachle considered a mere throwaway. This engine became the CAE Model 140, later designated the Model 141. About 1951, the Air Force needed an air starter for its J47 turbojet engines. The competition was between CAE and Garrett, and the CAE Model 141, given the military designation of MA-1A, was chosen. Starting in 1952, CAE built 1,076 MA-1/1A units.

10. Frank Marsh, Toledo, Ohio, to Rick Leyes, unpublished Turboméca engine data, 9 August 1988, National Air and Space Museum Archives, Washington, D.C.

11. Bachle to Leyes, 9 August 1988.

In mid-1952, following an intense CAE marketing effort, the Marboré II was selected for the Ryan Firebee Remotely Piloted Vehicle (RPV) target drone and given its military designation of J69. In December 1952, the Cessna Aircraft Co. won an Air Force competition for a jet trainer with their proposal of a T-37 aircraft powered by two Marboré II turbojet engines.[12] Selection of the J69 for use in the Firebee and the T-37 gave CAE the toehold it needed to become a legitimate contender in the aircraft gas turbine marketplace.

Turbojet Engine Development

THE J69 ENGINE

The J69 became the centerpiece of CAE's turbojet engine development effort. Not only was the J69 series of engines highly successful as a result of its use in the Ryan Firebee and the Cessna T-37, but two other highly successful turbojet engines were derived from it. One was the J100 engine, used in advanced versions of the Firebee, and the other was the J402 series of engines used in the McDonnell Douglas Harpoon missile and the Beech MQM-107 Streaker target drone.

The J69 engine had a single-stage centrifugal compressor, a through-flow annular combustor and a single-stage, axial-flow turbine. Two models of the engine were produced initially. One was the 920-lb-thrust J69-T-9 used in the Cessna T-37 trainer aircraft. The other was the J69-T-19, a 1,000-lb-thrust version of the engine used in the Ryan Firebee target drone. The J69-T-19's higher thrust was obtained by raising its turbine temperature, at the cost of shortening engine life. Although engine life and reliability were critical for the T-37 trainer application, shortening engine life to increase performance was an acceptable tradeoff for the remotely piloted Firebee.

A unique feature of the J69 engine was its slinger-type fuel injector, which was an integral part of the engine's main rotor shaft assembly. This feature, developed by Turboméca, was subsequently used in all of CAE's engines. Fuel was supplied at a moderate pressure into the hollow main rotor shaft. Centrifugal force generated by the shaft's rotation caused the fuel to spread out in a thin layer on the inner wall of the shaft. The fuel then flowed out through radial holes in the wall of the rotor shaft to the surface of the shaft where it was "slung off," atomized by the rotating shaft's tangential velocity. The velocity of the fuel thus thrown off into the annular combustor was equivalent to an injection pres-

12. "U.S.A.F. Trainer," *Flight, 66* (October 29, 1954): 650.

sure well in excess of 2,000-lb-per-square inch, eliminating any need for a high-pressure pump. Because the injector was an integral part of the engine rotor shaft, fuel atomization varied with engine speed only and was unaffected by fuel flow rate.[13]

The initial J69s were very similar to the Marboré II, but later models differed significantly in internal design and had vastly improved performance. Demands for increased thrust and higher-altitude operation required major aerodynamic, thermodynamic, and mechanical changes. The J69 family included man-rated engines, high-altitude engines, supersonic engines, and missile engines. Thrust grew from 880 lb for the J69-T-2 to 1,920 lb for the YJ69-T-406, and ultimately to 2,700 lb for the J100-CA-100, a derivative of the J69 that was used on a high-altitude version of the Firebee photo reconnaissance vehicle. The J69-T-25A model engine and spare parts for it were still being produced in the 1990s.[14]

FIREBEE APPLICATION

By early 1947, the Pilotless Aircraft Branch of the Air Force had established the performance specifications for a jet powered target aircraft. The aircraft, designated the Q-2, was to be capable of 521 kts at 15,000 ft and have a service ceiling of 40,000 ft.

Thirty-one companies responded to the first request for quotation, but after analysis by the Air Force, no proposal was accepted. The project was then opened for rebid with a due date of January 1948, to which 18 manufacturers responded with 14 designs. In August 1948, the Ryan Aeronautical Company was awarded the contract for the Q-2 subsonic pilotless, jet-powered training target, given the name Firebee.

In keeping with its procurement policy of contracting separately for airframe and engines, the Air Force selected the Fredric Flader Co.'s XJ55 turbojet engine to power the Firebee. The Flader engine, the design of which was based on

13. H. C. Maskey and F. X. Marsh, "The Annular Combustion Chamber With Centrifugal Fuel Injection," (SAE Paper No. 444C presented at the Automotive Engineering Congress, Detroit, Michigan, 8-12 January 1962), 2-4.

14. Eli Benstein, telephone interview by William Fleming, 25 July 1990, Teledyne CAE Interviews, National Air and Space Museum Archives, Washington, D.C. The Air Force told CAE that, if the company retained the J69 designation for new engines for the Firebee and T-37, it could continue to develop them using product improvement funds, instead of having to compete for new engine development funds. This decision resulted in the long-lived retention of the J69 designation.

Teledyne Ryan Firebee II (BQM-34E/F) supersonic drone powered by Teledyne
CAE YJ69-T-406 turbojet engine. Photograph courtesy of Teledyne CAE.

use of a supersonic axial-flow compressor, encountered difficulty in achieving
its stated performance. After numerous engine tests and ground runs in the Q-2,
the Flader contract was terminated and the Fairchild J44 designated as the replacement engine. The first successful flight of the XQ-2, powered by the
Fairchild J44, occurred in the spring of 1951 following its launch from a Boeing
B-17 at Holloman Air Force Base, N.M.[15]

Immediately after acquiring its license in mid-1951 to produce the Turbo-
méca engines in the U.S., CAE began searching for promising markets. The
Ryan Firebee was recognized as a potential opportunity for use of the Marboré
II engine. Development of the J44, the engine against which CAE had to compete, had been sponsored by the Navy. So, Carl Bachle of CAE approached the
Navy and explained the advantages of using the Marboré II in the Firebee.[16]
However, the key Navy engineer whom Bachle had to convince was a great
proponent of the J44, and he advised Bachle that CAE should not try to upset
the Navy's plans for use of its J44 engine in the Firebee.

15. William Wagner, *Lightning Bugs* (Fallbrook, Cal: Armed Forces Journal International in
    cooperation with Aero Publishers, Inc., 1982), 86-88.

16. Bachle to Leyes, 9 August 1988.

Bachle and others from CAE then took their Marboré II proposal to Ryan. They emphasized the engine's high fuel efficiency, resulting from its high pressure ratio and compressor efficiency. The chief engineer of Ryan immediately recognized that the Marboré II would give the Firebee about an hour and a half on target compared to the three-quarters of an hour provided by the J44. Consequently, Ryan went to the Navy and said that it wanted the Marboré II.[17]

By mid-1952, following a year of uncertainty, CAE was established as a contender when the Marboré II was designated to power some of the Air Force and Navy versions of the Firebee. Other versions continued to be powered by the Fairchild J44. In applying the Marboré II to the Firebee, CAE developed a short-life version of the engine, the J69-T-19, by increasing the turbine temperature and introducing various compressor modifications to raise the thrust from 880 to 1,000 lb.

The J69 was both more efficient and had a higher altitude capability than Fairchild's J44. The J69's higher compressor pressure ratio gave it superior fuel economy. Due to the J44's relatively tight compressor stall margin, which became tighter at altitude, the J44 surged at altitudes below the J69's operating ceiling with a resulting over-temperature condition. As a result, the J69 soon won favor, and by 1958, the Navy had terminated orders for the J44.

By late 1954, the Army, Navy, and Air Force were using models of the Firebee, designated, respectively, the XM-21, the KDA-1, and the Q-2A, and by 1958, pilots of the three services had found the Firebee to be a realistic threat simulator.[18] Canada joined in 1957 with an order for 30 KDA-4 Firebees for cold-weather testing. Between 1954 and 1958, nearly 1,380 Firebee target drones were produced.

In 1959, CAE began delivering an uprated model, the J69-T-29. With the addition of an axial-flow compressor stage ahead of the centrifugal compressor to increase both engine airflow and pressure ratio, CAE increased the thrust from 1,000 to 1,700 lb. Powered by this improved engine, a larger, faster, and higher-flying version of the Firebee, the Q-2C (later redesignated the BQM-34A), attained a speed of Mach 0.97 in level flight and a ceiling of over 60,000 ft. So advanced was the Q-2C that, in its early flights, chase pilots had difficulty staying with it.[19]

By April 1987, 5,218 improved Firebees had been produced, nearly all powered by the J69-T-29. (Some of the last Firebees produced used the General Electric

17. Bachle to Leyes, 9 August 1988.

18. Wagner, *Lightning Bugs*, 89-90.

19. Wagner, *Lightning Bugs*, 10, 89-90.

Teledyne CAE J69-T-29 turbojet engine used in Ryan Firebee target and reconnaissance drones. Photograph courtesy of Teledyne CAE.

J85-GE-7). By January 1987, this subsonic version of the Firebee had flown more
than 32,000 flights in every conceivable climatic and combat environment.[20]

Meanwhile, with the introduction of supersonic combat aircraft during the
1960s, the need emerged for a supersonic target drone. A major step in advancing the state of the art of pilotless aircraft was taken in 1966 when the Navy
funded development of a supersonic version of the Firebee.[21] Shortly after
Ryan started work on this newest member of the family, CAE began modifying
the J69-T-41A, used in a photo reconnaissance version of the Firebee, to fly su-
personically. The principal modification was a change in compressor materials
to permit the engine to be operated at the elevated inlet temperatures encountered at Mach 1.1 at sea level and up to Mach 1.5 at 60,000 feet.

Using this supersonic version of the J69 engine, designated the YJ69-T-406,
the BQM-34E Firebee II successfully completed its flight test program in
1970.[22] Production of the BQM-34E Firebee IIs began in 1970 and continued

20. *Jane's All the World's Aircraft 1987-88* (London: Jane's Publishing Company Limited, 1987),
860.

21. Wagner, *Lightning Bugs*, 91.

22. Wagner, *Lightning Bugs*, 91.

until 1980, when 269 had been delivered to the Navy and Air Force.[23] Meanwhile, from 1968 through 1976, 271 J69-T-406 engines were delivered by CAE to power these aircraft.[24]

PHOTO RECONNAISSANCE APPLICATION

By the late 1950s, the technology existed to use pilotless vehicles for photo reconnaissance; however, the military was reluctant to exploit it. This was not surprising as piloted high-altitude reconnaissance, using the Lockheed U-2 airplane, had already been developed and proven. Further, there were those in the military who felt that having a pilot in the loop was essential to good photo reconnaissance, and that the embryonic pilotless technology would not lead to a viable alternative. It took two highly publicized and embarrassing incidents to allow the pilotless photo reconnaissance program to materialize. Even so, the program was faced with an on-again, off-again existence for about three years. Leading to full acceptance of the unmanned approach was the need for extensive photo reconnaissance over Southeast Asia during the Vietnam War, combined with the increasing capability of enemy missiles to destroy high-flying aircraft.

By 1960, Ryan had recognized the potential of the Firebee for use as an unmanned photo reconnaissance aircraft. Early that year, Robert R. Schwanhausser, Ryan's project engineer for target drones, presented a briefing to top Air Force officials at the Pentagon on the potential of the Firebee to serve as a reconnaissance aircraft, eliminating the need to risk pilots in covert reconnaissance flights over unfriendly nations.

Following the briefing, Ryan began discussing a proposal for a reconnaissance drone powered by the J69-T-29 engine with officials at the Pentagon. On May 1, 1960, Francis Gary Powers' U-2 was shot down over Russia. Two months later, a Boeing RB-47 and its crew were also shot down by the Russians. A week later, on July 8, 1960, Ryan was awarded a contract by the Air Force to begin feasibility studies and demonstrations of its proposed reconnaissance drone. However, this work was terminated only four months later, in November 1960, following Vice President Richard M. Nixon's loss in his bid for the presidency. (It is customary in the case of a change of administration following an election loss that the outgoing administration does not approve any new de-

23. *Jane's All the World's Aircraft 1979-80* (London: Jane's Publishing Company, 1979), 640.

24. Teledyne CAE, "J69 Engine Delivery History," unpublished tabulation of J69 engine deliveries, 9 August 1988, National Air and Space Museum Archives, Washington, D.C.

velopment effort that will leave the incoming administration with a new long-term project. Hence funding for a follow-on effort to develop and produce a number of Firebee drone aircraft was held up by Dr. Harold Brown, Director of Defense Research and Engineering, thereby effectively killing the reconnaissance drone program.[25])

Not until 15 months later, following an intensive marketing effort by Ryan, did the new administration approve the program. On February 2, 1962, Ryan was awarded a contract by the Air Force for four Q-2C Special Purpose Aircraft (SPA) modified to photo reconnaissance configuration and incorporating features providing low vulnerability to attack. The program was given the code name "Fire Fly," and the aircraft were designated Model 147 SPAs. By early 1963, Fire Fly had been compromised and was succeeded by the name "Lightning Bug."

By late summer 1962, 147A aircraft launched from C-130s based at the Eglin Air Force Base had completed a number test flights. In November 1962, during the Cuban missile crisis, the 147s were deployed to be available to fly over Cuba. Although they were not used over Cuba at that time, out of the crisis came a contract for Ryan to proceed with development of the 147B, the first true reconnaissance drone. The downing of Maj. Rudolph Anderson, Jr., in a U-2 over Cuba, helped stimulate this go-ahead. The 147B model, with the J69-T-29 engine, was the first high-altitude drone. It had a much larger wing than the 147A, giving it a ceiling of 62,500 feet compared with 52,500 feet for the earlier model, and increasing the range from 1,227 to 1,680 nautical miles.[26]

Flight tests of the 147B were conducted through the first half of 1964. Then, with the Tonkin Gulf incident on August 4, 1964, a limited number of model 147B Lightning Bugs were deployed to Okinawa and then to South Vietnam for reconnaissance flights over southeast China, the Chinese island of Hainan, and North Vietnam. This activity marked the beginning of extensive pilotless aircraft reconnaissance activity in Southeast Asia during the Vietnam War, which had a major impact on J69 engine development and production requirements.

Early in 1965, after only five months of operational experience with the 147B, Ryan began cutting metal for the uprated 147G model. The principal change was the new 1,920-lb-thrust J69-T-41A engine, which produced 220 more pounds of thrust than the J69-T-29 used in the 147B model. The J69-T-41A achieved higher thrust by using a high-pressure-ratio, higher airflow

25. Wagner, *Lightning Bugs*, 2, 10-15.

26. Wagner, *Lightning Bugs*, 23, 42.

transonic compressor stage ahead of the centrifugal compressor, in place of the axial-flow stage used in the J69-T-29, a modification that raised the ceiling of the engine and the 147G aircraft in which it was used. By July 1965, the first Firebee 147G was delivered; it became operational three months later, replacing the 147Bs, which had flown a total of 78 missions. Through 1976, a total of 612 J69-T-41A engines were delivered for use in the Firebee 147G.[27]

Acceptance of pilotless aircraft for aerial reconnaissance was slow in coming. Ryan's biggest sales problem always was that nearly every customer "wore a pair of wings on his chest," i.e., was either a pilot or an advocate of piloted aircraft. RPVs were viewed as a competitive threat, both financially and programmatically. Another factor: reconnaissance drones and piloted U-2s were under the same command, which meant that there was direct funding competition between them. Technical factors also influenced the military's resistance. Reconnaissance drones represented a new technology, with consequent development problems. Operationally, the drone was less flexible than the piloted system—a pilot could make visual observations and change reconnaissance targets as appropriate—and recovery of the drones was difficult and unpredictable.

However, acceptance gradually emerged. RPV reconnaissance could be performed without hazard to human life. Shortly after surface-to-air missiles (SAMs) began to be deployed around Hanoi, the Air Force scheduled a dual mission in which a Firebee drone was to fly over a SAM site while a U-2 flew off to the side to witness what happened. As the U-2 pilot watched from his standoff position, he saw the drone shot down. After returning to base, the pilot reportedly commented, "From now on, you guys can have that mission."[28]

THE J100 ENGINE FOR HIGHER ALTITUDES

By the mid-1960s, the Air Force wanted to raise the altitude capability of the Firebee to as high as 75,000 feet to reduce still further its vulnerability to ground and air attack. The increased capability also raised the flight altitude above the level at which contrails form.

It had been found during target practice operations with the Firebee target drones that attacking aircraft pilots and ground missile crews were using the Firebee contrails to help locate and track the drones. Since the reconnaissance drones used in Vietnam operated at the same altitudes as the target drones, their contrails likewise contributed to their vulnerability to attack.

27. Teledyne CAE, "J69 Engine Delivery History."

28. Wagner, *Lightning Bugs*, 99.

The requirement for increased altitude capability led to development of the 2,700-lb-thrust J100-CA-100 engine. Derived from the J69 family, this engine was a missilized version of the CAE Model 356-27 developed earlier for piloted aircraft. The 356-27 was intended for the upgraded light strike fighter version of the T-37 trainer, unsuccessfully proposed by Cessna in the counterinsurgency (COIN) program competition. Soon after Cessna lost that competition, a need emerged for an improved Firebee engine that had the thrust and altitude performance capabilities embodied in the 356-27 design.[29]

Although the J100-CA-100 was derived from the J69 family, the two engines shared no parts in common. While its predecessor, the J69-T-41A, had a single transonic axial-flow stage ahead of the centrifugal compressor stage, the J100 employed two transonic stages. Also, a two-stage turbine replaced the single-stage J69-T-41A turbine. These modifications provided the J100 with a 40% thrust increase over its 1,920-lb-thrust predecessor. The additional transonic stage increased the compressor pressure ratio sufficiently to enable the Firebee to operate at altitudes in excess of 75,000 feet. During test runs, the J100 was operated at simulated altitudes as high as 90,000 feet.[30]

The J100 engine was remarkable in two respects. It was the first engine in which CAE used titanium in place of aluminum or magnesium for the inlet duct and the axial and centrifugal compressor housings. Use of titanium contributed to the engine's high thrust-to-weight ratio: 6.4, or 16% higher than that of the J69-T-41A. The J100 was also the highest thrust engine ever built by CAE.

Delivery of the J100 for the Model 147T Firebee began in 1968, and, by 1972, when production was terminated, 96 engines had been delivered.[31] Operational flights of the 147T over Southeast Asia began in the spring of 1969.[32]

FIREBEE SUMMARY

The target drone version of the Firebee, which was powered principally by CAE's J69 engines, and the photo reconnaissance version of the Firebee powered by the J69 and J100 engines, were major technical and financial successes for both Ryan and CAE. By April 1987, 6,498 subsonic Firebees and 269 supersonic

29. Frank Marsh and Harley Greenburg, interview by Rick Leyes, 9 August 1988, Teledyne CAE Interviews, transcript, National Air and Space Museum Archives, Washington, D.C.

30. Marsh and Greenburg to Leyes, 9 August 1988.

31. Teledyne CAE, "Engine Shipments," unpublished tabulation of Teledyne CAE engine shipments, 9 August 1988, National Air and Space Museum Archives, Washington, D.C.

32. Wagner, *Lightning Bugs*, 166.

Firebees had been produced. From 1954 through August 1995, 6,972 J69s were produced for target and reconnaissance versions of the Firebee. This included 18 J69-T-6, 27 J69-T-17, 634 J69-T-19, 4,924 J69-T-29, 1,098 J69-T-41/A/B, and 271 J69-T-406 engines. From 1968 until 1972, 96 J100-CA-100 engines were produced for the Firebee. These engine deliveries represented more than two-thirds of the J69s and all of the J100s that CAE had produced.[33]

Production of the J69 for the Firebee ended in December 1994, although the Firebee itself remained in production. As a system cost-reduction measure, the Navy had decided that subsequent Firebees would be powered by excess General Electric J85 engines made available by the Air Force and modified for the drone application. Nevertheless, a few J69s continued to be produced for foreign sales, and spare parts production continued to support the Firebee and T-37 trainer aircraft.[34]

There were two principal requirements that drove the Firebee program and thereby created the market for the J69 and J100 engines. The target drone version of the Firebee was driven by a practical and economic requirement for training jet pilots and ground-to-air missile crews. By the early 1950s, the remotely piloted target vehicle had become a realistic, elusive, and economic target. The photo reconnaissance Firebee was driven by the requirement for reconnaissance over hostile territory without endangering human life. The need for extensive photo reconnaissance during the Vietnam War, accompanied by the increasing capability of SAMs to destroy high-flying aircraft, led to extensive Firebee use.

The J69- and J100-powered photo reconnaissance versions of the Firebee flew a total of 3,435 operational sorties over Southeast Asia between 1964 and 1975. Because the vehicles were designed for parachute recovery, some flew a number of missions; four of the vehicles flew more than 45 missions each.[35]

T-37 TRAINER APPLICATION

During the latter half of the 1940s, the Air Force and Navy were developing new turbojet-powered fighter aircraft to replace slower piston engine fighters in their operational squadrons. With turbojet-powered fighters came the need for jet-powered training aircraft. In the late 1940s and into the 1950s, the military satisfied this need with two-place versions of several of its new jet fighters.

33. Teledyne CAE, "Engine Shipments."

34. James R. Apel, Toledo, Ohio to Rick Leyes, letter, 26 May 1995, National Air and Space Museum Archives, Washington, D.C.

35. Wagner, *Lightning Bugs*, 213.

However, the growing complexity of these aircraft, as well as the multiple versions of modified aircraft being used as trainers, made them expensive to manufacture, operate, and maintain. The services realized that a new, less-expensive means was needed to train their jet combat pilots.

In 1952, the Air Force released a request for proposals for a jet-powered trainer to the aircraft industry. Competition among the aircraft manufacturers was keen because a development contract for the trainer could eventually lead to large production orders.

In early 1951, Bachle, supported by a Continental attorney, had negotiated the licensing agreement with Turboméca, personally making commitments on behalf of Continental. As Bachle recalled, soon afterward, he was intensely seeking customers for the Turboméca engines in an effort to get off the hook for the licensing commitment he had made. CAE recognized the planned jet trainer as a potential candidate for the Marboré II turbojet, which had already been man-rated by Turboméca. In late 1951, Bachle visited the Cessna Aircraft Co., which had begun design of a turbojet-powered entry for the trainer competition. He spent a great deal of time with the project engineer for Cessna's trainer, ultimately convincing him to use the Marboré engine. These visits occurred at the same time that Bachle was also trying to convince the Navy and Ryan to use the Marboré II in the Firebee.[36]

A number of the entries in the trainer competition were single-engine designs. Both turbojet- and turboprop-powered versions were proposed. Engines other than CAE's Marboré II used in the trainer proposals were the Fairchild J44, a small Allison engine, and a British Mamba which was to have been built by Curtiss-Wright.[37] In addition to Cessna's use of the Marboré II in its entry, the Temco Aircraft Corporation also submitted a twin-engine design using the Marboré II.

In December 1952, the Cessna Aircraft Company's Model 318 was selected. The Cessna trainer, given the military designation T-37, was to be powered by two CAE Marboré II engines.[38] The Cessna design was unusual for an American trainer in that it provided side-by-side seating for the instructor and student pilot.[39] In CAE's view, factors favoring Cessna's selection in the competition were the Marboré II's superior fuel economy and its good service-life record, established during its earlier use in Europe.[40]

---

36. Bachle to Leyes, 9 August 1988.

37. "Our First Jet Lightplane," *Flight Magazine*, (February 1953): 22.

38. "U.S.A.F. Trainer," 650.

39. "Our First Jet Lightplane," 22-23.

40. Eli Benstein, interview by Rick Leyes, 9 August 1988, Teledyne CAE Interviews, transcript, National Air and Space Museum Archives, Washington, D.C.

Cessna's selection to develop the T-37 trainer was followed by two years of intensive effort at CAE to uprate the performance of the Marboré II and to fabricate the first engines. In order to meet the military specification for time required to clear a 50-ft obstacle after takeoff, the T-37's engines were each required to produce a minimum rated thrust of 920 lb. However, the Marboré II engine, given the military designation of J69-T-2, was rated at only 880 lb thrust. CAE believed that obtaining a five percent increase in thrust would be a relatively straightforward task. However, CAE had not taken into account that, in France, engines were rated on the basis of the average thrust for all engines produced, with a tolerance of plus or minus two to three percent. In the U.S., engines were rated on the basis of minimum guaranteed thrust for all engines produced. Consequently, to achieve 920 lb thrust on minimum thrust engines meant raising the average thrust by about eight percent.

More than a year passed before CAE had its first J69 engine running; nearly half of this time was consumed in converting the engine drawings from metric to English units of measurement and translating the notes on the Turboméca drawings, the remainder in fabricating and assembling the first engines. Another year was required to uprate the engine to 920 lb thrust.

The Air Force required other changes to make the engine acceptable. The fuel control was completely redesigned so that starting and acceleration of the engine were automatically controlled in response to throttle movement. This control was considered advanced for its time.

Uprating the engine's thrust, CAE found that the turbine temperature could not be raised because the engine's hot-end materials were unable to withstand higher-temperature operation and still provide the required engine life. The thrust increase itself was ultimately obtained through a series of painstaking engineering refinements and polishing of the design to increase component efficiencies. An example of such refinements was a decrease in labyrinth seal clearances to reduce air leakage. Another was a structural modification to correct a vibration problem in the inducer.[41]

A prototype J69 was first flown in the T-37 trainer on October 12, 1954 by Bob Hagan, Cessna's Chief Test Pilot. The first prototype engines flown were J69-T-15s. These engines were built by Turboméca and Americanized by Continental with the installation of all AN fittings, lines, etc., and the use of a ratiometer-type fuel control built by the Marquart Co.[42] Delivery of the first

---

41. Eli Benstein, telephone interview by Rick Leyes and William Fleming, 2 April 1990, Teledyne CAE Interviews, tape recording, National Air and Space Museum Archives, Washington, D.C.

42. "AN" refers to Army-Navy standard specifications for engine fittings, lines, etc.

920-lb-thrust production engines, designated the J69-T-9, began in 1955; 1,189 J69-T-9 engines were produced through 1959.

In 1954, CAE moved into a new research and engineering facility in Detroit, built to handle both engine development and production. However, rapidly expanding production soon made it necessary to look for additional space. In the spring of 1955, CAE took over the idle government-owned plant in Toledo, Ohio; on the following September 16, the first engine rolled off the new J69 production line at the Toledo plant.

However, even with J69 production at both Detroit and Toledo plants, all was not well with the program. Engine costs had exceeded the contract amount, and deliveries were behind schedule. Cessna was storing airplanes while waiting for engines. In 1957, Ray Powers, who had been Vice-President for Operations at Studebaker-Packard, joined CAE as Vice-President and Assistant General Manager at Toledo. Powers was placed in charge of production and, by 1958, had consolidated all J69 production in Toledo. By putting the right people in the right places, he soon had J69 production on schedule.[43]

Shortly after the T-37A became operational, another problem was encountered with the J69. On occasion, the engine experienced compressor surge during high-altitude maneuvers. The cause was a combination of inlet flow distortion and a relatively tight compressor surge margin that became tighter still at high altitudes. The T-37 had an S-shaped inlet ahead of each engine. In addition, a large nose cone on the front of the engine housed the starter generator. The combination of the duct shape and the presence of the nose cone produced an asymmetric airflow profile entering the engine. This flow distortion, combined with the tight compressor surge margin, resulted in a tendency for the engine to surge at altitudes of 20,000 to 25,000 ft. Of course, surge during training flights was most undesirable, because it was accompanied by an explosion-like bang "that scared the heck out of the young pilots."[44]

The Air Force lived with this situation briefly before deciding that it was intolerable. In 1955, CAE was directed to correct the surge problem and to increase the thrust. CAE undertook an extensive redesign of the J69-T-9 engine. Drawing on the inducer and diffuser technology of the J69-T-19 Firebee target drone engine, CAE redesigned the J69-T-9 compressor and increased its size. The turbine was redesigned to match the new compressor. The "new" engine, the J69-T-25, had an adequate compressor surge margin, a thrust increase from 920 to 1,025 lb, and it weighed 6 lb less than its predecessor. Describing the ex-

43. Wagner, *Continental! Its Motors and Its People*, 155-156.

44. Benstein to Leyes and Fleming, 2 April 1990.

Cessna T-37A military jet trainer powered by Teledyne CAE J69 engines. National Air and Space Museum, Smithsonian Institution (SI Neg. No. 95-1059).

tensive redesign, Eli Benstein commented, "We jacked up the J69 nameplate and installed a new engine under it."[45]

Production of the J69-T-25 for the T-37B began in 1959. By the time it ended in 1976, 1,804 engines had been produced. About 1,500 J69-T-25 engines still existed in 1990, powering the T-37B as well as foreign military aircraft.[46] As many as 1,400 of these engines, 1,300 operational and 100 spares, kept 638 T-37s flying at the Air Force Air Training Command bases. Production of spare parts to service these engines and keep them flight-ready continued to be an important segment of CAE's business through the 1980s. In fact, the manufacturing of J69-T-25 spare parts had been a primary function of CAE's Toledo plant from the early 1970s.[47]

In the early 1980s, it appeared that the T-37 might be approaching the end of its lifetime as an Air Force trainer. In July 1982, Fairchild was selected by the Air Force to develop a next-generation trainer. Designated the T-46A, it used the Garrett F109-GA-100 turbofan engine. Cessna's unsuccessful entry in the competition was to be powered by a new turbofan engine proposed by CAE. The

45. Benstein to Leyes and Fleming, 2 April 1990.

46. Benstein to Leyes and Fleming, 2 April 1990.

47. Hans Due, telephone interview by Rick Leyes and William Fleming, 2 April 1990, Teledyne CAE Interviews, tape recording, National Air and Space Museum Archives, Washington, D.C.

design of this new engine would draw on technology developed by CAE during its participation in the Air Force Advanced Turbine Engine Gas Generator (ATEGG) program. Although Cessna and CAE lost the T-46A competition, they came out winners in the long run. In March 1986, the Air Force discontinued the T-46 program in favor of reengineering and modifying the existing fleet of Cessna T-37 aircraft.

Although the T37 was an old airplane in the Air Force inventory and the highest-flying-time aircraft that any U.S. military service had ever flown, it was still doing a good job of pilot training. However, the T-37 was scheduled to be replaced during the late 1990s by a new Joint Primary Aircraft Training System trainer aircraft. Hans Due, J69 Program Manager, projected that the T-37 would likely continue to be flown by the Air Force for another 10 to 15 years before it was replaced. He said that as long as the airplane flew, CAE would continue building J69 spare parts and any additional engines that were needed.[48]

The J69-T-25 engine powered the T-37B for many years. No significant changes were made to the engine; it performed well and had developed a record of high reliability. According to CAE, by 1990 it had become the lowest-maintenance-cost engine in the Air Force inventory, largely because it could be easily dropped from the fuselage of the T-37 and taken to a hangar for servicing.[49] Another achievement for engine and aircraft: by 1990, the T-37B had become one of the safest aircraft in the Air Force inventory.[50]

T-37 SUMMARY

The T-37 trainer and the J69s that powered it were technical and financial successes for both Cessna and CAE. From 1955 until 1976, 3,016 J69 engines were produced for the T-37, nearly one-third of all J69 engine models manufactured. This included 23 J69-T-2, 1,189 J69-T-9, and 1,804 J69-T-25/A engines. Production of these engines, plus the long years of spare parts production and continued Air Force funding for J69 component improvement, comprised approximately one-third of CAE's total business over the years that the engine was used in the T-37. CAE anticipated that its production of J69-T-25 spare parts would

48. Due to Leyes and Fleming, 2 April 1990.

49. Benstein to Leyes and Fleming, 2 April 1990.

50. Lt. Col. James Tothacer, Chief of Fighter Trainer Branch, Headquarters Air Force Inspection and Safety Center, Norton Air Force Base, Cal., telephone interview by William Fleming, 16 July 1990, Teledyne CAE Interviews, National Air and Space Museum Archives, Washington, D.C.

continue through the 1990s and into the 2000s. Measured by the revenues it produced, the J69 family was CAE's most successful engine series.

THE J402 ENGINE

Some 20 years after CAE's entry into the aircraft gas turbine business with the highly successful J69 engine family, the company began development of what became its second most important engine family, the J402 series. A derivative of the J69 engine family, the J402 was an expendable short-life turbojet designed for use in short-range missiles. Its design was based on simplicity, low cost, and the Navy's requirement for zero maintenance after extended on-the-shelf storage. These requirements were new to the turbine engine industry, and the J402 was considered a unique engine. Development of the technology needed to produce such an engine was driven by the Navy's zero maintenance requirements for its new Harpoon anti-ship missile. As a result, the J402 became the first turbojet engine ever designed for long-term storage without any maintenance.

In 1972, following a competitive evaluation, CAE received a contract from the Navy to develop the J402-CA-400 (CAE Model 370) engine to power the McDonnell Douglas Harpoon (AGM-84A), an anti-ship cruise missile that could be launched from surface ships, aircraft, and submarines. It was launched and initially powered by a rocket booster stage, which was jettisoned after burnout. The J402 sustainer engine then powered the missile for the balance of the brief flight.

The contract award for the J402-CA-400 engine had been preceded by an extensive series of cruise missile studies by the Navy during the late 1960s, during which both solid-rocket and expendable turbojet propulsion systems were evaluated for the sustainer stage. The governing propulsion system requirements for the studies included:

- Low acquisition cost,
- Significant volume and diameter constraints, and
- A "wooden round" requirement, meaning no maintenance.[51]

The Navy decided in the early 1970s to use a turbojet sustainer engine to power the missile. The advantage of the turbojet over the rocket as a sustainer

---

51. Dennis E. Barbeau, "Versatile, Low Cost Turbojet Propulsion for Unmanned Vehicles," (Unnumbered paper presented at the Fourth International Conference on Remotely Piloted Vehicles Sponsored Jointly by the Royal Aeronautical Society and the University of Bristol, Bristol, U.K., 9-11 April 1984), 1.

engine was that the turbojet-powered missile carried only the turbojet's fuel, since its oxidizer was provided by the air passing through the engine. A rocket-powered missile carried both fuel and oxidizer. Hence, in a missile of given size and weight, the turbojet offered increased range.

The unique requirement of the Harpoon missile was that it be treated as a "wooden round" of ammunition. A wooden round was a hypothetical system that had 100% reliability, infinite shelf life, and required absolutely no special storage, maintenance, or handling treatment. Applied to missiles, the wooden round concept in the late 1960s and early 1970s had been almost exclusively the province of solid propellants.[52]

Translating the wooden round requirement to the J402, the engine was required to be capable of "on-the-shelf" storage for up to five years in a fully fueled and ready-to-go Harpoon missile. The engine would not be tested or inspected for several years while it was stored in such places as an Azroc launcher on shipboard or was flown on a number of missions under the wing of a Lockheed P-3 or Boeing B-52. The engine would be expected to have at least a 99% reliability to start, to accelerate to maximum power in 4 to 5 seconds, and to operate satisfactorily during the powered-cruise portion of the mission.

In addition to the wooden round requirement, there were several others imposed by the Harpoon: the engine had to fit into a missile slightly over one foot in diameter; its weight was to be about 100 lb; and it had to have a thrust-to-weight ratio of approximately five, higher than many other engines of the time. The engine also needed to have a specific fuel consumption in the same class as other small turbojet engines. Required in large quantities, the Harpoon had to be affordable; low cost was a mandatory design factor. The challenge facing CAE: could such an engine be made to work?

The 660-lb-thrust J402-CA-400 developed for the Harpoon weighed only 101 pounds for a thrust-to-weight ratio of 6.6, and had an sfc of 1.2. The J402 engine design was based on technology embodied in the J69-T-406 engine used in the supersonic version of the Firebee target vehicle. The J69-T-406 technology was transferred by aerodynamically scaling the engine down to 32% of its airflow capacity. Accordingly, the thrust was scaled from 1,920 to 660 lb, the major diameter from 22.3 to 12.5 inches, and the length from 44.8 to 29.2 inches.

52. Raymond Smith, "Development of the First Wooden Round Turbine Engine," figures and notes used in unpublished presentation at Society of Automotive Engineering Aerospace Vehicle Conference, Washington, D.C., 8-10 June 1987, National Air and Space Museum Archives, Washington, D.C. Much of the information on the J402 engine in this history was obtained from this reference.

Teledyne CAE J402-CA-400 turbojet engine used in the McDonnell Douglas AGM-84A Harpoon cruise missile. Photograph courtesy of Teledyne CAE.

The J402-CA-400 was a single-shaft, fixed-geometry turbojet with a sin-gle-stage transonic axial-compressor stage followed by a single centrifugal stage. The engine had an annular slinger-type combustor and single-stage axial turbine. About 20% of the airflow passed through hollow turbine inlet nozzle vanes be-fore entering the combustor. The regenerative cooling thus provided allowed the engine to operate at high turbine inlet temperatures without cooling losses. The ability to operate at elevated turbine temperatures contributed to the en-gine's high thrust-to-weight ratio. Engine life was reported to be one hour at rated thrust.[53]

To minimize cost and improve reliability, the number of parts was reduced, as was machining time, through greater use of castings, and the subsystems were simplified. No machined airfoils were used in the engine, investment castings were numerous, and assemblies were brazed into a single piece to reduce the part count. This resulted in the J402 rotor system including only 16 parts com-pared to 149 for the parent J69-T-406.

Bearing lubrication was provided by a self-contained and non–circulating sys-tem. The front ball bearing was contained in a small closed-sump unit requiring no pump or oil tank, and the rear roller bearing was grease packed. Extensive ef-fort was placed on development of the lube system for the forward bearing,

53. Frank Marsh, Toledo, Ohio, to Rick Leyes, unpublished note, December 1988, National Air and Space Museum Archives, Washington, D.C.

McDonnell Douglas Harpoon anti-ship missile, powered by a
Teledyne CAE J402-CA-400 turbojet engine, breaking the water
surface after being launched from a submarine. National Air and
Space Museum, Smithsonian Institution (SI Neg. No. 95-1057).

with nearly 80 configurations run on a test rig. The most difficult lube system
requirement to satisfy was that for an engine start after a minus 65 degrees F cold
soak, followed by acceleration to full speed in slightly less than eight seconds.[54]

There were no gears in the engine. The fuel pump was driven directly off the
rear of the engine rotor shaft and the alternator off the front of the rotor shaft,
so all three turned at 41,200 rpm.

The J402-CA-400 was required to be stored for periods up to five years without the benefit of any maintenance. To achieve this storage life in humid and
marine environments, materials having inherent corrosion resistance were used.
The compressor and turbine rotors and the combustor housing were fabricated

54. Harley Greenburg, interview by Rick Leyes, 9 August 1988, Teledyne CAE Interviews,
transcript, National Air and Space Museum Archives, Washington, D.C.

from corrosion-resistant steel alloys. The already good corrosion resistance of aluminum castings used for the air inlet and compressor casings was further enhanced by applying an electro-coat polyester coating.

To achieve the required starting reliability, the J402-CA-400 used pyrotechnic devices to accelerate the engine rotor from rest and to ignition rpm. This approach to engine starting was considered by CAE to be a first in the industry.[55] The rotor was accelerated by two ammonium-nitrate cartridges, the hot gas from which impinged on the centrifugal compressor to accelerate the engine to about 35% of rated rpm in two seconds. Ignition was provided by two magnesium-teflon igniter cartridges recessed into the combustor housing. All four cartridges were ignited simultaneously. By 1988, 422 Harpoons had been fired. On only two occasions were engine-related problems encountered, demonstrating a 99.5% reliability for the propulsion system.[56]

The designers were concerned that a conventional type of hydromechanical fuel control might not function properly after a lengthy storage period in the types of environment to which it would be subjected. They decided to use a full authority electronic analog fuel control. According to Dr. Raymond Smith, CAE Scientific Engineer, this was the first full authority electronic control to be used on a production jet engine.[57] Operation of this control proved to be highly successful even after being subjected to the engine's storage environment.

The test procedure for qualifying the J402-CA-400 engine was patterned after that used for solid rockets (which also operated only a few minutes at maximum thrust). CAE's Frank Marsh commented that the Navy's selection of Al Rezetta, a solid rocket engineer, to manage the program had a major influence on the design and success of the qualification program.[58] The philosophy he brought into the reliability program was that of running multiple tests of short duration in a variety of environmental conditions. In a 30-engine qualification program, each engine was required to start once and run for its short mission life. As a number of problems were identified and design changes required, the program was expanded to include three additional engines.[59]

55. Smith, "Development of the First Wooden Round Turbine Engine."

56. Greenburg to Leyes, 9 August 1988.

57. Smith, "Development of the First Wooden Round Turbine Engine."

58. Frank Marsh, interview by Rick Leyes, 9 August 1988, Teledyne CAE Interviews, transcript, National Air and Space Museum Archives, Washington, D.C.

59. D. E. Barbeau, "A Family of Small, Low Cost Turbojet Engines for Short Life Applications," (ASME Paper No. 81-GT-205 presented at the Gas Turbine Conference and Products Show, Houston, Texas, 9–12 March 1981), 6.

Field operations with the Harpoon missile demonstrated the high reliability of the J402-CA-400 engine. Of 131 Harpoon flights up to July 1980, only two confirmed engine failures and one possible were experienced, producing an engine reliability of 0.977 to 0.992.[60] In 25 launches between 1982 and 1987, there were no system failures. During the March 1986 Gulf of Sidra incident, Navy A-6 Intruders used air-launched Harpoons to sink two Libyan warships, a Combattante II missile boat, and a Nanuchka II missile corvette.

Combat reliability of the engine after several years of on-the-shelf storage was demonstrated during subsequent military operations in 1987 and 1988. Harley Greenburg, Director of Product Support and Improvement, described the highlights of those operations. During that period, 10 Harpoon missiles were launched and 10 hits were achieved. The engines used in these missiles were up to seven and one-half years old. Up to 1988, the oldest J402 engine fired in a Harpoon missile was eleven and one-half years old. To evaluate reliability on a continuing basis, three engines were selected at random by the Navy each year and tested at the Trenton Naval Air Propulsion Center. The engines were then disassembled and inspected with a formal report written on each.[61]

The J402-CA-400 engine was originally qualified in August 1974. The first production engine was delivered on April 30, 1975, and by March 1991, 6,068 engines had been delivered for use in the Harpoon missile. By that time, the Harpoon had become the most widely deployed missile in the Navy. It was also in the inventory of the Air Force and most of the free-world's naval forces. It was anticipated by CAE that J402 production would continue well into the 1990s with total quantities approaching 10,000.[62]

J402 TARGET VEHICLE APPLICATION

During the early 1970s, the J402-CA-700 (CAE Model 372-2A) engine was selected to power the Beech MQM-107 Streaker Variable-Speed Training Target (VSTT) Vehicle that was to be developed for the Army Missile Command. The mission of the MQM-107 was to tow a variety of gunnery banners and targets for missile evaluation and launch crew training and to serve as an aerial target for air-defense systems. Several modifications of the J402 were required for this application. To accommodate the Time Between Overhaul requirement of 15

---

60. Barbeau, "A Family of Small, Low Cost Turbojet Engines for Short Life Applications," 7.

61. Greenburg to Leyes, 9 August 1988.

62. Robert H. Snook, Toledo, Ohio, to Rick Leyes, letter, 27 September 1989, National Air and Space Museum Archives, Washington, D.C.

hours, turbine temperature was lowered and the maximum engine speed was reduced from 41,200 to 40,400 rpm. To extend the bearing life from one hour to
15 hours, the previously grease-packed rear bearing was lubricated by spraying it
with a small amount of fuel. The fuel used to cool the bearing was discharged
into the exhaust gases. A reusable start and ignition system was installed to provide a multiple-start capability. The direct-drive alternator located in the nose
cone of the J402-CA-700 was determined to be a cost-effective way to provide
electrical power required in the MQM-107. The turbine also was redesigned to
improve its performance for the longer-life mission of the J402-CA-700.[63] The
engine was rated at 640 lb thrust, slightly less than that of the J402-CA-400. Production of the J402-CA-700 began in 1975 and continued until 1984, by which
time 522 engines had been produced.[64]

The J402-CA-702 (CAE Model 373-8B), used in the MQM-107D, was a
company-funded growth version intended for use in missiles and unmanned
target vehicles. CAE initiated design of the engine in the mid-1970s after studies
had confirmed that adding an additional axial compressor stage to the J402
could increase the engine's thrust by nearly 50% without increasing the engine
diameter and with only a four-inch increase in length. The single-stage turbine
was redesigned to accommodate the higher work output required to drive the
two axial and single centrifugal compressor stages. The improved version of the
J402 engine was rated at 970 lb thrust and was capable of operating at altitudes
above 60,000 feet.

On October 24, 1985, Beech was awarded a contract by the Army for the
MQM-107D, an improved version of the Streaker, and, in the same year,
the J402-CA-702 was qualified for use in the MQM-107D. From 1986, when
production began, through March 1991, 454 J402-CA-702 engines were manufactured.[65]

SHORT LIFE ENGINES SUMMARY

The Navy's decision in 1970 to use an expendable short-life turbojet engine instead of a solid rocket in its Harpoon anti-ship missile produced engine requirements that resulted in a revolutionary change in turbojet design philosophy.

63. Marsh to Leyes, 9 August 1988.

64. Teledyne CAE, "J402-CA-700 Engine S/N's by Order," unpublished tabulation, 9 August
1988, National Air and Space Museum Archives, Washington, D.C.

65. Robert H. Snook, Toledo, Ohio, to Rick Leyes, letter and manuscript/data review, 1 May
1991, National Air and Space Museum Archives, Washington, D.C.

While companies independently developed their capabilities to design and produce small engines that embodied simplified fabrication concepts, reduced cost, and high reliability, CAE was unique in combining these design innovations in maintenance-free, short-life engines. By the late 1980s, CAE had established itself as a leading developer and producer of small, short-life turbojet engines.

Turboshaft and Turboprop Engine Development

In mid-1951, at the same time that CAE was converting Turboméca's Marboré II turbojet for use in the Firebee and T-37 trainer, the company was beginning to convert and market the Turboméca Artouste I and Artouste II turboshafts. These engines, respectively designated T51-T-1 and T51-T-3 by the military, were followed by three additional turboshaft engines, the T72, T65, and T67. In 1978, another division of Teledyne, the Teledyne Continental Motors (TCM) Aircraft Products Division, started development of its first turboprop engine, the TP-500. Unfortunately for CAE and TCM, none of the turboshaft engines ever reached production.

T51 ENGINE FAMILY

The T51 family comprised two engine models: the 280-shp XT51-T-1 (CAE Model 210) based on the Artouste I, and the 425-shp XT51-T-3 (CAE Model 220-2) based on the Artouste II. The XT51-T-1 was a simple single-shaft engine with a centrifugal compressor, annular slinger-type combustor, and two-stage turbine. The XT51-T-3 was a higher-airflow version of the XT51-T-1. A propeller gear box was developed for the XT51-T-1 engine, which was offered in turboprop and turboshaft versions.

CAE's marketing of these engines, beginning in 1951, was focused largely on the potential of these engines to replace piston engines in aircraft and helicopters. However, conceived by Turboméca in the late 1940s, the Artouste I and II had compressor pressure ratios of only about 4:1, and their specific fuel consumption was approximately twice that of the piston engine. Consequently, use of the XT51-T-1 and XT51-T-3 was limited to several experimental aircraft.[66]

The first of the T51 family to receive funding by the military was the 280-shp turboprop version of the XT51-T-1, flown in 1954 in an experimental

---

66. Ray Smith, interview by Rick Leyes, 9 August 1988, Teledyne CAE Interviews, transcript, National Air and Space Museum Archives, Washington, D.C.

model of Cessna's L-19 Bird Dog observation/reconnaissance airplane. The standard version of this airplane, the Army L-19A, was powered by a Continental 213 hp O-470-11 six-cylinder, horizontally-opposed, air-cooled engine. As part of an Army program to investigate the benefits of using small gas turbines in such aircraft, both CAE's XT51-T-1 and Boeing's XT50 were installed.[67]

The turboprop engine was viewed as having several advantages over the piston engine. It was projected that the turboprop would weigh less, simplify engine installation, eliminate cooling problems, and almost completely suppress vibration. In addition, the turbine engine could operate on a variety of fuels, including diesel fuel and any grades of automotive or aircraft fuels. Operation of the XL-19 with the XT51-T-1 and the XT50 confirmed these advantages. However, neither of these gas turbines could compete with the piston engine's superior fuel economy and flight endurance.

The 280-shp turboshaft version of the XT51-T-1 first flew in 1954 in an experimental version of the Bell H-13 helicopter. The operational version of the H-13, used commercially and by the Air Force, was powered by the Lycoming 260-hp VO-435 horizontally-opposed, air-cooled engine.[68] The 425-shp XT51-T-3 was used experimentally in the Sikorsky XH-39 helicopter, an experimental turboshaft-powered version of the Sikorsky S-52 utility helicopter developed for Air Force and the Navy. Mounted on top of the fuselage, the XT51-T-3 replaced the S-52's 245-hp Lycoming O-425-1 piston engine. Powered by the XT51-T-3, the prototype XH-39 set a helicopter world speed record of 156 mph on August 29, 1954, and a helicopter altitude record of 24,500 feet on the following October 17.[69]

The turboshaft engine enjoyed the same advantages over the piston engine in the helicopter as it had in fixed-wing aircraft: low engine weight; simplified installation; and elimination of cooling problems and engine vibration. The gas turbine offered another important advantage in helicopter installations. Its small size, compared to piston engines, and its configuration permitted the gas turbine to be mounted above the fuselage, leaving the cabin space near the center of lift available for cargo and passengers. This altered the design of helicopters once the turboshaft had achieved suitable fuel economies.

Although the XT51 family of turboshaft engines had only limited experimental use, they made an important contribution in that their early experimen-

---

67. *Jane's All the World's Aircraft 1955-56* (London: Jane's All the World's Publishing Company Limited, 1955), 231.

68. *Jane's All the World's Aircraft 1955-56*, 220.

69. *Jane's All the World's Aircraft, 1955-56*, 314-315.

tal applications demonstrated the advantages of small gas turbines in light aircraft and helicopters. The next step necessary to realize their potential was to improve fuel economy.

T72-T-2 ENGINE

In 1959, with a view to expanding its product line, CAE initiated development of a turboprop engine in the 500-shp class with a fuel economy competitive with that of a piston engine. The development of this engine, the T72-T-2 (CAE Model 217-5), was company funded.

The T72 design contemplated a two-spool, free turbine mechanical arrangement, the development of which could be time consuming without the further complication of designing new aerodynamics into the engine. Consequently, to minimize development time and company investment, CAE decided to scale the engine aerodynamically from previously developed CAE and Turboméca engines.

The T72 engine had a single-stage axial compressor followed by a single centrifugal stage, an annular slinger-type combustor, a two-stage gas generator turbine, and a single-stage free turbine driving a front power takeoff. The axial compressor was a CAE transonic design with a pressure ratio of 1.7. The gas generator and power turbines were also derived from earlier CAE engine designs. The centrifugal compressor and the annular slinger-type combustor were derived from Turboméca designs. Using this scaling approach, the T72 passed its first Preliminary Flight Rating Test (PFRT) in early 1961, only 20 months after the start of design.[70]

The T72 was initially intended for use in general aviation aircraft. However, in 1960, the Marines expressed interest in a turboshaft-powered Assault Support Helicopter (ASH) to carry a six-man squad. The T72 development objectives were altered to accommodate this new application, and the engine was actually built as a 525-shp turboshaft to power the ASH. To demonstrate the capability of its new engine, CAE successfully flew the T72-T-2 in the Republic Lark (license-built Sud-Aviation Alouette II) helicopter six months after the engine passed its PFRT.

By December 1960, the Navy, the contracting agent for the Marines, released a formal request for the proposed ASH. However, soon after the T72 passed its

70. I. W. Nichool, "The T65 and T72 Shaft Turbines," (SAE Paper No. 624A presented at the Society of Automotive Engineers Automotive Engineering Congress and Exposition, Detroit, Michigan, 14-18 January 1963), 1-3.

PFRT, the ASH was canceled by the Navy, because the Marine Corps' need for an assault support helicopter was too urgent to wait for a new design.[71] Consequently, a decision was made to buy an "off-the-shelf" helicopter. It was announced in March 1962 that the already operational Army Bell HU-1B was selected in place of the ASH.[72]

Although the T72 was in the same power class as the Pratt & Whitney Canada PT6, it never competed in the commercial commuter aircraft market with the PT6 because CAE concentrated its marketing on military applications. With the ASH canceled, and the T72 too small for medium military helicopters and too large for light helicopters, CAE was unable to find any military applications for it. Its development was discontinued.

T65-T-1 ENGINE

About the time the T72 was discontinued, a series of events occurred that eventually led to development of the T65 engine. By 1961, CAE's sales had dropped to just over half those of two years earlier. Contributing to this drop were reduced Air Force requirements for T-37s and Firebee target drones, a slower-than-anticipated buildup in Navy Firebee target drone production, and smaller Air Force orders for CAE's MA-1A gas turbine-powered air generators.

Eli Benstein, at that time Director of Advanced Technology and later CAE's Chief Scientist, and two of his associates, Doug Oliver, Manager of Advanced Structural Design, and Marvin Bennett, Director of Marketing, were instructed to get out and acquire new business wherever they could find it. A promising opportunity opened in connection with the 250-shp Allison T63, chosen to power the Army's Hughes 500 Light Observation Helicopter (LOH). CAE had not participated in that engine competition. But, by 1961, Allison's T63 was in trouble. Compressor efficiency was low, turbine blade tips were rubbing, turbine wheels had failed, and compressor inducer blade failure had occurred.[73]

During a drive from Detroit to Dayton, Eli Benstein and his boss, Marv Bennett, were discussing Allison's problems with the T63. Benstein observed

71. Raymond Smith and Frank Marsh, interview by Rick Leyes, 9 August 1988, Teledyne CAE Interviews, transcript, National Air and Space Museum Archives, Washington, D.C.

72. "Navy to Purchase Assault Helicopters from Bell for Marine Corps," Department of Defense Office of Public Affairs News Release No. 320-62, 2 March 1962, Washington, D.C.

73. David A. Anderton, "Allison's T63 Turbojet Engine Can Grow to Meet Army Light Helicopter Requirements," *American Aviation*, Vol. 27, No. 11 (April 1964), 59-61.

that there ought to be a competitor for Allison's T63 engine. He recalled that, as they drove, they developed the concept for such a competitor engine, a 300-shp engine scaled to one-half the size of the 600-shp T72.

Soon afterward they approached Ray Bittner, in the contracting branch at Wright Field, with their concept for a competing engine for the ailing T63. Bittner reportedly thought it was a good idea, and CAE followed up by talking about it to other people in the military.[74] Later, a request for proposal was issued for the development of an alternate engine to power the Hughes 500 Light Observation Helicopter. It was possible that CAE's marketing efforts influenced the Army's decision. CAE competed against Boeing and won the alternate engine competition.

CAE decided to scale down the Turboméca Astazou II engine, from which the T72 had been derived. The design point of the T65 was 305 shp to permit flat rating at 250 shp under sea-level, standard day conditions. CAE knew that a geometric scale-down of the Astazou engine would run well the first time out, if they didn't botch the mechanical and structural design. As a result of this design approach, the T65 completed its 50-hour PFRT only 16 months after design was started.[75]

The engine ran well. However, in 1963, CAE was unexpectedly confronted with a new Air Force requirement that wound up imposing a one-year delay in the engine's development schedule. The Air Force wanted an automatic fuel flow cutoff added to the engine to prevent the engine from overspeeding and self destructing in the event the helicopter rotor shaft became decoupled from the engine. Without such an override, the free turbine would accelerate very rapidly, resulting in its destruction in about one and one-half seconds. Allison's T63 had satisfied this requirement earlier.

There was debate within CAE as to whether the redesign should be performed prior to requesting a T65 production order from the Air Force or should CAE immediately press the Air Force to place a production order for the T65 as a second-source engine. This second alternative would allow CAE to accomplish the redesign work while the engine was being tooled up, allowing production to begin as soon as the redesign was completed. CAE's engineering department had a very strong influence over how the company was operated. Ed Hulbert, the Vice President of Engineering, decided to proceed with redesign because, in his view, it was really the right thing to do.[76]

---

74. Benstein to Leyes, 9 August 1988.

75. Nichol, "The T65 and T72 Shaft Turbines," 1-2.

76. Marsh to Leyes, 9 August 1988.

Teledyne CAE T65 turboshaft engine that competed unsuccessfully against Allison's T63 as an alternate engine for the Army's Light Observation Helicopter. Photograph courtesy of Teledyne CAE.

During the ensuing one-year delay to modify CAE's T65 design, Allison received a follow-on production contract for its T63 that included full funding for production tooling. When the T65 redesign work was completed, the military argued that it had already paid for Allison's production tooling and therefore would not fund a second set of tooling for CAE. To be cost competitive in a toe-to-toe competition with Allison, CAE would have to expend approximately $900,000 for T65 tooling. Unwilling to risk such an investment, CAE decided to terminate its effort to compete with Allison's T63.[77] Reflecting on the decision to proceed with the T65 redesign, Frank Marsh said that, ". . . from marketing's standpoint, it was probably a mistake because, if we hadn't spent a year doing that to make the engine technically better, we might not have lost the competition. We might have been it."[78] Marsh felt that the redesign could have been done later, enabling CAE to be in there early enough to compete with Allison for the production tooling funds. Some at CAE, in retrospect, suspected that an Air Force objective all along had been to spur Allison to solve the T63 problems that it faced.

77. Benstein and Marsh to Leyes, 9 August 1988.

78. Marsh to Leyes, 9 August 1988.

Although unsuccessful in the Hughes 500 helicopter competition, CAE felt that it could successfully compete with Allison in other markets for the T65 engine. In 1964, CAE had received an offer from the Bell Aircraft Co. for a contract to deliver 100 T65 engines. Reportedly, Bell was uneasy with the T63 engine it was using, and Bart Kelley, Bell's Vice President of Engineering, had called CAE to make the offer and express a desire that CAE tool up and put the T65 into production.[79] However, an order for only 100 engines would not begin to cover the cost of the tooling. CAE had to decide whether the company should risk its own funds for production tooling in the hope that sufficient future orders would be received to cover the costs. After a series of studies, a decision was made that the company should not make this investment.

The decision was made by Continental's corporate management in Muskegon. Continental had for years been a piston engine manufacturer. Gas turbines were regarded by some in the company hierarchy as an offshoot of the company's primary business. Continental's choice lay between funding T65 production tooling at CAE's Toledo plant or providing funding for Continental's Aircraft Products Division in Mobile, Alabama, to develop and produce the Tiara, a new series of piston engines. Continental opted to fund the Tiara engine series.

The Tiara engines were to have been wonder engines, but they were commercially unsuccessful. Their fuel consumption was high, and by the time they were introduced in the late 1960s, fuel economy had become an important consideration. Further, the Tiara's performance was not significantly better than existing engines. Airframers found it difficult to justify the engine's certification costs for installation in their aircraft. The combination of poor fuel economy and the problem with marketing the engine led the Aircraft Products Division to discontinue production of the engine in the late 1970s.[80]

TURBOMÉCA REVISITED

In 1961, about the same time that CAE began design of the T65 by scaling the Turboméca Astazou II engine to the necessary size, CAE returned to Turboméca in France for technology to be used in the design of a new twin-turboshaft engine, the T67. Among the engines that Turboméca had developed

79. Marsh to Leyes, 9 August 1988.

80. William Brogdon, telephone interview by Rick Leyes, 18 July 1990, Teledyne Continental Motors Aircraft Products Division Interviews, National Air and Space Museum Archives, Washington, D.C.

was the Astazou X, an engine with a high-efficiency compressor having two transonic axial stages followed by a single centrifugal stage. To attain the level of fuel economy desired in the T67, CAE needed the high efficiency and increased pressure ratio of the Astazou X.

In 1961, CAE obtained from Turboméca the sales and manufacturing rights for the Astazou X.[81] Subsequent application of the Astazou X compressor design by CAE produced a fuel economy for the T67 twinpack engine that was nearly a 20% improvement over that for the earlier T72-T-1.

T67-T-1 ENGINE

Developed with company funding, the T67 was seen as an engine that would be used in military and commercial fixed-wing aircraft and helicopters. Design of the engine, begun in 1963, was based on the use of two CAE Model 217-10 engines connected by a common gearbox to a single power-output drive. The CAE Model 217-10 engine was a new design using the two transonic axial-compressor stages and the centrifugal-compressor stage of the Astazou X. The T67-T-1 (CAE Model 217A), a 1,540-shp twinpack engine, was designed with overrunning clutches that allowed either twin- or single-power-unit operation in flight, as well as automatic single-power-unit operation if one unit failed.

The T67-T-1 completed its 60-hour PFRT in 1964. In 1965, the engine was flown experimentally on a Bell UH-1D Iroquois utility helicopter. However, its use in the Iroquois was only experimental, as the Lycoming T53 turboshaft had already been selected to power the production version of the Iroquois.

By 1967, the T67 had been uprated to 1,600 shp, and it was offered by CAE in the competition for an engine to power the Bell UH-1 twin-engine helicopter. CAE's competition was Pratt & Whitney Canada with its twin version of the PT6 turboshaft engine (military designation T400).

CAE put a major effort into this competition, hoping to recover its investment in developing the engine and funding most of the PFRT. However, Pratt & Whitney Canada's twin PT6 was selected. The U.S. military version of the UH-1 was designated UH-1N and the Canadian military version CUH-1N. CAE viewed the success of Pratt & Whitney Canada in this competition as being due to several factors. CAE acknowledged that Pratt & Whitney Canada had a good proposal. Further, the basic PT6 was a mature engine that had been in commercial used for several years. Although regulations required U.S. mili-

81. Benstein and Marsh to Leyes, 9 August 1988.

tary aircraft and their components to be made domestically, Pratt & Whitney Canada formed a subsidiary to build their engine in West Virginia, an economically depressed area.[82]

In addition to these factors, the Canadian government had an interest in acquiring a twinpack version of the Bell UH-1 helicopter powered by the Canadian-built twin PT6, and provided financial support to Pratt & Whitney Canada for development of the engine's twin gearbox and the tooling to produce it. With this support, Pratt and Whitney Canada was able to submit a substantially lower bid price than could CAE. The price differential was due mainly to the production tooling costs included in CAE's bid. With the loss of this competition and with no other application of the T67 in sight, work on the T67 was discontinued. This decision marked the end of turboshaft development by CAE.[83]

TP-500 ENGINE

Although Teledyne CAE discontinued development of turboshaft engines in the late 1960s, one additional turboprop engine was developed later by another division of Teledyne, the TCM Aircraft Products Division, in Mobile, Alabama.[84] In 1978, TCM began development of the 500-shp TP-500 turboprop engine. TCM's principal products were a series of four- and six-cylinder horizontally-opposed, air-cooled engines for single- and twin-engine aircraft. These engines, ranging from 100- to 520-cubic-inches displacement, had been in production for a number of years.

In the late 1970s, TCM perceived that a possible market niche existed for a turboprop engine at the top end of the six-cylinder turbosupercharged engine power spectrum as a low cost replacement for piston engines. Although the gas turbine field was new to TCM, several engineers in the company had come from turbine engine manufacturers. Drawing on this limited expertise, a small engineering staff was assembled to perform a market survey. The survey indicated that a 500-shp turboprop would indeed be attractive as a replacement for six-cylinder turbosupercharged engines in six- to eight-passenger, twin-engine business aircraft. The survey also indicated that penetrating this market would require development of a low-cost engine. Cost became the primary driver in the TP-500 design.

82. Benstein to Fleming, 25 July 1990.

83. Benstein and Marsh to Leyes, 9 August 1988.

84. In 1972, Teledyne Continental Motors was split into three divisions, one of which was the TCM Aircrafts Products Division that specialized in aircraft piston engines.

According to Kevin Brane, TP-500 Project Manager, a number of factors led TCM to enter the gas turbine field.[85] The company had reached the peak displacement of piston engines that it could market without making a major new investment, and the company's sales had begun to fall. It observed that turboprop manufacturers had entered the commuter aircraft market and were doing well. Further, the turboprop offered an opportunity to diversify TCM's product line by taking a step in the direction of higher-power engines for business and commuter aircraft, and the turboprop also offered a reliable and attractive engine for the retrofit market.

Preliminary design of the TP-500 began in June 1978. Because of its limited gas turbine expertise, TCM believed that the most expedient way to enter the gas turbine business would be to contract out the preliminary design work. TCM awarded a contract for the engine's preliminary design to Noel Penny Turbines, a small research and development company in Great Britain. TCM eventually developed a TP-500 engineering staff of 40 to 45 people, including three Noel Penny personnel who spent about a year assisting in the initial development work.

The TP-500 was a single-shaft turboprop with a single-stage centrifugal compressor, reverse-flow annular combustor, two-stage turbine, and two-stage planetary gearbox. The main consideration leading to selection of the single-shaft configuration was the driving design objective of low cost, which meant minimizing the number of engine parts and avoiding mechanical complexity. Also, the single-shaft arrangement was considered simpler for retrofit installations. Weight was not an important factor in the design, because in the retrofit market the piston engines being replaced were relatively heavy compared to the gas turbine.

Although there was little interface with CAE engineers during development of the TP-500, CAE did provide technical assistance when called upon, shared its data and technical expertise, and made technical inputs (and was reimbursed). CAE's willingness to provide assistance reflected that the TP-500 was no threat to CAE's business. The TP-500 was intended for the general aviation market; CAE was building small turbojet engines for the military market.

Component development testing began with a compressor rig test in 1979. Tests of other components culminated in the first flight of the engine in September 1982 on a Piper Cheyenne flying testbed, which had a TP-500 on the

85. Kevin Brane, interview by Rick Leyes, 2 November 1989, Teledyne Continental Motors Aircraft Products Division Interviews, tape recording, National Air and Space Museum Archives, Washington, D.C.

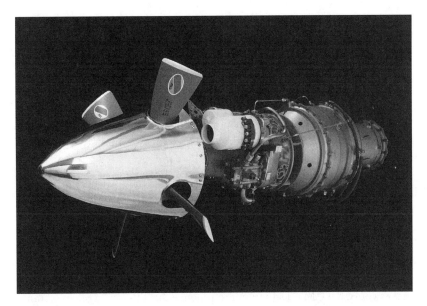

This Teledyne Continental Motors TP-500 turboprop engine was developed for use in general aviation aircraft. Photograph courtesy of Teledyne TCM Aircraft Products Division.

left side and a Pratt & Whitney Canada PT6 on the right. The first production-configuration engine was delivered for flight testing in March 1988, and by the following October, the engine had received its FAA Type Certificate.

Several applications were projected for the engine. One of the first was the Beech Baron, a twin-engine business aircraft in the 5,500-lb gross weight class. Later, Piper showed interest in using the engine to power its twin-engine Navaho, a business aircraft in the 6,500-lb gross weight class. Some consideration was also given to using the engine in the single-engine Piper Malibu. Principal competitors were Garrett's TPE331, Pratt & Whitney Canada's PT6, Allison's Model 250-series engines, and Textron Lycoming's LTP101, all in the same power class. Although this array of competitors might appear formidable, TCM felt that the TP-500 could effectively compete with these engines in the retrofit market on the basis of its low cost.

By May 1989, TCM had spent several million dollars in developing the TP-500 and obtaining its FAA Type Certification. Production configuration engines had been delivered for flight testing, and the engine was ready to be produced once production tooling was in place. However, at that time, the project was suddenly placed on hold by the company, and the TP-500 engineering staff was reassigned. Ironically, this action resulted from the success of a TCM ground-based gas turbine engine for the Army and Air Force—the RGT-3600—that used facilities and manpower assembled for the TP-500.

In 1983, TCM began developing the RGT-3600. This engine was a 450-shp class regenerative gas turbine designed to supply power for a variety of ground based applications. In 1987, TCM received an $8.7 million contract from the Army for a generator set to provide power for its Patriot missile system, and in 1988, a $48 million contract from the Air Force for a ground power generator system. Both units were based on the RGT-3600.

With these two major military contracts, TCM found itself stretched to the limit. TCM was also concerned that the retrofit market might not pay for TP-500 production tooling. A decision was made to hold TP-500 production until sufficient customers were found and concentrate all company gas turbine expertise on the RGT-3600. In early 1989, engineering effort on the TP-500 was halted and reassigned to the RGT-3600.

With his project on hold, Kevin Brane observed that the ball was in the marketeer's court.[86] TCM's decision not to produce the TP-500 engine was similar to CAE's earlier decision not to produce its T65 engine, in favor of placing its efforts on more promising market areas.

TURBOSHAFT ENGINES SUMMARY

CAE's experience in the turboshaft marketplace was disappointing to say the least. There had been a history of clever and innovative engineering by CAE in satisfactorily developing its several turboshaft engines. However, the company had been unsuccessful in marketing those engines for a variety of reasons.

Near the end of the 1960s, the company redirected its business approach. It concluded that its major market was unmanned special purpose aircraft such as reconnaissance drones, target vehicles, missiles, and decoys. Although CAE was interested in developing engines for manned aircraft other than trainers, its resources were too limited at that time to pursue both unmanned and manned applications. Accordingly, the company focused its major efforts on unmanned applications, which were believed to offer the greater near-term business potential. By that time, CAE had begun development of its first turbofan engine for unmanned applications.[87]

86. Brane to Leyes, 2 November 1989.

87. Michael L. Yaffee, "Small Gas Turbines Developed," *Aviation Week & Space Technology*, Vol. 93, No. 18 (November 2, 1970): 40-43.

Turbofan Engine Development

By 1967, CAE had begun to expand into the field of turbofan engines, leading to what might be typified as a "bad news-good news" scenario. The bad news was that CAE was unsuccessful in competing for military contracts to produce the two turbofans that it developed, and was unsuccessful in marketing and obtaining contracts for a Turboméca turbofan licensed for U.S. production. The good news was that as a result of CAE's work in developing its two turbofan engines, the company secured a very profitable turbofan engine production contract extending over a number of years. However, this production contract was for its primary competitor's engine, the Williams F107.

CAE MODEL 472 ENGINE

In 1967, CAE began design studies of the Model 472 turbofan engine, later given the military designation of F106-CA-100. Development of this engine was undertaken as a company funded program. The engine was intended to compete with the Williams F107-WR-100 engine in the Subsonic Cruise Armed Decoy (SCAD) missile competition.

Boeing, Lockheed, Beech, and Northrop were engaged in design studies for the SCAD missile. A major issue among the airframe designers was the level of engine thrust required during the cruise mode. The Beech design required a relatively large engine; Boeing's called for a very small engine. This disparity in thrust requirement among the competing airframe designers left CAE in the difficult position of having to select an engine size and hope that it was in the right ball park.

CAE's initial specification was for an engine having slightly more than 600 lb thrust with some capacity for growth. Most of the competing airframe designers called for an engine of only 500 lb thrust, and Boeing wanted an even smaller engine. However, when the military specifications for the engine were finally issued, CAE's engine fit them well. Later, when Boeing went through its first structural analysis, the weight of the missile rose, and it turned out that CAE's thrust selection of 600 lb was right for Boeing too.

The Model 472 engine was a CAE design except for the annular combustor with its slinger-type fuel injector, the latter a variation of the Turboméca design for a two-spool engine. It incorporated a manifold slinger injector designed by CAE and adopted for use in all of CAE's engines. The Model 472 was the first two-spool engine fully designed by CAE. Earlier CAE two-spool turboshaft engines had been derivatives of Turboméca designs. The low pressure spool consisted of two fan stages driven by a two-stage turbine. The high pressure spool was composed of three axial-flow stages followed by a centrifugal-flow stage driven by a two-stage turbine. The Model 472 was a compact engine, which could best be illustrated by comparing it with CAE's J402, a turbojet of equivalent thrust used in the Harpoon missile. The diameter and length of the two engines were virtually the same. However, instead of the J402's three stages of rotating machinery, the Model 472 had 10 stages in the same space.

Using company funds, CAE completed the design, fabrication, and performance testing of the engine core by early 1972. At that time, the Air Force held an industry-wide competition for the development of the cruise missile engine. Two contracts were awarded on May 31, 1972, one to CAE and the other to Williams International.[88] The two contracts called for delivery of prototype demonstration engines in six months for testing at the Air Force Arnold Engineering Development Center at Tullahoma, Tenn. This gave CAE six months to develop the low pressure spool and deliver the demonstration engines. The engines were delivered to Tullahoma in November 1972, and the competitive runoff with the Williams F107 was conducted until February 1973.

There were two aspects of the competition considered by the Air Force final evaluation. One was engine hardware and performance validation in simulated altitude tests at Tullahoma. The other was the PERT-type (Program Evaluation and Review Technique) critical-path schedule network, a cost-schedule control system, a configuration management system, and an integrated logistics support system.[89]

CAE's engine performed well during the simulated altitude tests. However, in February 1973, following evaluation of the engine test results and the proposed management control systems, the Williams F107 was declared the winner. At that time, further work on the F106 (Model 472) engine was discontinued. However, it was soon revived in modified form as the Model 471-11DX engine.

CAE MODEL 471-11DX ENGINE

In December 1972, the Navy had awarded study contracts for its Sea-Launched Cruise Missile (SLCM) to Vought, General Dynamics, Boeing, Lockheed, and McDonnell Douglas. In January 1974, Vought and General Dynamics were se-

88. At that time, the CAE Model 472 engine was given its F106-CA-100 military designation.

89. Robert Anderson and Frank Marsh, interview by Rick Leyes, 9 August 1988, Teledyne CAE Interviews, transcript, National Air and Space Museum Archives, Washington, D.C.

Teledyne CAE Model 472 (military designation F106-CA-100) turbofan engine.
Photograph courtesy of Teledyne CAE.

lected by the Naval Air Systems Command to develop and demonstrate competing prototype missiles. The Vought BGM-110 missile was to be powered by the CAE Model 471-11DX engine and the General Dynamics BGM-109 missile by the Williams F107-WR-100.

The Model 471-11DX was a modification and refinement of the F106-CA-100 engine. The thrust was increased from the 600- to the 650-lb class, and the compressor was further developed to improve its efficiency. The engine was structurally strengthened to withstand 100g-plus forces from nuclear depth charges, compared to a 15g capability for the F106. Accessories were rearranged from a chin position for the F106 to the top of the engine, giving the Model 471-11DX a circular outline more suitable for torpedo tube launching. One stage of the two-stage low pressure turbine driving the fan was eliminated. To improve combustion efficiency with the high-density, 140,000-Btu/ gallon hydrocarbon fuel, selected by the Navy to maximize the missile's range, the hole pattern controlling airflow and mixing in the combustor was modified.

CAE began running a complete Model 471-11DX engine incorporating these modifications in May 1975. The engine was designed for a 50-hour life as was the F106 from which it was derived. Cast components were used extensively throughout the engine. For example, each fan, compressor, and turbine rotor stage had integrally cast blades and disk.[90] In late 1975, two Model 471-DX engines were shipped to the Air Force Arnold Engineering Development Center (AEDC) for integration tests in Vought's BGM-110 missile. The main objective of these tests, conducted in the AEDC 16-ft transonic wind tunnel, was to determine engine inlet compatibility and to measure vehicle drag under powered conditions.[91]

A fly-off competition between the Vought and General Dynamics missiles was held by the Navy in February 1976. The General Dynamics BGM-109 made two successful launches demonstrating transition from underwater launch to in-flight glide. Through no fault of CAE's engine, both of the Vought BGM-110 demonstration launches ended in failure. As a result of these failures, and a $4 to $5 million cost overrun announced by Vought at that time, the Navy stopped all further work on the BGM-110 and discontinued support for the 471-11DX. On March 17, 1976, the Navy awarded General Dynamics the airframe contract for the F107-powered BGM-109, designated the Tomahawk missile. Two months later, the Navy named Williams the winner of the engine contract.[92]

There was one last brief attempt to market the Model 471-11DX engine to the military. By 1976, the Williams F107 had the inside track as the favored engine for all cruise missiles. However, a decision had been made by the military that there would be a second manufacturing source for all cruise missile components. Consideration was given to an industry-wide competition to develop a second engine. CAE approached the military offering the Model 471-11DX. But there were two factors working against CAE. One was the military's recognition that parallel development of two engines would be an expensive proposition. At that time, further development work was still needed on both the F107 and 471-DX engines. The other factor was that after being approached by the military, Williams had agreed to give the military the rights to license another company to manufacture the F107. This agreement was reached with the provision that Williams would be guaranteed a large percentage of the first production order. These

90. "Cruise Missile Engine Design Pushed," *Aviation Week & Space Technology*, Vol. 103, No. 1 (July 7, 1975): 41-43.

91. "Vought Ends BGM-110 Cruise Missile Tests," *Aviation Week & Space Technology*, Vol. 104, No. 1 (January 19, 1976): 39.

92. Kenneth P. Werrell, *The Evolution of the Cruise Missile* (Maxwell Air Force Base: Air University Press, 1985), 155.

two factors sounded the death knell for the Model 471-11DX engine. Early in 1977, Williams was given the go-ahead to complete development of the F107 engine for use in both the Navy and Air Force versions of the cruise missile.

F107 ENGINE PRODUCTION

Following the contract award to Williams International in 1977 for completion of F107 development, the company held an industry-wide competition for a second-source producer of the engine. This was held in conformance with Air Force requirements for the selection of second-source contractors. Bids were submitted to Williams by several companies including CAE, Garrett, and Sunstrand. The competition was monitored by the Joint Service Cruise Missile Projects Office, and in September 1978, CAE was selected.

Although the second-source production approach for the F107 engine was ultimately successful from the viewpoint of Frank Marsh, a CAE Vice President in charge of the company's F107 manufacturing program, developing the expertise and implementing the procedures leading to CAE's production of the engine was difficult.[93] Many of the problems encountered initially were people and attitude related. Although the top management of both companies agreed to share information and to work together, convincing the people at the working level to talk openly to each other was a problem. In the opinion of those at CAE, the Williams engineers found it psychologically difficult to give their data freely to a prime competitor.

It took three years for CAE to set up its production program. In the first year, CAE made only a few engine parts. During the second year, Williams brought a former government employee into the company and placed him in charge of the licensing activity. Under his direction, the information exchange process improved. By the end of the second year, CAE was building a number of parts and had put into place the production processes and tooling for building them. In the third year, CAE built its first three engines, which were run successfully through a qualification and substantiation program.[94]

CAE's first production F107 engine was shipped from its Toledo plant in October 1982. CAE found that its prices were higher than those of Williams, and, as a result, CAE's engine orders decreased compared to Williams's orders. In view of the higher costs and the limitations in production capacity at its Toledo plant, CAE decided that unless F107 production was moved elsewhere, the company would find itself out of the business.

93. Marsh to Leyes, 10 August 1988.
94. Marsh to Leyes, 10 August 1988.

After obtaining a guarantee of sorts on a future production share if its prices were acceptable, CAE established a new plant in Gainesville, Georgia. Using automated storage and retrieval of manufacturing materials, as well as computer-controlled machinery and management information systems, this plant was designed as a modern and efficient manufacturing facility. CAE funded the purchase of the building and installation of the tooling itself. Although development of the Gainesville plant was stimulated by the need for more cost effective F107 production, other CAE engines were also built there.[95] The Gainesville plant was completed in 1986, and production of both the F107 and the J402-CA-702 moved there from Toledo. The Toledo plant was retained as the focal point for CAE's technology and engine development activities, as well as for production of the J69 engine, J69 spare parts, and the J402-CA-400 Harpoon engine.

The F107 production program was financially beneficial to CAE, and it kept the company active in the turbofan engine field for a number of years. However, in December 1991, CAE delivered the last of 861 F107s. In 1989, believing that the requirement for Tomahawk missile production was nearing its end, the Navy would no longer support dual-source production by both CAE and Williams. A competition was held to select one source for the last two planned engine production lots. Williams was selected, and CAE production of the F107 was scheduled for termination. Then, came the Persian Gulf War in 1991, during which the Tomahawk missile was highly effective. Production of the Tomahawk weapon system, and its F107 engine manufactured by Williams, was extended indefinitely.[96]

LARZAC 04 ENGINE

During 1973, about the same time that work was beginning on the 471-11DX, CAE obtained a license to manufacture another Turboméca engine in the U.S. This engine was the 3,000-lb-thrust Larzac 04 turbofan developed by France's two turbine engine manufacturers, Turboméca and SNECMA, in a joint entity known as Groupement Turboméca-SNECMA. This engine, first run in France in 1969, was used in the Alpha Jet light trainer aircraft developed and produced by Dassault-Breguet of France and Dornier of West Germany. The Larzac engine first flew in the Alpha Jet 01 on October 26, 1973, and by January 1975, the engine had completed its 150-hour endurance test.

95. Marsh to Leyes, 10 August 1988.

96. Apel to Leyes, 26 May 1995.

The Larzac 04, the CAE version of which was designated the Model 490-4, was a two-spool low-bypass engine featuring modular design. A two-stage fan on the low pressure spool, providing both bypass air and air to the high pressure spool, was driven by a single-stage uncooled turbine. The high pressure spool consisted of four axial stages driven by a single-stage turbine with air-cooled blades and vanes. The combustor was annular with pre-vaporizing fuel tubes. The engine cycle was optimized to produce high performance at low altitudes and high-subsonic flight speeds.[97]

CAE had entered into the licensing agreement with Groupement Turboméca-SNECMA with the initial objective of using the engine for missiles and commercial aircraft. About 1980, subsequent to CAE's unsuccessful efforts to find a commercial application for the Larzac engine, the Navy issued a Request for Proposal for its VTX advanced trainer, later designated the T-45. Lockheed, with licensing rights to manufacture the Alpha Jet trainer in the U.S., formed a team with Dassault-Breguet and Dornier in the competition. Licensed to produce the Larzac engine in the U.S., CAE supported Lockheed's team.[98] Six aircraft designs were submitted in the competition. The eventual winner, announced in the fall of 1981, was the international team headed by McDonnell Douglas and including British Aerospace as the principal subcontractor. They had offered a modified version of the British Aerospace Hawk trainer powered by the Rolls-Royce-Turboméca Adour engine.[99]

CAE's license agreement to manufacture the Larzac engine was allowed to expire about 1985, following a decade of unsuccessful efforts to market the engine in the U.S. At that time, CAE was required to purge its files of all data on the engine and attest to the fact that it had done so as part of the completion of the agreement with Groupement Turboméca-SNECMA.[100]

TURBOFAN ENGINES SUMMARY

The inability of CAE's turbofans to compete successfully in the marketplace was technically and financially disappointing. Nevertheless, these development efforts paid off by providing the technical skills and experience that won the

---

97. Teledyne CAE, *"NATO Calls it the Engine for Alpha Jet,"* Teledyne CAE Brochure, unnumbered, undated (circa mid-1970s).

98. Smith to Leyes, 9 August 1988.

99. *Jane's All the World's Aircraft 1982-83* (London: Jane's Publishing Company, Limited, 1982), 253, 417.

100. Smith to Leyes, 9 August 1988.

F107 second-source production contract. This contract led to a modern, semi-automated production facility used to produce all of CAE's engines except the J69 and the J402-CA-400.

Demonstration Engines

In 1961, CAE began work leading to an Air Force contract for a lightweight lift engine for vertical take-off and landing (VTOL) aircraft. Four years of work gave CAE visibility within the Air Force, enabling the company to begin an Air Force-sponsored technology development and demonstration program. Twenty years of Air Force and Navy technology demonstration work followed.

XLJ95 LIFT ENGINE

Development of CAE's first demonstration engine, the XLJ95, later designated the ATEGG-365, stemmed from the 1961 assignment given to Eli Benstein, Doug Oliver, and Marvin Bennett to go get some new business. At that time, the aircraft industry, supported by the military, was developing and demonstrating a variety of vertical take-off mechanisms, including a number of experimental aircraft. Driving this work was the international environment in Europe during the early 1960s. The Cold War was at its height and Soviet forces, supported by the Warsaw Pact countries, were sitting on the NATO front with the ability to destroy NATO's fixed bases. One of NATO's potential responses was to disperse its fighter aircraft out onto the roads and fields, places requiring vertical takeoff. NATO was looking for fighters and other types of aircraft that could take off vertically from dispersed sites of this sort. A multiple number of dispersed sites would be less vulnerable to destruction under attack than a large airport with a concentration of aircraft.

In their efforts to develop vertical take-off aircraft in support of NATO's requirements, airframers were confronted with a new and difficult propulsion problem. To take off vertically required anywhere from 50 to 250% more thrust than was needed to fly horizontally. If the main engine were sized to provide the lift for vertical takeoff, it operated at a very low percentage of maximum power during horizontal flight. For almost any engine, this meant that operation was well off the design point and specific fuel consumption was very high.

Two approaches were taken to alleviate this problem. One was to size the main engine in the conventional manner for efficient horizontal flight and to augment its thrust, by such means as afterburning, to provide the vertical lift needed for takeoff. This approach required either redirecting the exhaust downward for takeoff and landing, or using a tail-sitter configuration. The other approach, known as the lift engine technique, was to use one or more high-thrust-to-weight-ratio, vertically-mounted engines to provide vertical lift during takeoff and landing, while using the main engine for horizontal flight.

CAE, opting for the lift engine approach, elected to design a high-thrust-to-weight-ratio engine. They believed that an engine in the 2,500-lb-thrust class could be designed with a thrust-to-weight ratio as high as 20:1. In recalling his marketing efforts, Benstein said, "We went around the country marketing the proposition that a 20:1 thrust-to-weight-ratio engine could be built for vertical takeoff, and that since its mission was about a minute and a half to two minutes for takeoff and landing, it was an economical thing to do."[101]

A key to obtaining high thrust-to-weight ratio was that, although the engine needed to be man rated, it did not have to run for thousands of hours and thus did not require all of the normal safety factors. However, a host of peripheral issues had to be considered in the design of the vertical lift engine. These included inlet flow distortion imposed on an engine in a vertical position, the number of engines required for reliability, and the thrust requirement on a hot day with one engine out. The challenge was to design an engine that was compact, very lightweight, and very tolerant of its environment.

CAE's first experimental lift engine, demonstrating a thrust-to-weight ratio of 12:1, operated horizontally; because of finishing constraints, it made limited use of lightweight materials and design techniques. However, as a step toward the 20:1 projection, it represented a major advance at that time. CAE acquainted the military with its results and revisited the airframers around the country.

Soon afterward, the Air Force at Wright Patterson Air Force Base held a competition for a lightweight lifting engine. CAE competed with Pratt & Whitney and General Electric for this contract. CAE won the competition, and in 1962 was awarded a contract to develop and demonstrate a lift engine in the 4,000-lb-thrust class that would fit inside a 55-gallon drum. The result was an engine that was three feet long, 18 inches in diameter, and designed to operate in a vertical position.

This original lift engine demonstrator program lasted from 1962 until 1966. The design challenge was to get the engine weight down and the thrust up. A conventional 4,000-lb-thrust engine at that time would have weighed about 1,000 lb. CAE's objective was an engine weight of 200 lb.

The XLJ95-T-1 engine had a five-stage axial-flow compressor with a pressure ratio of about 6:1, a slinger-type combustor, a single-stage cooled turbine and a very short cusped jet nozzle. Steps taken to minimize engine weight in-

101. Benstein to Leyes, 9 August 1988.

Teledyne CAE XLJ95 demonstration vertical lift engine which achieved a 22:1 thrust-to-weight ratio. Photograph courtesy of Teledyne CAE.

cluded the use of very thin structural members designed to satisfy the lift engine operating scenario. The lubrication system was self contained, eliminating the need for a gear box, pumps, scavenge lines, and fittings. A lightweight fuel control was developed that was adequate for the short periods during which the engine would be operating at sea level only.

Testing was conducted with the engine mounted vertically in a facility built in the tower of one of the company's buildings in Detroit. The engine was operated for about 200 hours with no catastrophic failure. By the end of the program in 1966, the technical feasibility of a lightweight lift engine had been successfully demonstrated. The engine had achieved a thrust-to-weight ratio of 22:1, higher than that of any other engine.

The success of the XLJ95 established the company's technical reputation with the Air Force. The XLJ95 served as the springboard that launched CAE into the Air Force's newly established engine technology demonstration program.

TWENTY YEARS OF TECHNOLOGY DEMONSTRATION

In the mid-1960s, the Air Force established the Light Weight Gas Generator (LWGG) program at Wright-Patterson Air Force Base (WPAFB) under the jurisdiction of the Power Plant Laboratory to develop and demonstrate technology advances in the gas generator section of small gas turbine engines. The gas generator was the basic engine core, the key to an engine's performance, to which other components could be added to produce a complete turbofan or turboshaft engine.

Contracts for technology demonstration brought with them two incentives for the engine companies. Such work enabled a company to enhance its technology data base using government funds, and demonstration hardware could evolve into new marketable products.

In 1969, the LWGG program evolved into the Advanced Turbine Engine Gas Generator (ATEGG) program. The program was restructured to include both large and small gas generators to demonstrate advanced propulsion technology for fighter, bomber, and large transport aircraft as well as for pilotless and trainer types of aircraft. By the mid-1970s, the Navy had joined the Air Force in sponsoring the renamed Joint Technology Demonstrator Engine (JTDE) program. Its scope was expanded to include the demonstration of complete turbofan and turboshaft engines. The military's overall objective in each of these programs was to develop technology that would be available on a non-proprietary basis for any company to apply in the development of a new engine. All of the engine technology demonstration activities carried a security classification. Availability of information on such topics as design details, development problems, and technology demonstrated was limited.

Following its successful XLJ95 demonstration, CAE was awarded a contract in 1967 under the LWGG program for its second-generation demonstration engine, designated the ATEGG-440. The objective was to demonstrate high pressure ratio and high turbine temperature technology in a core typical of that of a 4,000-lb-thrust turbofan. The application of this technology would be for engines used in advanced trainers and remotely piloted vehicles.

The design objective of the ATEGG-440 was to achieve an efficient machine having the maximum possible pressure ratio and turbine inlet temperature. The resulting gas generator had a single-spool rotor; a multi-stage, variable-geometry, axial-flow compressor followed by a centrifugal stage; a slinger-type combustor; and a single-stage cooled turbine. This was CAE's first venture into variable compressor geometry. According to CAE, although the specifics were still classified in 1990, the resulting compressor produced a higher pressure ratio from a single spool than anybody in the small engine industry had previously achieved.[102] An additional major accomplishment was that the engine's cooled

102. Benstein to Leyes, 9 August 1988.

turbine blades were very small, approximately the size of a thumb nail. The ATEGG-440 was first run in January 1970 and was tested for 72 hours, during which all of the program objectives were achieved.[103]

Following that success, CAE proposed to the WPAFB Power Plant Laboratory that the company be allowed to undertake a more conservative and practical design for their next demonstration engine, the ATEGG-555. What CAE wanted to do for its third-generation demonstration engine was to develop a core that had a more reasonable pressure ratio, but still operated at a high turbine temperature. CAE proposed to develop a core applicable to a 2,000-lb-thrust turbofan for future commuter or advanced trainer aircraft, while also providing technology for missile and remotely piloted vehicle engines. CAE received a contract to proceed with the ATEGG-555 in the early 1970s. The new engine was much simpler in design than the ATEGG-440. The engine first ran in February 1974, and 586 hours of testing had been accumulated by 1976, when the contract ended.[104]

Based on the success of the ATEGG-555 and some intensive engineering marketing, CAE received its first contract under the JTDE program for development and demonstration of the JTDE-455, the company's fourth-generation demonstration engine. The objective of this contract was to use the ATEGG-555 core and develop a fan and power turbine for it, thereby producing a complete turbofan engine. It was intended that this demonstration engine would lead to a new family of durable, low-cost engines for use in cruise missiles, trainers, and liaison aircraft until the end of the century.

By this time, the WPAFB Power Plant Laboratory had become sensitive to the ultimate cost of producing engines. The theme throughout the development of the JTDE-455 was to provide improved performance with lower production costs. Emphasis was placed on such factors as reducing the number of compressor stages and using cost-cutting materials. The JTDE-455 was first run in October 1977. Tested for 242 hours, it demonstrated its design objectives.[105]

During the early 1980s, two more demonstration engines were developed by CAE. One was the ATEGG-585, another fourth-generation turbofan core. The other was the ATEGG-589, an improved fifth-generation turbofan core. The ATEGG-585 first ran in December 1982, but only for 10 hours of test time. The

103. Teledyne CAE, "ATEGG/JTDE - 20 Years of Technology Development," unpublished chart, National Air and Space Museum Archives, Washington, D.C.

104. Teledyne CAE, "ATEGG/JTDE - 20 Years of Technology Development."

105. Teledyne CAE, "ATEGG/JTDE - 20 Years of Technology Development."

ATEGG-589, first run in May 1986, was tested only 22 hours. Emphasis contin-ued to be placed on increasing performance while seeking the optimum balance between cost and performance in the application of new technology. Subse-quently, the number of compressor stages required to achieve a given pressure ratio was reduced by a factor of two.

A CAE derivative of the technology demonstration program was its paper Model CAE-444, a 1,500-lb-thrust turbofan proposed as the power plant in Cessna's unsuccessful entry in the Next Generation Trainer (NGT) competition held by the Air Force. The winner was the Fairchild Republic Company's entry powered by the Garrett F109-GA-100 engine. The proposed Model CAE-444 was never built.

During the ATEGG and JTDE programs, CAE had been competing with General Electric, Pratt & Whitney, and Allison for program funds. All three of those companies, being manufacturers of large engines, were working on tech-nologies applicable to fighter, bomber, and military transport aircraft engines. Over the years, these applications had been of higher priority to the Air Force than were the trainer, target drone, and remotely piloted vehicle applications of CAE's demonstration engines. Because CAE was working with small engines, its engineering effort's funding was small relative to that of the large engine companies. In addition, CAE was making a number of innovative engineering advances. These two factors permitted CAE to compete for program funds for as long as it did.

However, in 1987, after 20 years of demonstration engine contracts, CAE was dropped from the JTDE program as a result of a military budget crunch. Allison had been dropped several years earlier, and Garrett had been brought into the program. When the budget crunch hit, the Air Force had to make a decision be-tween Garrett and CAE. CAE was perceived as oriented toward missiles and re-motely piloted vehicles, and not competitive in man-rated applications. Garrett, on the other hand, had been on the winning team in the Next Generation Trainer competition with its F109 engine. Other Garrett engines were being used in a variety of aircraft. Thus, by 1987, it became obvious to CAE that the military perceived Garrett's advanced technology work as more supportive of manned aircraft technology needs than its own, and also that there wasn't going to be enough money to support both companies.

TECHNOLOGY DEMONSTRATION SUMMARY

The 20 years CAE spent developing demonstration engines provided an oppor-tunity for the company to develop a wealth of technical know-how on improving the performance and efficiency of small engines. However, this effort was a mixed blessing for the company. CAE's Chief Scientist, Eli Benstein, observed that, "For 20 years, it was an interesting program, but it represented a schizophrenia on our part."[106] He considered it super technology, but also observed that it took away from the company's ability to focus on what might have been a more fruitful marketplace for its engineering capabilities.

Because there had been a significant sharing of program costs with the military, the technology demonstration program had absorbed 40 to 80% of the company's budget for Independent Research and Development (IRAD). For example, the Air Force insisted that no component be run on an engine until it had been tested. However, the Air Force would not pay for component tests, which left the funding of such activities to the participating contractors. In addition, company funding had to maintain the caliber of engineering staff required to solve the many high-tech problems that arose.[107]

Over the years, there had been disagreements within the company over how its IRAD money should be spent. This siphoning off of IRAD funds detracted from the company's ability to invest in developing marketable products. However, one thing that once again stood out clearly was that the views of the engineers, rather than those of the marketeers and business developers, were dominant in CAE's decision making.

Post-ATEGG Market Development

In 1985, at the time when CAE's participation in ATEGG was coming to a close, the overall financial picture of the company was good. Production of the F107 and J402 engines was profitable. The J69 also continued to be an important part of the company's business with a several-year horizon being projected for continued delivery of J69-T-29s and J69-T-41As for the Firebee and for continued supply of J69 spare parts for T-37 trainer engines.

With the likelihood of being dropped from the ATEGG program, decisions were made to direct the company's discretionary funding, previously used to support ATEGG work, into the development of new marketable products. This meant a substantial increase in the budget for new product development and an

---

106. Benstein to Leyes, 9 August 1988.

107. Benstein to Leyes, 9 August 1988.

important transition of emphasis from development of new technology to new products that would expand CAE's market base.

Despite past debates over IRAD spending priorities, new market opportunities had not existed for the application of CAE's technology and engineering capabilities in the small, missilized gas turbine engines that it had become expert at building. But, by 1985, such opportunities were beginning to emerge in the form of requirements for miniature turbojet engines to power small tactical missiles, RPVs, and aerial target vehicles. In response, CAE designed a new family of small missilized engines.

THE MINIATURE ENGINE SERIES

During the two decades prior to 1985, the Air Force had learned that it could not send pilots into some arenas of aerial combat and had recognized that pilotless aircraft such as drones, reconnaissance vehicles, short-range and stand-off missiles, and decoys were important adjuncts to its piloted aircraft. In addition, a growing demand for greater range and flight endurance than could be achieved with existing or proposed rocket-powered vehicles meant that turbojet propulsion should be used. As a result, by 1985, the military began releasing requirements for miniature turbojet engines that could power small tactical missiles. At that time, no turbojets of this type existed.

The stimulus for the development of these engines came from the Fiber Optic Guided Missile (FOG-M) program, a technology program that had been conducted in-house by the Army Missile Command (MICOM) at Huntsville, Alabama, during the early to mid-1980s. The FOG-M was a small anti-tank and anti-helicopter missile equipped with a television-camera seeker and a two-way fiber optic guidance link connected to a ground-control console in a concealed position near the launch site. As the missile began searching for a target following launch, it unreeled the fiber optic "wire" connected to the ground-based control console. Monitoring the image transmitted by the on-board television camera, the controller could maneuver the missile, lock it onto a target, and then command it to attack.

Several rocket-powered variants of the FOG-M system were tested by MICOM in the early 1980s. Although the solid-rocket sustainer powering the missile offered low cost and high-speed dash performance, it lacked range and endurance. The turbojet offered increased range, throttleability for loiter and dash capability, and smoke-free operation to reduce visibility. Although it had long been technically feasible to power small expendable vehicles with turbojet or turbofan engines, they had been considered prohibitively expensive. Nevertheless, the increasing need for longer range and increased endurance generated strong interest in the use of these engines. In 1985, MICOM initiated a program to develop a series of miniature turbojets suitable for powering small tactical missiles. That program and the engines developed under it are described in Chapter 11.

In response to the requirements of that program, CAE began developing a series of little turbojet engines. Beginning in early 1986, CAE started the design and development of a family of small turbojet engines with 40 to 350 lb thrust designated as its 300 series.

To make its little engines cost effective for expendable missile applications, the prime design objective for CAE's 300 series engines was minimum cost. For the engines to be acceptable, it was necessary that their cost be reduced to between one-half and one-tenth that of current small gas turbine engines. Acceptable cost goals could best be achieved with a clean-sheet design concentrating on the fundamentals of low-cost manufacturing and design:

- Low parts count,
- Conventional manufacturing processes,
- Minimum significant engineering requirements (dimensional, process, etc.), and
- Simple manufacturing procedures.[108]

To achieve such sizeable cost reductions without sacrificing performance, CAE turned to the technology base that it had developed during its demonstration engine programs. The result was a low-cost, simple, back-to-basics configuration for the 300-series engines, like a little J69 but much simpler. For example, the 40-lb-thrust Model 305-4 engine had only 10 major parts and required only 15 manufacturing operations.[109]

Each of the first four 300 series engines consisted of a single-shaft rotor with a single centrifugal compressor stage, a slinger-type combustor, and a single-

108. Emanuel Papanddreas, "SCAT: A Small Low Cost Turbojet for Missiles and RPVs," (Paper No. AIA-88-3249 presented at the AIAA/ASME/SAE/ASEE 24th Joint Propulsion Conference, 11–13 July 1988, Boston, Massachusetts), 1–2.

109. Teledyne CAE, "Teledyne CAE's Compact Engine Series," Teledyne CAE Brochure unnumbered, undated (circa 1980s).

stage turbine. The fifth engine, the 320-2, added a transonic stage ahead of the centrifugal compressor. The small size and light weight of these engines are illustrated in the following table:[110]

Model Number	305-4	305-7E	312-1	320-1	320-2
Thrust (pounds)	40	90	177	240	350
Diameter (inches)	6.6	6.6	8.3	9.9	9.9
Length (inches)	9.9	10.7	12.6	17.19	17.19
Weight (pounds)	12	19	34	50	58

Design and fabrication of these engines was accomplished in a relatively short time. Design of the Model 305-4 engine began in January 1986, and it first ran just seven months later. Two of these engines were delivered to MICOM in February 1987. A Model 305-4 engine was also flown in May 1987 in a one-tenth-scale MiG-27 target vehicle. Design of the Model 305-7E, 312-1, and 320-1 engines was started in the latter part of 1986, and by early 1987, each of these engines had been run on a test stand. Design of the last engine in the series, the 320-2, was started in 1987, and it first ran in late 1989.

A design objective for the initial 305, 312, and 320 engine models was to be able to achieve a 100% growth in thrust within the existing engine frame size. For example, the 305-4 was a relatively low-technology engine, using a simple turbosupercharger-type of compressor design. The 305-7E followed with a state-of-the-art compressor having a higher airflow and pressure ratio. The thrust was more than doubled and the specific fuel consumption was lowered by 21%. The 312-1 was a new engine design with a larger frame size. It used advanced component technology from the ATEGG program, including a backward curved single-stage centrifugal compressor. The 320-1 was a 1.2 dimensional scale-up of the 312-1. The 320-2, with the same frame size as the 320-1, derived a 50% thrust growth by an increase in turbine operating temperature and by the addition of a transonic stage ahead of the centrifugal compressor. The transonic stage increased the airflow by 35%.[111]

110. Teledyne CAE, "Teledyne CAE's Compact Engine Series," "Model 305-7E Turbojet Engine," "Model 312 Turbojet Engine," "Model 320-1 Turbojet Engine," and "Model 320-2 Turbojet Engine," Teledyne CAE Brochures, unnumbered, undated (circa 1980s).

111. Eli Benstein, telephone interview by William Fleming, 23 April 1990, Teledyne CAE Interviews, National Air and Space Museum Archives, Washington, D.C.

Left to right are Teledyne CAE Models 305, 312, and 320 miniature turbojet engines developed for use in small tactical missiles, RPVs, and small aerial target aircraft. Photograph courtesy of Teledyne CAE.

During 1988, a competition was held by the Army Missile Command for the Non-Line-of-Sight (NLOS) missile and its power plant. CAE entered its 90-lb-thrust Model 305-7E engine against Williams International and Sunstrand Turbomach. The winner was Williams International, on the basis of minimum cost. However, the NLOS program became a victim of "peace dividend" budget actions, and the program was canceled in 1991.

Meanwhile, CAE offered a derivative of its Model 312-1 engine, the Model 318, in the Ground-Launched Tacit Rainbow (GLTR) missile competition. The GLTR was intended to destroy both enemy aircraft and ground defenses. The Raytheon/McDonnell Douglas team was selected as the prime supplier for the GLTR, powered by CAE's Model 318 engine. Earlier, Northrop had been selected as the prime supplier of the air-launched version of the Tacit Rainbow, powered by the Williams WR36 engine. The Raytheon/McDonnell Douglas team was also selected to develop a second-source version of the air-launched Tacit Rainbow, to be powered by the CAE Model 318 engine.[112] However, in 1991, the Tacit Rainbow program was terminated by the military.

In 1989 and 1990, CAE began development of several other miniature engines for both MICOM programs and other missile, RPV, and aerial target applications that evolved during the early 1990s. For example, the Model 312-5 (J700-CA-400) was selected to power the Improved Tactical Air-Launched De-

112. Benstein to Fleming, 23 April 1990.

coy (ITALD) and the Model 305-7E and 312-2 engines were being considered for a variety of U.S. and foreign missile programs.

MODEL 382-10 ENGINE

Development of a somewhat larger engine in CAE's 300 engine series, the Model 382-10, rated at 980 lb thrust, began in 1985. An unusual aspect was its classification as a "bypass turbojet" with a low-bypass ratio of only about 0.25. The engine had a fan stage followed by a mixed-flow compressor, slinger-type combustor, and a single-stage turbine. The compressor and fan design drew on CAE's earlier advanced technology work, while the turbine was similar to that used in the Model 373-8B engine. A major purpose of the bypass feature was to cool the engine skin, which greatly simplified missile engine installation. The engine, first run in 1989, was selected for use in the Teledyne Ryan model 350 air-launched missile, developed under the Navy's Mid-Range Unmanned Aerial Vehicle (MRUAV) program.[113]

Overview

STATUS OF THE CORPORATION

In 1961, the Ryan Aeronautical Company began a gradual acquisition of Continental Motors stock. By October 1965, holding just over 50% of Continental's stock, Ryan assumed full control of Continental and its subsidiaries, including CAE.[114]

By 1968, the technical and financial success of the combined Ryan-Continental group had grown to the extent that the parent Ryan Aeronautical Company had become a prime target for acquisition. In September 1968, Teledyne, Inc., a broadly-based advanced technology firm headquartered in Los Angeles, made an offering to buy all of Ryan's outstanding stock. The offering was accepted by Ryan, and the acquisition by Teledyne was executed on January 2, 1969.

In March 1969, Teledyne restructured Continental Motors Corporation, giving it control of three divisions, two of which were the Continental Motors Division (piston engines) and the Continental Aviation and Engineering Division

113. Benstein to Fleming, 23 April 1990.

114. Wagner, *Continental! Its Motors and Its People,* 169-173.

(gas turbine engines). Then, in December 1969, Continental Motors Corporation was merged into Teledyne, and the former Continental Motors Division became Teledyne Continental Motors and the Continental Aviation and Engineering Division became Teledyne CAE. In February 1972, TCM was further split into three units, one of which was the TCM Aircraft Products Division (aircraft piston engines).

Thus, by the 1980s, the various businesses that had been built up and acquired over the years by the original Continental Motors Corporation had been restructured as five distinct divisions of Teledyne. They were the TCM Aircraft Products Division; the TCM Industrial Products Division; the TCM General Products Division; Teledyne Wisconsin Motor; and Teledyne CAE.[115]

Then, in July 1993, Teledyne restructured the corporation in response to the changing economic climate and the drawdown of the defense budget. Among other changes, three of Teledyne's operating units were combined to form the Teledyne Ryan Aeronautical Company. These three units were Teledyne Ryan in San Diego, Teledyne McCormick Selph in Hollister, California, and Teledyne CAE in Toledo, Ohio, all of which remained in their previous locations supported by a headquarters in San Diego.[116]

In 1986, CAE constructed a new semiautomated manufacturing plant in Gainesville, Georgia, to which a major portion of CAE's engine production was moved. The engineering and management staff remained at the Toledo plant, devoted basically to engine design and development, component and engine development testing, and production of J69 and Harpoon missile engines. The Gainesville plant was used to manufacture all other CAE engines, including the F107 and J402-CA-702. The Gainesville plant continued to operate until 1991 when it was closed following completion of F107 engine production. At that time, all remaining production was moved to Toledo.

Between the late 1950s and the late 1980s, employment at CAE had decreased. In 1958, CAE had employed 2,000 people, and by 1989, there were 640 employees.

THE COMPANY'S MARKETS AND INFLUENCING FACTORS

CAE had a history of rapid initial business growth followed by subsequent decline. As a result, it was among the smallest of the seven major manufacturers of small gas turbine engines for aircraft. From 1955 to 1976, CAE produced man-rated J69 turbojets for the Cessna T-37 trainer aircraft. Unmanned versions of

115. Wagner, *Continental! Its Motors and Its People*, 169-203.

116. Apel to Leyes, 26 May 1995.

Teledyne CAE facility in Toledo, Ohio. Photograph courtesy of Teledyne CAE.

the J69 were also produced from 1954 through 1994 to power target drones and reconnaissance vehicles. In 1972, CAE began development of the first wooden round turbojet engine for the Harpoon missle.The company had only a minor share of the turbofan market. After failing to win a competition with its F106 turbofan engine in the mid-1970s, the company produced the Williams F107 turbofan as a subcontractor. Starting in the early 1970s, CAE concentrated its efforts on the development and production of low-cost turbojets for tactical missiles and target drones. Beginning in the mid-1980s, development began on miniature low-cost turbojets engines for small tactical missiles, RPVs, and aerial target vehicles.

CONTRIBUTIONS

CAE was a company that served only the military marketplace. The J69 turbojet, used in the T-37 military trainer aircraft and in Firebee target drones and reconnaissance vehicles, was CAE's most important engine. That engine contributed to the success and longevity of the T-37 military trainer aircraft. It also gave the Firebee the capability to conduct reconnaissance flights over hostile territory and to become a successful target vehicle for use in training military personnel. The J402 initiated the use of turbojets in tactical missiles.

LYCOMING | 5

Company Origin and Early Development

The roots of Lycoming lie in its predecessor company, the Demorest Fashion and Sewing Machine Company, founded in 1845 by Mademoiselle Demorest as a metal fabricating company. In 1889, the Demorest factory was moved to the present Lycoming factory site in Williamsport, Pennsylvania. By the early 1890s, the company was also producing high-quality bicycles. Shortly after the turn of the century, the sewing machine and bicycle business slackened, and the plant was transformed into a general foundry and machine shop under the name Lycoming Foundry and Machine Company.[1] (The new name came from that of a small stream, Lycoming Creek, located just west of the company's Williamsport plant.[2]) The company manufactured duplicating machines, cup vending machines, typewriters, gas irons, platen printing presses, and button sewing attachments.[3]

In the spring of 1909, the management of the Lycoming Foundry and Machine Co. decided to seek engine manufacturing business from the emerging au-

1. A. E. Light, "The Evolution of Reciprocating Engines at Lycoming," (Unnumbered AIAA paper presented at the meeting of the Dayton Section of the American Institute of Aeronautics and Astronautics, Dayton, Ohio, 24 April 1985), 9-1.

2. A. E. Light, "The Evolution of Reciprocating Engines at Lycoming," 9-1.

3. "Local Plant's Engines Honored in U.S. Automotive History," *Williamsport (Pennsylvania) Sun-Gazette*, April 1972.

tomobile industry. In 1910, the company delivered its first gasoline engine, of the purchaser's design, to the Velie Motor Vehicle Co. Its business grew and by 1915, Lycoming began to market a four-cylinder, two-bearing engine of its own design, the Lycoming Four, to power the Dort automobile. Soon the Lycoming Four was powering as many as 3,000 moderate-priced cars and trucks.

To meet the Army's demands during World War I, Lycoming turned out 15,000 Lycoming Fours in 1917. Contracts were received to deliver more than 60,000 of these engines during 1919 and 1920.[4]

By 1920, the company had grown to 2,000 employees. That year the company was reorganized and refinanced, and its name was changed to Lycoming Motors Corp. In 1924, the company's name was again changed to the Lycoming Manufacturing Corp.

Lycoming's automobile engine business thrived throughout the 1920s. In 1925, Lycoming accounted for 15 percent of all automotive engines produced in the U.S. These sales included an extensive export business with engine shipments to automobile companies in Europe, Japan, Australia, and Argentina. In 1926, with its workforce of 2,000 people and a payroll of $3 million, Lycoming realized $7 million in revenue from sales.

In the 1920s, an array of new engines was introduced by Lycoming. By the close of 1928, full lines of four- and six-cylinder automobile and truck engines were being manufactured in addition to a series of straight-eight engines. In fact, by that time, Lycoming had become the largest producer of straight-eight automobile engines in the world. In all, Lycoming ultimately produced 57 types of automotive and truck engines. A number of these engines were high-performance power plants designed for such stylish and expensive cars as the Auburn, Duesenberg, and Cord.[5]

In 1927, Lycoming became a subsidiary of the Auburn Automobile Company when Auburn acquired 62.5 percent of Lycoming's capital stock. The President of Auburn was E. L. Cord. In 1929, Cord established the Cord Corporation as a holding and management company for the properties that he and his associates controlled. At that time, the Cord Corporation acquired a large share of Auburn stock and the remaining 37.5 percent of Lycoming common stock. With this move, Auburn and its Lycoming subsidiary became holdings of the Cord Corporation.[6]

4. "Local Plant's Engines Honored in U.S. Automotive History."

5. "Lycoming Engines Had Big Role in Golden Age of Luxury Car," *Williamsport Sun-Gazette*, April 27, 1972.

6. "The Cord Corporation," *Aero Digest*, Vol. XXIII, No. 5 (November 1933): 22-29.

Although the strength of the company's business through the 1920s and into the 1930s was mainly in its automotive engine sales, during the 1920s, Lycoming had expanded its engine markets to include marine, agricultural, and industrial applications. Lycoming's marine inboard power plants were supplied to such boat manufacturers as Elco, Penn-Yan, Wheeler, and Horace Dodge. In addition, an improved 30-hp, liquid-cooled engine introduced in the late 1920s was used on the John Deere combine and as an industrial standby power plant for electric generators.[7]

During the 1930s, Lycoming's automotive engine business slackened as automobile companies increasingly built their own engines. By the end of the 1930s, Lycoming had ended its production of automobile engines.

In January 1928, E. L. Cord, who at that time was Chairman of the Board of Lycoming Manufacturing Corp., made the decision that Lycoming should enter the aviation business. At that time, Lycoming engineers began designing what was to be a low-cost radial aircraft engine. Unfortunately, the low-cost concept did not produce a reliable engine, and it was soon replaced by a more conventional design, which established Lycoming in the aircraft engine business. This engine was the R-680, a nine-cylinder, 215-hp, air-cooled radial engine. The R-680 made its maiden flight on April 13, 1929, in a Travel Air biplane. The engine became an immediate success and found its way into the first scheduled airliners. It was also used in military training and spotting aircraft before and during World War II. Eventually 25,000 of these engines were built.[8]

During the 1930s, in response to the emergence of the general aviation industry, Lycoming turned its attention to building engines that would meet the needs of this new and growing market. The Cord Corporation had acquired control of the Aviation Corporation in 1933 following a stormy battle that had begun the previous year. With a reorganization of the corporation's holdings in 1935, the Aviation Manufacturing Corporation was formed as a subsidiary of the Aviation Corporation (Avco).[9] Lycoming, with the Stinson and Vultee Aircraft companies, became subsidiaries of Avco. Lycoming then became known as Avco Lycoming.

In 1938, Lycoming introduced the O-145, a four-cylinder, 55-hp, horizontally-opposed, air-cooled engine. In the years that followed, Lycoming developed a series of four-, six-, and eight-cylinder engines of the same type.

7. "Lycoming Engines Had Big Role in Golden Age of Luxury Car."

8. A. E. Light, "The Evolution of Reciprocating Engines at Lycoming," 9-2.

9. *Avco Corporation–The First Fifty Years* (Greenwich, Connecticut: Avco Corporation, 1979), 14-19, 95.

Ranging in size from 100 to 450 hp, these engines were widely used in private and corporate planes built by such companies as Piper, Taylorcraft, Beech, and Ryan.[10] It was a 75-hp version of the Lycoming O-145 that powered the world's first successful helicopter, built and flown by Igor Sikorsky.[11] By the 1950s, nearly 50 models of Lycoming's horizontally-opposed aircraft engines had been certificated. By 1985, this number had increased to more than 600 engine models covered by 43 individual type certificates.[12] Meanwhile, in 1932, Lycoming had begun design studies of high-powered, liquid-cooled aircraft engines. In cooperation with the Army Air Corps, a 12-cylinder, horizontally-opposed, liquid-cooled engine rated at 1,200 hp was developed. Readied for endurance testing by the end of 1937, this engine (the O-1230) was introduced as a "flat" motor designed for wing installation.[13]

In response to the need for more power, design work began during the summer of 1939 on the XH-2470. This engine was a 24-cylinder, 2,400-hp, liquid-cooled engine with the cylinders arranged in the form of an H. First run in January 1940, the engine reportedly passed a development test in January 1942. The Navy intended to use this engine in its new Curtiss F14C fighter and the Army in the new Vultee P-54 fighter. Performance of the engine appeared so promising that, without awaiting construction of an airplane in which flight tests could be made, the Navy, in May 1942, ordered 100 engines, 50 of which were for the Army. Delivery was to begin in 1943. On the basis of this order, Lycoming set up a factory in Toledo exclusively to produce the H-2470. However, before the engine went into production, the Navy canceled its contract because it believed that the engine would not be available in time for use in the war. During 1943, the engine was extensively flight tested in the Army's XP-54 fighter. By the time these tests were completed, it had become clear to the Army also that the H-2470 could not possibly be ready in time for use in the war. Before the end of 1943, the Army also dropped the project.[14]

Near the end of 1943, Lycoming began designing the world's largest and most powerful reciprocating engine for aircraft use. This engine, the XR-7755, was a 36-cylinder, liquid-cooled, radial engine capable of delivering 5,000 hp.

10. "Lycoming in Aviation in Late 1920s," *Williamsport Sun-Gazette*, May 1972.

11. "Lycoming Has First Ranking Among Producers of Aircraft Engines," *Williamsport Sun-Gazette*, May 10, 1972.

12. A. E. Light, "The Evolution of Reciprocating Engines at Lycoming," 9-4.

13. A. E. Light, "The Evolution of Reciprocating Engines at Lycoming," 9-2.

14. Robert Schlaifer and S. D. Heron, *Development of Aircraft Engines, Development of Aviation Fuels* (Boston, Mass.: Graduate School of Business Administration Harvard University, 1950), 292-293.

Although the engine's design and performance goals were met, the XR-7755 was overtaken in 1945 by the development of the turbojet engine, and work on the XR-7755 was discontinued. Only two of these engines were ever built,[15] one of which is in the National Air and Space Museum in Washington, D.C.

Following World War II, the market for large piston engines began to disappear. Accordingly, Lycoming discontinued work on such engines and turned its attention to the further development and manufacture of its small piston engines for single- and twin-engine executive and utility aircraft. Thus in 1951, when Lycoming entered the field of aircraft gas turbines, the company was primarily a developer and manufacturer of small piston engines for the general aviation market.[16]

When the Korean War broke out in 1950, the government stepped up its orders for small aircraft and helicopters. To support this effort, Lycoming was awarded a contract to build Curtiss-Wright's nine-cylinder, R-1820 radial engine. Lycoming also received a contract to manufacture components of the General Electric J47 turbojet engine. In February 1951, Lycoming took over the old Chance-Vought plant in Stratford, Connecticut from the Air Force and, shortly afterward, the Bridgeport-Lycoming division was established.[17]

Anselm Franz

In 1950, Anselm Franz, a German jet engine pioneer during World War II who had come to the U.S. soon after the War, took the first steps in almost single-handedly laying the foundation for Lycoming's entry into the aircraft gas turbine business. Franz believed that Lycoming should develop, under his direction, a gas turbine engine in the medium-power range with an output of 600 to 1,000 shp. This power range had been largely neglected by the aircraft gas turbine industry up to that time. He viewed the development and production of this engine as an effort that would parallel Lycoming's piston engine business.[18]

Anselm Franz had completed his studies in mechanical engineering at the Technical University in Graz, Germany, in 1924. While Franz was still studying for his degree, the head of the Department of Turbomachinery at the university

---

15. A. E. Light, "The Evolution of Reciprocating Engines at Lycoming," 9-2, 9-3.

16. The Williamsport plant became a division of Lycoming, and in the 1990s, it was still manufacturing horizontally opposed reciprocating engines for general aviation aircraft.

17. *Avco Corporation—The First Fifty Years*, 40-41.

18. Anselm Franz, *From Jets to Tanks* (Stratford, Connecticut: Avco Lycoming Stratford Division, undated), 44.

offered him a job as "scientific helper." His acceptance of that offer proved to be a fateful event; from that point on throughout his life, he was associated with turbomachinery. After spending several years developing hydraulic torque converters, he entered the field of aviation in 1936 when he joined the Junkers Engine Development Division of the Junkers Company.[19]

While working on exhaust jet propulsion for reciprocating engines, Franz became interested in designing a jet engine. Because of the pressure on the Junkers Engine Development Division to improve and further develop their piston engines for Germany's military aircraft, Franz was forced to proceed with his jet engine work in a manner that would not disturb any of the high-priority, piston engine programs. Thus, he built a new independent organization within Junkers for the development of jet engines.

In June 1939, Junkers received a government contract for development of the Jumo 004 axial-flow turbojet engine. The award of this contract occurred two months before the world's first flight of a gas turbine engine, Dr. Hans von Ohain's HeS-3B, in the Heinkel He 178. Franz directed the design, development, and production of the Jumo 004 engine, which became the world's first successful axial-flow turbojet engine and the first mass produced. The Jumo 004 first ran in October 1940, and on July 18, 1942, the Messerschmitt Me 262 V3, powered by two Junkers Jumo 004A engines, made its initial flight. By the end of World War II, more than 6,000 Jumo 004 engines had been built.[20]

Following the war, Franz was brought to the U.S. under "Operation Paperclip" and was assigned to work as a consultant at the Wright-Patterson Air Force Base in Dayton, Ohio.[21] Before he left Germany, he had counseled several of his key engineers to elect likewise to come to the U.S. He told them that, after an appropriate period of time, he expected to start his own engine company in the U.S. and that there would be a position in it for them. While working for the Air Force, Franz principally served as a consultant for companies in the U.S. that had begun to develop gas turbine engines.

This exposure enabled him to survey the state of the U.S. gas turbine industry. He felt that it would not make sense to develop large engines, as General Electric, Pratt & Whitney, and Allison already had that area well covered. At the other extreme, Boeing was developing small gas turbines of 100 to 200 shp. However, no real effort was being made at that time to develop gas turbines in

19. Franz, *From Jets to Tanks*, 15–16.

20. Franz, *From Jets to Tanks*, 13–38.

21. Operation Paperclip was an activity whereby many German engineers and scientists were brought to the U.S. and Britain immediately following World War II.

the 600- to 1,000-shp range, a size that could have applications as a stationary power plant, a boat engine, a helicopter engine, and possibly as a small jet engine in the form of a ducted fan. Here was a niche he might exploit.[22]

During his stay at Wright Field, Franz evaluated the companies that he might approach. One was the Wright Aeronautical Corporation. He had an offer from the president of the company, but he concluded that it would not be the right step for him. He would have been a member of the engineering department reporting to the Vice President of Engineering, which he felt would not give him the free hand he desired. He decided that what he was looking for was a piston engine company with no gas turbine experience. Of course, Lycoming was such a company. He wrote to Victor Emanuel, President of the Avco Corporation, of which Lycoming was an operating division, and Emanuel put him in touch with Avco Lycoming President S. B. Withington.

Near the end of 1950, Withington flew to Dayton to meet with Franz. Franz recalled that as the two men met, Withington said, "Now, today, or never." In the discussion that followed, they reached agreement that Franz was to join the company, report to Withington, and have full authority to set up Lycoming's gas turbine development effort. Withington remarked, "But, of course, with authority goes responsibility, and if the whole thing blows up, then you (Franz) will get fired and not me." A deal was struck, and Franz joined Lycoming in January 1951.[23]

Arriving at Lycoming's Williamsport plant in January of 1951, Franz began assembling a team to help him design a medium-power, gas turbine engine and find a market for it. Cliff Pfleegor was the first hired. Pfleegor was working for Lycoming at that time as an ignition systems specialist on piston engines. Although he had no gas turbine experience, he was familiar with the Williamsport plant and the people working there.

Heinz Moellmann was the first gas turbine engineer hired. Moellmann had worked for Franz at Junkers, and, like him, had been brought from Germany after World War II by Operation Paperclip. Under Franz at Lycoming, he eventually would become Chief Engineer.

The remaining members of the team provided a cross section of expertise. Carl Schwanbeck was a test engineer who had worked for Pratt & Whitney. Earl Feese was a draftsman who came from Westinghouse. Salvitore Straniti was an

---

22. Anselm Franz, interview by Rick Leyes and William Fleming, 14 March 1989, Textron Lycoming Interviews, transcript, National Air and Space Museum Archives, Washington, D.C.

23. Franz to Leyes and Fleming, 14 March 1989.

engineering graduate fresh out of college. Ladislaw Srogi was an experienced and talented mechanical designer. Eugene Clark was a creative mechanical designer who came from Solar. Rita Border was the secretary, and Lois Lloyd was hired to perform the computations for the group.

This team of ten people was supported by two technical consultants. One was Dr. H. Adenstedt, a materials expert working at Wright Field who had been responsible for the Jumo 004 materials design at Junkers. The other consultant was Dr. F. Bielitz, a mechanical stress and vibration expert. It was this team that developed the initial conceptual designs and prepared the proposal to the Air Force that led to the award of Lycoming's first aircraft gas turbine development contract.[24]

In 1952, Avco Lycoming President Withington told Franz that he could stay at Williamsport or he could relocate to the Stratford, Connecticut, plant, where Lycoming was manufacturing Wright R-1820 piston engines and J47 turbojet engine components. Franz recalled that Withington told him, "You can go here or there, anyplace you want to. But the East Coast is beautiful, and it's the best part of the United States. And here (in Williamsport) you are a little bit in the woods." With that, Withington took Franz to Stratford and showed him the plant.

The R-1820 engine and J47 component manufacturing activities occupied only a small portion of the vast Stratford plant, a facility that had been used during World War II to build the Chance-Vought Corsair fighter airplane. Impressed that the plant would be ample in size for the growing business he envisioned, and realizing the advantages of being physically separated from Lycoming's entrenched piston engine activities in Williamsport, Franz opted for Stratford. Soon after Lycoming was awarded its initial gas turbine contract in mid-1952, Franz and his team moved to Stratford.[25]

According to Walt Schrader, who joined the team in 1953 as a turbine designer, "A very informal management process was used at Lycoming during the early period of T53 and T55 engine design and development." This process, which was most effective, was possible only because of the relatively limited size of the engineering staff at that time. Nearly all emphasis was placed on technical matters with little concern being given to organizational functions and responsibilities. As Schrader observed, "Anselm Franz knew everybody, and he knew what

24. Heinz Moellmann, interview by Rick Leyes and William Fleming, 12 October 1989, Textron Lycoming Interviews, transcript, National Air and Space Museum Archives, Washington, D.C.

25. Franz to Leyes and Fleming, 14 March 1989.

Some members of the Lycoming T53 engineering staff depicted with T53-L-10 turboprop engine. Left to right: Dr. Heinrich Adenstedt, Heinz Moellmann, Siegfried Decher, Dr. Anselm Franz, Dr. Friedrick Bielitz, Hans Berkner, and Kenneth Moan. Photograph courtesy of Textron Lycoming.

they were doing. Further, Anselm Franz personally reviewed and approved each design."[26]

Following award of the initial contract by the Air Force, the Army's contracting agency, Lycoming's gas turbine business at Stratford grew steadily.[27] This business growth was accompanied by a progressive expansion in production facilities. Test facilities were also developed to enable Lycoming to perform its necessary component and engine development and demonstration testing, and, by 1958, Lycoming had more than 30 test cells at Stratford.

Serving as Vice President and Assistant General Manager of Avco Lycoming, Franz continued in charge of gas turbine engine development and manufacturing at the Stratford plant from 1951 until his retirement in 1968. By that time, the number of employees had grown to over 10,000. Having had the responsi-

26. Walt Schrader, interview by Rick Leyes and William Fleming, 10 October 1989, Textron Lycoming Interviews, transcript, National Air and Space Museum Archives, Washington, D.C.

27. The Air Force Air Materiel Command at Wright-Patterson Air Force Base in Dayton, Ohio, served as the procurement agency for the Army in awarding and technically managing its early contracts for gas turbine engines in view of its technical expertise in that field.

bility for establishing and leading the development and growth of Lycoming's gas turbine business, Franz had indeed achieved his objective of having his "own" gas turbine company in America.

Following his retirement, Franz continued as late as 1990, at the age of 90, to be active as a consultant at Lycoming. He reflected that he not only worked with the engineers on their design problems, but he often dropped by the company president's office to offer his opinions on subjects of company management and operations.[28]

The Gas Turbine and the Army's Airmobility Concept

Lycoming's entry into the aircraft gas turbine business was related to the Army's need for gas turbine powered helicopters to replace its piston-powered models. This need was driven in part by the Army's airmobility concepts developed in the late 1950s.

At the time the U.S. entered the Korean War in 1950, the Army's role in aviation was limited to the use of light fixed-wing aircraft and helicopters for artillery spotting, liaison work, and expediting ground combat operations in forward areas of the battlefield. The Air Force was given the responsibility for all close combat and logistical support for the Army. The Korean War tested these arrangements and neither service found them satisfactory. The Army objected to the Air Force's inability to supply close air support when needed, and there was continuing disagreement concerning the use of multi-purpose jet fighters versus single-purpose, lightweight, fixed-wing aircraft for close air support. Attempts within the Department of Defense to resolve these disagreements culminated in a Memorandum of Understanding signed by Army Secretary Frank Pace and Air Force Secretary Thomas K. Finletter on November 4, 1952. This agreement raised the weight limit on Army fixed-wing aircraft, eliminated any weight limit for Army helicopters, and gave the Army the added functions of topographic survey and limited medical evacuation, including battlefield pick-up of casualties.[29]

---

28. Franz to Leyes and Fleming, 14 March 1989.

29. Richard G. Davis, *The 31 Initiatives: A Study in Air Force–Army Cooperation* (Washington, D.C.: Office of Air Force History, United States Air Force, 1987), 5-12. The brief history of Army aviation presented on the following pages was obtained from this reference document.

In Korea, the Army's reconnaissance and liaison functions expanded to include aeromedical evacuation and reconnaissance mainly using small, two-place Bell H-13 helicopters. In addition, rudimentary troop and logistics movements also began using the Sikorsky H-19 and Piasecki H-21 helicopters and the de Havilland L-20 fixed-wing aircraft.[30] It was missions such as these in Korea that first shaped the helicopter's role in support of ground operations.

After the Korean War, the Army continued to acquire large numbers of helicopters, at that time still piston engine powered, and to increase its mobility with transport helicopters. During the mid-1950s, The Army Aviation School at Ft. (then Camp) Rucker in Alabama began to test new mobility concepts, which included "sky cavalry" experiments involving the deployment of helicopter transported troops for scouting, raiding, and delaying roles.[31] General Hamilton Howze, first director of the Army Aviation Directorate, established in 1956, visualized airmobile formations led by air cavalry with integrated aerial firepower performing traditional Army missions of attack, reconnaissance, seizing terrain, and breakthrough.[32] Although the piston engine-powered helicopter performed poorly in these exercises, experimentation continued. These field exercises led to most of the strategies and tactics adopted by the Army at the beginning of the 1960s and, in the process, shaped the role of turbine power in Army aviation.[33]

During the early 1950s, the Army had begun to recognize the potential advantages offered by gas turbine-powered helicopters over the piston engine helicopters then in service. With a view to improving the performance and capabilities of its helicopter fleet and thereby implementing its air cavalry concepts, the Army became a key sponsor of small turbine engine development. One of the Army's first steps was the award of the contract to Lycoming in July 1952 to develop a 600-shp aircraft gas turbine engine. That engine, the T53, later powered the Army's Bell UH-1 Iroquois utility helicopter, Grumman OV-1 Mohawk high-performance observation plane, and Bell AH-1 Huey-Cobra attack helicopter.

In 1955, Bell was awarded a contract by the Army for development of the T53-powered H-40 as a helicopter suitable for front-line evacuation of casualties, general utility missions, and as a utility trainer. Production of helicopters

30. John Todd, "U.S. Army Helicopters," *Rotor and Wing International*, Vol. 25, No. 5 (May 1991): 124.

31. Davis, *The 31 Initiatives: A Study in Air Force–Army Cooperation*, 12-13.

32. Todd, "U.S. Army Helicopters," 124.

33. Davis, *The 31 Initiatives: A Study in Air Force–Army Cooperation*, 12-15.

Bell UH-1 Iroquois "Huey" helicopters powered by Lycoming T53 turboshaft engines. Photograph courtesy of Textron Lycoming.

(designated HU-1, later UH-1 "Huey") began in 1959. The Huey emerged as the backbone of Army Aviation and was a key to expansion of the Army's aviation role. From 1962 onward, thousands of these craft took to the air daily during the Vietnam War to perform a variety of tasks. Squadrons of medevac Hueys flew into areas of combat to pick up badly wounded soldiers and rush them to hospital facilities. Other Hueys were designed to carry troops into combat and as gunships.

Concurrent with development of the Huey, Grumman was awarded a contract to develop the fixed-wing Mohawk observation plane powered by two T53 turboprops. Deliveries of the Mohawk began in 1959, and it too was used extensively in Vietnam.

It was in 1954 that work began at Lycoming on the T55 engine for the Army. Twice as powerful as the T53, the T55 was the Army's first step in seeking a turbine-powered replacement for its piston-type logistics aircraft developed in the early 1950s. The aircraft replacement was the twin-turbine Boeing Vertol CH-47 Chinook, which was powered by two T55s and had a lifting capability of two to three tons. The initial contract for five Chinooks was awarded in June 1959. The addition of the Huey, Mohawk, and Chinook to the Army's inventory markedly expanded its close ground support and logistics capabilities.

Boeing-Vertol CH-47 Chinook medium-heavy lift helicopter that provided support
for heavy logistics and artillery movement during the Vietnam War. It was powered
by two Lycoming T55 turboshaft engines. Photograph courtesy of Textron
Lycoming.

The Army's role in aviation was further expanded following a special study
requested by Secretary of Defense Robert S. McNamara in 1962. That study
recommended that certain observation, utility, and cargo helicopters should
carry light automatic antipersonnel weapons. It also recommended that attack
versions of these helicopters should have an antitank capability and carry large
stores of ammunition. The Air Force opposed such recommendations, doubting
the ability of the Army's attack helicopters or Mohawk fixed-wing aircraft to
survive in high density combat and questioning their cost effectiveness compared to Air Force fighter aircraft. Nevertheless, following release of the 1962
study recommendations, the Army deployed an armed version of the Huey helicopter in Vietnam.

The issue of armed attack helicopters came to a head in early 1965 when the
Army awarded contracts for the program definition phase of an advanced attack
helicopter. The recommendations of the 1962 study, followed by combat oper-

ations in Vietnam, had driven the armed helicopter concept forward. With large numbers of troop and supply helicopters crowding the airspace over landing zones during helicopter assaults, Air Force jets found it difficult to coordinate and fly missions to suppress hostile ground fire. Consequently, the helicopters had to carry their own means of self defense to keep enemy heads down.

The war in Vietnam settled the armed helicopter issue once and for all. On September 7, 1965, after nearly a year of combat supported by an armed version of the Huey helicopter, the first Army attack helicopter—the prototype AH-1 HueyCobra—made its initial flight. Four days later, Secretary McNamara informed the Air Force Secretary, Eugene M. Zuckert, that any aircraft operating in the battle zone should be armed. By November 1967, the HueyCobra gunship became operational in Vietnam.[34]

Meanwhile, on April 6, 1966, the Chiefs of Staff of the Air Force and Army, Generals John P. McConnell and Harold K. Johnson, signed an agreement dividing responsibility for certain aircraft between the two services. The Army agreed to transfer the approximately 160 de Havilland CV-2B Caribou fixed-wing transports that it had acquired to the Air Force and to relinquish any future claims to fixed-wing, tactical-airlift aircraft. In exchange, the Air Force abandoned all claims to helicopters for airlift, fire support, and supply of Army forces.[35] This was a continuing initiative that further expanded Army aviation, particularly in terms of rotor aircraft.

It was in this environment that the T53 and T55 engines came into being. The part played by these engines was to provide the improved helicopter performance and operating capabilities that made possible the Army's new mode of air-ground combat and airmobility operations proven in Vietnam.

The success of the Army's airmobility concept and improved close air support, made possible by the gas turbine engine, resulted in a growth in the number of Army helicopters from about 2,600 in 1960, all powered by piston engines, to a fleet of nearly 10,000 helicopters by 1970, most gas turbine powered. Including its small fixed-wing aircraft, by 1970, the Army had a total of about 12,000 aircraft in its inventory and 24,000 aviator pilots on active duty. At that time, the Army had more pilots on active duty than did the Air Force.[36] By 1991, the Army had the world's third largest aviation force, after the Air Force and the Soviet Union's aviation forces.[37]

34. Davis, *The 31 Initiatives: A Study in Air Force–Army Cooperation*, 15-21.

35. Davis, *The 31 Initiatives: A Study in Air Force–Army Cooperation*, 15-21.

36. Davis, *The 31 Initiatives: A Study in Air Force–Army Cooperation*, 21-23.

37. Todd, "U.S. Army Helicopters," 124.

Bell AH-1 HueyCobra attack helicopter powered by Lycoming's T53 turboshaft engine. Photograph courtesy of Textron Lycoming.

Medium-Power Engine Family

The T53, Lycoming's first gas turbine engine, was the star in the company's crown. It was followed by the development of a number of other successful engines covering a range of power levels above and below that of the T53. Lycoming's engineers chose to group their engines into four separate families, the medium-power, high-power, low-power, and vehicular engine families. Most, and in some cases all, of the engines in a given family were derivatives of the initial turboprop or turboshaft engine developed in that family. Because of the similarity in size and the mechanical and aerodynamic interrelationships among the engines in each family, the ensuing historical account of Lycoming's gas turbine engines has been organized by family, beginning with the medium-power family, which included the T53.

DESIGNING THE T53

Lycoming's first step in the gas turbine business was to lay down a series of conceptual designs in response to evolving requirements of the Army and Air Force for turboprop and turboshaft engines. Dr. Franz filled the role of aerodynamics expert on the design team and was instrumental in determining the configuration of these early conceptual designs. The mid-power-range engine that Franz set out to design was to be substantially smaller and more versatile than conventional jet engines of that time. These were single-purpose power plants

The Lycoming T53 turboshaft engine. Photograph courtesy of Textron Lycoming.

used mainly for high-speed flight in military combat and bomber aircraft. In contrast, Franz viewed his smaller engine as being sufficiently versatile to serve a variety of applications, including as a turboshaft engine for helicopters and as a turboprop or possibly a turbofan for fixed-wing aircraft. He also envisioned its use as a stationary, marine, or vehicular power plant.

Franz believed that the major use of his engine would be in helicopters and small observation aircraft for the Army. Because such machines were likely to be operated and maintained at forward positions, they would be serviced by relatively unskilled personnel and would be confronted with operation in a sandy and dirty environment. These conditions dictated a rugged engine that could be serviced and maintained easily. As Franz expressed it, "such an engine has to be really a tough engine that can take abuse."[38]

With these requirements for ruggedness, versatility, and maintainability, he recognized that his engine could not be mechanically scaled down from large engines, but would require an original design of its own. The resulting mechanical arrangement that he conceived for the T53 was a free turbine engine with a front power takeoff. The gas generator core of the engine had a five-stage, ax-

38. Franz to Leyes and Fleming, 14 March 1989.

ial-flow compressor followed by a centrifugal compressor stage, both closely coupled to the turbine that drove them. A concentric power turbine shaft passed through the gas generator rotor to the front power takeoff, and a reverse-flow, annular combustor was wrapped around the engine's turbine assembly.

This arrangement offered a short, compact, and rugged engine package. A short engine meant a short concentric power shaft, minimizing shaft vibration problems. The mechanical arrangement of the T53 was a highly successful design, and it set the style for nearly all of Lycoming's gas turbines. In addition, the T53 with its close-coupled compressor turbine arrangement, concentric shafts, and front power takeoff set a precedent for front-drive turboshaft engines. The arrangement was widely adopted by the small gas turbine industry in America while the close-coupled compressor turbine arrangement with wrap-around combustors was employed selectively. Charles Kuintzle, Vice President of Business Strategies at Textron Lycoming, considered the T53 design an innovative step, expressing the view that:

> The T53 engine really was not only one of the first turbine engines for a heli-
> copter, but it moved turbine configurations in a direction which was far
> different from where it had been before, and it did that by necessity. The kind
> of things that were being studied and the problems we had to address came
> down to the question, how can you effectively build a shaft engine where the
> shaft comes out through the front of the engine.[39]

The T53's compact, front-power-takeoff configuration made the engine attractive for a variety of applications. Franz considered the front power drive as a must because it offered the optimum design for turboshaft, turboprop, and turbofan applications. In addition, the torque converter characteristic of the free turbine eliminated the need for a heavy clutch, which was required in piston engine-powered helicopters and ground vehicles. The engine's arrangement made for structural ruggedness. In addition, using a centrifugal stage for the high pressure portion of the compressor avoided the use of small, delicate blades that would have been required in the high pressure part of a small all-axial-flow compressor.

Based on experience with the T53 and T55 engines, Lycoming came to view the folded or reverse-flow combustion chamber design as a major safety feature,

39. Charles Kuintzle, interview by Rick Leyes and William Fleming, 12 October 1989,
    Textron Lycoming Interviews, transcript, National Air and Space Museum Archives,
    Washington, D.C.

since it provided four layers of steel shielding around the gas generator turbine wheels. Turbine blade failures that occurred during the development of the T53 and T55 were either contained within the engine or the failed blades were ejected out the tailpipe.[40]

The external combustion chamber arrangement, together with the engine's modular design, also led to ease of maintenance. The location of the engine's mounting points even made it possible to remove the combustor from an installed engine. Removal of the combustor provided ready access to all hot engine parts for maintenance and for easy removal and replacement of engine components.[41]

The T53 design stemmed from intensive study and proposal activity during 1951 and the first half of 1952 that produced several concepts. The objective of this activity was to define an optimum engine that could meet the often opposing requirements of the military and the airframe designers and that could open a wide market. One of Lycoming's first conceptual designs was a 350-shp engine having a single centrifugal compressor with a rotating diffuser and a two-stage turbine. A rotating diffuser was chosen because it would allow a higher pressure ratio from a single compressor stage and would raise compressor efficiency by several percentage points, improving fuel economy.

A plan for the development of this engine, designated the LT-X, was proposed to the Air Force in late 1951. However, following discussions with the Air Force and representatives from the helicopter industry, Franz realized that helicopters being proposed by the airframers required a power level higher than 350 shp and still better fuel economy. Another round of parametric studies was conducted covering a variety of engine configurations that would produce a power output of 600 shp. To this end, the compressor pressure ratio was raised from about 4 to 6 for better fuel economy, and the airflow was raised to increase power output.[42]

Engines with two different compressor configurations were designed and analyzed. One had two centrifugal stages, and the other had a multi-stage axial compressor followed by a centrifugal stage. Both configurations were proposed to the military. The use of two centrifugal stages had the advantage of making the engine more rugged and requiring fewer parts. However, the diameter of the engine increased because the accessories were to be mounted around the centri-

---

40. Lycoming Division of Avco Corporation, "Lycoming Gas Turbines," *AHS Newsletter*, (July 1960): 7.

41. Franz to Leyes and Fleming, 14 March 1989.

42. Heinz Moellmann, personal notes, 12 April 1992, National Air and Space Museum Archives, Washington, D.C.

fugal stages to provide easier access for maintenance. In the case of the axial-centrifugal compressor arrangement, ultimately used for the T53, the basic over-all engine diameter was retained, the accessories being wrapped around the smaller-diameter axial compressor. The result was a neater package than the two-stage, centrifugal-compressor configuration. In addition, this engine turned out to be lighter in weight and to have improved fuel economy.[43]

Both single-shaft and dual-shaft configurations were considered in Lycoming's design studies. Initially, the Army, the Air Force, and some aircraft and other engine manufacturers favored a single-shaft engine over a dual-shaft engine with a free power turbine. The single-shaft version was simpler, had fewer bearings, and was believed to respond more quickly to demands for power. A rapid power response was a factor of particular importance to helicopter pilots.

The military dropped its preference for the single-shaft engine when it was discovered that, following a flame-out in a single-shaft turboprop aircraft, the high inertia of the compressor rotor produced a large braking action, quickly reducing propeller speed and causing a sudden negative thrust. Special control devices were required to avoid such dangerous conditions. When multi-engine helicopters were introduced, single-shaft engines were considered unsuitable, for reasons of safety, in case of loss of power in one engine. In addition, it was demonstrated that the power response of the free turbine engine could be made essentially the same as that of the single-shaft engine by taking advantage of the inherent ability of the free power turbine to increase torque when reducing rotating speed, and by paying special attention to the design of the compressor rotor and control system.

Other advantages of the free turbine design were smaller diameter of the gas generator components and higher turbine efficiency. These advantages came from the designer's freedom to select the optimum rotating speeds for the power turbine and the gas generator rotor individually. Operating the gas generator compressor and turbine at higher rotating speeds than the free power turbine led to improved overall efficiencies and a smaller engine diameter.[44]

THE FIRST PROPOSAL

A Request for Proposal (RFP) for a turboprop engine in the 600-shp class was issued on April 4, 1952, by the Air Force Air Materiel Command, serving as the Army's procurement agency. The proposal called for a rugged, simple, single-

43. Moellmann to Leyes and Fleming, 12 October 1989.

44. Moellman to Leyes and Fleming, 12 October 1989.

shaft design, turboprop engine of 500-700 shp. Alternate free turbine designs were also requested. It was interesting to note that in the RFP, the statement, "A centrifugal type of compressor is desired for this engine" had been crossed out, and typed after it was the replacement statement which read, "Simplicity and ruggedness of the compressor is desired."[45] This change came about because Lycoming had made the Air Force and Army aware of the advantages of the axial-centrifugal compressor design.

In May 1952, Lycoming responded to this RFP by submitting design proposals for two single-shaft, 600-shp, turboprop engines, one with two centrifugal compressors and one with an axial-centrifugal compressor arrangement. These engines could be converted to turboshafts by removing the propeller gearbox. Lycoming also submitted 600-shp, dual-shaft, free turbine designs for each of the two compressor arrangements.[46] Lycoming's designs were evaluated at Wright Field in competition with approximately 12 other designs submitted by several other companies.

In July 1952, Lycoming received a contract to develop a 600-shp, free turbine, turboshaft engine, designated the LTC1. This engine had a five-stage axial compressor in front of a centrifugal compressor stage, which was driven by a single-stage, gas generator turbine. A single-stage, free power turbine drove a front power takeoff. The combustor was a reverse-flow, annular design wrapped around the turbine assembly.[47]

DEVELOPMENT OF THE T53

Not long after detailed design began, the power requirement dictated by helicopter designers increased. In response, by the time the engine passed its 150-hour qualification test, its output had been raised to 860 shp, with a pressure ratio of 6:1 and an airflow of 10 lb per second. This engine, the Lycoming model LTC1B (military designation T53-L-1), was the first gas turbine engine produced by the company.

As soon as Lycoming received the contract to develop the engine, it scrambled to hire design and development personnel. Franz assembled a team with groups to handle engine structural design, compressor and turbine aerodynamic

45. Headquarters, Air Materiel Command, Wright-Patterson Air Force Base, "Request for Proposal 345609," Dayton, Ohio, 4 April 1952.

46. Moellmann to Leyes and Fleming, 12 October 1989.

47. Avco Lycoming Division, "T53 Historical Background," Report No. 1628.5.15, undated (circa 1974), 10.

design, and shafting and gearing design, and he added an expert in manufacturing. Walt Schrader, head of Lycoming's advanced engine and preliminary design groups prior to his retirement in the mid-1980s, related that when he arrived at Lycoming in 1953 there were 50 to 55 people on the T53 team. The senior engineers on the team had worked on gas turbines in Germany during World War II. With less than five years' experience each, the remaining engineers were relatively green.[48]

One of the first tasks of this team was to conduct a critical design review and analysis of all engine components. From the beginning, special attention had been given to potential design problems in the free power turbine shafting with its front power drive. A major mechanical design problem was the low natural frequency of the relatively long, slender, power turbine shaft, its small diameter dictated by bearing design requirements. The natural frequency of the shaft fell within the engine's operating speed range, which would produce shaft vibration. In a solution that became known as the "pregnant worm," design team expert on mechanical stress and vibration Dr. F. Belitz increased the diameter of the shaft between the front and rear bearings to approximately twice that at the bearings. The thin-walled, hollow, enlarged portion stiffened the shaft enough to raise its natural frequency above its rotational speed range.[49]

During contract negotiations, the military considered the control system one of the most important items of the engine development program. This concern stemmed from difficulties experienced with early jet engine control systems, including their vulnerability to the dirt and ice often encountered in JP4 fuel. The principal design requirements were lightweight, small size, ruggedness, and the ability to provide a rapid power response and to handle dirty fuel. Considerable effort was exerted on developing a satisfactory fuel control for the T53.

After evaluating proposals from several fuel control manufacturers, Lycoming selected the Chandler Evans Company (CECO) as the control vendor. The type of system developed was a hydraulic control. Such a system was considered more adaptable and accurate than competing pneumatic controls. Electronic controls being used on some jet engines at that time were heavy and not yet considered to be sufficiently reliable.

An intensive product improvement program reduced the control's size and weight, increased its fuel flow capacity by 70 percent, and simplified the system to ease manufacturing and maintenance and to improve its reliability. According to Heinz Moellmann, Bell Helicopter Senior Vice President of

48. Schrader to Leyes and Fleming, 10 October 1989.

49. Moellmann to Leyes and Fleming, 12 October 1989.

Engineering Bartram Kelley summarized the success of this design program by stating that, "The T53 control was the yardstick for gas turbine engine controls."[50]

Parallel to engine design and mock-up activities, preparations for component and engine testing began with the acquisition of a test rig in a small facility near the Stratford plant to conduct combustor experiments. A turbine test facility was then assembled in a building behind the Stratford plant that had been used by Pratt & Whitney for combustor testing during the late 1940s. There, a couple of old vacuum pumps were augmented by a steam-driven, diesel locomotive supercharger to supply air; an electric heater; and a high-speed water brake connected to the torque arm of an old Toledo scale, read backward through a telescope. Compressor tests were conducted at the Williamsport plant where there was an electric motor of sufficient size to drive the compressor.[51]

Although these test facilities were improvised, component testing progressed well, essentially verifying the performance goals. The first test run of the LTC-1 engine occurred in December 1954, followed by the first test run of the T53-L-1 a month later. The T53-L-1 passed its 50-hour PFRT in July 1956 and was first flown in a modified Kaman HOK-1 helicopter in September 1956. A second T53-L-1 engine was flown a month later in the new Bell XH-40 helicopter, subsequently put into production as the HU-1/UH-1 Iroquois.

The T53-L-1A passed its 150-hour Model Qualification Test (MQT) in April 1958, with the first production engine delivered in March 1959.[52] However, prior to the MQT a number of problems had been encountered during ground testing of the engine, some a consequence of cost-reduction efforts. The principal engine parts at fault were the labyrinth seals and the main reduction gear. In addition, a problem with vibration of the main rotor shaft and a variety of minor problems had to be solved to ready the engine for its MQT.

During early test runs, the T53 was found to be an oil guzzler, using about one-fourth that of the engine fuel consumption. This oil loss was caused by leakage through the engine's labyrinth seals. Modification of the labyrinth seals to reduce their clearances was impractical because of the small size of the engine. The oil loss was stopped by installation of custom-designed, compact, positive seals that could tolerate high temperature, pressure, and rubbing speed.

From time to time during test runs, power turbine blades came off. Because high-temperature vibration sensors were not available, solving the problem be-

50. Moellmann, 12 April 1992.

51. Schrader to Leyes and Fleming, 10 October 1989.

52. Avco Lycoming Division, "T53 Historical Background," 10.

came a veritable detective job. The source of the blade failures was ultimately traced to the spur gears in the reduction gearing. The vibrating frequency of the spur gear meshing was transmitted through the gear train and in turn through the power turbine shaft and the turbine disk into the turbine blades, creating a resonant frequency that under certain conditions caused the blades to fail.

The potential for vibration excitation had been recognized when the reduction gearing was designed. Thus, a special "hunting" gear had been designed into the system in the hope that the magnitude of the excitation forces would be sufficiently low to avoid blade vibration. Replacing the spur gears with more expensive helical gears solved the problem. The solution to this vibration problem was picked up throughout the industry.

The vibration problem in the gas generator rotor shaft was traced to the curvic coupling used between the axial compressor stages. Mating gear teeth machined into the faces of the compressor disks to transmit the torque from one disk to the next deformed under load. The specified machining tolerances of the gear teeth were sufficiently broad to produce a nonuniform contact pressure among the curvic coupling elements as the engine rotated. This resulted in flexing where the contact pressure was greatest, lowering the natural frequency of the rotor assembly to within the engine's operating speed range. The problem was solved by tightening the machining tolerances to produce more uniform contact pressures between the curvic couplings, thereby stiffening the connection between the compressor stages and raising the natural frequency of the rotor above the operating speed range.[53]

Ground testing of the T53 included 30 hours of anti-icing runs at Mt. Washington, New Hampshire, in addition to cold-weather and high-altitude tests at Wright Field. It was reported from these tests that it was not possible to "flame out" the engine at any altitude below 35,000 feet and that the engine could be started without difficulty at -65°F.[54]

EARLY T53 MARKETING

On February 27, 1956, Lycoming publicly announced details of the XT53-L-1 engine.[55] Concurrent with development of the T53, Lycoming had begun

53. Schrader to Leyes and Fleming, 10 October 1989.

54. "Shaft Turbines by Lycoming," *The Aeroplane*, XCV, No. 2458 (October 10, 1958): 561.

55. Avco Lycoming Division, "XT53 News Release," Avco Lycoming Press Release, 27 February 1956.

looking for possible applications for the engine. Lycoming's initial contact was with Charles Kaman, President of Kaman Aircraft. He became interested in the engine and agreed to provide one of his helicopters as a testbed for it. About the same time, the Air Force had released an RFP to the industry for a new helicopter to be used by the Army. The size of this helicopter was such that the T53 could power it. Since the Bell Aircraft Co. was planning to respond to the RFP, discussions soon began between Lycoming and Bell concerning the use of the T53.

At that time, the military was divided on whether this new helicopter should use a gas turbine or a piston engine. Consequently, Bell designed two versions of its helicopter. The Bell engineers recognized the advantages offered by the gas turbine: its light weight and the fact that it could be mounted above the fuselage, freeing up cargo space that the in-fuselage piston engine installation would otherwise occupy. However, they were concerned about the gas turbine's lower fuel economy. Bell's design studies indicated that for flights of greater than one and one-half hours, the weight of the gas turbine and its fuel exceeded that of piston engine installations, thus reducing payload capability. Lycoming acknowledged this shortcoming of the gas turbine, but pointed out that whereas the piston engine had reached maturity, the small gas turbine was a new engine type that could expect to see improvements in fuel economy as its design matured.

In addition, Lycoming engineers noted a number of other advantages offered by the gas turbine. One was improved flight safety. Given the high power-to-weight ratio of the gas turbine, helicopter designers would be inclined to put more power into a gas turbine powered helicopter than into its piston-powered counterpart. More power would increase the power margin, and thus safety, during flight operations. Another advantage was the potentially longer life of gas turbine engines. For piston engines, the Time Between Overhaul (TBO) was less than 1,000 hours. However, longer TBOs were being obtained with large gas turbines in the fleets of military aircraft flying at that time, and the potential was for even longer TBOs. The gas turbine also operated smoothly, without the objectionable vibration characteristic of the piston engine.[56]

Bell submitted a proposal to develop a T53-powered helicopter, the Model 204 (military designation XH-40). On February 23, 1955, it won the Army competition.

Early flight tests of the T53-powered helicopter revealed torsional vibration between the engine and the helicopter rotor. A low-frequency oscillation of the

56. Moellman to Leyes and Fleming, 12 October 1989.

helicopter rotor at about three cycles per second originated from the power turbine. The rotor in the Bell helicopter had about five times the effective inertia of the power turbine, and it rotated at a speed of only about 200 rpm compared to the 20,000 rpm of the power turbine. This rotor was connected to the engine by a gear system with no damping provided between the engine and rotor.

During a visit to Bell by Lycoming engineers, Heinz Moellmann observed the oscillation while seated with the Huey's pilot in the cockpit. Moellmann requested that a fuel manifold pressure gauge be installed in the cockpit so that he could see if the engine fuel pressure was oscillating. With the gauge installed, the pilot tried to excite the oscillation but couldn't. The air bubble in the line between the fuel manifold and the pressure gauge was sufficient to dampen out the oscillation. The Lycoming engineers returned to the drawing board and quickly designed a fuel control damping system that eliminated the oscillation problem.[57]

Meanwhile, Lycoming had been working with the Grumman Aircraft Co. on the use of a turboprop version of the T53 in a light observation aircraft that Grumman was designing as its entry in a competition being held by the Army. The design competition for this aircraft was won by Grumman with its T53-powered OV-1 Mohawk; flight testing began in April 1959.[58]

EVOLUTION OF THE T53

To support the Huey and Mohawk, both turboshaft and turboprop versions of the T53 were developed. Initial deliveries of the 860-shp T53-L-1 turboshaft for the Huey began in March 1959. During the four years that followed, three progressively updated turboshaft models of the T53 reached production. One was the T53-L-5 and the other two were the T53-L-9 and T53-L-11.[59]

Turboprop and turboshaft models of the engine were developed in parallel. As improvements were introduced and the power output was uprated on the turboshaft models, the same improvements were put into the turboprop versions.[60] The first T53 turboprop, the 960-shp T53-L-3, was delivered in September 1959 for the Grumman OV-1 Mohawk. This engine was the turboprop version of the T53-L-5 turboshaft. In 1962, production began on an uprated

57. Moellman to Leyes and Fleming, 12 October 1989.

58. *Jane's All the World's Aircraft 1959-60* (London: Sampson Low, Marston & Company, Ltd., 1959), 305.

59. Avco Lycoming Division, "T53 Historical Background," 10-14.

60. Schrader to Leyes and Fleming, 10 October 1989.

T53 turboprop, the 1,100-shp T53-L-7, which was the turboprop version of the T53-L-9.

Up to that point, the progressive increase in power output of the T53 had been obtained mainly by increasing turbine temperature. These temperature increases were made possible by improved cooling of the turbine inlet guide vanes, improved turbine materials, and ceramic coating of the combustor liner. Mechanical design changes were also made to improve interchangeability of parts, simplifying logistics in mixed fleet operations, and to increase ruggedness of the engine structure. Other improvements included redesign of helical reduction gearing to accommodate higher torque and modification of the fuel control to improve engine operating characteristics.

By the late 1950s, the T53 was approaching the limit of its available power output. Any significant increase would require a major aerodynamic design change. Both Bell and Grumman continued to press for more power. Consequently, in 1959, Lycoming undertook a major redesign of the engine, which became the 1,400-shp T53-L-13. A complete T53-L-13 engine was first run in late 1960.

The power output of the T53-L-13 was nearly double that of the T53-L-1, first run in 1955. This gain, which came within five years, was achieved within the original engine frame size. The T53-L-13 incorporated five basic design improvements. The first two stages of the five-stage, axial-flow compressor were replaced by transonic stages having variable inlet guide vanes. The transonic stage design increased airflow and pressure ratio, and the variable inlet guide vanes provided good compressor stage matching over a wide speed range and a good compressor stall margin. A second stage was added to the gas generator turbine to improve turbine efficiency. Turbine blade cooling was introduced in the first stage of the gas generator turbine to allow an increase in turbine operating temperature. A second free turbine stage was added to supply the higher power to the output shaft.[61] Finally, there were improvements in the combustor involving the introduction of improved heat and corrosion resistant alloys to provide longer life, and the use of atomizing fuel nozzles to enable operation with alternate fuels. Previous T53 engines had used vaporizing-type fuel injectors, which had to be tailored to the specific fuel used in the engine.[62]

The T53-L-13 was first manufactured in 1966. Among the military applications of the T53-L-13 were the Bell UH-1H Huey and Bell AH-1G HueyCobra. A commercial version of this engine, the T5313B, was also produced for use in

61. Avco Lycoming Division, "T53 Historical Background," 11-15.

62. Schrader to Leyes and Fleming, 10 October 1989.

the Bell Model 205A commercial utility helicopter derived from the UH-1H Huey.

A year after completing the major design changes incorporated in the T53-L-13, its turboprop companion, the T53-L-15, was introduced for use on the Grumman Mohawk. However, the power output of that engine was limited to 1,160 shp to avoid exceeding the mechanical limits of the T53-L-7 propeller reduction gearing. A 1,400-shp turboprop version of the T53, the T53-L-701, was introduced later following redesign of the propeller reduction gearing to accommodate the full-power output capability of the engine.

The last in the series of T53s to be produced by 1990 was the T53-L-703, used in both the Bell AH-1Q and AH-1S HueyCobra. The design of this engine was modified for improved durability and long life. In addition, the turbine operating temperature was increased to raise the power output to 1,550 shp. The thermodynamic power output of this engine was 1,800 shp.[63] This engine began production in 1978.

FIELD EXPERIENCE DURING THE VIETNAM WAR

The T53 proved to be very durable during its extensive field operations in Vietnam. As might be expected, there were a number of minor nuts and bolts types of problems encountered, but all were solved quickly and effectively. The one major technical problem encountered by users of the T53 in Vietnam was erosion of the internal engine parts from ingestion of sand. The larger T55 engine used in the Chinook transport helicopter shared this problem. However, the problem was most severe with the T53 because the Huey and HueyCobra operated and landed at unprepared field sites to provide support for ground troops.

The sand ingestion problem in Vietnam was first encountered as the helicopters began operating out of fields and off the beaches and had been particularly bad on the beach near Danang during the early part of the Vietnam War. Waiting in formation, a squad of helicopters with whirling rotors kicked up tons of sand and dust. Sand that entered the engine peened over the leading edges of the first-stage compressor blades and eroded the rest of the compressor and the turbine, increasing blade tip clearance in the process. The result was a rapid and intolerable drop in power output and fuel efficiency, and before long, the compressor stall margin was reduced to the point that the engine would surge. In addition, engine life expectancy was drastically reduced.[64]

63. Avco Lycoming Division, "T53 Historical Background," 16-17.

64. Schrader to Leyes and Fleming, 10 October 1989.

Meanwhile, between 1958 and 1963, foreign object damage had been investigated in 11 engines that had been subjected to sand and dust environments in both field and test cell programs. Data accumulated in these tests showed that the T53 was exceptionally tolerant to sand and dust. Then, in 1963 and 1964, the Army 11th Air Assault Division at Ft. Benning, Georgia, conducted operations using the newly developed airmobility concept in which multiple helicopters operated extensively in tight formation over unprepared landing sites. These operations established a new perspective on what constituted a rigorous foreign object environment and underscored the need for inlet protection. As a result of these operations and related experience in Vietnam, foreign object damage mushroomed into a major problem.

Bell Helicopter Co. quickly designed a barrier filter that provided some immediate protection for the engine. Lycoming's response was to undertake a crash program to develop a suitable inlet particle separator and to make internal improvements to reduce the engine's vulnerability to sand and dust. Studies suggested that an inertial-type particle separator appeared to offer the most promising solution. The design objective was to provide a means of separating out sand and other foreign particles without having too great an effect on engine performance due to inlet pressure loss. Designed to be installed in the aircraft as an integral part of the engine, the separator had catching buckets that had to be emptied manually between flights. Following development tests of the separator at Lycoming, field evaluation tests by the Army at Ft. Rucker, Alabama, during 1965 and 1966 indicated that engine life in a sand and dust environment was increased by a factor of three with the separator installed. In May 1967, the separator was released for production, and by the end of 1968, more than 6,700 units had been produced for use in the field.

Work proceeded simultaneously to increase the engine's tolerance to ingestion of sand and dust. The most significant of these improvements was the development of replaceable stainless steel inserts to protect the magnesium axial compressor housing from erosion. The incorporation of Bell's foreign object damage screen together with Lycoming's inlet particle separator and internal engine changes improved engine life in a sand and dust environment by a factor of six. As a result, the mean time between depot return for engine erosion in Vietnam rose from 1,870 hours prior to separator installation to over 9,500 hours after installation of the separator and steel compressor case inserts.[65]

65. H. D. Connors and J. P. Murphy, "Gas Turbine Sand and Dust Effects and Protection Methods," (SAE Paper No. 700705 presented at the SAE Combined National Farm Construction, Industrial Machinery and Powerplant Meetings, Milwaukee, Wisconsin, 14-17 September 1970), 1-10.

Throughout the Vietnam War, Lycoming's capability to produce T53s was strained to its limits. Due to the military's increasing demand for more engines, there was a tremendous push to get engines out the door. As a result, a substantial amount of production support was brought to bear to solve production problems and to develop procedures that helped marginal engines pass their acceptance tests.[66]

In early 1968, a critical shortage of T53 and T55 engines began to develop as a result of increased operations in Vietnam. In fact, it became "normal" for replacement engines to be unavailable. The solution developed by the military was to initiate a dedicated air transportation system using C-141 cargo aircraft. Fully loading the C-141s with engines on each flight, unserviceable engines were flown to the Corpus Christi Army Maintenance Depot in Texas and serviced engines were returned from that point to Vietnam. Within a few months, one-way transport time for engines between Vietnam and Corpus Christi had been reduced from 10 days to 33 hours.[67]

Dave Knobloch, who became manager of the T53 and T55 programs in 1969, described a subsequent step taken to further reduce engine transportation time and to release the C-141s for other critical transport tasks. U.S.S. *Corpus Christi,* a small aircraft carrier, was modified to contain a complete machine shop, an engine disassembly and assembly area, and engine test cells to service both T53s and T55s. With its facilities manned by trained personnel, the ship lay off the Vietnam coast, and helicopters flew engines out to it where they were overhauled, tested, and returned to service.[68]

T53 SUMMARY

The T53 was one of the first gas turbines of its power class to be completely designed and developed in the U.S. It turned out to be the right engine at the right time. The T53 was one of the first small gas turbine engines to be developed as a multiple function power plant. It was used as both a turboshaft and a turboprop engine. The front-drive, concentric-shaft arrangement of the T53 set a design precedent for turboshaft engines that became a widely accepted standard throughout the industry in America. As one of the first turboshaft engines mass

66. Schrader to Leyes and Fleming, 10 October 1989.

67. "Turbine Engine Supply Management," *United States Army Aviation Digest,* Vol. 18, No. 4 (April 1972): 22-23.

68. David Knobloch, interview by Rick Leyes and William Fleming, 11 October 1989, Textron Lycoming Interviews, transcript, National Air and Space Museum Archives, Washington, D.C.

produced in the U.S. for helicopters, the T53 gave Lycoming its start in the aircraft gas turbine business and was the key to the company's continuing growth and success in that business. The T53 was also an important part in the expansion of Army aviation's role that occurred during the Vietnam War with the development of a whole new means of air–ground warfare made possible by the gas turbine-powered helicopter.

The importance of the T53 to Lycoming is illustrated by the following. Of the 26,210 aircraft gas turbine engines manufactured by Lycoming from 1959 through the end of 1989 (including 1,151 license-built engines), 18,959, or nearly three-fourths of them, were T53 turboshaft and turboprop engines. Of these, more than 10,000 were produced during the Vietnam War years for the Bell Huey and HueyCobra helicopters. In addition, 1,286 turboprop versions of the engine were produced between 1959 and 1980 for use in the Army's twin-engine Mohawk, which saw service in Vietnam. Although the T53 was the first gas turbine manufactured by Lycoming, its progressive uprating gave the T53 a production lifetime that extended over many years.[69]

The T53 made a major contribution to the Huey's operating flexibility and performance. The capability of this helicopter allowed the Army to extend its infantry attack and combat operations to include integrated ground and air operations. By 1965, the use of the T53-powered Huey to evacuate wounded from the battlefield had helped to keep the death rate down to less than two percent of the wounded.[70] The turbine engine helicopter, with its great power, its reliability, and its smaller requirement for maintenance was the technological turning point that made the concept of airmobility a reality.[71]

High-Power Engine Family

ORIGIN AND DESIGN OF THE T55 ENGINE

Development of the T55, the first and most prominent member of the high-power engine family, started in 1954 soon after work had begun on the T53. Both turboshaft and turboprop versions of the T55 were developed. However,

---

69. Textron Lycoming, "Original Engine Shipments by Year of Manufacture," 23 July 1990, unpublished graph, National Air and Space Museum Archives, Washington, D.C.

70. *Avco Corporation - The First Fifty Years*, 59.

71. Col. Gene Gurney, *Vietnam: The War in the Air* (New York, N.Y.: Crown Publishers, Inc., 1985), 37-38.

only a few prototype turboprop engines were built for limited experimental use in two prototype aircraft.

In April 1954, nearly two years after design of the T53 had begun, Lycoming received a contract from the Army and Air Force to develop the T55 as a turbo-prop engine.[72] Initially conceived by the Army as a turboprop that would power a future medium transport aircraft, the T55 was to be a 1,500-shp engine capable of substantial future growth in power output.[73] By 1957, engine tests had proven that the mechanical design concept was sound and the engine had passed its 50-hour qualification test at a power rating of 1,500 shp. However, the specific fuel consumption was higher than the original goal, and there were no firm applications in sight for the engine.[74] Meanwhile, in 1956, the Army announced that it planned to replace the piston-powered H-37 transport helicopter with a gas turbine-powered model.[75] As a result, in 1957, Lycoming redirected its work on the T55 to develop the T55-L-3 turboshaft version of the engine for use in the H-37 replacement.[76]

The T55 was to be a small, rugged, lightweight engine, highly reliable and easy to maintain. The T55 was virtually identical to the T53 in physical size and in its basic arrangement of major engine components. Nevertheless, aerodynamically and mechanically it was a completely new design. As in the T53, the T55's compressor, a mixed axial/centrifugal design, and turbine were close-coupled. A reverse-flow combustor was wrapped around the hot section of the engine. In appearance, the two engines bore a striking similarity. However, with an engine volume only slightly larger than that of the T53, the T55 produced almost twice as much power.[77]

The major design differences between the two engines were that the T55 had seven axial-flow compressor stages instead of five, and the T55 compressor was designed to deliver twice the airflow. The T55 had a single-stage gas generator turbine and a two-stage power turbine that rotated in opposite directions in

72. Avco Lycoming Division, "Appendix I, Historical Background–T55 Engine Series," Report No. 1755.5.36, undated (circa 1976), 69.

73. Thomas Dickey, interview by Rick Leyes and William Fleming, 10 October 1989, Textron Lycoming Interviews, transcript, National Air and Space Museum Archives, Washington, D.C.

74. Erwin J. Bulban, "Lycoming T55 Cleared for Airframe Use," *Aviation Week*, Vol. 71, No. 6 (August 10, 1959): 87-94.

75. Gurney, *Vietnam: The War in the Air*, 48.

76. Bulban, "Lycoming T55 Cleared for Airframe Use," 94.

77. Moellmann to Leyes and Fleming, 12 October 1989.

The Lycoming T55 turboshaft engine. Photograph courtesy of Textron Lycoming.

place of the T53's single-stage co-rotating turbines. Although design and development of the T55 trailed that of the T53 by about two years, as new information was acquired and lessons were learned on one program, those data were passed to the other program.[78]

The first T55 model, the T55-L-1 turboprop, was run initially in February 1955 and was flight qualified in 1957 at 1,500 shp. One engine was delivered to the Navy in September 1962 for use in its Short Airfield Tactical Support development program.[79] A later turboprop model, the T55-L-9, was used during the early 1960s in a prototype turboprop version of the North American P-51 Mustang. Beginning in 1963, the engine was used briefly in the North American YAT-28E prototype conversion of the piston engine-powered T-28 trainer aircraft. Conclusion of the flight tests of these prototype aircraft ended the brief lifetime of the turboprop versions of the T55.[80]

78. Dickey to Leyes and Fleming, 10 October 1989.

79. Avco Lycoming Division, "Historical Background–T55 Engine Series," 69, 72-73.

80. Dickey to Leyes and Fleming, 10 October 1989.

Soon after the Army made its decision to replace the H-37, Lycoming developed the first T-55 turboshaft model, the 1,900-shp T55-L-3. This engine incorporated internal gearing suitable for helicopter use. Only prototype and test versions of this engine were built. However, it was on the basis of the T55-L-3's performance that the T55 engine was selected in July 1959 for the Boeing Vertol Model 114 transport helicopter, later designated the CH-47A. The initial engine model used in the CH-47A was the T55-L-5 uprated to 2,200 shp. Following its selection for the Chinook, the T55 had been uprated to 2,200 shp.[81]

THE CHINOOK AND ITS ROLE IN AIRMOBILITY

In the evolution of airmobility, the Huey helicopter became the cornerstone and the Chinook became an important building block. In late 1956, the Army announced plans to replace its piston-powered H-37 transport helicopter with a new turbine-powered version. In September 1958, following a design competition for the H-37 replacement, a joint Army-Air Force source selection board recommended that the Army procure the Vertol medium transport helicopter. The Army ultimately settled on a larger version, the Vertol Chinook, as its standard medium transport helicopter.[82]

The Boeing Vertol to CH-47 Chinook originally evolved from the Vertol Model 107, a twin-rotor transport helicopter powered by two engines. With the Army's decision in 1956 to replace the Sikorsky H-37, Vertol began designing its YHC-1A (military version of the Model 107). Prototype versions of this helicopter, powered by two Lycoming 860-shp T53-L-1 engines, began flying in April 1958. Soon afterward, Boeing acquired Vertol and continued development of the YHC-1A. The production version of the YHC-1A, later designated the CH-46 and powered by two 1,250-shp General Electric T58 engines, was selected in 1961 by the Marine Corps. In June 1959, the Army selected as its transport helicopter a derivative version of the YHC-1A, the YHC-1B (Model 114). The production version of the YHC-1B was designated the CH-47A Chinook. The Chinook was an extensively redesigned and enlarged version of the YHC-1A powered by two Lycoming 2,200 shp T55-L-5 engines.[83]

81. Avco Lycoming Division, "Historical Background–T55 Engine Series," 72-73.

82. Gurney, *Vietnam: The War in the Air*, 48.

83. *Jane's All the World's Aircraft 1962-63* (London: Sampson Low, Marston and Company, Limited, 1962), 188-189.

By February 1966, a total of 161 Chinooks had been delivered to the Army. When the 1st Cavalry Division arrived in Vietnam in 1965, it brought its Chinook battalion. Although considered an assault troop carrier, the Chinook's highest priority mission in Vietnam became transporting artillery batteries to positions that were inaccessible by other means, and resupplying them with ammunition. The Chinook soon proved so valuable for artillery movement and heavy logistics that it was seldom used for assault troop transport. The Army's decision in 1958 to move to the large version of the Chinook helicopter proved to be indisputably sound, and it produced for the Army a capability it did not previously have.[84]

GROWTH AND EVOLUTION OF THE T55

From 1959 to 1978, the power output of the T55 turboshaft engine was progressively raised from 1,900 to 3,750 shp. This growth in power output was in response to the continuing demand for more power to increase the lifting capability of the Army's Chinook transport helicopter.

The first T55 production turboshaft model, the 2,200 shp T55-L-5, was initially delivered for use in the Chinook CH-47A in February 1961. Within the engine, the free power turbine was connected directly to the power output shaft with no internal gearing. In addition, an automatic interstage compressor air bleed was incorporated to avoid compressor stall during rapid acceleration. The increase in power output from 1,900 shp for the T55-L-3 prototype model to 2,200 shp for the T55-L-5 was achieved with a number of refinements. The most prominent of these was the use of improved turbine materials, allowing operation at higher turbine inlet temperatures.[85] A total of 146 T55-L-5s were produced between 1960 and 1963 for use in the CH-47A.[86]

The first 2,650 shp T55-L-7 was delivered in July 1963 for the Chinook CH-47B. This engine was an uprated version of the T55-L-5 with compressor and combustor modifications.[87] The most significant design change was the introduction of atomizing fuel nozzles to replace the vaporizing T-cane injectors, eliminating hot spots in the gas stream at the combustor outlet and providing a

84. Gurney, *Vietnam: The War in the Air*, 48.

85. Avco Lycoming Division, "Historical Background–T55 Engine Series," 69, 73.

86. Textron Lycoming, Table of aircraft gas turbine engines delivered through 1989, unpublished production numbers, National Air and Space Museum Archives, Washington, D.C.

87. Avco Lycoming Division, "Historical Background–T55 Engine Series," 73.

more uniform temperature profile.[88] The atomizing fuel nozzles reduced the temperature spread between the hottest and coolest spots at the combustor outlet by as much as 100°F., permitting a corresponding increase in average turbine inlet temperature, which pushed the power output up from 2,250 to 2,650 shp.[89] This series of T55 engines was used extensively in Vietnam, with a total of 1,471 T55-L-7 engines being delivered between 1963 and 1968 for use in the CH-47B model of the Chinook.[90]

The Army continued pressing for more power to increase the Chinook's lifting capability. However, in its existing configuration, the T55 was approaching the limit of its power output capability. Major design changes were required to obtain further significant power increases. The first of these redesigned engines, the 3,750 shp T55-L-11, was delivered in August 1968. Principal changes were variable compressor inlet guide vanes, transonic rotor blades in the first three compressor stages, aerodynamically and mechanically redesigned turbines with a second stage added to the core turbine, and a redesigned airframe gearbox. The aerodynamic redesign of the compressor increased both airflow and pressure ratio, major factors contributing to the engine's increased power. The aerodynamic redesign of the turbine was made to match it to the higher engine flow rate and pressure ratio, and redesign of the airframe's gearbox accommodated the higher power output.

By the early 1970s, service and in-flight problems with the T55-L-11 in Vietnam led to the recall of the engine from field service and an extensive mechanical redesign to improve reliability, maintainability, and durability. The compressor design was also modified. The variable inlet guide vanes were removed and wide-chord blades were used in the first and second axial compressor stages. The resulting engine was the 3,750 shp T55-L-712 used in the Chinook CH-47D.[91] Production of this engine began in 1978, and by the end of 1989, 849 T55-L-712 engines had been manufactured.[92]

---

88. A T-cane injector is a small, hollow T fabricated from circular tubes and installed at the forward end of the combustor. An atomizing fuel nozzle injects fuel into the T-cane from its base. Heated by the hot combustion gases surrounding it, the T-cane vaporizes the fuel as it flows rearward down the leg of the T and out the two arms of the T into the combustor where it is burned.

89. Knobloch to Leyes and Fleming, 11 October 1989.

90. Textron Lycoming, Table of aircraft gas turbine engines delivered through 1989.

91. Avco Lycoming Division, "Historical Background–T55 Engine Series," 73–76.

92. Textron Lycoming, Table of aircraft gas turbine engines delivered through 1989.

In early 1990, Lycoming announced that the T55-L-714 rated at 4,867 shp had been qualified for use in the Army's MH-47E Special Operations Forces' deep penetration helicopter. Major improvements included cooled gas generator turbine blades, improved durability of the gas generator turbine assembly, improved tip clearance control, and a Full Authority Digital Engine Control (FADEC). Production of the engine began in late 1990.[93]

COMMERCIAL AND FOREIGN MILITARY APPLICATIONS

A commercial version of the T55-L-712, the AL5512, was produced between 1979 and 1985 and used in the commercial version of the Chinook, the Boeing Model 234. A number of Model 234 helicopters were produced to support logging and fire-fighting operations in the northwest United States and oil rig operations in the North Sea and Alaska, at one point averaging over 100,000 passenger miles per day.[94] Forty-four AL5512 engines were manufactured for Model 234 helicopters.[95]

Meanwhile, in the early 1970s, the T55-L-7C had been uprated for use in three models of the Bell 214 helicopter, the 214A and 214C military models and the 214B commercial model.[96] The Bell Model 214 was an improved version of the T53-powered UH-1H Huey Plus helicopter and had a 50 percent greater lifting capability than the Huey Plus. In 1972, Bell announced that it had received an order from the Army for 287 Model 214As, to be acquired by Iran through the U.S. government.[97] Flat rated at 2,250 shp, 433 LTC4B-8D engines were produced between 1973 and 1977 for use in these helicopters. The commercial version of the Model 214B was introduced in 1975 for agricultural and fire-fighting purposes. Eighty-eight T5508D engines, also flat rated at 2,250 shp, were produced for use in the 214B.

T55 FIELD EXPERIENCE IN VIETNAM

The T55-L-7 was initially fielded in the Chinook CH-47B in Vietnam. Only minor problems were encountered with the engine itself. However, as had been

93. Textron Lycoming, "T53/T55 Engine Modernization and Growth for the 21st Century," *Textron Lycoming Press Release*, 4 February 1990, 4.

94. Knobloch to Leyes and Fleming, 11 October 1989.

95. Textron Lycoming, Table of aircraft gas turbines delivered through 1989.

96. Knobloch to Leyes and Fleming, 11 October 1989.

97. *Jane's All the World's Aircraft 1973-74* (London: Sampson and Low, Marston & Company Ltd., 1973), 265.

the case with the T53, erosion of the compressor rotor blades due to the ingestion of sand became a major problem. As initially deployed, the inlets of these engines were unprotected except for a heavy screen ahead of the compressor to prevent the ingestion of birds and other large objects.

The Army took two routes in solving the problem on the two engines. In the Huey and HueyCobra, the T53 was submerged within the fuselage. Thus, a velocity type of inertial separator was developed and installed in the aircraft as an integral part of the engine assembly. This separator incorporated catching buckets that could be removed and emptied. In contrast, the T55 was mounted above the Chinook fuselage, and its external installation made design of a sand separator a difficult problem. Screens were installed at the engine inlet to keep sand and other foreign materials out of the inlet. As might be expected, the T55's inlet screens were less effective than the T53's inertial separator. This was made clear by the T55's shorter TBO.

Lycoming fielded a completely redesigned 3,750-shp T55-L-11 in the Chinook CH-47C in 1969. With the Army's urgent need for more power to increase the Chinook's lifting capability during the rapid Vietnam buildup, the T55-L-11 had been rushed into service with less than 3,000 hours of testing. Although the tests had been relatively problem free, a number of problems emerged in service. These became so serious that the engine was withdrawn from service in Vietnam.[98]

A number of compressor blade problems occurred during operation of the T55-L-11 in Vietnam that remained with the engine throughout its service there. First-stage compressor blades suffered fatigue failures; the source of the blade excitation was at first a mystery. Tests revealed that improper inlet guide vane adjustment during field maintenance induced a flutter mode in the first stage blades.

Stress corrosion also caused first-stage blade failures. The material used in the first-stage blades was AM350, an alloy then new to the industry. Later, it was found that if AM350 was not heat treated properly, carbides that formed in the grain boundaries caused rotor blade fatigue failure in high-stress areas. The extent of this problem was such that Lycoming removed AM350 from the first compressor stage.

A more serious problem for the T55-L-11 in Vietnam was a progressive failure of the power turbine second-stage rotor blades. These long blades had interlocking blade tip shrouds that wore out due to rubbing, leaving the blades free standing. Without blade tip support, a vibratory resonance occurred that eventually produced blade failure. At times, sufficient torsional force was generated to tear

98. Knobloch to Leyes and Fleming, 11 October 1989.

loose the power turbine and shaft, taking the tailpipe with them. The power turbine second stage was located aft of the combustor and had only a light shroud around it, insufficient to withstand the ballistic force of a four-inch-long turbine blade. Blades that failed went through the shroud and into the aircraft.

Lycoming and the Army viewed these turbine failures as extremely dangerous. In some instances, the "shrapnel" damaged the hydraulic system, forcing the aircraft to land. After about 18 months of service in Vietnam, the engines were withdrawn from operation and replaced by the earlier model T55-L-7, which continued to provide effective service throughout the war.[99]

Lycoming conducted an extensive Service Revealed Difficulty program after the recall of the T55-L-11. An important lesson learned from the experience with this engine in Vietnam was the hazard of releasing an engine for field operations before it had completed an extensive development test program.

T55 SUMMARY

From the aspect of Lycoming's business volume, the T55 engine became an important part of the company's history and was second in importance only to the T53. From 1960 through the end of 1989, 4,233 T55s (including 191 license-built engines) were produced. More than half of these T55 engines were produced during the Vietnam War years. Further, like the T53, the T55 had a long production life.

During the Vietnam War, the T55 contributed to establishing a triad of Lycoming-powered helicopters that provided the Army with airmobility. The T55-powered Chinook's large capacity for movement and resupply of artillery and materiel provided an essential complement to the assault-troop-movement and air-attack capabilities of the T53-powered Huey and HueyCobra.

INTRODUCING THE HIGH-BYPASS TURBOFAN—THE PLF1A-2

In the early 1960s, Lycoming developed the PLF1A-2, the first high-bypass turbofan engine built and successfully tested in America and possibly in the world.[100] However, the concept of bypass engines was not new. Nearly three decades earlier, in 1935, Sir Frank Whittle in England obtained his patent for a bypass engine. With funding support by the British Government, in 1952, Rolls-Royce began development of the low-bypass ratio Conway engine, which became the first production bypass type of turbojet. With a bypass ratio of 1:1, the Conway first

---

99. Knobloch to Leyes and Fleming, 11 October 1989.

100. Avco Lycoming PLF1A-2 Accession Memorandum, National Air and Space Museum, NASM Number 7742, Catalogue Number 1989-0042, 17 April 1989.

The Lycoming PLF1A-2 turbofan engine, the first high-bypass turbofan built and successfully tested in America and possibly in the world. This engine is in the National Air and Space Museum collection. National Air and Space Museum photo by Carolyn Russo, Smithsonian Institution (SI Neg. No. 92-5808).

ran in 1953 and completed its type test in 1955. It was put into production soon afterward and was used in the Douglas DC-8-40, the Boeing 707-420, the Vickers VC-10 and Super VC-10 airliners, and the Handley Page Victor B. Mk.2 bomber and aerial tanker.[101]

Although Pratt & Whitney had started design work on its 5:1-bypass ratio JT9D[102] and General Electric on its 8:1-bypass ratio GE1/6 engine,[103] both for use in large commercial transport aircraft, it was Lycoming's PLF1A-2 that, in 1964, became the first U.S. engine to demonstrate high-bypass, turbofan engine technology.

The concept of a high-bypass turbofan engine at Lycoming evolved from a series of conceptual studies begun in the mid-1950s.[104] These studies were di-

101. "Rolls-Royce Takes Wraps Off Conway," *American Aviation*, Vol. 19, No. 8 (September 12, 1955): 38-40.

102. Quarterly JT9D Status Report, Pratt & Whitney Aircraft Co., 1 August 1975.

103. General Electric Co., *Seven Decades of Progress* (Fallbrook, Cal.: Aero Publishers, Inc., 1979), 148, 149.

104. Dickey to Leyes and Fleming, 10 October 1989.

rected toward making the turbojet engine more competitive with the turbo-prop in thrust and fuel economy at takeoff and in low-speed flight.

By the first quarter of 1962, the T55-L-7 was selected as the core of a proto-type turbofan. An engine was fabricated during 1963, and in February 1964, testing of the PLF1A-2 began.[105] Development feasibility tests were successfully completed in 1965. The engine had a thrust of 4,320 lb and a bypass ratio of 6.1:1.

TESTING THE PLF1A

Although only two PLF1A-2 engines were built, they were used in tests over several years. Following completion of development feasibility tests on the first engine in 1965, the second engine was sent to the Air Force's Arnold Engi-neering Development Center (AEDC) for altitude testing.[106] During 1965, tests were performed to determine the fan and engine performance of the PLF1A-2 at altitudes up to 40,000 feet and flight Mach numbers up to 0.8.[107] In January 1966, a PLF1A-2 was sent to Siebel in West Germany for tests re-lated to a V/STOL application. In March 1967, an engine was sent to the NASA Lewis Research Center for noise testing. Finally, one of the engines was sent to Wright–Patterson Air Force Base for use in a classified research program which began in September 1968.[108] In 1989, Lycoming donated the only re-maining PLF1A-2 engine to the National Air and Space Museum in Washing-ton, D.C.[109]

The development feasibility testing of the PLF1A-2 at Lycoming in 1964 and 1965 produced results that correlated well with original design predictions.[110] A few mechanical problems were encountered during the tests, but they were readily solved. Testing of the engine at the NASA Lewis Research Center in its

105. Paul L. Lovington, Avco Lycoming Textron, to Rick Leyes, letter, 23 February 1987, National Air and Space Museum Archives, Washington, D.C.

106. Lovington to Leyes, 23 February 1987.

107. E. E. Turner and D. W. Jones, "Altitude Performance of the Lycoming PLF1A-2 Prototype Turbofan Engine," (AEDC Paper No. AEDC-TR-65-259, Arnold Engineering Development Center, Air Force Systems Command, Arnold Air Force Station, Tennessee, February 1968).

108. Lovington to Leyes, 23 February 1987.

109. Avco Lycoming PLF1A-2 Accession Memorandum.

110. Avco Lycoming, "Proposal for Development of the Lycoming PLF1B-2 High-Bypass Turbofan Engine. . . ," (Proposal No. 3000.1, to the Chief, Bureau of Naval Weapons, Department of the Navy, Washington, D.C., 18 December 1964), 1-4.

Quiet Engine Program revealed that the fan produced a high noise level, a major source of which was the fan's relatively high tip speed. A significant factor learned from tests that dealt with minimizing noise was the importance of the number of rotor and stator blades in the fan and the axial spacing between the rotor and stator.[111]

THE SEARCH FOR APPLICATIONS

The PLF1A-2 had been developed and funded by Lycoming as an experimental prototype engine to demonstrate high-bypass engine technology and thereby position the company to market and produce turbofan engines. Concurrently, Lycoming began developing and marketing a production version, the PLF1B-2. By 1963, two German companies had undertaken design studies of small transport aircraft designed around the PLF1B-2. One aircraft was the Weser Flugzeugbau WFG-614 and the other was the SIAT 311A. The SIAT 311A was to have almost STOL performance.[112] Whether due to skepticism of the high-bypass turbofan within the aircraft industry or to other reasons, neither of these commercial applications, nor any others, materialized for Lycoming. In the mid-1960s, the marketplace was just not ready to accept this new breed of engine, even though its feasibility had been demonstrated. Lacking industry support, Lycoming decided that the investment to develop the engine for the commercial market was not justifiable.[113]

Looking for possible military applications for the engine, in 1964, Lycoming submitted an unsolicited proposal to the Navy's Bureau of Naval Weapons to develop the PLF1B-2 engine. This proposal supported a North American Aviation concept for a small, carrier-based, Anti-Submarine Warfare (ASW) aircraft, the VSX, using PLF1B-2 engines in an advanced ASW version of the North American T-39 series of aircraft.[114] Continued sales efforts by Lycoming with supporting studies from North American Aviation confirming Navy in-house studies led to an urgent industry-wide competition sponsored by the Navy in mid-1966 for development of the power plant for a small ASW aircraft, the VSX (S-3A). Unfortunately, Lycoming was forced to submit a "no-bid" because T53 and T55 program commitments at that time in support of the Vietnam War effort had taxed the company's facilities and engineering manpower to its

---

111. Schrader to Leyes and Fleming, 10 October 1989.

112. "Lycoming Turbofans," *Flight International*, Vol. 84, No. 2836 (July 18, 1963): 97.

113. Lovington to Leyes, 23 February 1987.

114. Avco Lycoming, "Proposal for the Development of the Lycoming PLF1B-2. . . ."

limits. As a result, a golden opportunity to establish early leadership in the small high-bypass turbofan engine marketplace was lost.[115]

PLF1A-2 SUMMARY

The PLF1A-2 opened the new era of high-bypass technology as the first U.S. engine to demonstrate the concept. In addition, a factor of importance to Lycoming was that its work on the PLF1A-2 and PLF1B-2 provided experience that led to the success of its first production turbofan engine, the ALF 502.

A HEAVY LIFT HELICOPTER ENGINE—THE LTC4V

By the late 1960s, the Army was developing plans for a new heavy lift helicopter as a successor to the Chinook. To implement these plans, the Vertol Division of Boeing was awarded a contract on May 11, 1971, for the first phase of the development of such a helicopter. This machine was to be capable of lifting as much as 20 tons of cargo. It was designed as a tandem-rotor vehicle with three engines geared together to drive the two rotors and was to meet the shore-based heavy lift requirements of all U.S. Services.[116] The Army was assigned responsibility for development of the helicopter with Navy participation in the program and with funding provided by both services.

By 1967, Lycoming had recognized that the heavy lift helicopter offered a potentially fruitful future turboshaft engine market. The power requirement of each engine was projected at that time to be in the 5,000- to 7,000-shp range. However, the T55, Lycoming's then-largest turboshaft engine, was approaching its potential limit near the 4,000-shp level. For Lycoming to enter the heavy lift helicopter marketplace meant the development of a new higher power engine. Lycoming used its own funds to undertake the development of the LTC4V, a new 5,000-shp turboshaft engine, and viewed this engine as a possible successor to the T55 in the large engine family.

The design of this engine departed from Lycoming's traditional dual-shaft, axial-centrifugal compressor arrangement with closely coupled turbines. The LTC4V was a three-spool engine with its three shafts running concentrically. The engine core was comprised of two axial-flow compressors mounted on separate rotors, each of which was driven by its own turbine. The power tur-

---

115. Lovington to Leyes, 23 February 1987.

116. *Jane's All The World's Aircraft 1972-73* (London: Sampson Low, Marston & Co., Ltd., 1972), 278-279.

bine shaft ran concentrically within the other two rotor shafts to a front power takeoff. One LTC4V engine was assembled and a second set of engine hardware was purchased. In 1968, the 5,000-shp LTC4V-1 was successfully demonstrated. At that time, Lycoming was in the process of scaling up the engine to 6,000- and 7,000-shp models with a larger compressor and a zero compressor stage. The larger compressor was tested on a compressor rig, but a complete higher-power version of the engine was never assembled.

The engine was relatively complicated and represented an advanced design for its day, producing its 5,000 shp with a specific fuel consumption of only 0.43, considered excellent at that time. With a weight of 590 lb, it had a power-to-weight ratio of 8.7, unheard of in those days. Three factors contributed to this ratio: all titanium construction, which made it an expensive engine; high pressure ratio; and an extremely high turbine temperature for the time.

During the 1968-1969 period when the engine was being tested, Lycoming submitted several proposals to the Army for its use in heavy lift helicopters, but received no positive response.[117] By 1973, Allison's 8,000-plus-shp XT701-AD-700 turboshaft had been selected for heavy lift helicopters.[118] Unsuccessful in penetrating this market, Lycoming discontinued further work on the LTC4V engine.[119] Then, in the mid-1970s, interest of the military services in the heavy lift helicopter disappeared, and all work on this class of helicopter was canceled.

ALF502 TURBOFAN ENGINE

Throughout most of the 1960s, Lycoming had to concentrate essentially all of its engineering and manufacturing resources on the T53 and T55 engines for use in Vietnam. Thus, it was not until late 1968 that Lycoming was able to turn its attention to the design of a turbofan engine for an emerging market. Studies begun at that time indicated that an engine of moderate pressure ratio and turbine temperature with a 6:1 bypass fan was very competitive with more ambitious high pressure ratio cycle engines. Moreover, such a turbofan engine could be constructed using the T55 core. The ability to derive this cycle using an existing turboshaft engine provided a decided advantage in the maturity of many

117. Aavo Anto, interview by Rick Leyes and William Fleming, 10 October 1989, Textron Lycoming Interviews, transcript, National Air and Space Museum Archives, Washington, D.C.

118. *Jane's All the World's Aircraft 1973-74*, 285.

119. Anto to Leyes and Fleming, 10 October 1989.

critical components. Also, cost effectiveness and low design risk were important considerations in the design of a new turbofan, and the T55 core offered obvious advantages in these areas.

In September 1969, two turbofans were offered to a few potentially interested airframe companies. One was the ALF301 based on the T53 and rated at 3,000 lb thrust. The other was the ALF501 based on the T55 and rated at 6,000 lb thrust. Both engines offered the same conservative design approach. By the summer of 1970, an internal decision was made to launch the development of fan component hardware for the ALF301. Simultaneously, there was growing interest in the larger T55-based turbofan for the Northrop A-9 prototype aircraft. Although the ALF301 fan module was successfully tested beginning in March 1971, the program had been already been eclipsed by work on the ALF502. Design of the ALF502A high-bypass-ratio turbofan had begun in the fall of 1970 in anticipation of the contract signed with Northrop in January 1971 for use of the engine to power its A-9A prototype aircraft.[120]

Developed under contract to the Air Force, the ALF502A had the military designation of YF102-LD-100 and was used in one of two aircraft entered in the Air Force experimental attack (AX) aircraft competition. Two YF102s powered Northrop's A-9A entry in the competition. The competing aircraft, Fairchild's A-10 entry, was powered by two General Electric TF34 engines.

The Lycoming engine developed for this aircraft competition was a 7,200-lb-thrust turbofan with a bypass ratio of 6.1:1. Like the PLF1A, this engine used the T55-L-7 as its core. The T55's two-stage power turbine drove a single fan stage and two axial-flow compressor stages mounted behind the fan on the fan rotor. The two compressor stages on the fan rotor were added in response to the continuing press by the Air Force for more thrust during the AX program. These stages served as booster stages for the T55 core, increasing both the engine's airflow and pressure ratio.[121]

The fan itself was a completely new design with emphasis on minimizing noise by using the optimum numbers of blades and vanes. It had no inlet guide vanes to prevent wakes and turbulence entering the fan rotor. An important design feature that minimized fan noise was a wide spacing between the fan rotor blades and the following cascade of stators. This design approach allowed the wakes from the rotor blades to dissipate sufficiently by the time they reached the stator to maintain a low blade-passing-frequency noise level (the

120. T. A. Dickey and E. R. Dobak, "The Evolution and Development Status of the ALF502 Turbofan Engine," (SAE Paper No. 720840 presented at the SAE National Aerospace Engineering and Manufacturing Meeting, San Diego, Cal., 2–5 October 1972), 1–5.

121. Schrader to Leyes and Fleming, 10 October 1989.

noise generated by aerodynamic interaction between the rotor wakes and the stators).

In addition, the engine had a reduction gear between the power turbines and fan. Use of the reduction gear allowed independent optimization of fan and power turbine rotating speeds. The reduction gear between the fan and power turbine permitted the turbine to turn at a sufficiently high rotating speed that the turbine blade-passing frequency, and the resultant frequency of the noise generated, was above the audible frequency range. In contrast, the power turbines of direct drive engines operated at lower rotating speeds than the ALF502 turbine in order to keep the fan tip speed low enough for good aerodynamic efficiency. Such lower rotating speeds generated frequencies that were in the audible range, and also led to the need for one or two additional turbine stages.[122]

The engine featured four easily removable and interchangeable modules for ease of maintenance. These modules were the fan module which contained the reduction gear, the gas generator module which housed the high pressure compressor and turbine, the combustor and power turbine module, and the accessory gearbox.[123] This engine first ran in June 1971, completed its Preliminary Flight Rating Test in March 1972, and was first flown in the Northrop A-9A in May 1972.[124]

Six YF102 engines were built and used in the Northrop A-9A in 238 flights totaling 650 flight hours.[125] During early flights of the YF102, an oil scavenging problem was encountered. The oil circulating through the accessory gearbox failed to drain properly; sloshing of the oil by the rotating gears made the oil overheat. The problem was solved during a six-month cut-and-try process that produced several minor modifications within the gearcase. There was no recurrence of the problem, and the engine was relatively trouble free during the AX fly-off.[126]

Although the engine performed successfully, the Northrop A-9A lost to Fairchild's A-10 in the competition. This loss ended the military's interest in the YF102. Later, NASA acquired all six of the YF102 engines that had been manufactured during the AX program to power its four-engine Quiet, Short-haul

122. Dickey to Leyes and Fleming, 10 October 1989.

123. A. G. Meyer, "Transmission Development of Textron Lycoming's Geared Fan Engine," (Paper No. 88 FTM 14 presented at the American Gear Manufacturers Association, Alexandria, Va., October 1988), 1.

124. Dickey and Dobak, "The Evolution and Development Status of the ALF502 Turbofan Engine," 6.

125. Meyer, "Transmission Development of Textron Lycoming's Geared Fan Engine," 2.

126. Dickey to Leyes and Fleming, 10 October 1989.

Research Aircraft (QSRA). Those engines were still in use in that aircraft as late as 1989.[127]

In 1971, Lycoming began developing a commercial version of the engine, designated the ALF502D. The principal changes made to convert the YF102 to a commercial engine were to remove one of the two booster compressor stages on the fan rotor and to lower the turbine operating temperature to provide the increased engine life necessary for commercial operations. As a result, the thrust of the engine was derated from 7,200 to 6,500 lb thrust.

Development testing of the ALF502D engine was performed in both ground and flight test facilities: extensive sea-level testing in Lycoming's test cells at the Stratford plant; altitude tests at the Air Force Arnold Engineering Development Center in Tullahoma, Tenn., in 1972.[128] By the mid-1970s, flight tests of the engine began using a surplus Navy carrier-based bomber, the North American AJ-2, which the company had acquired. Flight test operation in the AJ-2 was performed by lowering the engine through the bomb bay doors during flight.[129]

Development of two additional commercial versions of the ALF502 series engines, the ALF502L and ALF502R, followed. An uprated version of the ALF502D, the ALF502L began development in 1977. The principal design change was to replace the second axial-flow booster stage on the fan rotor shaft. Three models of the ALF502L, the L-2, L-2A, and L-3, were produced, each rated at 7,500 lb thrust.

Development of three models of the ALF502R, the ALF502R-3, R-3A, and R-5, began in 1978. These were derated versions of the ALF502L with one of the axial-flow booster stages removed from the fan rotor shaft. The R-3 was rated at 6,700 lb thrust, and the other two models were rated at 6,970 lb thrust. A fourth model, the ALF502R-6 followed with the second axial-flow booster stage replaced on the fan rotor. The R-6 was rated at 7,500 lb thrust.

ALF502 COMMERCIAL APPLICATIONS

Drawing on the successful operation of the YF102 engine during flight testing of the Northrop A-9A, Lycoming found a commercial market for the ALF502.

---

127. Schrader to Leyes and Fleming, 10 October 1989.

128. Michael Cusick, "Avco Lycoming's ALF502 High Bypass Fan Engine," (SAE Paper No. 810618 presented at the SAE Business Aircraft Meeting and Exposition, Wichita, Kans., 7-10 April 1981), 2.

129. Dickey and Dobak, "The Evolution and Development Status of the ALF502 Turbofan Engine," 10.

The 6,500-lb-thrust Lycoming ALF502D turbofan engine. Photograph courtesy of Textron Lycoming.

The first customer was Dassault in France, which had developed the Falcon 30, a twin-engine turbofan aircraft powered by two 6,500-lb-thrust ALF502D engines and capable of carrying 30 passengers. Intended for regional transport operations, the Falcon 30 was a derivative of the Dassault's Falcon 20 twin-engine business jet, powered by two General Electric CF700 engines. A prototype of the Falcon 30 first flew in May 1973, and by the end of 1974, had logged more than 105 flying hours.[130] However, after about 270 hours of flight testing, Dassault discontinued further development of the Falcon 30, having lost confidence in the market for an aircraft in its size category.[131]

130. *Jane's All The World's Aircraft 1974-75*, 65-66.

131. Dickey to Leyes and Fleming, 10 October 1989.

The next customer found for the ALF502D was Canadair. Earlier, Lycoming had interested William Lear, Sr., in the engine. Lear designed an advanced-technology, long-range, executive transport, the LearStar 600, that had two rear-mounted ALF502Ds. The airplane was designed to carry 14 passengers over a nonstop distance of 5,000 miles or to carry 30 passengers as a commuter airliner over a series of shorter distances without refueling. The design cruising speed of the LearStar 600 was 600 mph, hence the chosen designation.[132]

Having already sold Learjet, the company that he had founded, and having insufficient capital to develop and build the airplane himself, Bill Lear approached Canadair as a prospective developer of the LearStar 600. In April 1976, Lear sold the aircraft's exclusive production rights to Canadair, Inc. Canadair drastically modified Lear's design and renamed it the Challenger 600. Dissatisfied with the manner in which Canadair had revamped his design, Lear severed all ties with the project shortly after consummating the production license agreement. By late 1976, Canadair had launched production and marketing of the CL-600 Challenger, an eight- to ten-passenger executive jet.[133]

The ALF502 had been considered as the power plant for the Challenger beginning with Lear's initial design studies of the airplane.[134] Although General Electric actively courted Canadair with the objective of getting its CF34 turbofan on the Challenger, the final nod went to the ALF502. Two factors influenced the choice of the 7,500-lb-thrust ALF502L-2 over the 8,650-lb-thrust CF34. One was the inability of GE to meet the production schedule set for the Challenger. The other was that an order for 25 Canadair 600s had been received from Federal Express specifying the ALF502 engine.[135]

The Challenger 600 first flew in November 1978 and proved to be successful. About 85 of these aircraft powered by ALF502L-2, L-2A, and L-3 engines were manufactured for use throughout the world.[136] However, design changes made during its development had resulted in an increase in aircraft weight from the

132. K. M. Molson and H. A. Taylor, *Canadian Aircraft Since 1909* (London: Putnam & Company Ltd., 1982), 152-153.

133. Donald J. Porter, *Learjets: The World's Executive Aircraft* (Blue Ridge Summit, Pa.: Tab Books, 1990), 31, 89.

134. *Jane's All the World's Aircraft 1979-80* (London: Macdonald & Jane's Publishing Group Limited, 1979), 18.

135. David M. North, "CF34 Upgrades Challenger Capabilities," *Aviation Week & Space Technology*, Vol. 118, No. 20 (May 16, 1983): 63.

136. "CL-600 Challenger," *Society of Experimental Test Pilots, Technical Review*, Vol. 14, No. 4 (1979): 138-144.

LearStar's estimated gross of 24,000 lb to a weight of 32,500 lb for the developed version of the Challenger 600. This weight increase led to a performance shortfall, primarily in range, from that guaranteed to the early customers.[137] The range shortfall resulted from limitations in fuel capacity because the aircraft had outgrown the 7,500-lb-thrust ALF502. In 1979, Canadair announced the Challenger 601, powered by two GE 8,970-lb-thrust CF34-1A engines and enlarged to provide additional fuel capacity for greater range. The first flight of the Challenger 601 was on April 10, 1982.[138]

The third and most prolific customer for the ALF502 was British Aerospace (BAe), which selected the engine for use in its BAe 146 regional transport aircraft. That airplane, initially powered by four ALF502R-3 engines, first flew in September 1981. BAe's use of the ALF502 was preceded by a decade long stop-and-go relationship with Lycoming.[139] This relationship began in 1970, when Hawker-Siddeley, the forerunner of BAe, and Beech Aircraft formed the Beechcraft Hawker Corporation, a Beech Aircraft subsidiary, to design, develop, manufacture, and market a family of next-generation corporate jet aircraft. One of the aircraft planned was a small business jet, the BH200.[140] The two partners became interested in Lycoming's turbofans, and the 2,600-lb-thrust ALF301 turbofan being proposed by Lycoming at that time was selected for use in the BH200. The ALF301 used the T53 core to drive a single-stage fan. Work on the ALF301 had progressed to aerodynamic tests on the fan when plans to build the BH200 were terminated, and Lycoming promptly discontinued further work on the ALF301. As a result, a complete model of the ALF301 engine was never assembled.[141]

In August 1973, Hawker-Siddeley announced plans for its HS 146 commuter transport, to be powered by four ALF502 engines. However, development of this airplane ceased within a few months when the British government nationalized the aircraft industry. In addition, confidence in the aviation marketplace was badly shaken by the energy crisis at that time. Nevertheless, research and design of the aircraft continued on a limited basis. It was not until 1978, following the establishment of BAe, that development of the airplane was resumed, and it was given the designation of BAe 146. Four 6,700-lb-thrust ALF502R-3 engines were selected to power the 71- to 93-passenger Series 100 BAe 146 air-

137. Molson and Taylor, *Canadian Aircraft Since 1909*, 153-154.

138. North, "CF34 Upgrades Challenger Capabilities," 63.

139. Dickey to Leyes and Fleming, 10 October 1989.

140. *Jane's All The World's Aircraft 1972-73*, 242.

141. Dickey to Leyes and Fleming, 10 October 1989.

British Aerospace BAe 146-100 regional transport aircraft powered by four 6,700-lb-thrust Lycoming ALF502R-3 turbofan engines. Photograph courtesy of Textron Lycoming.

craft, which began scheduled operations in 1983 as the first of the new generation of regional jet aircraft. The engine was later uprated to 6,970 lb thrust for use on the enlarged 82- to 109-passenger BAe 146 Series 200 transport.[142] By that time, 819 ALF502R engines had been manufactured for the BAe 146.

In 1983, when the BAe 146 was introduced into commercial use, and during the years that followed, it was advertised as the world's quietest jetliner. The ALF502 had been designed to achieve low noise levels, minimizing both fan- and turbine-generated noise.

ALF502 FIELD OPERATING EXPERIENCE

Although the ALF502 ultimately developed a good in-service record with low engine removal and in-flight shutdown rates, it was troubled by a variety of growing pains following its introduction. An early problem was a vibratory res-

142. *Jane's All The World's Aircraft 1983-84* (London: Jane's Publishing Company Limited, 1983), 256-257.

onance in a large bell gear in the main reduction gearbox. This gear would occasionally crack and fail, thus requiring in-flight engine shut down. It was found that the bell gear was ringing like a bell at a resonant frequency of approximately 4,000 Hz, which is a musical tone fairly high on the musical scale. A mechanical damper installed in the bell gear solved the problem.[143]

Another early problem was flutter of the first stage in the core compressor resulting in fatigue failure of the blades. A similar but not identical problem had been encountered on the T55. Following extensive investigation, it was discovered that the blade failures resulted from a phenomenon known as "stall-flutter" which occurred at an abnormal setting of the variable inlet guide vanes. This phenomena had been studied in propellers, but little published data for gas turbine blades were available at the time. The solution was to install mid-span mechanical dampers between the first-stage blades in addition to establishing a closely monitored procedure for setting inlet guide vane schedules.[144] Much later, the first three stages of the compressor were redesigned using an entirely different airfoil, and the vibration dampers were removed.

After the engine had begun commercial service on the BAe 146, rapid wear was encountered in the reduction gears, a problem never encountered in Canadair's Challenger 600. A variety of causes were possible, so a shotgun approach was taken, and all possible corrective actions were pursued in parallel. The major change was a modification to accomodate differential expansion between the magnesium frame and the reduction gear carrier. In any event, after this and other corrective actions were taken, the problem completely disappeared.

A major problem that still existed at the end of the 1980s was that the ALF502 was an expensive engine to run because airline operators had to replace parts in the turbine section of the engine more often than was cost effective. Lycoming engineers began a long-term effort in the late 1980s to redesign many of the hot parts of the engine to increase their life.

Although hot section parts had to be replaced more often than desired, the ALF502 offered a low engine removal rate of about 1 per 5,000 hours. The engine also had an extremely good in-flight shutdown rate of about 1 in 20,000 hours. Thus, by the end of the 1980s, the ALF502 had demonstrated its success as a commercial engine, powering 151 BAe 146 aircraft used by 29 operators on five continents.[145]

143. Dickey to Leyes and Fleming, 10 October 1989.

144. Dickey and Dobak, "The Evolution and Development of the ALF502 Turbofan Engine," 8-9.

145. Dickey to Leyes and Fleming, 10 October 1989.

NEW TURBOFANS—THE 500 SERIES

In September 1988, Lycoming announced plans for its new LF500 Series of turbofan engines based on the ALF502R.[146] The LF507 was the first in the company's LF500 family of commercial turbofans, intended to cover the 7,000-to 14,000-lb-thrust range. Two models were produced, the LF507-1H and LF507-1F, both flat rated at 7,000 lb thrust to 74 degrees F.[147] The first of these engines, the LF507-1H was certificated in October 1991, and the second model, the LF507-1F, in March 1992.

Derived from the ALF502, the LF507 used the ALF502 core and had a single-stage fan with a two-stage, low pressure compressor and two-stage power turbine.[148] Major emphasis was placed on reliability and durability.[149] The basic difference between the two LF507 models was that the -1H had a hydro-mechanical control, while the following -1F was equipped with a FADEC.[150] During early 1991, a prototype -1H engine underwent testing, and in May 1991, a BAe 146-300 powered by four LF507-1H engines equipped with a hydromechanical control made its maiden flight.

The LF507-1F was initially used in the BAe RJ70, a 70-passenger regional transport introduced as an "optimized" derivative of the BAe 146. The RJ70 was basically a 4,000-lb-lighter BAe 146-100 with 12 fewer seats. Much of the weight reduction was in fuel with a resulting reduction in maximum range from 1,300 to 800 miles. Using the LF507, this airplane was expected to offer operators improved performance and lower operating costs than the ALF502-powered BAe 146-100.[151] The LF107-1F was also used in the BAe 146 family of regional transports.

By mid-1991, Lycoming had begun building the first production LF507-1F engines, and the first four production engines for the RJ70 were delivered in

146. *Jane's All The World's Aircraft 1989-90* (Coulsdon, Surrey: Jane's Information Group Limited, 1989), 748-749.

147. "Advanced Textron Lycoming LF500 Turbofans Pass First Four-Engine Flight With Flying Colors," *Textron Lycoming Press Information*, 30 April 1991, 1-2.

148. "Textron Joins Battle for EMB-145 Power," *Flight International*, Vol. 136, No. 4184 (September 30, 1989): 21.

149. "Textron Lycoming Building Larger Engines on ALF502 Core," *Air Transport World*, (October 1988): 82-85.

150. "Textron Joins Battle for EMB-145 Power," 21.

151. "Textron Lycoming Begins Building of First Production LF507," *Commuter Regional Airline News International*, Vol. 4, No. 33 (August 19, 1991).

March 1992.[152] The initial order for 20 BAe RJ70 regional jets powered by the LF507 was placed by Business Express in late 1991.[153] By August 1996, 48 LF107-1H engines and 270 LF107-1F engines had been produced.

Lycoming announced the 500 Series common core engine in February 1994. The objective of that program was to combine the benefits of experience on the LF507 engine with an infusion of selected technology developed and demonstrated by Lycoming.[154] AlliedSignal's (the company that acquired Lycoming in October 1994) common core engine was intended to position the company to have a family of power plants available for the regional aircraft market in the late 1990s.[155]

The common core engine incorporated the LF507's seven axial-stage and single centrifugal-stage compressor configuration with the diameter of the first three axial stages increased by about ten percent. Computational fluid dynamics was used to upgrade the aerodynamic design of the compressor. In addition, the single centrifugal stage used an improved efficiency "lean-back" impeller. These compressor design changes boosted airflow by 23 percent and overall engine pressure ratio by 63 percent. A new short-residence-time combustor was fabricated from a proprietary Lycoming material called Lycolite®, an open-core honeycomb material featuring serpentine paths for cooling air, developed by Lycoming under the Integrated High Performance Turbine Engine Technology (IHPTET) initiative. Lycoming engineers estimated that Lycolite® reduced wall cooling air requirements by 60 percent compared to conventional film cooling techniques. In addition, the Lycolite® short-residence-time combustor improved the temperature uniformity of the flow stream, thus reducing the likelihood of hot spots in the turbine. The two-stage turbine configuration of the LF507 was retained with the application of advanced cooling techniques to boost the turbine's temperature capability by as much as 250°F above that of the LF507, with turbine metal temperatures remaining equal to or below those in the LF507.[156]

152. "Textron Lycoming Begins Building of First Production LF507."

153. "Business Express Launches RJ70 With Order for 20 BAe Aircraft," *Aviation Week & Space Technology*, Vol. 136, No. 1 (January 6, 1992): 25.

154. "Textron Lycoming's New Common Core Engine Combines Demonstrated Technology With Proven Performance," *Textron Lycoming Press Information*, February 1994, 1-3.

155. Stanley W. Kandebo, "Common Core Tests Target Technology Validation," *Aviation Week & Space Technology*, Vol. 142, No. 1 (January 2, 1995): 51.

156. Stanley W. Kandebo, "Lycoming to Test New Engine Core," *Aviation Week & Space Technology*, Vol. 140, No. 10 (March 7, 1994): 32-33.

The common core engine was sized for turboprops in the 7,500-shp class and turbofans producing 12,000 to 14,000 lb thrust. Lycoming put its initial efforts into the LP512 turboprop member of the 500 Series family to match the development schedules of the de Havilland Dash 8-400 and ATR82 regional transports. It was planned that the LP512 would have an initial rating of 7,500 shp, but could be uprated to 11,000 shp. Lycoming planned to sell the LP512 on the basis of its weight. With a close-coupled gearbox mounted directly to the front of the engine casing and its short core, the LP512 could fit into a shorter, more slender cowling than the Allison AE2100 and require less mounting structure. Lycoming estimated that the LP512 would save nearly 1,000 lb over the AE2100 in a twin-engine aircraft.[157] In early 1995, AlliedSignal was continuing development of the LP512.

A PARTNERSHIP WITH GENERAL ELECTRIC—THE GLC38/T407

In December 1987, Lycoming and General Electric formed a partnership to further develop, market, and produce GE's GE27 engine. General Electric had developed the GE27 for the Army as a demonstrator turboshaft engine. With the GE/Lycoming partnership, the engine became the GLC38, for GE/Lycoming Commercial engine. The military version of the engine, used on the Lockheed P-7A Long-Range Air ASW Capable Aircraft (LRAACA), was the T407-GE-400.

The events that led to formation of this partnership occurred over about four years. In March 1983, GE was awarded a contract to develop a Modern Technology Demonstrator Engine (MTDE) in the 5,000-shp class for the Army. This engine, the GE27, had a five-stage, axial-flow compressor followed by a single centrifugal stage, a short annular combustor, a two-stage, air-cooled gas generator turbine, and a three-stage free power turbine. The engine incorporated the latest aerodynamic, materials, and cooling technology, giving it a power-to-weight ratio of about 6.5. The GE27 engine core first ran in December 1983, and the first full engine run was made in late 1984.[158] The engine completed a successful demonstration program, exceeding all of its goals.

In 1985, GE offered the GE27 in the engine competition held for the Bell-Boeing V-22 Osprey tilt-rotor aircraft. Winning the competition, the Allison Gas Turbine Division of GM was awarded a contract in December 1985 for its

157. Bill Sweetman, "New Power for Regionals," *Interavia*, Vol. 49, No. 583 (October 1994): 17.

158. *Jane's All the World's Aircraft 1987-88* (London: Jane's Publishing Company Ltd., 1988), 955.

6,000-shp T406-AD-400 turboshaft.[159] GE was left with a successful demonstrator engine in which the company had made a heavy investment and for which no ready application existed. The company also was faced with a further large expense if it were to perform the additional development work to carry the engine through certification for the commercial market.

GE visited a number of engine companies in the U.S. and Europe to discover any interest in a partnership to complete development of the engine and to market and produce it. As part of this effort, GE made its initial contact with Lycoming. GE was looking for a partner to share the GE27's development and marketing. Lycoming realized the advantage of having a new, larger and more modern turboshaft engine in its product line to augment the T55 series.

Lycoming also believed that an arrangement with GE could offer some long-term advantages in the tank gas turbine engine market as well. Lycoming had already captured the existing main battle tank engine market with its AGT1500 vehicular turboshaft engine, and a next generation version, GE's LV100 Advanced Integrated Propulsion System (AIPS), was viewed as a potential follow-on. At that time, GE had a contract with the Army to develop AIPS as a complete propulsion system for a tank, including the engine, transmission, final drives, hydraulic system, cooling system, electrical system, and fuel system. Lycoming knew that it could make significant contributions to this program and thereby aim toward capturing the future tank engine market. If it succeeded, the two companies could dominate the tank turbine engine world well into the following century.[160]

It was this series of events and business logic that led to the formation of the General Electric/Lycoming revenue sharing agreement in 1988. General Electric brought with it two other partners to participate on the GLC38 program. One was Ruston Gas Turbines Ltd., a manufacturer of industrial gas turbines located in Lincoln, England. Ruston was to manufacture many of the rotating parts, including compressor disks and turbine rotors. The other partner was Bendix, which was to provide the fuel control. Lycoming's role in the partnership was to mechanically redesign portions of the engine to improve its producibility and increase its lifetime in order to meet the requirements of the commercial market, thereby making it a commercially viable engine. Once manufacturing began, Lycoming was to be responsible for manufacturing approximately 30 percent of the engine as shipped. The partnership was managed by a joint program office

159. *Jane's All the World's Aircraft 1987-88*, 374.

160. Kuintzle to Leyes and Fleming, 12 October 1989.

with a General Electric director and a Lycoming deputy director. Providing the focal point between the two companies, these two people administered the work being carried on by their respective companies.

The initial application for the engine came as a result of General Electric's efforts to interest Lockheed in its GE27 engine prior to formation of the partnership. Following formation of the partnership, Lockheed selected the military version of the engine, the T407-GE-400 (GE38), to power its entry in the LRAACA competition to replace the Navy's Lockheed P-3. In October 1988, Lockheed won the competition and received a contract to develop the P-7A powered by four T407 engines, ultimately producing 5,600 shp each. The T407 proposal submitted to Lockheed by the General Electric-Lycoming partnership was for a complete turboprop propulsion system including engine, propeller gearbox, propeller, and nacelle. As a result, SPECO of Cleveland was added as a fourth partner with the responsibility for manufacturing the propeller reduction gearbox. Unfortunately, in July of 1990, the P-7A program was terminated after Lockheed failed to reach an accommodation with the Navy regarding projected program cost overruns.[161]

In 1988, the commercial turboprop version of the engine, the GLC38, was offered for use in the twin-engine Saab 2000, a new regional transport seating up to 50 passengers. The version of the GLC38 offered for the Saab 2000 was derated to 3,200 shp, which would have resulted in outstanding fuel economy and reliability.[162] In 1989, the competing Allison GMA 2100 was selected to power the Saab 2000.[163]

By late 1989, the mechanical redesign of the engine had been completed. Lycoming was building its first test engine, and parts manufactured by Lycoming were being shipped to General Electric to enable them to build their first test engine. Concurrently, GE was supplying their parts to Lycoming for a second engine to be built and tested at the Stratford plant. The Lycoming engine went on test on December 26, 1989 and the GE engine in early 1990. Test of the GLC38 at both General Electric and Lycoming were completed and met all performance projections.[164] Data on the T407-GE-400 and GLC38 engines are included in the GE data charts.

161. James Lunny, interview by Rick Leyes and William Fleming, 10 October 1989, Textron Lycoming Interviews, transcript, National Air and Space Museum Archives, Washington, D.C.

162. *Jane's All The World's Aircraft 1989-90*, 735-736.

163. "Allison Engine Chosen for Saab 2000; Sweden Certifies Saab 340B Transport," *Aviation Week & Space Technology*, Vol. 131, No. 3 (July 17, 1989): 29.

164. Lunny to Leyes and Fleming, 10 October 1989.

HIGH-POWER ENGINES SUMMARY

The key engine in the high-power family was the T55 turboshaft. Success of this engine resulted from three successive actions by the Army: 1) the 1956 decision to replace its piston engine-powered air transport helicopter with a gas turbine-powered model; 2) the selection of the T55 to power that helicopter, the Chinook; and 3) the extensive deployment of the Chinook in Vietnam and its continued use after the war. A total of 4,233 T55s were produced between 1960 and the end of 1989, representing 16 percent of Lycoming's total aircraft gas turbine production to that time. The Army's Chinook transport used 3,668 of these engines.

The T55 also served as the core for the two turbofan engines in the high-power engine family, the PLF1A and the ALF502. One of the more significant contributions of the T55 as part of the large engine family was that, using it as the gas generator, Lycoming produced America's first high-bypass turbofan, the PLF1A-2. That accomplishment led to the subsequent design and development of one of the world's quietest turbofan engines, the ALF502. As a quiet and fuel efficient engine, the ALF502 was used worldwide in the Canadair Challenger CL-600 executive aircraft and the BAe 146 regional jet. Between 1972 and the end of 1989, a total of 6 YF102s and 1,019 ALF502 D/L/R engines were built. The ALF502 spawned a new family of engines, the Series 500, which initially featured the LF507 commercial turbofan. Success of the LF507 was followed by the common core engine, an advanced technology engine which was developed as the core for a family of turboprop and turbofan engines.

Low-Power Engine Family

In the mid-1960s, Lycoming began shaping its technology development program around components and demonstrator engines in the 5-to-10 lb-per-second airflow size. This program began with work on advanced compressor designs, small reverse-flow combustors with improved mixing and temperature profiles, and high work single-stage turbines.[165] In 1968, this component technology led to the development of Lycoming's first engine in its low-power engine family, the PLT32 demonstrator engine. From that engine evolved the LT101 series of turboshaft and turboprop engines and the ALF101 turbofan.

165. Avco Lycoming Division, "Avco Lycoming's LTS/LTP 101 Series Turboshaft/Turboprop Engines," July 1979, 1.

Two follow-on demonstrator engines, the PLT34 and PLT34A, were developed during the 1970s. The PLT34A led to the development of the T800 as a Lycoming and Pratt & Whitney joint venture, formed in 1984. In 1989, Lycoming began development of the PLT210 demonstrator engine. That engine was designed to serve for a decade or more as the vehicle on which new technical concepts and component technology would be demonstrated.

THE FIRST SMALL DEMONSTRATOR ENGINE—THE PLT32

The PLT32, development of which began in 1968 using Lycoming funds, was designed as a small, simplified, advanced technology gas generator core in the 600 shp class. Its purpose was to incorporate and demonstrate the technology that had evolved from earlier component technology development and testing. The PLT32 was a single-shaft gas generator core having a single axial-flow compressor stage followed by a centrifugal-flow stage driven by a single-stage turbine. A reverse-flow combustor was wrapped around the hot section of the engine.

The PLT32 demonstrated technology advances in compressors, combustors, and turbines for small simplified engines, leading to subsequent development of the PLT34 and PLT34A demonstrator engines. In addition, the technology demonstrated in this small, simplified gas generator allowed Lycoming to spin off from it the LT101 series of engines in the early 1970s.[166]

THE LT101 SERIES

Development of the 600-shp-class LT101 series engines began in 1972. These engines were designed as simplified, compact, rugged, modern-technology, low-cost engines for civil applications. The gas turbine section was similar to the PLT32 with one axial and one centrifugal compressor stage, a single-stage turbine and a reverse-flow annular combustor. A single-stage free power turbine mounted on a concentric shaft extending forward through the gas generator section drove a front power takeoff. The engine was designed with four basic modules: the engine inlet and particle separator module, the gas generator module, the combustor and power turbine module, and the reduction gear and accessory gearbox module. This design provided quick and easy access to hot engine parts for field inspection.[167]

166. Kuintzle to Leyes and Fleming, 12 October 1989.

167. A. Myers and E. Pease, "Lycoming's LTS101 - Low Cost Turbine Power in the 600 hp Class," (SAE Paper No. 730911 presented at the National Aerospace Engineering and Manufacturing Meeting, Los Angeles, California, 16-18 October 1973), 8-9.

Three gearbox designs were developed for use on the various LT101 engine models. Front-drive gearboxes provided output speeds of 1,925, 6,000 or 9,545 rpm. The 6,000-rpm gearbox had both forward and aft drives. Also, the engines were provided with either scroll or radial inlets.

Simplicity, low cost, and the attainment of low fuel consumption during part-load operation were emphasized, leading to selection of a moderate turbine inlet temperature that would not require turbine cooling. As a result, the optimum cycle pressure ratio for the engine was approximately 8.4. The emphasis on low cost led to the extensive use of low-cost investment casting of components throughout the engine in place of welded, brazed, or fabricated assemblies.[168]

Although the initial intent was that aircraft use of the engine would be solely for helicopters, both turboshaft and turboprop versions of the engine were developed and produced. The primary objective was to design and develop a low-cost industrial engine that would also comply with aircraft standards.[169] Nevertheless, Lycoming did not push the turboprop version of the engine, focusing instead on marketing and producing the LTS101 turboshaft models for helicopters.[170] Between 1973 and the end of 1989, a total of 1,839 LTS101 turboshaft and 154 LTP101 turboprop engines were made.

The engine was intended to have a manufacturing cost approaching those of piston engines. Early in its production cycle, the LTS101 achieved a cost of $40 per horsepower, which was considerably less than for its turbine engine competitors. In comparison, however, mature piston engines in the same power class were being manufactured at costs of $25 to $35 per horsepower.[171] In the fixed-wing general aviation field, a major attraction of the piston engine was its low initial cost. However, studies performed by Lycoming suggested that with a manufacturing cost of approximately $40 per horsepower, the LTP101 might have a lower life cycle cost than the piston engine for many fixed-wing applications.[172]

The principal turboprop models of the engine were the 620 equivalent shp (eshp) LTP101-600A-1A and the uprated 700 eshp LTP-700A-1A. These en-

168. J. Moore and E. Pease, "Lycoming's LTS101 Engine Design," (SAE Paper No. 740165 presented at the Automotive Engineering Congress, Detroit, Michigan, 25 February - 1 March 1974), 1-12.

169. Myers and Pease, "Lycoming's LTS101 Engine Design," 1.

170. Kuintzle to Leyes and Fleming, 12 October 1989.

171. Myers and Pease, "Lycoming's LTS101—Low Cost Turbine Power in the 600 hp Class," 8-9.

172. M. Bentele and J. Laborde, "Evolution of Small Turboshaft Engines," *Society of Automotive Engineers Transactions*, Vol. 81 (1972), Paper 720830.

Lycoming LTS101 turboshaft engine. Photograph courtesy of Textron Lycoming.

gines were used in a number of small fixed-wing aircraft, several of which were conversions of existing aircraft from piston to gas turbine engines. Among the aircraft converted from piston engines to the LTP101 were the Piaggio P.166, Page Ag-Cat, Snow Thrush, and Cessna 421.

The LTS101-600, LTS101-650, and LTS101-750 series, in the 550- to 700-shp range, powered various models of Bell, Aerospatiale, and MBB/Kawasaki private and commercial helicopters, and air ambulance and Air National Guard helicopters. In the early 1980s, Aerospatiale, a major user of the engine, introduced its LTS101-powered AS350D AStar helicopter. In 1983, the Coast Guard began acquisition of a fleet of LTS101-powered Aerospatiale HH-65 Dolphins for air-rescue and drug-interdiction missions.

LT101 MANAGEMENT MOVES

Responsibility for the management and manufacture of the LT101 engines underwent two major moves. Around 1980, the Stratford facility faced a major demand for increased output. New products were being introduced to its existing product line. Insufficient manufacturing capability and program management resources existed at Stratford to handle the development and manufacturing of the LT101, ALF502, and the AGT1500 on top of the normal military production base. The decision was made to move the LT101 engine program to the Williamsport plant. Williamsport had excess management and manufacturing capacity as well as facilities capable of testing LT101-size engines. Lycoming selected the LT101 as the engine to go to Williamsport because its size and its commercial applications were more closely related to Williamsport's piston engine experience than were those of Lycoming's other two new gas turbines.

Williamsport continued development of the engine and began production by 1980. However, following early engine problems and a series of LTS101 in-flight engine failures, Lycoming's management concluded that Williamsport lacked the technical and managerial expertise for the implementation of the program. In 1987, the engineering and program management responsibility and customer support were returned to Stratford. Nevertheless, the Williamsport plant retained responsibility for LTS101 and LTP101 engine assembly, overhaul, and testing.[173]

LT101 OPERATIONAL EXPERIENCE

Lycoming's goal to manufacture low-cost components to secure a niche in the small turbine engine marketplace led to a troubled operational history for the LT101 from the mid-1980s until 1990. The engine suffered from power turbine blade cracking and experienced numerous in-flight shutdowns caused by failure of power turbine blades. As a result of the turbine blade cracking, the engine had a Time Between Overhaul rate of less than 600 hours compared to its initially designed 2,400 hours.[174] In fact, FAA Service Difficulty Reports published between 1984 and mid-1990 revealed 154 instances of LTS101 power turbine blade cracks in civilian helicopters. The Coast Guard's experience with the turbine wheel had not been much better.[175]

173. Kuintzle to Leyes and Fleming, 12 October 1989.

174. "Lycoming to Pay U.S. $17.9 Million in LTS101 Performance Settlement," *Aviation Week & Space Technology*, Vol. 133, No. 3 (July 16, 1990): 24-25.

175. "LTS101 Turbine Blade Failures Probed," *Aviation Equipment Maintenance*, Vol. 9, No. 12 (December 1990): 6-8.

Soon after the Coast Guard's first encounter with power turbine blade crack-ing, Lycoming introduced engineering and manufacturing changes in an effort to improve the turbine's reliability. In 1987, a new power turbine wheel was in-troduced. However, the process of integral casting of disk and blades was re-tained. Temperature control during manufacturing of the turbine wheel was held to closer tolerances. However, blade cracking and turbine failures contin-ued with operating times to crack detection that ranged from as low as 41 hours to a high of 1,595 hours.[176]

In 1988, after more than 100,000 hours of test cell operation, Lycoming engi-neers duplicated the turbine's field behavior and diagnosed the cause of the problem. They traced turbine blade cracking and failure to the one-piece, in-vestment cast construction of the power turbine. The power turbine contained 27 blades cast together with the turbine disk as an integral unit, referred to as a blisk. The blades were subject to cracking because the blade and disk portions of the turbine wheel had different rates of thermal expansion, producing stress-fatigue cracks in the trailing edge root area of the blades. After the cracks had progressed sufficiently, turbine blade failure followed.[177]

The power turbine was redesigned with conventional, inserted blades in the disk in lieu of the single-piece casting. Production of this new turbine began in 1989, and in February 1990, the Coast Guard began retrofitting its LTS101 en-gines. By mid-1990, with 55 percent of the Coast Guard's LTS101 fleet retro-fitted, no complaints had been logged with the new wheel. The Coast Guard completed the retrofit in its fleet of 96 HH-65As by September 1990.

Meanwhile, the Coast Guard had initiated a program to resolve the power plant problem by re-engining the HH-65A. An HH-65A was modified to ac-cept two Allison-Garrett LHTEC 800 power plants originally developed for the Army's LHX helicopter, and a flight test program of the T800 powered HH-65A began with the first flight with T800s in August 1991.[178]

However, by early 1991, the Coast Guard announced that it was satisfied with the performance of the modified LTS101 engine in its HH-65A helicopters and had ended plans to seek a replacement engine. Although the scheduled flight tests of the T800 equipped HH-65A began in August 1991, those tests were ter-minated following a 50-hour flight test program and plans to certificate the en-gine in the HH-65A were scrapped. According to Lycoming LTS101 Program

176. "LTS101 Turbine Blade Failures Probed," 6-8.

177. "Lycoming to Pay U.S. $17.9 Million in LTS101 Performance Settlement," 24-25.

178. "T800-Powered Dolphin Flies, But Future Funding Off," *Helicopter News*, Vol. 17, No. 22 (November 1, 1991): 5.

Manager Ward Hemmenway, by the spring of 1991, the Coast Guard was achieving readiness rates of 75 percent with its LTS101 powered HH-65As. That rate exceeded the Coast Guard's goal of 71 percent availability and greatly exceeded the 59 percent rate of a year earlier.[179]

By early 1994, buoyed by the apparent field success of the redesigned power turbine as well as other modifications to the LTS101, Lycoming was offering modification packages to owners of unretrofitted engines. Called LTS101 Plus, the package was offered in two parts: the Plus I package was a basic set of improvements to the power turbine, compressor rotor, bearings, and oil seals. The Plus II package included the Plus I improvements and a new rear bearing support housing, oil-wetted fuel pump splines, and electronic engine overspeed protection. Both packages were offered as new engines.[180]

In early 1996, following its acquisition of Lycoming, AlliedSignal announced that it would invest some $15 million in the LT101 engine to improve its reliability and boost customer satisfaction. By that time, more than 1,650 of the LT101 engines were operational. LT101 program manager Robert L. Miller said, "We need to ensure that the fleet reliability of the LT101 series is at a level commensurate with the rest of AlliedSignal Engine's products."[181] The LT101's nominal unscheduled removal rate was about 1.5 per 1,000 operating hours. AlliedSignal's goal was to reduce that rate to 0.4.

Initial efforts to improve reliability centered on the power turbine to eliminate a blade shift problem. Interactions between the rivets that held the turbine blades in place and the blades and turbine disk could allow the blades to move, unbalancing the wheel. Plans were to install redesigned rivets to eliminate the problem.

Other reliability improvements centered on integrating technologies developed for the LTS101-750 Plus 2 engines into the entire LT101 fleet over a 12- to 18-month period. In early 1996, only 100 "Plus 2" engines were in service, operating primarily in the MBB/Kawasaki BK117 helicopter. Two additional improvements were a low-coke combustor and a flexible fuel manifold. The new combustor liner was expected to eliminate turbine blade erosion caused by carbon buildup, cut visible smoke, and increase liner life. The flexible manifold

179. "Coast Guard Drops HH-65A Reengining," *Aviation Week & Space Technology*, Vol. 134, No. 18 (May 6, 1991): 68.

180. "Textron Rolls Out LTS101 Mod Packages," *Helicopter News*, Vol. 20, No. 4 (February 18, 1994).

181. "AlliedSignal Commits to LT101 Improvements," *Aviation Week & Space Technology*, Vol. 144, No. 11 (March 11, 1996): 70.

was expected to eliminate repetitive inspections required with the current rigid design.

AlliedSignal also implemented substantial reductions in the cost of engine parts. The cost of engine components was cut by 2 percent, the cost of the top 20 percent of high-usage parts was cut by 23 to 79 percent, and the average cost of LT101 standard parts was cut by 80 percent. According to LT101 program manager Robert Miller, AlliedSignal expected to benefit by these price reductions, "because the parts pricing is an element of getting the improvements fielded. The lower cost of parts provides the customer with an incentive to improve engine reliability and that benefits the LT101 program."[182]

LT101 SUMMARY

Lycoming was ultimately successful in developing the LT101 as a simplified low-cost engine. However, the emphasis that had been placed on simplicity and low cost in designing the engine were factors that led to in-flight turbine failures resulting in a poor record of engine reliability. Once the reason for the problem was correctly diagnosed, the solution was straightforward. However, by the time the problem had been solved, a negative image had evolved which was difficult to overcome. By mid-1991, as a result of a $50 million investment by Lycoming, all product improvements that the company considered necessary had been completed, resulting in significantly improved engine reliability and accompanied by improved customer relations and satisfaction. In addition, following its acquisition of Lycoming, AlliedSignal embarked on a program to further improve engine reliability.

ALF101—LYCOMING'S FIRST SMALL TURBOFAN

The ALF101 was a small, simplified turbofan using for the engine core the LT101 gas generator modified to include a second axial-compressor stage, and a single-stage fan driven by the LT101 single-stage power turbine. Unlike the LT101, the ALF101 had two axial-compressor stages ahead of the single-stage centrifugal compressor. The ALF101 produced 1,610 lb thrust, and with its bypass ratio of 8.5, it had a specific fuel consumption of only 0.36.

Development of the ALF101 began in 1976 in response to NASA's interest in testing and demonstrating small, quiet turbofans with low exhaust emissions.

182. "AlliedSignal Commits to LT101 Improvements," 70.

This work was part of NASA's Quiet Clean General Aviation Turbofan (QCGAT) program, in which noise, emission, and performance tests were to be run on small turbofans at the Lewis Research Center in Cleveland. Lycoming proposed the ALF101 to NASA and received a contract for its development. Lycoming built two ALF101 engines, tested them at the Stratford plant, and in 1979, sent them to Lewis for performance testing. Reportedly, for budgetary reasons, NASA dropped its plans to also conduct noise tests at the laboratory and instead used noise data obtained by Lycoming during the earlier tests of the engine at the plant.[183]

Engine emissions were found to be extremely low and the noise level was below the program objective. In fact, Walt Schrader, who as Director of Lycoming's Advanced Projects in the late 1970s was in charge of the ALF101 program, recalled that the noise generated by the engine was so low that a person standing 50 feet from an ALF101 when it was operating at full power could be heard when talking at a conversational level. Its low noise level was not surprising since the fan was a scaled-down version of the ALF502's quiet fan. Key steps to minimize noise generation were selecting the number of fan and stator blades to avoid noise generating frequencies and locating the stator blades a sufficient distance behind the fan to allow the rotor blade wakes to dissipate before reaching the stator vanes.[184]

During engine development, Lycoming had been trying to find customers for the engine in the commercial market, but without success. Although the ALF101 had been successfully demonstrated as a clean, quiet, fuel-efficient general aviation engine, it was too small for any of the aircraft applications at that time. Consequently, following conclusion of the performance tests at NASA in 1979, work on the engine was discontinued.[185]

PLT34 AND PLT34A DEMONSTRATOR ENGINES

The PLT34 and PLT34A were demonstrator engines developed as part of two separate technology development programs sponsored by the military. The PLT34, a gas turbine core, was developed under a contract extending from 1971 to 1974 as part of the military's Small Turbine Advanced Gas Generator (STAGG) program. The objective of the STAGG program was to demonstrate

183. Schrader to Leyes and Fleming, 10 October 1989.
184. Schrader to Leyes and Fleming, 10 October 1989.
185. Dickey to Leyes and Fleming, 10 October 1989.

advanced technology component interaction, i.e., compressor, combustor, and turbine elements operating in a real engine environment.

The PLT34A was a follow-on development under a contract extending from 1977 to 1982 as part of the military's Advanced Technology Demonstrator Engines (ATDE) program. The objective of this program was to demonstrate a complete propulsion system for an advanced helicopter using the PLT34 core gas generator together with a power turbine, a gearbox to drive accessories, an electronic fuel control, and inlet particle separator.

Design of the PLT34 engine began under the STAGG program for the purpose of developing an 825-shp, advanced technology, gas turbine core. The STAGG competition was divided among four companies. Garrett and Williams were each successful in winning contracts for the development of core engines in the auxiliary power unit class of 1.5- to 2-lb-per-second-airflow size. Pratt & Whitney's Government Products Division in Florida won a contract to develop a 3.5-lb-per-second core demonstrator, and Lycoming won a contract for its 5-lb-per-second PLT34 core demonstrator.

According to Aavo Anto, who had been associated with advanced technology engines for 25 years at Lycoming and who was Project Manager of PLT34 Advanced Engines, the primary objective of the STAGG program was to push the state of the art in small axial compressor and turbine technology.[186] Consequently, the PLT34 compressor used high pressure ratio, axial compressor technology that reduced the number of compressor stages to only three, i.e. two axial stages and one centrifugal stage. In contrast, a compressor of conventional design would have required eight or nine stages to achieve the same pressure ratio. The two axial stages were designed for extremely high pressure ratios, in this case approaching 1.8 to 1.9 per stage. The second stage was fully supersonic from hub to tip. This compressor represented Lycoming's first use of a fully sonic compressor stage. Together with the centrifugal stage, an overall compressor pressure ratio of 13.5 was obtained. Achievement of high pressure ratios in the axial stages meant high rotational speeds. As a result, the engine ran at 59,000 rpm, nearly 1,000 revolutions per second.

The PLT34 turbine had two stages. In designing the first stage, the challenge was to scale down the materials and casting technology used successfully in large engines to a blade design of only a half-inch in height in a turbine assembly no more than six inches in diameter. This challenge was met by using thin-wall castings containing cooling passages for the first-stage turbine rotor blades and directionally solidified metallurgy for the second-stage blades. Turbine nozzles

186. Anto to Leyes and Fleming, 14 March 1989.

were integrally cast containing cooling passages in the vanes. Turbine rotor inlet temperatures ran in excess of 2,200°F.

The second-stage turbine, uncooled, was designed as a dual alloy turbine: an outer ring containing the turbine blades was cast separately from the turbine hub. The hub, which used powdered metallurgy, was hot pressed into the outer ring. The result was a turbine disk made from an alloy with fairly fine grain structure that provided the cyclic life requirements of the turbine disk, and a cast alloy rim with large grain structure that provided the best compromise for stress life and stress rupture. This turbine was Lycoming's first attempt at application of the dual alloy principal.

The PLT34 first ran in December 1973. All of the testing was conducted at the Stratford plant in Lycoming's sea-level test facilities. In 1975, the engine successfully achieved durability objectives and demonstrated acceleration, cold engine starting, and multi-fuel capability. These objectives reflected requirements for a flexible engine capable of operating in Army helicopters of the future.[187]

Development of the PLT34A began in 1977, following the award to Lycoming of one of two ATDE contracts. Under one contract, Lycoming developed the PLT34A. The other contract went to Allison for development of a twin-centrifugal compressor engine. General Electric, Garrett, and other companies in the competition were not successful in winning contracts.

Using the PLT34 as its core, the PLT34A added a single-stage, high-work, high-speed, 30,000-rpm power turbine, a high-efficiency inlet particle separator, a digital electronic fuel control, and new high-speed engine accessories. The engine first ran in November 1979. By 1982, Lycoming had successfully demonstrated the engine's durability and achieved its performance objectives. Having fulfilled the contract objectives, work on the PLT34A ATDE engine came to an end at that time.[188]

APW34/T800 ENGINE

With the conclusion of work on the PLT34A engine, Lycoming began positioning itself for its next major competition. This was the Army's T800, the engine designated to power a new light experimental helicopter, the LHX. Accordingly, in 1984, Lycoming began design of its AL34 based on its PLT34A engine. The AL34 was designed as a 1,100-shp engine that included three fundamental changes from the PLT34A design. The first was to scale up the engine

---

187. Anto to Leyes and Fleming, 10 October 1989.

188. Anto to Leyes and Fleming, 10 October 1989.

to increase its airflow. The second was the use of single-crystal blade casting technology in the core turbine in order to sustain an increase in turbine inlet temperature. By that time, single-crystal technology for small, air-cooled turbine blades had become available. The third change was the addition of a second power turbine stage, which allowed a reduction in the speed of the power output shaft from 30,000 to 23,000 rpm.

The RFP for the T800 engine released by the Army in 1984 contained a then-unique requirement that each proposer consist of a team. The Army's team concept was based on the idea that the team could be separated at some point in the production cycle, allowing for two competing producers of the engine. In other words, the teaming arrangement was to be a marriage designed for divorce. Consequently, before responding to the RFP, partners had to be selected and joint-venture agreements had to be drawn up with "divorce" in mind.

Lycoming elected to team with Pratt & Whitney's Government Products Division in West Palm Beach, Florida, which worked with the other Pratt & Whitney divisions in Canada and at East Hartford, Connecticut. The main attraction in teaming with Pratt & Whitney was their turbine materials and cooling technology, which had been demonstrated in P&W's F100 fighter engine. That engine was running with turbine inlet temperatures several hundred degrees hotter than were small turbine engines at that time. Another factor attracting Lycoming to P&W was that its Canadian affiliate, Pratt & Whitney Canada, was recognized as an industry leader in centrifugal compressor design. Basically, the corporate elements of the P&W team possessed technologies that complemented one another. Lycoming provided its high pressure ratio axial compressor technology, its ability to build extremely small high-speed accessories, and its digital fuel control systems. Pratt & Whitney contributed its knowledge of high-temperature alloys for turbines, turbine blade cooling technology, and centrifugal compressor design expertise.

Two other teams had entered the competition. Garrett had teamed with Allison and General Electric had a leader-follower arrangement with Williams International. The Lycoming-Pratt & Whitney team received one contract and the Garrett-Allison team received the other for competing versions of the T800 engine. Both teams were to develop and demonstrate their engines, after which one team would be selected to produce its T800 engine for the Army's future LHX helicopter.

Upon completion of the Lycoming-Pratt & Whitney teaming negotiations, redesign began on the AL34 to incorporate Pratt & Whitney technology. The engine was the APW34 with the military designation of T800-APW-800. The overall configuration and layout of the APW34 was the same as that of the AL34. The P&W Florida division designed the high pressure turbine and the inlet particle separator while Pratt & Whitney Canada designed the centrifugal compressor and power turbine.

An important feature of the engine's mechanical design was integration of its accessories to minimize their size and the elimination of external fuel and oil lines. A key lesson learned in Vietnam was that engine accessories and fuel and oil lines were vulnerable to small-arms fire during low-altitude operations. To minimize their vulnerability, integrated accessories were located on top of the engine to shield them from ground fire and to provide easy access for servicing. With the exception of the main fuel line from the fuel tank to the engine, there were no external fuel or oil lines. External oil lines were replaced by internal coring though the accessory gearbox structure.

One of the most difficult design problems that faced the Lycoming-P&W team was design of the inlet particle separator (IPS). A key hazard in helicopter engines was foreign object damage resulting from ice breaking off of the engine inlet surfaces, and sand and dust kicked up by the rotors during takeoff and landing. Avoidance of such damage required an IPS sufficiently effective to capture at least 95 percent of the particles entering it. An additional requirement was that particles passing through the separator and into the engine should only be very fine, talcum-powder-sized dust, which would not harm the compressor and turbine.

Eleven engines were built for development testing and demonstration purposes. Approximately half of the engines were tested by Lycoming at its Stratford plant, and the remainder were tested by the P&W Government Products Division. Pratt & Whitney Canada also tested two engines in altitude test chambers at the Canadian Research Laboratory in Ottawa.

The Lycoming-P&W engine performed well and met its design objectives. However, when the Army made the final selection in late 1988, the Garrett-Allison team was the winner. According to the debriefing given by the Army following the selection, characteristics of the two engines were similar, and their power-to-weight ratios were very close. The deciding factors appeared to be production engine price and the support guarantees that the two companies offered to the military over a potential 20-year span of production.[189]

Following its loss in the T800 competition, Lycoming unsuccessfully sought to find a partner interested in sharing the cost and risk involved in developing the

189. Anto to Leyes and Fleming, 10 October 1989.

engine for possible commercial use. Lycoming believed that the APW34/T800 had been a fertile technology proving ground and that its technology would ultimately find its way into Lycoming's future military and commercial engines.[190]

PLT210 DEMONSTRATOR ENGINE

Work on the PLT210 demonstrator engine began in 1987 to provide a focus for the company's current and future work on compressors, combustors, and turbines. The decision was made to concentrate the company's component technology efforts on one size class, that of 10-lb-per-second of airflow, and to integrate and demonstrate this advanced technology in a gas generator of the same size.

There were several reasons for selecting a size of 10 lb per second. By that time, the two engines that had been Lycoming's biggest sellers were the T53 and the AGT1500,[191] which had airflows of 10 to 12 lb per second. It was expected that the company's future engines would be only moderately larger or smaller than these two engines. Thus, only a limited amount of scaling would be required to apply demonstrator engine technology to the design of new engines. The 10-lb-per-second size was also convenient for component and engine fabrication, testing, and handling. Lycoming's machine tools could easily fabricate components of that size, the existing test facilities had ample drive power and air handling capacity to operate that size equipment, and hardware of that size could be picked up and handled easily without the use of heavy lifting equipment.

The PLT210 was designed to be a gas generator that could be used to demonstrate Lycoming's new technology as it was developed through the 1990s and into the 2000s. Harold Grady, who directed Lycoming's Advanced Engineering Department, observed that, "The PLT210 is not so much an engine as it is a technology program."[192] The engine was designed to be changed with time and altered to demonstrate technology advances. The engine built in 1990 would be significantly different from the one built in 2000. At any point, technology

---

190. Anto to Leyes and Fleming, 10 October 1989.

191. The AGT1500 was a small gas turbine engine developed and manufactured by Lycoming for use in the U.S. Army's M1 "Abrams" Main Battle Tank.

192. Harold Grady, interview by Rick Leyes and William Fleming, 11 October 1989, Textron Lycoming Interviews, transcript, National Air and Space Museum Archives, Washington, D.C.

could be picked out of the PLT210 technology stream for use in designing a new engine or uprating an existing one.

Work on the PLT210 began with company funding. However, there was a congruence between the technology work on that engine and the military's tri-service technology paths. In the late 1980s, the military services established the IHPTET program. That program dealt with technology for the full spectrum of engines from small helicopter size engines to the huge engines for supersonic fighters. One of the IHPTET subprograms, the Joint Technology Advanced Gas Generator (JTAGG) program, was aimed primarily at Army requirements for propulsion. In 1989, Lycoming competed for and was awarded a technology demonstration contract under that program. This contract, to extend from 1989 to 1994, conformed with the company's plans for the PLT210 and provided funds to support much of the engine's technology demonstration work.

The contract had several objectives, requiring the demonstration of significant improvements in fuel economy and thrust-to-weight ratio superior to that of current engines. In addition, a demonstration of engine lifetime was required. Basically, the Army was looking for a smaller, lighter, more efficient, and more powerful engine for its future helicopters.

By 1989, Lycoming was testing PLT210 components, and the first complete gas generator ran successfully in July 1991. Details of the engine and its components remained classified at that time. Advanced Engineering Director Harold Grady characterized the PLT210 program as exciting and as an activity that he believed would give the engineers who had developed new concepts and advanced components the satisfaction of seeing their hardware working in a complete engine. He observed that as a result the program had generated a good deal of enthusiasm among the engineering staff and had acted as a stimulus to their creativity.[193]

LOW-POWER ENGINES SUMMARY

The key aspect of the low-power engine family was the new technology that emerged from its small demonstrator engines. This advanced technology had allowed Lycoming to extend its portfolio of engines to include the small LTS/LTP101 turboshaft/turboprop engines as well as the T800 turboshaft engine. In addition, new technology produced by the demonstrator engine program found its way into other Lycoming engines. By the end of the 1980s, the LTS/LTP101 series had contributed only minimally to company revenues due

193. Grady to Leyes and Fleming, 11 October 1989.

to its reliability record. Nevertheless, in 1990 the LTS/LTP101 series were the only production engines in Lycoming's low-power engine family.

Vehicular Engine Family

The story of Lycoming would be incomplete without a brief description of its vehicular engine family, in which there were three similar engines. The first and by far most successful was the AGT1500 turboshaft engine developed for use in the Army's M1 Main Battle Tank. That engine was followed by the model TF15, which was a modification of the AGT1500 and designed for use in a number of marine applications. The third engine was the PLT27 turboshaft, a flight-weight version of the AGT1500 that was developed for use in helicopters.

AGT1500 TANK ENGINE

As early as 1951, Dr. Anselm Franz and S. B. Withington visited the Detroit Tank Arsenal to present a proposal for development of a gas turbine engine for a tank application. Their early attempts to convince the Army to use gas turbines in tanks were unsuccessful. However, in 1963, Lycoming succeeded in obtaining an Army contract to develop a tank turbine engine, the AGT1500.[194] The first complete AGT1500 engine ran in November 1966.

Rated at 1,500 shp, the AGT1500 was a three-shaft engine with contra-rotating gas generator shafts and a free power turbine shaft. The low pressure compressor had five axial stages that were driven by a single-stage turbine. The high pressure compressor had four axial stages followed by a centrifugal stage, which were driven by a single-stage turbine. A two-stage power turbine drove a rear-power takeoff through a reduction gearbox. Mounted to the engine inlet was a self-cleaning filter system to clean sand and foreign objects from the air before it entered the engine.

A recuperator recovered waste heat from the exhaust gases leaving the turbine to improve the engine's cycle efficiency and thus its fuel economy. Air discharged from the high pressure compressor passed through the recuperator where it was heated by the exhaust gases and was then returned to the engine where it entered the engine's single-can combustor. The recuperator was wrapped around the power turbine and gearbox; after the exhaust gases had passed through it to be cooled, they were discharged radially upward.

194. Franz, *From Jets to Tanks*, 69-72.

In 1976, the AGT1500 was selected to power a prototype version of a new Army tank, the XM1, being developed by Chrysler Corporation, intended as a replacement for the M60. At the Aberdeen Proving Grounds in Maryland, the Army conducted a set of competitive tests between two versions of XM1 prototype tanks, one powered by the AGT1500 and the other by a diesel engine built by Continental Motors. The AGT1500-powered version of the tank was selected. With the success of the AGT1500 in this competition, Lycoming captured the tank turbine engine market.[195]

By 1979, the AGT1500 was in volume production for use in the Army's 60-ton M1 *Abrams* Main Battle Tank, then manufactured by Chrysler Corporation and later by General Dynamics. In the wake of the large-volume production of the M1 tanks that followed, sales of the AGT1500 grew, and, in 1989, accounted for 44 percent of Lycoming's nearly $1 billion revenue.[196]

The AGT1500 offered a number of advantages over diesel engines previously used as tank power plants. It produced lower levels of noise and vibration than the diesel engine, reducing crew fatigue and contributing to their comfort. It was smaller, freeing substantial space for additional fuel and other purposes. Light, modular in construction, with top-mounted accessories, it was easier to maintain than the diesel engine. In Army comparative tests with the diesel-powered model, the AGT1500 version had faster acceleration and greater mobility and maneuverability than the diesel version, especially over rough terrain.[197]

The AGT1500 enjoyed a successful record and met with full acceptance in the M1 tank. In fact, both the engine and the tank reportedly operated beyond expectations during Operation Desert Storm in the 1991 conflict with Iraq.[198] As a result of its success, the production rate grew, reaching 990 engines in 1989. By the end of that year, AGT1500 engines had accumulated more than six million operating hours in the field.

The AGT1500 member of Lycoming's vehicular engine family was a major contributor to Lycoming's overall gas turbine sales, accounting for 44 percent of Lycoming's total 1989 revenues. By the end of 1989, a total of 7,677 AGT1500 engines had been manufactured.

195. Frantz, *From Jets to Tanks*, 69–72.

196. Textron Lycoming, "1989 Lycoming Sales Split," unpublished graph, National Air and Space Museum Archives, Washington., D.C.

197. Avco Corporation, *The First Fifty Years*, 89–90.

198. Thomas Strom, "Vehicular & Small Turbomachines Committee," *Global Gas Turbine News* (July/August 1992): 15.

In 1986, the Army initiated an improvement program for the AGT1500 with a planned introduction of an improved version in the mid-1990s. By 1990, a number of improvements were being incorporated in the engine to make it more efficient and to meet customer requirements. These and other planned improvements to the AGT1500 engine were intended to help Lycoming maintain its unique position in the tank turbine marketplace.

PLT27—A NEW HELICOPTER ENGINE

In 1965, soon after development of the AGT1500 had begun, Lycoming decided to develop a flight-weight version of the engine for use in helicopters. At that time, requirements were emerging from the Army and Navy for helicopters requiring engines in the 1,500- to 2,000-shp class. Since the AGT1500 was in that power class and had good fuel economy, Lycoming felt that a flight-weight version of the engine would be a good candidate to satisfy the military's requirement.

Accordingly, using company funds, Lycoming began development of the PLT27 in 1965 as a 2,050-shp turboshaft. With a bit of trimming and adaptation of the AGT1500 to the typical Lycoming reverse-flow wrap-around combustor configuration, a lightweight aircraft engine was produced. The PLT27 used the twin-spool gas generator system and power turbines of the AGT1500. The high pressure turbine was modified slightly to rematch it for operation without the recuperator. Both front and rear power takeoffs were developed for the engine.

The Army's main requirements at that time were for an engine that could be serviced easily, that could be disassembled and maintained in the field with only a small number of tools, and that was combat resistant. Accordingly, the PLT27 broke into three easily disassembled modules: a power turbine section, a gas turbine section, and an accessory gearbox. Accessories were top mounted both for ease of maintenance and to be shielded from ground fire. Serviceability of the engine was demonstrated by having two high-school students with 52 hours of training disassemble and reassemble the engine using only five standard tools.[199]

A complete PLT27 engine first ran in December 1969. Considerable ground testing of the engine followed, and the engine completed a demonstration preliminary flight rating test. With a specific fuel consumption of only 0.45, it was one of the most fuel efficient turboshaft engines in the country at that time.

The engine was marketed by Lycoming in three major military competitions. The first was the Army's Utility Tactical Transport Aircraft System (UTTAS)

199. Lunny to Leyes and Fleming, 10 October 1989.

competition, in which the PLT27's competitors were the General Electric GE12 engine and the Pratt & Whitney Canada ST9. In December 1971, the GE12, given the military designation of T700, was selected over the PLT27 and ST9. In August 1972, T700-powered versions of the Sikorsky and the Boeing Vertol UTTAS entries were selected by the Army to compete in a fly-off of prototype models. In December 1976, the T700-powered Sikorsky YUH-60A was selected as the UTTAS helicopter.

The second competition was for the Army's Advanced Attack Helicopter (AAH) program. The PLT27 again lost to the GE T700. In June 1973, T700-powered helicopters offered by Hughes and Bell were selected by the Army for a prototype AAH fly-off, and in December 1976, the Hughes YAH-64A was announced the winner.

In 1977, the PLT27 was involved in its third and final competition when it was offered as the power plant for the Light Airborne Multi-Purpose System (LAMPS) III helicopter. The Department of Defense decided at that time to develop five classes of helicopters with a single helicopter in each class to be powered by a given engine. These helicopters were to be used by all the military services. However, to meet its specialized requirements, the Navy wanted an engine of its own in its version of the LAMPS III rather than one developed primarily by the Army. During the LAMPS competition, the Navy received permission to solicit a competitive bid for an alternate power plant. It was at that time that Lycoming appeared with its PLT27 engine as a competitor to the GE T700. Sikorsky won the LAMPS competition with a T700-powered variant of its S-70L, later designated the SH-60B Seahawk. Developed under contract to the Navy, the primary missions of the Seahawk were anti-submarine warfare and anti-ship surveillance and targeting. Secondary missions included search and rescue, medical evacuation, and vertical replenishment.

Between 1971 and 1977, the PLT27 lost three times in competition with the General Electric GE12/T700. As a result, the T700 had become well entrenched as the 2,000-shp class engine in both Army and Navy helicopter programs. Work on the PLT27 continued only briefly following the LAMPS loss, after which further development and marketing efforts were discontinued.[200]

VEHICULAR ENGINES SUMMARY

The AGT1500 member of Lycoming's vehicular engine family was a major contributor to Lycoming's overall gas turbine sales. Further, it was the AGT1500

---

200. Lunny to Leyes and Fleming, 10 October 1989.

that allowed Lycoming to capture the tank turbine market as the uncontested supplier of tank turbines to the Army.

Expanded Manufacturing and Service Capacity

About 1980, Lycoming faced a heavy demand on its manufacturing capacity at the Stratford plant. The company projected a growing demand for replacement parts, maintenance, and repair of its engines. To satisfy this future demand, it determined that an additional facility should be created to manufacture spare parts. That decision led to the establishment of the Textron Turbine Services plant in Greer, South Carolina, as a parts manufacturing facility, which became operational in 1981.

The 200,000-square-foot Greer facility helped to unload the Stratford plant and also to establish itself as a low-cost producer of parts such as turbine, compressor, and fan disks, and all turbine blade castings. Parts built at Greer were either delivered to Stratford for use in assembling engines or were shipped as spare parts directly to customers.[201]

As part of Textron Turbine Services, an Engine Repair Center was established at Greer consisting of a team of technicians and engineers dedicated to the maintenance, repair, and overhaul of a part of the LT101 fleet of engines. Another element of Textron Turbine Services was a facility in Hatfield, England, that specialized in repair and overhaul of Lycoming ALF502 turbofans. These service operations were part of a long-term Lycoming strategy aimed at capturing a portion of the component/engine repair and overhaul market.

In addition to its Greer facility, in early 1987, Lycoming opened a 136,000-square-foot Customer Support Center (CSC) near its Stratford plant to perform commercial depot maintenance on ALF502, T53, and T55 engines. The development and implementation of new repair technologies by the Center led to more cost effective repairs and a reduction in customer operating costs. Robert Gustafson, Lycoming's Vice President-Customer Operations, estimated that the number of engines repaired at the Center would grow to as many as 250 per year by 1992.[202] The CSC was also the focal point for the company's commercial and military customer-service and product-support operations. Concen-

201. Kuintzle to Leyes and Fleming, 12 October 1989.

202. "Avco Lycoming Business Plan Stresses Quality and Service," *Aviation Week & Space Technology)*, Vol. 126, No. 22 (June 1, 1987): 79-81.

trated in a single location were the important aftermarket functions of spare parts, maintenance, service engineering, warranty administration, and technical publications. In addition, CSC was the head of Lycoming's customer communications network, spanning seven continents and connected to a team of professionals ready to go anywhere to support Lycoming customers.

Overview

STATUS OF THE CORPORATION

In 1985, Avco Lycoming was acquired by Textron, Inc. and became Textron Lycoming. By the early 1990s, in addition to the plant in Greer, South Carolina, and the service facility in Hatfield, England, Lycoming had 1.86 million square feet of factory and office space at its Stratford, Connecticut, plant, and within that 96-acre plant site were its engineering and engine manufacturing facilities, 35 engine test cells, and 32 component test cells. The nearby Customer Support Center provided commercial depot maintenance on ALF502, T53, and T55 engines.

The number of employees in the company peaked at a level just over 10,000 in 1968 as production of T53 and T55 engines was approaching its maximum. Employment then began declining in the late 1960s, reflecting reductions in military orders for those engines. With the drop in sales that followed, employment plummeted, reaching a level of less than 3,000 by 1976. With growing orders for its newly developed LTS101 and ALF502 engines during the 1970s and 1980s, the employment level crept higher, holding between 4,000 and 5,000 employees through the 1980s.[203]

Following the drop in T53 and T55 sales and revenues during the 1970s, Lycoming's revenue turned upward again, largely on the success of its AGT1500 tank turbine engine and the ALF502. From 1980 to 1986, the company's revenue from the sale of aircraft, helicopter, and tank turbine engines grew from $185 million to $946 million.[204] In 1990, total company sales were $993 million. However, by 1993, Lycoming's reported sales had dropped to $620 million and the staff numbered approximately 2,900.

203. Textron Lycoming, "Manpower History," undated (circa 1990), unpublished graph, National Air and Space Museum Archives, Washington, D.C.

204. "Avco Lycoming Business Plan Stresses Quality and Service," *Aviation Week & Space Technology*, Vol. 126, No. 22 (June 1, 1987): 79-81.

From 1951 until after its acquisition by AlliedSignal, Lycoming gas turbine engines were primarily developed and manufactured at this facility in Stratford, Connecticut. Photograph courtesy of AlliedSignal.

THE COMPANY'S MARKETS AND INFLUENCING FACTORS

Over the years, beginning in 1959, Lycoming produced turboshaft engines for use in helicopters. It was the major supplier of turboshaft engines for Army medium and medium-heavy lift military helicopters through the late 1970s. In the early 1970s, Lycoming entered the civil marketplace with the development and production of turbofan engines for executive and regional transport aircraft. The company's military market was significantly expanded, beginning in the late 1970s, with production of a turboshaft engine for powering Army tanks. Annual deliveries of Lycoming T53 and T55 engines increased from 76 in 1959 to a peak of 2,239 in 1970. However, production of these engines fell off in the early 1970s because of reduced Army helicopter acquisition for the Vietnam War. Production from the mid-1970s to the end of 1989 ranged from a high of 1,093 in 1975 to a low of 277 in 1989, with an average of 473 engines per year. During the last half of the 1980s, ALF502 turbofans accounted for nearly 30 percent of the aircraft gas turbine deliveries.[205]

The success of its T53 and T55 turboshaft engines in the Army's medium and medium-heavy lift helicopters confined Lycoming's business to the military

205. Textron Lycoming, "Original Engine Shipments by Year of Manufacture."

marketplace for nearly 20 years. The decline in the Army's demand for T53 and T55 engines following the 1970 production peak finally freed up sufficient resources to allow the company to concentrate on developing engines for commercial applications. As a result, during the subsequent 20-year period, there was a dramatic shift in the sales split among the markets served by Lycoming's engines. With the entry of the ALF502 into the regional jet market in the early 1970s, followed by the LTS101 turboshaft used in both commercial and military helicopters, and the growth in use of the T53 and T55 in commercial versions of the Huey and Chinook helicopters, Lycoming's commercial sales began to grow. In 1989, 46 percent of Lycoming's sales were to the commercial market and 54 percent to military applications. However, more than 80 percent of the military sales in 1989 were for the AGT1500 tank engine, with the remainder being for T53s and T55s.[206] During the period from 1970 to 1989, Lycoming's gas turbine sales for military aircraft shrank nearly by a factor of ten.

Some 1,025 ALF502s were manufactured between 1972 and the end of 1989 for use on regional jet transport and business aircraft. By the late 1980s, the ALF502 accounted for nearly 30 percent of the company's annual deliveries of aircraft gas turbines. In 1988, Lycoming announced plans for the 500 Series of engines that initially included the LF507, and, by late 1991, the first production LF507 turbofan engines had been delivered for use in the new British Aerospace RJ70 regional jet transport.

The LTS101, introduced in the mid-1970s as a simple, low cost, and economical turboshaft engine for use in light helicopters and small fixed-wing aircraft, suffered from an unfortunate history of poor in-flight reliability due to turbine failures that gave the engine a negative image within the helicopter industry. Nevertheless, 1,839 LTS101 engines were produced between 1973 and the end of 1989. By 1990, the turbine problem had been solved, earlier models of the engine were being retrofitted, reliability had improved, and production of the engine was proceeding.

During the 1980s, it became more difficult for Lycoming and others to do business with the military. Growing budget uncertainties led to uncertainty in program schedules and allowable costs. Once a company entered a competition, it never knew whether the announced program would be carried out on schedule with manufacturing orders for an end product. This uncertainty had a strong effect on Lycoming's business strategies. A great deal of contingency planning and tradeoff evaluation were required to select those programs with the highest probability of success. Also, there was a desire to participate in as

206. Textron Lycoming, "1989 Lycoming Sales Split."

many program competitions as possible to maximize the possibility for obtaining new contracts. That desire, accompanied by the uncertainty over which programs might go to fruition, led Lycoming and others to teaming arrangements that would permit company resources to be spread over a number of programs.

Growing competition among the engine companies during the 1980s became a source of difficulty. From the mid-1950s until the late 1970s, the aircraft gas turbine industry had flourished as aircraft applications for new engines grew. However, by the 1980s, the engine business and its aircraft market had matured. The number of new applications had decreased. There were too few new applications and too many competitors. In addition, the cost of fully funding development of a new engine was approaching the limits of an individual engine company's ability to afford, particularly with the possibility of limited production orders and the amortization of costs for engine development and production tooling. This environment led to the formation of Lycoming's 1987 partnership with GE to develop, market, and produce a commercial version of GE's GE27 (T407-GE-400) turboshaft engine, the GLC38 turboprop.

In the early 1990s, Lycoming was poised to continue pursuing the business strategy that it had established in the late 1960s, broadening its business base in the civilian aircraft markets. In 1989, Lycoming's president, John Myers, stated that the company intended to make the general aviation and commercial airline industries the two dominant segments of its business in the future. In view of the shrinking military market, the company hoped to increase its civilian sales from slightly less than half to about two-thirds of its total revenue.[207] Nevertheless, Charles Kuintzle, a Senior Vice President of Lycoming, observed that the company needed both military and commercial business, as the two complemented one another.[208]

A major transition in the corporation's history took place on May 12, 1994, when AlliedSignal, Inc., announced that it had signed a memorandum of understanding to acquire the Lycoming Turbine Engine Division of Textron, Inc., for approximately $375 million in cash plus the assumption of certain liabilities.[209] The acquisition became official on October 31, 1994. At that time, the Lycoming Turbine Engine Division employed approximately 2,900 employees. The Ly-

---

207. "Textron Turbine Engine Group Seeks More Commercial Business," *Aviation Daily*, Vol. 298, No. 3 (October 4, 1989): 27.

208. Kuintzle to Leyes and Fleming, 12 October 1989.

209. "AlliedSignal To Acquire Textron's Lycoming Turbine Engine Division," *Textron News Release*, 12 May 1994, 1-3.

coming acquisition was described as an expansion of AlliedSignal's engine product line into new markets—primarily for 50- to 100-seat regional aircraft and light, medium, and heavy helicopters—beyond those of its 6,000-employee, Phoenix-based AlliedSignal Engines facility.[210]

The acquisition brought together into one company (AlliedSignal Engines) the small gas turbine aircraft engine business of AlliedSignal's Garrett Engine Division and the aircraft and tank turbine engine business of Textron's Lycoming Turbine Engine Division. According to AlliedSignal officials, one of the primary reasons for the purchase was to acquire Lycoming's common core engine, the new gas generator for the company's 500 Series of turboprop and turbofan power plants. Jim Robinson, Vice President and General Manager of Business Operations for AlliedSignal Engines commented, "We first thought of a joint venture with Lycoming, (to get their technology), but marriages often don't go well."[211] In the acquisition announcement, James F. Hardymon, Textron Chairman and Chief Executive Officer, said, "The consolidation of Textron Lycoming with AlliedSignal's engines operation is a good transaction for Textron and Allied-Signal. It will create an efficient player in the small turbine engine business."[212]

In late 1995, AlliedSignal announced plans to end all operations at the Stratford plant and move its work to the AlliedSignal Engines facility in Phoenix, Arizona.[213] AlliedSignal completed its operations in Stratford and turned the facility over to the Army in September 1998. Anselm Franz's 1950 dream of having an engine plant of his own had come to fruition with the success and growth of Lycoming in the small gas turbine aircraft engine business. Some 45 years after it was founded, that company faded from existence with its integration into AlliedSignal Engines. It took along as its legacy the engines and technology it had developed.

CONTRIBUTIONS

Lycoming and its products made important contributions to military operations and civilian air transportation. The most significant was that made by the T53 as

210. "AlliedSignal Completes Acquisition of Textron's Lycoming Turbine Engines," *AlliedSignal News Release*, 31 October 1994, 1–2.

211. Stanley W. Kandebo, "AlliedSignal Completes Lycoming Acquisition," *Aviation Week & Space Technology*, Vol. 141, No. 19 (November 7, 1994): 35.

212. "AlliedSignal to Acquire Textron's Lycoming Turbine Engine Division," *Textron News Release*, 12 May 1994, 1.

213. "News Breaks - AlliedSignal Aerospace," *Aviation Week & Space Technology*, Vol. 143, No. 14 (October 9, 1995): 21.

the power plant for the Huey, the Army's first operational turbine-powered helicopter, and the HueyCobra. During the Vietnam War, these helicopters played a key role in revolutionizing Army infantry operations by contributing to the success of the Army's airmobility. The T55-powered Chinook helicopter also played a major role in airmobility operations during the Vietnam War. The AGT1500 tank turbine engine contributed importantly to Army ground combat operations. As the power plant in the Army's main battle tank, it made possible a new generation of battle tanks with faster acceleration and greater mobility than earlier piston engine-powered tanks. Lycoming's principal contribution to civil air transport was made by the ALF502 engine. Low noise level and good fuel economy were hallmarks of that engine which was first used in the four-engine British Aerospace BAe 146 jetliner, one of the world's first regional jet transport aircraft.

GENERAL ELECTRIC SMALL AIRCRAFT ENGINES | 6

Early Company History

The Edison General Electric Company was formed in 1889 to consolidate the separate but interlocking electrical companies controlled by Thomas A. Edison. The parent company of Edison General Electric was the Edison Electric Light Co., founded by Edison in 1878. Late in 1879, Edison had invented and made patent application for a carbon filament lamp—the first practical and economical electric light for domestic use.[1] Over the next decade, through the development and sales of the incandescent electric light, electric lamps and appliances, and power generation equipment and distribution systems, the Edison companies grew and consolidated to become the preeminent electric company in the United States.

One of the chief competitors to the Edison General Electric Company was the Thomson-Houston Company. Established in 1883 in Lynn, Mass., Thomson-Houston originally manufactured arc dynamos and arc lights based on the patents of Elihu Thomson and Edwin Houston. Thomson-Houston moved into the incandescent lamp business under a license arrangement with a company that held non-Edison incandescent light patents. With the acquisition of other businesses in the electric industry, Thomson-Houston grew rapidly; by 1889, it was second to the Edison group in business volume.

1. Matthew Josephson, *Edison* (New York: McGraw-Hill Book Company, Inc., 1959), 221-222.

In 1885, the Edison company began patent infringement litigation against companies owning competing incandescent lamp patents. By this time, Thomson-Houston and Westinghouse Electric and Manufacturing Co., the third major player in the electric industry, owned most of the non-Edison patents. However, in 1891, Edison won a key court case against the United States Electric Lighting Company, a subsidiary of Westinghouse, upholding the validity of Edison's original lamp patent.[2] Edison was granted a legal monopoly on the production of incandescent lamps until the patent expired in 1894.

This patent litigation had been costly to all parties. Edison had the dominant patents for the incandescent lamp business, a direct current electrical generation and distribution system, and a growing electric traction business for powering street cars and other vehicles. Thomson-Houston was a leader in the arc light business, alternating current power systems, and electric street railways. As business grew, it became more difficult for either company to produce electrical installations or certain products relying solely on their own patents or technology. Combining the patents and products of each company would put the merged company into a dominant position in the electrical industry. In 1892, the two companies amalgamated to become the General Electric Company.[3] GE became the largest electrical trust in the U.S.[4] Headquarters was established in Schenectady, N.Y., the site of the former headquarters of Edison General Electric.

GE DEVELOPS STEAM TURBINES

Within a decade after Edison had first demonstrated his light bulb in 1879, electric lighting and power had become significant technological elements in American society.[5] The development of electrical generating and distribution systems toward the end of the 19th century created a demand for electricity on a large and economical scale. By linking water- or steam-powered turbines to generators, electrical power could be produced far more efficiently than with conven-

---

2. Arthur A. Bright, Jr., *The Electric-Lamp Industry: Technological Change and Economic Development from 1800 to 1947* (New York: Arno Press, 1972), 87-88.

3. *Century of Progress: The General Electric Story 1876-1978*, ed. Bernard Gorowitz (Schenectady, New York: Hall of History Foundation, 1981), 6.

4. For purposes of brevity, the General Electric Company is abbreviated GE. General Electric Aircraft Engines is also abbreviated GE. The distinction between the General Electric Company, which is the corporate parent, and General Electric Aircraft Engines, a major industrial segment or division of GE, will be clear in the text.

5. Bernard S. Finn et al., *Edison: Lighting a Revolution, The Beginning of Electric Power* (Washington, D.C.: Smithsonian Institution, 1979), 16.

tional reciprocating engines. Steam turbines had higher power-to-weight ratios, were more compact, and had fewer parts than reciprocating engines.[6]

In 1896, Charles Curtis had patented a velocity-staged, steam-driven turbine.[7] In 1897, Curtis joined GE in Schenectady as a consulting engineer to develop the commercial potential of the steam turbine. Working with Curtis, GE engineer William Emmet redesigned and improved the turbine. In 1903, the first Curtis-Emmet steam-driven turbines were put into commercial operation at the Chicago Edison Company. The most powerful steam turbine generator of its time, it required only one-tenth the space, weighed one-eighth, and cost one-third as much as a piston engine generator of equivalent power. The turbine business expanded rapidly, and the biggest selling point was economy in space.[8] The electrical capacity of plants could be expanded without the need to expand existing buildings. By the late 1960s, steam turbine generators had long accounted for more than 80% of all of the electrical power generated in the U.S.[9]

DR. SANFORD MOSS

While GE was developing steam turbines, Cornell University doctoral candidate Sanford Moss was conducting gas turbine research. In the course of his experiments, he became the first person in the U.S. to power a turbine with combustor gas. The initial turbine wheel used by Moss was a De Laval steam turbine. Compressed air was provided by a reciprocating steam engine compressor and was fed to a continuous combustion chamber.[10] Because the power required for the compressor was more than that produced by the turbine, Moss considered the experiment a failure.[11]

6. Leonard S. Reich, *The Making of American Industrial Research: Science and Business at GE and Bell, 1876-1926* (Cambridge: Cambridge University Press, 1985), 56.

7. Edward W. Constant III, *The Origins of the Turbojet Revolution* (Baltimore: The Johns Hopkins University Press, 1980), 79.

8. John W. Hammond, *Men and Volts: The Story of General Electric* (New York: J. B. Lippincott Company, 1941), 283.

9. General Electric, *The Story of the Turbine* (September 1969), 5.

10. Sanford A. Moss, "The Gas Turbine: an "Internal Combustion" Prime Mover" (Ph.D. diss., Cornell University, 1903). Moss's gas turbine is described in Chapter 1 and 6 of his dissertation.

11. Sanford A. Moss, "Gas Turbines and Turbosuperchargers," *Transactions of the American Society of Mechanical Engineers Vol. 66* (New York: American Society of Mechanical Engineers, 1944), 353.

When Moss graduated, he went to work in June 1903 at GE in the Steam Turbine Department at Lynn. With others, Moss was able to get into operation a number of gas turbine models, both at Schenectady and Lynn.[12] These gas turbine engine experiments ended in 1907; low turbine and compressor efficiency and material temperature limitations resulted in low power output and high fuel consumption compared to other internal combustion engines.

However, as part of this work, the first research in the U.S. on centrifugal compressors began.[13] In a 1904 patent application, Moss established, probably for the first time, the relationship between centrifugal compressor design and the velocity and volume of compressible fluid flow.[14] GE used Moss's theory to develop successful centrifugal compressors, and the first centrifugal compressor machine was put into commercial service at the Lynn plant in 1906. In time, an extensive centrifugal compressor business grew. GE compressors supplied pressurized air for such applications as blast furnaces, pneumatic tubes and conveyor systems, and gas and oil furnace air supply. With the commercial success of centrifugal compressors assured, Moss established the Turbine Research Department (later the Thomson Laboratory) within the Lynn Steam Turbine Department.

In November 1917, National Advisory Committee for Aeronautics (NACA) Chairman Dr. William F. Durand wrote to GE President E. W. Rice and asked the company, and Moss in particular, to undertake the development of a turbosupercharger for the Liberty engine.[15] Durand, a Cornell faculty member when Moss was a student there, knew of his turbine research. Durand was also familiar with GE's extensive business in steam turbines and in centrifugal compressors.

Through a World War I cooperative arrangement to exchange information among the Allies, NACA had received drawings of a turbosupercharger system designed by Auguste Rateau, a French engineer who had successfully operated his system on a mountain in France. GE based the general plan of its turbosupercharger on the Rateau design, but the details of the design were based entirely on the years of experience of the engineers of the turbine and centrifugal compressor departments at GE.[16] The resulting turbosupercharger, the first one in the U.S., was built in Lynn. It was delivered to the War De-

---

12. Moss, "Gas Turbines and Turbosuperchargers," 353.

13. Sanford A. Moss, *Superchargers for Aviation* (New York: National Aeronautics Council, Inc., 1944), 13-14.

14. Moss, "Gas Turbines and Turbosuperchargers," 355.

15. Moss, "Gas Turbines and Turbosuperchargers," 362-363.

16. Moss, *Superchargers for Aviation*, 32.

partment's Airplane Engineering Division at McCook Field in Dayton, Ohio on June 4, 1918, and shortly thereafter it was mounted on the front of a Liberty engine.

Preliminary turbosupercharger testing was conducted during the summer of 1918, indicating the need for altitude testing. Between September 10 and October 7, Moss and McCook Field engineer C. P. Grimes directed the testing at Pikes Peak, Colorado.

Preliminary calibration had yielded an engine rating of 354 hp. At the mountain's summit, the engine delivered 230 hp without supercharging and 352 hp with supercharging. Summarizing their work, Moss and Grimes stated: "Tests of the Turbo-Supercharger at the Summit showed that it was able to raise the MEP (mean effective pressure) of the Liberty Motor to practically the value obtained at sea level. It was further possible to exceed this value (to the limiting point of pre-ignition) corresponding to density, at a considerable distance below sea level."[17]

Moss and Grimes had conclusively demonstrated that the GE turbosupercharger enabled an engine to maintain rated power at altitude by supplying compressed air to the intake manifold at sea-level atmospheric pressure. Following the successful Pikes Peak test and the World War I Armistice, on July 12, 1919, flight tests began with the turbosupercharged Liberty engine installed on a Packard-built LePere aircraft. Tests showed that engine horsepower was "greatly improved at high altitudes and the speed and ceiling of the particular airplane used have been remarkably increased."[18]

From 1919 to about 1930, flight tests of the turbosupercharger concentrated on setting altitude records. After that period, more effort was devoted toward improvement of performance at normal operating heights. This work was conducted by the Army Air Corps, NACA, various aviation companies, and GE's supercharger department with the assistance of the turbine and other GE de-

17. C. P. Grimes and Dr. S. A. Moss, "Test of Turbo Supercharger made at Pikes Peak in conjunction with 12 Cylinder Liberty Aeroplane Motor," *War Department, Bureau of Aircraft Production, Airplane Engineering Division, Experimental Engineering, McCook Field, Dayton, Ohio Report Ser. No. 498* (September and October 1918): 2. See also: W. A. Reichle, "Report on Test of the General Electric Company Turbo-supercharger for the 12-Cylinder Liberty Engine," *War Department, Bureau of Aircraft Production, Airplane Engineering Division, Experimental Engineering, McCook Field, Dayton, Ohio, Engine Report No. 172 Ser. No. 588* (12 December 1918).

18. Edward T. Jones, "Power Plant Section Test Report No. 64 Report of Flight Test on Moss Supercharger," *War Department, Air Service, Engineering Division, McCook Field, Dayton, Ohio Report Ser. No. 1071* (25 November 1919): 7.

partments.[19] In 1937, a major order for turbosupercharger from the Army Air Corps resulted in the formation of the GE Lynn Supercharger Department. Production skyrocketed with the start of World War II; many thousands were made by GE and its licensees for aircraft such as the Boeing B-17, Lockheed P-38, Republic P-47, and Consolidated B-24.[20]

In 1940, Moss and the Army Air Corps were awarded the coveted Collier Trophy for "outstanding success in high-altitude flying by the development of the turbo-supercharger."[21] The turbosupercharger gave American bombers excellent high-altitude performance, helping to make possible strategic bombing tactics that ultimately contributed to the success of the war effort. An important turbosupercharger contribution to both civil and military aviation was over-the-weather flight, which improved safety and passenger comfort. It is ironic that the turbosupercharger, originally a gas turbine by-product, became both a reason for GE's selection to enter the jet age and a means by which the company could accelerate its gas turbine development.

GE Enters the Jet Age

The history of GE's early jet engine development has been well documented and only a broad-brush review of it is necessary here.[22] The first substantive efforts took place from the mid-1930s to the early 1940s when certain people at GE's Lynn and Schenectady divisions conducted analytical and marketing studies of gas turbine engines.[23]

19. Moss, *Superchargers for Aviation*, 34.

20. John Anderson Miller, *Men and Volts at War: The Story of General Electric in World War II* (New York: McGraw-Hill Book Company, Inc., 1947), 76.

21. *The Aircraft Year Book for 1942*, ed. Howard Mingos (New York: Aeronautical Chamber of Commerce of America, Inc.), 252.

22. The following books are excellent sources of information: Robert Schlaifer and S. D. Heron, *Development of Aircraft Engines, Development of Aviation Fuels* (Boston: Harvard University, 1950); William R. Travers, *The General Electric Aircraft Engine Story* (Unpublished manuscript, 1978), National Air and Space Museum Library, Washington, D.C.; William R. Travers, *Engine Men* (Unpublished manuscript, 1986); General Electric Company, *Seven Decades of Progress: A Heritage of Aircraft Turbine Technology* (Fallbrook, Cal.: Aero Publishers, 1979); General Electric Company, *Eight Decades of Progress: A Heritage of Aircraft Turbine Technology* (Cincinnati: The Hennegan Co., 1990).

23. *Eight Decades of Progress*, 28, 38. See also Schlaifer and Heron, 442, 445.

In 1939, Sanford Moss proposed a gas turbine-driven propeller (turboprop) for aircraft propulsion to GE management. Moss's study was forwarded to Dale Streid, then a young GE engineer, for review. Streid made a simple graph comparing the specific fuel consumption (sfc) of the turboprop power plant and a jet propulsion gas turbine (turbojet) at the 1,700°F turbine inlet temperature proposed by Moss. The curves crossed at about 450 mph because the propeller efficiency fell off drastically at this speed. Streid concluded that if a gas turbine were to be developed, it should be a turbojet rather than a turboprop. However, Streid was not knowledgeable enough about the aviation industry to recommend that GE invest in any kind of development in this field. Streid also questioned the ability of the gas turbine to operate at this temperature. On the basis of Streid's review, GE management declined to invest in Moss's proposal. Upset, Moss invited the "young upstart" to Lynn where Streid was shown turbosupercharger operating at 1,700°F; Moss, however, couldn't refute Streid's conclusion about jet propulsion.[24]

Glenn B. Warren, Chief Turbine Design Engineer in Schenectady, also had a long-standing interest in gas turbine engines. In 1919, as an undergraduate student at the University of Wisconsin, Warren had conducted successful experiments on gas turbine combustion chambers and nozzles.[25] After joining GE in 1919, he proposed the development of gas turbines for power production, but neither the state of the art of metallurgy nor compressors was sufficiently advanced to permit this.

However, GE's Engineering Advisory Committee approved a resumption of work on a gas turbine engine in 1937. This decision was the result of a combination of factors, including: Warren's and Moss's early work and continuing interest in the gas turbine; knowledge of high-temperature materials derived largely from GE's steam turbine and supercharger work; and the development of an efficient axial-flow compressor by Brown-Boveri, which could not be patented and which they had publicly disclosed.[26]

In response to a request by the Navy for the fabrication of a few components for a shipboard gas turbine unit, a program of gas turbine studies and component testing was initiated under the overall direction of Managing Engineer A. R.

24. Dale D. Streid, Cincinnati, Ohio, to Rick Leyes, letters, 20 February 1992 and 1 April 1992, National Air and Space Museum Archives, Washington, D.C.

25. Glenn B. Warren, "An Experimental Study of Gas Turbine Chambers and Nozzles," *The Wisconsin Engineer* Vol. 26, No. 1 (October 1921): 1.

26. Glenn B. Warren, Phoenix, Arizona, to Arthur I. Strang, letter, 5 January 1965, National Air and Space Museum Archives, Washington, D.C.

Smith and the engineering leadership of Warren and Designing Engineer Alan Howard. As part of this work, a component test was performed on a small-scale turbine using high-pressure air supplied by a reciprocating compressor. In 1937, Smith visited the British Thomson-Houston Company (an offspring of the original U.S. company) and saw an aircraft gas turbine engine. This was the first British turbojet engine, invented by Flight Lieutenant Frank Whittle.[27]

Another outgrowth of this investigation was a study presented by Alan Howard at the February 1941 GE Engineering Council meeting at which he recommended development of a 4,000 shp axial-flow gas turbine engine for locomotive, industrial, or marine use.[28] The minutes of the meeting indicated that GE had previously conducted studies on aircraft gas turbines at the request of Wright Field, and also had made a mathematical analysis of jet propulsion of the type proposed by Frank Whittle and of an aircraft gas turbine with a propeller.

In addition, several external events took place early in 1941 that pushed GE into the gas turbine business. In January 1941, a committee of the National Academy of Sciences had recommended to the Navy the purchase and testing of a gas turbine for marine propulsion.[29] Shortly thereafter, GE's Schenectady division took a Navy contract for the development of a gas turbine power plant for PT boats that included a four-stage axial-flow compressor, a combustion chamber, and a single-stage turbine.

In February 1941, as a result of intelligence about new German power plant developments, General H. H. Arnold, Chief of the Air Corps, asked Vannevar Bush, Chairman of both the National Defense Research Committee (NDRC) and the NACA, to form a committee to study jet propulsion.[30] Subsequently, W. F. Durand was chosen to head the Special Committee on Jet Propulsion. Industrial representatives on the committee were GE's Schenectady Steam Turbine Division, Westinghouse, and Allis-Chalmers, all chosen for their turbine manufacturing experience and because they, unlike conventional engine manufacturers, presumably would not be opposed to unorthodox power plant developments. The committee decided that each of the manufacturers should proceed with a

27. Frank Whittle, an officer of the Royal Air Force and also a principal of Power Jets Ltd. which sponsored his engine, tested his first turbojet engine on April 12, 1937.

28. General Electric, "Minutes of the Meeting of Engineering Council, Schenectady, February 24, 1941," unpublished photocopy, National Air and Space Museum Archives, Washington, D.C.

29. The Committee on Gas Turbines Appointed by the National Academy of Sciences, *An Investigation of the Possibilities of the Gas Turbine for Marine Propulsion* (Washington, D.C.: United States Navy Department Bureau of Ships, January 1941), 2.

30. Virginia P. Dawson, *Engines and Innovation: Lewis Laboratory and American Propulsion Technology* (NASA SP-4306) (Washington, D.C.: NASA, 1991), 46.

design of its own. On July 7, 1941, under the direction of Alan Howard, GE started design in Schenectady of an axial-flow turboprop engine, the TG-100 (T31). In late 1941, GE received an order from the Army for this engine.[31]

The event that most quickly accelerated GE into turbine engine development was the transfer of Frank Whittle's turbojet technology. In the spring of 1941, General Arnold had visited England, learned about the Whittle turbojet engine, and witnessed taxiing tests and short flights of the Gloster aircraft that it powered.[32] Realizing the potential of the engine, Arnold set in motion the transfer of that engine technology to the U.S. GE was selected to develop an engine based on the Whittle W.1.X.[33]

On September 4, 1941, GE received a briefing and reports on the Whittle turbojet engine; on October 4, it received the Whittle-designed Power Jets W.1.X experimental engine and production drawings.[34] The project was placed under the direction of Donald F. "Truly" Warner at GE's Lynn plant. With authorization from the Army, GE made mechanical design changes, redrew the engine completely to conform to American and company production standards, made original designs for missing components, and integrated into the new engine the technology gained from its turbosupercharger experience. The new engine, known as the GE Type I, was first tested on March 18, 1942 and ran successfully on April 18. On October 1, 1942, the Bell XP-59A, powered by two GE I-A engines, flew for the first time, becoming America's first jet aircraft.[35]

Aircraft turbine engine work at GE's Lynn and Schenectady divisions proceeded on separate tracks although there was some interchange of component technology.[36] The focus at Schenectady was on the axial flow TG-100 turbo-

31. Alan Howard and C. J. Walker, "An Aircraft Gas Turbine for Propeller Drive," *Mechanical Engineering* (October 1947): 827.

32. H. H. Arnold, *Global Mission* (New York: Harper and Brothers, 1949), 242.

33. Historical Office, Air Technical Service Command, Wright Field, "Case History of Whittle Engine" (Secret Report), October 1944, Historical Study No. 93, Office of History, Headquarters Air Force Logistics Command, Wright-Patterson Air Force Base, Dayton, Ohio.

34. The production drawings that were received were incomplete in some important respects, particularly the lack of an automatic control system. The controls for the Type I-A were based on patents of D. F. Warner.

35. David M. Carpenter, *Flame Powered* (Lynn, Mass.: Jet Pioneers of America, 1992), 33. On this day, test pilot Robert M. Stanley flew the XP-59A about 25' off the ground. On the following day, Stanley made the second flight to altitude.

36. For example, A. J. Nerad of GE's Schenectady Research Laboratory designed the TG-100 combustor, and the I-14 engine utilized this new combustion liner.

prop engine, first ground tested on May 15, 1943, without a propeller, and on May 15, 1945, with a propeller. On December 21, 1945, the TG-100 made its initial flight in a Consolidated Vultee XP-81, becoming the first turboprop to be flown in the U.S.

At Lynn, the I-14 and I-16 (J31) centrifugal-flow engines, with higher power ratings than the I-A, were developed. In January 1943, the Army met with GE to discuss the possibility of a 3,000–4,000-lb-thrust turbojet engine, and in March asked GE to submit a proposal for such an engine. Subsequently, GE proposed the I-40 (J33), a centrifugal-flow engine that could be produced in a short time, as well as the axial-flow TG-180 (J35), which had the promise of greater performance but required a longer development time. The TG-180 project was undertaken by Schenectady and the I-40 by Lynn. First off the block, the I-40 was tested at Lynn on January 9, 1944, and went into service in 1945 in America's first operational jet fighter, the Lockheed P-80A. The TG-180 first ran on April 23, 1944, and was flight tested in February 1946 in a Republic XP-84.

However, the supercharger division at Lynn would soon become the center of GE's turbine engine activities, while the steam turbine division at Schenectady would phase out. Both divisions made significant contributions to the development of the gas turbine engine, but those that were derived from GE's long history in the development of steam turbine engines were generally lesser known. For example, major and fundamental discoveries were made in connection with resonant wheel vibration and the vibration of long turbine buckets.[37] Solutions to these vibration problems were published by ASME and were used in the design of turbojet engines by GE and other companies during and after World War II. Another invention patented by Glenn Warren in the 1920s was the variable reaction-impulse design for long bucket stages. This invention resulted in greater efficiency for the long turbine blades that characterized early GE turbine engines. Steam turbine, axial-flow compressor development work, which was undertaken in Schenectady, was integrated into the first GE axial-flow gas turbine engines.

THE B-1 EXPENDABLE TURBOJET

During the course of this work at Lynn and Schenectady, GE also developed a little-known, small, expendable turbojet, designated the B-1, for use in a short-range cruise missile. The B-1 was never part of GE's mainstream gas turbine activities, and it never successfully powered the missile. However, it was an inter-

37. Warren to Strang, 5 January 1965.

esting engineering initiative that used existing turbosupercharger technology to develop an essentially "off-the-shelf" small gas turbine engine.

In late 1943, Eugene Stoeckley, the engineer in charge of GE's Turbosupercharger Development Test Facilities at Lynn announced, "We have been able to make a Type B turbosupercharger operate in 'bootstrap'."[38] Bootstrap was a term used to describe the test condition in which a turbosupercharger was able essentially to operate itself. In this demonstration, the discharge air from the supercharger's compressor was fed into a combustor to produce the hot gas needed to drive the turbine. Sufficient energy was developed by the turbine to drive the compressor. This engineering development was significant for two reasons: it provided a technique that could greatly reduce the production testing cost of turbosupercharger; and it duplicated what happened in a turbojet engine, leading to modification of the Type B-31 turbosupercharger to make the B-1 turbojet engine.

As a reaction to the terrifying experiences of the British with the German V-1 buzz bomb, by late 1943, the Army Air Forces began development of a similar type of weapon. The Army selected Northrop to design and develop a flying wing "Power Bomb," designated the JB-1, as a ground-launched missile that would carry two 2,000-lb demolition bombs. It was to have a cruising speed of approximately 400 mph and a pre-programmed guidance system that would guide the missile with reasonable accuracy to a target approximately 200 miles away.

The Power Bomb was to be powered by two B-1 turbojets, each producing 400 lb thrust. This engine, stemming from the "Operation Bootstrap" demonstration, was conceived by Russ Hall, an engineer at the Lynn plant. The B-1 was a converted Type B-31 turbosupercharger of the kind used in the Boeing B-29 Superfortress.[39] Development began in the spring of 1944; conversion required a complete aerodynamic redesign of the turbine rotor and stator blades and the development of a combustor.

A piloted-glider test model of Northrop's JB-1 was flown in 1944 at the Army Air Forces flight test facility located at Muroc, Cal., to check out the control system. On December 7, 1944, a ground-launched version of the missile, the JB-1A, powered by two B-1 engines, made its initial test flight at the Eglin test range in Florida. The JB-1A climbed rapidly following launch, stalled, and then crashed 400 yards from the launch point. The JB-1 program encountered a num-

---

38. William Travers, Marietta, Georgia, to Bernard Maggin, letter, 25 July 1988, National Air and Space Museum Archives, Washington, D.C, 1.

39. Travers, *The General Electric Aircraft Engine Story*, I-3 - I-4.

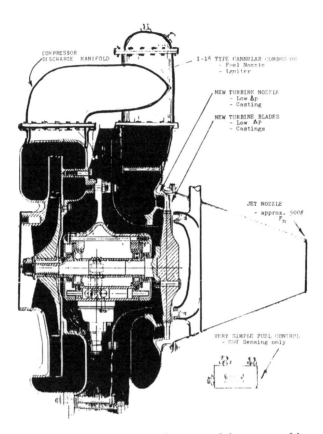

GE Type B-1 turbojet engine that powered the unsuccessful
Northrop JB-1A "Power Bomb." Drawing courtesy of William Travers.

ber of difficulties, among which was the fact that the B-1 engine produced only
half of the expected 400 lb thrust, and difficulty was encountered in obtaining a
sufficient number of test engines. Consequently, the B-1 engine was dropped
from the program and Northrop fitted a pulse jet to the JB-1 airframe.[40]

New Directions after the War

At the end of the war, GE made a number of organizational changes with
respect to its turbine engine business. On July 31, 1945, the Aircraft Gas Tur-

40. Kenneth P. Werrell, *The Evolution of the Cruise Missile* (Maxwell Air Force Base, Alabama:
Air University Press, 1985), 68–69.

bine Division was created at Lynn, an organizational change which included the turbosupercharger business. Subsequently, engineering responsibility for both of Schenectady's turbine engine projects, the TG-100 and the J35, was transferred to Lynn. The Steam Turbine Division at Schenectady henceforth would concentrate on its traditional business of industrial and marine turbines.

The end of the war almost ended GE's engine production program. To meet wartime needs, the production of J33 and J35 engines had been licensed to divisions of General Motors. In 1944, a Navy Defense Plant Corporation facility operated by GE in Syracuse, N.Y., had been selected for the additional production of Type I-16 and J33 engines. However, GE was not able to meet the production schedule, a problem that handicapped the Lockheed P-80 program. The Allison Division of General Motors was selected by the Army Air Forces in June 1944 for additional licensed production of the J33.[41] In 1945, another General Motors Division, Chevrolet, in Tonawanda, N.Y., was selected for primary production of the J35.

With the end of the war in August 1945, the Army Air Forces re-evaluated its jet engine production program. For economic and other reasons, the Army Air Forces turned over all postwar production of the J33 to Allison in September 1945.[42] Subsequently, GE's Syracuse plant was closed. In January 1946, the Army Air Forces decided to have General Motors continue production of 1,200 J35s and specified that GE manufacture only 100 J35 engines.[43] In 1947, J35 production was turned over to Allison when Chevrolet began postwar automobile production. Thus, GE lost an important production opportunity for two

41. Historical Office, Wright Field, "Summary of Case History of Turbo-Jet Engine J33 (I-40) Series" (Former Secret Report), Historical Study 95, Office of History, Headquarters Air Force Logistical Command, Wright Patterson Air Force Base, 4.

42. "Summary of Case History of Turbo-Jet Engine J33 (I-40) Series," 4-5. The Army Air Forces also selected Allison for postwar I-40 production because it considered Allison's production performance on the I-40 engine more satisfactory than that of GE, Allison's quote was lower than GE's quote, and because the Army Air Forces considered that it would be in the best interest of the Government to insure participation in the postwar aircraft program by old-line engine manufacturers such as Allison.

43. Historical Office, Air Materiel Command, Wright Field, "Case History of Turbo-Jet Engine J35 (TG-180) Series," August 1947 (Former Secret Report), Historical Study No. 96, Office of History, Headquarters Air Force Logistics Command, Wright-Patterson Air Force Base, Dayton, Ohio, 15-16. The reasons for this decision included the fact that General Motors offered a price reduction on a significant portion of the engines to be produced, the cost of canceling the contract with GM would be high, and GE did not have adequate production facilities to produce the quantity of engines required within the deadline.

engines that it had developed, a major blow that created uncertainty about the company's future role, if any, in the engine business.

In this period shortly after the war, some in GE wanted to remain in the engine business, and some wanted to get out of it. Aircraft jet engines were perceived as a high-risk investment, not part of GE's traditional product base. On the other hand, engineers who had designed and developed these successful engines were enthusiastic and wanted to remain in this promising new field. They were extremely discouraged when GE lost the J33 and J35 production contracts, and were concerned about their jobs.[44] Charles E. Wilson, President of General Electric, decided to play hard ball. Reacting to the January 1946 Army Air Forces decision that allowed General Motors to continue production of almost all J35 engines, Wilson indicated to the Assistant Chief of Air Staff-4, Washington, that his company would not remain in the J35 program with a contract for only 100 engines—or in the jet engine development program—unless some production was included.[45]

Between January and March 1946, turbine engine engineers at the newly created Aircraft Gas Turbine Division at Lynn, determined to stay in the business, had put together a postwar plan that included new engine development, test facilities, and production capability. To compensate for the loss of the J33 and J35 production run, a major element of the plan was the development of a new engine, more powerful and fuel efficient than the J35, but within the same frame size.[46] For them, the new engine was a call to action to stay in the engine business. GE proposed this engine to the Army Air Forces as a J35 "product improvement," if it could receive a production commitment.[47]

For their part, Army Air Forces authorities responsible for the jet engine program had become apprehensive about possible curtailment of GE's development work after the war ended, when it was realized that engine production cutbacks threatened GE. Substantial portions of GE's jet engine development programs had been borne by urgent, wartime production contracts, and because development expenses were not production-related, it would be difficult for GE to obtain reimbursement. This situation had been aggravated by the decision to give General Motors most of the J35 production contract and a decision early in

---

44. William R. Travers, Marietta, Georgia, to Rick Leyes, letter, 24 March 1992, National Air and Space Museum Archives, Washington, D.C.

45. "Case History of Turbo-Jet Engine J35 (TG-180) Series," 16.

46. *Eight Decades of Progress*, 61.

47. Jacob Neufeld, *Air Force Jet Engine Development: A Brief History* (Washington, D.C.: Office of Air Force History, 1990), 10.

1946 by the Army Air Forces not to provide GE with money for J35 development facilities. The Army Air Forces saw its future as a leader in jet aircraft
threatened. GE had done so much of the pioneer work in jet engines that without its continued research, development in the U.S. would receive a decided
setback.[48]

The Army Air Forces were also particularly interested in having the TG-181
(GE's designation for the improved TG-180, later changed to TG-190) developed. GE allowed that it would be willing to invest in R&D facilities at its own
expense, if it received a production order large enough to justify this cost. This
offer, the unexpected availability of fiscal year 1946 funds, and the fact that the
postwar military aircraft fleet would depend on promising new turbine engines
such as the TG-190 clearly influenced an Army Air Forces decision to procure
production of the new GE engine.

On May 24, 1946, a contract was prepared to cover the procurement of the
TG-190. Although only a portion of the money was obligated at the time, the
contract covered 25 million dollars (the same amount as the remaining fiscal
year funds) of engine procurement. This was a deviation from normal engine
procurement procedures, which required a contract for a few experimental test
engines, a satisfactory completion of the 150-hour model test, and then orders
for production engines. The TG-190 (later J47) engine as well as GE's continued participation in jet engine development work was considered sufficiently
important by the Army Air Forces to bend the rules.

For GE, and for the Lynn Division, the significance of this contract was that
the J47 engine program established the company as a permanent engine manufacturer. The Army Air Forces had gained the advantage of having two capable
companies (GE and Allison) develop and produce the most advanced models of
American-designed jet engines.[49]

The J47 engine itself set a design standard for the use of all-axial-flow compressors in GE's engines for many years. A disagreement had existed among GE
engineers between those supporting centrifugal-flow and those supporting axial-flow designs, with strong arguments on each side, and GE had designed both.
Because the axial-flow engine offered higher pressure ratios and thus more

48. Ruth E. Vorce, "Case History of the J-47 Turbojet Engine" (Former Secret Report), July
    1950, Historical Study No. 97, Office of History, Headquarters Air Force Logistics
    Command, Wright-Patterson Air Force Base, Dayton, Ohio, 1.

49. Ethel M. DeHaven, "Case History of the J47 Turbojet Engine January 1950–April 1953"
    (Former Secret Report), September 1953, Historical Study No. 98, Office of History,
    Headquarters Air Force Logistics Command, Wright-Patterson Air Force Base, Dayton,
    Ohio, 8.

power, better performance, and had other important advantages, GE dropped the all-centrifugal-flow engine design.

The J47 was first tested on June 21, 1947, and went into production at Lynn in 1948. The engine had been selected to power numerous military aircraft, and it became clear that orders would exceed Lynn's production capability. An Air Force-owned facility in Lockland, (renamed Evendale in 1953), Oh., which had produced Wright Aeronautical piston engines, was recommended to GE by the Air Force as an additional site. It was conveniently located near a large city (Cincinnati) and the Air Force Power Plant Laboratory at Wright Field in Dayton. By the summer of 1948, the Air Force had made available money for the reactivation of the plant. On February 28, 1949, the new GE plant shipped its first production J47 engine.

Between 1949 and 1950, GE moved its Aircraft Gas Turbine Division to Lockland. Most of GE's aircraft engine engineers went, too, but a small group remained in Lynn. With the outbreak of the Korean War in June 1950, J47 production, cut back as a result of reduced defense spending, resumed and increased. GE expanded its operation at Lockland, investing in a new engineering and production center north of the complex, purchasing privately-owned buildings on the site, and planning for the eventual purchase of the leased government property at Lockland. J47 production continued at Lynn as well, and Packard and Studebaker were licensed for production. When production ended in 1956, more than 35,000 J47s had been manufactured, making it the most mass-produced turbojet engine in history.[50]

The Start of the Small Aircraft Engine Department

The decision by GE to enter the small gas turbine aircraft engine business was shaped by several converging influences: corporate management's policy of decentralizing the decision making process; a committee study that showed a potential market for small gas turbine engines; the easing of J47 production with the end of the Korean War; winning a Navy contract for a helicopter engine; and the reluctance of some Lynn engineers to move to Evendale.

Ralph J. Cordiner, elected GE President in 1950, established a new organizational structure in February 1951. GE's volume of business had grown rapidly, and the corporation's highly centralized decision making process was no longer suitable for its size. Cordiner's plan was to put the responsibility and authority

---

50. Robert B. Meyer, Jr., ed., "Classic Turbine Engines: J47 Launches General Electric as a Major Jet Engine Manufacturer," *Casting About 2* (1986): 12-15.

for decision making closer to the scene of the problem.[51] Individual operating departments organized around a specific product or market area were created and given the responsibility, authority, and accountability for their niches. Each department had profit and loss, pricing, and budget planning responsibilities.

As Vice President and General Manager of GE's Aircraft Gas Turbine Division, Cramer W. "Jim" LaPierre had the responsibility to carry out Cordiner's objective by determining if GE's engine business should remain as a single business department or be further decentralized. In the spring of 1953, LaPierre assigned Jack Parker to chair a committee to look into this. Parker, who retired as GE Vice Chairman and member of the Board of Directors, recalled:

> The purpose of the exercise was to determine where we could best put our engineering and manufacturing talent to work in the engine field. One of the parts of the field that, up to that point, had been really untouched was the small gas turbine business . . . We thought that there was a big void there. We thought that gas turbines of lower power could be very profitably and efficiently employed in a lot of aircraft as well as for other uses, particularly in the military, but also branching out into civilian use.[52]

LaPierre brought in executive Marion Kellogg to devise a new organizational structure. Kellogg proposed five departments: research and development, jet engines, manufacturing, small engines, and aircraft accessory turbines. With the end of the Korean War in July 1953 and the subsequent easing of J47 production, LaPierre was in a position to make organizational changes consistent with GE's decentralization policy.[53] LaPierre established the Small Aircraft Engine Department in Lynn on October 1, 1953. The charter limited the new department to engines of 5,000 lb take-off thrust or 5,000 shp. LaPierre selected Parker to head it up.

As Cordiner's decentralization policy evolved in 1952, the Lynn Division was manufacturing J47 jet engines and small accessory components for jet engines. Encouraged by the new decentralization policy, the engineers and managers remaining at Lynn had an interest in developing new products, including engines.

Exploring new opportunities, the Lynn division soon received a request to bid on a 600-shp turbine engine. The Lynn division responded with a proposal

51. Ralph J. Cordiner, *New Frontiers for Professional Managers* (New York: McGraw Hill Book Co., 1956), 47-48.

52. Jack Parker, interview by Rick Leyes, 3 December 1991, GE Interviews, transcript, National Air and Space Museum Archives, Washington, D.C.

53. "Getting Along on a Lean Diet," *Business Week*, No. 1289 (May 15, 1954): 132-134.

to the Air Force on May 5, 1952. GE lost this bid to Lycoming, which went on to develop its famous T53 engine. However, the Lynn engineers soon found another opportunity.

As a result of efforts by Marketing Manager H. M. "Hank" Wales, GE was allowed to bid on a small turboshaft engine for the Navy. Supported by Accessory Turbine Manager John Turner and Engineering Manager Bill Meckley, Project Engineering Manager G. W. "Bill" Lawson and engineer Len Heurlin headed a team of draftsmen and put together a proposal. The Navy wanted an 800-shp engine with an sfc of 0.70 lb/hp/hr and weight of 0.5 lb/hp. On December 31, 1952, GE submitted its design, an axial-flow, free turbine turboshaft engine weighing 400 lb. The compressor was based on a scaled-down J47 engine compressor.

Competing against Lycoming, Continental, and Fairchild, GE was awarded the contract for the XT58 in July 1953. With the decentralization of the Aircraft Gas Turbine Division and the formation of the Small Aircraft Engine Department (SAED) on October 15, 1953, the XT58 became the department's project and first product. To meet the criteria for a department, SAED's income would be derived from J47 sales while the XT58 project was germinating. GE was in the small engine business.

The T58 Turboshaft Engine

At the time the department was formed, Jack Parker was appointed General Manager of the SAED. Parker's Engineering Manager was Ed Woll. Harold T. Hokanson was selected as the first T58 Project Manager. The original T58 design team was directed by Hokanson's Design Engineering Manager, Bill Lawson. Among those in Lawson's engineering subsection were Floyd W. Heglund supervising engine design and D. C. "Dave" Prince supervising thermodynamic design. Weight and material control were under O. A. DesMaisons and evaluation engineering was supervised by H. J. Baier. The Engine Control Manager was John W. Jacobson who reported to Hokanson.

As the design study on the XT58 continued in the fall of 1953, a helicopter mission analysis, and later, a market study were made to determine the type and size of engines that would be required in five years.[54] The mission analysis revealed that an 800-shp engine was not powerful enough to accomplish some military missions. Also, several design reviews conducted by Evendale's Devel-

54. James S. Siegal, "T58 Project History," an unpublished GE draft history, 28 March 1958, National Air and Space Museum Archives, Washington, D.C., I-9 - I-10.

Early GE T58 turboshaft engine and some of the program participants. From left:
L. Asher, L. Callahan, J. Turner, J. Parker, E. Woll, and H. Hokanson. Photograph
courtesy of GE.

opment Department showed that engine weight could be reduced and power
increased by incorporating less conservative, state-of-the-art, large turbine engine technology. In part, the weight reduction drive was motivated by overweight problems with other GE engines, such as the J73 and XJ53. From a market viewpoint, a higher power rating was more desirable in order to exceed that
of the Lycoming T53 by a sufficient margin. In February 1954, the XT58 design
team started over with a design goal of a 1,050 shp engine weighing 250 lb
without gearbox.

The prototype engine configuration, finalized in March 1954, had an eight-
stage axial-flow compressor, annular combustor, two-stage gas generator turbine, and single-stage free power turbine driving a reduction gear at the rear of
the engine. The first three compressor stator stages were variable to provide increased stall margin during starting. In essence, the engine was an opposed-shaft,
rear-drive design. The gas generator consisted of the compressor, combustor,
and two-stage turbine. Downstream, the single-stage power turbine drove the
reduction gearing. There was no mechanical connection between the gas generator and power turbine.

The weight and sfc of the engine were selected to meet the typical mission
requirements for a light cargo helicopter with a 200-nautical-mile range. GE's
objective for the XT58 was to replace the piston engine in the helicopter. According to Ed Woll, who at the time was SAED's Engineering Manager:

> Our challenge was to push the piston engine out of helicopters. Obsolete
> them. If we didn't obsolete them, we didn't have a market. It's fundamental

that piston engine fuel consumption was better than a gas turbine. But on the other hand, the piston engine was heavier, and you've got some limitations where you can mount the engine. So we felt that we had some definite advantages with the gas turbines for helicopters.[55]

The design rationale for the compressor was driven by weight and sfc. The need for good fuel economy resulted in the selection of an axial-flow compressor with moderately high pressure ratio.[56] The compressor was an extremely lightweight, transonic design with an 8.3:1 total pressure ratio rise in eight stages—highly loaded for the time. Fewer compressor stages meant less weight and cost, and axial-flow compressors with higher stage loadings had been demonstrated by NACA.[57] Very thin rotor disks were fastened together by explosive rivets.

The annular combustor was the first to be installed in any GE engine. The small size of the annular combustor made it possible for the first time to test an entire combustor. For earlier, larger jet engines, a limited airflow test capacity dictated single can tests and consequently a can-type combustion system. The annular combustor would eliminate six or seven very small diameter (3 in) combustion cans as well as the operational and manufacturing problems associated with them. The annular combustor promised simplicity and longer life, more nearly uniform distribution of discharge temperatures, less hot metal surface to be cooled, and lighter and stronger structural elements. The responsibility for the design of the combustor was assigned to the Engineering Section's Combustor Unit under the supervision of J. A. "John" Benson.

Large jet engine experience was adapted to the T58 gas generator turbine. A lightweight, short-chord turbine was used. To maintain basic reliability and long life, a moderate turbine inlet temperature was chosen.[58] The decision to use a rear-end drive was made to avoid two unknown areas: the difficult design of the long, slender, high-speed shaft that would be necessary to connect the power turbine in the rear of the engine with the output drive at the front; and the difficult design of the bearings of the gas generator, which would have to be enlarged to

55. Ed Woll, interview by Rick Leyes and William Fleming, 20 August 1991, GE Interviews, transcript, National Air and Space Museum Archives, Washington, D.C.

56. H. T. Hokanson, "The T58 - A New Heritage for Helicopter Powerplants," (AHS Paper No. A-56-12-03-7000 presented at the American Helicopter Society Annual Forum 1956 Proceedings), 28.

57. Hokanson, "The T58," 29. Also, Siegal, "T58 Project History," II-5.

58. Robert H. Cushman, "GE Bids for Helicopter Market with T58," *Aviation Week & Space Technology*, Vol. 64, No. 23 (June 4, 1956): 60.

unprecedented diameter and peripheral speed in order to accommodate the forward drive shaft penetrating the center of the engine. The advantages of the free turbine configuration included better part-load fuel economies and a power output that could be supplied at a speed independent of the power level needed (i.e., for a given power setting, a wide range of rotor speeds was possible).[59]

The decision to start over on the XT58 in order to increase its power to 1,050 shp and reduce its weight to 250 lb left SAED with the uncomfortable task of proposing the now more costly new design to the Navy. The Navy was approached on March 8, 1954. While startled by the proposal, the Navy agreed that the mission analysis work was good and that a power increase was justified, but critically questioned the advanced design.[60] Following this meeting, a series of new proposals, discussions, and negotiations ensued. In June 1954, the Navy gave preliminary approval to a 1,050-shp engine.

On March 11, 1955, the prototype XT58 engine went to test but would not self sustain when the electric starter was disengaged. By additional tuning and by removing the power turbine, the engine did run as a gas generator on April 5, but performance was low. The aerodynamic performance of two prototype engines was poor and component test compressor data indicated that only a major redesign could improve performance.[61] Another problem was power turbine bucket failures due to vibration. Dampers were developed which shifted the bucket frequency and solved the problem.[62]

The most challenging problems were the compressor performance and mechanical design deficiencies. The design concept of using high-shear explosive rivets for the compressor rotor stage assembly proved inadequate for maintaining mechanical stability throughout the operating range of the engine and resulted in unacceptable levels of vibration. First-stage compressor rotor blade vibratory stress fatigue failures also required a solution.

Various iterations of the compressor failed to solve the aero-mechanical problems completely. The original engine had gone to test without the benefit of compressor component testing because of a tight development schedule and budget constraints. The original compressor had been built without interstage instrumentation due to the compact nature of the compressor, the relatively

59. "Small Gas Turbines for Helicopters," *Aviation Age*, Vol. 23, No. 2 (February 1955): 51.

60. Siegal, "T58 Project History," I-12.

61. Siegal, "T58 Project History," I-19.

62. F. W. Heglund, "Design Analysis of the General Electric T58 Engine," (GE Paper No. GEA-6816 presented at the ASME, First International Gas Turbine Symposium, Washington, D.C., March 1958), 6-7.

large size of the instrumentation available, and a lack of high-speed slip ring capability. An interim research compressor, which was "stretched out" for instrumentation, confirmed that the required airflow and pressure ratio were not within the performance capabilities of the aerodynamic design.

A new compressor design was begun in June 1955 that included significant aerodynamic and mechanical changes. To improve compressor performance, stage loading was reduced and two compressor stages were added. The resulting 10 stages produced the required 8.3:1 pressure ratio. Variable inlet guide vanes (IGVs) were installed to provide adequate stall margin throughout the operating speed range and new airfoil designs were used for the aft compressor stages. The major mechanical design change was the introduction of a "one-piece smooth-spool compressor rotor" for the aft eight stages.[63] This new design used circumferential dovetails for fastening the airfoils to the rotor, and most significantly for the engine's performance, facilitated the elimination of shrouds on the stator vanes. The one-piece smooth-spool rotor provided exceptional mechanical stability to the rotor throughout the operating range and outstanding control of the blade tip clearances. Although the weight of the rotor was more than that of the unacceptable riveted rotor, it was lighter than other state-of-the-art concepts. The new design was called the C-8 compressor.

Under the general direction of Ed Woll, Manager of Engineering, and Hal Hokanson, T58 Project Manager, major contributions were made by Frank Lenherr to the compressor aerodynamics, Don Berkey to the preliminary design, and Art Adinolfi to the mechanical design with his design team, including Ed Movsesian and Tom Hull. Design drafting was led by Lou Lechthaler and manufacturing process design by Ted Ferrer. Bill Lawson, Floyd Heglund, Frank Pickering, Gerry Sonder, and Arnie Brooks also made significant contributions to the effort.

In January 1956, testing began on the XT58 with the C-8 compressor installed, and on February 2, the guaranteed military rating of 1,050 shp was surpassed. This was a critical turning point for the XT58 development program, which had been on the verge of cancellation by the Navy for lack of a successful compressor. The weight-saving compressor design had been stretched too far. Too much performance was expected from too few stages, and the lightweight compressor was unstable at high rotor speeds. The redesigned smooth-spool compressor became a key GE technology, the basis for the GE T64 and TF34 engine compressors. The smooth-spool drum rotor was also incorporated

---

63. The smooth spool has been given various names over time, including solid drum, solid spool, monolithic rotor, spool rotor, drum rotor, integral spool, etc.

in the GE1 demonstrator engine in the company's large engine activity, and became a standard feature in all subsequent GE aircraft engines.

T58 GROUND, FLIGHT, AND FURTHER DEVELOPMENT TESTING

Evaluation of potential flight test vehicles for the T58 had begun in 1954. A survey of available helicopters indicated that only two current models were able to absorb full power of two turbine engines and yet be capable of flight on a single engine. These were the Sikorsky S-58 and Piasecki (later Vertol) H-21. While either airframe would have been satisfactory as a flight test vehicle, consideration of Navy plans and possible future business made the S-58 installation more desirable.[64] In October 1955, Sikorsky received a contract from BuAer for modification of two S-58 airframes for T58 engines.

A rotor test stand was built at the GE Flight Test Center in Schenectady and tests began in April 1956. The stand was equipped with a Sikorsky S-58 rotor and transmission system to observe engine and control system response to the rotor. Among other things, these tests refined and demonstrated the stability of the control when subject to throttle bursts. They also demonstrated that there was a minimum power differential (i.e. split in the power load) between engines under increasing load in a dual-engine installation.[65]

After a number of problems, the XT58 engine passed its 50-hour PFRT on August 4, 1956. On January 30, 1957, the first hovering flight took place at the Sikorsky plant in Bridgeport, Connecticut in a Sikorsky HSS-1F (S-58) modified to take two T58 engines. In March 1957, preliminary anti-icing, altitude-performance, and cold-weather-starting tests began at an Air Force cold-weather site atop Mount Washington in New Hampshire. In September 1957, the first flight of a Vertol H-21D, powered by two T58 engines, occurred.

Several noteworthy development problems occurred during the 150-hour Model Qualification Test (MQT) program. After extended testing, the rear stages in several compressors began to exhibit a fatigue condition in the dovetail, causing at least one blade failure.[66] The problem was traced to a torsional vibration of the blade that resulted in fatigue cracks at the neck of the dovetail blade

---

64. Siegal, "T58 Project History," I-22.

65. N. N. Davis, "Test Experience with the T58 Engine," (GE Paper No. GEA-6666 (4M) 6-57 presented at the American Helicopter Society Forum, Washington, D.C., 9 May 1957), 4-5.

66. The blades in the smooth-spool rotor had circumferential dovetails. They were assembled through a loading slot and slid around the rotor circumferentially.

attachment. The rotor blade dovetails strengthened, the vibration characteristics were altered and the problem solved.[67]

As a result of inadequate (non-uniform) temperature distribution, early combustor liners showed a gradual deterioration resulting in collapsed structure and hot spots.[68] The solution was found in an annular flow splitter ring that fit between the compressor diffuser and the combustor inner liner; this insured a positive distribution of air over the outer and inner liner. Other combustor changes were made as well.

The helicopter engine control presented a complex design problem. At the time, helicopter pilots had a high workload (constant retrimming of the controls was necessary) because rotor efficiency was highest at a single speed, and power required to hold this rotor speed varied with even slight changes in flight conditions. A constant-speed rotor and engine control requiring minimum pilot effort was highly desirable. The design problem was compounded by an "elastic" rotor system that made it difficult for the engine control to sense the actual speed and load of the rotor blades.[69] Finally, the fact that the T58 was a free turbine engine, which permitted the selection of a power turbine speed independent of the gas generator speed, complicated the control problem.

Hamilton-Standard and GE jointly developed an engine control under the direction of John Jacobson, SAED Controls Engineering Manager. Computer studies and the rotor test stand tests helped to reduce the problems. The controls first shipped with flight test engines had insufficient stability margin and did not permit consistent automatic engine starting.[70] These controls were improved substantially as a result of accelerated endurance cyclic bench testing, and the T58-GE-2 passed the MQT in September 1957.

PINCH AND ROLL MANUFACTURING DEVELOPMENT

Paralleling development testing of the T58 was the development of appropriate manufacturing methods for small gas turbine engines. The change from manufacturing large engines to small engines required the development of new pro-

67. N. N. Davis, "T58 in Flight," (ASME Paper No. 57-A-290 presented at the ASME Annual Meeting, New York, N.Y., 1-6 December, 1957), 2-3. This paper was also titled "T58 Testing in Factory, Field and Flight."

68. N. N. Davis, "T58 in Flight," 3-4.

69. Factors contributing to this elasticity included rotor blade flexibility and rotor blade lag hinge attachments, rotor transmission and speed reduction gear backlash, and drive shaft twist.

70. N. N. Davis, "T58 in Flight," 4.

cesses because small engine parts, such as compressor and turbine blades, had to be made with much greater precision than those of conventional large jet engines. The challenge was to hold the tolerances for correct airfoil shape and clearance control to maintain efficiency. At the time, skeptics believed strongly that efficiency of a small engine would be low and manufacturing costs high.

In many instances, J47 manufacturing methods could not be used because T58 parts were too small. For example, J47 compressor blades could be precision forged. However, the very thin T58 blades had extremely close tolerances; die forging yielded poor dimensional control, high rejection rates, and high die wear.[71]

Searching for an alternative, GE engineers first turned to Cincinnati Hydrotels and Dekel die sinkers to produce machined compressor blades, which then required skilled hand finishing. While this was an improvement, the cost was high and the process far from practical for the production quantities required. The solution was found in a "pinch and roll" process put into operation first in Ludlow then Rutland, Vermont. In this process, a series of dies first pinched a slug of material at the root of the blade and then rolled from the root to the tip to get the airfoil. William Paille, Gene Belli, and Frank Flowler of Lynn's manufacturing engineering group were credited with inventing this patented GE process.[72]

NEW AIRFRAME APPLICATIONS

With the success of the HSS-1F testbed helicopter in hand, Sikorsky petitioned the Navy for a contract to design a new airframe around the T58. With the improved performance potential of a gas turbine-powered helicopter, it made sense to design a new helicopter around the power plant rather than go into production with a conversion of an old piston engine airframe. In August 1957, Sikorsky received a contract from the Navy for the HSS-2 (commercial designation S-61) anti-submarine helicopter. The first flight of the prototype was on March 11, 1959, powered by two 1,050-shp T58-GE-6 turboshaft engines. The HSS-2 (redesignated SH-3 in 1962) Sea King went into Navy service in 1961. The SH-3A broke a number of speed records shortly after its service introduction—it was the first helicopter to break the 200-mph barrier.[73] An Air Force

71. L. W. Waitt, "Manufacturing Small Engines," (ASME Paper No. 57-F-20 for presentation at a joint session of the Gas Turbine Power and Aviation Divisions at the ASME Fall Meeting, Hartford, Conn., 23-25 September 1957), 2.

72. "Roll Forming Gets an Assist," *Steel*, Vol. 142, No. 22 (June 2, 1958): 70.

73. Norman Polmar and Floyd D. Kennedy, Jr., *Military Helicopters of the World: Military Rotary-Wing Aircraft Since 1917* (Annapolis: Naval Institute Press, 1981), 300.

variant, known as the HH-3E Jolly Green Giant (powered by 1,400-shp T58-GE-5 engines) was used extensively during the Vietnam War to rescue downed American flyers deep within North Vietnam.

In 1956, the Navy had a requirement for a long-range utility helicopter capable of operating from small ships. Competitors included Sikorsky and Kaman Aircraft. The Kaman HU2K-1 Seasprite (civilian designation K-20) was selected making it the second T58 application. The first flight of the HU2K-1 was July 2, 1959, powered by a single 1,050-shp T58-GE-6 engine. Starting in 1967, the HU2K-1 helicopters were converted to twin T58 engines. In the early 1970s, LAMPS I (Light Airborne Multi-Purpose System) program, the HU2K-1 helicopters were reconfigured for anti-submarine warfare and redesignated SH-2. Having lost the Navy utility helicopter competition, Sikorsky proceeded with a commercial version of its entry known as the S-62. The first flight of this single-engine T58-GE-6 helicopter was on May 14, 1958. In 1962, the Coast Guard ordered these helicopters for search and rescue operations. Originally designated the HU2S Sea Guard (later the HH-52), this helicopter was equipped with a 1,250-shp T58-GE-8 engine. The HH-52A had a watertight boat hull fuselage for water landings.

The T58 had first flown in Vertol's testbed turbine helicopter, the H-21D, in 1957; however, there was no procurement of this aircraft by any service. In the late 1950s, Vertol (later acquired by Boeing) had developed the Model 107 (YHC-1A military designation) for evaluation by the Army as a medium lift troop transport. While the Army elected to develop the Lycoming T55-powered Vertol Model 114 (YHC-1B, later CH-47) instead, the Marine Corps decided to procure the YHC-1A in quantity as the HRB-1 (later designated the CH-46A) Sea Knight. A tandem-rotor design, the Model 107 was originally powered by two T58-GE-8 engines. The H-46 was used by the Marines as a medium assault helicopter and the Navy as a Vertical On-board Delivery (VOD) aircraft to transport munitions and supplies from auxiliary ships to warships.

Although the T58 was not GE's first commercial engine, it did contribute to the company's early commercial marketing experience with helicopters.[74] On July 1, 1959, the CT58-100-1 (commercial version of T58-GE-6) became the first U.S. gas turbine engine to be type certified by the FAA for commercial

---

74. GE's first certificated engine was the TG-190B, a commercial derivative of the J47 turbojet engine; however, no commercial application was found for this engine. GE's first commercial engine which found an application (the Convair 880 airliner) was the CJ805-3, a commercial derivative of the J79 turbojet; the CJ805-3 was certificated on September 9, 1958. GE's first commercial experience for turbine aircraft engines can be traced back at least as far as the early 1950s when a small planning team began working with airframe manufacturers and airliners to establish aircraft requirements.

helicopter use. Within six months, all three U.S. helicopter airlines had ordered CT58-powered fleets. Los Angeles Airways and Chicago Helicopter Airways purchased Sikorsky S-61s and New York Airways purchased Vertol 107s, Model IIs.

On December 21, 1960, Los Angeles Airways put into service the first U.S. turbine helicopter to go into scheduled service, a leased Sikorsky S-62.[75] The success of this operation led to the formation of a new helicopter airline, San Francisco & Oakland Helicopter Airlines, which began service in the San Francisco Bay area on June 1, 1961, with leased Sikorsky S-62s.[76] On March 1, 1962, Los Angeles Airways inaugurated the world's first multi-engined, turbine-powered, transport helicopter service with the Sikorsky S-61L, a 28-passenger helicopter specifically designed for all-weather airline use. For Los Angeles Airways, the higher speed and higher capacity S-61 had a much higher productivity (in terms of seat or ton-miles) than did the company's earlier piston engine Sikorsky S-55 helicopter.[77] The first scheduled service of New York Airways took place on July 1, 1962, with Vertol 107s.

EARLY SERVICE PROBLEMS

For the Navy, an important problem that showed up early in service was engine power deterioration and corrosion related to salt water ingestion. The anti-submarine mission of the Sikorsky HSS-2 included long-duration hovering near the surface of the sea. Engineering studies showed that at low altitude rotor downwash agitated the surface of the ocean creating salt spray which, under certain wind conditions, would be circulated up and into the engine inlet.[78] Rough, irregular salt deposits on the first blade and vane stages changed the airfoil sections and reduced airflow through the compressor causing power loss. To remove the salt, a fresh-water wash followed by a water-displacement oil spray procedure was developed. To solve the corrosion problem caused by residual salt, new component materials, platings, and high temperature paints

75. William H. Gregory, "S-62 Utilized for Turbine Familiarization," *Aviation Week*, Vol. 74, No. 14 (April 3, 1961): 40.

76. R. E. G. Davies, *Airlines of the United States Since 1914* (Washington, D.C.: Smithsonian Institution Press, 1982), 475-476.

77. Clarence M. Belinn, "The First Two Years of Operational Experience with the Sikorsky S-61," *Journal of the Royal Aeronautical Society*, Vol. 69 (March 1965): 190.

78. Ted F. Stirgwolt, "Salt Water Ingestion by Gas Turbine Engines," (Paper No. 24344 presented at the American Helicopter Society, Inc. Seventeenth Annual National Forum, Washington, D.C., 5 May 1961), 1-3.

Los Angeles Airways Sikorsky S-62 helicopter powered by a GE T58 turboshaft engine. This was the first U.S. turbine helicopter to be put into scheduled service. National Air and Space Museum, Smithsonian Institution (SI Neg. No. 92-15619).

were developed, and lubrication applied in appropriate areas for added protection.[79]

Commercial service also led to T58 engine changes. Commuter operations servicing hub airports and outlying communities had short flights and high daily utilization rates with more engine starts and more rapid engine-hour buildup than the military services. The teething problems of the new T58 engine showed up rapidly in both the Sikorsky S-61 and the Boeing Vertol Model 107.

In the case of Boeing Vertol Model 107, a low-frequency vibration caused by the rotor fed into the engine, causing a vertical bending mode in the engine frame. The bending amplitude increased with increased airspeeds. When the limit was exceeded, excessive stresses were generated within the power turbine bearing support assembly, ultimately resulting in failure of the assembly. The solution was the inclusion of some structural improvements in the helicopter along with a new forged-flange bearing-support assembly and a heavier-wall

79. David A. Anderton, "Service Experience Leads to T58 Changes," *Aviation Week & Space Technology*, Vol. 78, No. 17 (April 29, 1963): 49.

power turbine section. A high-frequency vibration problem was generated by the engine power turbine, the high-speed shaft assembly connecting the engine and transmission, and the high-speed input gear assembly within the transmission. This vibration resulted in rapid wear of engine mount bearings, cracking of engine power turbine casings, malfunctions of the engine-mounted components, and structural fatigue of the airframe. The solution was found in maintaining rigid control of the balance and runout of the entire high-speed drive train. Similar problems and solutions occurred with the Sikorsky S-61.[80]

Additional service problems cropped up. Early records of engine malfunctions showed that 54% were caused by the fuel control system, another major problem.[81] Numerous modifications ultimately provided the basis for a new control system. A premature wear rate of the splined fuel pump drive shaft was fixed with design and material changes in the spline and its drive gear. First-stage compressor stalls encountered during rotor engagement under certain atmospheric conditions were eliminated by reducing backlash in vane mechanisms and by adjusting vane tracking. Enroute power deterioration was reduced by preventing rotor head lubrication from entering the engine and with compressor wash procedures.

Commercial operation substantially improved the reliability and performance of the T58 for both military and civil use. Problems, particularly those which were a function of time, resulted in fixes which were integrated into military engines before they showed up in service. Military spare parts requirements were more easily determined on the basis of commercial experience, and repair technicians knew where to focus their maintenance efforts.

In the 1960s, as T58-powered helicopters went into service during the Vietnam War, serious cases of engine damage due to erosion occurred. Erosion was shown to be the major cause of premature turbine engine removal in combat operations.[82] Erosion test programs determined that sand concentrations experienced in actual helicopter operations were far in excess of the military specifications. The main problems encountered by the T58 in Vietnam were power deterioration, loss of deceleration stall margin, compressor blade and vane erosion, and deposition of laterite (volcanic ash) on blades, vanes, and the

80. J. H. Yost, "Turbo Shaft Engine Installation Problems in Helicopters," (ASME Paper No. 64 WA/GTP-12 presented at the Winter Annual Meeting, New York, N.Y., 29 November - 4 December 1964), 11.

81. David A. Anderton, "Service Experience Leads to T58 Changes," 48.

82. G. C. Rapp and S. H. Rosenthal, "Problems and Solutions for Sand Environment Operation of Helicopter Gas Turbines," (ASME Paper No. 68-GT-37 presented at the Gas Turbine Conference and Products Show, Washington, D.C., 17-21 March 1968), 2.

variable geometry mechanism. Installation of previously developed devices including Sikorsky CH-53 barrier filters and Sikorsky H-3 anti-ice deflectors helped reduce sand ingestion. Dust covers for the variable geometry activation rings and levers reduced contamination in that area. These devices and new maintenance checks and procedures significantly reduced stalls and premature engine removals during Vietnam service.

MAJOR T58 GROWTH STEPS

The YT58-GE-2 was the prototype engine in the T58 engine series, and a derivative engine, the T58-GE-6, followed in short order. Beyond these two engine models, there were four major engine growth steps in the T58 series. In chronological order, the four growth step military engines were the: T58-GE-8; T58-GE-5 and -10; T58-GE-14; T58-GE-16. Commercial derivatives were made of some of these engines.

The first 1,024-shp YT58-GE-2 engine was shipped in October 1956 and the last in August 1958. First shipped in July 1958, the T58-GE-6 was similar to the -2 model, but had additional installation flexibility including power takeoff in any one of three positions either forward or aft, optional main reduction gear, three-position exhaust, torque sensor, and qualification on JP-4 and JP-5 fuels. A total of 161 of these engines had been delivered by September 1960. The commercial version of the -2 and -6 engines was the CT58-100, and 71 were shipped.

The first growth step came with the introduction of the T58-GE-8 series, which passed its MQT in July 1960. With a moderate increase in turbine temperature, the power was raised to 1,250 shp. The first production engine in this series was delivered in November 1960, and by June 1968 when the last was delivered, 1,803 engines had been manufactured. Derivative military versions in this series included the T58-GE-1 rated at 1,300 shp and the T58-GE-3 rated at 1,325 shp. The commercial derivative of the -8 was the CT58-GE-110.

The next growth step consisted of the T58-GE-5 and -10 engines, both rated at 1,400 shp and qualified for service in Vietnam. The compressor first-stage rotor and stator were changed to increase airflow by 8%, air leakage at the 10th-stage compressor seal was reduced, and the compressor generally was cleaned up to reduce parasitic drag. Power turbine inlet temperature was raised approximately 20°F. A new, more reliable integrated (electrical/hydromechanical) fuel control was introduced that improved load sharing in twin engine installations, responded more quickly to over-temperature conditions, and reduced power drop by sensing the collective rotor pitch angle changes.

Between January 1966 and March 1980, 607 of the -5 version engines were delivered; between 1966 and 1984, 1,507 engines in the -10 series were delivered. The commercial version of the -10 engine was known as the CT58-140, and 352 were delivered in this series. Other military engines derived from the -5 engine, included the T58-GE-400B, T58-GE-100, and T58-GE-402, all of which had a small step growth increase to 1,500 shp. The -400 series had electrical overspeed protection, improved hot-section material, and a higher power turbine temperature. The -100 had improved hot section material, improved power turbine blade-tip-to-shroud clearances, and a higher power turbine temperature. The -402 was a conversion of the Navy's T58-GE-10 engine.

The third growth step was the YT58-GE-14, a non-production prototype engine. The -14 engine was important in as much as it introduced a two-stage power turbine, incorporated into the fourth growth step model, which did go into production. This fourth and last major growth step in the T58 series was the T58-GE-16 rated at 1,870 shp. This model had air-cooled gas generator turbine blades, a second-stage power turbine, and higher turbine temperatures. The air-cooled turbine was derived from technology developed in GE's large TF39 engine. The -16 engine passed its MQT in March 1970, and the first production engine was delivered in June 1974. The last of 742 -16 engines was delivered in March 1984, ending the 28-year T58 engine production run in Lynn.

From its initial power rating of 1,050 shp (T58-GE-6), the T58 grew to 1,870 shp (T58-GE-16), a 78% increase in power. All of this was made possible with the normal methods of increasing airflow and temperature.

LICENSED PRODUCTS AND OTHER APPLICATIONS

In March 1958, a license was granted to de Havilland Engine Co. Ltd. (later Bristol Siddeley Engines Ltd. and then Rolls-Royce Ltd.) in Great Britain to manufacture the T58 in Europe. Under this license, the T58 was known as the Gnome turboshaft engine. The engine was also licensed to Ishikawajima Harima Heavy Industries in Japan in December 1960 and Klöckner-Humboldt-Deutz in Germany in March 1963.

In late 1959, as part of a move to diversify, GE studied non-aircraft applications for its gas turbines. The small gas turbine, such as the T58, generally was not competitive with the diesel engine. However, for some applications, its small size and weight were attractive. Known as the LM100, the derivative T58 engine was adapted for emergency electrical power for telephone companies, pipeline pumping, off-highway vehicles, experimental hydrofoil boats (e.g. H.S. *Denison*), and air cushion vehicles such as the Bell SK-5.

Unusual applications for the T58 included the Bell X-22A VTOL aircraft, a tilting-duct VTOL research aircraft powered by four YT58-GE-8D engines that first flew in March 1966. Another experimental application was the T58-GE-8 powered Piasecki 16H-1A Pathfinder II high-speed research helicopter which first flew with the GE engine installed in November 1965. An Army and Navy sponsored test vehicle, the Pathfinder reached a speed of 225 mph.

As a matter of record, GE drew up proposals for the T58 as both a turbojet and turboprop engine. The turbojet version was known as the SJ109 and the turboprop version as the ST105 (military designation YT58-GE-4). Neither version found an airframe application.

T58 SUMMARY

Sponsored by the Navy, the T58 was GE's first small turbine engine. By the early 1960s, the T58 had become very successful and was one of the bread and butter production engine programs at Lynn. More than 6,000 T58 and CT58 engines and engine kits were shipped by GE and a considerable number of additional engines were built by licensees, although the exact number of license-built T58 engines is not known. The T58/CT58 had a 28 year production run from October 1956 until March 1984. The peak production years for the T58 were between 1961 and 1970.

The T58 was primarily a military helicopter engine. By production numbers, the most important military T58 engine models were the T58-GE-5, -8, -10 and -16, which powered a number of applications. As a commercial engine, the primary CT58 helicopter applications were the Sikorsky S-61 and S-62 and the Boeing Vertol 107. The most popular commercial engine was the CT58-140. The CT58 had the distinction of being the first gas turbine helicopter engine certificated by the FAA. It also powered the first U.S. turbine helicopter to be put into scheduled airline service. In both commercial and military applications, the most important CT58/T58 competitor was the Lycoming T53.

The T58 was the first high-performance, axial-flow gas turbine for helicopter use.[83] According to retired Vice President and Chief Engineer Frank Pickering, "I would consider the major technical breakthrough was taking an axial-flow compressor and building a successful, efficient compressor in that size, not only

83. Edward Woll and Fred F. Enrich, "Some Key Decision Points in the Development of the Modern Aircraft Gas Turbine," *The Leading Edge* (Fall 1989): 13.

from a design standpoint, but learning how to manufacture it, and manufacture it in a repeatable kind of way."[84]

Floyd Heglund, who became T58 Engine Design Manager in October 1956, later summarized some additional features of the engine which he believed had made it successful:

> First off, I think the thing that made it mechanically a sound engine was the integral (compressor) spool and the use of curvic couplings (between the gas generator turbines), which were new at the time . . . We (then) had something that was stable, reliable, and repeatable . . . On the aerodynamic side, it used some variable stators to get it through the starting range, and that was different. The other thing that was a big plus for the engine was the rear-end drive . . . (because) it lent itself, configuration-wise, well on the airframe and (because) it didn't open up that (problem) area of trying to bring a shaft right up through (the T58).[85]

The T58 was one of the first-generation turbine engines that significantly improved the utility of the helicopter. The Korean War had demonstrated certain shortcomings of the helicopter, largely traceable to its reciprocating power plant.[86] Helicopters with high power demands had to carry heavy piston engines (with a 1:1 power-to-weight ratio) that limited their payloads. Also, the relatively poor reliability of piston engines added to the difficulty of remote field maintenance. Constantly trimming power to set rotor speed, pilots had their hands full flying such helicopters. The small turboshaft engine included such advantages as a much higher power-to-weight ratio; improved payload; a constant-speed control that reduced pilot workload; freedom from vibration; good cold-start capability; the ability to run at sustained high power levels without reduced engine life; and higher air speed.

The J85 Turbojet Engine

In fall 1953, SAED General Manager Jack Parker and Engineering Manager Ed Woll decided to add two more engines to the department's product line in addi-

84. Frank Pickering, interview by Rick Leyes and William Fleming, 20 August 1991, GE Interviews, transcript, National Air and Space Museum Archives, Washington, D.C.

85. Floyd Heglund, interview by Rick Leyes and William Fleming, 21 August 1991, GE Interviews, transcript, National Air and Space Museum Archives, Washington, D.C.

86. David C. Gerry, "Gas Turbine Powerplants for Helicopters," *U.S. Army Aviation Digest*, (July 1959): 23.

tion to the T58.[87] They concluded that they were going to need a 2,500-shp turboshaft engine and a 2,500-lb-thrust turbojet engine—the T64 and the J85. Neither engine had a specific application, but a look around the industry suggested some possibilities. More power and a lower sfc would be needed for a heavy helicopter, and a turboshaft engine about twice the size of the T58 should do it. A small turbojet might be suitable for an executive jet, military trainer, or, with an afterburner, a small fighter. While there was no clear requirement for the J85, the Air Force was in the process of defining a high thrust-to-weight demonstrator engine.

As part of an advanced design study for the turbojet, Chief Draftsman Louis Lechthaler was assigned to do three 2,500-lb-thrust engine layouts at 5:1, 7:1, and 12:1 pressure ratios. Lechthaler was assisted by engineers Frank Lenherr on the compressor, John Benson on the combustor, and Art Kohn on the turbines. The primary design objectives were high thrust-to-weight ratio and low cost.

Woll then took the drawings to the Air Force at Wright-Patterson Air Force Base and to the Air Force headquarters in Washington, D.C. for review and to determine if there was an interest in the engine. The consensus was encouragement of the 7:1-pressure-ratio engine. In 1954, a letter-type proposal was submitted to the Air Force by GE for a 2,500-lb-thrust engine weighing 250 lb (10:1 thrust-to-weight ratio). The proposal engine was designated X-104 by GE. About the same time that GE had decided to broaden its small engine product base, the Air Force, in an unrelated initiative, had developed a need for a small, high-performance aircraft engine. Air Force studies indicated that a considerable weight saving could be achieved with the design of a new engine for future use in trainer aircraft.[88] For example, it would be an advantage to have an engine that weighed less than the Continental J69 used in the Cessna XT-37 trainer aircraft. A further substantiation of a need for a high thrust-to-weight ratio engine in the 2,000-lb-thrust class became apparent when proposals for an XQ-4 supersonic target drone were evaluated. The only engine that could meet the schedule was the XJ81-WE-1, the Rolls-Royce RB93 Soar, licensed by Westinghouse and undergoing development modifications. However, the Air Force preferred to have an American-built engine available as a backup when the XQ-4 went into production. The XJ81 was approved as an interim power

87. Woll to Leyes and Fleming, 20 August 1991.

88. "Presentation on the XJ83-R-1 and XJ85-GE-1 Engines" (Confidential Report), undated, Office of History, Headquarters Air Force Logistics Command, Wright-Patterson Air Force Base, Dayton, Ohio.

plant for the XQ-4, and a requirement was established for a 2,000-lb-thrust American engine for follow-on Q-4 production.

To minimize the development risk of this follow-on engine, the Air Force decided to pursue the development of an engine with two contractors having different approaches and designs. An RFP was prepared based on an earlier proposal by Project Engineer William Downs of the Power Plant Laboratory at Wright-Patterson Air Force Base, who had suggested the possibility of an exploratory, 10:1 thrust-to-weight gas turbine generator demonstrator for VTOL applications.[89] The primary goal of the new engine sought by the Air Force was low specific weight (engine weight per pound of thrust). The Air Force was also very interested in having a fully American small jet engine. The Air Force felt that the lack of such engines had led to the use of partially developed foreign engines that had not provided an entirely satisfactory solution to the high thrust-to-weight goal that the Air Force had in mind.[90] Awards were made to Fairchild for the XJ83-R-1 and to GE for its XJ85-GE-1. The XJ85 was designated the MX2273 Project by the Air Force.

On November 28, 1954, GE signed an Air Force development contract to build and demonstrate two lightweight engines with a 10:1 thrust-to-weight ratio. In January 1955, GE formed the J85 project. Joseph C. Buechel was appointed Project Manager and Sherman Crites became Manager of Engine Design. Art Adinolfi was placed in charge of mechanical design. Engineer Sherman Rosenthal was responsible for the thermodynamic performance, aerodynamicist Frank Lenherr was responsible for the compressor aerodynamic design, John Benson provided the combustor design, and engineer Art Kohn designed the turbine. Clem Gunn handled applications and installations. Engineer Clarence Danforth contributed his experience in aeroelasticity to the compressor design. John Jacobson directed the controls activity.

Originally, the J85 was designed for a decoy missile application. It was a simple single-spool, seven-stage axial-flow engine with a straight-through-flow annular combustor and a two-stage turbine. The engine was unique in that it was a two-main-bearing turbojet that had both an overhung compressor and an overhung turbine. Bearing loads were transmitted to the central main frame section. The output was expected to be 2,000 lb for a dry weight under 300 lb. The extreme shortness of the engine (slightly over 36") was the key to its light weight.

89. William Downs, taped interview by Rick Leyes and William Fleming, 20 November 1991, National Air and Space Museum Archives, Washington, D.C.

90. "Presentation on the XJ83-R-1 and XJ85-GE-1 Engines."

The original J85 compressor was based on a NACA-Cleveland experimental four-stage transonic compressor designed by engineer Ray Standahar. The derivative J85 seven-stage compressor had a pressure ratio of 7:1.

Design and component development were begun in February 1955, and the first component tests were started in November. When the engine was first tested, the highly loaded compressor prevented it from accelerating to a speed beyond that provided by the starter. A simple, quick fix incorporated bleed valves to unload the compressor, allowing the engine to self-sustain for the first time in January 1956. However, there were still stall problems, and the engine would not come up to design speed.

REDIRECTED J85 PROGRAM AND AFTERBURNER DEVELOPMENT

In the meantime, what had begun as an Air Force gas generator demonstrator program became a systems requirement for an aircraft weapons system. In February 1956, the Air Force selected McDonnell for the GAM-72 Quail program, an air-launched decoy missile designed to protect Boeing B-52 bombers. In June, GE was selected as the engine contractor for the GAM-72. The engine to be developed for the GAM-72 was a YJ85-GE-3.

Also in 1956, a second application for the J85 was secured on the Northrop T-38A Talon trainer aircraft. For GE, Northrop's interest in the J85 engine was a significant turning point in the J85 program. This second, man-rated engine application had a large market potential. It also offered the opportunity to develop an afterburner for the engine.

In the early 1950s, Northrop had investigated a next generation, lightweight supersonic fighter. In December 1952, preliminary design was begun on the N-102 Fang which resulted in an aircraft mock-up built around a single GE J79 turbojet engine. While no procurement was made, the N-102 established a design precedent for lightweight jet fighters. In 1954, Northrop studies in Europe and Asia suggested the need for a small, lightweight, low-cost fighter with high thrust-to-weight engines and a low attrition rate for both plane and pilot.[91] In 1954, in conjunction with the N-102 project, Northrop became interested in GE's small J85 engine, under development for missiles. A safety-package of two could be installed and still weigh less than a single larger engine. When GE said that an afterburning version of the engine could be produced that would increase the J85's power as much as 40%, Northrop embarked on the design of the N-156 lightweight, supersonic fighter.

91. Ted Coleman, *Jack Northrop and the Flying Wing: The Story Behind the Stealth Bomber* (New York: Paragon House, 1988), 201-202.

GE XJ85 turbojet engine prototype No. 001. This Air Force development program was called Project MX2273. Photograph courtesy of GE.

In January 1955, Northrop proposed to the government the N-156F as a single-place fighter and the N-156T as a two-place trainer.[92] The Air Force concluded that the N-156T would meet its specifications for a new TZ trainer aircraft and issued a letter of intent for prototype trainers on June 15, 1956. This led to the T-38, the first supersonic two-place trainer for the Air Force, important because it was the first Air Force jet trainer with performance characteristics similar to contemporary operational aircraft. The aircraft was also attractive as a trainer because it was simple, easy to maintain, and low cost.

Responding to Northrop's request, GE began preliminary design work on an afterburner version of the J85 for the T-38. Engineering Manager Ed Woll was

92. Fred Anderson, *Northrop: An Aeronautical History* (Los Angeles: Northrop Corporation, 1976), 175.

Northrop T-38 Talon supersonic trainer aircraft powered by twin GE J85 engines. Photograph courtesy of GE.

the person who had first told Northrop engineers that an afterburner could be developed for the J85. In order to allow the afterburner to fit within the Northrop aircraft without lengthening it and to keep the weight down, Woll decided to make the afterburner one foot shorter than that used on GE's J79 engine.[93] Woll assigned Bill Lawson to make the first designs of the J85 afterburner. Engineers John Benson and Ev Waters as well as GE engineering consultant Mike O'Brien were also involved in the project. Engineer Hank Schnitzer designed the variable jet nozzle and the mechanical nozzle activation system for the afterburner.[94]

In 1956, GE proposed to the Air Force (Proposal SJ110) its afterburning engine. The proposed engine had a seven-stage compressor and was rated at 2,450 lb thrust dry. Soon afterward, the Air Force awarded GE a contract for development, designating the engine XJ85-GE-5. To power the prototype YT-38, an interim engine was developed, the YJ85-GE-1, a non-afterburning engine rated at 2,180 lb thrust. It had a seven-stage axial-flow compressor, annular combustor, and two-stage turbine.

In addition to the GAM-72 and the YT-38, other airframe applications appeared promising as well. The design of the North American T-39 Sabreliner, intended as a trainer and utility jet aircraft, had started early in 1956; the proto-

93. Woll to Leyes and Fleming, 20 August 1991.

94. Later, the J85-GE-21 engine used a hydraulically actuated nozzle system rather than a mechanically actuated system.

type version would be powered with a J85 engine. Still pending was the production contract for the XQ-4 target drone, which had originally motivated the Air Force to sponsor the development of the high thrust-to-weight ratio engine. With these applications in hand, the J85 had a promising future, if a difficult compressor problem could be resolved.

SOLVING THE COMPRESSOR PROBLEM

Fairchild's J83 was a competitive engine that could exclude the J85 from at least one potential application, if the J85 could not be made to work properly. The J85 program was in trouble because the compressor had not been performing properly, and it was not clear what was causing the problem. Numerous experiments with modified compressor blades had not worked.

In a moment of inspiration, Ed Woll had the compressor blades removed from a J85 engine that had not been operating satisfactorily, put on a comparator, and checked against the drawings. The J85 had been designed with high-aspect ratio (long, narrow) compressor blades. Their resulting aerodynamic sensitivity meant that a correct airfoil shape was especially critical. When checked against the blueprints, it was discovered that the blades did not have the proper shape. The J85 compressor blades had been manufactured in the same manner as GE's large engine blades, by forging. In this process, the extra metal flashing left on the airfoil was removed by a cropping die. However, this process also chopped off the airfoil leading and trailing edges, leaving them blunt.

The removed compressor blades were turned over to tool and die makers who made leading- and trailing-edge guillotine gages (to check the airfoil contour) and hand-benched (dressed) the blades to the exact profiles specified by the blueprints. The blades were re-installed in the seven-stage compressor of the engine that had not been operating satisfactorily. Much, but not all, of the predicted performance and stall margin was gained back. Thus, part of the compressor performance problem was identified as a manufacturing problem. While the forging process had been satisfactory for GE's large engine compressor blades, it clearly was not accurate enough for small engine blades. Vendors were found who could accurately machine the blades. Later, the pinch and roll process developed in the T58 program was used to manufacture J85 compressor blades.

Unfortunately, by the time this fundamental problem had been discovered, it was late into the development program. By the summer of 1957, it was clear that even the improved seven-stage compressor was not going to solve the remaining compressor performance problem. The J85 team had run out of development time and had production applications facing it. The safest decision was to make some compromises in the design, lower the risk, and proceed with a more conservative approach. It was critical that the compressor produce the required airflow and have sufficient stall margin. "This was," according to Ed Woll, "the kind of a decision you have to make when you are on the firing line."[95] Thus, in July, GE announced the planned change to an eight-stage compressor J85.

In September, Harold T. Hokanson was appointed J85 Project Manager, and proceeded with the development of the eight-stage compressor. Work continued on the seven-stage J85 as an interim flight engine to power prototype missiles and aircraft.

The eight-stage-compressor engine for the GAM-72 was known as the J85-GE-7 and for the Northrop T-38, the J85-GE-5. A significant mechanical design change made to the basic engine was the addition of a front frame at the engine inlet that incorporated variable inlet guide vanes, an anti-icing system, and a support structure for a new forward compressor rotor bearing.[96] The new eight-stage compressor design included dovetail rotor blade fasteners for all rotor blades other than those in the first stage, which continued to use pinned-root retention. The design also included a reduction in stator blade and rotor blade tip clearances, made possible by the stabilizing effect of the new front bearing, and more accurate manufacture of the compressor blade airfoils. The engine retained the annular main combustor and the two-stage overhung turbine. The new -5 engine had an afterburner that consisted of a diffuser, pilot burners using four full spray bars, and the main burner with 12 spray bars and a unique through-flow flame holder. The afterburner also included a variable-area convergent-divergent exhaust nozzle to accommodate the large variation in volumetric flow between afterburning and non-afterburning power settings and to completely expand the exhaust stream in supersonic flight conditions. An afterburner fuel control system was introduced to modulate the afterburner fuel flow in response to throttle lever demand and to modulate the exhaust nozzle area. The success of these modifications proved to be a critical turning point in the J85 program. The GE design team referred to this engine as the eight-stage, three-bearing machine.

95. Ed Woll, taped interview by Rick Leyes, 24 November 1992, National Air and Space Museum Archives, Washington, D.C.

96. In this case, had the compressor been overhung (i.e. with the bearing downstream of the compressor), the additional stage could have caused problems under certain conditions. An out-of-balance compressor rotor, high G maneuvers, and foreign object ingestion could have caused the blades to rub against the case and possibly break.

About the time GE was close to solving the difficult J85 compressor problem, its competition with the Fairchild J83 engine also came to an end. The J83 powered the prototype XSM-73 (WS-123A) Bull Goose (also known as Blue Goose), an intercontinental range surface-launched decoy designed to protect B-47 and B-52 bombers. The Fairchild Bull Goose had been designed to accept either the Fairchild J83 or the GE J85 engine. Because the J83 and the J85 were being developed for similar applications, strong competition existed at this time between the two engine programs. Though each program had technical problems, the Air Force dropped the J83 program late in 1958: it could not afford both programs; the J85 could power both the Goose and Quail missiles; the J85 was lighter and smaller; and the J85 afterburning engine had been selected to power a prime aircraft system, the T-38.[97] In December 1958, the Air Force canceled the Goose due to budgetary pressures and because the missile could not simulate a B-52 on enemy radar. After the Air Force dropped the J83 program, work on the J83 was terminated, and Fairchild went out of the jet engine business.

THE FIRST J85 ENGINE MODELS

Two interim J85 engine models were developed, one for a decoy missile (-3), and the other man-rated (-1). These were followed by two production models, the -7 for the decoy missile and the -5 for piloted applications. Both piloted and pilotless engines used the same basic compressor, combustor, and turbine. However, they had different accessory gearboxes, and they passed different type tests. The -5 engine program also included the development of an afterburner.

The interim decoy missile engine was the YJ85-GE-3. It had a seven-stage compressor, and although originally quoted at 2,450 lb thrust, was derated to 2,250 lb during the qualification program. The design was started in April 1956, and the first engine test took place in December 1956. The YJ85-GE-3 passed its 15-hour PFRT in March 1958, and the J85-GE-3 passed its 15-hour QT in November 1958. The first successful powered flight of the McDonnell GAM-72, powered by the -3 engine, was in August 1958.[98] This was also the first flight of the J85 engine.

An engine start problem had caused early GAM-72 test drops from B-52s to fail. Engineer Fred O. MacFee, Jr. and his team traced the problem to air cur-

97. Nick Constantine, "The J85 Story," unpublished, undated paper, National Air and Space Museum Archives, Washington, D.C.

98. Werrell, *The Evolution of the Cruise Missile*, 125.

rents within the B-52 bomb bay that had caused the compressor to rotate in the reverse direction. The energy needed to start the engine was being used to stop the reverse rotation. The problem was remedied, and the GAM-72s started successfully. MacFee became J85 Project Manager in 1959.

The YJ85-GE-1, the interim aircraft engine, had a seven-stage compressor and was rated at 2,180 lb thrust. The design was started in January 1955, and the first successful engine test was in January 1956. The development proposal was submitted by GE on February 28, 1957. The YJ85-GE-1 passed its 50-hour PFRT in August 1958, but was not carried through a 150-hour QT. On September 16, 1958, the prototype North American T-39 Sabreliner (NA-246 UTX) flew with J85-GE-3 engines installed.[99] This was the first manned flight of a J85 engine. The J85-GE-3 installed on the Sabreliner was a -1 engine with a larger accessory gearbox than the missile version. On April 10, 1959, the prototype Northrop YT-38 flew for the first time powered by YJ85-GE-1 non-afterburning engines. On July 30, 1959, the prototype Northrop N-156F flew for the first time also powered by YJ85-GE-1 engines.

The flight of the Northrop YT-38 marked another critical turning point in the J85 program. Because of the difficult compressor problem, the Air Force had been on the verge of canceling the J85 program. While many of the compressor stall problems had been resolved before the J85 engines had been shipped to Northrop, the engine showed little or no stall margin when it was installed in the YT-38. Conferring with SAED General Manager Gerhard Neumann, Ed Woll picked April 10, 1959, for the YT-38 flight date without knowing the cause of the stalls. Neumann then persuaded the Air Force to wait until the April 10 cut-off deadline. In January 1959, an intensive effort was launched to resolve the stall problem. Ed Spear, a GE technical representative at the Northrop facility, and his ground crew, joined the Woll team in search of the solution.

The investigation was conducted with the engine installed in the airplane with different airframe configurations. The stalls were traced to one major problem: hot exhaust gases were drawn forward between the engine and the fuselage and into the engine inlet. The hot gases expanded the compressor casing, opening up the rotor tip-to-case clearance, and caused the engine to stall while it was running on the ground at high speed. In addition to various engine and airframe modifications, the primary fix was to install suck-in (vacuum breaker) doors in

---

99. The Pratt & Whitney JT12 turbojet engine was eventually chosen for the production T-39 because it had a higher initial thrust rating than the GE J85 and because the Air Force had requested design changes which resulted in a heavier aircraft.

the fuselage near the compressor face which prevented the recirculation of hot air during ground running. The deadline was barely met as test pilot Lew Nelson followed Woll's request to "just get air under the wheels."[100]

Relatively few -3 and -1 interim engines were manufactured. The first XJ85-GE-3 engine was shipped by GE in the last six months of 1957. Production of 38 YJ85-GE-3 engines procured by the Air Force was completed in March 1959, and production of 61 J85-GE-3 engines under Air Force contract was completed in April 1960. A total of 105 J85-GE-3 engines of all models were manufactured by GE. In July 1959, delivery was completed of all 12 YJ85-GE-1 engines procured by the Air Force for use in two Northrop YT-38s and two Northrop N156 aircraft plus four spare engines.

The J85-GE-7 was the production missile engine for the McDonnell GAM-72A Quail. The -7 engine had an eight-stage compressor and was rated at 2,450 lb thrust. Design effort on the -7 engine started in June 1957 and the first engine went to test in June 1958. The -7 passed its first official 15-hour QT in January 1959. In March 1959, -7 flight tests began on a Convair F-102A testbed aircraft. The engine was mounted on a fully-instrumented, retractable pod lowered from the F-102 missile bay. Simulated altitude tests were conducted at the Arnold Engineering Development Center in Tullahoma, Tenn., and climatic and altitude tests were conducted at the Wright Air Development Center in Dayton, Oh. The -7 engine completed a second 15-hour MQT in October 1959, and the first engine shipments were in November. The first free flight of the -7 engine powered GAM-72A was in March 1960; by February 1961, one Quail-equipped B-52 squadron was operational. A total of 577 T58-GE-7 engines were shipped between 1959 and 1962 for the GAM-72.

The first afterburning J85 production engine, available for either aircraft or missile installation, was the J85-GE-5. The -5 engine had an eight-stage compressor and was rated at 3,600 lb thrust with afterburning. The -5 design started in July 1956, and the first engine to test was in January 1959. Altitude and afterburner military qualification tests were conducted at the Arnold Engineering Development Center. The YJ85-GE-5 completed its 50-hour PFRT on October 3, 1959. The first flight of XJ85-GE-5 was completed in September 1959 in the Convair F-102A testbed aircraft. The first YJ85-GE-5 engines for the T-38 program were shipped on October 31, 1959, and the first flight in a Northrop YT-38 with this engine was in November 1959. The -5 engine completed its 150-hour MQT in July 1960.

---

100. Woll to Leyes and Fleming, 20 August 1991.

The first J85-GE-5 production engine was shipped in December 1960, and production was completed in 1971 when 2,825 -5 engines (of all models including 68 YJ85-GE-5 engines) had been shipped. This constituted one of the largest production runs of J85 engine models, resulting primarily from the use of the -5, -5A, and -5B engines in the Northrop T-38A trainer, one of the two primary J85 airframe applications. The -5 engine was also the first major growth step in the J85 engine series.

The non-afterburning version of the -5 engine had a variety of experimental and other limited applications. The X-14A, powered by two such engines, was a research vehicle built by Bell and tested by NASA for the purpose of exploring the problems associated with VTOL aircraft. It made significant technical and pilot-training contributions to VTOL programs.[101] The -5A engine powered the Q-4B, a supersonic target drone built by the Radioplane Division of Northrop. Other experimental applications for the -5 engine included boost power on a de Havilland Otter and Kaman UH-2C helicopter. An important engine application for the -5 engine, described in detail later, was to power lift fans in the Ryan XV-5A VTOL aircraft.

Two non-afterburning engines that were derived from the J85-GE-5 included the J85/J2 and the J85-CAN-40, each rated at 2,850 lb thrust. As a result of the escalating war in Vietnam, the Air Force became interested in the immediate acquisition of counter-insurgency aircraft. Cessna responded with the J85/J2 powered YAT-37D attack aircraft, a derivative of the Air Force T-37 jet trainer. Eight J85/J2 engines were built by GE. The J85-CAN-40 was manufactured under license by the Orenda Engine Division of Hawker-Siddeley Canada Ltd., in Malton, Ont., to power the Canadair CL-41R Tutor aircraft.

The most important accomplishment associated with the J85-GE-5 engine was the afterburner, which had been proposed originally to the Air Force in 1956 and had first been tested as a component late in 1957. The afterburner was qualified as part of the -5 engine in 1960. For the time, the J85 afterburner development was a unique technical advancement for small gas turbine engines.

MAJOR J85 GROWTH STEPS

The major J85 growth steps were the J85-GE-5, -13, -15, and -21 engine models. The addition of the afterburner distinguished the -5 engine as the first major

101. Mike Rodgers, *VTOL Military Research Aircraft* (Somerset, England: Haynes, 1989), 201. A later version of the aircraft, the X-14B, used J85-GE-19 engines.

growth step in the J85 series. The second major growth step was the -13 engine, accomplished largely with increased turbine temperature. The third step was the -15 engine, incorporating an improved turbine. The final major model was the -21 engine, with an additional compressor stage to increase airflow. There were numerous derivative engines from each of these major models.

The -5 engine production was started in 1960 for the Northrop T-38A. The -13 engine program, which started in 1961, evolved in response to the Northrop N-156F, which eventually became the F-5A. The -15 and -21 models were more powerful engines, primarily for later versions of the F-5 fighter. Thus, in a broader sense, the roots of the -13, -15, and -21 can be traced to the start of Northrop's N-156F program.

Though the Air Force originally was not interested in a lightweight fighter, Northrop decided to build its N-156F Freedom Fighter early in 1958. U.S. defense officials had become interested in buying a new airplane more suitable for the defense requirements of friendly nations. With their shortage of highly skilled technicians, less complete maintenance facilities, and lower budgets, a less complex and costly aircraft for these nations would be desirable. In May 1958, the Air Force, acting on behalf of the Department of Defense, authorized Northrop to build prototype N-156F aircraft.

On July 30, 1959, the prototype N-156F flew powered by YJ85-GE-1 engines. This and another prototype aircraft were later powered by YJ85-GE-5 engines when they became available. On April 25, 1962, the Department of Defense announced the selection of the N-156F as a defensive aircraft for favored nations under the Military Assistance Program (MAP). In October 1962, Northrop was authorized to produce single- and two-seat versions of the aircraft, redesignated F-5A and F-5B respectively. The prototype YF-5A flew on July 31, 1963, powered by J85-GE-13 engines.

The J85-GE-13, a re-rated version of the J85-GE-5, was selected originally to power the Northrop F-5A supersonic fighter. To modify the -5 engine, turbine inlet temperature was increased to produce 2,720 lb thrust dry and 4,080 lb with afterburner. The -13 design effort was started in March 1961, and the engine completed its 150-hour MQT in late summer of 1963. The -13A, license-built by Alfa-Romeo, powered the Aeritalia G.91Y aircraft. A total of 2,156 J85-GE-13 engines were built by GE, making it one of the major production models in the J85 series.

A derivative, non-afterburning version of the J85-GE-13 was the J85-GE-4, rated at 2,950 lb thrust. The -4 and -4A powered the Rockwell T-2C Buckeye jet trainer and the Teledyne Ryan MQM-34D Mod II Firebee drone. The -4A was also used for boost power on the Rockwell OV-10BZ aircraft. GE built 740 J85-GE-4 engines of all models. Another non-afterburning derivative of the J85-GE-13 was the J85/J4, rated at 2,950 lb thrust, which powered the Canadair CL-41G. Thirty were built.

The -13 engine was followed by the J85-GE-15 rated at 2,925 lb thrust dry and 4,300 lb with afterburner. The increased thrust came from an improved turbine design and higher turbine inlet temperature. The initial applications were for the Northrop F-5A-15 and the license-built Canadair CF-5A/B fighters; design began in February 1964, and the engine passed its 150-hour MQT in May 1966. A license was granted to Orenda Engine Division in Canada to manufacture the -15 engine. GE built four YJ85-GE-15 engines; the number of Orenda-built J85-GE-15 engines is not known.

Other models included the J85-GE-17 and J85-GE-19. The -17 was a non-afterburning derivative of the J85-GE-5 engine. Rated at 2,850 lb thrust, it was used for take-off and climb boost on the Fairchild C-123H/K and the Fairchild AC-119K aircraft. The -17A was used on the Cessna A-37A/B attack aircraft, and the -17B on the Saab 105 attack/reconnaissance aircraft. GE manufactured 2,058 J85-GE-17 engines of all models. The J85-GE-19, rated at 3,015 lb thrust, was a special vertically-operating J85 engine used on the Lockheed XV-4B Hummingbird II, a direct jet-lift VTOL research aircraft. The -19 engine was also used on the NASA-sponsored Bell X-14B VTOL research aircraft. Only 12 J85-GE-19 engines were manufactured.

The fourth and final major model in the J85 series was the J85-GE-21. The start of this program was in February 1962, when Ed Woll began a growth program for the J85. Engineer Art Adamson and Aerodynamicist Frank Lenherr began the original growth studies, and later engineers Brian Rowe, Lee Fischer, and Paul Setze were responsible for the design that went into production. George Eddy was the Design Manager during the major production and service phases.

The J85-GE-21 was rated at 3,500 lb thrust dry and 5,000 lb thrust with afterburning. The 23% increase in thrust over the J85-GE-13 was achieved by increasing airflow from 44 to 52.5 lb/sec with the addition of a zero-stage compressor—for a total of nine compressor stages—and by increasing turbine inlet temperature by 56°F.[102] Variable stators in the first three stages of the compressor were substituted for the original bleed system, designed as a quick

---

102. "GE Tests Advanced J85 Turbojet Engine," *Aviation Week & Space Technology*, Vol. 91, No. 7 (August 18, 1969): 58. T4 was also increased 56°F.

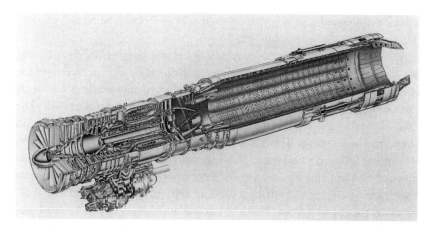

GE J85-GE-21 afterburning turbojet engine cutaway. Photograph courtesy of GE.

fix for the compressor start problem. To keep the weight increase to a minimum, both the compressor rotor and blades were made of titanium instead of
the steel in earlier models. Mid-span dampers were used on the first two compressor stages to resist vibratory stress. The compressor was a smooth-spool design of the type originally developed for the T58 as compared to the J85
eight-stage, stacked-rotor compressor, in which each disk was connected to its
neighbor with studs and spacers. This improved rotor stability and meant that
all blades were individually removable without rotor disassembly (with the exception of the first stage, eight-stage J85 engines required rotor removal for
blade replacement).

J85-GE-21 development was conducted in two phases. The initial development phase started in February 1962 and ended in 1969. This program was a
low-level effort using Component Improvement Program (CIP) funds. The
first zero-stage compressor test was in July 1963. The first -21 engine was tested
in December 1965, and the 60-hour PFRT was completed in November 1967.
The first flight was on March 28, 1969 in a Northrop YF-5B-21, a GE-modified
Northrop F-5B named after GE's engine.[103] The second development phase
took place from January 1971 until June 1972. The Air Force approved the
MQT in May 1972, and the first production engine was shipped the same

103. The YF-5B-21 program was largely a GE initiative. Northrop was concerned that a higher
    powered F-5 posed a potential conflict with another one of their aircraft programs known
    as the P530 Cobra.

month. GE shipped 3,483 J85-GE-21 engines of all models, the highest production total of any J85 engine model.

A few of the major in-service problems of the J85, resolved in the -21 engine, included an afterburner liner cracking problem, caused by high deflection of the liner support hangers in the vicinity of the variable exhaust nozzle (VEN) actuator mount ring; and stresses induced by restrained axial thermal growth. The solution was to add two axial slip joints in the liner to provide flexibility to relieve the high stresses. The VEN had two primary problems, high deflection at high-flight-speed conditions, and high friction. The nozzle housing was stiffened by changing from a cylindrical to a polygonal shape and the nozzle leaves were strengthened to reduce bending deflection. Nozzle friction was reduced by the use of carbon "lipstick" lubrication of the VEN leaf roller bearings and applying dry-film lubricant to all of the nozzle sliding surfaces. Engine stall and flameout problems following high-altitude, high angle-of-attack afterburner light-offs, were fixed with the introduction of a compressor discharge pressure lag tube and dump system. Number two bearing failures were fixed by strengthening the bearing cages. Compressor 4th and 6th stage instability problems resulting in blade failure were solved by re-cambering the airfoils.[104] The longevity and operation of afterburner components and the main fuel and afterburner control were some of the additional problem areas addressed.

In the fall of 1969, the Department of Defense opened a competition for an aircraft that would be a relatively low-cost successor to the F-5A for its allies under the Foreign Military Sales program. Emphasis was placed on an air-superiority as opposed to tactical fighter role; the new aircraft would counter late-model Soviet MiG-21 fighters. Northrop entered the F-5A-21 powered by the high thrust J85-GE-21 engine, and on November 20, 1970, the Air Force announced Northrop the winner of the International Fighter Aircraft competition. In 1971, the aircraft was redesignated F-5E Tiger II. The first F-5E made its initial flight on August 11, 1972.

A highly successful aircraft, the F-5 was flown by some 23 countries by 1976. In the U.S., F-5E and F-5F aircraft were used as aggressor aircraft in air combat maneuvering training in the Air Force and in the Navy Fighter Weapons ("Top Gun") School. Production of the F-5 series ended in 1987, when 3,805 F-5/T-38 aircraft had been produced. As of 1991, F-5 aircraft remained in service with 31 air forces. It was the popularity of the F-5E and F-5F that accounted for the large production run of the J85-GE-21 engine model.

---

104. George W. Eddy, Lynn, Massachusetts, to Ed Woll, letter 19 August 1992, National Air and Space Museum Archives, Washington, D.C.

THE LIFT FAN

The J85's combination of high thrust and light weight made it an ideal power plant for V/STOL aircraft. The VTOL application with the greatest impact on GE was the XV-5A program. In the late 1950s, the Army was interested in a fast VTOL aircraft suitable for battle-zone reconnaissance and capable of operating from forward positions in unprepared terrain. Among those studied in the early 1960s were the Lockheed XV-4A, the Hawker-Siddeley XV-6A (P.1127), and the Ryan XV-5A, each with a different type of propulsion system.

In May 1957, GE received a study contract from Transportation Research Command (TRECOM) of the Army for V/STOL propulsion, specifically for lift-fan propulsion.[105] Among the numerous types of systems being explored for VTOL propulsion including direct lift, vectored thrust, rotating ducts, and augmented jet ejectors, GE's lift fan, conceived by Pete Kappus, had an advantage because it had a very high augmentation (lift-to-thrust) ratio. The fan gave a vertical lift of about three times the thrust of the basic engine in horizontal flight.[106] This meant that a smaller engine could be used for both vertical and forward flight, thus saving installed engine and fuel weight. Building on its steam turbine and aft fan experience, GE designed a tip turbine-driven lightweight fan.

In May 1959, GE received a development contract for the fan, the first of which went to test in November 1959. In May 1960, GE submitted to TRECOM a proposal for a research aircraft and flight program to test the lift fan. On November 15, 1961, GE received an award to build two XV-5A aircraft. In an unusual situation, GE, an engine company, became the prime contractor for the aircraft system. Ryan Aeronautical, an airframe company, was selected as a sub-contractor to build the two aircraft; Ryan selected Republic Aviation, another airframer, for the flight test program. Wind tunnel tests were conducted by NASA's Ames Research Center. GE's Project Manager was Art Adamson. The work was carried out at GE's Evendale plant. Key engineering work on the project was done by a team which included Lee Jensen, Brian Rowe, Joe Rowe, Ted Stirgwolt, John Simon, Cal Conliffe, and Bob Thorp.

Propulsion for the Ryan XV-5A was provided by two GE X353-5 convertible lift fan system units, each consisting of a lift fan, a diverter valve, and a J85-GE-5 engine (without afterburners). The -5 engines served as gas generators for the vertical phase and conventional turbojets for horizontal flight. The

105. Travers, *Engine Men*, 224.

106. David A. Anderton, "GE Projecting Lift-Fan Aircraft Designs," *Aviation Week & Space Technology*, Vol. 76, No. 3 (January 15, 1962): 55.

diverter valve directed exhaust gas flow from the turbojets to the lift fans, which were installed in both wings of the XV-5A and to a smaller fan (GE Type X376) in the fuselage nose, used primarily for pitch control. For transition from hover to horizontal flight, exit louvres directed the fan efflux aft.

The first flight of the Ryan XV-5A was on May 26, 1964, at Edwards Air Force Base. While the XV-5A was successfully tested for some period of time, one aircraft was lost during a demonstration flight in April 1965 for the Army. The Army accepted two aircraft on January 27, 1965, and conducted extended flight tests from November 1965 to September 1966. The loss of one XV-5A and the Army's loss of its close air support role (including battlefield surveillance) to the Air Force—which was not interested in the Army-developed XV-5A—were factors in the eventuality that the aircraft concept was not pursued for an operational program. The second aircraft (modified and called the XV-5B) was used for flight testing by NASA. Of more lasting value to GE was the fact that an 80-inch-cruise-fan project conducted in the early 1960s, using the same tip-driven turbine technology, helped establish the company as an early leader in the development of compressor stages typical of the fan component in high-bypass turbofans.

J85 SUMMARY

GE shipped more than 12,000 J85 engines of all models, making it a real "cash ringer" for the company. More J85 engines were produced than any other small GE engine. In the 1960s, the J85 emerged as GE's lead production engine at Lynn, giving rise to the nickname "Turbotown" for the Small Aircraft Engine Department. In the 1970s, the J85 reached its peak production and became GE's Aircraft Engine Group's mass production leader. Production ended in 1988, 34 years after the engine was designed. While the J85 had a large number of applications, it principally powered the Northorp T-38 Talon and F-5 Freedom Fighter. The T-38 was the first supersonic trainer, and the F-5 was a low-cost fighter that was used by many nations from the 1960s through the 1980s.

From a technical viewpoint, while the original 10:1 thrust-to-weight design objective was never met in any J85 production engine, the J85 had the highest thrust-to-weight ratio of any production engine built for its time. For example, the early 2,250-lb-thrust J85-GE-3 for the McDonnell GAM-72 Quail had a thrust-to-weight ratio of 8.68:1 and the last version, the 5,000-lb-thrust J85-GE-21 for the Northrop F-5E/F, had a 7.3:1 ratio. Of significance is the fact that the J85 remained one of the highest thrust-to-weight ratio engines in the world for many years after it was conceived.

The J85 was notable for its combustor and afterburner technology. The heat release per unit volume per atmosphere of the very compact J85 combustor was, in fact, superior to any other GE engine for some time. The J85 was also unique because it was the first U.S. small turbine engine to go into production with an afterburner. By shortening the length of the afterburner to less than that of the J79 engine afterburner, GE was able to increase the ratio of the heat release per unit volume per atmosphere. The J85 also introduced, for the first time on a GE engine, a now-standard afterburner technology: a variable exhaust nozzle using flaps to vary the exhaust nozzle throat area through a continuous range to accommodate the considerable changes in volumetric flow associated with the gas temperature rise in the afterburner. Earlier GE engines, such as the J47, had used an eyelid-type of movable clam shell mechanism as a variable exhaust nozzle. Finally, afterburner flame holder technology was improved in the J85 by replacing standard V-gutters with U-shaped bluff bodies with through-flow fuel/air scoops. This permitted the bluff bodies to act more like actual combustors than simple flame holders.

THE CJ610 AND CF700 BIZ JET ENGINES

J85-derived engines, known as the GE CJ610 and CF700, powered the majority of the first generation of business jet aircraft during the 1960s and early 1970s. In July 1964, the CF700 engine became the first small turbofan engine to be certificated in the U.S. Building on a substantial J85 production base for military aircraft, it was logical for GE to seek civilian aircraft applications for its successful, high-performance engine.

By the late 1950s and early 1960s, airlines were rapidly replacing their piston-powered airliners with jet aircraft. However, large corporations were generally still using piston-powered World War II conversions, light twins, and small airliners to transport top executives. Market surveys forecasted a tremendous market potential for business and utility aircraft.[107] Predictions suggested that, by 1970, the majority of all multi-engine business aircraft would be turbine-powered.[108] In fact, by the late 1950s and early 1960s, some very large corporate operators had already begun to switch from piston engine aircraft to turboprop aircraft then available, including the Vickers Viscount, Grumman Gulfstream G I, and Fairchild Friendship F-27. The reasons for this were the

---

107. Harvey J. Nozick, "Business Aircraft of the Future," *Business & Commercial Aviation*, (January 1961): 26.

108. Nozick, 27.

greater speed, comfort (less vibration), availability (less downtime due to main-tenance and overhaul), higher resale value, all- (i.e. over-the-) weather perfor-mance, and prestige of turbine aircraft.

Within this predicted corporate market was a need for smaller high-performance aircraft with speeds equivalent to jet airliners. The first U.S. entries into this field were jets developed for the Air Force. On August 1, 1956, the Air Force announced that it would purchase a privately developed, flight-proven aircraft meeting its Utility Trainer Experimental (UTX) and Utility Cargo-Transport Experimental (UCX) specifications. In response to the UTX require-ment, North American developed its prototype NA-246 UTX design (later commercially designated Sabreliner). Lockheed and McDonnell vied for the UCX with Lockheed developing its prototype L329 (later commercially desig-nated JetStar) and McDonnell the Model 119 (later Model 220). However, be-cause of budget cutbacks, neither the UTX nor UCX aircraft could be acquired by the Air Force when planned. In October 1958, the Air Force did place an initial production order for the North American Sabreliner, later designated T-39A. In June 1960, the Air Force also placed a small initial order for Lockheed JetStars, given the military designation C-140 and powered with P&WC J60 (JT12) turbojet engines. Seeking a civilian market, all three companies cer-tificated their aircraft in the early 1960s. Also seeking to enter this executive jet niche were companies such as Bill Lear's Swiss American Aviation Corporation; Dassault; Aero Commander; and Israel Aircraft Industries, which eventually ac-quired Aero Commander.

On September 16, 1958, GE J85s powered the prototype Sabreliner on its first flight. However, GE lost the early Sabreliner business when Pratt & Whit-ney J60s were selected for the military production version. The J60 had a higher thrust rating than the J85, which proved critical when Air Force-ordered design changes resulted in a heavier production aircraft.[109] Despite this setback, GE had additional opportunities as a number of other small jets were in design to meet the executive transport market.

About June 1959, GE's Small Aircraft Engine Department decided to enter the business jet market and to develop a fan engine with company money. In November 1959, a project was set up for the CF700 aft fan commercial engine, funded by GE. The project eventually also assumed responsibility for the civil version of the non-afterburning J85 turbojet engine known as the CJ610.

109. Kenneth W. Hamilton, "Development of the Gas Turbine Engine Installation in the North American Aviation Sabreliner," (SAE Paper No. 650363 presented at the Business Aircraft Conference, Wichita, Kansas, 6–8 May 1965), 2.

Announced in 1959, the 4,000-lb-thrust CF700-1 was a subsonic version of the J85 mated to a bypass aft fan unit, designed for use in light and medium weight corporate or utility transports.[110] In May 1960, the CF700 ran for the first time. Also, in May 1960, GE announced the CJ610 engine, a slightly modified J85. The development and application of the CJ610 was undertaken within the CF700 Project, whose project manager was Jim N. Krebs.[111] By 1961, GE was developing the 4,200-lb-thrust CF700-2B, 2,850-lb-thrust CJ610-1, and 2,400-lb-thrust CJ610-2B for executive jets.[112] The CF700 completed its official 150-hour endurance testing in April 1961, but formal military approval and civil certification were not sought until aircraft applications were more immediate. The CF700 completed MQT requirements in May 1964 and FAA certification on July 1, 1964.

The application that launched the CF700 program was the McDonnell 119/220. At the same time that McDonnell was competing for the Air Force navigation trainer in 1959, it was also marketing this aircraft as an executive transport. The aircraft was then flying with two Curtiss-Wright Orpheus turbojets. McDonnell had negotiated with Pan American World Airways, Inc., the sale or lease of 170 Model 119B aircraft with four CF700s. In May 1961, McDonnell considered the engineering work required to install GE CF700-1 engines, but decided against using them pending a Pan Am order. Negotiations between McDonnell and Pan Am foundered, and McDonnell dropped the Model 119/220 project in 1965.

SELLING THE CJ610

GE continued to search for commercial applications for the CF700 program. A visit in 1960 to Aero Commander in Bethany, Okla. proved fruitful, but in an unexpected way. The company planned to install Pratt & Whitney JT12 engines in their new business jet design called the Jet Commander. GE Engine Specialist Ron Krape and CF700 Project Manager Jim Krebs convinced the company to consider another alternative. When it was clear that a turbofan engine the size of the CF700 was too large for the aircraft, Krebs later recalled saying to Aero

110. "News Digest," *Aviation Week & Space Technology*, Vol. 71, No. 11 (September 14, 1959): 39. The decision to certificate the CJ610 and CF700 engine for commercial aircraft use was made by GE in either June or July 1959.

111. "Jet Engine Designed for Business Aircraft," *General Electric Jet Times*, 1967, 5.

112. Andy Keil, "GE Eyes CJ610, CF700 Executive Sales," *Aviation Week & Space Technology*, Vol. 74, No. 16 (April 17, 1961): 131.

Commander's Vice-President of Engineering Research and Development Ted Smith, "Well, let's talk about this because it wouldn't be that big of a deal for us to take the fan off our engine."[113] Subsequently, this was agreed to, and on December 6, 1961, a contract for 100 CJ610 engines was signed. Aero Design and Engineering had become the first CJ610 customer and the Jet Commander the first application.

The engine designed for the Jet Commander was the CJ610-1, the first engine in the CJ610 series, derived from the J85-GE-5. Rated at 2,850 lb thrust, the CJ610-1 engine became the first small turbojet engine to be certificated in the U.S. (on December 6, 1961), and it powered the Jet Commander 1121 on its maiden flight on January 27, 1963.

The second CJ610 customer was Bill Lear. In 1959, Lear's Swiss American Aviation Corporation based in Geneva, Switzerland (later Wichita, Kan.) had begun the first designs of what would become the Learjet.[114] The Learjet was intended to be a very high-performance, high-altitude, relatively low-priced executive jet seating up to seven passengers. It was originally designed around a Continental J69 engine. Top level GE engineering teams met with Lear and his engineers in Switzerland in 1961 to determine the feasibility of installing CJ610 engines and to convince Lear to buy them instead of J69s. One GE sales pitch to Lear was that the dry J85 would "make a sports car out of your airplane."[115] Actually, Bill Lear preferred to think of his airplane as a "Cadillac." Advantages of the CJ610s over J69s on the Learjet were improved performance and extended range.[116] On April 6, 1963, Lear signed the initial engine contract with GE for 10 engines. On October 7, 1963, the prototype Learjet Model 23 flew for the first time, powered by CJ610-1 engines. In November 1963, 86 engines were put on order for the Learjet.

The next engine in the CJ610 series was the CJ610-4, installed on an improved Learjet. Like the CJ610-1, the -4 engine was derived from the J85-GE-5. The -4 engine differed from the CJ610-1 in accessory gearbox location, and weighed 4% less. In May 1964, the -4 engine was certificated at 2,850 lb thrust.

---

113. Jim Krebs, interview by Rick Leyes and William Fleming, 21 August 1991, GE Interviews, transcript, National Air and Space Museum Archives, Washington, D.C.

114. Originally, the aircraft was known as the Lear Jet. The name was changed to a single word in 1969.

115. Travers, *Engine Men*, 186.

116. Cecil Brownlow, "Overseas Firms Plan Executive Transports," *Aviation Week & Space Technology*, Vol. 74, No. 10 (March 6, 1961): 89.

It was installed on the Learjet Model 24, a higher gross weight aircraft certificated to FAA Transport Category standards for charter or commuter airline use in addition to executive transport. The Learjet Model 24 first flew on July 31, 1964.

Certificated in June 1966, the 2,950-lb-thrust CJ610-6 engine, developed from the -4 engine, provided a 3.5% thrust increase over the -4 engine and a 2% cruise sfc reduction. The Learjet Model 25, a stretched version for eight passengers, was certificated in 1967 with the CJ610-6. Later, Learjet Models 24B/C/D as well as Models 25B/C received their certification with the -6 engine as well.

The CJ610-8A, rated at 2,950 lb thrust, was certificated in April 1977 for use on Learjet Century III Models 24 E/F and 25 D/F/G Models. The -8A engine was designed for operation up to 51,000 ft to give better economy and over-the-weather capability for these high-altitude Learjet models. The aircraft was actually flown to 55,000 feet. The main differences were a longer-life turbine and a turbine nozzle area change. The -8A engine was also used on the high-altitude, longer-range Learjet Model 28 and 29 Models ("Longhorn" series).

The fast, glamorous Learjet became the best selling jet in the world for many years. By the end of 1981, the Gates Learjet Corporation had led the industry in the total number of business jet aircraft delivered for 17 consecutive years.[117] Learjet then accounted for 25% of the world's business jet fleet. Of these Learjets, the Model 20 Series was CJ610 powered. The Learjet Model 20 Series, including the Models 23, 24, 25, 28, and 29, was the largest single CJ610 airframe application. After the Garrett TFE731 turbofan-powered Learjet Models 35 and 36 were introduced in 1973, the Model 30 Series and later models began to replace the first generation Learjets. By 1975, the number of Learjet Model 30 series delivered exceeded the number of Learjet Model 20 series delivered. As a result, the CJ610 eventually lost ground to the TFE731. However, as a turbojet, the CJ610 performed efficiently at a higher altitude than did the turbofan TF731; thus, for a period of time, a limited demand continued for the very high altitude Century III Learjets.

The National Air and Space Museum has in its collection N802LJ, the second Learjet built and the production prototype of this famous line of business jets. The Learjet, as represented by this prototype, is historically significant because it

117. Donald J. Porter, *Learjets: The World's Executive Aircraft* (Blue Ridge Summit, Pa.: Tab Books, 1990), 34. On April 10, 1967, The Gates Rubber Co. took controlling interest in Lear Jet Industries. In 1969, Lear Jet Industries became Gates Learjet Corporation, and the airplane was redesignated the Gates Learjet.

was first U.S. jet aircraft designed and produced specifically for business and corporate operation.

Several foreign applications were found for the CJ610. In 1962, Hamburger Flugzeugbau (HFB) made a decision to substitute CJ610 engines for Pratt & Whitney JT12 engines on its HFB 320 (Hansa) executive transport.[118] Ed Woll had visited the management of HFB, which he had not met previously, offered them a deal that "they couldn't refuse," and consummated the arrangement on a handshake.[119] Any further consideration of the Pratt & Whitney engine was subsequently dropped.

Germany's first civil jet, the Hansa 320, an unusual design with forward-swept wings, flew for the first time on April 21, 1964, with CJ610-1 engines. Starting in 1968, Hansa 320s were equipped with 2,950-lb-thrust CJ610-5 engines. The -5 engine, developed from the CJ610-1, had improved take-off thrust and reduced fuel consumption; it was certificated in June 1966. The final version of this aircraft was powered by 3,100-lb-thrust CJ610-9 engines, certificated in January 1968. Only 45 Hansa jets were built, making this a relatively limited application for CJ610 engines.

In 1967, Israel Aircraft Industries (IAI) acquired all production and marketing rights for the North American Rockwell Corporation (formerly Aero Commander) Jet Commander executive jet. Renamed the Commodore Jet, the 1121B was powered by 2,950-lb-thrust CJ610-5 engines. As a result of modifications and improvements, the Model 1121 evolved into the Model 1123 Westwind in the early 1970s. The Westwind was powered by 3,100-lb-thrust CJ610-9 engines.

Thus, in the early 1960s, a number of applications had been found for the CJ610. The CF700 development program, which had been delayed when McDonnell did not select its engine for the Model 119/220, was still looking for a home.

THE CF700 MAKES A COMEBACK

Pan American still wanted to get into the executive jet market and had been searching for an airplane that was different from all of the others. A turbo-fan-powered jet would be an ideal candidate, and GE had the prototype engine

118. "GE CJ610-2B Turbojets Chosen To Power Hamburger HFB.320," *Aviation Week & Space Technology*, Vol. 77, No. 6 (August 6, 1962): 109.

119. Ed Woll, Boca Grande, Florida, to Rick Leyes, letter, February 1993, National Air and Space Museum Archives, Washington, D.C.

which could make it possible. Pan Am had first attempted to strike a deal with McDonnell and then with de Havilland, but nothing resulted from the negotiations. Pan Am next turned to Dassault. Dassault was willing to accept the price, quick delivery schedule, and stringent performance specifications that Pan Am required for its executive jet, provided that an engine manufacturer would also accept Pan Am's requirements. This was the opportunity that GE had been looking for. In April 1963, the pace of the CF700 program picked up again as a result of negotiations for the use of the engine on Dassault's Mystére 20 jet.

GE's General Manager Ed Woll and Marketing Manager Fred Brown commuted across the Atlantic to both sell the GE engine and negotiate the difficult contract conditions set by Dassault. Driven by Pan Am's stringent requirements, Dassault wanted severe penalties for failure to deliver on time and meet performance specifications; they also wanted a low introductory price for the engine. At Dassault, negotiations started in the afternoon and continued on until late at night, day after day. With the traditional French holiday month of August looming as a deadline, both sides realized that it was now or never. With compromises on both sides, an acceptable deal was struck in the early morning hours of August 1. Dassault employees then went on vacation, and GE went to work to meet its first key contractual commitment—the delivery of two flight test engines by the end of the year.[120]

Also, in August 1963, Pan Am formally created its Business Jets Division to diversify and to sell business jet aircraft.[121] This company had selected the CF700-powered Mystére 20 as its sales product and had renamed the Americanized version the Fan Jet Falcon. This selection breathed new life into the CF700 program, and in late September, drawings for the new CF700 were released for manufacturing. At the end of 1963, only five months after signing the contract, GE met its first commitment by airshipping the two flight test engines to Dassault as promised.

The first production engine was shipped to Dassault in March 1964. The 4,200-lb-thrust CF700-2B became the first small turbofan engine certificated in the U.S. on July 1, 1964.[122] The first flight of the CF700 engine was on the Falcon on July 10, 1964. The Falcon thus became the first turbofan-powered business aircraft.

One problem of note occurred during early Falcon flight tests. The first engine nacelle design had separate inlets for the fan and the engine core. It was dis-

120. Woll to Leyes, February 1993.

121. "Business jets and the Fan Jet Falcon," *Interavia*, Vol. 22 (May 1967): 767.

122. The certification of the CF700-2B was renewed in July 1963.

covered during flight tests that this configuration led to a serious installation drag problem and a subsequent shortfall in aircraft range. A team of GE installation aerodynamicists led by John Kutney and Dassault engineers worked together to redesign the nacelle so that the engine core and the fan had a common inlet. Additional wind tunnel tests revealed that repositioning the new nacelle relative to the rear of the fuselage would realize a further decrease in drag, enabling the airplane to meet its original design specifications.

In the U.S., the Falcon 20's primary customer was Federal Express, which modified the 10-seat executive jet to a cargo configuration and used it for its overnight air express service. The Federal Express Falcon 20 inaugurated the world's first airline devoted exclusively to air express. The first of these aircraft, N8FE, now part of the National Air and Space Museum collection, was powered by two GE CF700-2D engines. In spite of the fact that the CF700 had not been designed to withstand the rigors of airline-type operation of 2,000 duty cycles a year, GE, under the direction of Art Adinolfi, implemented a service and support program that Fred Smith, the head of Federal Express, called outstanding and a major contributor to the success of Federal Express.

In July 1965, the CF700-2C, was certificated at 4,125 lb thrust. It was installed on the initial production version of the Falcon 20 and on the Falcon 20 Series C. The -2C engine was flat rated for additional hot day performance. This was followed by the CF700-2D, rated at 4,250 lb thrust and certificated in January 1968. The increased performance -2D engine was a modified -2C engine with a redesigned compressor turbine having a higher thermodynamic efficiency. Improved materials were also introduced in other components. The -2D engine had a 3% take-off and 4% climb-thrust increase as well as a 3% sfc improvement compared to the -2C engine. The Falcon 20 Series D was certificated with the -2D engine. In October 1969, the -2D2 engine, rated at 4,315 lb thrust, was certificated. The -2D2 engine had a 1.5% take-off thrust increase compared to the -2D engine and incorporated a new-design tailpipe. The Falcon 20 Series E and F were certificated with the -2D2 engine.

In 1971, Rockwell International made a decision to use the -2D2 engine on its Sabre Model 75A business jet aircraft. Originally, the Model 75 had been designed for the Garrett ATF3 engine. When this engine was not available in time, the Pratt & Whitney JT12 was substituted. With the JT12, the Model 75 was underpowered. The installation of the CF700-2D2 allowed the Sabre Model 75A to meet its design performance goals.[123] The CF700-2D2 improved engine

123. Richard N. Aarons, "The Sabre 75A," *Business & Commercial Aviation*, Vol. 33, No. 3 (September 1973): 66-68.

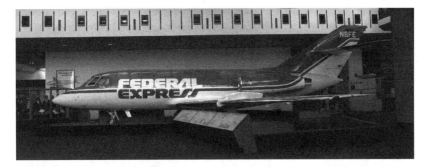

GE CF700-powered Dassault Falcon 20. The Federal Express Falcon inaugurated the first airline devoted exclusively to air express. This aircraft is in the National Air and Space Museum's collection. National Air and Space Museum photo by Mark Avino, Smithsonian Institution (SI Neg. No. 84-1542-3).

was certificated in May 1975 at 4,500 lb thrust and was also used on the Rockwell Sabre 75A. Ironically, GE had finally succeeded in getting back some of this business after powering the prototype Sabreliner many years earlier.

An interesting application for the CF700 was the Bell Lunar Landing Research and Lunar Landing Training Vehicles, powered in part by a 4,200-lb-thrust CF700-2V turbofan engine, and designed to provide realistic moon landing simulations for the Apollo Program. The turbofan engine, modified for vertical operation and installed on a gimbal mounting behind the cockpit, was controlled automatically and provided lift equal to five-sixths of the vehicle's gross weight. Thus, the engine counteracted five-sixths of the Earth's gravity, the remaining one-sixth being comparable to gravity on the Moon. The first flight of the Lunar Landing Research Vehicle was on October 30, 1964.

The development funding for the CF700 that was used in the Lunar Landing Trainer was provided by a 1963 Air Force contract. Subsequently, a military qualification program was conducted in parallel with the civilian CF700 certification program. A 50-hour vertical operation test was completed in 1963. The CF700 was given the military designation TF37-GE-1.

CONVERTING A MILITARY ENGINE INTO COMMERCIAL ENGINES

From a technical perspective, both the CJ610 and CF700 used the J85 gas generator, fuel control, and gearbox as a building block. The principal changes on the CF700 gas generator were a lower turbine inlet temperature, varying gearbox accessory drive locations, incorporation of a fuel heater, and a different lubrication system. On the CF700 engine, a fan component was attached to the rear flange of the turbine frame. The fan rotor was aerodynamically coupled to the gas generator rotor. The inner airfoil portion of the fan rotor blades formed a turbine driven by exhaust gases from the gas generator. The outer airfoil formed a compressor or a fan. Bypass air was drawn into the engine by this fan, compressed, and discharged through the exit guide vanes. The exhaust gas from the two streams mixed in a converging tailpipe.[124]

The arrangement of compressor blades at the tip of the fan and turbine buckets at the root inspired GE's nomenclature for them, i.e. "bluckets."[125] The aft fan was based on the same aerodynamic design as the GE CJ805-23 aft turbofan engine used on the Convair 990. It was determined that a bypass ratio of approximately two would give the best overall performance for the intended applications. In comparison to the CJ610, the addition of a fan to the CF700 resulted in a 45% increase in take-off thrust, a reduction in cruise sfc by 12–14%, and a reduction in the take-off noise level by one-half for a weight increase of under 300 lb.[126] This improved performance resulted in a range increase of 15–20%, a 20% decrease in field (take-off) length, and doubled commercial airport capability (i.e. more runways were available due to a shorter runway requirement).[127]

The original CF700 preliminary design was directed by Project Manager Jim Krebs and the original detail design was supervised by Manager of CF700 Engineering Brian Rowe, who led a design team that included Art Adinolfi, Mel Bobo, George Eddy, Bob Hufnagel, and Bob Smuland. The top management team of Gerhard Neumann, Fred Garry, and Ed Woll provided supervisory support and guidance to the program. With the exception of a slightly choked turbine problem, many of the design problems of the aft fan component had been worked out in the CJ805-23 program.[128] The aft fan arrangement was chosen because it was the fastest, least risky, and least costly way to make a fan

---

124. Brian H. Rowe, "The Design and Development of the CJ610 Turbojet and the CF700 Turbofan Engines for Use in Business Jet Aircraft," (SAE Paper No. 650381 presented at the Business Aircraft Conference, Wichita, Kansas, 6-8 May, 1965), 136.

125. David A. Brown, "GE Seeks Company Pilot Views in Defining CF700 Engine Loan, Overhaul Programs," *Aviation Week & Space Technology*, Vol. 80, No. 14 (April 6, 1964): 101.

126. "CF700: General Electric's Baby Aft Fan Engine," *Flight International*, Vol. 86, No. 2910 (December 17, 1964): 1051.

127. Robert I. Stanfield, "General Electric's Executive Turbines," *Flying*, Vol. 68, No. 5 (May 1961): 86.

128. In a slightly choked turbine, the gas flow through the turbine is close to the turbine's capacity limit due to the high Mach number of the gas flow as a result of a smaller than required effective turbine area.

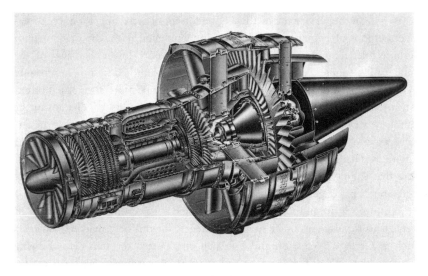

Cutaway of GE CF700 turbofan engine. Photograph courtesy of GE.

engine. The CJ610 was a single-shaft engine, and there was no way to run an LP (low pressure) shaft through the middle of the engine. By following the CJ805-23 precedent, an aft fan package could be "bolted on" and thereby transform the single-shaft engine into a moderate-bypass turbofan. While it was a significant improvement over the straight turbojet from which it was derived, the performance of the aft fan was limited by the fact that it did not supercharge the compressor as did a front fan, in most cases.

A focus of the development and testing program was to adapt the military J85 engine to commercial operation. In addition to standard component and endurance tests, the engine met other FAA certification requirements including bird ingestion, overspeed, and low-cycle fatigue tests. However, unexpected low-cycle fatigue problems did follow the engine into service. While it was originally thought that the number of duty cycles on executive jets would be less than those on military jets, the reverse turned out to be true. Executive fleet leader programs were established whereby engines subjected to accelerated use were closely monitored to uncover problems such as this. Furthermore, new methods of predicting rotor parts' fatigue life by numerical computation and experimentation were developed.[129]

129. Frederick W. Garry, "Designing Business Jet Aircraft Engines for Increased Service Life," (SAE Paper No. 670234 presented at the Business Aircraft Conference, Wichita, Kansas, 5-7 April 1967), 19-33.

To extend the service life of the engine for commercial operation, conservative methods of engine operation were implemented. First, the gas generator turbine inlet temperature was set approximately 60°F lower than that of the military J85 engine. Secondly, CJ610 power was monitored by Engine Pressure Ratio (EPR), rotor speeds, and exhaust tailpipe temperature as opposed to exhaust temperature (T_5) and gas generator speed on the J85.[130] The advantage of this new system, which gave a direct measurement of thrust, was that the engine was not taken to maximum temperature at each takeoff.[131] Finally, the engine was flat rated to higher standard day temperatures (86°F) in order to ensure that engine performance would be met at various airport density altitude and aircraft passenger load configurations.

As more CJ610 and CF700 engines entered service on executive jets, GE needed to develop customer support strategies for this new market. Managers such as Gerhard Neumann and Ed Woll emphasized the necessity of being number one in customer service and support. Accordingly, headquarters for small commercial engine customer services was transferred from Evendale to Lynn, where Art Adinolfi was made manager with responsibility for the CT58, CJ610, and CF700 product lines. Operator seminars were held to discuss problems, receive owner and operator feedback, provide information on product improvements, and define loan and overhaul programs. An emergency service phone line with the number (216) GE CARES was set up for 24-hour service. For operator convenience, GE service and overhaul facilities, technical service representatives, and distributors were located around the world. A warranty protection program was established, the TBO of the engine was increased significantly over time, and engine upgrades were made available.

Another example of good customer rapport was a quick response technique developed by then project manager for the CJ610/CF700 Program, Art Adinolfi. Whenever there was an engine problem, Adinolfi communicated immediately by telegram with the senior executive on the flight:

> Aware of your unfortunate experience. We apologize for any inconvenience that it may have caused you and your associates. Please be assured that upper management at General Electric is taking a real hard look at this problem, and we'll be back shortly to tell you what went wrong and what we're going to do to fix it.[132]

130. The same system was used for the CF700, except that fan speed was also monitored.

131. Rowe, "CJ610 and CF700 Engines," 142.

132. Art Adinolfi, interview by Rick Leyes and William Fleming, 22 August 1991, GE Interviews, transcript, National Air and Space Museum Archives, Washington, D.C.

THE RISE AND FALL OF GE'S BIZ JET MARKET

Although costly in terms of available resources, product support strategies such as these were effective in helping sell GE small commercial engines. GE promoted both the CJ610 and CF700 on the basis of a large military J85 production base, extensive military flight hours, the highest thrust-to-weight ratio available in their respective classes, maintainability, fuel economy, and competitive pricing. In less than two years after entering service, the combined applications for the CJ610 and CF700 engines accounted for over 50% of the total turbojet/turbofan-powered business aircraft in use.[133] Sales for the GE CJ610 outpaced competitor turbojet engines such as the Pratt & Whitney JT12 and Bristol Siddely Viper 20.

However, by the late 1960s and early 1970s, increasing environmental concern, an Arab fuel embargo, and the desire for better range had shifted consumer interest toward turbofan-powered aircraft. GE responded by placing increased emphasis on noise suppression, emission reduction, and better fuel consumption. About the same time, airframers such as Lear, Lockheed, and Dassault were considering upgrading and re-engining their executive jets. GE seriously studied the possibility of a new dual-rotor front-fan engine designated the T64/F1 but referred to informally as the "CF64," the core of which was common to the T64 turboshaft engines. While a demonstration engine was tested, a decision was made not to proceed with further development.

The CF64 would have been an expensive undertaking and did not have a military development and production base to underwrite it. There were also competing priorities within GE, among them the TF39, J79, CF6, and GE4 programs. According to retired GE Vice Chairman Jack Parker, "To be perfectly honest, I felt that we had to use our engineering talent where it would be best for the company. Some of the big engines were selling for a million dollars a copy, and the small engines were selling for $150,000 a copy. There was about as much engineering on each, so we opted to go for the large sizes."[134] To this, retired Vice President and Group Executive Gerhard Neumann added:

> The maintenance requirement (for business jets) is so dramatic . . . There are not enough sales. One guy buys one airplane or two airplanes . . . a company maybe three airplanes and there are only six engines sold. You have to send a guy down there. It isn't worth it . . . We ought to give the customer all the

133. General Electric, "CJ610 Jet Power for Business Jet Aircraft," GE Brochure No. FPD-122-5-67 (3000), May 1967.

134. Parker to Leyes, 3 December 1991.

service he wants, but I could not promise the buyer of business jet engines the support that I (would) like to give them, because we just didn't have enough people for that . . .. So we deliberately let the business jet go.[135]

Thus, in the 1970s, GE relinquished its leading position in the business jet market and concentrated its commercial engine resources on its large engine programs because of the high support requirement relative to the number of business jet engines sold. Between its introduction in 1973 and the end of 1977, Garrett's TFE731 captured more than 45% of the small to medium business jet market. Much of Garrett's success was at the expense of GE.[136] Pratt & Whitney's JT15D was also a strong competitor. While GE continued to maintain and produce CJ610 and CF700 engines, it did not actively seek to get back into this market until it later introduced the CF34 turbofan engine.

CJ610/CF700 SUMMARY

The production line for the CF700 closed at the end of 1981 after 18 years of continuous production and some 1,165 CF700 engines. CJ610 production continued until 1982, when 2,059 CJ610 engines had been manufactured. While some doubt exists about these production numbers, the combined total of CJ610 and CF700 engines (of all models) manufactured was 3,224.

The two top production years for the CJ610 were 1965 (299 engines) and 1966 (369 engines); after that time, yearly production declined significantly, ranging from a high of 124 (1975) to a low of 32 (1977). In part, the decline was due to competition from other fanjet engines. The CJ610-6 was the most popular engine model, with 808 CJ610-6 engines manufactured for the Learjet Model 24B/C/D and the Learjet Models 25/B/C.

For the CF700, the top three production years were 1966 (128 engines), 1967 (154 engines), and 1968 (133 engines). Thereafter, production varied from a low of 22 in 1970 to a high of 102 in 1979. The most popular engine model was the CF700-2C, of which 406 were manufactured for the Falcon 20 and Falcon 20 Series C.

The CJ610/CF700 engines powered the majority of the first generation business jet aircraft during the 1960s and early 1970s. For their time, the CJ610 and

135. Gerhard Neumann, interview by Rick Leyes and William Fleming, 22 August 1991, GE Interviews, transcript, National Air and Space Museum Archives, Washington, D.C.

136. J. Philip Geddes, "Giants Battle in US Small Turbine Market," *Interavia*, Vol. 33 (March 1978): 180.

CF700 had the highest thrust-to-weight ratio of any small commercial turbine engine in service.[137] They were, respectively, the first small turbojet and turbofan engines certificated in the U.S. The CJ610's most important application was the Learjet Model 20 and the CF700's was the Dassault Falcon 20. Due to competing priorities, GE elected in the 1970s to let go of its leading position in the business jet aircraft market.

The T64 Turboshaft/Turboprop Engine

As a result of the decision that had been made in the fall of 1953 by SAED General Manager Jack Parker and Engineering Manager Ed Woll to add two more engines to the department's product line, preliminary design work on a 2,500 shp turboshaft engine was begun. There was no specific application for the engine, but investigation revealed several possibilities including a heavy lift helicopter and turboprop aircraft. A principal objective was to supersede the performance capability of the large Pratt & Whitney and Wright radial engines then installed in such applications. It was decided that the new engine should have the following characteristics: an sfc lower than the T58; more than double the power of the T58; and a better cycle with about 12:1 pressure ratio—a turboshaft engine about twice the size of the T58.[138]

The first layout of the new turboshaft engine was made by Louis Lechthaler late in 1953 at SAED's headquarters, then in an office building in North Station in Boston. Frank Lenherr was responsible for the compressor, John Benson and Jack Pierce were responsible for the combustor, and Art Kohn handled the turbine. Jack Parker and Ed Woll defined the key parameters of the engine and how far the design would be stretched. Preliminary design work continued in 1954.

The T58 program had given GE an entree with the Navy, which had a tradition of buying many of their engines from Pratt & Whitney. Now the Navy was a potential customer for GE's new turboshaft design. The responsibility for the design, the proposal, and the strategy to sell the engine was in the engineering department. The key element of the strategy was to put the new engine's advanced technology compressor on test at the time of the proposal. If it could

---

137. This is true comparing the CJ610 to the Pratt & Whitney JT12A, Turboméca Gabizo and the Bristol Siddely Viper 20 turbojet engines and by comparing the CF700 to the Turboméca Aubisque turbofan engine.

138. Travers, *Engine Men*, 197.

demonstrate the feasibility of the compressor, GE knew that it would have an important sales point and that Navy concern about an advanced technology compressor would be reduced. Accordingly, with its own R&D money, GE decided to design, build, and put a compressor on the test stand that was specifically directed to help the company win a Navy competition, should it take place. GE then submitted an unsolicited proposal that was successful in inspiring the Navy to establish a formal competition.

After the Navy issued a formal RFP, Aeronautical Engineer Dennis Edkins put together a proposal for the turboshaft engine. At the time the proposal was submitted in December 1956, the new compressor had not yet been tested. When the first T64 compressor runs were made in 1957 and test results came in, GE began to send to the Navy on a regular basis actual and predicted compressor maps based on the data. Ironically, on one occasion Ed Woll happened to be in the Navy office in Washington when the results were phoned in, and he put the test points on the predicted compressor map that had been submitted to the Navy. This data was crucial in establishing the credibility of GE's proposal.

Competing against Lycoming, GE was awarded a $58.5 million U.S. Navy Bureau of Aeronautics engine development contract in April 1957. When the award was announced publicly on May 8, the engine, then called the T64, was to be developed as both a turboshaft and turboprop engine.[139] SAED's engineering organization directed the new program, and later Ed Woll took over the T64 and directed its early development until he became the General Manager of SAED.

The T64 was designed as a free turbine "convertible" (i.e., turboshaft or turboprop) engine. As such, it was GE's first bi-functional power producer. The compressor was a 14-stage axial-flow type with a pressure ratio of 12.6:1 and an airflow of 24.5 lb/sec. The inlet guide vanes and first four stages of stator vanes were variable to assure surge-free characteristics during starting and partial-power operation. The compressor rotor was originally designed with 14 individual disks. However, early in the program it was redesigned to include three forward disks joined by curvic couplings followed by two smooth, integral spools, comprising eight and three stages, respectively. The engine had a straight-through-flow annular combustor. It had a two-stage gas generator turbine and two-stage power turbine, the latter mechanically independent of the gas generator turbine. The 3:1 power-to-weight ratio engine was designed for the 2,600-shp class.

139. General Electric, "Multi-Million Dollar Contract Awarded General Electric to Develop New T64 Engine," GE Brochure or Press Release No. GED-3437 (.5M) 5-57, May 1957.

The primary design goal was low specific fuel consumption—-the 1957 model specification called for a guaranteed sfc of .506—with weight a secondary factor.[140] This goal was driven by the fact that the primary competitor for the T64 was still the piston engine. Once the temperature limits of the new engine were established, GE determined that it could boost the compressor pressure ratio. A key factor required to achieve the low sfc goal was a high pressure ratio compressor, and the decision was made to stretch the ratio to better than 12:1.[141] The decision later proved sound because the high pressure ratio, high-capacity compressor also provided the T64 with extra performance margin.

Conventional wisdom suggested that a three-stage gas generator turbine should drive the high pressure ratio, 14-stage compressor. For an appreciable period, GE pursued a 2,500-shp study with five turbine stages (three-stage gas generator and two-stage power turbines).[142] However, engine cost, length, and weight concerns suggested that a two-stage gas generator turbine would be better. Studies indicated that it was a risky decision, but technically feasible. When the engine ran, the two-stage turbine performed well, vindicating the decision. The T64 was the first GE engine to have such a high pressure ratio compressor driven by a two-stage turbine. This established a new GE benchmark in turbine loading. Among the features that made this achievement possible were: GE's first film-cooled stage-one turbine nozzle; GE's first application of abradable honeycomb labyrinth seal stators and turbine blade shrouds (to control high pressure turbine tip clearance and air leakage); and cast turbine blades with hollow tips to provide the required thickness to the tip airfoil sections without compromising centrifugal stresses at the blade roots.[143]

The T64 was also GE's first front-drive, free turbine engine. The front drive was necessitated by GE's desire to develop a turboprop version of the engine and by the fact that airframers preferred a front drive to make the engine easier to install in helicopters. Ed Woll commented, "It (the front-drive engine) was a departure for us. Now, we had more confidence that we could do this job. We knew that we could do the compressor, combustor, turbine, and so forth. We

140. Cecil Brownlow, "T64 Designed in Three Basic Versions," *Aviation Week & Space Technology*, Vol. 70, No. 23 (June 8, 1959): 61.

141. Woll to Leyes and Fleming, 20 August 1991.

142. W. T. Gunston, "T64: Design Philosophy Behind GE's New Shaft Turbine," *Flight*, Vol. 79, No. 2075 (January 13, 1961): 63.

143. Edward Woll, Fredric Ehrich, and William M. Meyer, "The T64: GE Aircraft Engines' Most Enduring Product," *The Leading Edge* (Winter 1994): 27.

added what we might call the green stuff, and we bored a hole through that engine."[144]

One of the major complications for GE was bringing a long, slender shaft through the center of the engine. According to Dr. Fred Ehrich, then Manager of T64 Engine Design:

> GE engines, since the time of the T58, had axial-flow combustors. It was a major deviation from the traditional style of small engines, which, for the most part in that era, and still to a large extent, had reverse-flow combustors, which had a larger diameter and allowed the turbine to be tucked in underneath the combustor. This made for a shorter engine, which in terms of rotor dynamics, rotor critical speeds, made it a much easier job of taking a long slender shaft through the engine. So no GE engine up to that time had a long slender shaft through it with a front drive. Of course, achieving a sound rotordynamic design with the superior axial-flow compressor was a basic ingredient not only for making the T64 a successful engine, but preparing GE for the era of front fans.[145]

There were two basic T64 configurations based on the same gas generator that could be converted from a turboshaft to turboprop engine with a minimum of effort and a maximum of parts interchangeability.[146] The first was a turboshaft engine, the models of which were the T64-GE-2 and -6. Both were originally rated at 2,650 shp. Later the -6 engine was uprated to 2,850 shp. Intended primarily for helicopter operation, the -2 had a 2.61:1 reduction ratio achieved by a offset helical gear driven by a high-speed pinion from the power turbine shaft. The -6 engine had no reduction gearing and was designed for VTOL aircraft.

The second T64 configuration was a turboprop engine with models T64-GE-4 and -8. The -4 engine was originally rated at 2,570 shp and the -8 at 2,700 shp. Later both models were uprated to 2,850 shp. A second-stage planetary reduction gear was added to the -2 gearbox, yielding a combined reduction ratio of 13.44:1 in order to achieve appropriate propeller speeds. The engine could be used as a turboprop with the offset prop shaft center line above (T64-GE-8) or below (T64-GE-4) the gas generator center line.[147]

144. Woll to Leyes and Fleming, 20 August 1991.

145. Dr. Fred Ehrich, interview by Rick Leyes and William Fleming, 22 August 1991.

146. George E. Behlmer, "Progress Report of General Electric's T64 Gas Turbine engine," *American Helicopter Society Newletter*, Vol. 6, No. 8 (August 1960).

147. Frederic F. Ehrich, "Design and Development Review of the T64 Turboprop/Turboshaft Engine," (ASME Paper No. 61-GTP-7 presented at the Gas Turbine Power Conference and Exhibit, Washington, D.C., 5-9 March 1961), 2-3.

GE T64-GE-6 turboshaft engine. Photograph courtesy of GE.

The original 1957 Navy contract was written to fund the T64 through its 150-hour qualification tests. The first gas generator run was in August 1958. The XT64-GE-4/-8 engine was first tested in January 1959, and the XT64-GE-2/-6 engine was first run in March 1959. First test runs were disappointing as performance was only 1,850 shp, considerably below the guaranteed performance specification of 2,650 shp. It was obvious to the engineers that they had a "big hole" in the engine because the components were not that inefficient.

Suspecting an air leakage problem, GE engineers literally "glued" together with RTV sealant all suspect engine flanges and seals and eliminated those which would not affect manufacturing or maintenance; some flanges were double bolted. Labyrinth seal and turbomachinery tip clearances were minimized. In April 1959, this attention to detail and fine tuning resulted in achieving 2,850 shp, exceeding the guaranteed performance. As a result, GE was able to manufacture engines with an average power in excess of its guarantee, and the difference was called "shipment margin." The sum of a variety of small air leaks had proved to be a significant part of the engine airflow causing loss of power, a problem characteristic of small turbine engines in general.

To assure high reliability of a new engine design in production, the Navy had specified that the development program for the new T64 engine design include

10,000 full-scale test hours at the time of the qualification test.[148] Included in this program were anti-icing tests, first conducted on the top of Mt. Washington during the winters of 1959 and 1960. In September 1960, the first environmental and altitude tests began at the Navy's Aeronautical Engineering Laboratory in Philadelphia. At GE, two test cells, large enough to accommodate the turboprop propeller, were built specifically for T64 testing.

At the GE test cells, the YT64-GE-2/-6 turboshaft engine, connected to a dynamometer, tested without incident, and completed its PFRT on schedule in April 1960 (PFRT approved in August 1960). However, the turboprop tests did not go as smoothly. A major problem, which delayed the turboprop engine program, was discovered after the first turboprop PFRTs were run in June and July 1960. Hairline cracks had developed near the mounting of the reduction gear's rear housing and a total gearbox redesign was implemented.[149] The offset reduction gear mounting, the dynamic loads imposed by a large propeller undergoing speed changes and reversals, and a thin wall casing designed to meet a challenging weight specification were all part of the problem. A heavier, thicker wall casing and other gearbox changes helped solved the problem.

Originally, the gearbox had been attached to the power section by two struts and the torque tube, through which the power shaft connected to the propeller reduction gear in an arrangement similar to that of the Allison T56. After the T56 encountered severe problems with a partially failed mount, possible failure modes in the T64 gearbox casting were analyzed, and GE engineers decided to add a third mounting pad to the new gearbox and an additional mounting strut. The three mount pads and the torque tube considerably strengthened the design. The modification eventually was stipulated by the FAA as a requirement for certification of turboprop engines. In June 1961, the YT64-GE-4/-8 turboprop completed its PFRT after a penalty run (PFRT approved in June 1961).

Flight tests were delayed pending completion of the turboprop PFRT. The decision to use the turboprop T64, rather than the turboshaft, for the first flight tests was based in part on the assumption that any further problems would involve the gearbox. When the Canadian government made available, at no cost, an aircraft of precisely the type best suited for the turboprop YT6-GE-4, a modified de Havilland DHC-4 Caribou (military CV-2A) and also offered to

---

148. This requirement was instituted to minimize the probability that a new engine design with insufficient exposure to full scale testing would "luck-out" and pass the standard 150-hour qualification test.

149. David H. Hoffman, "GE Sees Variety of Uses for T64 Engine," *Aviation Week & Space Technology*, Vol. 75, No. 18 (October 30, 1961): 63.

invest some money in the test program, the Navy and GE accepted the offer.[150] The first flight of the modified DHC-4 transport, and thus the first flight of the T64, was on September 22, 1961.

The official 150-hour endurance qualification testing had been initiated in July 1961. At the 147-hour point of the 150-hour qualification test, a first-stage compressor blade separated. Analysis showed that there had been a major oil leak in a test cell system. After ingesting oil fumes over a long period, the engine inlet had become contaminated with oil residue. The partial blockage of the inlet significantly reduced the airflow into the engine, causing the first-stage compressor blades to swing to a high angle of incidence, in turn inducing aerodynamic instability (stall flutter) of the blades and consequent failure. The compressor blade was redesigned for more stall flutter incidence margin in the event of airflow blockage. During outdoor engine testing near Lynn's salt-water Saugus River, the high pressure turbine inlet nozzle and blades encountered a hot corrosion problem. The turbine blade alloys at high temperature in the presence of sea salt and residual sulfur in the engine fuel resulted in a chemical reaction known as sulfidation (called "green rot" by the engineers), which caused severe metal disintegration. In part, the remedial measure was to apply aluminum oxide coatings. High-temperature coating for corrosion resistance was pioneered by GE for the first time on the T64 HP (high pressure) turbine and nozzle.

Early testing revealed another weak area of the original design. For ease of maintenance, the Navy had specified a combustor casing and liner with horizontal flanges that could be split apart for maintenance without having to remove the HP turbine from the engine. Not only did the split combustor prove easy to remove, but it became the prime reason for removal because it failed prematurely. A decision was made to return to the stronger unsplit design of the type used on other GE engines.

With ample funding and no immediate application, the Navy wanted and got a high-quality engine. Engine components were redesigned and retested to the point of near-perfection. The slightest cracks in components had to be fixed. In charge of the program for the Navy was Jack Horan. GE engineers referred to their tough taskmaster as "No Crack Jack." The 150-hour endurance qualification testing was completed for the T64-GE-2 in April 1962, for the T64-GE-6 in June 1963, and for the T64-GE-4/-8 in October 1963.

In January 1962, the T64 had been selected for the Vought-Hiller-Ryan XC-142A Tri-Service, tilt-wing VTOL transport, and special tests were con-

---

150. Hoffman, "T64 Engine," 64. Because GE was going out of the country, Ed Woll took
    GE's proposition to the Secretary of the Navy, and he approved the project immediately.

ducted in association with this program. The T64 had been designed to operate from a -45 degree attitude, useful in helicopter towing operations, to a +100 degree attitude, appropriate to tilt-wing and propeller applications.[151] The T64 was, in fact, the first GE engine designed to operate in a vertical position. Proof tests of this attitude capability were conducted at GE's test facilities in Peebles, Oh., between October 1962 and April 1963, and the official attitude qualification tests were conducted between July 1963 and September 1963.

After all of the Aeronautical Engine Laboratory environmental, altitude, and anti-icing tests on the T64 were completed, the original T64 development program came to an end in December 1964, and the T64 received its full model qualification the same month. The final cost of the development program, started in 1957, was $63 million dollars.

EARLY T64 TURBOSHAFTS AND TURBOPROPS

Numerous applications were sought for the T64 engine. Early heavy lift helicopter prospects included the Sikorsky S-60, Vertol HC-1B, and Sikorsky S-65. Examples of turboprop re-engining prospects were the Grumman S2F Tracker, Grumman SA-16 Albatross, the Grumman Gulfstream, and the Fairchild F-27 aircraft. The primary competitor turbine engine was the Lycoming T55, especially for Army helicopter applications. To a lesser extent, the Pratt & Whitney JFTD12 (military T73) was a competitor for heavy lift helicopters. A direct competitor, particularly for several turboprop transport aircraft, was the Rolls-Royce Dart. The T64 was also bracketed on the high side by the Allison T56, which was not an immediate competitor, but became one as the T64 grew in size. With the exception of the Sikorsky S-65, none of these possible airframe applications materialized, in part because of these competitor engines.

In September 1961, the T64-GE-1 was selected for the Vought-Hiller-Ryan XC-142 and was uprated for that application from 2,850 to 3,080 shp. The XC-142 was an experimental four-engine, tiltwing aircraft, built in response to a February 1961 Department of Defense RFP for a VTOL transport aircraft to meet the needs of the Air Force, Army, and Navy. The first conventional flight of this aircraft took place on September 29, 1964, and the first transition flight

151. F. Ehrich, J. Shackford, and H. Glessner, "Design and Development of the T64 Turboprop/Turboshaft Engine for Operation in V/STOL Attitude," (AIAA Paper No. 64-174 presented at the AIAA General Aviation Aircraft Design and Operations Meeting, Wichita, Kansas, 25-27 May 1964), 230.

occurred on January 11, 1965. The product of a considerable expenditure of money on V/STOL aircraft, the XC-142 was expected to determine the future military use of V/STOL transport aircraft.[152] However, only five XC-142 aircraft were built, and no production orders were received.

The XC-142 program gave the T64 a start, and the uprated (3,080 shp) T64-GE-1 engine in the XC-142 became the first T64 production engine (and the first growth step of the T64). The increase from 2,850 to 3,080 shp was achieved primarily with an increase in the power turbine inlet temperature to 1,180°F. Like the T64-GE-6, the -1 engine had no reduction gearing. The first of 38 production engines was delivered in June 1963 and the last in February 1967.

In November 1961, shortly after being selected for the XC-142, a second T64 application materialized, the Hughes Model 385 (XV-9A) hot-cycle research helicopter, funded by the U.S. Army Transportation Research Command. Hot exhaust gas from two T64-GE-6 engines was ducted to the rotor blades and discharged at their tips. The purpose of this interesting concept was to reduce weight and complexity of helicopter shaft drive systems. The biggest payoff was believed to be a much improved payload/(gross) weight ratio for heavy lift helicopters.[153] The first flight of the only aircraft built was on November 5, 1964. The aircraft successfully demonstrated the feasibility of the propulsion system, but the aircraft, constructed only as a testbed, had poor handling and stability characteristics.[154]

The first significant T64 application was the Sikorsky CH-53 Sea Stallion (Sikorsky S-65), powered with two 2,850 shp T64-GE-6 (later 3,080 shp -6B) engines, which provided a genuine production base for the T64. On August 27, 1962, it was announced that the CH-53A was selected by the Navy as a heavy assault transport helicopter for the Marine Corps. GE had campaigned with Sikorsky to install two T64 engines in their heavy helicopter instead of three Lycoming T55 engines studied originally. The first flight of the new helicopter was on October 14, 1964, and production deliveries began in mid-1966. The first T64-GE-6 production engine was delivered in October 1963 and the last in December 1967, for a total of 392 -6/-6A/-6Bs, almost all installed in the CH-53A and B models.

152. Erwin J. Bulban, XC-142A Expected to Define Military Role," *Aviation Week & Space Technology*, Vol. 80, No. 25 (June 22, 1964): 16.

153. "Hughes XV-9A to Verify Hot Cycle System," *Aviation Week & Space Technology*, Vol. 80, No. 25 (June 22, 1964): 60.

154. Polmar and Kennedy, *Military Helicopters of the World*, 230.

As a direct result of the urgent need for an aircrew rescue aircraft during the Vietnam War, the Sikorsky S-65 was introduced into the Air Force.[155] Known as the HH-53B "Super Jolly," it was powered by 3,080 shp T64-GE-3 engines. As a variation of the T64-GE-1 (uprated), the -3 turboshaft engine was also a first growth step T64 engine.

By the mid-1960s, several applications had been found for the T64 turboprop engine. The first of these, the de Havilland DHC-5 Buffalo, had been selected by the Army and was equipped with T64-GE-10 engines. Early in May 1962, the Army had submitted RFPs for a new STOL tactical transport aircraft. De Havilland Canada won the competition with a derivative version of the Caribou (the DHC-5 Buffalo). Flight experience for this aircraft was obtained using a testbed de Havilland DHC-4 (military designation CV-2) Caribou fitted with T64-GE-4 turboprop engines. These engines nearly doubled the installed power of the aircraft over that with piston engines.[156] The first flight of the DHC-5 was on April 9, 1964, and delivery to the Army for evaluation began in April 1965. The Army version of the DHC-5 was designated CV-7A. Four Buffalos for the Army were delivered between April and September 1965; they were the only ones ordered because of a change in military policy (i.e roles and missions).[157] The transport mission was assigned to the Air Force, which already had the C-130.

In 1966, the Army agreed to transfer its fixed-wing transports (CV-2 Caribou) and tactical airlift role to the Air Force. The Army CV-7A Buffalos were redesignated C-8As by the Air Force and used for a variety of non-military purposes, especially research. As a result of this change in policy, GE lost a much-sought-after sale to the Army. However, GE's effort was not in vain as the Buffalo program opened the door for future Army sales.

Like the prototype turboprop T64-GE-8 engine described earlier, the 2,850 shp T64-GE-10 engine used on the DHC-5 Buffalo was a T64-GE-6 with a 13.44:1 ratio propeller reduction gearbox installed above the gas generator centerline. The -10 engine incorporated a modified fuel control into the T64-GE-8 engine to provide for propeller reversing through the power lever.[158] Between

155. Gordon Swanborough and Peter M. Bowers, *United States Military Aircraft Since 1908* (London: Putnam Aeronautical Books, 1989), 561.

156. T. R. Nettleton, "Handling Qualities Research in the Development of a STOL Utility Transport Aircraft," *Canadian Aeronautics and Space Journal* (March 1966): 93.

157. K. M. Molson and H. A. Taylor, *Canadian Aircraft Since 1909* (London: Putnam & Co. Ltd., 1982), 284.

158. David A. Brown, "Turbine-Powered Buffalo Built for Short, Rough Fields," *Aviation Week & Space Technology*, Vol. 80, No. 21 (May 25, 1964): 84.

December 1963 and October 1966, 25 T64-GE-10 engines were built for the U.S. versions of the DHC-5.

The Canadian military version of the DHC-5A Buffalo, the CC-115, was powered by 2,970 shp CT64-820-1 (military designation T64-GE-14) turboprop engines. GE had certificated the CT64-820-1 engine in conjunction with the Buffalo program in order to help sell the aircraft in the civilian market. Certificated and first delivered in January 1967, the CT64-820-1 was a derivative T64-GE-10 with FAA required modifications and a unique accessory gearbox optimized for the Buffalo aircraft. Between January 1967 and April 1972, 153 CT64-820-1 engines were produced. Later models of the Buffalo were powered by growth T64 engines.

Other early turboprop applications were found for the T64. A T64 manufacturing agreement was established with Ishikawajima-Harima Heavy Industries (IHI) in Japan. Certain T64 components were manufactured by IHI, and the company assembled the engine. IHI-manufactured T64-IHI-10 engines rated at 2,765 shp were used to power the Kawasaki P2J twin-engine anti-submarine patrol plane. The prototype aircraft, a converted Lockheed P2V-7, Neptune (Kawasaki GK-210) flew for the first time on July 21, 1966. In January 1966, another Japanese company, Shin Meiwa, was awarded a contract to develop an anti-submarine flying boat for the Japan Maritime Self-Defense Force. The prototype of this aircraft, known as the PS-1 (Shin Meiwa SS-2), flew for the first time on October 5, 1967. The PS-1 was originally powered by four T64-IHI-10 turboprop engines. The ASW mission of the PS-1 required it to repeatedly land on the sea and dip its sonar. As the pilot controlled the position of the flying boat with its engines, the propellers would clip waves causing the T64s to ingest salt water. The resulting severe engine corrosion proved to be a difficult problem for GE and IHI to solve.

In 1993, the T64-IHI-10J engine was selected to re-engine the Nihon Kokuki YS-11E commuter aircraft. The T64-IHI-10J was rated at 2,605 shp.

TURBOSHAFT GROWTH VERSIONS OF THE T64

Shortly after the original T64 development program (i.e. T64-GE-2/-4/-6/-8) had completed its MQT in December 1964, the Navy had issued a contract to GE to improve the corrosion resistance of the engine for operation in a salt water environment. T64-GE-6 testbed gas generators incorporating all-titanium compressors designed for a new growth version of the T64 were built and tested. The major components of the compressor, including the case, stator vanes, rotor disks, and rotor blades, were made of titanium alloy more corrosion resistant than the original steel, and lighter. The small titanium blades could not be manufactured by the pinch and roll process, but GE's Thompson Laboratory, headed by Ralph Patsfall, found an alternative solution involving precision chemical milling of oversize forgings to a final size. The all-titanium compressor was the first of its type built by GE for the Navy, and it would become an important component of the new engine then under development.

The new engine model—the T64-GE-12—was the second T64 growth step. Flat rated at 3,400 shp, the -12 first ran on November 5, 1965, in a development program headed by T64 Project Manager Leonard R. Heurlin. The higher power of the T64-GE-12 was achieved by a 12% higher compressor airflow and an increased compressor pressure ratio and turbine inlet temperature. Increased airflow was accomplished by allowing the inlet guide vanes and variable stator vanes to swing open farther. The gas generator turbine rpm was raised in order to accommodate the increased airflow and pressure ratio. To permit the higher turbine inlet temperature, the first-stage turbine rotor blades were air cooled. This was GE's first use of air-cooled turbine rotor blades on a small turbine engine. Cooling air entered the blade through a slot in the bottom of the blade and passed into a plenum chamber and then into five vertical passages cast inside the blade. (The T64-GE-6 engine used solid turbine blades). Segmented first- and second-stage turbine nozzles, whose vanes could be removed, replaced welded nozzles used on the original T64. In addition to the film cooling introduced on the T64-GE-6 (air was discharged in the trailing edge region of the pressure side of the first-stage nozzle), film cooling on the first-stage turbine nozzles was improved by the addition of holes on the leading edge. Another major design change was to replace the conventional thimble hole and louvre design combustor with a stepped-strip or shingle combustor that improved film cooling of the liner.[159] The shingle combustor was an important transitional configuration between the early louvered shells and later machined ring components.

A 1965 technical review of the T64 engine written by Brian Rowe uncovered problems with the T64-GE-6 that needed to be addressed in the T64-GE-12.[160] A few of the more important problems were a less than desirable stall margin at idle, a limited-life combustor, limited first-stage nozzle life from hot corrosion and cracking, inadequate bearing life, and very low carbon seal life.

The stall margin problem was addressed with the opening of the compressor vanes. The new combustor design was intended to accommodate the higher

159. Michael L. Yaffee, "GE Completing T64-12 Engine Assembly," *Aviation Week & Space Technology*, Vol. 83, No. 19 (November 8, 1965): 53.

160. Brian H. Rowe, *T64 Technical Review* (Lynn, Mass.: General Electric, 1965).

temperature requirements of the -12 engine as well as to improve combustor operational life. The segmented turbine nozzle design in the -12 engine was driven by the need to reduce cracking with easier maintenance a secondary benefit. Introduction of leading edge film cooling on the nozzle helped reduce the corrosion problem. Miscellaneous fixes were introduced to solve the bearing and carbon seal problems.

By late 1965, GE's T64-GE-12 engine, developed under Navy contract, had been selected to power the Army's Advanced Aerial Fire Support System (AAFSS), initiated by the U.S. Army Materiel Command in August 1964. The Army urgently needed heavily armed helicopter gunships during the Vietnam War. Under the Army's airmobility concept, the deployment of troop transport helicopters into hostile areas required a weapon system to perform the escort function and protect the column from enemy ground fire.[161]

GE had proposed two versions of its T64-GE-12 to the Army: the T64-12/AAFSS, flat rated at 3,400 shp with a military maximum rating of 3,695 shp; and the T64/S4B, also flat rated at 3,400 shp, but with a military maximum of 3,925 shp. The Army selected the higher-powered version, designated T64-GE-16. The -16, a third growth step in the T64 series, incorporated the same technical features as the -12, but operated at slightly higher temperatures at its maximum rating.

In November 1965, Lockheed was declared the winner of the AAFSS, and on March 23, 1966, the company received an engineering development contract for the Army's AH-56A Cheyenne. Powered by one flat-rated 3,400-shp T64-GE-16, the Cheyenne was a high-speed, two-seat compound attack helicopter with a rigid rotor system and a pusher propeller in the tail. Ten were ordered for development and flight testing. The first flight of the Cheyenne was in September 1967. On January 7, 1968, the Army ordered 375 AH-56 helicopters with initial deployment to Vietnam anticipated to occur no later than 1970.

For GE, the Cheyenne program was a great opportunity, with the promise of a very substantial production order, and it was GE's first significant entree with the Army. The Army's development of the Lycoming T55 made it the incumbent engine in that branch of the armed forces. The Army's predisposition to use the T55 was reinforced by early Navy service problems with GE's T58 engine and the fact that the T64 was a Navy-sponsored engine.

---

161. P. W. Theriault, "Status Report on Design Development of the AH-56A Cheyenne," (AHS Paper No. P68-201 presented at the Annual Forum Proceedings, Washington, D.C., 8-10 May 1968), 1.

Unfortunately for GE, the AH-56 production contract was canceled on May 19, 1969. Significant cost overruns on the program, technical problems with the helicopter, and a fatal crash of a preproduction test aircraft were contributing factors. On August 9, 1972, the entire AH-56 program was canceled, and a new competition for an Advanced Attack Helicopter (AAH) began. GE's hope for a major T64 production run was dashed. Only 13 T64-GE-12s, selected for the original AAFSS program, were built between November 1968 and January 1969. Only 34 T64-GE-16 engines, for the AH-56 flight test and production models, were delivered, the first in April 1967, one month before the AH-56 roll out, and the last in June 1969, one month after the AH-56 production contract was canceled.

While neither the T64-GE-12 nor the T64-GE-16 engine programs resulted in significant production, related engine models, known as the T64-GE-7 and T64-GE-413, did. The -7/-413 engines were part of the third major T64 growth step along with the -16 engine and had the same 3,925-shp maximum rating as the T64-GE-16. Performance growth for the -7/-16/-413 engines was achieved by increasing power turbine inlet temperature to 1,325°F. The -7 engine passed its MQT in December 1967 (simultaneously with the -12), and the -413 engine passed its MQT in March 1968. Follow-on models were the -7A, derived from the -7, and the -413A, derived from the -413 and -7A. The engines in the -7/-413 series were used to upgrade the performance of various models of Sikorsky CH-53 and HH-53 helicopters.

The CH-53 and HH-53 upgrade program was good business for the T64 program. A total of 322 engines and 110 kits of -7/-7A models were delivered between December 1967 and March 1979.[162] For the T64-GE-413/-413A program, 517 engines were delivered between February 1969 and July 1973.

The fourth major growth version of the T64 engine was again directly tied to an important military helicopter requirement. In 1973, the Sikorsky S-65A was chosen by the Navy as part of a Heavy Lift Helicopter competition. The fact that the original S-65 design had a provision for the addition of a third engine was a factor in the selection. The Navy/Marine Corps specification called for a multipurpose, shipboard-compatible helicopter capable of lifting 16 tons. The intent of the Navy was to use the helicopter for vertical on-board delivery operations, mine countermeasures, and removal of battle-damaged aircraft from carrier decks. The Marines needed the helicopter for amphibious assault, to airlift com-

---

162. A "kit" is a full or nearly full set of engine parts, subassemblies, and modules that are shipped to another location (generally to a licensee) for final engine assembly and acceptance testing after addition of any of the licensee's manufactured parts.

bat equipment, and to retrieve tactical aircraft without disassembly. The prototype, known as the YCH-53E and powered by three 4,380-shp T64-GE-415
engines, first flew on March 1, 1974. Deliveries of the CH-53E to the Marine
Corps started on June 16, 1981. The formidable CH-53E Super Stallion (Sikorsky S-80), powered by three 4,380-shp T64-GE-416 engines, became the largest
and most powerful helicopter produced outside of the Soviet Union. A variation
of this helicopter, known as the MH-53E Sea Dragon, was used for airborne
mine countermeasures.

The 4,380-shp T64-GE-415 and T64-GE-416 growth engines, derived from
the predecessor T64-GE-7A/-413As, achieved their improved power primarily
by increasing power turbine inlet temperature to 1,410°F. Better turbine cooling and an improved combustion liner accommodated the higher temperature.
A derivative of the -416, the -416A incorporated further turbine improvements.

The -415 engine was also used on the Sikorsky CH-53E and the RH-53D.
The -415 engine passed its MQT in September 1973, and between that month
and November 1975, 24 engines were produced. The -416 and the -416A engines were used on the CH-53E and MH-53E. The -416 engine passed its
MQT in October 1978. Between January 1980 and 1989, 527 T64-GE-416 production engines were delivered. The MQT for the -416A was passed in December 1988. T64-GE-416A production started in December 1988; the engine continued in production with 121 engines delivered by December 1991.

A further T64 growth model was the T64-GE-419, rated at 4,750 shp, and
developed to provide the MH-53E with increased power for hot-day, one-
engine-inoperative capability. Performance was increased via improved hot-end
materials and a higher turbine temperature. The engine was similar to the
T64-GE-416 with an improved turbine and an integral fuel/oil heat exchanger.
Design of the -419 engine started in January 1988, and the first engine was run
in June 1989. The engine was qualified in September 1991.

TURBOPROP GROWTH VERSIONS OF THE T64

T64 turboprop performance growth and technology improvements paralleled
that of the T64 turboshaft engine. These engines incorporated earlier turboprop
technology as well as that from the third T64 growth step, the T64-GE-7/
-16/-413. The two primary applications were later models of the de Havilland
DHC-5 Buffalo and the Fiat G.222 aircraft.

Some de Havilland DHC-5As, originally powered with 2,970-shp CT64-
820-1 engines, underwent a field modification which upgraded their engines.
Also rated at 2,970 shp, the conversion engine was called the CT64-820-3. It in-

Sikorsky MH-53E Sea Dragon helicopter powered by three T64 turboshaft engines.
Photograph courtesy of GE.

cluded a film-cooled, segmented first-stage nozzle and improved combustor; it
also incorporated T64-GE-7/-413 technology. All CT64-820-3s were field
conversion engines, thus they were not production engines per se. The
DHC-5B was an aircraft model proposed with a GE T64/P4C engine.[163] Later
DHC-5D/-5E aircraft were powered by 3,133-shp CT64-820-4 engines. Certificated in November 1974, the CT64-820-4 engines incorporated hot-day and
altitude performance improvements and numerous other upgraded features.
Production of the DHC-5 ended in December 1986 with a total of 123 aircraft
of all versions built. Sold in relatively small quantities to many countries around
the world, the DHC-5's T64-engine support proved to be a difficult logistical
problem for GE.

In the spring of 1963, Fiat was awarded a research contract by the Italian Air
Force for a military transport designated the G.222. This aircraft flew for the first
time on July 18, 1970, and was powered by CT64-820-2 engines. The first delivery of a production aircraft was in November 1976. The G.222 was powered by
two 3,400-shp T64/P4D engines produced by Fiat under a GE partial license.
GE had combined its interest in marketing the engine in Italy with its need for a
T64 gearbox vendor by selecting Fiat to manufacture the gearbox. In 1969, Fiat
became part of Aeritalia, which in turn became part of Alenia in 1990. The prin-

---

163. The CT64-P4C was a proposed (i.e. paper) engine, but was never manufactured.

cipal customer of the Alenia G.222 was the Italian Air Force; in August 1990, the U.S. Air Force ordered it, and gave it the designation C-27A Spartan.

The T64/P4D passed its MQT in November 1974. The T64/P4D was derived by combining components of the T64-GE-10 and CT64-GE-820 turboprop gearbox with the power producing section of the T64-GE-7/-413. The compressor casing was stiffened as required for turboprop maneuver loads and the fuel control was modified for the propeller reversing function. The gearbox was also uprated.

T64 SUMMARY

By 1993, the T64 had been in production longer than any other GE engine. Between 1963 and November 1991, more than 2,600 T64 and CT64 engines and T64 kits were produced at GE's Lynn plant. Additional T64s were produced under license by MTU in Germany, IHI in Japan, and Fiat in Italy.[164] The most important production years of the T64 were between 1966 and 1975, which generally coincided with the production of the T64's primary application, namely the A through G Models of the Sikorsky CH/HH-53 helicopter. Within this time frame, two production peaks occurred. The first was between 1966 and 1967 (386 total engines for these two years) and consisted primarily of T64-GE-6 series engines for the CH-53A helicopter. A second major peak occurred between 1969 and 1973 (1,062 total engines for these five years), composed largely of T64-GE-7 series and T64-GE-413 series engines for upgraded models of the CH-53 and HH-53 helicopters. Annual production of the T64 continued at a slower but steady pace from the mid-1970s and through the 1980s. Between 1980 and 1988, the T64-GE-416 was the primary T64 production engine, used largely on the CH-53E and MH-53E helicopters.

When the T64 was developed in the late 1950s, it had the highest pressure ratio yet chosen for a single compressor rotor turboshaft engine. The high pressure ratio combined with progressively higher turbine temperatures resulted in a very low sfc, which was significantly better than its competitors at the time. With power ratings close to 5,000 shp in the latest production engines, the T64 had nearly doubled its original guaranteed performance rating of 2,650 shp. With such improvements and additional state-of-the-art technology incorporated over the years, the T64 remained competitive in the 1990s and was the only one of GE's engines from the 1950s and 1960s still in production.

---

164. Another licensee was Bristol Siddeley, later Rolls-Royce. However, this license was never exercised.

A number of T64 technical achievements were used in other GE engines. But the most important legacy of the T64 was the fact that it contributed further to the development of the very heavy lift helicopter, in particular the various models of the Sikorsky H-53. The longevity of the T64 program was attributable to a good engine design, a thorough development program, and the success of the CH-53 helicopter. In part, Jack Horan's contribution to the success of the T64 program needs to be acknowledged. According to William M. Meyer, Manager of the T64/TF34 Programs:

> Jack Horan was a dedicated civil employee who had technical responsibility back in the 1960s and 70s for the Navy's turboshaft/turboprop gas turbine engines, which included the T64. Jack demanded technical perfection. This engine over the years has consistently demonstrated high reliability in its original and growth models. Credit for this good experience must be shared with Jack because of the unwavering position he took in his drive for perfection and his insistence on operational excellence. We are grateful for his guidance.[165]

Mostly, the T64 was a Navy and Marine corps helicopter engine. Relatively few CT64s were sold, the de Havilland DHC-5 being the primary application. The T64 program was complicated with a large number of engine models developed for relatively few major airframe applications. Compared to other GE engine programs, the total number of engines produces over the years was relatively small. In the early 1960s, considerable competition existed for heavy lift helicopters and turboprop applications. Had some of the potential opportunities materialized for the T64, such as the Cheyenne program, the outcome might well have been different.

The TF34 High-Bypass Fan

The TF34 engine was developed for the Navy by GE for an anti-submarine warfare (ASW) aircraft originally known as the VSX and later as the S-3A. In the late 1950s and early 1960s, the Navy conducted numerous in-house studies for a new carrier-based ASW aircraft to eventually upgrade or replace its piston-powered Grumman S-2 Tracker aircraft. Fixed-wing aircraft, helicopters, and VTOL/STOL aircraft were examined; for these aircraft, various engine alternatives were also studied. On February 23, 1965, the Navy issued a Tentative

165. William M. Meyer, Lynn, Massachusetts, to Rick Leyes, letter, 25 May 1993, National Air and Space Museum Archives, Washington, D.C.

Specific Operating Requirement for a new carrier-based anti-submarine airplane. Subsequent studies supported a recommendation for a conventional, twin turbofan engine, carrier-based aircraft.

The Navy had investigated several advanced turbine engine technology programs and had sponsored some of its own. Among the technologies considered was the relatively new high-bypass turbofan engine. The TF39 high-bypass turbofan engine, which GE had proposed for the Air Force-sponsored Lockheed C-5A large transport aircraft program, was an example of an available turbine technology; in the proper power class, this technology was suitable for the VSX. Also under consideration was the Navy-sponsored Allison T78 regenerative turboprop. In October 1965, the Navy narrowed all VSX engine candidates to an 8-9,000-lb-thrust turbofan engine or a 3,500-shp turboprop.

At the same time, the Navy proposed that an Advanced Engine Development Program be financed to apply the technology being developed to the specific size engine required by the VSX aircraft. By this time, the Office of the Secretary of Defense required that a new engine have an application in an already approved airplane program before giving approval for its development.[166] The practical effect of this policy was that engine development independent of a weapon system, i.e., an aircraft application, ceased. Because the engine development had to precede the VSX by a year or more and because such a program officially could not be launched prior to an approval of the VSX, the Advanced Engine Development Program was conceived.

In December 1965, the Navy recommended that this Advanced Engine Development Program be restricted to new turbofan designs. An important reason for this was to conserve industry and Navy financial resources. While the turbofan was the preferred engine for the VSX, turboprop engines were still competitive. The GE T64-GE-12 turboshaft engine and the Allison T78-A-2 regenerative turboprop engine, both Navy programs, were in a relatively advanced state of development and could be modified or improved to meet the anticipated requirements of the VSX, if they were later shown to be better candidates. As it later turned out, turboprop candidates were eliminated completely because they could not meet the VSX mission requirements and aircraft speeds.[167]

166. A. Stuart Atkinson and Charles C. McClelland, "Rationale for a Military Aircraft Engine Development Program," *Astronautics and Aeronautics*, Vol. 7, No. 9 (September 1969), 35.

167. "TF34 Engine Program" and "Chronology of VSX Engine Program (GFE)" unpublished, undated papers, National Air and Space Museum Archives, Washington, D.C. These papers were believed to be written by the Propulsion Division of the Naval Air Systems Command about 1973.

In May 1966, the Navy formally issued to the engine industry a Request for Quotation for a 14-month-long Advanced Engine Development Program. Proposals were received from the United Aircraft Corporation, Curtiss-Wright Corporation, General Motors Corporation (Allison Division), and GE. The GE proposal was put together under the direction of Manager of Engineering Brian Rowe. A preliminary design and proposal team led by Art Adamson laid out the design of an engine on the basis of the T64 gas generator with a high-bypass fan.[168] This design was derived from the T64/F1, generally referred to as the "CF64" demonstrator engine. Designer and engineer Eric Doorly contributed significantly to the original design of the new engine.

On January 3, 1967, the Navy inaugurated the program by selecting GE and General Motors (Allison Division) engine designs from the competitors. Included in the program were a design, an engine mock-up, a development program proposal, and a liaison with airframe contractors. In November 1967, the Secretary of Defense initiated the VSX program. In January 1968, GE's team, including among others Bob Ingraham, Dave Gerry, and George Behlmer, submitted a proposal to the Navy. The gas generator compressor was approximately 10% larger in dimensional scale than that of the T64 and handled approximately 20% more airflow. On January 13, 1968, a source selection plan for the VSX engine development program was set in motion. Competing against General Motors's engine (Allison TF32), GE received a contract to develop the TF34-GE-2 engine on April 16, 1968.

The Navy engine development proposal and contract included several different features. Instead of the cost-plus-fixed-fee arrangement that had been customary for full-scale engine development, the contract was a fixed-price, incentive-fee type. The incentives were based on PFRT schedule, and engine durability as demonstrated by a Post Endurance Test run after the normal MQT. GE would lose contract money if the engine were late for contract milestones, but would earn financial incentives for being ahead of schedule. An interesting feature included a proposal based on a specification for a production engine configuration (as compared to additional separate specifications for ground test and flight test engines). Guaranteed installed-engine (nacelle) performance was another feature of note.[169] Another unique feature of the Navy proposal request was that the criteria employed to compare the relative performance of competing designs consisted of the minimum combined engine and fuel weight required to

---

168. Travers, *Engine Men*, 321.

169. Atkinson and McClelland, "Rationale for a Military Aircraft Engine Development Program," 40-41.

perform specified missions. This allowed maximum flexibility for the engine designer in making trade-offs between engine weight and fuel consumption.

On August 4, 1969, Lockheed (teamed with LTV Aerospace Corporation) was declared the winner of the VSX competition and the designation S-3A assigned. The Lockheed S-3A Viking was a compact, high-wing, twin-engine aircraft. The primary mission of the aircraft was to find and destroy enemy submarines, achievement of which required both high subsonic flight speed (in order to minimize the time to the target area) and long endurance over the target area. The key technology required to satisfy both of these requirements was the high-bypass turbofan configuration.

The TF34-GE-2 was a 9,275-lb-thrust dual-rotor, front-fan engine.[170] It had a single-stage fan with a bypass ratio of 6.2, a pressure ratio of 1.5:1, and airflow of 338 lb/sec. A 14-stage axial-flow compressor on the HP shaft produced a 14:1 pressure ratio rise and a total core airflow of 47 lb/sec. Variable inlet guide vanes and variable stator vanes were incorporated in the early compressor stages. The engine had an annular combustor with a low pressure fuel injection system and machined liner. An air-cooled, two-stage HP turbine and a four-stage LP turbine with tip-shrouded blades for high component efficiency completed the basic engine.

Between 1966 and 1967, GE's Aircraft Systems Analysis Unit at Lynn had performed detailed unofficial aircraft mission studies to refine the engine design.[171] When the Navy officially specified five individual missions for the ASW aircraft to demonstrate the capabilities of the selected engine cycle and to provide a basis for competitive engine evaluation, these specifications and GE's earlier studies were used together to determine the optimum cycle. While higher temperatures resulted in more performance, a more conservative 2,200°F turbine inlet temperature limit was defined to meet a design objective of high reliability at the beginning of service.[172] For the temperature selected, the 14-stage compressor resulted in the best sfc levels and best weight for the defined missions. The bypass

170. J. E. Worsham, "New Turbofan Engines - F101 and TF34," (SAE Paper No. 720841 presented at the National Aerospace Engineering and Manufacturing Meeting, San Diego, Cal., 2-5 October, 1972), 1-2.

171. P. Niederer and R. E. Houlihan, "Design Considerations of the TF34-GE-2 High Bypass Turbofan Engine," (SAE Paper No. 690688 presented at the National Aeronautic and Space Engineering and Manufacturing Meeting, Los Angeles, Cal., 6-10 October, 1969), 4.

172. The temperature was determined after considerable internal debate among GE engineers. With the technology in hand, some argued for higher temperatures. Others felt that the Navy was not comfortable with higher temperatures and wanted a more conservative value. In the end, the opinion of the latter group prevailed.

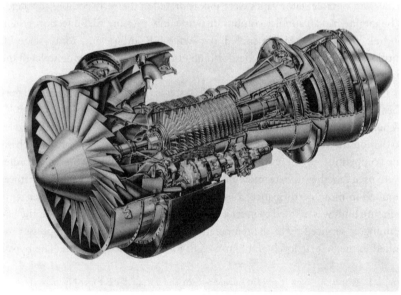

TOP: GE's TF34 turbofan engine was originally developed for the Lockheed S-3A Viking anti-submarine warfare aircraft. Photograph courtesy of GE.

BOTTOM: Cutaway of GE TF34 turbofan engine. Photograph courtesy of GE.

ratio chosen was a compromise between sea level missions favoring a high ratio and the penalty associated with a higher ratio during high-altitude missions.

The TF34 was designed to be a "low-risk" engine to meet severe reliability requirements in the ASW role within a short time frame. To do so, it used proven technologies of the GE's T64, TF39, and GE1 engines. The fan used important technology first demonstrated during GE tests of the "CF64" engine, specifically, a low aspect ratio (wide-chord), unshrouded fan. It was designed for good efficiency at 10% above flight idle to accommodate the important loiter portion of the aircraft's mission. The HP compressor was approximately a 1.20-flow scale of the T64 compressor, and the HP turbine was also based on T64 technology. The LP turbine was based on TF39 technology. The original combustor layout was derived from a GE12 combustor that had resulted from a Curtiss-Wright subcontract.

In April 1969, the TF34-GE-2 ran for the first time, one month ahead of schedule. The first major problem in the program came late in 1970, when it became clear that GE could not meet its specification weight of 1,260 lb with acceptable durability within the time scale required for the S-3A program. The Navy granted a deviation allowing a weight increase to 1,421 lb in exchange for a reduction in program cost to the Navy.[173] The weight of the TF34-GE-2 ultimately reached 1,478 lb. GE staged a comeback when it completed the PFRT on March 1, 1971, ahead of schedule and qualifying for an incentive award. At the same time that the GE development tests were taking place, the Naval Air Propulsion Test Center conducted altitude tests from February to August 1971. From September 1971 to June 1972, the Navy conducted official inlet distortion, anti-ice, rocket gas, climatic, air starts, and altitude performance and transients qualification tests.

The first flight of a YTF34 engine took place on January 22, 1971, at GE's Flight Test Center in California; the engine was installed in an S-3A nacelle and flown on the wing of a B-47 testbed. Flight testing was conducted for 330 hours through September 1972. Major problems that were found during this testing included: reduced starter assist air-start envelope; reduced windmill air-start capability; altitude compressor stall; engine rollback with S-3A loads; and main fuel control "G" load sensitivity.[174] Specification deviations were later permitted for a reduction in the air-start envelope.[175] The inability of the engine to carry

173. "TF34 Engine Program," 10-11.

174. "TF34 Engine Program," 11-12.

175. The "air start envelope" is the speed and altitude boundary within which an engine can be started. "Engine rollback" refers to an uncontrollable reduction in engine speed and thrust.

the required engine bleed and accessory drive loads at idle throttle setting above approximately 8,000 feet was solved by incorporating a high-idle reset on the fuel control.[176] The compressor stall problem was solved by control schedule changes for the variable compressor stators and adjustment of the compressor operating line. Manager of Engineering Frank Pickering worked toward solving the difficult performance margin and weight problems.

The first flight of the YS-3A aircraft, powered with YTF34 engines, was on January 21, 1972. During this same month, the Navy examined engines which had completed 150-hour endurance tests and reviewed problems impacting the program. The troubles were altitude stall problems, sea level stall problems, unacceptable combustor and turbine durability, and combustor coking and flashback.[177]

The last proved to be the most challenging to solve. The combustor for the TF34 had originally been proposed as a vaporizing type with mushroom-shaped injectors, as in the GE12. Very early in the development, a new type of carbureting injector under development for the F101 engine in Evendale was adopted. The injector was shaped like a scroll and vaporized the fuel without the application of heat (as was the case in the vaporizing system). The final fuel/air mixing took place in the shear layer between counter-swirlers placed in the combustor dome around each injector. However, the long residence time of the fuel in the injectors caused them to coke up, resulting in flashback of the flame into the injector. The corrective action was to lengthen the fuel tubes, delivering fuel deeper into the injectors and shortening the residence time of the fuel in the scroll to preclude coking. Solving the problem caused a two-month delay, but the scheduled MQT was met.

The TF34-GE-2 successfully completed all qualification test requirements on August 25, 1972. On October 17, 1972, the engine successfully completed a 150-hour Post Endurance Test, earning GE an incentive award for being ahead of schedule. Except for the minor weight increase and changes in the starting envelope, the program met all performance, schedule, and cost requirements of the development contract.

For flight testing, GE had received a support contract for two XTF34 ground test engines, 28 YTF34 engines, and 4 TF34 engines. Production authorization for the TF34-GE-2 was given on March 31, 1971, so GE could secure long-lead parts. The first production engine was delivered in August 1972. The -2 engine

176. Burt D. O'Laughlin, E. Lloyd Graham, and John Cristiansen, "S-3A Development Tests," (AIAA Paper No. 73-778 presented at the AIAA 5th Aircraft Design, Flight Test and Operations Meeting, St. Louis, Missouri, 6-8 August 1973), 4.

177. "TF34 Engine Program," 12.

became fleet operational in February 1974 when the first S-3A deliveries were made to a Navy training unit.

Another significant problem that occurred early in service resulted from a non-specified engine duty cycle. The aircraft, and subsequently, the engine was designed for long-endurance patrol missions. This meant that the engine was expected to be subject to fewer throttle transients and to be operated for extended periods at flight idle, conditions that were relatively easy on the engine. However, to prepare pilots for landing the new aircraft on carriers, the S-3A was assigned in 1974 to a training school where it was subjected to numerous takeoffs and landings as well as formation flying. More engine cycles and throttle transients than expected led to high-temperature engine problems. The original impingement-cooled first-stage turbine nozzle diaphragms had to be replaced prematurely. The substitution of a nozzle diaphragm with a film-cooled leading edge resolved the problem.

Starting in September 1972, TF34-GE-2 problems identified after the engine went into service were corrected. A new engine, designated TF34-GE-400A, resulted, replacing the -2 with a more durable and maintainable engine. Rated at 9,275 lb thrust, it incorporated a modified fuel control and configuration changes, as well as different external piping and an adaptive control system for optimizing accessory power extraction. The -400A also incorporated a simple subsystem to avoid engine stalls that might be encountered when the engine ingested exhaust gases from rockets fired from launching positions near the engine inlets. When a rocket was fired, the subsystem automatically closed the compressor's replaced stator vanes to temporarily increase its stall margin. The first -400 production engine was delivered in December 1974. S-3As and later S-3Bs were delivered with the -400 engine.

THE TF34-GE-100

In March 1967, the Air Force issued an RFP for design studies of a pure close air support aircraft based on experience gained during the Vietnam War. The Air Force was looking for an aircraft which would embody the best characteristics of the Douglas AD-1 Skyraider and the yet-to-fly Vought A-7D Corsair II, be cheaper to produce than either, and have STOL capability from rough fields.[178] In May 1970, RFPs were issued for a competitive prototype development of an A-X aircraft. A unique aspect of this RFP was the fact that design-to-cost was the only requirement. Unlike most RFPs, which established

178. Lou Drendel, *A-10 Warthog in Action* (Carrollton, Texas: Squadron/Signal Publications, 1981), 4.

performance requirements, this RFP instead set performance goals to satisfy specified aircraft characteristics. These goals were: responsiveness, lethality, survivability, and simplicity.[179]

On December 18, 1970, Fairchild Republic Co. and Northrop Corp. were selected to participate in a competitive prototype aircraft evaluation, each receiving contracts to develop two prototype aircraft. The Northrop entry was designated YA-9 and the Fairchild aircraft YA-10. The engines were not government furnished equipment; the choice was up to the contractors. In 1970, Fairchild selected the GE TF34-GE-100, and in 1971, Northrop contracted with Lycoming to provide its 7,200-lb-thrust YF102-LD-100.

The TF34 was already developed and qualified to Navy standards. Re-engineering for cost reduction was the primary objective of the TF34-GE-100 program, and Fairchild and GE worked closely to find ways to do it. A review of the Air Force and Navy requirements provided many cost savings; certain Navy requirements were not necessary for the close air support mission. JP5 fuel heaters and anti-icing equipment were unnecessary. Engine weight was less critical to the Air Force, and costs were saved by substituting steel parts for titanium and eliminating special weight reduction machining.[180]

The TF34-GE-100 engine had the same basic design features and performance capability as the earlier TF34-GE-2, and the same aerodynamics and thermodynamics. That the -100 was rated at 9,065 lb thrust as compared to the 9,275-lb-thrust -2 engine was a result of their different installation and performance requirements. In contrast to the -2 engine, designed for under-the-wing pylon mounting with a short fan duct, the -100 engine was side mounted and had a long fan duct. Critical design features were the same for both models, permitting a 90% parts commonality.

The prototype YA-10, powered by YTF34 flight engines, flew for the first time on May 10, 1972. An Air Force fly-off competition began in October 1972, and on January 18, 1973, the YA-10 won the A-X competition. The fact that Fairchild had selected an already developed engine and that the price of the engine could be reduced further was one of the factors which gave the A-10 a significant edge in the competition.[181] On March 1, 1973, GE was awarded a contract for the development of TF34-GE-100 engines to power the A-10A.

179. C. W. Adams and U. A. Hinders, "Design-to-Cost for the A-10 Close Air Support Aircraft," (AIAA Paper No. 74-963 presented at the AIAA 6th Aircraft Design, Flight Test and Operations Meeting, Los Angeles, California, 12-14 August 1974).

180. Adams and Hinders, "A-10 Close Air Support Aircraft."

181. Adams and Hinders, "A-10 Close Air Support Aircraft."

The effort to win the A-10 application for the TF34 engine was conducted under the direction of GE Project Managers Lou Tomasetti and Steve Chamberlin.

With two major applications, full-scale production of the TF34 had begun in 1972. The first TF34-GE-100 to test was in July 1973, and it passed its MQT in October 1974. The first pre-production YA-10A flew on February 15, 1975, with production TF34-GE-100 engines installed. The first production A-10As were delivered in March 1976. Officially designated Thunderbolt II, the A-10A was universally known by the nickname "Warthog."

Two development problems of note occurred with the TF34-GE-100 during early A-10 flight testing. Hot gas and powder residue from the A-10's GAU-8/A 30mm cannon was ingested by the engines, leaving a residue on the fan and compressor blades that accumulated and gradually reduced the stall margin of the engines until compressor stall developed.[182] Muzzle deflectors, airframe modifications, and regular engine washes ultimately resolved the problem. The other problem was an interruption of oil flow to the engine roller bearings caused by negative-G pop-up maneuvers, solved by installing an accumulator in the oil pressure system that provided oil during negative-G maneuvers.

Though satisfactorily resolved, the early gas ingestion/engine stall problem was one of several issues that left some lingering doubts about the A-10's combat effectiveness.[183] When put into combat in 1991 during the Persian Gulf War's Operation Desert Storm, however, the TF34-powered A-10 more than proved its worth. Primarily attacking frontline infantry positions, 144 A-10s flew almost 8,100 sorties. Warthog pilots claimed approximately 1,000 tanks, 2,000 other vehicles, 1,200 artillery pieces, and two helicopters shot down in air-to-air engagements. No other aircraft received more attention during this war than the A-10.[184]

THE CF34

In the 1970s, GE had allowed its leading position in the business jet market to decline in order to concentrate on large engine programs and because business

---

182. Robert R. Ropelewski, "A-10 Engine, Gun Tested in Gas Ingestion," *Aviation Week & Space Technology*, Vol. 112, No. 8 (February 25, 1980): 40.

183. While designed as a rugged, damage-tolerant, and highly maneuverable aircraft, the A-10 was a relatively slow jet. Thus, there was concern about its ability to evade intensive flak and missile fire.

184. Richard P. Hallion, *Storm over Iraq* (Washington, D.C.: Smithsonian Institution Press, 1991), 210-211.

jet engines required so much support compared to the number sold. However, by the mid- to late-1970s, changing circumstances led GE to reconsider its decision, and the company began to look at possible applications for a commercial derivative of its TF34 engine.

On April 7, 1976, design work started on what eventually became the Canadair CL-600 Challenger business and commuter jet. This jet, originally designed by William P. Lear and known as the LearStar 600, was conceived as an ultra-long-range, very-high-speed business aircraft.[185] Major design changes were made by Canadair, resulting in, among other things, a wide-body configuration and substantial weight increases. Increasing the aircraft gross weight meant that the original Lycoming ALF502D engine chosen for the aircraft no longer had the necessary performance capability. Canadair then evaluated a proposed uprated Lycoming ALF502L engine and a commercial derivative of the GE TF34 engine known as the CF34. A decision was made to install the ALF502L, a lower thrust engine than the CF34, when Federal Express, which had placed 25 aircraft orders, specified the Lycoming engine. Another factor in the choice was the inability of GE to meet what later proved to be an over optimistic schedule set for the Challenger.[186]

By the time the Challenger received its FAA certification in November 1980, the aircraft weight had continued to grow, in part because of certification requirements, and performance objectives fell short of projected figures. Despite its uprating, the 7,500-lb-thrust ALF502L engine was not able to compensate for the additional increase in aircraft weight. Canadair was also hampered because Lycoming was having difficulty meeting its scheduled delivery dates for the ALF502L engines.[187]

Canadair needed more power to solve the performance problems with the Challenger. The company was also interested in stretching the Challenger to compete with the new Gulfstream American Gulfstream III corporate jet. In the long run, the big market was commercial airlines, and an enlarged jet would help pave the way toward that possibility as well. GE's CF34, too large for Bill Lear's original design, had the growth capability that the Challenger needed to

185. K. M. Molson and H. A. Taylor, *Canadian Aircraft since 1909* (London: Putnam & Co. Ltd., 1982), 152-153.

186. David M. North, "CF34 Upgrades Challenger Capabilities," Vol. 118, No. 20 *Aviation Week & Space Technology,* (May 16, 1983): 63.

187. Marc Grangier, "The Canadair Challenger Goes into Service," *Interavia,* Vol. 36 (April 1981): 333-334.

solve its immediate performance problems. It also had ample performance for a stretched Challenger design.

From its perspective, GE saw airframe applications for CJ610 and CF700 engines dwindling and an increasing market share for new production and retrofit programs going to Garrett's TFE731 and Pratt & Whitney Canada's JT15D engine.[188] GE had earlier concluded that it needed a bigger market for the TF34, and a decision was made to re-enter the executive aircraft market by seeking the Challenger application.[189] As GE saw it, there would be no real competition for the CF34 in this particular case; furthermore, the Challenger was the only commercial aircraft that was really compatible with the CF34.[190]

In April 1978, GE once again proposed to Canadair the installation of its engine (the CF34-1A) in the Challenger. GE's sales pitch took pride in its product by offering a "technically superior engine with fuel consumption performance in the class of GE's CF6, which powers the most modern wide-body aircraft in airline service."[191] GE also emphasized that the CF34 would provide exceptional reliability because of the maturity of the military TF34 engine program and the engine's established production base. For Canadair, GE highlighted the fact that the CF34 could readily power all versions of the Challenger being studied, provide the longer ranges needed, and still allow for future growth. The clincher was GE's willingness to provide engineering support, test engines and facilities, and financial resources to support a development and certification program.

In short order, Canadair made several major decisions. First, on July 3, 1979, the company announced the Challenger E (for Extended), which had a lengthened fuselage, greater payload and range, and a new power plant—the 9,140-lb--thrust CF34-1A engine. Renamed the Challenger 601, the prototype aircraft made its first flight on April 10, 1982. Second, in March 1980, Canadair and GE jointly undertook a development program to certificate the original Challenger CL-600 with a CF34-1A engine. Canadair subsequently offered its customers the choice of the GE CF34 or the Lycoming ALF502 for the standard Challenger. While this competitive offering was standard practice for the airline in-

188. "GE Keys Civil Engine Efforts to CF34, CT7," *Aviation Week & Space Technology*, Vol. 111, No. 1 (July 2, 1979): 74.

189. Neumann to Leyes and Fleming, 22 August 1991.

190. Adinolfi to Leyes and Fleming, 22 August 1991.

191. General Electric, "Executive Summary: General Electric CF34 Turbofan Engine for the Canadair Challenger," Proposal CF34-478-001, April 14, 1978, National Air and Space Museum Archives, Washington, D.C, 1.

dustry, it was a precedent for the business aircraft world. The Challenger was also the first civilian application of the TF34 engine. The replacement of the Lycoming ALF502L engines with the higher-thrust GE CF34s transformed the Challenger into the corporate aircraft that Canadair had originally promised its customers in the late 1970s.[192]

The CF34 had been announced originally by GE in April 1976. It was essentially the same engine as the TF34-GE-100 except for external configuration differences required for individual installation and FAA certification. Some distinguishing features of the civilian derivative were a Kevlar fan blade containment ring, more fan exit guide vanes to reduce noise, and fire shields.[193] The engine was flat rated and targeted for the 7,000-8,000-lb-thrust class of aircraft.

The CF34 was initially aimed at the general aviation market.[194] As it turned out, however, the original certification and production schedule was paced to coincide with an established military airframe requirement. This was the Coast Guard's Medium Range Surveillance (MRS) aircraft program, for which GE supported a Lockheed Georgia Company JetStar proposal. However, in January 1977, the Coast Guard was authorized to select the low bidder, which was the Garrett-ATF3-powered Dassault Falcon 20G. The CF34 certification schedule was then delayed until the Challenger application was assured. While a substantial portion of the FAA testing had been accomplished during the Navy and the Air Force TF34 qualification programs, additional FAA testing for the CF34 began in June 1980. The first engine run was in 1981. The CF34 engine, rated at 9,140 lb thrust with Automatic Power Reserve (APR), received its FAA certification in August 1982. The CF34 engine entered commercial service on the Challenger 601 in November 1983.

Next, a CF34-3A engine, flat rated to 70°F for better climb and hot-day take-off performance, was introduced on the Challenger 601-3A, which first flew on September 28, 1986. The -3A engine was rated at 9,220 lb thrust with APR. At the time that it was introduced, the CF34-3A-powered Challenger had the lowest sfc of any large-cabin intercontinental corporate jet, and was the first of such jets to be permitted unrestricted operation at airports with evening noise curfews, such as Washington National Airport in Washington, D.C.

192. North, "CF34 Upgrades Challenger Capabilities," 63.

193. David Velupillai, "The Light Turbofans," *Flight International*, Vol. 119, No. 3751 (March 28, 1981): 924.

194. Eric Falk, "GE Aims New CF34 Turbofan at General/Commercial Aviation Markets," GE Press Release AEG-976-67, September 14, 1976.

Powered by GE CF34 turbofan engines, the Canadair Challenger 601 was a large cabin, intercontinental corporate jet. Photograph courtesy of GE.

On March 31, 1989, Canadair announced the CL-601 RJ Regional Jet. A stretched version of the Challenger, the 50-passenger CL-601 RJ was intended primarily for commuter airline operation. GE CF34-3A1 engines, rated at 9,220 lb thrust with APR, had been selected in 1988 for the Regional Jet. The prototype Regional Jet flew on May 10, 1991. Bruce Gordon was the CF34 Project Manager at the time.

Design of the CF34-3A1 had started in 1987. Early in the development program, GE had decided that the performance requirements and ratings of the CF34-3A1 Regional Jet engine would be identical to those of the CF34-3A business jet engine. In order to meet the more demanding requirements of airline use, GE focused on improving engine durability and over-all maintainability.[195]

Material improvements were introduced in the fan disk and fan compressor spool for better fatigue strength and improved low-cycle fatigue.[196] The introduction of directionally solidified HP turbine blades and more modern serpentine cooling in the first-stage HP turbine blades were modifications made to improve component life. Additional durability changes included: material improvements in other compressor and turbine rotating components; a new non-metallic abradable coating on the compressor rotor and stator; new ignition exciters; and improved seals and lubrication system modifications. In part, such

195. H. Lowey, "GE's CF34 Engine for Business and Regional Jets," (AIAA Paper No. 90-2041 presented at the AIAA/SAE/ASME/ASEE 26th Joint Propulsion Conference, Orlando, Florida, 16-18 July 1990).

196. John Morris, "Adapting the CF34 for Airline Use," GE Press Release No. GEAE-91-33L, October 16, 1991.

improvements were based on experience with GE's large airline engines, the CF6 and CFM56. To improve maintainability, external components were relocated to make them more accessible, the fuel heater and oil cooler were combined into a single unit, and lockwire was eliminated. An airline on-condition maintenance program replaced maintenance based on TBO intervals.

The first CF34-3A1 engine run was in 1988, and the first production engine was delivered in 1990. The -3A1 first flew on the Regional Jet on May 10, 1991. The -3A1 engine received its FAA certification on July 24, 1991. An improved version of this engine, known as the CF34-3B, was under development in 1993 for the Challenger 604 jet. The CF34-3B was designed for a 7% higher thrust capability, improved hot-day/high-altitude performance, and 3% lower fuel consumption. The objective of the -3B was to provide more thrust without raising the engine temperature much by using a higher efficiency compressor and increasing the airflow through the engine. The -3B was flat rated at 9,220 lb take-off thrust (with APR) to 86°F. The first flight of the Challenger 604 prototype with -3B engines installed was on March 17, 1995, and the -3B was FAA certified in May 1995. A related development, the CF34-3B1, a -3B with airline ratings, was certificated concurrently with the -3B.

In February 1995, a major upgrade of the CF34 engine was selected for the 70-seat Canadair CRJ-X regional jet. Known as the CF34-8C, this model had an initial take-off rating of 13,000 lb thrust with growth potential to 18,000 lb thrust. The engine incorporated a new, higher pressure ratio and larger diameter (45.6" v. 44") fan with wide fan blades. Replacing the original 14-stage high pressure compressor was a new 11-stage compressor, seven stages of which were derived from GE's F414 fighter aircraft engine. Technology from GE's F414 and GE90 engines was used in a multi-hole combustor. The engine retained its two-stage high pressure turbine, but used single crystal airfoils which were stronger and more resistant to thermal fatigue and corrosion. A four-stage low pressure turbine had a new aerodynamic design for higher efficiency and a reduced parts count including fewer airfoils. A dual FADEC control was part of the design.[197] The preliminary design phase of the -8C engine was completed in May 1995, and component testing began early in 1996. Ishikawajima-Harima Heavy Industries was a revenue sharing participant in the -8C program and Kawasaki Heavy Industries was a subcontractor for manufacturing the gearbox.

As a power plant for an early regional jet, the CF34 turbofan brought jet performance and passenger comfort to the commuter airline industry, hitherto pri-

197. David Hughes, "CF34-8C to Power new Regional Jet," *Aviation Week & Space Technology*, Vol. 142, No. 7 (February 13, 1995): 70.

marily served by turboprop aircraft. For GE, the Regional Jet program provided renewed vigor for the CF34 program. At Lynn, GE invested $3.4 million in a CF34 production facility, which opened in April 1991.

OTHER TF34/CF34 VENTURES

In 1970, the TF34 had been selected as the engine of the eight-engine Boeing Airborne Warning and Control System (AWACS). However, early 1973, TF34 propulsion for the E-3A AWACS was canceled. In 1974, the TF34-GE-400A was selected to provide auxiliary (thrust) power for the Sikorsky S-72 Rotor System Research Aircraft (RSRA), jointly developed for NASA and the Army. The RSRA was a test vehicle that could be variously configured to fly as a pure helicopter, a pure airplane, or a compound helicopter. It was used to evaluate a variety of rotor and propulsion systems. It first flew as a compound helicopter (with the TF34 engines) on April 10, 1978.

In 1976, a modified version of the TF34-GE-100 was selected to power a second-generation Boeing Compass Cope High Altitude Long Endurance (HALE) RPV. Designated YQM-94A by the Air Force, the reconnaissance/surveillance RPV was intended to cruise at heights above 55,000 feet for more than 20 hours. However, in July 1977, the Air Force terminated this program because the potential payloads under development for the RPV might not mature in the form or time envisaged.

In 1984, the CF34 engine was selected by American Aviation Industries in Van Nuys, Calif., for a Lockheed JetStar retrofit program called the FanStar. GE was responsible for the design and implementation of the engine conversion portion of the program, the purpose of which was to meet FAA noise requirements and to increase fuel efficiency. The Fanstar flew successfully on September 15, 1985, but the program was not pursued because of poor business prospects for it.

Several interesting studies were conducted with TF34 engines. In 1981, NASA awarded GE a $5.6 million contract for the design, fabrication, and testing of a TF34 convertible turbofan/turboshaft aircraft engine. The objective of the program was to explore technology for integrating turbofan and turboshaft engines into a single engine.[198] It was anticipated that such an engine would be capable of powering a high-speed rotorcraft in vertical and horizontal flight up

---

198. Ken Atchison and John M. Shaw, "NASA Awards $5.6-Million Contract for Aviation Engine Program," *NASA News*, Press Release No. 81-64, May 15, 1981. DARPA also funded this research project.

to approximately .85 Mach number.[199] The demonstration was successful, but the concept was superseded by the tilt-rotor approach to V/STOL propulsion, as in the Bell Boeing V-22. Another NASA-sponsored study was conducted to establish a preliminary design definition of a quiet nacelle for a series of turbofan engines based on the TF34 turbofan engine.[200] This nacelle concept was not applied to any TF34 installations; the engine was never challenged by low-noise requirements because it was a relatively small engine and had the highest bypass ratio of any engine in its size class.

A non-aircraft derivative of the TF34 was the LM500 marine and industrial turbine. Developed in cooperation with Fiat Avio of Italy, the LM500 was a simple-cycle, two-shaft gas turbine in the 6,000-shp class. It was used for pipeline pumping applications and for Danish Standard Flex300 patrol boats.

TF34/CF34 SUMMARY

When it was introduced, the TF34 had the highest thrust-to-weight ratio and lowest sfc of any engine in its thrust class.[201] Production for the TF34 began in 1970. While some doubt exists about the production figures, by November 14, 1991, approximately 2,540 TF34 and CF34 engines had been produced at GE's Lynn manufacturing facility. Of these engines, 2,126 engines were TF34s and 414 were CF34s. The primary TF34 production years were between 1974 and 1982. The annual production rates varied from 106 engines in 1974 to 196 engines in 1982. The first CF34 engines were manufactured in 1981. TF34 production was completed in 1983, and CF34 production continued at a low level, but steady pace since then.

The volume production years (1974-1982) of the TF34 were directly related to the S-3A and A-10 production run. A total of 187 S-3A aircraft were produced between 1972 and 1978. The first production A-10As were delivered in March 1976, and the last of 713 aircraft for the Air Force was delivered in March 1984. The start of the CF34 production run in 1981 coincided with the Challenger 600 and 601 programs.

199. A. I. Bellin and A. Brooks, "Status Report: DARPA/NASA Convertible Turbofan/Turboshaft Engine Program," (ASME Paper No. P83-GT-196, 3 March 1983), 1.

200. D. P. Edkins, R. Hirschkron, and R. Lee, "TF34 Turbofan Quiet Engine Study," (NASA No. CR-120914, GE No. GE AEG N72-26691, and Contract No. NAS 3-14338, 1972), 3.

201. General Electric, "TF34 High Bypass Turbofan: Power for the U.S. Air Force A-10," GE Brochure No. AEG-1033(5/75), 4.

The TF34 was developed as a military engine and primarily produced military applications. Following a GE marketing tradition, it was successfully parlayed into a commercial engine. The TF34/CF34 was notable because it was the first small GE engine to incorporate high-bypass fan technology and go into production. It was a pioneer high-bypass fan engine for small combat, executive, and airline aircraft.

The T700 Turboshaft Engine

In the 1960s, the helicopter became an important weapon and tool for U.S. military forces during the Vietnam War, used for close air support, troop deployment, logistics, medical evacuation, search and rescue, reconnaissance, and aircraft recovery operations. The Army was the principal user of helicopters, deploying them in great numbers. The most widely used helicopter was the Bell UH-1 Iroquois (Huey) utility helicopter, powered by a single Lycoming T53 turboshaft engine. During the course of the war, a clear need began to develop for a future utility helicopter with more payload and performance. Such a helicopter would require a more powerful, twin-engine installation. Furthermore, a twin-engine installation, with single-engine-flight capability, would provide a greater safety margin.

Also during the Vietnam war, turboshaft engines of all makes and models exhibited a specific pattern of problems resulting from an extremely hostile operating environment, difficult and primitive maintenance conditions, and the need for improved engine design. Low service life, high spare parts requirements (including large numbers of replacement engines), high and costly fuel consumption, complex and frequent maintenance procedures, and vulnerability to ground fire were among the variety of problems that had become the norm. While retrofit airframe/engine modifications were made that helped significantly, a better solution was desired. Not only was there a need for a new utility helicopter engine with improved performance, but also one with significantly better reliability, maintainability, survivability, and fuel consumption. For the pilot, another needed improvement was a reduction in engine-related work load so that he could concentrate on flying the mission rather than having to divert his attention frequently to monitor and adjust the engine.

In the mid-1960s, the Army Concept Formulation Studies of a future replacement for the UH-1 helicopter began. Aircraft system characteristics and requirements were defined. Eventually, they became the basis for the Utility Tactical Transport Aircraft System (UTTAS). The UTTAS requirement was issued to the helicopter industry late in 1971. The UTTAS helicopter would be powered by twin engines and would be required to have the capability to lift a crew of three plus a combat-equipped infantry squad of 11 men out of ground effect at a 4,000-ft altitude on a 95-degree F day. The combination of helicopter engine difficulties during the Vietnam War and the evolving need for a new helicopter laid the groundwork for Army specifications for a new turboshaft engine. From this environment came the GE12 and later the GE T700 turboshaft engines.

THE GE12

In the early 1950s, the Air Force had sponsored turboshaft engine design competitions and developments for the Army that resulted in the Lycoming T53 and T55 engines. At the time, each branch of the military service had its own dedicated supplier. Lycoming had become the Army's supplier of helicopter engines, but GE had a continuing interest in penetrating this market. In particular, Ed Woll mounted a concentrated effort to get GE more involved with the Army and address its engine needs. His reasoning was that the Army was the principal user of helicopters and deployed them in great numbers.[202]

GE first attempted to sell its T58 engine to the Army as an uprated replacement for the T53 engine. This proposal was studied by the Army, but was not implemented, because, in part, of the complexity of logistically supporting both engines internationally. While GE's initial effort did not succeed, it did arouse Army interest in GE's more technologically advanced engines. In 1962, the de Havilland Buffalo (designated CV-7A), powered by GE T64 engines, was selected by the Army; in 1966, the Army selected for its AAFSS helicopter gunship program a Lockheed helicopter (designated AH-65 Cheyenne), also powered by the T64. Unfortunately for GE, the Buffalo program ended in 1965 and the AH-56 in 1969. Nonetheless, GE's efforts to cultivate a relationship with the Army showed steady progress over time.

Indicative of the progress was a visit in the 1960s to GE by General Robert R. Williams, then the commanding general of the U.S. Army Aviation Center and Commandant of the U.S. Army Aviation School (and shortly later Director of Army Aviation). General Williams informed GE of the Army's need for a helicopter engine in the 1,500-shp class. In a discussion with Woll, Williams indicated that some uncertainty existed about the size of the engine required for a

---

202. Ed Woll, Boca Grande, Florida, to Rick Leyes, letter, 16 March 1994, National Air and Space Museum Archives, Washington, D.C.

demonstration program. Army studies had shown that 1,500-shp would give about 10-15% margin. Williams was interested in knowing whether GE would prefer to scale up a demonstrator engine for a production engine, if the engine were too small, or to scale it down, if it were too large. In view of the margin that existed, Woll recommended to Williams that the Army stay at the 1,500-shp size. Williams was also interested in knowing if GE would compete for a demonstration contract, which Woll assured him GE would do.[203]

In the early 1960s, long-range planners in the Army had begun to look at future helicopter engine possibilities. For the first time, the Army had been given the responsibility to manage an engine demonstration program on its own without Air Force program initiation. The Army decided to sponsor engine studies from a broad spectrum of engine manufacturers. Between May 1964 and August 1967, the Army awarded a series of industry-wide contracts for analysis, research, design, component design and experimental tests, and compatibility studies for advancement in the state of the art of turboshaft machinery.[204] Specifically, the purpose of these studies was to design and develop a 1,500-shp-class gas turbine aircraft engine that was lighter and consumed less fuel than existing engines in that power range. The results from these contracts provided assurance that a lightweight 1,500-shp-class engine was possible.

In 1967, the Army issued an RFP for a 1,500-shp-class demonstrator engine, and GE responded with a proposal for an engine known as the GE12. The GE12 design was conceived by Proposal Team Leader Art Adamson and a design team led by Art Adinolfi. Included among those involved during this phase were: Les King, who did the compressor aerodynamics; Bob Warren and Jerry Kean, who worked on the mechanical areas; and Anthony L. "Tony" Rodes, who was GE's Washington representative and who worked closely with the Army. After the contract for development was awarded to GE, the concept was further refined and developed by many of the same team under the leadership of the first GE12 Project Manager, Eric Doorly, and GE12 Engine Design Manager Jerry Kean. Later, Doorly left GE and Arnie Brooks became the second GE12 Project Manager, reporting to Len Heurlin. Kean also left the GE12 development project later and was replaced by Bob Mattheson.

In August 1967, GE and Pratt & Whitney were awarded contracts for a parallel effort to design, fabricate, and test an advanced technology 1,500-shp-class

203. Woll to Leyes, 16 March 1994.

204. M. C. Duffles and S. M. Cote, "T700 Engine Program History and Status," Document No. 109515, Propulsion Division, Naval Air Systems Command, Department of the Navy, Washington, D.C., 31 July 1990, 9.

free turbine demonstrator engine. Other competitors had included Allison, Teledyne CAE, and Lycoming. The demonstrator engines were sponsored by the U.S. Army Aviation Materiel Laboratories (USAAVLABS) in Ft. Eustis, Va. Pratt & Whitney's engine was designated ST9.

While the power rating of the demonstrator was not directly specified by the Army, the rated gas generator airflow was set at 10 lb/sec. This flow size would provide a demonstrator in the 1,500-shp class, providing the projected cycle parameters and efficiencies were achieved. Lower and higher power parameters were considered because of uncertainty about the final UTTAS helicopter configuration. The initial objectives of the program were to demonstrate a reduction of 25% in sfc and an increase of 40% in power/weight ratio relative to operational engines in the same power class, such as the Lycoming T53, GE T58, and Pratt & Whitney Canada PT6. For the time, the 25% reduction in sfc was a major technical stretch. Both engines used significantly higher pressure ratios than existing small turboshaft engines.[205] Important additional design considerations were ease of maintenance, high reliability, and low vulnerability.

The GE12 was a 1,500-shp front-drive, free turbine turboshaft engine. It had an axial-centrifugal compressor, annular combustor, two-stage HP turbine, and two-stage LP turbine. The basic front drive, concentric-shaft configuration was specified by the Army based on their preference for the T53-type of layout. The complete engine was constructed of major sub-assemblies to facilitate field maintenance through modular replacement or replacement of individual components.

The compressor, consisting of five axial stages followed by a single centrifugal stage, provided a high 17:1 pressure ratio. With a straight-through-flow configuration, the annular combustor was compact and short. It used a vaporizing fuel system, eliminating the need for atomizing fuel nozzles.[206] The HP turbine was air-cooled and operated at a temperature of 2,250°F, similar to the TF34. The HP turbine blades were of short radial height, which made air cooling a difficult technology for the time. The LP turbine was designed to operate efficiently at part-power levels, especially at 30 to 60% of military rated power; the turbine blades had tip shrouds. The GE12's combination of increased compressor pressure ratio and higher turbine inlet temperature, both of which were

205. "Outlook Cloudy for GE12, ST9 Program Despite Added Funding," *Aviation Week & Space Technology*, Vol. 91, No. 20 (November 17, 1969): 32.

206. "GE12 to Demonstrate Advanced Design, Component Technology," *Aviation Week & Space Technology*, Vol. 91, No. 17 (October 27, 1969): 26.

greater than existing small turboshaft engines, contributed to significant improvements in engine performance, size, and weight.

GE introduced an interesting combination of technology on the GE12. One example was the axial-centrifugal compressor, the first that had been used in a GE engine. For the sake of durability, an axial-centrifugal compressor was chosen because the axial blades could not go below a certain size. Low aspect ratio, wide-chord blades with thick leading edge airfoils were used in the axial compressor. The first for such use in a small GE engine, they were made of steel and designed for ruggedness and erosion resistance. In contrast to the T58, the GE12 had 80% fewer compressor airfoils and 40% fewer stages, but the overall compressor pressure ratio had approximately doubled, a testimonial to advanced aerodynamics technology. GE had not worked on centrifugal compressors since its 1940s-era J33 turbojet engine. GE retirees were brought in to consult with the compressor designers and company archives were searched for centrifugal compressor data. The centrifugal impeller was characterized by backswept vanes at the impeller tip.

During the preliminary design of the GE12, design tradeoff studies were made of both vaporizing and atomizing combustor designs. The requirement to have light weight, high-heat-release rate, contamination tolerance, low-frame radiation emissivity, and low-smoke operation led to the selection of a vaporizing combustor design. In the vaporizing system, the fuel was injected into a vaporizer exposed to the combustion zone. The heat of combustion then induced vaporization of the unburned injected fuel as it was fed into the combustion zone for burning. The vaporizing combustor with a candy-cane injector geometry can be traced to the Armstrong Siddeley Sapphire turbojet engine licensed by Curtiss-Wright of Woodridge, N.J. Curtiss-Wright undertook further development and evolved a vaporizing system with a mushroom-shaped injector. The design, development, and fabrication of the original GE12 combustor was subcontracted by GE to Curtiss-Wright. The GE12 retained the mushroom-type vaporizing combustor throughout the course of its development, although it was eventually redesigned by GE for use in the T700 without the vaporizing mushroom system.

An important aspect of the combustor was that it was a 5,000-hour-life, machined-ring design. In the late 1960s, a significant advance was made with the introduction of combustion liners machined from rolled and welded bar stock rings. The primary design feature of the machined ring combustor was greater strength and rigidity for dimensional control, resulting in longer life, more consistent pattern-factor control, and uniform, annular film cooling. In contrast, contemporary sheet metal combustors had less dimensional stability, were prone to cracking as a result of their punched cooling slots, and generally had a shorter life. The initial cost of the machined ring combustor was high, but the lifetime cost was intended to be low because it lasted much longer and because fewer replacement parts were needed. This idea was initially difficult for GE to sell to the Army because of the higher initial cost and because of the Army's skepticism that the combustor would, in fact, last 5,000 hours. Their skepticism stemmed from the fact that the typical stamped and fabricated sheet metal combustor had a 500-to 1,200-hour life expectancy.

The original demonstrator program, begun in 1967, was a two year program, at the end of which it was intended that either the GE12 or the Pratt & Whitney ST9 would be selected to power the Army's UTTAS helicopter. By 1969, however, no engine development money was yet available. Because the UTTAS program had not yet been approved, the Army could not commit to, nor had the money for, an engine development and production program. In order to keep the engine program viable in the interim, the Army managed to partially fund both companies for additional work on the demonstrator engines (through September 30, 1971 when the program ended). For its part, GE provided most of the funds to keep the program going after the original two-year contract ended.

During the GE12 demonstrator program, the high pressure ratio, high tip speed axial and centrifugal compressors proved to be difficult technologies. When first tested, the GE12 would not run because the compressor was highly loaded and had insufficient bleed. Holes drilled in the rear compressor casing were a temporary fix. Also, considerable compressor component work needed to be done in order to match the flow capability of the axial and centrifugal compressors.

The first run of the gas generator was in December 1968, and the first of two engines to test was in 1969. In June 1970, GE announced that the GE12 had successfully completed its official 10-hour demonstrator test.[207] Relatively early in the demonstrator program, both the GE12 and the ST9 engines had shown that the feasibility of advanced technology engines in the 1,500-shp class was possible. However, the ST9 had come much closer to meeting the Army's maintainability objective, giving it a competitive edge. Pratt & Whitney had announced that the ST9 could be disassembled with only three commonly available tools.[208]

Reflecting on this problem, former T700 Project Manager William J Crawford III said, "The real challenge that we felt we had was in the maintainability

---

207. Eric Falk and Ernie Stepp, "GE Demonstrator Engine Completes Official Test," GE Press Release AEG-670-37, June 1970.

208. "Outlook Cloudy for GE12, ST9 Program," 32.

side, because our engine fell short of what we believed we could achieve at that point in time and what it would take to surpass the competition."[209] The design of the GE12 had been driven primarily by the difficult-to-achieve performance objectives, and less attention had been paid to the maintainability objectives. GE12 engine disassembly and component removal was time consuming and required the use of special tools. Previously, maintainability had never been a major design requirement or issue. If an engine had been maintainable, it was more a matter of happenstance. Maintenance was costly for the Army, and it was time consuming. GE soon learned that the Army meant business on the maintainability objective. The Army decided to assess a heavy time penalty for the use of special tools for a given maintenance task.

Not only did GE have to improve the maintainability of the GE12 demonstrator engine, but it also had to overcome the reputation of its T58 engine, considered difficult to maintain. For example, a T58 fuel control change on some installations required removing the entire engine and other components for access. Reinstalling the fuel control required sequential, interdependent adjustments. The whole process could take up to a week and a half. One objective in redesigning the GE12 was to guarantee that the fuel control could be replaced, with no adjustments and no special tools, in 15 minutes.

During the follow-on contracts that the Army issued to keep both engine programs alive, GE assembled a special task force from its Lynn and Cincinnati plants in order to determine how to make design changes to meet the maintainability objective. GE determined that it could actually calculate maintainability; the calculations were verified with demonstrations. This led to guaranteed component and module replacement times. Improved maintainability was facilitated by a new engine design that required only 10 tools for field maintenance.

GE also undertook an important effort to redesign the GE12's axial-centrifugal compressor when it learned that the Army would count parts in the evaluation. The competitor ST9 engine, with its dual centrifugal compressor, had far fewer parts than the GE12, whose axial compressor had individually removable blades. In response, GE adopted the blisk (bladed disk) idea. In the blisk, the compressor rotor and blades were machined from a single forging. The blisk concept was thoroughly tested on the GE12's first and second compressor stages. Blisks were subsequently incorporated in all five axial stages. The compressor parts count was reduced from about 300 parts to 11 parts. The blisk construction also benefitted the design by permitting a very low radius ratio of the

209. William J. Crawford III, interview by Rick Leyes and William Fleming, 21 August 1991, GE Interviews, transcript, National Air and Space Museum Archives, Washington, D.C.

first stage compressor blades and by eliminating leakage through the disk/blade assembly. The blisk concept was made possible by GE manufacturing engineers under the direction of H. William "Bill" Lindsay.

Another issue that GE wrestled with during this period was whether or not to make an inlet particle separator integral or separate. There were advantages and disadvantages to either approach, and it was not clear to GE which way the Army wanted to go. Previously, inertial particle separators had been developed as airframe-mounted components or as add-on engine components.[210] The purpose of the separator was to reduce foreign object damage (FOD) to the engine compressor and other internal components. The magnitude of the FOD problem was considerable. During the Vietnam War, sand and foreign object damage accounted for 40-60% of all unscheduled helicopter engine removals.[211]

On the basis of separator experience and studies with T58 and T64 engines, GE proposed the use of an integral separator that would operate all of the time the engine was running. The primary reason for an integral separator was GE's belief that the engine's axial-centrifugal compressor could not survive a sand-ingestion test if the Army tested the engine without the separator. There were other important reasons as well. A separator could be most efficiently designed as part of an engine. An integral design would give the engine company final responsibility for engine performance. Performance was critically dependent on a successful installation. Experience had shown that after separators had been removed several times for cleaning or for an engine overhaul, they did not fit well and leaks developed. Also, separators were generally turned off by pilots once they reached altitude and not always turned on again for landing.

In the inlet particle separator (IPS) designed by GE, a series of non-rotating swirl vanes set up an outward rotation of inlet air, centrifuging out higher-mass particles, which were collected in an outer annular scroll. Engine air was drawn into the compressor through an inner annulus containing a set of static de-swirl vanes. The scavenge air was drawn out by a separate engine-mounted blower and then discharged overboard.[212]

210. For example, in 1966, Allison had contributed to the design, fabrication, and testing of an airframe-mounted Donaldson inertial separator to protect its T63 engine. In 1968, Lycoming had qualified an add-on inertial particle separator for its T53 engine. Neither of these were integral inlet particle separators.

211. Michael L. Yaffee, "T700 Aims at Low Combat Maintenance," *Aviation Week & Space Technology*, Vol. 100, No. 4 (January 28, 1974): 45.

212. William J. Crawford, "The T700-GE-700 Turboshaft Engine Program," (SAE Paper No. 730917 presented at the National Aerospace Engineering and Manufacturing Meeting, Los Angeles, California, 16-18 October 1973), 3.

In addition to providing about 90% protection against FOD, the IPS performed several other useful functions. The ejected air was available to cool exhaust system IR suppressors. The ingested air was used for air-oil cooling, thus avoiding the use of a separate oil cooler. The frame provided oil tank storage and acted as an engine front frame, a mount for the accessory gearbox, and provided the main engine front mounts. The primary disadvantages were increased engine weight and inlet pressure drop. The IPS also added to engine cost.

Just prior to GE's proposal for a full-scale engine development program, the integral IPS design was validated by sand erosion tests that showed a totally unexpected 2% rise in efficiency. Examination showed that the centrifugal compressor was eroding slightly and creating its own backsweep. This was also interesting because it was contrary to the generally accepted notion that centrifugal compressors were more tolerant of sand and dust than axial compressors. In this case, the centrifugal compressor eroded before the axial one did.

The last major design change addressed at the conclusion of the GE12 demonstrator program and before GE's proposal for a full-scale development program (leading to the T700) was a redesign of the GE12 power turbine in order to change the power turbine speed. The ST9 and the GE12 had significantly different power turbine speeds, and the Army decided that it needed to have a common speed in order to run a helicopter competition. A compromise speed was negotiated: Pratt & Whitney raised their original speed, and GE lowered theirs.[213]

By the end of the GE12 demonstrator program in September 1971, the two engines in the program had accumulated 453 hours of full-scale engine development testing, and some 11,833 hours of component and accessory testing. The sfc and weight-reduction objectives of the GE12 program were met in 1971.

In 1970, authority had been given to the Army to proceed with engineering development of the UTTAS. In 1971, the Army issued an RFP for a 1,500-shp UTTAS power plant and a competition was held. The RFP called for among other things: 20-30% lower sfc (as compared to operational engines); integral inlet particle separator; reduced logistic support; improved survivability; 37-50% reduction in maintenance manhours (as compared to operational engines); 5,000-hour design life; low-cycle fatigue design and test requirements. Etched in stone was an Army requirement for .520 sfc at 900 shp (60% power) in order

213. William J. Crawford III, Cotuit, Massachusetts, to A. Stuart Atkinson, letter, 23 November 1992, National Air and Space Museum Archives, Washington, D.C.

to minimize fuel consumption at helicopter cruise conditions. In order to reduce pilot workload, the control system requirements included automatic overspeed and overtemperature protection, constant power turbine speed governing, and torque matching in multiple engine installations.[214]

Through 1970 and early 1971, before the Army RFP was formally issued, GE and the Army had many detailed reviews on specification and development program requirements. These discussions, led by Bill Crawford for GE, enabled the GE team to be well prepared when the RFP was formally issued, and no RFP deviations were necessary.

Major efforts in 1971 included writing the proposal, completing the GE12 component improvement tests that substantiated the proposal claims, and finalizing all of the design changes from the demonstrator to the final production engine design. In September 1971, the GE12 production engine and development proposal was submitted in response to the RFP.

The team working many seven-day weeks throughout 1971 on the pre-proposal and post-proposals included Bob Turnbull (Engineering Manager), Bill Crawford (Project Manager), Arnie Brooks (Army Technology Coordinator), Gerry Donohie (Applications), Tony Rodes (Marketing), Pete Kastrinelis (Integrated Logistics Support), Carl Hollenbeck and John Wellborn (Requirements), Barry Weinstein (Technology), Everett Bishop and O. D. Taylor (Lubrication Systems and Structures), Les King and Dave Klassen (Compressor), Martin Ray (Combustor), Lou Bevilacqua (Turbine), Jack Moulton and Joe Curran (Controls), Floyd Heglund (Development Coordination), Michel Compagnon (Performance), Frank Gallagher (Contracts), and Zack Boyages (Finance).

The GE12 proposal occurred during difficult times at GE. The U.S. Supersonic Transport (SST) and its GE4 engine had been canceled in March 1971. Earlier, in 1970, GE had lost major competitions to Pratt & Whitney's F100 engine for the F-15 and the F401 for the F-14 fighter aircraft programs. Layoffs were heavy and a win was important to the company as well as the GE12 team.

THE T700-GE-700

In December 1971, the GE12 engine was selected over Pratt & Whitney's ST9 and a private-venture development, the Lycoming PLT 27, an aircraft derivative

214. Charles C. Crawford, Jr., and William J. Crawford, III, "Aircraft Turbine Engine Development—Current Practices and New Priorities," (AGARD Paper No. CP-302 presented at the 57th Specialists' Meeting of the AGARD Propulsion and Energetics Panel, Held at the Ecole Nationale Supérieure de l'Aéronautique et de l'Espace (ENSAE), Toulouse, France, 11–14 May 1981), 3-1.

of a demonstrator tank engine of the same general size. For the next several months, the GE12 team was kept together while contract discussions with the Army continued and the UTTAS program solidified. On March 6, 1972, a contract was awarded for the design, development, and qualification of the engine then designated T700-GE-700. The contract also included support of the UTTAS competitive airframe manufacturers.

The T700 team was headed by Bill Crawford, T700 Project Manager, who was supported by Bob Turnbull, T700 Engineering Manager. Key technical advisors were Fred Garry, Manager of Military Projects and Ed Woll, Manager of Engineering. A number of the engineers on the T700 team had also been key persons on the GE12 project. Members of the mechanical design team included Lou Zirin, Bob Mattheson, and Everett Bishop. Lou Zirin had originally sketched out the novel integral IPS. Included on the compressor team were Les King, Dave Klassen, Arnie Brooks, and Steve Chamberlin. Lou Bevilacqua was initially involved in the turbine design. John Wellborn, Carl Hallenbeck, Bob Binford, and Lou Scott were members of the controls team. Joe Curran was responsible for integration of the engine control with the helicopter. Zack Jennings, Herb Katz, and Jack Moulton devised a no-adjustment torque sensor for the engine. Pete Kastrinelis was the Integrated Logistics Support Manager, Zack Boyages was responsible for finance, and Frank Gallagher was the contract manager who did most of the contract negotiating with the Army. Other important contributors included Tony Rodes, Floyd Heglund, Pete Chipouras, Michel Compagnon, Tom Milne, and Jerry Donohie.

The T700 engine incorporated major design changes that occurred during or subsequent to the GE12 demonstrator program, among them the integral IPS, compressor blisks, a redesigned power turbine, and improved fuel injectors. GE eliminated the original mushroom-type vaporizing tubes used throughout the GE12 development, which were subject to carbon deposition and whose life expectancy was uncertain. Under the leadership of Ralph Sneedon, a new fuel injection system, termed the "central injector," was introduced, which introduced fuel at a moderate pressure and partially atomized into the center of a pair of concentric counter-swirlers. The shear layer at the interface of the counter-swirling air streams accomplished the final atomization and vaporization necessary for effective combustion. This type of injection system set a design precedent for future GE engines, both large and small. In addition, the compressor was flared (inlet diameter increased) slightly to get additional airflow, and the centrifugal compressor diffuser was redesigned from a conical-throat to a parallel-throat configuration. The vertical and horizontal dimensions and the external configuration of the T700 were determined in part by the requirements of

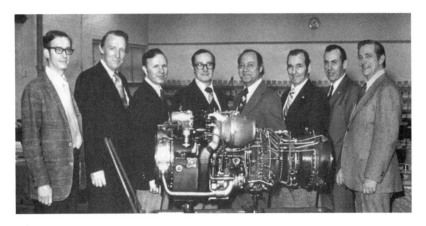

Some of the early GE T700 engineering team members. From left: Michael Compagnon, Floyd Heglund, Dave Klassen, Bob Turnbull, Lou Zirin, Tom Milne, Jack Moulton, and Ev Bishop. Photograph courtesy of GE.

the Boeing and Sikorsky UTTAS helicopter candidates. Of special note was the fact that the T700 engine modifications from the GE12 incorporated design features that made it much more maintainable, reliable, and survivable than operational turboshaft engines.

The T700 was designed for 0.24 hr maintenance manhours per engine operating hour for all maintenance levels including corrective, preventive, and servicing actions. This was four times better than then operational 1,400- to 1,500-shp engines and was based in part on increased engine reliability.[215] The engine had a design life of 5,000 hours and a low-cycle fatigue life of 15,000 cycles.

The basic engine design also contributed to lower maintenance times. For quick replacement of major subsystems, the engine consisted of only four modules: accessories and control; the cold section (integral IPS, compressor, and diffuser); the hot section (combustor and HP turbine); and the power turbine, each interchangeable between engines of the same model.

Other steps were taken to reduce maintenance time and improve survivability. External plumbing was reduced by coring passages for internal routing of lubrication and air lines. Items requiring frequent servicing were grouped on the accessory module. The engine control and accessory module was top mounted in the candidate UTTAS aircraft to improve service access and to re-

215. Crawford, "The T700-GE-700 Turboshaft Engine Program," 7.

duce vulnerability to enemy ground fire. The engine control required no rigging or adjustment when put into service. An emergency lubrication system was capable of sustaining the engine for at least six minutes at high power. Lockwire connectors were eliminated in favor of self-locking nuts. Instead of electrical clamps, "broomstick" type spring clips were used to retain electrical harnesses. A wrench-arc system was used to tighten hydraulic fittings: to avoid the need for a torque wrench, fittings were snugged tight and then turned one nut flat further to prevent leakage.

Another maintenance-friendly, time saving design feature that later emerged from the development program was the T700 variable-geometry axial compressor that required no adjustment at assembly, no engine "tuning" to get more performance. This was made possible by a variable geometry system that used identical, non-adjustable actuating levers for all three variable compressor stages.

To improve survivability, a low pressure, suction fuel transfer system was incorporated into the design. Typically, helicopters used fuel lines that were pressurized from the tank to the engine. In combat, a bullet through a high pressure fuel line would cause fuel to spray on hot engine components causing in-flight fires. By substituting a low pressure fuel system, which used suction feed from the fuel tank, this problem would be virtually eliminated. Fuel line punctures would result in air inflow leaks rather than external fuel leaks.

The T700 engine was designed for the pilot as well as the mechanic. Turboshaft engines of the 1960s demanded a significant amount of attention by pilots. Starting procedures were complicated and required critical timing to avoid engine over-temperatures. Pilots had to monitor the engine to insure proper torque matching. Fuel control design had focused on the needs of the engine rather than optimization of the handling characteristics of the airframe. In contrast, the T700 control system was designed to be a simple system requiring a low level of pilot attention.[216] The T700 engine monitored itself and automatically controlled power through cyclic and pitch stick movements, allowing the pilot to concentrate on flying the mission.[217] The control had automatic engine starting, automatic temperature limiting, speed limiting, power turbine overspeed limiting, torque limiting, torque matching, and isochronous rotor speed governing.

216. J. J. Curran, "T700 Fuel and Control System," (Paper presented at the 29th Annual Forum of the American Helicopter Society, May 1973), 2.

217. Joseph F. Wansong, "T700 Engine—Designed for the Pilot and Mechanic," (Paper No. A-83-39-10-C000 presented at the 39th Annual Forum of the American Helicopter Society, 9-11 May 1983), 105.

STARTING T700 DEVELOPMENT

The T700 Full-Scale Engineering Development Program began in March 1972 upon the award of the development contract. The Army told GE that they had been selected because of the great attention GE had paid to maintainability and the many good engine design features, but mainly because the Army had great confidence that GE could do the job that had been promised.[218] George Kovacich was the Army's T700 program manager from St. Louis and Nick Kailos was the Army's T700 technical leader from Ft. Eustis. Their confidence was severely tested the following year when GE ran into two major technical problems. The first was a very large performance deficiency, and the second was a major axial compressor failure on two early engines. Each of these problems took approximately one year each to understand and fix. The major design changes that had been made in the transition from the GE12 to the T700 were suspected as contributors to the problems.

The first T700-GE-700 engine went on test February 27, 1973. The first time up, the engine produced only a little over 900-shp. After trimming to compensate for the poor component efficiencies and opening up the power turbine area, the engine was still deficient by approximately 15% in both shp and sfc. This shocked both GE and the Army. The performance deficiency showed up in almost every engine component, but especially in the power turbine which was approximately nine points in efficiency below requirement. No one had ever seen a power turbine down that much; it made everyone question the test cell torque measurement.

Ed Woll, who had then become Vice President of Engineering, joined Bob Turnbull to solve this problem. A "Group 1" task force was set up with full authority for any action within GE. Task force meetings were held in Bob Turnbull's office every morning at 8 a.m. with experts brought in as needed, particularly Marty Hemsworth, Gene Stoeckly, Mel Bobo, and Don Berkey in addition to the T700 team. A feature of these meetings was that all data from the last 24 hours of testing and analysis was reviewed and an updated action plan was generated each morning. This included making any decisions on design changes or tooling since the Project and Manufacturing people were also present. The task force reduced inlet separator losses, improved both axial and centrifugal compressor efficiencies, improved compressor matching, reduced combustor pressure loss, reduced HP turbine internal leakage and clearances, completely re-

218. Robert C. Turnbull, Cincinnati, Ohio, to Edward Woll, letter, 16 March 1994, National Air and Space Museum Archives, Washington, D.C.

designed the flowpath between the HP and LP turbines, redesigned the LP turbine blading, and incorporated LP turbine casing cooling.[219]

While GE was trying to resolve the performance problem, the Army convened an outside panel of experts to consider the problem. The outside panel concluded that GE's corrective actions were valid. The Army, which had considered launching a competitive engine, gave GE a grace period to solve the problem.

Each T700 engine went to test incorporating those improvements that were available at the time. The engine's ease of maintainability helped greatly. For example, three power turbine configurations were tested in one 24-hour period. Step by step, the performance improved from the initial 900 shp to over 1,600 shp during the first year. This steady improvement in performance maintained both Army and GE management's confidence in the T700 team. Finally, in November 1973, the T700 was able to demonstrate its required power.

The second major development problem involved compressor stages three, four, and five. Early in the development program, two T700 engines suffered major axial compressor failure just after reaching full power on their initial break-in run. Shortly after stabilizing at full power, a severe vibration spike occurred and output shp immediately dropped to zero. Engine inspection showed that every stage three, four, and five compressor blade had broken off identically at the base of the disk. Fortunately, the compressor casing did show very slight evidence of multiple per-revolution rub patterns that approximated the various stages' first flex frequencies. Based on this, the cause for the blade failures was concluded to be rub-induced vibration at first-flex frequency that overstressed the blade roots. To eliminate this problem, the blades in all three stages were beefed up approximately 15% near their bases and heavy break-in rubbing was avoided.[220]

A third problem of lesser concern to both GE and the Army than the performance and compressor problems was the fact that the engine was overweight and exceeded its original 400-lb weight specification by 15 lb. In part, this was due to the fact that GE had originally proposed a magnesium gearbox in order to meet the RFP specifications. However, after the contract had been awarded, the Army, thinking of possible future Navy use of the engine, requested a material change from magnesium to aluminum, which, although heavier, had significantly superior resistance to salt water corrosion. The Army and the airframers

219. Turnbull to Woll, 16 March 1994.

220. Turnbull to Woll, 16 March 1994.

agreed to a 15-lb increase in the weight specification as long as engine power was increased to compensate for it.

As the important development problems were put to bed, T700 tests proceeded normally. Numerous accelerated environmental tests were conducted on the T700 engine prior to field testing. They included: salt, ice, water, sand, and bird ingestion tests; extreme temperature testing; and low-cycle fatigue testing. The T700 PFRT testing was completed on July 6, 1974, and PFRT approval was obtained on October 11, 1974, both ahead of contract schedule.

UTTAS/AAH GROUND AND FLIGHT TESTING

After selecting GE as the UTTAS engine manufacturer in March 1972, the Army awarded the UTTAS airframe contract in August 1972 to two competing helicopter manufacturers. Shortly after the UTTAS selection, the T64-powered Cheyenne attack helicopter program was canceled for technical reasons. The Army then developed a new program to replace the ill-fated Cheyenne, the Advanced Attack Helicopter (AAH), and created a new competition with five airframe manufacturers bidding. The T700 was one of the candidate engines. Although not uniquely specified by the Army, the majority of the bidders selected the T700 as their preferred power plant. In June 1973, the Army selected Hughes and Bell as the AAH competitors, both of whom had specified the T700 as their primary choice of propulsion. Thus, a second major victory was won by the T700 team.

The Army's UTTAS RFP had been circulated to the helicopter industry late in 1971. On August 30, 1972, Sikorsky Aircraft Division of United Aircraft Corporation, and Vertol Division of the Boeing Company were each awarded a Basic Engineering Development (BED) contract to build three production prototype flight aircraft and one Ground Test Vehicle (GTV). Both competing models were powered by government-developed and -furnished T700-GE-700 engines.

The primary missions of the UTTAS helicopter included tactical transport of troops, supplies, and equipment in combat and combat support operations.[221] Secondary missions included combat support and service tasks—medical evacuation; transportation of command, maintenance, and medical personnel; and transportation of supplies and equipment. The UTTAS helicopter could carry both internal and external loads.

221. Robert A. Wolfe, "Army UTTAS Program," (SAE Paper No. 770952 presented at the SAE Aerospace Meeting, Los Angeles, 14-17 November 1977), 2.

The UTTAS GTV testing began in June 1974 and continued until March 1976. Virtually all GTV testing was done with XT700 engines, the first of which was delivered on March 29, 1974. The first YT700 flight test engine was delivered on July 31, 1974. The UTTAS flight testing began in September 1974 and finished in March 1976. All flight operations used YT700 engines.[222]

The first flight of the first prototype Sikorsky YUH-60A—and the T700 engine—was on October 17, 1974. The Boeing Vertol YUH-61A first flew on November 29, 1974. Competitive tests started in March 1976, and on December 23, 1976, the Sikorsky YUH-60A—later named Blackhawk—was selected for production. GE received its first production contract to provide engines for the Black Hawk. The first production UH-60A later flew on October 17, 1978, and was delivered to the Army for pilot training in April 1979.

The second major T700 application was the AH-64A Apache helicopter. In November 1972, following the cancellation of the Lockheed AH-56 Cheyenne helicopter program, the Army's AAH Task Force issued an RFP for an anti-tank helicopter capable of operating day or night, in adverse weather, and with great accuracy.[223] On June 22, 1973, competitive engineering development contracts were awarded to Hughes Helicopters and the Bell Helicopter Company, both of whom had selected the T700. Each company was to build a GTV, a static test airframe, and two flying prototypes. The AAH GTV testing began in June 1975 and ended in May 1976. The AAH flight testing started in September 1975 and concluded in May 1976. The Hughes YAH-64A flew for the first time on September 30, 1975, and the Bell YAH-63A flew on October 1, 1975. Government competitive testing started in June 1976 and finished on September 30, 1976. On December 10, 1976, the Hughes YAH-64A was announced the winner.

One hundred T700 engines were built for the UTTAS and AAH field (i.e. ground and flight) test programs, which took place between 1974 and 1976. Of these, 18 were XT700 engines, used in the four GTVs. The other 82 YT700 engines were used to support the 12 flight test helicopters. The XT700 and YT700 engines jointly accumulated over 10,700 hours operating in the field.

The T700 engine had demonstrated very high reliability during flight testing. There were no occurrences of engine-caused in-flight power loss or in-flight shutdowns in the first 200 flights and 10,000 hours of field operations. The

222. Charles H. Greene, George J. Kovacich, and W. J. Crawford, III, "T700 Flight Test Experience on UTTAS and AAH," (SAE Paper No. 760934 presented at the Aerospace Engineering and Manufacturing Meeting Town and Country, San Diego, Cal., 29 November–2 December 1976), 3-4.

223. René J. Francillon, *McDonnell Douglas Aircraft since 1920: Volume II* (London: Putnam Aeronautical Books, 1990): 317-318.

Top: This GE XT700-GE-700 turboshaft engine was one of the experimental engines used in the 1974 UTTAS competition and is in National Air and Space Museum's collection. National Air and Space Museum photo by Carolyn Russo, Smithsonian Institution (SI Neg. No. 93-4183-5).

Bottom: The business end of a GE T700-powered Hughes AH-64 Apache attack helicopter. Photograph courtesy of GE.

T700 had an unscheduled engine removal rate of 1.73 engines per 1,000 operating hours, which was a significant achievement so early in the flight test program. Throughout both the UTTAS and AHH programs, with careful and sometimes painful engine/airframe integration effort, all four candidate aircraft used the same T700 configuration.

T700 MQTS AND MATURITY PROGRAM

Official Model Qualification Tests were conducted in parallel with the helicopter test programs. Shortly after the UTTAS and AAH flight testing programs had ended and before the government competitive testing, the T700 engine completed two 150-hour Model Qualification endurance tests. The 150-hour MQTs qualified the engine to two fuel and lubrication specifications. In the first 150-hour MQT, completed on January 22, 1976, all test objectives were met or bettered with the exception of the fuel pump, which required a reconfiguration and a separate supplemental test. The second endurance test was finished on February 21, 1976, and again, all test objectives were met or bettered, and the tests were completed ahead of the original contract schedule. Most of the 300 test hours were conducted between maximum continuous and intermediate rated power, which was the equivalent of more than 3,000 hours of typical field operation. This was done to accelerate engine life testing.

The conclusion of the 150-hour MQTs marked the end of the T700 Full-Scale Engineering Development Program. There were 14 T700 engines in the factory development program. From the time the first engine was run in February 1973 until the end of the first quarter of 1976 when the MQTs were completed, more than 8,200 full-scale factory engine test hours were accumulated. Technically, the T700 engine development program had been a success, and almost all prime item development specifications had been met or exceeded. On May 13, 1976, GE was cited by Army Materiel Development and Readiness Command for technical excellence in the development of the T700. The engine had completed every major test on the first run and there had been no program delays due to engine failure. The development program was completed within the 56 million dollar contract target cost in a four-year time period of over 40% total inflation.

More than a year before the scheduled T700 MQTs, it became clear that delays in the UTTAS program meant that the T700 development program would be completed long before production engines were needed. To correct residual problems uncovered in the field and factory programs and to further improve the T700 engine, a "maturity program" was conceived by the Army, which awarded GE a contract for it in March 1975.

The principal objective of the program was to further "debug" the engine prior to production by additional intensive testing. The program was divided into two major parts—factory and field. The factory program included two 1,000-hour accelerated endurance tests, a 1,500-hour mission cycle test (later called Accelerated Simulated Mission Endurance Test (ASMET), a low-cycle fatigue test (3,500 cycles), fixing service-revealed difficulties, and qualification of the production parts list. The field program, led by an Integrated Logistics Support Management Team: supported accelerated helicopter testing; updated engines to the latest configuration; overhauled and repaired engines; and performed high-time engine inspections.

The Maturity Program resolved problems that were exposed during the competitive testing of the helicopters. The most important of these were a redesign of the No. 3 bearing and the power take-off assembly. Other improvements included a better hydromechanical control unit (fuel control), a new high-pressure fuel pump, a lower sensitivity chip detector, new power turbine seals, and an improved airseal to reduce the electrical control unit operating temperatures.[224] The field program found ways to decrease field maintenance as much as 85%. For example, highly repetitive procedures such as SOAP (Spectrometric Oil Analysis Program) sampling and changing oil filters were among the Army's most time consuming maintenance tasks. SOAP samples were eliminated and the oil filter capacity was increased to extend its service life.

Performance improvements also resulted from the Maturity Program. Compressor stall margin was improved more than 30% and power turbine efficiency was improved. All in-flight engine shutdown and flight abort problems discovered during competitive testing of the helicopters were corrected. In summary, more than 50 major improvements resulted from the Maturity Program and were incorporated into the first production engine.

Official MQT testing was completed in March 1976. The conclusion of the Maturity Program coincided with MQT approval, which came in March 1978. The first T700 production engine was also delivered in March 1978. By the time that this first production engine was delivered, 42,000 engine development test hours had been accomplished, including 24,000 hours of flight testing in five helicopter systems.[225] Starting in May 1978, a series of Component Improvement

224. K. F. Koon, "The T700—A Case Study in Reliability Growth," (GE Paper No. 243 presented at the Institute of Environmental Sciences, Cambridge, Massachusetts, 8 March 1988).

225. R. A. Dangelmaier, "T700: Modern Development Test Techniques, Lessons Learned and Results," (AIAA Paper No. P82-1183 presented at the AIAA/SAE/ASME 18th Joint Propulsion Conference, Cleveland, Ohio, 21-23 June 1982), 2.

Program contracts followed the T700 qualification. The purpose of these contracts was to correct service revealed difficulties, to enhance the reliability of the engine, and to reduce life cycle costs.

In part, the type of testing conducted in the T700 Maturity and Component Improvement Programs reflected the broader need for more realistic engine testing. By the mid-1970s, it was realized that actual flight mission usage contained more cyclic content and somewhat more time at maximum temperature than development testing then contained.[226] Cyclic content was increased by initiating the accelerated mission and the simulated mission endurance tests in the T700 testing program. The more realistic testing in the T700 program paid off in the form of early maturity and high reliability early in the engine's lifetime. A 1982 comparison between the T58 and the T700 showed that the T700 had a significantly more favorable operating cost, readiness, and mission completion rate.[227] The T700 engine in its third year of operation was more mature than the T58 at 20 years.[228]

The highly innovative T700 engine had gotten off to an excellent technical start and had two major production applications, both with the same model engine. There was one continuing concern on the part of the Army, however. GE had not met the design-to-cost goal for the T700, and the engine was high in cost. GE had made a considerable effort to comply with the design-to-cost provisions that were added to the original development contract in November 1973. GE argued that ownership cost constituted approximately 80% of any system's life cycle cost, and that the design of the T700 sought to address this issue by reduction of maintenance, spare parts, and fuel cost due to lower sfc. Also, a cost control program kept the original development contract within 8% of its 1972 target cost despite high inflation.[229]

226. R. C. Turnbull, "Recent General Electric Engine Development Testing for Improved Service Life," (SAE Paper No. 0148-7191/78/1127-0990 presented at the Aerospace Meeting Town and Country, San Diego, California, 27-30 November 1978), 7.

227. R. G. Ruegg, "Helicopter Engine Development: New Standards for the '80s," (AHS Preprint No. RWP-20 presented at the Rotary Wing Propulsion System Specialist Meeting Sponsored by the Southeast Region of the American Helicopter Society, Williamsburg, Virginia, 16-18 November 1982).

228. W. J. Wriggins, "The T700 Engine for the Eighties and Nineties," (GE Paper No. AEBG-5/82-2330L presented at the International Helicopter Symposium, Bückenberg, Germany, 21 May 1982).

229. William J. Crawford III, "The Army's T700 Engine: A Design to Cost Perspective," (GE Paper No. AEG-2/77-1415LR presented at the American Institute of Industrial Engineers, 1977), 9-13.

Between 1978 and 1989, 2,661 T700-GE-700 engines were produced. The T700-GE-700 engine had an intermediate rating of 1,536 shp, an sfc of 0.469, and a weight of 415 lb.[230] It was used on the Sikorsky UH-60A Black Hawk and Hughes AH-64 Apache.

T700 FIRST GROWTH STEP: T700-GE-401/-701

The T700-GE-401 and the T70-GE-701 engines were the first growth step beyond the T700-GE-700 engine, with 10% more power than the baseline -700 engine. The -401 was a Navy engine, and the -701 was an Army engine derived from the -401. The -401 program started in 1978, and the -701 program in 1980.

In the 1970s, the Navy proposed the LAMPS (Light Airborne Multi-Purpose System) Mk III as a replacement for its ASW helicopter, the Kaman SH-2 Seasprite. On September 1, 1977, after a competitive source selection, the Navy selected a derivative Sikorsky S-70L for its LAMPS Mk III. Later designated SH-60B Seahawk, the Sikorsky helicopter was a UH-60A Black Hawk modified for shipboard compatibility and naval missions. The primary missions of the LAMPS Mk III were anti-submarine warfare and anti-ship surveillance and targeting; secondary missions included search and rescue, medical evacuation, and vertical replenishment. On February 28, 1978, full-scale development of the SH-60B was authorized.

After a competition with the Lycoming PLT 27 engine, GE was awarded a contract for the development of the T700-GE-401 engine for the SH-60B on February 28, 1978. This contract provided for the fabrication of 14 T700-GE-401 engines for test and evaluation and performance of all full-scale development engineering and support requirements. Like other turboshaft engines, the T700 engine had been developed by one branch of the military service and then adapted to meet the needs of other services. The Army's olive-drab T700 was now wearing Navy blue.[231]

The -401 engine was a "navalized" version of the T700 engine and was the first T700 growth step. With an intermediate rating of 1,690 shp, the -401 was a

---

230. The "intermediate rating" is the maximum power available for 30 minutes of continuous operation.

231. The T700 was developed from the beginning as a multiple service use engine. However, single-engine performance for long range was more critical to the Navy than it was to the Army. The Navy got permission for a competitive engine bid for an engine with more power than the T700-GE-700 and with other special Navy requirements. After a rigorous competition, the source selection authority picked the improved performance T700-GE-401.

10% growth step in power relative to the -700 engine. Growth was achieved by a 3% increase in airflow and by a 50-degree F increase in turbine temperature. The engine was given an anti-corrosion treatment for use in the Navy environment. A primerless ignition reduced cost and improved reliability by simplifying the fuel system and the combustor. Development testing had shown that primer fuel nozzles, used for starting, were not necessary. The -401 engine also had a one-engine-inoperative contingency power rating of 1,723 shp, which could be enabled by pilot demand.

This first growth step was accomplished primarily by the increase in turbine temperature. For the higher gas temperature, cooling improvements were made in the combustor, the first-stage gas generator turbine blades and shrouds (materials and cooling), and third-stage nozzle. Rather than flaring the compressor, the increased airflow was made possible by reducing the radius ratio of the first-stage blisk at the hub to open the annulus. Changes in the axial and centrifugal compressor resulted in a one point efficiency improvement.

In addition to performance-oriented changes, the new model engine incorporated components that represented substantial cost and efficiency improvements in the manufacturing process. Single-piece castings replaced earlier, labor-intensive fabricated components. The new cast components, all interchangeable with original parts, included the front-mounted swirl frame, midframe, integral third-stage nozzle and duct, exhaust frame, and bearing support structures.[232]

The first -401 engine test took place during May 1978. The PFRT was completed in May 1979. The first YSH-60B helicopter, powered by a -401 engine, flew on December 12, 1979. Numerous qualification tests took place between 1979 and 1981. The Navy subjected the engine to a new 300-hour MQT, which replaced the earlier 150-hour MQT. An investigation during the first 300-hour -401 MQT revealed a performance loss due to first-stage turbine blade distress. A follow-on 300-hour MQT was successfully conducted in May 1981 with an improved first-stage turbine blade/cooling flow configuration. This test also incorporated a new low-stress gas turbine rotor, redesigned because the accelerated simulated mission endurance and low-cycle fatigue tests showed the original rotor did not quite meet the 5,000-hour design life.

The first production -401 engine was delivered in 1982. The first production SH-60B helicopter flew on February 11, 1983, and deliveries began shortly after

232. Keith F. Mordoff, "GE Cuts T700 Fabrication Costs with Expanded Use of Castings," *Aviation Week & Space Technology*, Vol. 122, No. 12 (March 25, 1985): 55.

that. Later, the -401 engine was selected for the Bell AH-1W SuperCobra, the Kaman SH-2G Super SeaSprite, the Sikorsky S-70B-2 Seahawk, and the Sikorsky VH-60 Blackhawk helicopters.

The T700-GE-701 was a derivative of the -401 engine developed in response to the Army's need for an additional 10% increase in power for the AH-64 Apache. It was essentially identical to the -401, without the Navy's anti-corrosion and fire resistant features, deleted because they were not required for the Army environment and their elimination reduced engine unit cost. The -701 engine shared most of the components, systems, and 90% of all parts with the -700 engine. It incorporated a transient droop fix for the Apache helicopter, which had experienced a rotor rpm loss during some aggressive maneuvers. The -701 engine had an intermediate engine rating of 1,690 shp; it also had an automatic contingency power rating of 1,723 shp under one-engine-inoperative conditions.

The -701 contract was awarded in August 1980, and the first engine went to test in June 1980. Because of the commonality between the -701, -401, and -700 engines, the only unique-to-the -701 engine tests were an 150-hour endurance test and an Electronic Control Unit (ECU) component qualification test (for the automatic contingency feature). Following the start of the -701 qualification program, a minor change was made in the centrifugal compressor configuration from the -401 baseline and a change was made in the overspeed and drain valve system; these changes required additional tests. The -701 MQT qualification acceptance was in February 1983.

Full production for the Hughes AH-64A Apache was authorized in March 1982.[233] Subsequently, a contract was awarded to GE in April 1982 for the first year production of -701 engines for the AH-64A Apache production program. Between 1985 and 1989, 1,618 T700-GE-701 engines were purchased by the Army for the Apache. The -701 engine had a major impact on the Apache, significantly increasing its acceleration and climb performance and its in- and out-of-ground-effect hover altitude.

THE SECOND GROWTH STEP: T700-GE-401C/-701C

The T700-GE-401C/-701C engines constituted the second growth step from the baseline T700-GE-700 engine. The intermediate power rating was in-

---

233. In 1984, Hughes Helicopters became a subsidiary of McDonnell Douglas, and the helicopter became the McDonnell Douglas AH-64A.

creased to 1,800 shp. Design and development for both engines started in the 1985-1986 time period.

The T700-GE-401C and -701C engines evolved from an integrated GE growth plan designed to concurrently develop and qualify two T700 military engines and two commercial turboprop versions of the T700 engine.[234] The need for this growth step became apparent in 1985-86 as a result of three developments. First, the weight and mission requirements were increasing for the Black Hawk and Seahawk helicopters. Second, a T700-GE-401 engine had been chosen for the EH Industries EH 101 helicopter, but more power was needed for a production engine. Thirdly, both the Saab 340 and the CASA Nurtanio CN-235 commuter aircraft were using turboprop versions of the T700 engine, but needed more power.[235]

The second-step growth program was designed to integrate selected, proven technology advances made in the 1980s to T700 and CT7 engines. Advances were applied in building block fashion to T700-GE-700, -401, and -701 military engines as well as the commercial CT7-2A engine. This simultaneous, integrated engine development significantly reduced the development time required for each new model and provided greater commonality of parts.

Design modifications throughout the -401 and -701 were made for the -401C and -701C engines. An important change in the centrifugal compressor diffuser was the incorporation of quasi-rectangular passages that increased efficiency and reduced losses. This innovation was developed and patented by GE as part of the Modern Technology Demonstrator Engines (GE27) program. Other compressor improvements included thickening the backwall of the centrifugal compressor for durability.

An improved cooling design (serpentine cooling) and material was used in the first- and second-stage HP turbine blades and nozzles in order to accommodate higher temperatures. To reduce clearances, ceramic shrouds were used on the HP turbine in the first GE application of this technology on a production engine. Material changes were also made in the third-stage power turbine blades.

These engines featured a digital electronic control as a primary system with a full-featured hydromechanical control as a backup. It provided cost and weight

---

234. The two commercial turboprop versions of the T700 engine referred to here were the CT7-6 and the CT7-9.

235. H. G. Donohie, "T700: Growing to Meet the Challenge," *Vertiflite*, Vol. 35, No. 3 (March/April 1989): 48.

advantages and added hot start prevention, engine history collection, and diagnostics capability. Transient droop improvements previously incorporated in the -701 engines for the Apache were also applied to the -401C and -701C engines to provide faster rotor speed response for both the Apache and Black Hawk helicopters.

The -401C and -701C engines had improved power-to-weight ratios and sfc over the -401 and -701. These engines provided increased gross weight margin, better one-engine inoperative performance, and improved hot and high performance for the UH-60 and SH-60 helicopters. An increase in the capability of the main rotor transmission of the Sikorsky H-60 series helicopters made it possible to accommodate the more powerful engine. Both the -401C and the -701C had an 1,800-shp intermediate rating and a 1,940-shp one-engine-inoperative contingency rating.

The first -401C/-701C engine to test was in mid-March 1987. The PFRT was initiated in mid-July 1987 and completed by the end of July. The flight test program on the -401C started on September 1, 1987, and was completed on September 30. In May 1988, the -401C/-701C completed its Qualification Test and received approval in the same month. Also in May 1988, GE was awarded a multiple-year contract to produce the -401C and -701C engines for the Black Hawk and Seahawk. GE had competed with the Rolls-Royce Turboméca RTM 322 engine for this major award. The first -401C engines were delivered to the Navy in 1988. The -401C was used in the Sikorsky SH-60B, SH-60F, HH-60H, and HH-60J helicopters. The -701C was used in the McDonnell AH-64A, C, and D helicopter models as well as the Sikorsky UH-60L, MH-60G, and MH-60K helicopter models.

T700 SUMMARY

From the early 1970s through the mid-1980s, the T700 became the engine of choice on nearly every major intermediate-size helicopter developed.[236] Low sfc and weight were critical T700 design specifications, however, very long component life, survivability, and maintainability were additional design specifications. Initial objectives of the program were to demonstrate a reduction of 25% in sfc and an increase of 40% in power/weight ratio relative to operational engines in the same power class. Together, these design specifications were a major step forward in turbine engine technology and set a precedent for future

236. Louis A. Bevilacqua, "T700/CT7 Engine Program," (GE Paper No. 203 presented at the
    U.S. National Aerospace Exposition '86, Beijing, China, 15-21 May 1986), 11.

Army turboshaft engine development. Furthermore, it was a highly innovative engine which incorporated an interesting combination of technology.

The design features of the T700 as well as a long and extensive development and testing program led to an engine that proved exceptionally reliable in service. By 1986, the T700 shop visit rate was competitive with modern high by-pass ratio engines powering wide-bodied civil airliners.[237] As of 1988, the T700 engine had a 99.8% availability and a mission completion rate of 99.7 to 99.9%, the highest in the industry.[238] This was unprecedented reliability for a small turboshaft engine.

During the time that the T700 was being developed, the Army reorganized its aviation maintenance units and reduced considerably the number of people based on the promise made by the T700 for reduced maintenance requirements. This resulted in significant savings to the Army. The T700 was one of a very few systems that, once fielded, did live up to its reliability, availability, and maintainability promises.[239] Early Army planning had considered procuring 50% spare engines for the T700 program based on 1960s experience. Actual service had demonstrated a need for about 15% spares, which provided a substantial savings in spare engine procurement.[240] Thus, the T700 established new standards for reliability and maintainability against which other helicopter engines were subsequently measured.

Future T700 growth plans called for specific step increases which would eventually place the engine in the 2,600-2,800 shp range. In 1991, GE initiated design of a T700 compressor with improved aerodynamics. On September 1993, tests were completed on the centrifugal portion of a new, increased airflow T700 compressor that was intended to boost the power of the engine by 15-20%. The increased power T700 would be known as the T700/T6E and was

237. H. G. Donohie and W. W. Rostron, "T700/CT7 Derivative Growth Engine Reliability Consistent with Longstanding Industry Traditions," (Paper No. 84 presented at the Twelfth European Rotorcraft Forum, Garmisch-Partenkirchen, Germany, 22-25 September 1986), 84-12. Shop Visit Rate was defined as the total number of engine removals per 1,000 engine flight hours requiring maintenance action at intermediate or depot level shops.

238. K. F. Koon and Art Nordstrom, "T700—The Result of Army Experience," *Vertiflite*, Vol. 34, No. 2 (March/April 1988): 47.

239. "Black Hawk Program Managers Discuss the T700: Maj. Gen. Richard D. Kenyon (ret.), Former Black Hawk Program Manager," *Lynn Headlines*, GE Aircraft Engines, May 9, 1988, 3.

240. Eugene E. Martin, "T700—A Program Designed for Early Maturity and Growth Potential," (Paper No. 85a presented at the Tenth European Rotorcraft Forum, the Hague, the Netherlands, 28-31 August 1984), 6.

targeted for heavy lift versions of the Black Hawk helicopter and the NH-90 helicopter.[241]

Commercializing the T700: The CT7 Turboshaft/Turboprop Engine

GE's decision to develop a commercial derivative of the T700 followed Gerhard Neumann's long-standing company policy that a military production base needed to be established first. This requirement was met in the early 1970s after the T700 had been selected for both the UTTAS and Apache helicopter programs. While the UTTAS competitive development program was underway, GE was approached by the Boeing Vertol Company and then by Sikorsky Aircraft Division who were interested in engines for commercial derivatives of their UTTAS helicopters, the Boeing Model 179 and Sikorsky S-70. Some T700 engines were subsequently sold to both companies for prototype commercial helicopter development. These sales, with Army approval, marked the beginning of the commercial T700 program.

On July 2, 1974, at the same time that the T700-GE-700 PFRT testing was nearly completed, GE applied to the FAA for certification of its commercial turboshaft engine. In September 1976, GE announced the introduction of its new commercial helicopter engine, the CT7 rated at 1,536 shp.[242] Changes in the CT7 from the T700-GE-700 reflected certification requirements such as fireproofing some external lines and conforming to commercial standards for fastener and locking devices. Additional small design changes were made in the oil and fuel filter, blower, and ECU.[243]

The next airframer with an interest in the T700 was Bell. During the 1970s, Iran, through the U.S. government, purchased from Bell a large number of Model 214 helicopters, upgraded derivatives of the UH-1. The Iranian helicopters needed still more power for hot day and high-altitude conditions, and twin T700 engines would fit the bill. GE agreed to develop and certificate a commer-

---

241. Michael O. Lavitt, "T700 Improvements," *Aviation Week & Space Technology*, Vol. 139, No. 12 (September 20, 1993): 19.

242. Eric Falk, "CT7 Description and Technical Design Data," GE Press Releases AEG-976-69, AEG-976-70, and AEG-976-71, September 1976.

243. James L. Nye, "An Advanced Technology Engine Family for General Aviation," (Paper No. 79-1161 presented at the AIAA/SAE/ASME 15th Joint Propulsion Conference, Las Vegas, Nevada, 18-20 June 1979), 3.

cial version of its T700 engine. A Bell-modified Model 214ST powered by 1,625-shp T700/T1C engines was tested in Iran in February 1977.[244] T700 Program Manager Bill Crawford had just completed the negotiations for the sale of 875 T700/T1C engines to Iran when all contracts with Iran were canceled because of the Islamic revolution in December 1978. Despite the loss of business, Bell decided to proceed with the development and certification of the Model 214ST, and GE continued with its commercial T700 program.

Thus, during the 1970s, the first commercial spinoffs of the T700 engine were gradually developed. GE's civilian certification program was conducted in parallel with the military T700 engine development program. Realizing that a broader T700 production base would reduce manufacturing costs and save money, the Army agreed to let GE develop with its own money a commercial T700 engine. Subsequently, the Army allowed GE to use some of the T700 qualification testing to satisfy FAA requirements. FAA certification testing was completed in April 1977. On June 29, 1977, the FAA issued a type certificate for the GE CT7-1. On September 6, 1978, the CT7-2, and T700/T1C engines received their FAA certification. All of these engines were derivatives of the T700-GE-700.

In November 1979, Bell decided to manufacture an initial series of 100 production Model 214ST helicopters with CT7-2 engines, rated at 1,625 shp. The CT7-2A, also rated at 1,625 shp, went into production for the Model 214ST. The CT7-2A was FAA certificated in May 1981 and entered commercial service with Petroleum Helicopter, Inc., on a Bell 214ST in 1982.[245] The Model 214ST was the first civilian production application of the CT7 series engine. Next, another variant designated CT7-2B, also rated at 1,625 shp, was approved for the Westland 30 Series 200 helicopter, which first flew on September 3, 1983. It was later used on the Westland 30 Series 300 as well. The CT7-2A and CT7-2B commercial engines were certificated and entered production during the same time period that the first growth step engines, the T700-GE-401/-701, were undergoing similar development. The CT7-2D (the commercial equivalent of the T700-GE-401/-701), rated at 1,623 shp, was FAA certificated in 1985 for the Sikorsky S-70C helicopter. This version featured additional corrosion protection for use in marine environments. The CT7-2D1 (the commercial equivalent of the T700-GE-401C/-701C) received FAA certification in

244. Alain J. Pelletier, *Bell Aircraft since 1935* (London: Putnam Aeronautical Books, 1992), 167.

245. R. E. Gaerttner, "The CT7 Turboshaft: Commercial Adaptation of a Military Turboshaft Engine," (GE Paper No. 86 presented at the International Helicopter Symposium, Hanover, Germany, 21-22 May 1982), 1.

June 1989 for the Sikorsky S-70. The 1,625-shp CT7-2C engine was also an optional Sikorsky S-70C power plant.

COMMUTER TURBOPROP ENGINES: CT7-5 AND CT7-7

An important trend of the late 1970s and early 1980s influenced GE to commercialize its T700 engine further and develop it as a turboprop engine. Airline deregulation in 1978 had led to a rapid increase in commuter airline traffic and the need for larger commuter aircraft.[246] Airframers including de Havilland, Embraer, Saab, and many others were pressing ahead with plans to enter this market. In 1979, GE carried out a study under contract with the NASA Lewis Research Center to identify advanced engines and technology for future 30- and 50-seat commuter turboprop aircraft.[247] In response to the need for new commuter aircraft, Vice President and Group Executive Officer Brian Rowe announced in November 1979 that a major effort had been launched at GE Lynn's Business and General Aviation Engine facility to provide new-technology, fuel-efficient aircraft engines for the commuter and general aviation industries.[248] Thus, GE publicized the development of the CT7 turboprop engine, which it said was being studied seriously by more than 10 airframers designing 30-passenger-size commuter aircraft. The CT7's fuel efficiency and low maintenance requirements would be ideal for commuter airline operators, GE felt.

Aggressively competing with other engine manufacturers for the 30-passenger-size commuter aircraft, but particularly with Pratt & Whitney Canada, GE sought applications for its turboprop engine, designated CT7-5 and CT7-7. GE's CT7 lost competitions against Pratt & Whitney's PT7 (later PW100) with de Havilland and Embraer, both of whom were long-standing Pratt & Whitney Canada customers. However, in June 1980, the 1,654-eshp CT7-5 was selected for the 35-passenger Saab-Fairchild SF340 commuter aircraft; this sale launched the CT7 turboprop engine program. On January 30,

---

246. See PW100 section of Pratt & Whitney Canada chapter for further details how deregulation affected the growth of the commuter aircraft market and the size of commuter aircraft.

247. Arnold Brooks and Robert Hirschkron, "A Review of Commuter Propulsion Technology," (GE Paper No. 74 presented at an SAE meeting, Savannah, Georgia, 25 May 1982). Also, R. Hirschkron and R. E. Warren, "Commuter Turboprop Propulsion Technology," (SAE Paper No. 801243 presented at the Aerospace Congress & Exposition, Los Angeles, California, 14 October 1980).

248. Robert N. Salvucci, "General Electric Launches Major Effort in Commuter and General Aviation Engine Programs," GE Press Release AEG-1179-131, November 1979.

Crossair was the launch customer for the Saab-Fairchild SF340 aircraft. This was the first commuter aircraft application for GE's CT7 turboprop engine. Photograph courtesy of GE.

1981, CASA-Nurtanio announced that the CT7 turboprop engine (CT7-7) had been selected to power its new 35-passenger Spanish-Indonesian commuter aircraft designated CN-235.[249] The 1,700 eshp CT7-7 became the second CT7 turboprop application.

GE had a compelling case for selling its turboprop engine as evidenced by its pitch to Saab-Fairchild.[250] The fuel economy of the CT7-5 was better than any other engine in the 1,500- to 2,000-shp class; lower fuel consumption meant lower direct operating costs, resulting in higher profits. The T700 had undergone a long development and testing period and was an operational engine in service on several important applications. Its excellent reliability and maturity was a matter of record. The engine had ongoing funding by the Army, Navy, and GE, which ensured quick resolution of operational problems. The engine was exceptionally maintainable which lent itself well to time-critical commuter maintenance requirements, reduced maintenance skill levels, and reduced cost. It was warranted and guaranteed to protect operators from unscheduled expenses. GE

249. Robert N. Salvucci, "CASA P.T. Nurtanio Announce Selection of GE Turboprop Engine for New Commuter/Utility Airliner," GE Press Release AEG-181-11, January 1981.

250. General Electric, "Executive Summary: CT7-5 Turboprop Engine for the Fairchild-Saab SF-3000," Proposal CT7-0480-002, 11 April 1980.

had considerable commercial and turboprop engine experience, and it had an international service and support organization. The major technical risk was the new propeller gearbox, built by Hamilton Standard to GE specifications.

The overall objective of the commercial turboprop program was to define a turboprop configuration responsive to the needs of the regional airline industry.[251] The turboprop performance was to be increased 20% over the baseline T700-GE-700. The CT7-5/-7 engines were a part of, and further expanded, the T700 engine's first-step growth development compared to the T700-GE-700. Other engines in this first-step growth included the CT7-2 and T700-GE-401/ -701. A further development of the CT7-2 and T700-GE-401/-701 engines, the CT7-5/-7 engines had increased compressor airflow and turbine temperature ratings, and consequently, more power.

The CT7-5/-7 incorporated technology from the T700-GE-401/-701 program. The three major changes to the T700 turboshaft engine incorporated in the CT7-5/-7 engine were: the addition of a propeller gearbox; component redesign to simplify or eliminate turboshaft features not required for turboprop applications; and component design to meet increased performance requirements. Performance was increased by increasing compressor airflow 8% and increasing turbine inlet temperature by 50°F compared to the first-step growth military turboshaft engines (T700-GE-401/-701).

The inlet particle separator was simplified to reduce inlet pressure loss by 60% (and thereby increase engine power) while still providing adequate inlet protection for commuter turboprop requirements. (While the turboshaft engine required inlet particle separation to cope with massive amounts of ingested sand and dust churned up by the helicopter down-wash from unprepared landing and take-off sites, the turboprop operated in a much more benign environment). Consequently, the high-intensity swirl flow in the turboshaft IPS was eliminated and the turboprop IPS was redesigned with axial-flow induction and with a modified flow path to provide for separation of ingested materials, but not fine sand and dust. To provide for the required 8% airflow increase in the CT7-5/-7 engine and to provide for future growth in later engine models, the compressor was designed for a 12% airflow improvement. The combustor, fuel injectors, and ignition system were the same as those used on the T700-GE-401/-701. Directionally solidified material of improved temperature capability was used for the gas generator turbine. The power turbine was not changed. The exhaust frame was modified to incorporate outlet guide vanes to decrease exhaust losses and an

251. J. D. Stewart, "CT7-5A/7/7E Turboprop Engine for Commuter Airline Service," (ASME/GTD Paper presented in Amsterdam, NL, June 1984), 3.

ejector plenum and system were added to provide for IPS bypass flow in lieu of the turboshaft IPS blower. The engine control system was altered to provide for the requirements of the turboprop engine. The propeller gearbox was a compound idler system chosen for high efficiency, low parts count, low heat rejection, and weight and cost advantages. An optional propeller brake could be mounted on the gearbox to lock the propeller and allow the engine to be run as an APU.[252]

The CT7 turboprop engine program was officially initiated in late 1980 with a launch order from Saab-Fairchild. In December 1981, a test program was started using a core engine for compressor development. The first full engine test was conducted in March 1982. A new turboprop engine/propeller test cell for development and certification was finished by April 1982 at Lynn. Flight testing was conducted at the GE Flight Test Center in Mojave, California, where a CT7-5A engine was mounted on a Grumman Gulfstream G1 turboprop aircraft. Flight testing began in 1982 and concluded in January 1983. Approximately 2,300 engine test hours were run on the turboprop configuration. FAA certification for the CT7-5A/-7/-7E engine models was awarded on August 29, 1983.[253]

On January 25, 1983, the Saab-Fairchild SF340 (powered by the CT7-5A) made its first flight, and on June 14, 1984, the SF340 went into commercial service with its launch customer, Crossair of Switzerland, followed in October by Comair of Cincinnati, Oh.[254] However, starting in September 1984, a series of engine malfunctions occurred which led to the grounding of the aircraft in Switzerland and the U.S.[255] Despite considerable promise, the CT7 ran into serious problems.

Some of the CT7-5A engine failures were traced to compressor second-stage blade failures. The fix was to increase compressor-blade-to-case clearance to prevent blade rubbing. Additionally, to improve vibratory stress margins in the second-stage blades, the blade angle of incidence was reduced, all of the blades

252. Stewart, *CT7-5A/7/7E*, 6-13.

253. J. D. Stewart, "CT7 1700 Horsepower Engine," (AIAA Paper No. AIAA-84-2229 presented at the AIAA/NASA General Aviation Technology Conference, Hampton, Virginia, 10-12 July 1984), 10.

254. Hans G. Andersson, *Saab Aircraft since 1937* (London: Putnam, 1989), 170-171.

255. "Swiss Ground SF-340 Aircraft after Crossair Engine Incidents," *Aviation Week & Space Technology*, Vol. 121, No. 11 (September 10, 1984): 36. "FAA Approves Fix for CT7 Engine," *Aviation Week & Space Technology*, Vol. 121, No. 14 (October 1, 1984): 34. "GE Modifying CT7 Second Stage Following Operational Failure," *Aviation Week & Space Technology*, Vol. 121, No. 20 (November 12, 1984): 32.

were shot-peened, and an alternative method of applying anti-corrosion coating was utilized (the orginal method had negated the conditioning effects of shot-peening). A number four engine bearing failure was traced to a faulty batch of roller bearing cages, which were then replaced. These were among the problems that led to the immediate grounding of the SF340, but premature turbine wear also became an expensive warranty problem for GE to fix. Several factors led to this problem.

First, the aircraft had grown in weight beyond its original design specificat-ions, and GE was required to develop engines that ran hotter to produce the necessary power. It was determined that the engines were being used at maxi-mum power (and thus at the higher temperatures) approximately 13% of the time. However, the engines had originally been designed for military applica-tions which only required maximum power about 2% of the time. Because this duty cycle had not been anticipated, the design life of some turbine parts, par-ticularly turbine shrouds, was rapidly exceeded, and they wore out prematurely. The higher than expected time at high turbine temperatures also caused turbine nozzle vane cracking, coking in the exhaust system, a friction-induced error in the torque sensor, and shorter bearing life. In short, all of the engine parts were worked harder than they had been designed for.

This resulted in unscheduled maintenance and loss of passenger revenue for airline operators and very high parts replacement costs for GE who had war-ranted the engines. For the short run, GE hired experienced airline pilots to show commuter pilots techniques to reduce engine wear. The permanent fix was to change the critical turbine parts with designs and materials that could tol-erate significantly higher temperatures.

Another airframe application was the Spanish-Indonesian CASA/IPTN CN-235 Series 10 which was powered by 1,700 shp CT7-7A engines. The CN-235 was a twin-turboprop regional and military transport aircraft. The first flight of the aircraft was on November 11, 1983. The aircraft became opera-tional in 1987 as a military transport, and on March 1, 1988, entered service as a commuter airliner.

SECOND GROWTH STEP: CT7-6 AND CT7-9

Four engines, the CT7-6, CT7-9, T700-GE-401C, and T700-GE-701C, consti-tuted the second major growth step of the T700 engine. The development pro-gram of these engines was completely integrated. All incorporated an improved efficiency centrifugal compressor and an improved HP turbine. Additional im-provements in the CT7-6 and CT7-9 included an increased airflow axial com-pressor and an improved efficiency LP turbine.

The need for this step two growth became apparent in 1985-86 when more power was needed for the Black Hawk and Seahawk helicopters, the European Helicopter Industries (EHI) EH 101 helicopter and other possible European helicopters, and the Saab 340 and CASA-Nurtanio CN-235 commuter aircraft. The CT7-6 was developed as result of GE studies showing that a 2,000-shp version of the T700 engine had an excellent market opportunity in Europe.[256] It found an application on the three-engine, multi-mission EH 101 helicopter. The development of the CT7-9 was driven by the increased power needs of the Saab 340B and the CN235-100.

The EHI EH 101 evolved from the twin-engine Westland WG.34, an ASW helicopter design selected in September 1978 to replace the British Royal Navy Sea King helicopter.[257] The WG.34 proof-of-concept test rig was powered by T700-GE-401 engines. A broadly similar requirement by the Italian Navy led to a June 1980 joint venture between the Westland and Agusta companies known as EHI Ltd. Subsequent market research resulted in a decision to develop naval, civil, and military versions of the helicopter. In January 1984, the British and Italian governments signed an agreement to proceed with the development of naval helicopters. Three T700-GE-401A engines powered the prototype EH 101 helicopter on October 9, 1987. On October 15, 1987, the Italian Ministry of Defense selected the CT7-6A to power the Italian Navy EH 101 prototype helicopters, replacing the T700-GE-401A engines previously destined for these helicopters. More power was needed for the EH 101 helicopter production engine, and the CT7-6 was developed to fill that need.

The CT7-6 was co-developed by Alfa Romeo Avio, Fiat Aviazione, and GE. GE had long-standing relationships with both Italian companies. Participation with Fiat included licensed production of J47, J79, and T64 models and co-development and co-production of LM500 and LM2500 power units. Alfa Romeo had licensed production of T58, Gnome, and J85 engine models. Both Italian companies had participated in CF6-80 and T700 engine programs. Co-development (50% GE and 50% Alfa/Fiat) of the CT7-6 started early in 1985.[258]

Alfa Romeo Avio, with the capability to test T700-GE-401A engines, was responsible for running the official 150-hour endurance test engine. Other Alfa

256. H. G. Donohie, "T700/CT7 Growth Engine for European Helicopters," (Paper No. 86 presented at the Eleventh European Rotorcraft Forum, London, England, 10-13 September 1985), 86-9.

257. Derek N. James, *Westland Aircraft since 1915* (London: Putnam Aeronautical Books, 1991), 434.

258. C. Massaro, R. Bonifacio, and R. Scenna, "Co-development of CT7-6 Engines—A Continued Tradition in Technology and Reliability," (Paper No. 94 presented at the Fourteenth European Rotorcraft Forum, Milano, Italy, 20-23 September 1988), 94-2 - 94-3.

Romeo responsibilities included modifying the turboshaft exhaust frame to retain T700/CT7 baseline engine length, preparation of the installation manual and engine mock-ups, Italian certification coordination, and providing engines for the Italian Navy's prototype EH 101. Fiat Aviazione redesigned the T700/CT7 turboshaft inlet frames to achieve compatibility with the larger-airflow axial compressor used in the CT7-6 and modified the diffuser case to maintain engine mount commonality with other T700/CT7 turboshaft models. Fiat performed diffuser case static load testing and generated field technical manuals. GE had responsibility for overall program integration and management as well as technical direction. GE was responsible for leading the design, development and FAA certification program, defining engine configuration, building certification and CT7-6 flight test engines, and conducting certification tests.

In addition to incorporating an improved-efficiency centrifugal compressor and an improved HP turbine, both introduced on the T700-GE-401C/-701C, the CT7-6 (and the CT7-9) included an axial compressor with 8% increased airflow, a more efficient LP turbine, and an increase in turbine temperature to 2,500° F, yielding a 29% growth in power over the baseline T700-GE-700 engine. The high-airflow axial compressor was a modification of the CT7-5 turboprop compressor with fine tuning in the front end (IGVs and stages one and two) to improve the efficiency, stall margin, and speed flow characteristics. Improvements in CT7-6 aerodynamic design involved changes in axial compressor blade and vane contours, centrifugal compressor diffuser shape, and LP turbine stator and rotor blade contours. Material improvements were made in the first- and second-stage HP turbine blades, the HP turbine shrouds and supporting structure, and the first-stage LP turbine blades. Turbulated, serpentine cooling was an additional improvement made in the HP turbine; this was approximately 30% more effective than single-pass radial cooling. The electronic control incorporated three-engine load sharing and an accessory drive-mode feature to meet unique EH 101 requirements.[259]

Both the CT7-6 and CT7-9 first went to test in May 1987. The CT7-6/-6A, rated at 2,000 shp, was FAA certificated on June 30, 1988; it was also certificated later by the Italian and British governments. Also in June 1988, GE delivered the first CT7-6 engines to EHI. The first flight of a CT7-6 powered EH 101 helicopter took place at Westland Helicopters, Ltd., on September 30, 1988, as part of the EH 101 flight test program (T700-GE-401A engines were used prior to

---

259. P. L. Kastrinelis and W. E. Lightfoot, "CT7-6: The Most Recent T700 Growth Derivative Engine," (ASME Paper No. 90-GT-241 presented at the Gas Turbine and Aeroengine Congress and Exposition, Brussels, Belgium, 11-14 June 1990), 2-4.

this to flight test pre-production EH 101 helicopters). The CT7-6A was identical to the CT7-6, except that it included additional corrosion protection (marinization) features to meet military requirements. The CT7-6A powered the prototype Italian Navy EH 101 on its first flight in April 1989. A military version of the CT7-6, the T700/T6A, was selected in September 1990 to power the Italian Navy's EH 101 and was qualified in 1991 by the Italian Ministry of Defense.

While GE was successful in winning the EH 101 application in Italy, it lost the EH 101 competition in England. GE had teamed with England's Ruston Gas Turbines, Ltd., to manufacture the CT7-6 engine for the EH 101. In June 1990, the Rolls-Royce RTM 322 was selected for the British Royal Navy Merlin (EH 101). GE came back, however, when in August 1991, it was announced that the CT7-6A1 won a competition against the RTM 322 for the EH 101 New Shipborne Aircraft (NSA) Program in Canada.[260] However, the Canadian EH 101 program's political unpopularity caused its cancellation in November 1993 when new prime minister Jean Chretien took office.[261]

Immediately following the certification of the CT7-6 turboshaft engine, the CT7-9 turboprop engine, flat rated at 1,750 shp (1,870 shp with APR), was certificated by the FAA in June 1988. Like the CT7-6 civil turboshaft engine, the CT7-9 civil turboprop engine featured the same improvements as the T700-GE-401C/-701C engines, but also added an improved axial compressor and upgraded power turbine. CT7-9 climb and cruise power was increased up to 15% and fuel consumption decreased up to 4% compared to the CT7-5A/-7 series, and a digital electronic control was introduced.

The first two production CT7-9 turboprop engines were shipped to Industri Pesawat Terbang Nusantara (IPTN), Bandung, Indonesia, in June 1988 for the CN-235 Series 100 aircraft built by CASA of Spain and IPTN. The 45-passenger CN-235 Series 100 was powered by CT7-9C engines rated at 1,870 shp (with APR). In September 1989, the Saab 340B was introduced with 1,870-shp (with APR) CT7-9B engines; this aircraft was intended primarily for hot and high conditions, and the CT7-9B engines gave better altitude performance and better climb and cruising speeds than CT7-5 series engines installed on earlier versions of this aircraft. On January 19, 1990, GE opened up new market opportunities when it signed contracts with the Czech and Slovak

---

260. John Morris, "Canada Chooses GE Engines for Proposed NSA Helicopter," GE Press Release GEAE-20L, 7 August 1991.

261. David Hughes, "Canada Defers Next Move on Helicopters," *Aviation Week & Space Technology*, Vol. 139, No. 21 (November 22, 1993): 96.

Federal Republic aviation industries to supply 1,940-shp (with APR) CT7-9D engines and support for the 40-passenger LET L610 G regional airliner in Western markets. GE had competed with Pratt & Whitney Canada for this business.[262] In January 1991, the first GE engines were delivered to LET, and the aircraft made its first flight on December 18, 1992.

By 1996, a two-step growth program that would provide significant power increases for the CT7 turboprop engine family was underway. The CT7-9+ was designed to offer 4% more power than the CT7-9B and hot and high take-off performance improvements, achieved primarily through the use of a new, advanced centrifugal compressor stage. A second model, the CT7-11, incorporated an advanced axial compressor with the centrifugal compressor stage in the CT7-9+, providing an additional 10% power increase to 2,000 shp.

CT7 SUMMARY

GE's decision to develop a commercial version of the T700 engine followed the engine's selection for the UTTAS and Apache helicopter programs, and the resulting T700 military production base which these programs established for GE. The first requirements for a commercial turboshaft engine evolved gradually during the early 1970s. Airline deregulation in 1978 resulted in a need for new commuter aircraft which subsequently led to the development of the CT7 turboprop engine in the early 1980s.

Of the 8,006 T700 and CT7 engines built by November 14, 1991, just 16% or 1,252 engines were in the commercial CT7 family. By November 14, 1991, the largest number of CT7 engines produced were the CT7-5 models and CT7-9 models, the majority for the Saab 340 commuter aircraft.

By 1991, the CT7 turboprop had demonstrated an in-service dispatch reliability rate of more than 99.9%.[263] The CT7 had found an important application on the popular Saab 340 commuter aircraft. In Europe, the Saab 340 had more than a 60% share of the 30-40-passenger aircraft market in 1991.[264]

The success of the CT7 as a commuter turboprop engine can be attributed to several important factors. The CT7's fuel consumption was approximately 10%

262. N. Isler and S. Morandi, "A Western Engine for an Eastern Aircraft; Re-Engining of CSFR LET 610 Turboprop Transport with the GE CT7-9," (AIAA Paper No. AIAA-91-1911 presented at the AIAA/SAE/ASME/ASEE 27th Joint Propulsion Conference, Sacramento, California, 24-26 June 1991), 1.

263. John Morris, "GE's CT7 Turboprop Engine Proves its Reliability Worldwide," GE Press Release GEAC-91-31L, 3 November 1991.

264. John Morris, "GE's Product Support Plays a Vital Role in CT7 Operations," GE Press Release GEAE-36L, 16 October 1991.

better than any other engine its class, a fact which saved customers money and made for lighter weight aircraft. The Army's 1960's design requirements for a lightweight, fuel-efficient helicopter engine, which had improved and grown over time, dovetailed nicely with 30-40-passenger commuter aircraft requirements. Also, airline deregulation resulted in a major expansion of commuter aircraft which really helped the small engine business, including the CT7 turboprop engine.

While CT7 turboprop engines outsold CT7 turboshaft engines, the combination of T700 and CT7 turboshaft engines was extremely successful. By 1993, the T700/CT7 family of turboshaft engines powered, or were options on, every mid-sized helicopter in the Western world requiring 1,600 to 2,000+ shp.[265] Having won four out of five contests between 1988 and May 1992, the T700/CT7 turboshaft engine was very successful in both civilian and military competitions against the RTM 322.[266]

The GE27, GE38, T407/GLC38, and CFE738 Engine Family

GE27

The Modern Technology Demonstrator Engines Program (MTDE) was the fourth in a series of technology demonstrator engine programs that began in 1967, sponsored by the U.S. Army's Aviation Applied Technology Directorate in Ft. Eustis, Va.[267] Of the competing engine candidates, including those of Garrett, Lycoming, and Allison, the GE27 and the Pratt & Whitney PW3005 were selected in 1983. The objective of the program was to develop a demonstrator turboshaft engine in the 5,000-shp class. The Army was interested in an engine for a new helicopter to replace the Boeing Vertol CH-47. The Navy also

265. International Gas Turbine Institute, American Society of Mechanical Engineers, *1993 Technology Report, Land Sea, and Air* (Atlanta, Georgia: International Gas Turbine Institute, 1993), 66.

266. Stanley W. Kandebo, "Emerging Helicopter Programs Spur New Rivalry Between GE, Rolls/Turboméca," *Aviation Week & Space Technology*, Vol. 136, No. 19 (May 11, 1992): 53.

267. Edward T. Johnson and David D. Klassen, "Developments in New Gas Turbine Engine Demonstrator Programs," (ASME Paper No. 86-GT-80 presented at the International Gas Turbine Conference and Exhibit, Dusseldorf, West Germany, 8-12 June 1986), 1-4. The preceeding programs were: the 1,500 hp Demonstrator Engine—1967; Small Turbine Advanced Gas Generator (STAGG)—1971; Advanced Technology Demonstrator Engines ATDE)—1977.

had an interest in the Army's MTDE program; it was looking for an engine for an aircraft to replace the Lockheed P-3 ASW aircraft and/or be retrofitted on the P-3 to give it enhanced capability. The Army, Navy, and Air Force had a combined interest in an engine for the Joint-Services VTOL Experimental (JVX), which sought to develop a multi-mission, medium lift VTOL aircraft (later known as the V-22 Osprey). The MTDE engine fit this application as well. The MTDE program was jointly funded by the Army and each winning engine company, and also had Navy support.

A noteworthy aspect of the MTDE program was that the major design requirements were given equal priority. They included reliability/durability, overhaul/support cost, acquisition cost, weight, survivability, and performance. For GE, a challenging MTDE design objective was the design-to-cost requirement. This meant that GE had to design an engine approximately four times larger than the T700 yet costing only 20% more, starting at the 250th production unit. An important performance specification was the altitude requirement, varying from 5,000 feet for the turboshaft application to 30,000 feet for the turboprop application. Component design compromises were made in order to balance the extremes of these flight envelopes.

The engine was developed under the direction of Program Manager Richard Hickok. Major preliminary design contributions in the design of the advanced axial-centrifugal compressor were made by Barry Weinstein, Manager of Engineering, Robert Neitzel, Manager of Preliminary Design, as well as Alex Bryans and Les King in the design of the compressor.

In anticipation of this demonstrator program, GE began its preliminary design work in the late 1970s. To broaden its product line, the preliminary design was laid out for turboshaft, turboprop, and turbofan applications.[268] A high pressure ratio compressor was selected on the basis of anticipated long-range, long-endurance turboprop or turbofan applications. A 22-23:1 compressor compression ratio was selected because it was probably the highest cycle achievable in a single rotor without difficult starting problems.[269] A quick way to demonstrate this technology was to take readily available T700 hardware and redesign the compressor. Late in 1982, GE ran a 23:1 pressure ratio compressor in a scaled size on a T700 engine.

The GE27 that was designed for the MTDE program had an inlet particle separator, compressor with five axial and one centrifugal stages, short through-

---

268. Jim Krebs, interview by Rick Leyes and William Fleming, 21 August 1991, GE Interviews, transcript, National Air and Space Museum Archives, Washington, D.C.

269. Woll to Leyes and Fleming, 20 August 1991.

flow annular combustor, two-stage air-cooled HP turbine, and three-stage LP turbine. A high compressor pressure ratio and high turbine temperatures were prerequisites for low sfc, an important objective. In describing the engine, Program Manager, Dick Hickok commented:

> The GE27's claim to fame was that it was to be the most modern technology that had been demonstrated in that size engine. In other words, you're talking a size of 27-28 lb-per-second (compressor airflow). It had a 22-23:1 pressure ratio compressor. No one else had done that. It had a centrifugal (compressor) in a size that no one else had done. It had single crystal (HP turbine) blades. It had a turbine temperature in the 2,400-2,500-degree class. That hadn't been done in an engine of this size in a total engine format. All those things (individually) had been done somewhere. What we did was demonstrate this (combination of) technology as a repertoire, so to speak . . . put it together in one engine and see how it worked. It worked successfully, probably one of the most successful (programs).[270]

The high pressure ratio axial-centrifugal compressor was selected after more than 300 trade-off studies had been carried out. The compressor technology was integrated from various engine and demonstrator programs. The axial-centrifugal compressor was derived from the T700/CT7 engine. Axial compressor technology was enhanced by the contributions of the Air Force High-Through-Flow Axial Compressor and GE25 axial compressor programs. Likewise, IR&D (Independent Research and Development) centrifugal compressor and GE26 axial-centrifugal compressor programs added to centrifugal compressor technology. An important reason for selecting a centrifugal compressor in this high pressure ratio, high flow compressor was to avoid numerous, very small and impractical axial blades at the back of the compressor.[271]

The GE27 was the first engine to demonstrate an efficient modern centrifugal compressor design in the 5,000- to 6,000-shp class. While the axial-centrifugal compressor configuration was an excellent choice for the requirements of the MTDE program, the design trade-off was a constraint on future growth to more than 7,500 shp or 7,500 lb thrust without significant changes in engine configuration or dimensions. Beyond that power level, the centrifugal

270. Dick Hickok, interview by Rick Leyes and William Fleming, 21 August 1991, GE Interviews, transcript, National Air and Space Museum Archives, Washington, D.C.

271. D. D. Klassen, "Configuration Selection and Technology Transition in 5,000-shp Class Engines," (AIAA Paper No. AIAA-83-1411 presented at the AIAA/SAE/ASME 19th Joint Propulsion Conference, 27-29 June 1983, Seattle, Washington.

compressor was a problem due to tip speed limitations at higher rpm and inherently lower centrifugal compressor efficiency compared to an axial compressor. Unfortunately, analysts and airframers would later cite the need for more growth capability as one reason for the selection of alternative engine choices when derivatives of the GE27 competed in various airframe competitions.

The core engine first ran in December 1983 and the full engine late in 1984. By 1987, after more than 4,000 hours of component testing and more than 800 hours of engine testing, four world records were set with the GE27, including: lowest sfc, highest power-to-weight ratio, highest specific power (ratio of power to airflow), and highest single-spool pressure ratio.[272] The GE27's 0.380 sfc was the lowest recorded sfc for a turboprop engine. Compared to the T700-GE-701 engine, the GE27 sfc and thermal efficiencies were 25% better and the power-to-weight ratio was improved by 62.5%.[273]

The JVX program was formally initiated in December 1981 and was run by the Army. In January 1983, the Department of Defense transferred responsibility of the JVX program to the Navy. In April 1983, the JVX airframe competition was won by a joint Bell Helicopter and Boeing Vertol design known as the Model 901 TiltRotor. As the new program manager, the Navy decided to re-evaluate engine candidates for the JVX aircraft and asked GE to consider proposing its T64, a shaft version of the TF34, and the GE27. GE had an internal competition among the candidate contenders and eventually decided to bid the GE27 for the JVX engine competition.

On December 24, 1985, Secretary of the Navy John Lehman announced that the Allison 501-M80C (later designated T406) engine had been selected over the GE27 and the Pratt & Whitney PW3005. Reasons for the selection were lower risk, established reliability and support system, and lower research and development costs.[274] Derived in part from an existing engine, the T406 had lower development costs and less risk. The T406 was also a larger engine, and had more power margin for hot-day and single-engine-out operation. With a

272. Michael J. Zoccoli and Kenneth P. Rusterholz, "An Update on the Development of the T407/GLC38 Modern Technology Gas Turbine Engine," (ASME Paper No. 92-GT-147 presented at the International Gas Turbine and Aeroengine Congress and Exposition, 1-4 June 1992, Cologne, Germany), 2.

273. Edward T. Johnson and Howard Lindsay, "Advanced Technology Programs for Small Turboshaft Engines: Past, Present, Future," (ASME Paper No. 90-GT-267 presented at the Gas Turbine and Aeroengine Congress and Exposition, 11-14 June 1990, Brussels, Belgium), 2.

274. "Allison Selected to Develop Engine for VTOL V-22 Osprey," *Aviation Week & Space Technology*, Vol. 124, No. 1 (January 6, 1986): 23.

lower compressor pressure ratio and turbine temperatures, the T406 had greater growth capability (without changing the dimensions or configuration of the engine) than did the GE27.

The V-22 program was primarily driven by a Marine Corps need for an assault aircraft that could quickly move troops and equipment from ship to shore. Because this was a relatively short-range mission, a case could be made for a power plant with modest fuel economy. The very low sfc that the GE27 had been designed to achieve was less critical in this context than it would have been for a longer-range aircraft. While the GE27 was an advanced technology engine compared to the T406, advanced technology was not the driving factor in the final selection. The formal engine contract was awarded to Allison in 1986, and a full-scale development program for the V-22 Osprey was approved by the Naval Air Systems Command on May 2, 1986.

GE38

The loss of the JVX engine competition came as a surprise and great disappointment to GE. Despite this setback, GE continued development of the engine knowing that it had an exceptional new gas generator core for advanced turbofan, turboprop, and turboshaft engines. Thus, the GE27 was superseded by the GE38, a commercial engine development effort.

The GE38 became a family of related shaft-power and turbofan engines. The GE38 turboshaft program had military (T407) and commercial (GLC38) turboprop engine variants. The turbofan engine program was called the CFE738. Cooperative programs for the development of these turboshaft and turbofan engines were explored and solidified. The work on both types of engines proceeded in parallel.

T407/GLC38

To reduce the development risk, GE backed off slightly on the performance of the GE38, which became the T407-GE-400. The sfc commitment was lowered by 3%. To share development costs, GE sought additional revenue-sharing partners. Potential applications included Lockheed P-3 aircraft, Boeing Vertol CH-47 helicopters, and Lockheed C-130 aircraft, all of which were sold worldwide. In 1988, after considering a number of possibilities, GE entered into revenue sharing partnership agreements for development of a turboprop engine with Textron Lycoming of Stratford, Conn., Ruston Gas Turbines of England, Bendix Controls Division of AlliedSignal located in South Bend, Ind., and later in 1989 Steel Products Engineering Company (SPECO) of Cleveland, Ohio.[275] This partnership was managed by a joint project office with a GE director and a Lycoming deputy director.

The immediate objective of the partnership was to develop a power plant for Lockheed's proposed P-3 replacement aircraft. Locking in a partnership was critical for GE for it had a self-imposed constraint that it could not compete for this engine development unless the financial resources (its own as well as its partners) had been committed. GE headed the partnership and had the lead in contact with Lockheed, as well as responsibility for certain compressor, combustor, and high pressure turbine parts. Lycoming was to participate in the management, development, and production of the engine. The Bendix role was to provide a Full-Authority Digital Electronic Control (FADEC) system. Ruston was to develop and produce HP and LP turbine parts. SPECO was responsible for the gearbox.

In October 1988, Lockheed was selected by the Navy to produce the Long-Range Air Anti-Submarine Warfare Capable Aircraft (LRAACA), the replacement for the Lockheed P-3. Late in 1988, Lockheed selected the GE38 over the Allison T406 for the LRAACA program. This win was consistent with GE's policy to have a military customer before a commercial customer, and it launched the GE38 program. Full-scale development began in late 1988. The GE38 turboprop engine was subsequently designated T407-GE-400 by the Navy.

The design objectives of the T407 engine included high output power per unit airflow, reliability from reduced parts count, ease of maintenance via extensive modularity, sfc levels that were as much as 25% below those of existing 5,000-6,000-shp engines. The immediate goals of the engine development program were to redesign the GE27 engine to a 10,000 hour/30,000 cycle life standard, and to deliver the first T407 engine to test in the fourth quarter of 1989.[276] The conversion of the engine from an Army technology demonstrator to a commercial product involved extensive revisions aimed at significant reduction in direct operating costs.[277]

---

275. This partnership was put together over time. The Lycoming/GE partnership dated from December 1987. Originally, Z-F of Germany was chosen as a partner to provide the propeller gearbox, but later, GE switched to SPECO when negotiations were not completed with Z-F.

276. Zoccoli and Rusterholz, "Development of the T407/GLC38," 1-3.

277. Michael J. Zoccoli and David D. Klassen, "T407/GLC38: A Modern Technology Powerplant," (ASME Paper No. 90-GT-242, June 1990).

The GE38/T407 used the GE27 core, with five axial and one centrifugal compressor stages, a short annular through-flow combustor, and two-stage air-cooled HP turbine. Additionally, the engine had a three-stage LP turbine and a front-drive output shaft. The turboprop version of the engine used a propeller gearbox with a two-stage reduction system based on T64 design with additional features incorporated from commuter airline experience. The compressor was in the 20:1-overall pressure ratio class at an airflow class of 30 lb/sec. The inlet guide vanes and first two stator stages were variable to improve part-power performance and stall margin. Engine control was achieved with dual redundant FADEC units, each separately housed and mounted on the engine. The engine was of modular design. It had a maximum performance rating of 5,660 shp and a 0.380 sfc.

GE was ultimately required by Lockheed to provide the total propulsion system including the propeller and engine nacelle and selected Hamilton-Standard to provide the propeller. In January 1989, the LRAACA was designated P-7A. On December 26, 1989 at Textron Lycoming, the T407 power section was run for the first time. In the early spring of 1990, the T407 flying testbed engine completed its first run at GE's field test site in Peebles, Oh. Lockheed and GE continued to work together until July 1990 when the P-7A program was terminated by the Navy. Lockheed could not meet the performance or schedule established in the original contract. In late July, Lockheed terminated GE. The T407 development program was put on hold at that time. Later, the T407 program was unsuccessfully revived for a competition for an updated, new production version of the Lockheed P-3 ASW aircraft known as the P-3 Orion 2.[278] The program was canceled in the face of a diminishing threat from the U.S.S.R. and a decreasing Department of Defense budget. In the meantime, work continued in support of commercial engine variants.

The commercial version of the T407 engine was the General Electric/Lycoming Commercial 38 (GLC38). Derated to extend its operational life, it was rated between 3,500 shp and 5,000 shp depending on the application. Turbine temperature was decreased, cooling air was increased, and engine speed was decreased. The T407 and GLC38 engines were part of the same development program. The partners were the same as for the GE38 program.

The GLC38 competed for several airframe applications, but was not successful. In July 1989, the Allison GMA 2100 was selected for the Saab 2000, a

---

278. Stanley W. Kandebo, "Allison GMA 3007 Turbofan Generates More Power, Uses Less Fuel than Predicted," *Aviation Week & Space Technology*, Vol. 135, No. 14 (October 7, 1991): 70.

50-passenger regional transport aircraft.[279] While it was generally assumed that the GE GLC38 would win this competition because GE was the engine supplier for the Saab 340, an independent analysis suggested that reasons for the GMA 2100 win included, among other things, a greater power growth capability and lower engine unit price.[280] The GLC38 also competed unsuccessfully for the 50-passenger IPTN N-250 commuter aircraft competition. GMA 2100 engines were selected for the N-250 in July 1990. According to IPTN, the GMA 2100's growth potential was a factor in the selection.[281] It was ironic to note that in competing against the GMA 2100, the GLC38 was actually competing against its earlier rival the T406, which used the power section of the GMA 2100. GE submitted a proposal based on the GLC38 for a competition for the de Havilland DHC-8 Dash 8 Series 400 in 1989, again in competition with the GMA2100. However, the program was not pursued by de Havilland at the time.

CFE738

In November 1984, GE and Garrett began to study the possibility of jointly developing a turbofan engine in the 5,600-lb-thrust class for corporate jets in the 30,000-lb-gross-take-off-weight class. Both companies had an interest in the midsize corporate business jet market. In June 1986, GE and Garrett announced a 50-50 joint venture to produce an engine known as the CFE738.[282] With a thermodynamic potential to grow to 7,000 lb thrust, the CFE738 filled a gap in each company's existing product line, specifically between the 5,400-lb-thrust Garrett ATF3-6 and the 9,000-lb- thrust GE CF34. GE's responsibility for the CFE738 was the GE27-based gas generator. Garrett was responsible for the low pressure fan system, including the three-stage LP turbine derived from its TFE731 turbofan engine. In June 1987, the Commercial Fan Engines (CFE) Company was formed by GE and Garrett to manage all phases of the development, manufacture, and marketing of an engine known as the CFE738. The joint venture was based on the successful CFM International arrangement be-

279. "Allison Engine Chosen for Saab 2000; Sweden Certifies Saab 340B Transport," *Aviation Week & Space Technology*, Vol. 131, No. 3 (July 17, 1989): 29.

280. "Another Major Victory for Allison," *The Power Letter*, Issue 95 (March 21, 1990): 1.

281. "Allison GMA 2100 Selected," *Aviation Week & Space Technology*, Vol. 133, No. 5 (July 30, 1990): 11.

282. "Joint Venture to Build Turbofan for Midsize Corporate Jets," *Aviation Week & Space Technology*, Vol. 124, No. 26 (June 30, 1986): 63.

tween GE and SNECMA.[283] Organizationally, it also included methodologies used by the Garrett-Allison LHTEC T800 engine program. The CFE program was GE's first partnership with an American company.

The CFE company resulted from the realization that the GE and Garrett engine projects were complementary and that each company had much to offer the other.[284] Garrett had been doing conceptual design work on its next-generation TFE731 engine designated TFE731-X, GE was exploring turbofan versions of its GE27, and both engines were in the same power class. By coming together as the CFE company, Garrett gained the GE core and GE gained some of Garrett's fan and low pressure turbine technology and the benefit of Garrett's dominant position in the business jet aircraft market.

Market studies conducted by the companies in 1985 indicated a demand for business jet aircraft with larger cabins, longer (3,500 nm) range, and the ability to cruise at higher Mach numbers.[285] By pooling resources, the companies could develop a challenger to Pratt & Whitney Canada's PW305 turbofan engine, a slightly smaller engine, but also in the 5,000-lb-thrust class. Although the CFE738 was originally intended as a pure business jet engine, design changes accommodated proposed regional jet airliners, such as the Shorts FJX.

The CFE738 was a 5,900-lb-thrust two-shaft turbofan engine. A single-stage fan was driven by a three-stage LP turbine. The gas generator consisted of a five-stage axial and single-stage centrifugal compressor, straight-through-flow annular combustor, and two-stage, air-cooled HP turbine. The CFE738 was the first small engine to operate at the high overall pressure ratios normally associated with 50,000-lb-thrust level engines.[286] With a relatively high hub pressure rise across the fan (1.55:1) plus a 23:1 core pressure rise, it operated at an overall pressure ratio (at altitude) of over 33:1. The ability of the CFE738 engine to operate successfully at such high pressure ratios was based on experience accumulated with T700/CT7 engines.

An interesting aspect of the CFE738 was that it was designed for on-airframe maintenance. It was also a business turbofan engine designed around airline-

283. Graham Warwick, "Small Turbines Step Forward," *Flight International*, Vol. 132, No. 4074 (August 8, 1987): 26. This was for the CFM56 turbofan engine.

284. Mark Lambert, "CFE738: The Turbofan for Tomorrow's Business Jets," *Interavia* (May 1988): 481.

285. "Two's Company for GE and Garrett," *Flight International*, Vol. 135, No. 4166 (May 27, 1989): 43.

286. David Moss and G. L. McCann, "CFE738: A Case for Joint Engine Development," (AIAA Paper No. 902522 presented at the AIAA/SAE/ASME/ASEE 26th Joint Propulsion Conference, Orlando, Florida, 16–18 July 1990), 3.

The CFE738 turbofan engine was developed jointly by GE and Garrett for the midsize business jet market. Photograph courtesy of GE.

level support technology. For instance, it had more borescope access than other contemporary business jet engines. It had a support system that allowed technicians to identify problems, download them into a laptop computer, and send them via fax or modem to CFE engineers for evaluation.

Full-scale engine development on the CFE738 began in May 1988. Eleven engines were built for the development program. The core for the prototype CFE738 was shipped to Garrett in March 1990 where it was mated to the rest of the engine. Testing of the complete engine began in May 1990. Ground tests were temporarily suspended in the summer of 1990 when a vibration problem in the engine's HP turbine led to a turbine blade separation. This problem was caused by a torsional vibration resonant with the first-stage turbine stator passing frequency. The problem was resolved by reducing the number of stator blades and raising the frequency of the rotor blades. Eleven engines were used in the development program. Flight testing began August 30, 1992, on Garrett's modified Boeing 720B test aircraft. During development, 4,000 hours of running time were logged, including 250 flight hours on the Boeing 720B and 300 hours on Dassault Falcon 2000 prototypes. The first production CFE738 engines were shipped in October 1993, and the CFE738-1 was FAA certificated on December 17, 1993.

A series of airframe competitions took place in the 1990s, and the CFE738 variously competed against Allison, Pratt & Whitney Canada, and Lycoming for them. In March 1990, the CFE738 and the Lycoming LF507 lost to Allison's GMA 3007 for the Embraer EMB-145 45-48 passenger regional jet. A launch customer was secured for CFE738 in April 1990 when it was selected by Avions Marcel Dassault-Breguet to power the Falcon 2000 business jet.[287] GE submitted the CFE738 and Pratt & Whitney Canada its PW305 for the Cessna Citation X business jet, but both candidates lost to Allison's GMA 3007 in October 1990. According to Cessna, the ability of the Allison engine to grow with the airplane was a major factor behind its choice.[288] Again, the gas generator of the GMA 3007 was the T406, an engine that had proved difficult for the GE38 family to beat in earlier competitions. In 1994, the CFE738 lost a competition for the IAI Galaxy business jet to the P&WC PW306A. Factors cited for P&WC's victory included lower engine weight and a simplified battery starting system versus the CFE738's requirement for pneumatic starting from an APU. This was considered an upset as the CFE738 already powered the Falcon 2000 while the PW306A was a proposed engine.[289]

For the Falcon 2000 eight-passenger business jet, the CFE738 was flat rated to provide 5,725 lb take-off thrust. The Falcon 2000 was essentially a Falcon 900 long-range trijet reworked into a smaller twin-engine configuration. It was a wide-body, medium-heavy business jet with a transcontinental range of 3,000 nm. The CFE738 was sized right for the Falcon 2000 application, and both GE and Garrett historically had done business with Dassault. Like its predecessor GE27, the CFE738 had a fuel efficiency unmatched by any other engine at the time. The CFE738-powered Falcon 2000 first flew on March 4, 1993.

THE GE27, GE38, T407/GLC38, AND CFE738 SUMMARY

Relative to other small turbine engine programs, the GE27/GE38 family did not win major airframe applications. Lack of success was largely determined by

287. "Allison, General Electric Secure Launch Customers for Latest Small Turbofan Engines," *Aviation Week & Space Technology*, Vol. 132, No. 22 (April 9, 1990): 22.

288. Nigel Moll, "Citation X is a 10," *Flying*, Vol. 118, No. 6 (June 1991): 108.

289. David Esler, "A new Player in the Medium-Thrust League," *Business & Commercial Aviation*, Vol. 75, No. 2 (August 1994): 84.

Allison's more powerful competitor engine. However, the GE27/GE38 engine family was noteworthy for its technical achievements, including the fact that it achieved a fuel economy unmatched by any other turbine engine. A successful derivative of the GE38 family was the CFE738 turbofan engine, a product of a joint venture between GE and Garrett. The CFE738 program was managed by the CFE Company, GE's first domestic partnership. The CFE738 successfully competed for the Dassault Falcon 2000, a wide-body, medium-heavy business jet with transcontinental range.

Overview

THE STATUS OF THE CORPORATION

GE contributed to the development of, built and tested in 1942 the first jet aircraft engine in the U.S. Subsequently, GE became the largest aircraft engine manufacturer in the world. GE passed Pratt & Whitney in jet engine sales around 1988. In 1990, the total revenue of GE's Aircraft Engine Business Group (AEBG) was $7.558 billion.[290] In that year, AEBG was the top earner among GE's diversified industry segments. Between 1986 and 1992, GE's aircraft engine business did well, consistently ranking number one or number two in revenue among GE's industry segments.[291] Total GE revenues for 1990 were $44.879 billion. With a year-end market value of $58 billion in 1989, GE was the second-largest American company.[292]

From its beginning in 1942 until 1988, GE Aircraft Engines grew to 14 plant sites with 13 million square feet of floor space and 38,000 employees. Of these, over 9,000 were in engineering. In Lynn, Mass., site of GE small engine manufacturing, employment peaked at about 13,000 in 1985. As a result of the worldwide economic recession that began in the early 1990s and GE restructuring efforts, Lynn employment declined to about 5,500 in 1993. In Evendale, Oh.,

---

290. General Electric Company, *1990 Annual Report* (Fairfield, Connecticut: General Electric Company, 1991), 33.

291. In 1988, General Electric Financial Services (GEFS) became a separate category in its annual report revenue and operating profit statement. Previously, financial services had been one of the GE's industry segments. While GE's aircraft engine revenue was consistently number one or number two among GE's industry segments between 1986 and 1992, this revenue was significantly less than that earned by GEFS as a result of the 1988 accounting change.

292. General Electric Company, *1989 Annual Report* (Fairfield, Connecticut: General Electric Company, 1990), 3.

where GE manufactured its large engines, 20,000 were employed in 1987; by 1993, employment had declined to 12,400.

GE's aircraft engine business was exclusively military until the late 1950s, although earlier commercial efforts had been made. GE had certificated its TG-190B, a derivative of the J47, on December 27, 1949 for commercial applications, but none were found. In the early 1950s, a small planning team began working with airframe manufacturers and airlines to establish aircraft requirements. GE's first successful commercial engine was the CJ805-3, which was certificated in September 1958 for the Convair 880 airliner. GE's commercial engine revenue remained a relatively small percentage of the company's engine business through the 1960s. Commercial engine sales began to increase in the 1970s. In 1989, commercial engine revenue exceeded military engine revenues. In that year, GE engine revenue was: commercial (48%); military (47%); marine and industrial (5%). This trend continued through 1992 when the breakdown was: commercial (52%); military (40%); marine and industrial (8%).

An export-oriented company, GE derived 35% of its total aircraft engine revenue from international business in 1988. Steadily increasing, the international portion of GE's total aircraft engine revenue was 49% in 1992. In 1992, GEAE exports were $3.6 billion, or 49% of the total value of GE exports. Of the $3.6 billion, sales were: Europe (42%); Pacific Basin (35%); Middle East and Africa (13%); Americas (10%).

As one of the big three aircraft engine manufacturers, GE had 40% of the entire "Free World" jet engine industry sales in 1988.[293] In 1993, GEAE and its French partner SNECMA, in their CFMI joint venture, had about 50% of the global large engine market. Large engines were consistently the largest portion by far of GE's engine business. In the small engine market segment, GE had nearly 30% of the total market. GE's Small Engine Operation was comparable to companies such as Lycoming, Garrett, and Allison.

Through the early 1990s, GE turbine engines were manufactured, assembled, and tested at two major centers. In Evendale, Oh., large commercial, military, and marine and industrial engines were manufactured. Since 1979, GEAE headquarters had been located in Evendale, also the largest GEAE facility. In Lynn resided the development and manufacturing of smaller engines for fighters, helicopters, commuter airliners and business jets. The F404, although generally considered a large military engine, was designed, developed and produced in Lynn. Small ma-

293. General Electric, "GE Aircraft Engines: An Overview," November 1988. The other two major engine manufacturers were Pratt & Whitney, with 30% of the total Free World jet engine sales, and Rolls-Royce with 10%.

GE's small aircraft engines are built in this factory in Lynn, Massachusetts. Photograph courtesy of GE.

rine and industrial engines were manufactured at Lynn as well. Satellite plants supporting the Evendale and Lynn facilities were located around North America, including a major outdoor test site at Peebles, Ohio; Flight Test Operations at Edwards Air Force Base, California; and additional flight testing at Mojave, California. GE engine service support facilities were located worldwide.

GE's aircraft engine research was conducted at both its Lynn and Evendale facilities. In general, advanced research and large commercial engine research was conducted at Evendale. Research primarily applicable to small engines, such as axial-centrifugal compressor research work, was conducted at Lynn. Research from each source was integrated into small and large engines wherever appropriate. A highly technical business with a continuing need for substantial reinvestment, GE Aircraft Engines returned 15% of its revenues for new product research and development.

THE COMPANY'S MARKETS AND INflUENCING FACTORS

When viewed as a whole, GE had a number of successful small turbine engine programs. Military engines constituted the majority of all small turbine engines manufactured by GE since its small engine business was formed in 1953. Accounting for a lesser portion of the business, small commercial engines emerged as an increasingly important presence in the 1960s. In addition to the typical reasons that account for successful products, GE's prosperous small engine business was in part a reflection of several key corporate strategies developed by top GE leaders, including Ralph Cordiner, Jim LaPierre, Jack Parker, Gerhard Neumann, Brian Rowe, and Ed Woll, especially in the small engine business.[294]

In 1951, Cordiner established a new GE organizational structure emphasizing decentralization. In carrying out Cordiner's objective, LaPierre and Parker were instrumental in establishing GE's small turbine business in 1953. Neumann, who held managerial roles from the mid-1950s until his retirement in 1979, instituted several important policies that had a strong impact on GE's small engine business including requiring a military production base before launching commercial engine programs, and implementing long-range strategic planning. Following Neumann's retirement, and a two-year period with Fred MacFee as Chief Executive, Rowe took over GE's engine business. Rowe's plan for the decade of the 1980s was to continue to build on GE's success and make specific improvements in technology, productivity, international markets, and customer support.

From its start in the jet engine business in 1941 until the 1960s, GE focused almost entirely on military engine development. Following that tradition and encouraging it, Neumann's policy was to let the military pay for engine development, secure a military production contract, get production tooling started, and then sell the engine commercially.[295] Recognizing the need for and the opportunity to expand in the commercial market, Parker and Neumann formally created in 1967 a Commercial Engine Projects operation. Included in this operation were the CJ610 and CF700 engines. By the late 1960s and early 1970s, GE had gained an important commercial presence. The "military before commercial" policy was used as both a practical means and a sales pitch to achieve this status.

Another important policy implemented by Neumann was long-range strategic planning. In October 1962, an Advanced Product Planning operation was created. Gradually, Neumann stretched the planning period to 20 years or more to account for long engine cycle life. Important objectives of long-range planning were to target and prepare for future military engine competitions, identify new airframe possibilities, and anticipate new technology required for future engines. For the engine business, one of the primary goals of long-range planning was to attract GE corporate resources. Engine development cost GE more than

---

294. Excellent design and performance, good timing and marketing, winning important competitions and airframe applications, and effective product support are typical reasons that account for successful turbine engine products.

295. Neumann to Leyes and Fleming, 22 August 1991.

any other product it manufactured, and GE's engine business had to compete for corporate investment against other GE businesses such as electronics and plastics. For that reason, long-range engine plans predicted return on investment and equity. Such plans worked, and the engine business got critically needed capital investment. According to former strategic planner Bill Rodenbaugh:

> I think that, at the time, if we had not had a good defendable strategic plan with a good track record, we would never have been able to attract the corporate resources necessary for us to grow to the point we got to. I believe the plan was as important as any other single element we had.[296]

In 1968, GE realigned its entire corporate structure into 10 major business segments, among them the Flight Propulsion Division, which became the Aircraft Engine Group. This formal recognition of the contributions and growth potential of the business proved to be important as GE's jet engine business became a top earner for the corporation.

During the decade of the 1980s, Brian Rowe presided over a major growth in the engine business from $1.8 billion in sales (and $81 million net income) in 1979 to more than $7 billion (and $650 million net income) in sales in 1990. Rowe's strategies to achieve this growth were based on continued investment in technology, further penetration of international markets, factory modernization to increase productivity, and continuous improvement in customer support.[297] In the early 1990s, Rowe was faced with new set of circumstances with the onset of an international recession which hit the aerospace especially hard. For Rowe, new priorities at GE included price reductions, reducing capacity, general cost cutting, significantly downsizing GE's workforce, greater use of computers for more efficiently carrying out development programs, and more cooperation between manufacturers in engine research and development projects.[298]

CONTRIBUTIONS

The primary thread of GE's turbine technology progressed from steam turbines through gas turbine experiments, turbo air compressors, turbosuperchargers,

296. Bill Rodenbaugh, interview by Rick Leyes and William Fleming, 22 August 1991, GE Interviews, transcript, National Air and Space Museum Archives, Washington, D.C.

297. Brian Rowe, interview by Rick Leyes, 15 January 1992, GE interviews, transcript, National Air and Space Museum Archives, Washington, D.C.

298. Johan Benson, "Face to Face with Brian Rowe," *Aerospace America* (March 1994): 6-8.

and finally gas turbine engines. In each case, the newer technology was built on knowledge gained from earlier development work. It was, however, the transfer of Frank Whittle's turbojet technology that was primarily responsible for GE's accelerated move into the jet engine business. Subsequently, GE became the first U.S. company to develop in part, build, and test a jet aircraft engine.

The company's turboshaft engines, which included the T58, T64, and T700, contributed significantly to the advancement of military utility, combat, and very heavy lift helicopters. The J85 turbojet made possible low cost, small jet fighter and supersonic trainer aircraft, and the CJ610/CF700 commercial derivatives powered the majority of the first generation of business jet aircraft. The TF34/CF34 was one of the first engines to introduce high-bypass fan technology into small combat as well as executive regional aircraft.

Small (and large) GE engines were characterized (in general) by their straight-through-flow designs, high compression ratio compressors using variable stators, compact combustors, and air-cooled turbine blades and nozzles that operated at very high temperature. Such technology contributed importantly to the success of GE engines.

WILLIAMS INTERNATIONAL | 7

Preface

The story of Williams International is in fact the story of Sam Williams himself. As its Chief Executive Officer, President, and Chairman of the Board, he piloted the course of the company since he founded it in 1954. A privately-held company, Williams did not have a hierarchy of bureaucracy to wade through to make decisions or start projects.[1] When he decided to do something, he could and did get started on the spot.

Early Years

While studying mechanical engineering at Purdue University, Sam Williams developed an interest in gas turbine engines. To him, engines were an exciting challenge. Imagining ways of making useful power attracted him; from childhood, he had been an inventor of sorts. As he learned more about aerodynamics and thermodynamics in college, he began to sketch and analyze turbine engines. Eventually, his interest in gas turbine engines led him to establish his own company, Williams Research.

1. William H. Gregory, "Innovation and Competition," *Aviation Week & Space Technology*, Vol. 123, No. 9 (September 2, 1985): 13.

383

EARLY EXPERIENCE AT CHRYSLER

After obtaining his Bachelor of Science degree in December 1942, Sam Williams joined the Chrysler Corporation Engineering Division. His first assignment was working on metallurgy and on piston engine development. As a result of his continuing interest in turbine engines, Williams became acquainted with the patent attorneys at Chrysler and began proposing various gas turbine engines. His associates became aware of his interest in the subject.

In 1944, Chrysler was awarded a contract by the Navy Bureau of Aeronautics to develop a 1,000 shp turboprop engine. The engine was intended for long-range, low-altitude, patrol and anti-submarine warfare aircraft. Because of the interest he had shown in turbine engines and his understanding of them, Sam Williams was selected to participate in the project. He contributed, laying out the engine configuration, performing thermodynamic cycle studies, and selecting the cycle pressure ratio. Gradually, an engineering and drafting team was assembled. The engine that they designed was designated the XT36-D-2.[2]

The Chrysler XT36 ran successfully on a test stand within two years of its conception, and according to Sam Williams, it had very good efficiencies for the time.[3] This 1,000-shp, single-shaft engine had an axial-flow compressor with a 6:1 pressure ratio, six can-type combustion chambers, a two-stage turbine, and a tube-bundle heat exchanger surrounding the engine to recover the waste heat. Sam Williams obtained a patent on the XT36 engine configuration.

Unfortunately, several years after World War II, the Navy's interest in further development of turboprop engines was superseded by its interest in jet engines, and it temporarily reduced its efforts in this area. Chrysler had spent $14 million under the Navy contract and had built a large organization to develop the engine. Nevertheless, the company elected to discontinue further work on aircraft gas turbine engine development after the War.

Not wanting to lose the expertise gained on this engine, XT36 project head George J. Huebner, Jr. successfully convinced Chrysler's management to put the core XT36 team, including Sam Williams, to work developing an automotive gas turbine engine. The resulting turbine engine had a single-stage centrifugal compressor with a 4:1 pressure ratio, a single can-type combustor, a sin-

2. XT36-D-2 was the Navy designation of the engine. Chrysler's designation was A-86A.

3. Sam B. Williams, interview by Rick Leyes, 7 September 1988, Williams International Interview, transcript, National Air and Space Museum Archives, Washington, D.C.

gle-stage turbine driving the compressor, a single separate power turbine, and a rotary heat exchanger.[4] The rotary heat exchanger was instrumental in reducing some important problems of using gas turbines in automobiles, namely high fuel consumption and wasted heat through the exhaust.[5] According to Williams, the general configuration of this engine was adopted by other automobile companies in their later development of automotive and truck gas turbine engines.[6]

The engine ran successfully in 1953 and was demonstrated in a 1954 Plymouth sports coupe. Sam Williams was one of the passengers in this car on its first demonstration run, which was cheered by coworkers as the car began moving forward under its own power.[7] On June 16, 1954, it was demonstrated publicly at the dedication of the Chrysler Engineering Proving Grounds at Chelsea, Mich. Sam Williams and George Huebner shared a patent on this engine configuration, and Sam Williams had various patents on its individual components.

FOUNDING WILLIAMS RESEARCH

As a result of his turbine engine experience at Chrysler, Sam Williams foresaw the possibility of aircraft, marine, and industrial small gas turbine engines. He believed that if an efficient, low-cost small turbine engine became available, applications and a market would follow. Going against conventional wisdom, which said that such engines would be too heavy and inefficient, late in 1954, he decided to take a chance and depart Chrysler to form his own organization. Soon thereafter, he obtained a contract with the Outboard Marine Corporation to develop a small gas turbine engine.

A lifetime goal of Sam Williams was to have his own company, and it seemed to him that the time had come to strike out on his own. Thus, on January 1, 1955, with $3,000 in savings and housed in a small rented warehouse in Bir-

4. The purpose of heat exchangers is to recover some of the heat energy that would normally be lost in the engine exhaust by transferring it to the compressed air entering the combustion chamber, thus reducing the amount of fuel required for heating in the combustion chamber.

5. "Chrysler Gas Turbine Licks Some Problems of Use as Auto Engine," *Wall Street Journal*, March 25, 1954, 1.

6. Williams to Leyes, 7 September 1988.

7. Don MacDonald, "Gas Turbines," *Motor Trend*, (August 1954): 16-17. Williams told Leyes on September 7, 1988 that he was one of the passengers in the car.

Left to Right: Sam B. Williams, Jack Benson, Harold Way, Ed Botting, and John Jones. Photograph courtesy of Williams International.

mingham, Mich., Sam Williams began formal operations of his new company under the name Williams Research Corporation.[8] His initial team was composed of four engineers whose talents he knew well.

Sam Williams regarded John Jones as "the best design engineer that he had met relative to design judgement and analytical capability."[9] Jones had established a fine reputation at Chrysler in a variety of design projects and would later become a key officer at Williams. Jack Benson had worked with Sam Williams at Chrysler as an experienced and practical designer. Before joining Williams, he had a successful engineering career at Ford, Bendix, and Chrysler. Harold Way, a manufacturing engineer, would work closely with Sam Williams on early designs and manufacturing processes. His prior engineering experience had been with Chrysler and Ex-Cello. Ed Botting "had a reputation for turning out drawings faster than any other designer at Chrysler," according to Sam Wil-

8. On April 26, 1981, the name Williams Research Corporation became Williams International Corporation. For purposes of brevity, the company will be referred to as Williams throughout the text.

9. Sam B. Williams, Walled Lake, Michigan, to Rick Leyes, 7 May 1991, editorial comments, National Air and Space Museum Archives, Washington, D.C.

liams.[10] He would do all of the design layout work and many of the detailed drawings of all of the early Williams engines.

Sam Williams was convinced that all of the advantages of big engines could be achieved in small engines as well. It was the general view of gas turbine engineers at that time that a small engine should be a miniature version of a large engine. Further, they believed that in scaling down a large engine, the component efficiencies would drop drastically, the fuel consumption would increase, and the power-to-weight ratio would be low. However, based on his automotive turbine experience at Chrysler, Sam Williams believed that if the design of small engines were approached correctly, good component efficiencies could be obtained.

Williams felt that others in the field had missed some fundamental points. Such people believed that in order to build a small engine, it was necessary to retain the same number of components when scaling down a large turbine engine. They thought that the cost to make all of the tiny parts would be prohibitive and that tolerances would be impossible to achieve. Also, they felt that it was impossible to obtain the rotor tip and seal clearances necessary to achieve good efficiencies. However, according to Sam Williams:

> I was convinced that we could overcome all of those things, that we could achieve the kinds of tolerances really quite easily that were needed to maintain good efficiency, that you could reduce the parts count very considerably on a small engine by combining the functions of many parts into one part, and because of the square cube law which says that (as engine size is reduced) power comes down as a square of the dimension (engine diameter) and the weight comes down as a cube of the dimension. It meant that you could achieve very good power-to-weight ratios when you make smaller power plants. We demonstrated all of these things in our early engines in this company.[11]

The WR1, The First Williams Engine

The common ancestor to all Williams engines was the WR1, a 75-shp regenerative free turbine turboshaft engine weighing only 75 lb that was designed for the Outboard Marine Corporation.[12] The design of this engine, which began in

---

10. Williams to Leyes, 7 May 1991.

11. Williams to Leyes, 8 September 1988.

12. Richard J. Mandle, "Twenty Year Evolution of the WR2/WR24 Series of Small Turbojet Engines," (SAE Paper No. 770998 presented at the Aerospace Meeting of the Society of Automotive Engineers, Los Angeles, Cal., 14-17 November 1977), 1.

1954, used the same thermodynamic cycle as automotive turbines, but was drastically different in configuration. To provide good fuel efficiency, this engine had a single metal rotary heat exchanger surrounding the combustor and turbine sections. Another unique feature of the engine was that engine accessories were driven directly off the main rotor shaft rather than through a gear box, eliminating the cost and weight of a gear train.

The engine was tested in a small runabout on the Detroit River. Outboard Marine then decided that a larger 150-hp engine would be more appropriate for its market. To meet this requirement, a larger engine using two ceramic rotary heat exchangers was designed. According to Sam Williams, this engine had the distinction of being the first turbine engine to run with ceramic heat exchangers, and its configuration became the accepted way of building automotive and truck turbines.[13]

Both of these marine turbine engines were patented by Sam Williams. After the 150-hp engine had been successfully operated in a boat, Outboard Marine decided to start its own development and manufacturing capability for marine engines. They engaged Williams as a consultant to aid in this process for about two years. During this time, they studied the risk of entering production and surveyed the intensive turbine development activities at the automotive companies. They concluded that it would be prudent to wait until the automotive companies made production decisions before taking on the challenge of a completely new type of engine and ceased their internal activities about 1964.[14]

Industrial, Automotive, and Auxiliary Power Applications

Williams was interested in developing gas turbine engines for a broad spectrum of applications. Although it was the development of engines for aircraft propulsion that ultimately became the big money maker, the company initially pursued the development of gas turbines for marine, industrial, and automotive applications. The work was conducted initially on behalf of other manufacturers through development contracts with Williams, but was curtailed as Williams's own developments for aircraft and aircraft auxiliary power turbines progressed.

13. Williams to Leyes, 7 September 1988.

14. Williams to Leyes, 7 May 1991.

By the late 1950s, with the ongoing and projected growth of its product line, Williams had reached the point where it needed more space. In 1959, prototype manufacturing was moved to a company-built plant in Walled Lake, Mich. There, the company could draw from the skilled labor pool, industrial resources, and subcontractors available in the Detroit area.

INDUSTRIAL GAS TURBINES

In 1956, Williams began development of an industrial gas turbine engine. At that time, Sam Williams successfully convinced the Waukesha Motor Co., a manufacturer of large industrial engines, that it should build and market a 400- to 500-shp industrial gas turbine engine. Several of the resulting T-400 engines developed by Williams were given trial applications. After field trials, the Waukesha Motor Company decided that it needed a larger engine. In 1971, the T-1500 (1,500 shp) was successfully run on a test stand after successful field trials in heavy vehicles and generator sets.

The T-1500 was an advanced industrial engine for the time, a three-shaft unit with 12:1 pressure ratio. It had a two-shaft gas generator that incorporated an axial compressor on one shaft and a centrifugal compressor on the other. This two-shaft axial-centrifugal compressor system (patented by Sam Williams) was later used in various Williams fanjet engines. All of the axial compressor rotors and stators as well as the three turbine rotors and stators were integrally bladed precision castings. Shortly after the initial running of the T-1500, the Waukesha Motor Company was sold, and its new owners were not interested in continuing the turbine activities.[15]

AUTOMOTIVE GAS TURBINES

Sam Williams had a continuing interest in automotive gas turbines. One of the company's early efforts in this area came in response to the Army's interest in turbine-powered land vehicles. Two 75-shp versions of the Williams WR8 automotive turboshaft engine were installed in jeeps and tested. However, this

---

15. According to Sam Williams, although large gas turbine engines had proved successful in industrial applications, by the late 1980s, small gas turbine engines were not yet economical in this field. Piston engines had become highly efficient and were more economical in the lower horsepower range. Nevertheless, Williams International continued to study the application of some of its auxiliary power units in industrial applications with the belief that technology advances could open up opportunities.

work did not progress beyond the experimental stage.[16] Later, as part of a proprietary program, Williams developed a 70-shp turboshaft engine for Volkswagen's first station wagon. The car was tested in Detroit and Germany. As a result of the tests, Volkswagen decided that it wanted a 150-hp production engine, which Williams also developed. These Volkswagen programs ran from 1962 to 1970. At that point, Volkswagen developed its own automotive turbine program and used Williams for consulting support.

Williams developed another automobile turbine engine, which in 1971 was installed in an American Motors Hornet. Working with a grant from the National Air Pollution Control Administration, the New York City Department of Air Resources used the car in an 18-month program to test emission levels, fuel economy, and performance. The car, driven around New York City, demonstrated significantly lower emissions than comparable, conventional reciprocating engine-powered cars. Williams also conducted turbine engine development programs on behalf of Toyota and GM.

While the automotive turbine engines developed by Williams showed initial promise, their cost and fuel consumption were not competitive with reciprocating gasoline or diesel engines. Although materials technology was advancing that could eventually permit higher operating temperatures and competitive fuel economies, the schedule of such basic materials advances could not be predicted with confidence. During the early 1970s, the automotive industry was under great pressure to improve fuel economy. The emphasis of its engine development work was redirected toward improving the efficiency of piston engines. The automotive companies curtailed most of their turbine work at Williams and within their own organizations, and the last sponsored automotive turbine work at Williams was concluded in 1975.

AUXILIARY POWER UNITS

Williams won a contract in 1963 for an auxiliary power unit (APU) to be used on the De Havilland Aircraft of Canada, Ltd.'s DHC-5 Buffalo Aircraft. This unit, the WR9-7C APU, provided bleed air for main engine starting and for cabin air conditioning, and also drove an electrical generator and hydraulic

---

16. Michael L. Yaffee, "Small Aircraft Turbofans Studied," *Aviation Week & Space Technology*, Vol. 92, No. 6 (February 9, 1970): 45. The 1.2 lb/hr/lb specific fuel consumption of the WR8 jeep engine was considered high for automotive applications. However, it was not prohibitive for the jeep's intended use as a lightweight compact airborne vehicle with limited mission time on the ground.

pump. Competing against Garrett and Solar in 1970, Williams won another APU contract for the Lockheed S-3A anti-submarine aircraft. This 150-shp unit, known as the WR27 APU, provided bleed air for starting the GE TF34 engines on the S-3A and for the aircraft's air cycle environmental control system. It also provided electrical power for ground use and in-flight emergency backup power for the main engine generators.

The industrial, automotive, and auxiliary power turbine engine programs furthered the development of the company's small engine technology, improved its ability to develop aircraft gas turbines, and helped to develop the company's engineering team. Of particular benefit was the emphasis on low-cost production processes in the automotive turbine engine programs. As a result, Williams developed manufacturing processes that enabled it to work toward low cost, quantity production. These processes were to pay off later in aircraft turbine engine design and production.

Jet No. 1 and the WR2 Series Engines

THE FIRST WILLIAMS TURBOJET

In 1957, Williams began the design and development of its first turbojet engine, Jet No. 1. The objective was to develop a simple and efficient engine that could be used to power small pilotless aircraft. Drawing on the initial experience and successful operation of the WR1 marine turbine engine, Jet No. 1 used the compressor, combustor and turbine of the WR1 gas generator section. Jet No. 1 had a single-stage centrifugal compressor, an annular combustor with centrifugally fed slinger-type fuel injection, and a single-stage axial turbine. A demonstrator engine, weighing only 23 lb and producing a thrust of approximately 60 lb, was successfully tested in 1957. The cost of this engine's development and demonstration was fully funded by Williams as a business venture.

Early in 1960, a photograph of the engine appeared in *Aviation Week & Space Technology*.[17] Shortly afterward, John Kerr, an engineer at Canadair Limited, sought out Williams to determine whether the engine might be suitable for a small aerial reconnaissance drone. Kerr's concept later became the Canadair CL-89 (AN/USD-501) high-performance battlefield reconnaissance drone. This short-range drone was fired from a mobile launcher with the aid of a booster

17. "23-lb Turbojet Develops 70 lb of Thrust," *Aviation Week & Space Technology*, Vol. 72, No. 3 (January 18, 1960): 126.

The publication of this Williams Jet No. 1 photograph ultimately led to the first aviation application of a Williams engine. Photograph courtesy of Williams International.

rocket. After reaching flight speed, the booster assembly fell away, and the sustainer turbojet took over to propel the drone for the remainder of its flight. This reconnaissance drone launched Williams into the aircraft engine business.

THE WR2 SERIES ENGINE

The serious interest shown by Canadair and the potential for its first aircraft application prompted Williams to redesign the engine.[18] This work began in 1960, and the engine was redesignated the WR2-1. Journal bearings were replaced

18. Mandel, "Evolution of the WR2/WR24 Series," 2.

with conventional ball and roller bearings to reduce the demand for oil flow. In order to minimize weight, a squirt type lubrication system was designed for a once-through, throw-away oil flow. This engine, built for test demonstration purposes, had no accessories or fuel control; the engine speed was regulated by a simple manually-operated fuel valve. The engine first ran at its design thrust of 70 lb in 1962. It was successfully tested under both high acceleration launch and simulated altitude conditions.

Between 1962 and 1963, the flight version of the engine was developed. This engine, designated WR2-2A, was uprated to 95 lb thrust by the introduction of several modest compressor improvements. To lower cost, the aluminum compressor and a one-piece multifunctional diffuser assembly were precision cast, reducing both the parts count and the machining operations required.

The WR2-2A was a single-speed engine designed to operate at maximum thrust and a constant rotating speed of 60,000 rpm. To avoid excessive exhaust gas temperatures during pre-launch runup on hot days, the engine was equipped with a simple spring-loaded two-position tail plug to vary the exhaust nozzle area. Held forward by the booster nose piece to increase the exhaust nozzle area during engine runup and launch, the tail plug was designed to pop out to the preset jet nozzle area immediately following booster separation.

One of the key design accomplishments was the development of the integrated governor and alternator assemblies. Both were mounted on the rotor shaft and operated at 60,000 rpm. The reason for designing these accessories in this manner was to achieve the simplest, lowest-cost, lightest-weight engine possible. The technical challenge was to make these accessories operate properly and reliably at the high rotational speed of 60,000 rpm. The resultant direct-drive alternator was believed to be the first of its kind used in a small turbojet.[19]

Flight tests—the first of any Williams engine—of the Canadair CL-89 were conducted in 1964 and 1965 at the Army Proving Grounds in Yuma, Ariz. Some 35 WR2-2A turbojet engines were built for testing. As the test program continued, the engine thrust was uprated first to 105 lb (WR2-4) and later to 115 lb (WR2-5). Introduced in 1968, the WR2-6 was the first production version of the WR2 series. It was rated at 125 lb thrust yet still weighed only 30 lb. Performance and starting improvements introduced in the WR2-6 were derivative technology developed in the WR24 engine, a parallel turbojet engine development program. Between 1968 and 1973, when production stopped, about 300 WR2-6 engines were produced for use in the CL-89.

19. Mandel, "Evolution of the WR2/WR24 Series," 5.

Williams WR2-6 turbojet engine used in the Canadair CL-89 reconnaissance drone. Photograph courtesy of Williams International.

HELICOPTER TIP-MOUNTED ENGINES

An interesting spin-off that occurred during the early years of WR2 engine development was the WR2-3 model, rated at 95 lb thrust, and its unusual application. The Hiller Aircraft Co. had developed a helicopter that was powered by tip-mounted ramjets; however, the ramjets had very poor fuel efficiency. As a result, in 1962, Hiller sought out Williams, knowing that the company's small WR2 turbojet engine had only one-fifth the fuel consumption of the ramjet being used.

During ground tests in 1964, a WR2-3 engine was mounted on a rotor blade tip and successfully operated despite the tremendous centrifugal forces. This concept was significant because a tip-mounted power plant eliminated the need for a heavy gearbox, and it produced no torque on the airframe, thus permitting a reduction in the size of the tailrotor, which was only needed for directional flight control.[20] However, in 1965 Hiller sold his company and interest in the project promptly faded. Consequently, the WR2-3 engine never reached production.

20. "Hiller Tests Rotor Tip-Mounted Turbojet," *Aviation Week & Space Technology*, Vol. 81, No. 6 (August 10, 1964): 77. Also, Yaffee, "Small Aircraft Turbofans Studied," 44-45.

THE VERTICAL ENGINE

Another interesting version of the WR1 marine engine was the WR3 helicopter turboshaft engine, a small 75-shp free turbine power plant, the development
of which began in 1957. This unit was known as the vertical engine because it
could be mounted with the rotor shaft in a vertical position. It used the WR2
compressor, combustor, and turbine as the gas generator section of the engine,
to which was added a free power turbine. The WR-3 weighed about 50 lb, had
a specific fuel consumption of 1.1 lb/hp/hr thrust and an airflow capacity of 1.4
lb/sec.[21] The Navy showed considerable interest in this engine for its one-man
helicopter program. However, Williams stopped its work on the engine when
the Navy's interest shifted toward larger helicopters.

The WR24 Series Target Drone Engine

While 300 WR2-6 engines were produced for the Canadair CL-89 reconnaissance drone, the WR24 engine series was Williams's first high-volume production engine. It was the existence of the WR2 engine that helped stimulate development of the Northrop Chukar (MQM-74) series of turbojet target drones,
initially powered by the WR24-6 engine. This engine was aerodynamically
similar to the WR2-6, and the weight and performance of the two were nearly
identical. However, the WR24-6 incorporated materials and accessory modifications necessary to satisfy the more stringent mission requirements of the
Chukar target drone.[22]

The Ventura Division of Northrop Corporation had been building small,
unmanned piston-powered KD2R-5 target drones for the Navy. Seeking a further application for his small turbojet, Sam Williams visited the Northrop Corporation in late 1964 to familiarize them with the WR2 engine and its use in
the Canadair CL-89. According to Sam Williams, following his visit, Northrop
saw the potential for a small jet-powered target aircraft that would be considerably faster and provide more realistic enemy aircraft simulation for Navy target
practice than existing target aircraft.[23] Thus, with its own funding, Northrop
built what became the MQM-74 Chukar target drone, considerably smaller and
less expensive than its earlier target aircraft.

21. Yaffee, "Small Aircraft Turbofans Studied," 45.

22. Mandel, "Evolution of the WR2/24 Series," 10.

23. Williams to Leyes, 7 September 1988.

This drone was precisely the type of application that Sam Williams had been seeking for his small turbojet engine. His belief had been that by developing a small jet engine, new applications would emerge. A small engine was the key to making a small pilotless version of a military aircraft that would function the same as the full-scale aircraft, but cost considerably less. Between 1968 and 1989, approximately 6,000 MQM-74s, BQM-74s, and upgraded models, all powered by the WR24 engine, were delivered to the Navy.

The mission requirements for the Northrop target drone were considerably more complex than those for the CL-89 Canadair reconnaissance drone. The Northrop drone was to have a one-hour flight duration compared to the 10-minute endurance of the CL-89. Also, the Northrop drone was to have an altitude capability of 20,000 ft and a variable flight speed compared to the 10,000-ft ceiling and constant flight speed of the CL-89 drone. Because this was a Navy target drone, an added challenge was the requirement that it be capable of reuse up to 20 times after sea water recovery.

To meet the salt water immersion requirements, the WR2 series magnesium compressor diffuser casting was replaced with anodized aluminum. Bearing materials and the turbine shaft were changed to stainless steel. Decontamination and maintenance procedures were also developed.

The one-hour flight endurance requirement made it impractical to use the squirt-lube, throw-away oil system of the WR2 series that consumed 0.5 lb per minute, or 5 lb for a typical 10-minute CL-89 flight. In an hour-long flight of the MQM-74, the oil consumed with such a system would equal the total engine weight. As a result, an oil mist lubrication system was developed that consumed only about one ounce of oil per hour under normal operating conditions.

Because variable air speed and maneuvering capability were required of the target drone, it was necessary to vary engine speed and thrust during flight. A simple fuel control system was developed that used the existing WR2 governor as a maximum speed limiter. Design of the fuel control was such that it could be adapted with minimum change to the basic engine configuration.[24]

During the course of its use in the several models of the Chukar target drone, the WR24 engine was progressively uprated. In 1968, the first flight model, the WR24-6, completed qualification tests at the Naval Air Engineering Center in Philadelphia, Pennsylvania and then entered production. The 121-lb-thrust WR24-6 (military designation YJ400-WR-400), weighing 30 lb, as did the WR2-6, powered the Northrop Chukar I (MQM-74A). This lightweight target

24. Mandel, "Evolution of the WR2/24 Series," 10-13.

Williams WR24-6 turbojet engine which powered the Northrop Chukar I (MQM-74A) target drone. Photograph courtesy of Williams International.

drone was used for anti-aircraft gunnery, surface-to-air and air-to-air missile training, and weapons system evaluation. By the time Chukar I production ended in 1973, about 2,200 WR24-6 engines had been produced for Chukar drones.

UPRATED ENGINES

In 1967, even before the development phase of the WR24-6 had begun to wind down, Williams took the initiative to finance the design of a higher thrust, zero-stage version of the engine.[25] The resultant prototype engine had a single axial compressor stage added ahead of the existing centrifugal compressor, thereby increasing the airflow by 27% and the compressor pressure ratio from 4.2 to 5.3. In addition, the turbine operating temperature was raised. These modifications increased the thrust of the prototype engine from 121 to 176 lb and reduced the specific fuel consumption from 1.25 to 1.2. These benefits were achieved in the prototype version of the engine with only a 2 lb weight penalty and a small increase in compressor length and diameter.

Two prototype engines were assembled in 1968, one of which was test flown on a conventional MQM-74A. This flight verified the engine's operational capability. Operational experience that was being obtained by that time with the

25. A zero stage is a single axial stage added ahead of the compressor to provide an increase in airflow and pressure ratio.

WR24-6 in the Chukar I began to influence the uprated engine's configuration. As a result, maintenance and logistics reports obtained through WR24-6 users had a major influence on the final design of the WR24. The principal design changes involved mounting the new axial compressor stage on a forward extension of the turbine shaft and reducing the number of bearings from four to two. This engine configuration was identified as the WR24-7 (military designation YJ400-WR-401). The production version of this engine was rated at 176 lb thrust and its weight increased by 44 lb, making it 14 lb heavier than the WR24-6.

Qualified in April 1973, the WR24-7 (YJ400-WR-401) was installed on the Chukar II (MQM-74C), a faster, improved, multiple-role version of the Chukar I. Between 1974 and 1990, 2,449 WR24-7 engines were produced for Chukar IIs.

In 1978, design was begun on the Chukar III (BQM-74C). The Chukar III target missile had air-launch capability and pre-programmed flight profiles. Early versions of the missile were powered by a Williams WR24-7A (YJ400-WR-402) turbojet, a slightly modified version of the WR24-7, having a rated thrust of 180 lb.[26] Between 1978 and 1984, 615 WR24-7A engines were produced. An uprated version of the WR24-7A, the WR24-7B (YJ400-WR-403) rated at 190 lb thrust, was also used in the Chukar III (BQM-74C). Between 1983 and 1990, 1,030 of these engines were produced. Beginning in 1986, the WJ24-8 (YJ400-WR-404) replaced the WR24-7A in the Chukar III target drone and the BQM-74E. Retaining approximately the same diameter and component layout as the WR24-7, the WJ24-8 produced about 36% more thrust and still retained significant growth potential. Weighing only 50 lb, the engine produced 240 lb thrust at 52,000 rpm. The growth in thrust was achieved by increasing the airflow and by introducing design improvements in the two compressor stages.[27] Between 1986 and early 1991, 318 WR24-8 engines were produced for the Chukar III and BQM-74E.

In Operation Desert Storm during the Persian Gulf War, the Navy BQM-74 target drones were used to simulate allied warplanes during early attacks on Iraqi air defenses. These drones were launched from deep in the Saudi Arabian desert against Baghdad and other well-defended targets. On September 29, 1995, the WR24-8 powered the first flight of a prototype Joint Standoff Weapon (JSOW). The missile was dropped from an F-4 test aircraft over the Naval Air

---

26. Richard A. Botzum, *50 Years of Target Drone Aircraft* (Newbury Park, Cal.: Publications Group of Northrop Corporation, 1985), 119-128.

27. "Powerplant Advances Tied to New Technologies," *Aviation Week & Space Technology*, Vol. 124, No. 17 (April 28, 1986): 111-112.

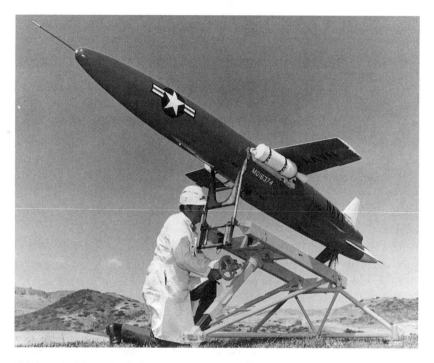

Northrop MQM-74B Chukar target drone powered by Williams WR24 series turbo-jet engine. Photograph courtesy of Williams International.

Weapons center at China Lake, Cal., following which it successfully completed an 11-minute test flight before being destroyed.[28]

WR19 Turbofan Development

With the success the company was having in developing new applications around its WR2 and WR24 series engines, Sam Williams felt that, ". . . the way to progress further in this business was to step out and develop some new engines that would bring to the small engine field the same kinds of performance, thrust-to-weight ratio, and so on that were being achieved in big engines. I felt that if we could develop a fanjet engine that performed like big fanjet engines,

28. David A. Fulghum, "JSOW Demonstrates Powered Flight," *Aviation Week & Space Technology*, Vol. 143, No. 15 (October 9, 1995): 62.

that (it) would open up still another market for us in long-range drones, missiles, and unmanned vehicles."[29] Subsequently, with its own money, the company initiated the WR19 turbofan engine in 1963, and in 1965, began development of that engine.

Sam Williams knew that development of a successful turbofan would require a high-cycle pressure ratio in order to achieve a low specific fuel consumption. The conventional ways of obtaining high pressure ratios in large engines were considered impractical for a small engine. For example, GE used variable geometry compressor stator blades in its single spool axial-flow J79 engine. That design required a large number of carefully machined parts, quite acceptable for large engines, but too expensive and complex for a small engine. In addition, because the blades in the last few compressor stages would be very tiny in a small engine, the task of obtaining the surface smoothness and low tip clearances needed for good efficiency would be both difficult and expensive. P&W had successfully developed its twin-spool J57 engine, which had two concentric shafts with an axial compressor on each shaft. Applied to a small engine, this approach also would result in very tiny blades in the last few compressor stages. In both cases, the task of aerodynamically matching the compressor stages to avoid compressor surge was a difficult one.

On the basis of the company's compressor design experience, the approach selected by Sam Williams for the WR19 was a two-shaft design with an axial-flow compressor and two fan stages on the LP (low pressure) shaft, followed by a centrifugal-flow compressor on the HP (high pressure) shaft. Use of the centrifugal-flow compressor on the HP shaft greatly simplified the mechanical design and fabrication of the HP compressor and kept down the parts count and the cost as well. This design achieved a high pressure ratio, originally 13 to 1 and subsequently substantially higher, which was imperative to obtain the fuel economy needed by the long-range missile application.

In addition to these advantages of his design, Sam Williams believed that the combination of an axial compressor followed by a centrifugal compressor on a separate shaft would provide much better aerodynamic matching than the two-shaft axial-compressor approach (used by P&W). A detailed engineering analysis of the concept supported this belief, showing that very good aerodynamic matching was possible. Both compressors would operate with a substantial stall margin and very close to their maximum efficiency across the entire speed range. This meant that good part-load and full-thrust efficiency were pos-

29. Williams to Leyes, 7 September 1988.

sible and that greater inlet flow distortion could be tolerated before encountering compressor stall. The WR19 compressor performed as predicted, and its design concept was patented by Sam Williams.

The WR19 engine first ran in August 1967, producing a thrust of 430 lb with a sfc of 0.66 lb/hr/lb thrust.[30] At that time, the WR19, which weighed only 67 lb, was the world's smallest turbofan engine. The engine had two fan stages, which supplied both bypass air and air to the core section of the engine, followed by a two-stage axial-flow compressor. The fan and compressor, both mounted on the LP shaft, were driven by a two-stage LP turbine. A single-stage centrifugal compressor on the HP shaft was driven by a single-stage axial HP turbine. The HP shaft rotated in the opposite direction of the LP shaft to minimize the gyroscopic effect. The engine used an annular combustor. The WR19's HP compressor and turbine and its annular combustor were adapted from the WR2 engine design.

The bypass flow and the hot core exhaust gases were allowed to mix downstream of the LP turbine and were discharged through a combined flow exhaust nozzle. Use of the combined flow exhaust nozzle resulted in an aerodynamic interaction between the HP and LP spools so that the bypass ratio, approximately 1:1, varied inversely with engine power level—high at low-power levels and low at maximum power. This variation in bypass ratio benefited sfc at cruise power and specific thrust at maximum power.[31]

Experimental Applications of the WR19 Engine

Beginning in 1963 and extending over a number of years, Williams began working with a number of aircraft companies and with the military on a variety of experimental applications for the WR19 engine. Although none of these experimental applications ever reached operational status, they illustrated the diversity of the company's efforts to open new markets for its small engines. An important technological contribution of this work, particularly that done during the early years, was the further development and demonstration of the WR19 engine in preparation for its most important future application, the cruise missile program.

30. T. K. Wills and E. P. Wise, "Development of a New Class of Engine—The Small Turbofan," (AIAA Paper No. 76-618 presented at the AAIA/SAE 12 Propulsion Conference, Palo Alto, California, 26-29 July 1976), 2.

31. Wills and Wise, "Development of a New Class of Engine," 2.

Williams publicity photograph comparing the world's smallest fanjet engine, the Williams WR19, to the then largest commercial fanjet engine, the Pratt & Whitney Aircraft JT9D. Photograph courtesy of Williams International.

JET FLYING BELT

In the early 1960s, the Bell Aerosystems Company was conducting successful experiments with its one man Rocket Belt, capable of a 21-second flight. During that time, Sam Williams became acquainted with the Bell people involved in this work, including Wendell Moore, who was the inventor of the Rocket Belt. Through this relationship, Moore learned that Williams had developed the WR19, a small turbofan engine of the size that could power the Rocket Belt.

Moore also recognized that using the WR19 instead of a rocket could signi-
ficantly increase the flight duration of the flying belt. Consequently, Bell and
Williams joined and endeavored to market what became known as the Jet Flying
Belt to the Army and to DOD's Advanced Research Projects Agency (ARPA).

On December 31, 1965, Bell was awarded a contract sponsored by ARPA to
develop a turbofan-powered flying belt for the Army. Bell then contracted with
Williams for the power plant. In March 1969, the WR19 completed its 50-hour
Preliminary Flight Rating Test (PFRT), and on April 7, 1969, the WR19 powered Jet Flying Belt, piloted by Robert Courter, a Bell test pilot, made its first
free flight.

The Jet Flying Belt had a gross take-off weight of 365 lb, which meant that it
had to be supported on a stand until the pilot throttled up the engine. Limited
to a six-gallon fuel load, the Belt's flight endurance was under ten minutes.
These limitations contributed to the fact that the Jet Flying Belt never became
operational for use by the Army and its development was discontinued and the
contract terminated in 1970.

In 1970, Bell granted Williams a license to manufacture, use, and sell the Jet
Flying Belt. However, Williams never used this license. The company concluded that the concept of strapping the unit on a person's back was not the optimum configuration. To obtain a useful range required a larger vehicle that
could carry more fuel. These conclusions led Sam Williams to conceive the idea
of a jet-powered, man-carrying platform.

WILLIAMS AERIAL SYSTEMS PLATFORM

In 1970, Williams proposed a small, man-carrying, jet-powered platform to the
Marine Corps. As a result of this proposal, the Marine Corps initiated the Small
Tactical Air Mobility Platform (STAMP) program in 1971. The Marine Corps
envisioned a small, rugged air vehicle that could operate in places inaccessible to
helicopters and motor vehicles. Such a vehicle could fly between tree trunks
and under the forest canopy and perform missions such as search and rescue, reconnaissance, and artillery fire direction. A contract for such a vehicle was
awarded to Williams in September 1972. The company then proceeded to develop a two-man flying platform, named the Williams Aerial Systems Platform
(WASP); it was promoted as having the agility of a wasp.

The WASP was powered by an uprated version of the WR19, known as the
WR19-9 BPR5 engine. It was rated at 670 lb thrust. The WR19-9 BPR5 utilized the WR19 configuration and technology, but in order to obtain the higher
thrust required, the engine core was scaled-up and a much larger single-stage

fan was installed on the LP shaft. The WASP I made its first flight (tethered) in December 1973 and made its first two-person flight (also tethered) in January 1974. These flights were made at sufficient heights to demonstrate that the vehicle could hover out of the influence of ground effect and could be controlled satisfactorily.[32] This successful demonstration completed the work required by Williams under the WASP I contract.

In September 1978, Williams received a contract under the Army Individual Lift Device (ILD) program in response to a Williams proposal to demonstrate its WASP II. The WASP II was a one-person air vehicle that used operator body movement and exhaust nozzle vanes for control. The WASP II was powered by a WR 19-7 turbofan engine rated at 570 lb thrust. This engine was a derated version of the WR 19-A7D. Additional accessory modifications and exhaust system modifications were made to fit the WASP II requirements. The first manned free-flight was made in April 1980. During the initial test program, speeds in excess of 45 mph, heights up to 60 feet, and flight endurance over 5 minutes were achieved.

Although these flights demonstrated that the vehicle could be flown safely, a number of problems were encountered. The Army concluded that the WASP II could be flown throughout its prescribed, limited flight envelope. However, it had problems with flight stability, was extremely sensitive to pilot position and movement, and was directionally unstable, all factors which contributed to high pilot workload. It also did not respond well to wind gusts above 5 knots.[33]

These problems were initially noted by an Army experimental test pilot. Subsequently, Williams incorporated a grid of yaw vanes in the exhaust duct, giving the pilot sufficient control authority to maintain stability in high crosswind gust loads. The modified vehicle was flown by the Williams test pilot to speeds of over 60 mph and for durations of 20 minutes. The vehicle was then cleared by the Army for flights by Army personnel. In a follow-on contract, three GI's without prior flight experience were taught to fly the WASP II at the Williams facility, and conducted demonstration flights at Fort Benning for the Army. The interest in the WASP II for highly specialized military applications continued for a number of years with Williams responding by submitting various proposals for full-scale development and production programs. None of these special ap-

32. Bernard Lindenbaum, "V/STOL Concepts and Developed Aircraft: Vol. I—A Historical Report (1940-1986)," (An unclassified Technical Report contracted by the Air Force Wright Aeronautical Laboratories and distributed by the Defense Technical Information Center), November 1986, 3-51.

33. Lindenbaum, "V/STOL Concepts," 3-55 – 3-58.

plications could attract the necessary funding. The last military contract ended in 1983.[34]

Following termination of the project by the Army, the WASP II was re-named the X-Jet by Williams. Williams continued to conduct flight tests and to improve the performance and stability of the WASP II in connection with its proposal efforts, but finally decided in 1987 to cease work on this project.

KAMAN SAVER

Another limited application of the WR19 was the Kaman KSA-100 Saver. The Saver was an experimental autogiro that was designed to be folded into a fighter ejection seat and could be deployed after ejection to provide the pilot with a means of escape from hostile territory. A 476-lb-thrust WR19-A2 engine pow-ered the prototype Saver. Development of the engine began in 1968 for use in the Kaman Saver, and the first test flights were completed in the spring of 1972, making the Saver the first turbofan-powered autogiro. The Saver never became operational. After the Vietnam War, the military lost interest in the project and support for it was discontinued.

WR19 Cruise Missile Applications

Concurrent with Williams's work on experimental applications for the WR19 engine during the late 1960s and early 1970s, the military began to support a parallel effort on the further development and the application of the WR19 en-gine. This effort involved application of the WR19 engine, given the military designation of F107, to the propulsion of long-range missiles.

During the late 1960s, the Air Force needed a small engine to power a missile having enough range that it could be used as a decoy or attack missile. At the same time, Williams was seeking additional applications for its WR19 engine. It was through Williams's work with Bell on the Jet Flying Belt program that the Air Force first became aware of the WR19 turbofan engine, with the thrust and fuel economy needed for a long-range missile.

Contract awards to Williams in 1969 to support Air Force cruise missile feasi-bility studies and to uprate the WR19 were pivotal for Williams. Derivatives of the WR19 growing out of these early contracts were eventually selected to power air-, sea-, submarine-, and ground-launched cruise missiles. The cruise

34. Williams to Leyes, 7 May 1991.

missile requirement for a small, efficient engine opened a major market for the uprated derivatives of the WR19. There were a number of uncertain years between the cruise missile studies that began in the late 1960s and the initial deployment of an operational cruise missile squadron in 1982. Nevertheless, this market was to dominate Williams's development and production efforts for more than two decades.

The story of Williams's involvement in this market was intimately intertwined with the evolution of the cruise missile program. Thus, the story of the F107 version of the WR19 is told in context with the history of the cruise missile.

SUBSONIC CRUISE ARMED DECOY PROGRAM

Toward the end of the 1960s, the Air Force sought a replacement for the McDonnell Quail (GAM-72), an air-launched decoy missile carried by the Boeing B-52 bomber for protection during penetration over hostile territory. However, Air Force contractor studies indicated that additional uses of cruise missiles were technically feasible and could provide increased flexibility to bombers as armed decoys or attack missiles.[35] Propulsion and guidance systems were considered critical areas in such missile programs.

Air Force contractor studies of a Subsonic Cruise Armed Decoy (SCAD) resulted in SCAD designs of many sizes, configurations, and capabilities. Many were designed around existing engines or scaled versions thereof, since few turbines in the 350- to 1,000-lb-thrust class existed.[36] Because of the prevailing feeling throughout the turbine engine industry that when an engine was scaled down, it would be inefficient and heavy, the Air Force was concerned that development of a cruise missile propulsion system would be time consuming and expensive.[37] In addition, for a cruise missile to achieve long range at low altitude, it would require a turbofan rather than a turbojet. Further, the engine would have to be very small to enable the missile to fit in the B-52 bomb bay.

During its cruise missile considerations, the Air Force discovered, through contacts at ARPA and Williams, that Williams already had a small turbofan engine running on the test stand. This engine, the WR19, was of the appropriate thrust level and had demonstrated the fuel economy required by the cruise missile. As a result, beginning in 1969 and continuing into 1972, Williams was called

35. Kenneth P. Werrell, *The Evolution of the Cruise Missile* (Maxwell Air Force Base: Air University Press, 1985), 144-146.

36. Williams to Leyes, 7 May 1991.

37. Williams to Leyes, 7 September 1988.

upon to support Air Force and airframe contractor study programs on the feasibility of what was to become the AGM-86A SCAD cruise decoy and missile concept.[38]

The WR19 engine was, at that time, the only engine available having the approximate thrust, size, and fuel economy needed for the SCAD cruise decoy/missile concept.[39] In April 1969, the Air Force awarded Williams a contract for preliminary design and preparation of engine specifications for an uprated version of the WR19 engine. A second contract was awarded in November 1969 for component development, and for design, construction, and testing of the engine. This "second generation" engine, designated the WR19-A2, with slightly modified accessories, was rated at 476 lb thrust, as compared with 430 lb for the original WR19 engine.

In 1970, the Air Force at Wright-Patterson AFB began an extensive in-house design study to determine the optimal vehicle configuration and propulsion system for the SCAD. Most system requirements were undefined at that time, so the objectives were to solidify the requirements and design with minimal B-52 modification and system cost. This design study was completed in early 1971.

The external SCAD configuration, weight, fuel volume, payload, drag, and engine performance requirements were defined. The engine study, designated "Minibrute," resulted in a small low-bypass ratio turbofan design in the 600-lb-thrust class. This design, resulting from extensive cycle iterations, could produce sufficient thrust at the high Mach No. sea level and altitude design points, yet have excellent sfc at part-power cruise (0.55 Mach) to maximize range. The Minibrute engine design was later used as the source of the engine performance, cycle, weight, and volume specifications for the Air Force SCAD RFP, which was released in February 1972.[40]

The AGM-86A SCAD was specified to be a decoy missile with the primary mission of providing a B-52 radar signature to assist B-52s in penetrating enemy defenses. A secondary mission of a limited number of armed SCAD missiles was to destroy enemy defenses. On May 31, 1972, the Air Force awarded two engine development contracts, one to Williams and the other to Teledyne CAE for the delivery of SCAD prototype demonstration engines.

Williams maintained the same WR19 basic engine with accessory changes to meet the specific installation requirements. This WR19-A7D engine was eventually designated as the F107-WR-100. In order to achieve the thrust and fuel

38. Wills and Wise, "Development of a New Class of Engine," 2.

39. Wills and Wise, "Development of a New Class of Engine," 2.

40. Williams to Leyes, 7 May 1991.

economy objectives, the overall engine pressure ratio of the WR19-A2 was raised from 7.6 to 14.5 and the turbine inlet temperature was raised from 2,125 to 2,306 degrees Rankine, while the bypass ratio was maintained near its optimum of 1:1. The increased pressure ratio was obtained primarily as a result of modifications in the HP spool. The rotating speed of the centrifugal compressor was increased, and the area of the HP turbine flow path was reduced in order to maintain the original core flow and bypass ratio. The higher operating temperature raised the LP rotor speed and with it the pressure ratios of both the fan and the LP compressor. Other design changes over the WR19-A2 were accessory system changes to provide for long-term storage, and to prevent oil and fuel spillage when the engines were stored in semi-inverted and inverted attitudes in the B-52 rotary launch rack.[41]

Development of the F107-WR-100 was not without its problems. Some problems that existed in the original WR19-2 needed to be solved, and new problems appeared. For example, the WR19-2 suffered from a marginal bearing cavity sealing and vent system, causing high oil consumption. This deficiency had been tolerable for the Jet Flying Belt application because its flights were only five to 10 minutes in duration. Better sealing and venting of the bearing cavities reduced the oil consumption considerably.

For mechanical simplicity, the thrust bearing located aft of the first fan was grease packed rather than oil lubricated in the F107-WR-100 engine. However, two opposing requirements were faced. One was starting and bringing the engine to full thrust in less than 10 seconds from a cold soak temperature associated with operation at 45,000 feet altitude. The other was the ability to maintain low bearing operating temperatures at low-altitude, hot-day, and high flight-speed conditions. Extensive testing was required to find a suitable grease and a way to adequately cool the bearing.

The increase in rotating speed of the F107-WR-100 HP rotor from 56,000 rpm to 64,000 rpm resulted in occasional shaft vibration. This problem was solved by eliminating various joint faces and integrating components with single piece castings and electron beam welding. In addition, extreme-cold high-altitude starts were made possible by extending start cartridge burn times and using an oxygen supplement system. The higher F107-WR-100 fan blade speed resulted in a vibration fatigue failure, a problem that was solved by increasing the blade chord and reducing the number of fan blades, thus altering the fan's natural frequencies.[42]

41. Wills and Wise, "Development of a New Class of Engine," 6.

42. Wills and Wise, "Development of a New Class of Engine," 8-11.

Williams F107-WR-100 turbofan engine. Photograph courtesy of Williams International.

A competitive runoff between the Williams F107-WR-100 and the Teledyne CAE F106-CA-100 engines was conducted from November 1972 until February 1973 at the Air Force Arnold Engineering Development Center in Tullahoma, Tenn. Williams won the competition, and Air Force support for further development of the F107-WR-100 engine was continued until June 1973. At that time, the SCAD program was abruptly canceled because of planning controversies within the Defense Department, which were also influenced by arms limitations planning considerations. Williams was put on a stop-work status.

The reasons for discontinuing the SCAD program were complex, and they were based to a significant extent on competing technologies and ideologies. Several versions of the SCAD missile, including an armed decoy, an unarmed decoy, and an armed attack (standoff) missile, were considered technically feasible. Of these possibilities, the Air Force was primarily interested in having an unarmed decoy, with an arming option, to assist B-52 penetration. From the viewpoint of the Air Force, using SCAD principally as a defensive weapon to protect B-52s was consistent with its historic reliance on the manned bomber as a strategic weapon. Also, as an evolving technology with limited accuracy, range, penetrability, and a small warhead, the missile was less useful as a standoff weapon. Most important, however, was the consideration that an air-launched cruise missile posed a potential threat to the Air Force's B-1 bomber program.[43]

Some in the Air Force feared that if SCAD's technical problems could be solved, the justification for a new penetrator bomber would be weakened by the potential of the standoff missile as a serious, less expensive alternative. However, this was also the reason why the Congress and the Office of the Secretary of Defense wanted the Air Force to include both the decoy and standoff versions of SCAD in its missile program. When the Air Force failed to propose a satisfactory plan for development of both options, Deputy Secretary of Defense David Packard canceled the SCAD program in June 1973.[44]

THE START OF THE AIR-LAUNCHED CRUISE MISSILE PROGRAM

About the same time that the SCAD program was terminated, Secretary of State Henry Kissinger wrote to Secretary of Defense William Clements to advise him of the utility of the strategic cruise missile as a bargaining chip in the SALT negotiations.[45] On July 20, 1973, SCAD was resurrected by the DOD as the Air-Launched Cruise Missile (ALCM) program. In December 1973, Deputy Secretary Packard specified that the ALCM program would make maximum use of the terminated SCAD airframe and engine development program.[46]

The ALCM was to be a small, winged missile capable of sustained subsonic flight following launch from a carrier aircraft. The missile was to be programmed for precision attack on a specific surface target and was to be guided by inertial and terrain comparison techniques. The ALCM was intended to give the launch aircraft standoff capability and to improve bomber penetration by destroying critical targets, thereby weakening enemy defenses.[47]

On February 1, 1974, the ALCM system development was begun by the Air Force with no change in engine requirements. Air Force support for continued development of the F107-WR-100 engine was resumed at that time. The engine's Preliminary Flight Rating Test began in August 1975. In addition, sea-

---

43. Werrell, *Evolution of the Cruise Missile*, 146-149.

44. Werrell, *Evolution of the Cruise Missile*, 149-150.

45. Werrell, *Evolution of the Cruise Missile*, 154.

46. Werrell, *Evolution of the Cruise Missile*, 156.

47. John W. R. Taylor and Kenneth Munson, eds., *Jane's All the World's Aircraft 1976-77* (London: Macdonald and Jane's Publishers Limited, 1976), 654.

level endurance tests were conducted at Williams and altitude performance tests were performed at the Arnold Engineering Development Center. During October 1975, both tests were completed.

THE SEA-LAUNCHED CRUISE MISSILE

During the early 1970s, while the Air Force was proceeding with the SCAD program, the Navy also became interested in cruise missiles. In January 1972, the Navy's Strategic Cruise Missile (SCM) program began.[48] The biggest boost for the Navy's program came shortly after the May 1972 signing of the Strategic Arms Limitations Treaty (SALT) agreement. At that time, Secretary of Defense Melvin R. Laird requested $1.3 billion from Congress for strategic weapons, including $20 million for the SCM program. In November 1972, after considering various methods of launching the missiles from submarines, the Navy decided on a cruise missile that could be launched from a torpedo tube. This missile was to become known as the Sea-Launched Cruise Missile (SLCM). The size and weight of this missile were constrained to the 21-inch diameter and 246-inch length of a submarine torpedo tube and the 4,200-lb weight capacity of the torpedo handling equipment.

In early 1974, the Department of Defense ordered the Air Force and the Navy to cooperate with each other in developing the key components of cruise missile technology.[49] The Air Force was to share its turbofan engine and high energy fuel technology and the Navy was to share its Terrain Contour Matching (TERCOM) guidance system. This arrangement, set forth in a formal program decision paper in December 1973, was approved by the DOD Defense Systems Acquisition Review Council (DSARC) in February 1974. In addition, the Navy was directed to conduct a fly-off to choose its SLCM contractor.

A competition was held among four airframe bidders to select the two companies that would engage in the fly-off. In spite of the fact that all four bidders selected the Williams engine, the Navy directed that one of the bidders, Vought Corporation, should use a competing Teledyne CAE engine.[50]

In March 1976, a Navy fly-off competition was held. The two competing systems were the Convair Division of General Dynamics BGM-109 SLCM powered by the Williams F107-WR-100 derivative engine, and the Vought Corporation BGM-110 SLCM powered by the Teledyne CAE CAE471-11DX

48. Werrell, *Evolution of the Cruise Missile*, 151.

49. Werrell, *Evolution of the Cruise Missile*, 154.

50. Williams to Leyes, 7 May 1991.

engine.[51] The success of the General Dynamics BGM-109 in this fly-off competition led to a BGM-109 airframe contract award by the Navy to General Dynamics on March 17, 1976. In May 1976, the Navy named Williams as the winner of the engine contract.[52]

The engine developed by Williams for the Navy's General Dynamics SLCM was the WR19-A7-1, a variant of the F107-WR-100 engine developed for the Air Force ALCM. The two engines had a parts commonality of as much as 80 to 90%. The major differences between the two engines was that the SLCM engine's accessory case was relocated to the top of the engine and the tailpipe design was changed to satisfy installation requirements. Because engine starts for the SLCM missile occurred at low altitude, the supplementary ignition oxygen system required for high-altitude starts of the ALCM's F107-WR-100 was not needed.

THE JOINT SERVICE CRUISE MISSILE PROGRAM

A number of events took place in 1975 and 1976 that led to unification of the Air Force and Navy cruise missile programs. With its selection of the General Dynamics BGM-109 in early 1976, the Navy program surged ahead. The Air Force cruise missile program had been suffering from internal and external problems. Internally, the Air Force had ranked the importance of the ALCM behind that of the B-1 bomber and the MX intercontinental ballistic missile. Externally, the House of Representatives, aware of these Air Force priorities, the progress being made on the Navy SLCM, and the cost savings that could result from having a single cruise missile program rather than a separate program for each service, deleted all money from the ALCM program budget in late 1975. Although the Senate later restored most of the money, the message to the Air Force was clear. To avoid having "a torpedo rammed up its bomb bay," the Air Force moved ahead on its development of the ALCM.[53]

On March 5, 1976, an Air Force B-52G bomber successfully launched the first ALCM free-flight test missile, a Boeing AGM-86A, over the White Sands Missile Range. This missile was powered by the Williams F107-WR-100 engine.

51. The Teledyne CAE CAE471-11DX engine was a variation of Teledyne CAE's
    F106-CA-100 that competed earlier in the SCAD program runoff with the Williams
    F107-WR-100 engine. For further details on this competition, refer to the Teledyne CAE
    chapter in this book.

52. Werrell, *Evolution of the Cruise Missile*, 155.

53. Werrell, *Evolution of the Cruise Missile*, 156.

The Navy subsequently passed the Air Force on June 5, 1976, by launching from a Navy A-6A aircraft the first fully-guided cruise missile, a General Dynamics BGM-109 SLCM powered by an early Williams F107-WR-400 engine.

As development of both weapon systems proceeded, a redefinition of the cruise missile program took place within the Department of Defense. As a result of subsequent recommendations by the DSARC in January 1977, a Joint Service Cruise Missiles Project Office was formed to stress commonality between the two programs. Full-scale engineering development was approved for both the SLCM and ALCM. The Air Force ALCM program was to concentrate on the long-range Boeing AGM-86B missile rather than the short-range AGM-86A. The Navy SLCM program was to concentrate on a long-range nuclear land-attack missile. In addition, an Air Force Ground-Launched Cruise Missile (GLCM) and a Navy anti-ship Tomahawk missile were approved for full-scale engineering development.

As both programs proceeded, an event occurred that gave the Air Force cruise missile program its biggest boost. On June 30, 1977, newly elected President Jimmy Carter canceled the B-1 bomber program. The cost of a B-1 bomber had risen from $39 million in 1969 to $84 million in 1975, and it had failed to meet its performance objectives. According to Secretary of Defense Harold S. Brown, "a B-1 force that would have equal capacity to B-52s with cruise missiles would have been about 40% more expensive."[54] Years later, Brown revealed that questions about the B-1 electronic system's capability to allow the bomber to penetrate enemy defenses, as well as existence of the then secret B-2 Stealth bomber alternative, also played a role in the B-1 program cancellation.[55] Clearly, with the cancellation of its top priority project, the Air Force was left with no alternative but to develop an effective cruise missile.

As a result of the B-1 cancellation, the Air Force cancelled the short range AGM-86A on September 30, 1977, and full-scale development began on the long-range Boeing AGM-86B. Early in 1977, Williams was contracted to complete development of the F107 turbofan engine as the propulsion system for both the Air Force AGM-86B ALCM and the Navy BGM-109 SLCM programs. Qualification testing for the engine began in October 1978.

With its selection in 1977, the F107 was firmly established as the sole engine in the cruise missile propulsion marketplace. Sam Williams had been proven correct in his belief that development of a small, efficient turbofan engine would

---

54. Werrell, *Evolution of the Cruise Missile*, 177.

55. George G. Wilson, "Radar-Evading B2 Bomber to Roll into Public View," *The Washington Post*, November 20, 1988.

open a new long-range missile market. Williams was about to embark on the second of its two most lucrative engine development and production efforts in the history of the company, the other being the WR24 for the Chukar target drone.

F107 ENGINE PRODUCTION

With the selection of the cruise missile engine now completed, the military turned its attention to how the F107 engine was to be produced. The ALCM FY1980-81 production plan called for a leader-follower approach to manufacturing. The military had decided that it wanted a second source on all components that went into the cruise missile. The military considered running a competition within the engine industry to develop a second engine. However, this approach would have required several years and been unduly expensive. The military then discussed with Williams a proposal to have the company license another manufacturer to build the F107 engine. Williams agreed to this approach and subsequently ran a competition among the engine companies to build its engine. The competition was monitored by the Joint Services Cruise Missile Project Office, and in September 1978, Teledyne CAE was chosen as the second source.

The agreement that Williams had with the military was that Williams and Teledyne CAE would each submit bids for production of the F107 engine in various quantities. The military would then decide the engine production split between the two companies. Williams was then responsible for contracting with Teledyne CAE to build its designated number of engines. Also under the agreement, Williams was required to provide Teledyne CAE with all of the know how necessary to produce the engine. In Sam Williams's opinion, this required a substantial transfer of manufacturing processes.[56] Because Williams was to warrant the engines built by Teledyne CAE, Williams had a vested interest in the quality of the Teledyne CAE engines and the success of its production program. Williams had the incentive to transfer the needed expertise in an effective manner.[57]

In preparation for its anticipated production of the F107 cruise missile engine, Williams made a major expansion in its Ogden, Utah, production facilities. At this location, remote from the parent plant, the company found a good labor market with the skills and work attitudes that it wanted. Administrative, engi-

---

56. Williams to Leyes, 7 September 1988.

57. Because this sort of technology transfer was a new and unnatural process among engine companies in the U.S., it met with some problems at the working level. Refer to the Teledyne CAE chapter of this book for discussion of these problems and how they were solved.

neering, and prototype development work remained at the company's Walled Lake, Mich. headquarters. That location, in the Detroit area, offered the technical skills, materials supply, and subcontractor capabilities that could best support development work.

A remote site was desirable for engine production so that production activities would not be interfered with by the distractions of product engineering and prototype development, activities with emphasis on product change, and the building of small quantities of engines or components. Williams wanted its Ogden plant to concentrate on production and to work toward achieving the highest possible quality product at the lowest possible price. The Ogden facilities had been producing WR24 series engines as well as auxiliary power units and became operational for producing F107 engines in June 1978.

CRUISE MISSILES BECOME OPERATIONAL: A CHRONOLOGY OF EVENTS

A number of versions of the cruise missile became operational for use as strategic and tactical weapons, some armed with nuclear and others with conventional warheads. Also, their configurations varied to permit air-, ground-, and sea-launch. With their deployment, cruise missiles became centerpiece issues of two important arms control treaties.

Between 1979 and early 1980, a fly-off competition between the Boeing AGM-86B ALCM powered by the F107-WR-101 and an air-launched version of the General Dynamics Tomahawk AGM-109 powered by a Williams F107-WR-102 was conducted by the Joint Services Cruise Missile Project Office. Both engines performed well for the fly-off with oil consumption improvements being the only significant development required during the program. On March 25, 1980, Secretary of the Air Force Hans Mark announced the Boeing AGM-86B as the winner.

The F107-WR-101 engine for the AGM-86B ALCM had a rated thrust in the 600-lb class. The first production engine was delivered in March 1981, three months ahead of schedule. In December 1982, the first Strategic Air Command B-52 squadron equipped with nuclear AGM-86B missiles became operational. By the time that the last production engine was delivered in March 1986, 1,926 of these engines had been produced (not including the 861 F107 engines produced by Teledyne CAE through 1991). Delivery of the last of 1,715 production AGM-86B ALCMs was completed in October 1986.[58]

58. Susan H. H. Young and John W. R. Taylor, ed. "Gallery of USAF Weapons," *Air Force Magazine*, Vol. 71, No. 5 (May 1988): 188.

Boeing B-52 bomber launching Boeing AGM-86B ALCM. Photograph courtesy of
Williams International.

In 1980, Williams had initiated a substantial uprating program for the F107.
This was accepted by the Air Force, and under Air Force contract, the engine
was designated as the F107-WR-103. In March 1982, Williams received a contract for full-scale development of the F107-WR-103 which, in April 1983, was
redesignated the F112-WR-100.

The Advanced Cruise Missile (ACM) was the successor to the AGM-86B
ALCM. In April 1983, the Convair Division of General Dynamics had been
selected to develop and manufacture the AGM-129A ACM to arm the B-52H
and B-1B.[59] The AGM-129A had improved range, accuracy, survivability, and
targeting flexibility as compared with the AGM-86B. It also embodied
low-observability (stealth) technology. The missile was powered by the F112-
WR-100 engine. The first production F112 engine was delivered in 1986, and

---

59. Young and Taylor, "Gallery of USAF Weapons," 188.

by early 1991, 575 F112 engines had been delivered for ACMs. Delivery of production AGM-129A missiles began in June 1990.

In 1983, the first General Dynamics BGM-109 SLCM Tomahawks were operationally deployed by the Navy. The SLCMs were able to carry out both land-attack and anti-ship missions. Some were designed for launch from submerged submarines and others by surface ships. The SLCMs were powered by Williams F107-WR-400 engines, rated in the 600-lb-thrust class.

The F107-WR-402, also rated in the 600-lb-thrust class, was an upgraded version of the F107-WR-400 developed for an improved Tomahawk missile. The government was interested in a demonstration test of an F107-WR-400 engine incorporating the design intent of the F107-WR-402. That demonstration test was authorized and conducted in 1986, using F112-WR-100 components in the first two fan stages and a modified intermediate-stage fan compressor. The test was completed successfully and was considered by Williams to be the first run date of the engine. In 1987, following the "configuration validation test," Williams received authorization to proceed with the design and development of the -402. The engine entered production in 1990.

An Air Force Ground-Launched Cruise Missile had been approved for development soon after establishment of the Joint Services Cruise Missile Project Office in 1977. Manufactured by General Dynamics, the GLCM was designated the BGM-109G. Like the SLCM, the GLCM was also powered by the F107-WR-400 engine. The GLCM, together with the Pershing II, was intended as a response to Soviet deployment of tactical nuclear weapons such as the SS-20 mobile ballistic missile in Europe.[60] The first BGM-109G GLCMs became operational in Europe in December 1983.

The missiles became the target of intense political controversy in Europe. They were also very expensive. Despite its controversial nature, however, the GLCM had some important attributes. It became an important issue during negotiations for the Intermediate-Range Nuclear Forces (INF) Treaty. Though the cruise missile issue was difficult for arms limitations negotiators to resolve, the GLCM was nonetheless an important factor in causing the Soviets to sign the INF treaty. Like other cruise missiles, the GLCM was highly accurate, hard to detect, hard to intercept, and could be launched in large quantities. With the signing of the INF Treaty on December 8, 1987, the United States and the Union of Soviet Socialist Republics agreed to eliminate their short- and intermedi-

60. *Jane's Weapon Systems 1986-87*, ed. Ronald T. Pretty (London: Jane's Publishing Company Ltd., 1986), 45.

ate-range GLCMs.[61] In exchange for eliminating the GLCMs and the Pershing IIs, the Soviets agreed to eliminate their SS-20s.

The INF Treaty, which eliminated a specific type of nuclear cruise missiles, was not seen as having a significant effect on Williams's overall cruise missile engine business.[62] Sam Williams believed that future wars would emphasize conventional weapons. Thus, conventionally armed cruise missiles, already an important part of the weapons arsenal, would probably become more important in the future.

The F107-WR-400, used in both the BGM-109 GLCM and the BGM-109G SLCM, was a variation of the F107-WR-100. The first production engine was delivered in 1982, and the last in 1990, with 2,614 produced.

In May 1990, the control of cruise missiles was once again the centerpiece issue of a U.S. and Soviet Union agreement on nuclear weapons. The Strategic Arms Treaty (START) was designed to reduce each side's long-range nuclear weapons by about 30%.[63] This historic agreement was the first attempt by the superpowers to achieve actual reduction in long-range nuclear weapons rather than merely capping existing levels.[64] As had been the case in previous arms control negotiations, cruise missiles proved the most difficult issue to resolve. According to Secretary of State James A. Baker III, the sea- and air-launched missiles had been "two of the most vexing problems we have faced" in the long running strategic arms negotiations.[65] That nuclear and conventionally armed cruise missiles were small in size and similar in shape, had long range, and could be air-, sea-, and land-launched were among the many factors that made negotiations complex.

The very qualities that made cruise missiles difficult to negotiate were the very heart of why they had become extremely effective weapon systems. In 1991, cruise missiles were first used in actual combat during the Persian Gulf

61. U.S. Department of State, *Treaty Between the United States of America and the Union of Soviet Socialist Republics on the Elimination of their Intermediate-Range and Shorter-Range Missiles, December 1987*, Dept. of State Publication 9555 (Washington, D.C.: United States Department of State, December 1987), 2-3.

62. Williams to Leyes, 7 September 1988.

63. *START: Treaty Between the United States of America and the Union of Soviety Socialist Republics on the Reduction and Limitation of Strategic Offensive Arms* (Washington, D.C.: United States Arms Control and Disarmament Agency, 1991).

64. Paul Mann, "Cruise Missile Accord Advances START Treaty," *Aviation Week & Space Technology*, Vol. 132, No. 22 (May 28, 1990): 19.

65. R. Jeffrey Smith, "Cruise Missiles Have Been 'Vexing' Problem in Talks," *The Washington Post*, May 20, 1990, A35.

War. During that engagement, the Navy fired 288 Tomahawk cruise missiles from 18 ships and submarines. Of these missiles, 85% hit their targets. No F107 engine problems were found to be the cause of missiles missing their targets. Williams International was congratulated by the Navy for the excellent reliability of their engines.[66]

SUMMARY OF THE WR19/F107 PROGRAM

Recognized as the world's smallest high efficiency fanjet engine at the time it was initially developed, the WR19/F107 was an important technical achievement in itself. The engine was one of the key technologies that made the cruise missile possible. The development of the cruise missile was controversial and heavily influenced by external, especially political, factors. But, once it was deployed operationally, the cruise missile became a major element of this nation's strategic defense.

The importance of the cruise missile was recognized on May 11, 1979, when Sam Williams received the 1978 Collier Trophy, the oldest and most prestigious aviation award given in the U.S. In awarding the trophy, the National Aeronautic Association cited Williams for the ". . . design and development of the world's smallest, high efficiency fanjet engine. The ingenious design resulted in a unique, lightweight, low cost and efficient engine which was one of the keys in proving the cruise missile concept."[67] The Association further stated that when the B-1 bomber program had been canceled, the cruise missile had been elevated in importance, becoming a crucial part of the Strategic Triad, the cornerstone of U.S. Defense. The selection and qualification of the Williams engine for both Air Force and Navy cruise missiles therefore made the engine one of the significant contributing technologies to this nation's strategic defense.

The Collier Trophy award was followed also in 1979 with the Federation Aeronautique Internationale award, which recognized the F107 engine's technical contribution to the advancement of aeronautics. At the time, the F107 engine represented a significant departure from what many people believed could be done with small engines. The F107 engine had the performance characteristics of larger engines. For instance, although the F107 engine was less than

---

66. Sam B. Williams, Walled Lake, Michigan, to Rick Leyes, letter, 15 August 1991, National Air and Space Museum Archives, Washington, D.C.

67. "Collier Trophy Winner Announced," National Aeronautic Association Press Release, 3 April 1979, 1-2.

1/20th the size of a Pratt & Whitney JT8D, a commercial airliner engine, the two engines had equivalent fuel economies and thrust-to-weight ratios.[68]

The cruise missile was a contentious, paradoxical issue in disarmament treaty negotiations. On one hand, negotiations were complicated by the missile's numerous operational configurations and difficulty in monitoring treaty compliance. On the other hand, the potential destructiveness of this weapon was an element that led to the successful negotiation of pioneering nuclear arms control treaties.

Cruise missile (and thus F107 engine) development was swept up in the dispute between competing military ideologies and technologies. While the WR19 had been demonstrated by the late 1960s, it's development was paced by the controversial cruise missile which was not operationally deployed until the early 1980s.

By 1990, over 4,000 cruise missile engines had been delivered to the Navy, Air Force, Boeing, General Dynamics, and McDonnell Douglas. In reflecting on this F107 program, Sam Williams took pride in some company achievements. The company had met its original design-to-cost goals for the engine. Every engine was delivered on or ahead of schedule. Also, according to Williams, the record of all flights, since the engine was qualified, was nearly flawless, and the reliability record exceeded the reliability goal set for the engine.[69]

Other Engine Developments

Several other engines were developed by Williams during the 1970s, the 1980s, and into the early 1990s, concurrently with the development and production of the WR24 and F107. Among these were engines for various military applications, including the WR34/WTS34 turboshaft series, the WR36-1 and P8300 turbojets, and three miniature turbojets designed for small tactical missile systems. Also, by the early 1990s, Williams successfully penetrated for the first time the civilian aircraft turbine engine market with its FJ44 turbofan engine and had begun development of a low-cost general aviation turbofan engine, the FJX.

68. L. Cruzen, "Cruise Missile Propulsion versus Commercial Airliner Propulsion—Different Challenges can Produce Similar Engine Cycles," (Paper No. AIAA-83-1176 presented at the AIAA/SAE/ASME 19th Joint Propulsion Conference, Seattle, Washington, 27-29 June 1983), 1. Although achieving performance parameters such as those demonstrated by the F107 engine was quite a feat, Williams was not alone. By 1978, Teledyne CAE had developed its competing F106 turbofan engine with comparable performance parameters. However, in the cruise missile engine competition, Williams had been the winner, and as such, was generally credited with this technology breakthrough.

69. Williams to Leyes, 7 September 1988.

WR34/WTS34 SERIES

Development of the WR34-15 turboshaft engine was initiated in 1979. The engine had been designed initially for a wide spectrum of applications. However, the primary application of this family of engines was the Canadair CL-227 Sentinel, a rotary-winged Remotely Piloted Vehicle (RPV) designed for reconnaissance, surveillance, and target acquisition. The first flight demonstrations of this RPV took place in Montreal in October and December 1981. This development vehicle was powered by a 32-shp Williams WR34-15-2. Starting in 1987, near-production versions of the CL-227, equipped with 51.5-shp WTS34-16 engines, began flight testing.

The WR34/WTS34 series engines were simple in design, with a single-stage centrifugal compressor, annular combustor, and single-stage radial inflow turbine. The maximum rated power output of the WTS34-16 was 50 shp at a rotating speed of 102,400 rpm. The engine design emphasized simplicity, ruggedness, simplified maintenance and overhaul procedures, and low manufacturing cost. The engine also had a multi-fuel and multi-lubricant capability.

WR36-1 ENGINE

The Williams WR36-1 turbojet engine, military designation F121-WR-100, powered the Northrop AGM-136A Air-Launched Tacit Rainbow anti-radiation missile. The AGM-136A program was a joint Air Force and Navy development program that started in 1981. The missile could be air launched and was designed as a low-cost anti-radar missile with the ability to loiter in the vicinity of target radars while searching for them, identifying their location, and destroying them. The AGM-136A carried enough fuel for more than 80 minutes of flight. The WR36-1 engine, development of which began in the early 1980s, was qualified in 1985, and by September 1988, four free-flight tests of the missile had been conducted.[70] The Tacit Rainbow contract with Northrop was terminated by the Air Force in September 1991.

P8300 TURBOJET ENGINE

The Williams P8300 turbojet engine was intended to provide an updated version of the Rockwell International AGM-130 glide bomb with a substantially increased range for a longer standoff capability. This glide bomb was launched by aircraft such as General Dynamics F-111s from low or medium altitudes. De-

---

70. "Pentagon to Review Funding, Scheduling for Tacit Rainbow," *Aviation Week & Space Technology*, Vol. 129, No. 12 (September 19, 1988): 64.

veloped during the late 1980s, the P8300 was a 1,000-lb-thrust class engine weighing 150 lb. Due to security classification, further details were not available.

MINIATURE TACTICAL WEAPON ENGINES

As discussed in Chapter 11 of this book, Williams developed three miniature turbojet engines during the late 1980s and early 1990s in support of the small tactical missile engine development program directed by the U.S. Army Missile Command (MICOM). These engines were the WJ119-2, the P8910, and the P9005.

The WJ119-2 Integrated Propulsion Module (IPM) was selected in 1989 as the sustainer engine for MICOM's Non-Line-of-Sight (NLOS) missile. That missile was an outgrowth of MICOM's earlier Fiber Optic Guided Missile (FOG-M) and had substantially greater range. The NLOS was a small anti-tank and anti-helicopter missile with a television camera in its nose. The missile was connected by a fine fiber optic wire to a ground-based control panel with a television console. By monitoring the television image, a controller could direct the missile to its target.

The WJ119-2 IPM consisted of a turbojet engine, inlet and exhaust ducts, fuel pump, fuel tank, start cartridge, ignitor, and controls. It also served as a structural element of the fuselage to which the fore and aft portions of the missile and the wings were attached. The 92-lb-thrust turbojet consisted of an integrally-cast six-stage axial-flow compressor, a reverse-flow annular combustor, and a single-stage axial-flow turbine.[71]

Development of the engine continued until January 1991 when the Army canceled development of the NLOS missile program due to fast-rising costs. The Army believed that the original bids for the program were too low.[72]

In 1989, Williams was awarded a contract for development of the 144-lb-thrust P8910 engine for the NLOS Technical Risk Reduction (TRR) program. The purpose of the NLOS TRR program was to reduce the technical risk of the NLOS program by developing and demonstrating alternate technology turbojet sustainer engines for the NLOS missile system that had the potential for reduced production cost.

71. Williams International, "WJ119 Integrated Propulsion Module," specification sheet, August 1991.

72. Barbara Starr, "FOG-M Shot Down by Rising Costs," *Jane's Defense Weekly*, Vol. 15, No. 1 (January 5, 1991): 14.

The P8910 engine developed by Williams was essentially a growth version of the WJ119-2 engine. The P8910 achieved a 42% increase in thrust over the WJ119-2 while maintaining the same engine envelope. That performance gain was achieved by increasing both compressor airflow and turbine operating temperatures.

Advanced tactical missile system concepts at MICOM led to the Long-Range Fiber Optic Missile (LONGFOG). With its greater range, the LONGFOG missile required a significantly larger sustainer engine than did the NLOS. The Williams P8910 engine satisfied LONGFOG's thrust requirement. Thus, development of the engine continued under the LONGFOG program following termination of the NLOS TRR program in 1991.

Success of MICOM's miniature engine development programs led to establishment of the Small Engine Applications Program (SENGAP) by the Defense Advanced Research Projects Agency (DARPA) in 1990. The objective of the SENGAP program was to develop and demonstrate low-cost, expendable turbojet engines that were only about 4 inches in diameter.

One of six SENGAP engine development contracts was awarded to Williams in 1990 for development of the 51-lb-thrust P9005 engine. That engine had a single-stage mixed-flow compressor, reverse-flow annular combustor, and single-stage axial turbine. At 4 inches in diameter and weighing 9.9 lb, the P9005 was among the world's smallest turbojet engines built by 1992.

THE FJ44 TURBOFAN ENGINE

Using the WR19-9 as a starting point, development of a turbofan engine known as the WR44 began in 1971. The engine was originally flight tested in a two-man VTOL system in 1973. It was chosen initially for the Foxjet International ST-600 business jet and for the American Jet Industries Hustler transport, neither of which was produced.

Using much of the technology and marketing studies for business jet applications of the WR19-9, Williams initiated work in the early 1980s on a larger power plant, known as the FJ44 fanjet engine. The FJ44, a 1,500-lb-thrust proof-of-concept engine, was designed as a power plant for lightweight (7,000-10,000-lb), four-to six-passenger business jet aircraft. This engine was first run in 1981.

It was the expertise derived from the F107 turbofan engine program that enabled Sam Williams to pursue his dream of developing a line of small turbofans for manned aircraft applications. According to Sam Williams, most of the business turbojet and turboprop aircraft that existed in the early 1980s were of old design using old technology.[73] Williams believed that development of a small turbofan engine based on the latest technology would permit the development of a small business jet aircraft that would have one-half the purchase cost and one-half the operating cost of current business jet aircraft. It would have the high Mach number capability and comfort of turbojet aircraft and the desirable characteristics of turboprop aircraft, including short field capability and affordable cost. Also, its fuel consumption would be about one-half that of existing business jets. It was his view that such an aircraft could replace the existing generation of small- and intermediate-sized turbojet and turboprop aircraft, which carried the majority of business passengers, thereby opening a substantial market for the advanced technology turbofan.

During 1983, the 1,500-lb-thrust proof-of-concept engine was tested extensively to study both performance and structural characteristics, including birdstrike and icing tests. When a market survey made at that time indicated that an 1,800-lb-thrust engine would be more suitable, a decision was made to uprate the FJ44 engine. Accordingly, a redesign of the engine was begun in 1983. The redesigned engine was first run in 1985, at which time it met both its 1,800 lb thrust and 0.44 sfc goals at lower than projected turbine operating temperatures.

The engine design emphasized simplicity together with low maintenance and purchase costs. According to Sam Williams, the FJ44 comprised the smallest number of aerodynamic elements that could produce the 3.3:1 bypass ratio needed for a modern business jet engine and the pressure ratio and thermodynamic characteristics needed to provide the necessary efficiency.[74] Epitomizing this design philosophy, the engine was assembled from approximately 700 parts, a total that Williams claimed was three to four times fewer than the parts count of other business jet engines.[75]

The first production version of the engine, the FJ44-1, was rated at 1,900 lb thrust and received FAA certification in March 1992. The engine included a single-piece fan milled from a solid billet of titanium and an intermediate axial-flow compressor stage on the LP shaft driven by a two-stage axial turbine. The HP core portion of the engine was composed of a titanium centrifugal compressor, an annular combustion chamber, and a single-stage axial turbine. The required high-efficiency components were made possible by the advanced technology design and development background derived from years of cruise

73. Williams to Leyes, 7 September 1988.

74. Williams to Leyes, 7 May 1991.

75. David Esler, "FJ44 Turbofan Earns Its Stripes," *Business & Commercial Aviation*, Vol.77, No.4 (October 1995): 56.

The first Williams engine to have a civil aircraft application was the FJ44 turbofan. Photograph courtesy of Williams International.

missile engine development. The FJ44 operated with conservative turbine temperatures, which contributed to low life cycle costs and built-in growth capacity.

Lacking a launch customer, Williams decided to stimulate interest in a new class of light business jets powered by the FJ44, by building and displaying a full-scale mock-up of a futuristic six-place aircraft called the V-Jet. Meanwhile, behind the scenes, Williams had been negotiating with a number of potential aircraft developers, including Cessna Aircraft, Swearingen Aircraft, and Burt Rutan's Scaled Composites, Inc. The outgrowth of these negotiations was that three aircraft, each powered by two FJ44s, were developed. The first of these was a proof-of-concept, all-composite aircraft called the Triumph that was built for Beech Aircraft by Burt Rutan of Scaled Composites, Inc. The initial flight of that aircraft was on July 12, 1988.[76] This flight was a significant event for Williams as the FJ44 was the company's first engine to power a civil aircraft. Although Rutan's Triumph displayed unusually stable flying characteristics, it was deemed to be unmarketable, as its performance barely exceeded that of contemporary turboprop business jets.[77]

---

76. William B. Scott, "New Beech/Composites Business Jet Makes First Flight," *Aviation Week & Space Technology*, Vol. 129, No. 3 (July 18, 1988): 24.

77. Esler, "FJ44 Turbofan Earns Its Stripes," 56.

The Cessna CitationJet powered by Williams FJ44 turbofan engines. Photograph courtesy of Williams International.

Announced in 1988, the second aircraft, designated SJ-30, was developed by Swearingen Engineering and Technology. This six-place business jet made its first flight on February 13, 1991. The third aircraft developed around the FJ44 was the Cessna CitationJet announced at the October 1989 National Business Aircraft Association show. In September 1990, Williams received a substantial FJ44 production order from Cessna Aircraft Company for the CitationJet. The entry-level, six-place business jet made its maiden flight on April 29, 1991.

For assistance in certificating, marketing, and servicing the company's first passenger-rated engine, Williams turned to Rolls-Royce in 1989 to make that company's long-life airliner technology and its worldwide marketing and customer support network available to the FJ44 program. Rolls-Royce joined Williams on the FJ44 program in November 1989, forming a separate company called Williams-Rolls to produce, market, and service the engine. Rolls-Royce held a 15% interest in the company. The engine was renamed the Williams-Rolls FJ44. Rolls-Royce participated in off-shore marketing and engine support activities. The engines were manufactured and underwent acceptance testing at the Williams Ogden, Utah production facilities. Rolls-Royce also participated in the design of the turbines and manufactured the turbine disks, turbine blades, and the LP turbine shaft for assembly in the Utah facility.[78]

The Williams-Rolls FJ44 engine was seen as the key to a new genre of entry-level business jets that would operate not only faster, but more efficiently than turboprop aircraft.[79] In Cessna's case, in order to recreate a true entry-level business jet as it did with the original Citation 500, a less expensive, more efficient engine was needed.[80] Most jet engine manufacturers had been concentrating on building more powerful, fuel-efficient, but expensive engines for longer-range business aircraft. It took a new player, Williams International, to invest in the type of lightweight, compact engine needed. Williams's teaming with Rolls-Royce on the FJ44 made possible aircraft such as the CitationJet and SJ-30.

During the early 1990s, a substantial market developed for the CitationJet. By late 1995, more than 100 CitationJets were in the hands of customers.[81] Many of these were companies that had been operating cabin-class aircraft powered by piston, turboprop, and older jet engines. A relatively low acquisition cost had priced the CitationJet below its slower turboprop competitors. The primary competitor was the Beechcraft King Air series of business aircraft. A measure of the CitationJet/FJ44 combination was a comparison of its performance to that of the progenitor of the Citation line, the Model 500 powered by the Pratt & Whitney Canada JT15D turbojet. According to Cessna, the CitationJet could fly 13% farther on 17% less fuel, and at mid-cruise weight, could cruise approximately 40% faster. Further, it accomplished this with a lower engine thrust output than the original Citation's engines. In fact, one CitationJet charter operator claimed that it was so economical that it offered jet transportation at the cost of a turboprop.[82]

The success of Sam Williams's concept of a low-cost, high-performance business aircraft built around the FJ44 turbofan had come to fruition with Cessna's production order. With the use of the FJ44 in entry-level business aircraft, the company had successfully penetrated the civilian aircraft turbine engine market for the first time.

78. Esler, "FJ44 Turbofan Earns Its Stripes," 56.

79. Barry Schiff, "The Triumph of Technology: The First of a New Breed- The Super-Efficient Business Jet," *AOPA PILOT*, Vol. 33, No. 6 (June 1990): 46.

80. J. Mac McClellan, "Play It Again, Cessna," *Flying*, Vol. No. 117, No. 1 (January 1990): 84.

81. Esler, "FJ44 Turbofan Earns Its Stripes," 56.

82. Esler, "FJ44 Turbofan Earns Its Stripes," 58, 60.

By 1991, Williams-Rolls was working to meet government requirements to allow the engine to be accepted into the military inventory. In 1993, Cessna entered the hotly contested Air Force/Navy Joint Primary Aircraft Trainer System (JPATS) competition with the Cessna 526, a derivative of the CitationJet that was powered by the Williams-Rolls F129 engine, a military version of the FJ44-1. In June 1995, the Pentagon announced that the competition had been won by the Beech PC.9 Mk.II turboprop powered by the P&WC PT6A-68.[83] Nevertheless, by late 1994, the Swedish Air Force had ordered 240 F129s as replacements for the aging Turboméca Aubisque RM9 in its fleet of Saab SK60 jet trainers. That order reflected the fact that pilot training throughout the world was demanding the low fuel and maintenance cost of a modern small turbofan.[84]

Another military application for the FJ44 was revealed on June 1, 1995, when the Department of Defense unveiled its previously top secret Tier III Minus "Darkstar" Unmanned Aerial Vehicle (UAV). The Darkstar was a stealthy battlefield reconnaissance vehicle of innovative design that was developed by the team of Lockheed Martin "Skunk Works" and the Boeing Defense and Space Group.[85]

In September 1995, Williams-Rolls announced the FJ44-2, an uprated variant of the FJ44. Already being run at its rated 2,300 lb thrust, the FJ44-2 had been selected to power the new Raytheon PD374 Premier 1 six-passenger light jet. The FJ44-2 was also selected for the Swearingen SJ30-2, a stretched version of the SJ30-1. According to Sam Williams, the key to the engine's uprated performance was introduction of a "wide sweep fan" that was somewhat larger in diameter than that of the FJ44-1. To generate a higher core flow, two more intermediate axial-flow compressor stages were added to the LP spool, for a total of three stages.[86]

It is interesting to note that while other small gas turbine aircraft engine companies developed turboprop engines as a part of their market basket of products, Williams did not. Williams' primary focus was on turbofans and turbojets with

---

83. Alton K. Marsh and Peter A. Bédell, "Beech/Pilatus Trainer Selected for $7 Billion Pentagon Contract," *AOPA PILOT*, Vol. 28, No. 8 (August 1995): 30.

84. "Twin-Engine Reliability for JPATS," *Aviation Week & Space Technology*, Vol 141, No. 15 (October 10, 1994): 14.

85. Jim Upton, "U.S. Defense Department Unveils Top Secret 'Darkstar'," *Atlantic Flyer*, (July 1995): B13.

86. Esler, "FJ44 Turbofan Earns Its Stripes," 57.

its turboshaft activities generally confined to airborne and ground APU applications. While Williams could have entered the business aircraft market with a turboprop engine, it chose the FJ44 instead. The reason for this focus of the company's efforts was a reflection of Sam Williams's personal philosophy, which as he put it was:

> Well, I guess I believe that turboprops are not the right way to go for speed and comfort and cost. We were accustomed to flying around with piston-powered propellers, and turbine-powered propeller was an alternative. Fanjet airplanes won out over turboprops in big sizes, and my view is that it's just a matter of time until they win out in small sizes. All the same reasons and parameters apply. You just have to make low cost, efficient small engines. Now, there are reasons for some turboprops, for cargo planes and things of that sort, but the interesting market to me is in the high-speed people transportation area.[87]

THE FJX TURBOFAN ENGINE

On December 16, 1996, Williams International received a $37.5 million matching grant contract from NASA under its General Aviation Program (GAP). The GAP's objective was for Williams to develop, within four years, a small, lightweight, ultra quiet, and fuel-efficient turbofan engine for four- to six-passenger, single- and twin-engine general aviation aircraft cruising at 200 knots or better. A major program goal was an order-of-magnitude cost reduction dropping turbine engine prices from their current hundreds of thousands to tens of thousands of dollars, making them cost-competitive with high-performance piston engines. Cost reduction was to be realized through reduced parts count, low-cost design techniques, and the development of advanced, automated, and high-volume production methodology.

A 550-lb-thrust, low-bypass ratio turbofan engine, designated FJX-1, was developed as an interim power plant. Two of these engines powered a testbed aircraft designed by Williams and built by Scaled Composites, known as the V-Jet II. First flight of the aircraft was on April 13, 1997. Under development was the FJX-2, a high-bypass, low velocity turbofan engine initially producing 700 lb static thrust with growth potential approaching 1,000 lb. Derived from the FJ44-1A, it was designed to weigh less than 100 lb.

87. Williams to Leyes, 7 September 1988.

Overview

STATUS OF THE CORPORATION

Williams International had a history of being able to aggressively capture limited segments of the small engine business in its sales to the military. Combined with that attribute was the company's emphasis on development in the small engine class. Its two greatest successes were the WR24 (J400) turbojet used in target drones and the WR19 (F107) turbofan used in cruise missiles. With these engines, Williams established itself in two important niches in the marketplace and secured a place alongside the other established small turbine engine manufacturers.

Williams began its engine development work in 1955 with four people working in a 3,200 square foot rented warehouse in Birmingham, Mich. By the late 1950s, needing more space, it moved to a company-built facility in Walled Lake, Mich. near Detroit. That facility grew to cover 287,000 sq ft on 69 acres and included the corporate offices, engineering design and development, limited manufacturing operations, and support functions. By the early 1970s, production orders for the J400 required additional manufacturing space. In the mid-1970s, a new manufacturing facility was established in Ogden, Utah. Production of the F107 engine required further expansion. The initial facility was closed, and a much larger new facility was constructed adjacent to the Ogden airport. The facility consisted of 185,000 sq ft of manufacturing area on 46 acres of land. By the late 1980s, the company had grown to 1,900 employees in its Michigan and Utah facilities.

Through early 1991, Williams had manufactured nearly 12,300 engines. Of these, over 6,600 were J400 turbojets for the Chukar target drone. In addition, there were 4,765 F107 and 575 F112 turbofan engines for cruise missiles.

THE COMPANY'S MARKETS AND INFLUENCING FACTORS

The initial product of the company was the WR1, a turboshaft engine developed during the 1950s for the Outboard Marine Corporation. It was the first turbine engine to power a boat and the first to use ceramic heat exchangers. Williams also developed turbine engines for experimental automobiles and APUs for aircraft. Though none of these early gas turbine engines were produced in large quantities, the pursuit of these various markets was consistent with Sam Williams's interest in developing small gas turbine engines for a variety of applications.

Williams International facility, Walled Lake, Michigan. Photograph courtesy of Williams International.

With the view that development of a very efficient small turbine engine could open other markets, Sam Williams designed the WR19. That engine was recognized as the world's smallest, high efficiency fanjet at the time that it was initially demonstrated. Early models of that engine were used in several experimental applications. Coincidentally, as these experiments were being conducted, the Air Force had a need for a long range air-launched missile and discovered that Williams had an engine of the required size, thrust, and fuel economy. There followed several years of controversy in the long-range missile program with a series of starts and stops intertwined with inter-service rivalry. Nevertheless, by 1976, the first of these missiles flew powered by the Williams F107 turbofan, which was an uprated version of the WR19. In 1982, the first long-range cruise missile was operationally deployed. With the selection of the Williams F107 engine over Teledyne CAE's F106 engine in the 1972–1973 cruise missile engine competition, the F107 became the only existing engine that could satisfy cruise missile power plant requirements. Thus, Williams totally captured this market.

Several other small engines were developed during the 1970s and 1980s for a variety of military applications, but by the early 1990s, none of these had found a production application. However, the FJ44 turbofan, developed during the 1980s for business jet aircraft, scored a major breakthrough for Williams. By 1992, the company had secured production orders for FJ44s to be used in the Cessna CitationJet. With this engine, the company had for the first time, successfully entered the civil aircraft turbine engine market. The FJ44 was the first Williams engine to be certificated for use in passenger-carrying aircraft.

CONTRIBUTIONS

For his many achievements in the turbine engine field, Sam Williams was awarded an Honorary Doctorate of Engineering by Purdue University in 1982. On December 9, 1988, the National Aeronautical Association honored him with the Wright Brothers Memorial Trophy. Sam Williams's personal contributions and that of his company can best be summarized in the citation accompanying this award which read:

> For significant and enduring contributions to aviation and national defense over a period of thirty-five years. His invention and production of small, light-weight gas turbine engines and his leadership in the introduction and processing of new technology and new materials have provided major impetus to U.S. aviation progress.[88]

To that honor was added the National Medal of Technology presented to Williams by President Clinton on October 18, 1995. That medal was given "for his unequaled achievements as a gifted inventor, tenacious entrepreneur, risk-taker and engineering genius in making the U.S.A. number one in small gas turbine engine technology and competitiveness, and for his phenomenal leadership in helping to revive the depressed United States general aviation business jet industry."[89]

Sam Williams possessed the remarkable ability to conceive, design, and successfully develop new small engine models that not only were available before there was a market for them, but also, in some cases, created new markets that had not previously existed. The predominant technical contribution that Williams made to the small gas turbine aircraft engine industry was the development of very small turbine engines with efficiencies equivalent to those of large engines. The most important product contribution was the development of the very small and efficient F107 turbofan engine used in cruise missiles. That engine was one of the key technologies that made possible the long-range cruise missile, an important element of military defense that directly affected U.S. foreign policy.

88. "Sam B. Williams of Williams International Chosen to Receive the Wright Brothers Memorial Trophy," National Aeronautic Association Press Release, 21 September 1988, 2.

89. "FJ44 Developer to Receive Medal at White House," *Business & Commercial Aviation Show News*, (September 28, 1995): 9.

PRATT & WHITNEY CANADA 8

Early Development and Growth

On November 29, 1928, Canadian Pratt & Whitney Aircraft Company, Ltd., was incorporated under the direction of its founding president James Young.[1] As a Vice President of the John Bertram and Pratt & Whitney tool companies in Dundas, Ont., Young was in a strategic position to start up a Canadian aircraft engine company for the Pratt & Whitney Aircraft Co. (P&WA) of Hartford, Conn., a new aircraft engine manufacturer.[2] P&WA had established an important foothold in Canada through use of its Wasp and Hornet engines in bushplanes and utility planes. The primary purpose in organizing the Canadian Pratt & Whitney Aircraft Company was to service and assemble the P&WA engines operating in Canada.[3] It was also to promote further sales. (Much later, as

---

1. Kenneth H. Sullivan and Larry Milberry, *Power: The Pratt & Whitney Canada Story* (Toronto: CANAV Books, 1989), 17-25.

2. The Pratt & Whitney Company of Hartford, Connecticut, a machine tool manufacturer, was a subsidiary of the New York holding company Niles-Bement-Pond. Niles-Bement-Pond had formed a link with John Bertram and Sons, a Canadian Tool Maker, which resulted in the formation of the Pratt & Whitney Company of Canada Limited, a small tool and gauge manufacturer. In 1925, the Pratt & Whitney Co. of Hartford financed the formation of the Pratt & Whitney Aircraft Company (P&WA). Also located in Hartford, P&WA designed, developed, and manufactured aircraft engines.

3. "The Canadian Pratt & Whitney Aircraft Co., Ltd.," *The Bee-Hive* (April 1930): 7.

433

Pratt & Whitney Canada Inc. (P&WC), it became an integrated manufacturer of its own gas turbine designs for worldwide sale).[4]

Young decided to locate the new company in Longueuil, on the south shore of the St. Lawrence River, opposite Montreal. With the Canadian Vickers aircraft factory and the Montreal airport nearby at St. Hubert, P&WC would be part of Canada's growing aviation industry. Montreal was a good source of labor and was the closest large Canadian city to Hartford. P&WC rented space in a factory manufacturing paper mill equipment. With 10 employees and a paid up capital of $85,100, the company started operation in February 1929.

Customers liked the convenience this new business offered for parts, service, and overhaul needs.[5] P&WA also benefitted by being able to avoid some engine import tariffs. Newly manufactured engine parts could could be shipped from Hartford to the Canadian factory and assembled there as complete engines. Existing tariffs heavily favored British engines, but engine parts were not taxed as heavily.

P&WC had strong sales during its first year of business, but with the 1929 stock market crash, its success was short lived. During the early years of the Depression, as engine sales dwindled, the company concentrated on spare parts sales and engine overhaul work principally for bushplane operators and the air freight business. For additional business, the company made an arrangement to sell, service, and manufacture Hamilton Standard propellers. With the establishment of Trans-Canada Air Lines in 1937 and its purchase of P&W-powered Lockheed 10A and 14 aircraft, P&WC picked up more business.

World War II dramatically reversed the effects of the Depression. From $488,000 in 1939, P&WC's sales rose to a peak of $8,557,000 in 1943, largely as a result of the British Commonwealth Air Training Plan, which established Canada as a major center for training pilots during the war. Early in the war, P&WC benefitted as the Canadian government placed orders for P&W aircraft engines through the company.[6] Most of these engines were used in the Noorduyn-built North American Harvard, Canada's standard training aircraft. P&WA manufactured the engines, and P&WC assembled and tested them.

4. The Corporation was incorporated on November 29, 1928 under the name Canadian Pratt & Whitney Aircraft Company, Limited. On December 11, 1962, the name was changed to United Aircraft of Canada Limited. On May 1, 1975, it was changed to Pratt & Whitney Aircraft of Canada Limited. On October 26, 1982, it became Pratt & Whitney Canada Inc. For purposes of simplification, the company is herein referred to as P&WC.

5. Sullivan and Milberry, *The Pratt & Whitney Canada Story*, 25.

6. Midway through the war, the Canadian government began placing orders for engines directly through the U.S. government rather than agents like P&WC or U.S. manufacturers.

Throughout the war, P&WC provided field service for the Royal Canadian Air Force and for the ferry service that transported aircraft from North America to Britain. While the majority of its business came from military contracts during the war, P&WC's commercial business was also an important segment, and the company reached its best-ever commercial sales in 1942. The war also gave the company its first manufacturing experience. In 1941, the Canadian government asked P&WC to establish a factory to manufacture Hamilton Standard propellers; thousands of propellers were built for Harvards and Consolidated Cansos.[7]

Toward the end of the war, sales dropped as rapidly as they had risen, and P&WC sought other business. Anticipating that many military aircraft would be converted to civilian use, P&WC began to plan and later purchase large numbers of surplus P&W engines and parts as they became available after the war. The engines were reconditioned and resold for use in commercial aircraft. In addition, P&WC overhauled engines for Canadair (formerly Canadian Vickers), which was converting to commercial use Douglas C-47s and C-54s, and Cansos. In another late-1940s venture, P&WC became an agent for Sikorsky Aircraft and marketed Sikorsky S-51 helicopters.

The postwar conversion to commercial business was successful. In the five years after the war, sales at P&WC totaled $15.8 million, more than 70% commercial. Despite this success, P&WC felt that in order to ensure its growth, it needed to change its overhaul and surplus-sale business to manufacturing. With the outbreak of the Korean war and the urgent need for more Canadian Harvard and U.S. Navy SNJ trainer engines, P&WC was finally able to start manufacturing engines. In a joint venture, P&WC spent its own money to purchase land and put up a factory. The Canadian government paid for the manufacturing equipment, and P&WA licensed P&WC to build Wasp engines. Production at the new plant began in May 1952, and the first Canadian-built Wasp was completed in December 1952.

A turning point for the company, the Wasp production program was the start of wider manufacturing operations. Initially, Wasp engines were produced exclusively for the Canadian government. To broaden its customer base and also lower the cost of the Wasp engines, P&WC first negotiated an agreement with the government to undertake other work in the new plant and to sell its products to additional users. It then negotiated an agreement with P&WA to supply Wasp spare parts to customers worldwide. This kept business levels steady when

7. The factory (Canadian Propellers Ltd.) was closed in 1945, and the charter of the company was surrendered to United Aircraft in 1947.

the Korean War ended in 1953. At this time, P&WA had begun to phase out production of spare piston engine parts and concentrate on turbine engine work. Thus, in 1954, it began to transfer to P&WC all tooling to make cylinders for R-985, R-1340, R-1830, and R-2000 engines. Also in 1954, P&WC won a licensing agreement to build Wright R-1820 engines for Royal Canadian Navy Grumman Tracker anti-submarine aircraft.

By the end of 1956, sales had passed $29 million and the company had developed an important piston engine and parts manufacturing program in its new factory. The company was involved in many different programs including helicopter sales and service, engine overhaul and service, and propeller and accessories sales and service. Building on this solid foundation, Ronald Riley, who had succeeded Young as President of P&WC in 1948, took the first cautious step in 1956 toward moving the company into the original design and development of gas turbine engines.

The Dirty Dozen: P&WC's First Turbine Engine Design Group

Studies carried out in the mid-1950s indicated that the piston engine spare parts business would decline in time.[8] Knowing that the company would eventually go out of business if it stayed with piston engines, it was important to determine what the long-term role of P&WC would be and in what direction it should move. Turbine engines looked promising, but the question was what aspect of this business would be most promising for P&WC. It was believed that the gas turbine engine would eventually have an impact on the light aircraft world. After discussions with its parent company, P&WC was authorized to build a team to design and develop small gas turbine engines. In 1956, P&WC began to hire a design team of gas turbine specialists.

At the time, Canada had only a few sources of turbine engine expertise, one being Orenda Engines, Ltd., in Malton, Ont., producer of engines for the CF-100 and F-86 aircraft, with a powerful engine, the Iroquois, under development. Orenda had experienced engine designers, but it was not clear that they would leave the growing Iroquois program to join a speculative venture. The Canadian government's National Research Council (NRC) in Ottawa also had people with gas turbine analytical expertise. Rolls-Royce of Canada, Ltd., near

8. C. B. Wrong, "The Story of Pratt & Whitney Aircraft of Canada," *Canadian Aeronautics and Space Journal*, Vol. 25, No. 2 (Second Quarter, 1979).

Some of the key individuals of the Pratt & Whitney Canada PT6 design and management team. Left to right: Gordon Hardy, Jim Rankin, Fernand DesRochers, Fred Glasspoole, Ken Elsworth, Allan Newland, Pete Peterson, Hugh Langshur, Jack Beauregard, Elvie Smith, Dick Guthrie, and Thor Stepenson. Photograph courtesy of P&WC.

Montreal, Que., overhauler of Rolls-Royce engines, was also a potential source. However, because not all of the needed skills were available in Canada, P&WC decided to seek gas turbine engineers in Britain as well.

Riley authorized a one-year budget of $100,000 to hire a design team. This meant that the person responsible for hiring the team, engineering manager Dick Guthrie, would have to find engineers to work for $6,000 or $7,000 per year. This constraint limited the search to relatively young engineers with limited experience in the field. Unlike some other companies that had entered the small gas turbine engine business, P&WC would not be able to hire well-established turbine engine engineers.

The first six people hired to start work on January 2, 1957 were Elvie Smith, Doug Millar, John Vrana, Pete Peterson, Allan Newland, and Jack Beauregard. Engineers Smith, Millar, and Vrana were hired from the NRC. Smith had conducted research work on anti-icing and afterburning in gas turbines. Millar was hired to do turbine design. Vrana was to become P&WC's expert in combustion chamber design. Peterson, Newland, and Beauregard had worked at

Orenda. Peterson was hired as an analytical engineer for controls and systems. Beauregard's responsibility was compressor design. Newland came as a mechanical designer.

By mid-June 1957, Guthrie had hired the remaining six. They were Ken Elsworth, Gordon Hardy, Fred Glasspoole, Fernand Desrochers, Arthur Goss, and Jim Rankin. Elsworth had worked at Bristol Aero-Engine, Ltd., in Britain on the Orpheus engine; he started in the design department at P&WC. Hardy was a project engineer at Blackburn & General Aircraft, Ltd., in Britain and came to P&WC as a designer. Desrochers was from Orenda and had been working in systems, design, and stress. Glasspoole was a draftsman already at P&WC. Rankin was a gear specialist from Leyland Motors of Longueuil. Goss's specialty was gearbox design; he had come from Rolls-Royce in the United Kingdom.

These 12, who formed the nucleus of P&WC's first turbine engine design group, were facetiously known within the company as the "Dirty Dozen." They joined Hugh Langshur, who was Chief Engineer of the existing engineering department at P&WC. In a logical progression starting in the 1930s and continuing in the 1940s, the company had built an industrial base on the sales and product support of P&WA-designed engines. In the 1940s, it had gained manufacturing experience, and in the early 1950s, it had acquired a permanent manufacturing capability of its own. Now, building on the stability of its piston engine business, P&WC was poised to establish its independence by designing its own engines, ultimately to complement those designed by its parent company.

The JT12

As the new team began looking for a project, coincidentally, in 1957, the Royal Canadian Air Force had announced its requirement for a small jet trainer. Canadair responded with its candidate aircraft, the CL-41 Tutor. Believing that they could come up with a design for the CL-41 engine, members of the team made some preliminary sketches, starting in July 1957, toward a proposed a 3,000-lb-thrust turbojet engine.[9]

As the team was going to P&WA in East Hartford to get training and familiarity with P&WA design practices, a decision was made to take the engine project along and design it there. Because some of the members of the team had

9. Robert I. Stanfield, "JT12 Design Geared to Permit Early Production Date," *Aviation Week*, Vol. 69, No. 16 (October 20, 1958): 59.

just arrived from England and for other reasons, not everyone had security clearance to enter the main plant. Some of the team worked in the Credit Union building located across the street from the plant. The P&WC team was not allowed to work on P&WA engine projects, so full time was devoted to the proposed 3,000-lb-thrust engine.

P&WA management organized several engineering teams to work on conceptual designs for the engine, each team to come up with its own design. The Canadian group designed a simple straight-through-flow axial-compressor engine which they designated the Design Study DS-3J. In the end, when the designs were critiqued, the Canadian "Credit Union Special" won. As P&WC's Vice President Design and Development Engineering Gordon Hardy recalled:

> The reason they picked ours is because we were young and relatively unsophisticated, and thus we made a very simple low-cost design, which would obviously work well. This caught the imagination of management. They didn't pick the other designs, which were more sophisticated and had a lot of clever stuff in them, which really wasn't needed, I guess.[10]

The DS-3J would be competing against the General Electric J85 and the Fairchild J83 for the CL-41 aircraft. Of these other two engines, the J85 was the principal contender. Compared to the DS-3J, the J85 was lighter, cheaper, and further ahead in development. P&WC felt that the DS-3J would be more rugged and easier to maintain and overhaul. In September 1957, P&WC decided to proceed with a jet engine that would be slightly larger (to give it a performance advantage) than the J83 or J85. This engine, the DS-4J, also would be designed for possible executive aircraft applications. Under the direction of a P&WA manager, the Canadian design team at East Hartford proceeded with a detailed design.

By the end of the year, the engine design was finished; however, as P&WC began to add up the numbers, the company realized that it did not have the development funds, the facilities, or the manpower to complete the engine. The rapid development schedule of the CL-41 would have forced P&WC to make immediate expenditures on facilities and new people. The DS-4J could not be financed from profits from the piston engine spare parts business because much of the money had been committed already to other areas of growth. Also, P&WC's assumption that research and development funding from the Canadian

---

10. Gordon Hardy, interview by Rick Leyes, 6 October 1988, Pratt & Whitney Canada Interviews, transcript, National Air and Space Museum Archives, Washington, D.C.

government would be available for the engine was not realized. For these reasons, in January 1958, the design was turned over to P&WA.

This engine became the JT12. The engine was a straight turbojet—a single-shaft engine with a nine-stage axial compressor, a cannular combustor, and a two-stage turbine. One model of this engine, the JT12A-6A had a 3,000-lb-thrust take-off rating, a 6.6:1 thrust-to-weight ratio, and 0.96 sfc.

The first JT12 engine ran on May 16, 1958, and a 50-hour endurance test was completed on August 16 of that year. In September 1959, the first engine was delivered to Canadair, and in January 1960, the CL-41 Tutor made its first flight. Shipment of production JT12 engines began in October 1960.

Eventually, the CL-41 won the Canadian trainer competition, but the JT12 was not selected for the aircraft. Instead, in what was considered to be a political decision, the Canadian government chose the J85 engine to be built under license in Toronto by Orenda.[11] The work would be split. The airframe would be built at Canadair in Montreal, and the engine would be built by Orenda. The decision came at a time when Orenda's major engine project, the Iroquois, had been lost following cancellation of the Avro Canada CF-105 Arrow fighter project by the Canadian government in February 1959.

Despite this loss, the JT12 program became successful. The principal applications of the JT12 were the North American Sabreliner and the Lockheed JetStar. The military version of this engine, known as the J60, powered the military versions of these aircraft, respectively designated the T-39 and the C-140. The J60 also powered the Fairchild SD-5 surveillance drone and the North American T-2B Buckeye, the U.S. Navy's first production jet trainer.

A turboshaft version, known as the JFTD12 (military T73), developed up to 4,800 shp and powered the twin-engine Sikorsky S-64 Skycrane (military CH-54 Tarhe). The JFTD12 used a JT12 gas generator with a two-stage free power turbine providing a rear drive. There were also industrial and marine versions of the JT12 engine. These engines, rated at 3,000 shp, were designated FT12. Production of the JT12/J60 engine series lasted for nearly two decades, during which 2,269 JT12 and J60 turbojet engines and 352 JFTD12 turboshaft engines were manufactured. With the completion of this engine's production in East Hartford, all of Pratt & Whitney's efforts in the small engine field reverted to P&WC.

It is noteworthy to mention that the JT12 was produced during a transitional era. It was part of the last generation of small pure jet engines developed for use on commercial and military aircraft. The turbofan era was beginning to emerge

---

11. Sullivan and Milberry, *The Pratt & Whitney Canada Story*, 119.

and would eventually replace the turbojet engine in many aircraft applications. For P&WC, perhaps the most important thing that came out of the program was that it gave the company's designers experience and credibility.

The PT6

In March 1958, following the transfer of responsibility for the JT12 to P&WA, the P&WC design team returned from East Hartford and were given instructions to look for a less expensive project. While the engineers worked on various design possibilities, turbine engine market studies were made. By September, the focus was on 250-500-shp engines. While some military possibilities existed, general aviation appeared most promising, particularly small twin-engine transports. A general aviation engine would cost less to develop and relatively few turbine engine companies had targeted this potential but undeveloped market.

Engineers Ken Sullivan and Elvie Smith of P&WC and P.J. Krones, a marketing representative from P&WA's installation engineering group, visited Piper, Beech, and Cessna among others to determine what their needs were.[12] They took with them some preliminary design work on a 350-shp engine. Piper was not interested because it felt that the higher fuel consumption of a turbine engine would make it unattractive in small business aircraft. Cessna, with considerable jet engine experience in its T-37 trainer, had an interest as long as the price could be kept to less than twice that of a reciprocating engine. Also, while Cessna had been visited by other turbine engine companies offering them lower horsepower engines, it was looking for a 400-shp engine for an upcoming aircraft.

Beech was the most interested candidate. In 1958, Beech was working on converting its military L-23F twin-engine transport into a commercial design to be known as the Model 65 Queen Air. This was exactly the type of airplane that the P&WC people felt should have turbines, and Beech was closest to doing that.[13] Upon their return to P&WC, Sullivan and Smith suggested that the company focus on a 450-shp engine with growth potential to 500 shp. Based on existing aircraft wing loadings, such an engine would provide direct operating costs comparable to piston engines of the same power, but at cruising speeds that were 50 mph faster.[14]

12. Sullivan and Milberry, *The Pratt & Whitney Canada Story*, 119.

13. Elvie Smith, interview by Rick Leyes, 6 October 1988, Pratt & Whitney Canada Interviews, transcript, National Air and Space Museum Archives, Washington, D.C.

14. Sullivan and Milberry, *The Pratt & Whitney Canada Story*, 120.

THE CONFIGURATION QUESTION

After the marketing studies had targeted the horsepower range, the next question that needed to be resolved was engine configuration. In establishing its design criteria, P&WC had decided that this had to be a low-cost, low-risk project. Also, the company had decided to select relatively conservative aerodynamic and mechanical design parameters because it lacked component test facilities. From a customer viewpoint, reliability, cost, fuel consumption, weight, and maintenance requirements were important considerations.

To meet these objectives, one of the questions focused on whether it should be a fixed-shaft or free turbine engine. There were advantages and disadvantages to each design. P&WC felt that a fixed-shaft engine would have fewer parts and thereby would be less costly to develop.[15] The fixed-shaft engine had an almost instantaneous response to throttle lever movement, an advantage for the pilot in emergency situations. However, the response rate of a free turbine would be adequate for what was normally required in service. The principal disadvantage of the fixed-shaft design was that the compressor, turbines, and drive shaft load were all committed to the same rotational speed, limiting the operational flexibility of the engine.

By comparison, because the gas producer section of the free turbine engine operated independently of the power turbine section and load, very high output power was available down to low output shaft speed and vice versa.[16] This "fluid torque converter" property was an advantage in helicopter operation during starting and run-up, and it eliminated a clutch requirement. In turboprop applications, the free turbine offered a wider range of engine power speeds in order to help reduce propeller noise both on the ground and in flight and to give more flexibility in selecting cruise speeds. The compressor could operate at the optimum design point for each flight condition while propeller speeds were independently controlled to deliver maximum efficiency. P&WC also felt that the output speed flexibility of the free turbine engine was important because it made the engine compatible with a wide variety of off-the-shelf propeller designs rather than more expensive custom-made designs.

The operational flexibility of the engine made the free turbine engine especially suitable for both turboshaft and turboprop applications. Because the P&WC designers were not too sure of the market at the time, an important

15. Sullivan and Milberry, *The Pratt & Whitney Canada Story*, 120.

16. R. H. McLachlan, "Handling Characteristics of Free Turbine Turboprop Engines," (Canadian Pratt & Whitney Aircraft Paper "SEM" 110, September 1963). For a given gas turbine power, the power turbine torque goes up as the power turbine speed goes down.

consideration was an engine that was adaptable to both markets.[17] Another advantage was that the free turbine required less starting power, which was helpful in cold weather. The biggest drawback of the free turbine was that it had a higher initial cost than did a fixed-shaft engine.

While generally leaning toward a free turbine engine because of the advantages they saw in the design, P&WC's decision to make the new engine a free turbine was largely based on Beech's preference for that type of engine.[18] In the event of an engine failure at high speed, Beech felt that the windmilling propeller would create a very large torque on a fixed-shaft engine because all of the internal components were attached to the single shaft. Absorbing this much power would cause extreme aerodynamic drag on one side of the aircraft and possibly cause structural failure of the tail. Because fewer internal engine components were attached directly to the propeller shaft on a free turbine engine, the smaller torque would result in less aerodynamic drag while the pilot was trying to feather the propeller. The decision to go with the free turbine engine was made in October 1958.

Another configuration decision was between choosing a concentric-shaft or an opposed-shaft design. In a concentric-shaft engine, a long power turbine shaft would pass through the compressor turbine shaft to the front of the engine where, through the reduction gearbox, it would drive the propeller. In this configuration, the engine could be designed to minimize the number of turns in the gas flow path. For example, the airflow could be delivered straight into the intake and exit straight out the rear of the engine. By routing the intake and exhaust airflow directly through, pressure and its performance losses could be minimized. An important design challenge with this configuration was avoiding power shaft vibration and the bearing problems associated with running a high-speed, relatively thin power shaft through the gas generator shaft.

From a mechanical viewpoint, the opposed-shaft engine envisioned by P&WC designers would be a much simpler engine.[19] In this design, the gas generator shaft would be at the rear of the engine, and the power turbine shaft would be "opposing" it at the front of the engine. As in the concentric-shaft engine, the gearbox would be at the front of the engine. However, in the opposed design, the air intake would be located at the rear of the engine instead of the front. Thus, the air would flow forward through the compressor, combustor, and the turbines, the exhaust gas discharging near the front of the engine.

17. Hardy to Leyes, 6 October 1988.

18. Smith to Leyes, 6 October 1988.

19. Hardy to Leyes, 6 October 1988.

This arrangement had some benefits, with lower compressor noise a result of the bending and reverse flow of the inlet air. Also, the inlet could be protected by a screen to reduce foreign object damage. Easy access to components requiring more frequent inspections or replacement was possible as well. For instance, the front half of the engine could be removed for a hot-section inspection, leaving the rear half on the wing.

The decision to go with the opposed-shaft design was a group decision of the designers. They felt that the concentric-shaft design was too much of a risk for their level of skill and experience.[20] In order to check the conclusions of the P&WC designers, engineers at P&WA were asked to design an engine with similar parameters. When they came up with a layout very similar to the opposed-shaft configuration, P&WC's design was given the go ahead. This design was known as the DS-10.

In its original configuration, the DS-10 was a free turbine, opposed-shaft engine. The gas generator spool, facing rearward, was a three-stage axial-flow and single-stage centrifugal-flow compressor driven by a single-stage turbine. The opposed, forward-facing power turbine shaft was driven by a single-stage turbine. The contrarotating power turbine drove the propeller shaft at the front of the engine through a reduction gearbox. The reduction gear had a single epicyclic stage for the turboshaft engine and an additional epicyclic stage for the turboprop.

The gas path of the DS-10 was unique. The air flowed forward through the compressor, combustor, and turbines. At the rear of the engine, intake air entered through a screened plenum, and then, via a radial-flow intake, flowed through the compressor. As the air exited the compressor, it passed through a conventional vaned radial diffuser and straightening vane assemblies. The air was then routed through an annular reverse-flow combustor. Exiting the combustor, the airflow again reversed, passing forward first through the gas generator turbine and then through the power turbine. Finally, the airflow exited through a double scroll exhaust unit near the front of the engine.

In December 1958, a decision was made that the engine would have a take-off rating of approximately 500-eshp, a design pressure ratio of 6:1 giving an esfc of .70 for a turbine inlet temperature of 2,170 degrees R.[21] Shortly thereafter, the designation DS-10 was changed to PT6. Actual design work on the PT6 started on December 30, 1958, when the design speed was established.

20. Hardy to Leyes, 6 October 1988.

21. Gordon Hardy, "Development of Gas Turbine Engines at Pratt & Whitney Canada," (CSME Paper No. 224 presented at the Canadian Society for Mechanical Engineering, Montreal, Quebec, June 1992), 17.

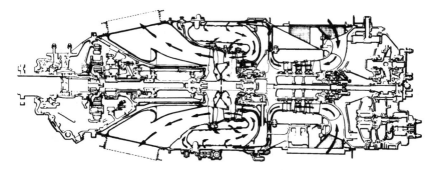

Cross-section of the P&WC PT6 engine highlighting its internal gas flow path. Air enters the compressor at the rear of the engine and the exhaust discharges at the front. Photograph courtesy of P&WC.

PT6 PRE-CERTIFICATION DEVELOPMENT PROGRAM

In January 1959, P&WC asked for and received $1.2 million from the Canadian government's Department of Defence Production. This money supplemented the P&WC investment required for a 30-month program to bring the engine to the 50-hour flight-qualification stage of the development program.

Detailed design work on the engine also started in January 1959. The aerodynamic design of the engine was completed by mid-May 1959. The opposed-shaft configuration had been chosen for mechanical simplicity. However, a concern was that the resulting turns and reverse flow in the airflow path were aerodynamically "dirty" and would result in pressure, and thus, performance losses. This problem was solved by judicious aerodynamic design, aided in some instances by model tests.[22] By designing for low velocity where turning was greatest, the engineers kept overall pressure loss to a value equal to that of a straight-through arrangement.[23] The low velocities were achieved by using large cross-sectional areas in the intake, combustor, and exhaust components.

In particular, information obtained from one set of experiments proved to be an invaluable guide to interpreting actual combustor behavior. A water flow vi-

22. J. C. Vrana, "Aerodynamics of the Gas Path in the PT6 Engine," (IAS Paper No. 60-96 presented at the CAI-IAS Joint Meeting, Montreal, Quebec, 17-18 October 1960).

23. E. L. Smith, "Development of the T-74 (PT6) Turboprop/Turboshaft Engine," (SAE Paper No. 624B presented at the Automotive Engineering Congress, Detroit, Michigan, 14-18 January 1963), 1.

sualization model was constructed in plastic and tested in the water tunnel at the National Research Council in Ottawa by Combustion Engineer John Vrana. Water carrying small metal particles was run through the model and a camera recorded the flow patterns.[24] Hot spots that would exist in an actual combustor were indicated by eddy currents.

In September 1959, a reevaluation of the PT6 began. It was determined that the aerodynamic and mechanical design required improvements and that the engine had structural problems. One concern was that the cast aluminum centrifugal compressor would not have the structural integrity required, and a decision was made to replace it with cast steel despite the extra weight and cost. Another problem was that the engine was 20% above its 250-lb design weight, and a component-by-component program was started to find ways to reduce the weight. These and other problems prompted the team to redesign the engine and redesignate it as the Mk.2. In the meantime, work continued on the original design then known as the Mk.1.

On November 16, 1959, the PT6 Mk.1 gas generator ran (unofficially) for the first time. On February 22, 1960, the complete Mk.1 engine ran, almost 14 months after full design launch. Four Mk.1 engines were built and run. Unfortunately, the best performance achieved (340 eshp at 0.91 esfc) by the Mk.1 design fell significantly short of the original design goals (500 eshp at 0.69 esfc). The main deficiencies were attributed to an inadequate compressor surge margin and poor turbine efficiency primarily due to leakage.[25] The engine also had mechanical limitations.

By July 1960, the Mk.2 gas generator was running, but with problems. The gas generator shaft developed uncontrolled vibration at relatively low speed (approximately 50% of rated gas generator rotor speed). A damped bearing mounting system for the thrust ball bearing was designed to solve the problem. Another redesign was required for the gas generator turbine stator vane mounts to prevent the vanes from collapsing into the rotor blades. Component problems such as these, combined with a cautious and frugal engineering team, resulted in slow progress.

P&WC management decided to reorganize the PT6 development program. Starting in November 1960 and for the next several months, the program was put under the technical direction of a six-person P&WA team. A move from a single-shift development program to an around-the-clock effort—a "make-them-and-break-them" approach—was ordered in order to uncover and solve

24. Sullivan and Milberry, *The Pratt & Whitney Canada Story*, 126.

25. Hardy, "Development of Gas Turbine Engines," 18.

problems more rapidly. Control of the program was consolidated under the development engineer. Meanwhile, tests necessary to get the engine into the air continued.

The complete Mk.2 turboprop and turboshaft engines ran in February 1961. The 50-hour test, which cleared the way for flight testing, was run in March. With the PT6 Mk.2 installed in the nose of a Beech 18 loaned by the RCAF, the first flight test took place on May 30, 1961. Tests were run on aircraft handling and performance, specific fuel consumption, propeller handling, and propeller drag.[26] Tests also included air starting, accessory loading, noise and vibration levels, and even in-flight propeller reversing. The aircraft sought out icing conditions to see how such conditions would affect engine performance. These pre-certification flight tests were conducted at altitudes as high as 25,000 ft.[27]

P&WC Canada was also gaining engine operational and installation experience from other aircraft and rotorcraft programs. In July 1961, a PT6 Mk.2-equipped Ten99 Hiller helicopter lifted off, marking the first time that a PT6 had powered an aircraft on its own. The Hiller Ten99 was a prototype helicopter developed for a U.S. Navy Assault Support Helicopter (ASH) program, later dropped. In February 1962, a Piasecki 16H-1 Pathfinder helicopter flew powered by a PT6B-2 turboshaft engine.[28] The Pathfinder was a privately financed high-speed research rotorcraft. The PT6 was also selected by Lockheed to power a high-performance, rigid-rotor research helicopter known as the XH-51A Aerogyro, which flew for the first time in November 1962. A Kaman K-1125 helicopter, powered by two PT6 engines, flew in April 1963. From May 1963 until July 1965, a specially modified de Havilland DHC-3 Otter powered by two PT6s and a General Electric J85 was flown in a STOL research program. A PT6-powered de Havilland DHC-2 Mk. 3 Turbo Beaver flew in December 1963.

One interesting problem surfaced during the testing of Lockheed XH-51A. Amusing to bystanders, but not to the pilot, was the "pogo stick" flight characteristic of the aircraft. A full-open or full-closed interstage bleed valve was used to bleed air off the axial-compressor stages at low power to prevent stall. It happened that the power at which the bleed valve would close was just about that required for the helicopter to lift off in ground effect. When the bleed valve

---

26. Sullivan and Milberry, *The Pratt & Whitney Canada Story*, 138.

27. "PT6 Exceeds its Estimated Performance," *Aviation Week & Space Technology*, Vol. 75, No. 19 (November 6, 1961): 74.

28. Norman Polmar and Floyd D. Kennedy, Jr., *Military Helicopters of the World: Military Rotary-Wing Aircraft Since 1917* (Annapolis: Naval Institute Press, 1981), 270.

closed, a spurt of power would cause the helicopter to take off abruptly. The rotor speed governor would then cut back the fuel causing the compressor to slow down. This would open the bleed valve causing the helicopter to lose power and drop rapidly. The cycle would then repeat itself. The solution was a proportional bleed valve.[29]

As experience was gained, in early 1962, it became clear to P&WC that an engine of 600 eshp would better serve their customers.[30] To accomplish this, several major aerodynamic changes in the PT6 Mk.2 were made including an increase of 13% in compressor air mass flow, an increase in pressure ratio from 5.8 to 6.25, and a 2 1/2 point increase in compressor efficiency. This performance was achieved in November 1962.

In addition to the Mk.2 gas generator mechanical problems, several others occurred during PT6 development testing. Developing the main reduction gear to a satisfactory level of reliability proved to be a difficult problem. Gear wear led to high vibration levels that threatened the integrity of the gearbox. A change in the pressure angle of the gear teeth, increased tooth face width, and closely held tolerances of tooth profiles and finishes during gear manufacturing were combined solutions that yielded consistent gearbox reliability.

A related mechanical problem occurred with the power turbine blades. Originally, the power turbine was designed with unshrouded blades. Intensive strain gauge tests were launched after a second-stage turbine blade failed when the engine program had about 1,600 hours total time.[31] These tests showed that high stresses were excited by reduction gearbox tooth meshing frequencies. The blades were shrouded, and close reduction gear tooth profiles were required to give smooth meshing. Locked shrouds on the blades provided damping and appropriate changes in natural frequencies.

By the end of 1963, the last year of the pre-certification development program, it was clear that significant technical progress had been made on the PT6. The prototype PT6A-1 engine had a take-off rating of 450 eshp and a 0.72 esfc. The non-flat-rated, take-off rating of the PT6A-6 certification engine was 610 eshp (550 shp), a 36% increase over the prototype. Fuel consumption had been lowered 11% to 0.64 esfc (0.65 sfc). In December 1963, the PT6 received its civil certification from both Canadian and U.S. government agencies. A total of more than 11,000 development running hours had been completed in addition

29. G. P. Peterson, interview by Rick Leyes, 7 October 1988, Pratt & Whitney Canada Interviews, transcript, National Air and Space Museum Archives, Washington, D.C.

30. Hardy, "Development of Gas Turbine Engines," 19.

31. Smith, "Development of the T-74 Engine," 5.

to over 1,000 hours in various prototype installations. Flight time at certification was just under 800 hours.[32]

The PT6 had become Canada's first turboprop engine. Eight engines had been built for the program, some turboprop and others turboshaft engines. The pre-certification development program had been financed primarily by P&WC with substantial support from the Canadian government. However, by the end of 1963, the engine program had yet to become financially self-sustaining.

SELLING THE PT6

Marketing efforts on the PT6 had started as soon as the design specifications had been written in November 1958. In 1959, numerous contacts were made with airframers around the world and the U.S. and Canadian governments. Initially, P&WC focused on marketing a turboprop engine, but potential opportunities for turboshaft engines developed.

In response to the 1960 U.S. Army Lightweight Observation Helicopter (LOH) program, P&WC, along with other small gas turbine engine manufacturers, took a serious interest in proposing a turboshaft engine for this very lucrative contract. P&WC offered a derated version of its PT6 engine carrying the military designation T74-P-2.[33] Of the 12 airframers that submitted proposals for the LOH competition, only Republic specified a PT6 for its helicopter. All others chose the 250-shp Allison T63. At 450 shp, the PT6 Mk.2 was really too big for the LOH as specified by the Army. Subsequently, when Bell, Hiller, and Hughes were selected as finalists in May 1961, P&WC was out of the LOH competition.

P&WC's interest in the turboshaft market was also stimulated by a U.S. Marine Corps requirement for the Assault Support Helicopter (ASH). In December 1960, the Navy requested that P&WC provide data on the PT6, funding estimates for a flight test program, and funding required for a 150-hour qualification test.[34] In October 1961, the U.S. Bureau of Naval Weapons issued a request for proposal to about a dozen contractors for the ASH. The PT6-powered Hiller Ten99 was one of the candidates for the ASH program. However, in March 1962, the Navy announced that it would buy Army HU-1B helicopters from the

32. J. P. Beauregard, "Turbine Powerplants for General Aviation," (AIAA Paper No. 64-176 presented at the AIAA General Aviation Aircraft Design and Operations Meeting, Wichita, Kansas, 25-27 May 1964).

33. Robert I. Stanfield, "Canada's Small Gas Turbine the PT6," *Flying*, Vol. 68, No. 4 (April 1961): 29.

34. Sullivan and Milberry, *The Pratt & Whitney Canada Story*, 142.

Bell Helicopter Company.[35] With minor modification, these "off-the-shelf" helicopters could more rapidly fill the Marine Corps' urgent need for an assault support helicopter than could an all-new design.

After these and other unsuccessful attempts to get into the U.S. military helicopter market, P&WC decided to redirect its efforts. As a Canadian manufacturer, despite the fact that it was U.S. owned, P&WC believed it was not going to win a new "clean-sheet-of-paper engine" contract from the U.S. military.[36] Allison had established an early lead on the light turbine engine military helicopter market and had been successful in cornering it. P&WC had to focus on other markets. The commercial fixed-wing market, the original target of the PT6 program, was still a potential opportunity, and P&WC decided to refocus on it.

By the spring of 1963, millions of dollars had been spent developing the engine, but no major company had yet come through with an order. However, continuing discussions with Beech Aircraft on the PT6, which dated from October 1958, were beginning to show promise. In 1958, Beech modified its military L-23F twin-engine transport into the commercial design known as the Model 65 Queen Air.[37] P&WC representatives who had seen the Model 65 during their market survey in 1958 were convinced that this was exactly the type of aircraft they were looking for.[38]

Beech had been interested in a turbine-powered aircraft and had worked with the French engine company Turboméca and the French government on such a project. As a result, a Turboméca Bastan-powered Beech Model 18 had flown in 1958, and a Turboméca Astazou-equipped Beech Baron flew in 1960. In 1962, Beech began an extensive design analysis of a turbine-powered executive aircraft. Beech considered a number of candidate turbine engines, including the Lycoming T53, Turboméca Bastan and Astazou, the Continental 217, and the P&WC PT6.[39]

In early 1961, Beech and P&WC agreed to combine their products in a test program. The companies set out to convince the U.S. Army to sponsor a

35. "Navy to Purchase Assault Helicopters from Bell for Marine Corps," Department of Defense Office of Public Affairs News Release No. 320-62, Washington, D.C., 2 March 1962.

36. Smith to Leyes, 6 October 1988. A "clean-sheet-of-paper engine" means an entirely new engine design. According to Smith, such an engine and its expensive development costs would be an additional factor which would make it difficult to sell the engine to the U.S. military.

37. Edward H. Phillips, *Beechcraft Staggerwing to Starship: An Illustrated History* (Eagan, Minnesota: Flying Books, 1987), 53.

38. Smith to Leyes, 6 October 1988.

39. J. C. Wilson, "Development of Turbine-Powered Pressurized Beech King Air," (SAE Paper No. 28037 presented at the Business Aircraft Conference, Wichita, Kansas, 6-8 May 1965), 19.

PT6-powered L-23F conversion and test program. The Army countered with an offer to sponsor a 100-hour flight test program if Beech and P&WC would sell for one dollar an L-23F with PT6 engines installed. The deal was accepted and the project moved ahead. On December 20, 1963, P&WC shipped its first two production engines, PT6A-6s, to Beech for installation on the L-23F conversion aircraft. This turbine-powered L-23F (Beech Model 87), designated NU-8F, first flew in May 1963.[40] Almost 10 months of flight testing followed before the aircraft was delivered to the Army in March 1964.

With increasing enthusiasm for a PT6-powered aircraft based on the acceptance of the L-23F program and experience derived from it, the looks and quietness of the aircraft, and the potential market, Beech decided that it would back a commercial turboprop aircraft. On July 14, 1963, Beech officially announced the King Air 90, a pressurized twin fitted with 550-shp PT6A-6 engines.[41] The King Air first flew on January 24, 1964.

P&WC sold 16 PT6A-6 engines to Beech for the King Air program, and P&WC had its first significant engine order. The King Air, the first small twin-turboprop airplane to be built in America, would become the most popular executive turboprop aircraft in the world. The P&WC PT6 was on its way to becoming one of the world's most popular turbine engines. Beech would also become P&WC's biggest customer.

There were many reasons for the success of the King Air program. According to Elvie Smith, retired P&WC Chairman of the Board:

> It was the right airplane at the right time, and it had Beech service and Beech quality behind it. It was promoted very effectively by Beech. Frankly, we were amazed when we learned the margins that the dealers had to work with. As Frank Hedrick, (Beech's Executive Vice-President) used to say: 'I want those guys to have the prospect of making so much money that they can't sleep at night for wanting to move those airplanes.' It was a very good, reliable, smooth, comfortable airplane. It had effective Beech product support. Olive Ann Beech (President of Beech Aircraft) for years and years used to read every customer complaint when it came in and used to check the answer that went out. They looked after the customer properly.[42]

40. In 1966, under Beech sponsorship NU-8F won a U.S. Army competition. This turned out to be an important contract for Beech. Designated the U-21, eventually 200 aircraft were delivered to the Army.

41. Edward H. Phillips, *Beechcraft Pursuit of Perfection: A History of Beechcraft Airplanes* (Eagan, Minnesota: Flying Books, 1992), 57.

42. Smith to Leyes, 6 October 1988.

First flight of the Beechcraft King Air No. 1 powered by two 550 shp P&WC
PT6A-6 engines. Photograph courtesy of P&WC.

Ken Sullivan, retired Senior Vice-President for Marketing and Product Support, analyzed the King Air and PT6 program:

> The King Air's selling points were quietness, no vibration, speed, and altitude
> capability. It could fly over weather. It was similar to the big transport jets
> flying at the time and just as glamorous. For its part, P&WC paid close atten-
> tion to customer service. King Airs carried top corporate executives who had
> no compunction about calling P&WC management if an engine problem
> caused them to miss an important meeting. As for the engine itself, it worked.
> The reputation of P&WA helped sell engines too because P&WC was an un-
> known engine company.[43]

While PT6-powered aircraft had good selling points, in the beginning, they
had an uphill struggle. They had to compete with well-established piston-
powered aircraft that had their own very good sales points. According to retired
Vice President Marketing Richard McLachlan, who was marketing the PT6 at
the time:

> It was tough in many ways because you could take a pressurized Queen Air with
> a piston engine and you could take a PT6, and the PT6 would not really bring
> any immediately obvious benefits to the airplane. The airplane would have es-

43. Kenneth H. Sullivan, interview by Rick Leyes, 5 October 1988, Pratt & Whitney Canada
Interviews, transcript, National Air and Space Museum Archives, Washington, D.C.

sentially the same speed, it would burn more fuel, which means you had to carry more fuel in the tanks for the same range. It was quieter and the engines would have longer lives, but these were not high sales points at that time. It had a significantly higher price, both the airplane and the engine. You wondered why anybody would ever buy a turbine engine. But as soon as they came on the market, the pressurized Queen Air with the piston engine disappeared completely, even though they reduced the price of the airplane to try to keep it going. The turboprop completely took over because everybody had to have one.[44]

While the glamorous era of the jet and a successful King Air marketing program were factors in the initial acceptance of the PT6, ultimately the engine had to sell itself within the dominant piston engine market. Important sales points were reliability, ability to use different kinds of fuel, and the ability to start in any kind of weather.[45] While a database had yet to be developed, theoretically a case could be made for lower life cycle costs. Despite the fact that the initial cost of turbine engines were higher than reciprocating engines, it was possible that long-term costs could be reduced or offset with other benefits.

For instance, turbine engines such as the PT6 had considerably better power-to-weight ratios than did piston engines. This meant bigger aircraft payloads, lower gross weight, and more speed. The earning capability of utility aircraft could be higher. Higher initial engine costs and fuel consumption could be offset. The net result, according to studies conducted by P&WC, would be higher cruising speeds at higher altitudes, with direct operating costs essentially identical to those of piston engine-powered aircraft.[46] For executive aircraft, the extra speed, quietness, and reliability would warrant the more expensive turbine engine. P&WC predicted a gradual takeover in the more expensive executive aircraft and more highly utilized utility aircraft markets.[47] This prediction proved to be accurate.

THE COMMUTER CONNECTION

While the initial success of the PT6 was tied to the King Air executive aircraft, an important factor in its acceptance and growth was its link to a new genera-

---

44. Richard H. McLachlan, interview by Rick Leyes, 7 October 1988, Pratt & Whitney Canada Interviews, transcript, National Air and Space Museum Archives, Washington, D.C.

45. McLachlan to Leyes, 7 October 1988.

46. Beauregard, "Turbine Powerplants for General Aviation."

47. Robert M. Sachs, "Small Propeller Turbines Compared with Other Power Plants in Low Speed Flight Applications," (AIAA Paper No. 64-799 presented at the AIAA/CASI Joint Meeting, Ottawa, Ontario, 26-27 October, 1964).

tion of commuter aircraft. In the 1960s, world air travel had increased dramatically largely because of the widespread use of the jet transport. Jets further increased the convenience of air travel while reducing costs.[48] Commuter airlines
shared in this growth. Between July 1964 and November 1968, the number of
scheduled air taxi operators (commuter companies) in the U.S. increased from
15 to 240.[49] Among the important factors contributing to commuter airline
growth were regulatory agency rule changes and the appearance of new and
efficient aircraft.

In 1959, the Civil Aeronautics Board (CAB), the federal agency that regulated airline passenger service, adopted a "Use it or Lose It" policy that said that
if a subsidized community could not generate at least five passenger boardings
per day, or 1,800 per year, the Board might allow the service to be abandoned.
As a result, some communities lost their certificated carriers in the 1960s. The
switch to jet aircraft by larger, certificated carriers had reduced the profitability
of short haul, low density routes. While such routes were unprofitable for larger
carriers, they presented opportunities for commuter airliners operating smaller
aircraft. In 1965, a CAB restriction preventing commuter airliners from operating on routes flown by certificated airlines was eliminated. Such regulatory rule
changes as these influenced the rapid growth of commuter airliners in the mid-
to late-1960s.

Another important factor that influenced and was influenced by this rapid
commuter growth was the appearance of new, more efficient aircraft types
that qualified under the 12,500 lb rule.[50] While the majority of the new aircraft that carried the bulk of the commuter traffic were piston powered, new
turbine-powered commuter aircraft became an increasingly important part
of this business. Two of the most widely used and enduring early turbine-
powered commuter aircraft were the de Havilland Twin Otter and the
Beech 99.

Originally designed as a STOL bushplane, the de Havilland Canada DHC-6
Twin Otter was certificated in August 1966 as a utility aircraft powered with
two PT6A-20 engines. However, the Twin Otter's ruggedness and short field
capability and the ability to haul 19 passengers made it an ideal candidate for the

48. Elizabeth E. Bailey, David R. Graham, and Daniel P. Kaplan, *Deregulating the Airlines*
   (Cambridge: The MIT Press, 1986), 17.

49. U.S. Federal Aviation Administration, *Scheduled Air Taxi Operators as of November 1968*
   (Washington, D.C.: Government Printing Office, 1969), 1-2.

50. Under Part 298 of the CAB economic regulations at this time, air taxi operators
   (commuters) could provide scheduled air service as long as they operated aircraft that did
   not have a gross take-off weight exceeding 12,500 lb.

commuter industry. By 1967, small commuter operations such as Pilgrim Airlines in New London, Conn., and Air Wisconsin in Appleton, Wis., were using Twin Otters. In 1968, Beech responded to this developing market by making available its Model 99, a stretched Queen Air made into a 15-passenger commuter airliner, and also powered by two PT6A-20 engines. The first Beech 99 went to Commuter Airlines in May 1968.

The Beech 99 and the Twin Otter contributed to the rapid growth of the commuter air travel. Largely as a result of their small turbine engines, aircraft such as these were able to operate at lower costs and carry more passengers than an earlier generation of piston-powered commuter aircraft (eg. the Beech 18). It has been argued that the development of the commuter industry as it is known today had to wait for a technological breakthrough in power plants for smaller aircraft. According to the FAA publication *Commuter Airlines and Federal Regulation: 1926- 1979*, the lightweight turboprop engine, principally the PT6, was responsible for this breakthrough.[51] While other small gas turbine engines eventually assumed an important role, the PT6, as used in the Beech 99 and Twin Otter, was the pioneer turboprop engine in the commuter aircraft industry.

EARLY OPERATIONAL EXPERIENCE IN COMMUTER AND EXECUTIVE AIRCRAFT

As the PT6 went into service, improvements were made in both commuter and executive aircraft applications. Time Between Overhaul (TBO) increased, methods of improving component longevity and durability were found, an icing problem was fixed, and other refinements made.

With the installation of the PT6 engine in commuter aircraft, the high utilization rate improved the engine more rapidly.[52] One important improvement was increased engine TBO. The PT6 entered service in 1964 with an initial TBO of 800 hours. Increases above the 800-hour mark were achieved in increments of 200 hours after evaluating a number of samples undergoing overhaul. Early Beech King Air operators typically flew 30 to 50 hours per month. However, commuter aircraft flew up to 300 hours per month. Early commuter operators started life with TBOs of 1,200 to 1,500 hours. While such limits were ac-

---

51. U.S. Federal Aviation Administration, *Commuter Airlines and Federal Regulation: 1926-1979* (Washington, D.C.: January 1980), 4-5.

52. F. R. Cowley, "The PT6 After 5 Million Hours," (SAE Paper No. 710385 presented at the National Business Aircraft Meeting, Wichita, Kansas, 24-26 March 1971), 5.

ceptable for corporate operators, they were inadequate for commuter airline operations.

Economic pressures on commuters caused them to press for rapid TBO increases, and their utilization rates provided the data necessary to convince the appropriate federal regulatory agencies to raise the limits.[53] By 1969, TBOs of more than 3,000 hours had been approved.[54] As the TBO increased over time, it could be shown that the operating cost per hour steadily declined.[55]

In 1968, as PT6-powered commuter aircraft were increasingly operated around the world, deteriorating performance problems occurred under certain conditions. Extensive corrosion in compressor sections following daily operation in salt water and heavily-polluted environments resulted in temperature limitations on engines. The solution was to incorporate a marinization (anti-corrosion) treatment in standard production and overhauled engines. Compressor washing procedures were also developed. Premature hot section deterioration, higher overhaul costs, and shorter engine life resulted from pushing the engines too hard. Experiments conducted with the operators showed that by flying at slightly reduced turbine inlet temperatures, hot section deterioration could be reduced with negligible impact on their schedules. This in turn reduced overhaul costs. Some premature hot section deterioration was traced to hot starts caused by undercharged batteries.[56] To alleviate this problem, the importance of good battery maintenance was stressed to operators.[57]

Engine icing first surfaced in executive aircraft service, and it proved both difficult and expensive to solve. Tests in the pre-certification development program had not revealed that engine icing was a problem. An alcohol anti-ice system met military specification icing conditions when tested at the National Research Council of Canada Engine Laboratory.[58] Flight tests in the Beech 18

53. C. B. Wrong, "Ten Years of PT6 Power for Commuter Airlines," (Pratt & Whitney Aircraft of Canada Ltd. Paper, Longueuil, Quebec, 10 May 1976), 3.

54. By 1990, the basic approved (DOT—Canada) TBO for most P&WC PT6 engine models was 3,500 hours. For larger PT6 turboprop engines in regional airline use where utilization was high, operators had, in general, (in conjunction with on-condition maintenance of the hot section made possible by a performance trend monitoring program) advanced their fleet TBO significantly beyond the 3,500 hour basic level.

55. Cowley, "The PT6 After 5 Million Hours," 6.

56. A hot start occurs when the engine starts and self-accelerates, but exhaust temperature exceeds prescribed limits. An undercharged battery would cause a slower than normal acceleration leading to higher than normal temperatures.

57. Wrong, "Ten Years of PT6 Power," 5-7.

58. Smith, "Development of the T-74 Engine," 5-6.

testbed and a King Air also indicated that a serviceable anti-icing system had been developed that met FAA standards.[59] Thus the PT6 went into executive aircraft service on the King Air with the certificated alcohol anti-ice system. However, soon after the PT6 had been put into service, anomalies began to occur in certain icing conditions, when enough ice would build up under the intake screen that a piece would break off and go through the engine.[60] The engine would "burp" and sometimes momentarily blow out. Occasionally, damage would be done to the compressor blades.

A major program was launched to solve the problem. The inlet guide vanes were deleted and the first-stage compressor blades were strengthened. These "ice choppers" helped, but did not resolve the icing accumulation problem. Engineer Pete Peterson conceived and tested an inertial particle separator that solved the problem.[61] The separator turned the intake air at an abrupt right angle before it reached the intake plenum. Supercooled water droplets and other particles over a certain mass exited the engine nacelle through a bypass duct. The inertial air separator was patented by P&WC in 1965, and retrofit modification kits were available for King Airs in 1966.

As solutions were found for the operational problems that cropped up, they were integrated into the next generation PT6 engines. Commuter operations were especially important in uncovering early PT6 problems, contributing to improvements in the engine, and rapidly increasing TBOs. Conversely, the PT6 engine contributed to commuter operations by lowering operating costs. In the larger picture, the shift to jet equipment, particularly turboprop and turbojet/fan aircraft, substantially reduced carrier costs.[62] By the end of the 1960s, the PT6 was well established as a pioneer turboprop commuter aircraft engine and was also well established as an executive aircraft engine.

THE SMALL PT6A TURBOPROP ENGINE FAMILY

Early important production models of the PT6A turboprop engine, certificated in the 1960s, included the PT6A-6 and the PT6A-20. A turboshaft engine certificated in May 1965 was the PT6B-9; it incorporated a single-stage reduction gear. The A-6, A-20, and the B-9 had the same maximum power and temperature capability. The A-20 differed from the A-6 and B-9 in one important

59. Sullivan and Milberry, *The Pratt & Whitney Canada Story*, 155.

60. Smith to Leyes, 6 October 1988.

61. Peterson to Leyes, 7 October 1988.

62. Bailey, Graham, and Kaplan, *Deregulating the Airlines*, 19.

respect. The A-20 had a maximum continuous power rating equal to its take-off rating.[63] The PT6A-27 turboprop engine was a growth model which was derived from and followed the A-20.

These engines were the first in a series that P&WC eventually classified as its small PT6A turboprop engine family (excluding the B-9 turboshaft engine), all broadly characterized by a single-stage power turbine. The first certificated engine in this family, the PT6A-6, has been described previously. Due to the number of highly specialized engine models in the small PT6A turboprop engine family, only the major models are highlighted.

The PT6A-20 was used on the Twin Otter, Beech 99, King Air 90 series, and other aircraft. Certificated in October 1965, it had a 550-shp take-off rating. Improvements over the PT6A-6 included a stronger, lighter, forged-titanium impeller with a significantly improved low-cycle fatigue life over earlier forged-steel impellers. A more durable inter-turbine temperature (ITT) system was developed to replace the original turbine inlet temperature (TIT) system. The PT6A-20 eventually replaced the original PT6A-6.

Demands for higher power in these aircraft led to the PT6A-27, which was certificated in December 1967 at a 680-shp take-off rating. Many significant changes were made in this model, including a larger diameter axial compressor and a new centrifugal compressor aerodynamic design, resulting in increased airflow and pressure ratio. To reduce cost, the impeller was designed to be "flank millable" instead of point-milled from the titanium forging. This approach was later built into the aerodynamic design system and was used on all P&WC centrifugal impellers. Also, the A-27 was the first P&WC engine to incorporate a pipe diffuser.[64] The pipe diffuser, replacing a multiple vane configuration, was invented by John Vrana during the company's centrifugal compressor research program and patented. It was a major step in increasing centrifugal compressor efficiency, especially as pressure ratios rose, leading to higher Mach numbers into the diffuser. While initial A-27 compressors had narrow-chord first-stage blades,

---

63. David A. Brown, "Growth Versions of PT6 Being Developed," *Aviation Week & Space Technology*, Vol. 81, No. 18 (November 2, 1964): 5.

64. This pipe diffuser straightened, diffused, and directed the airflow to the combustor through a series of expanding, tapered bores drilled in a structural annular ring, which discharged the flow into horn-shaped tubes which diffused and turned the flow to the correct conditions for entry into the combustor. Previous P&WC diffusers were of the multiple vane (also called radial vane and radial cascade) type. The multiple vane diffuser sandwiched a number of airfoil vanes between two parallel walls. The pipe diffuser reduced losses and improved compressor efficiency.

later engines of this model incorporated wide-chord blades that improved resistance to foreign object damage.

Other components of the A-27 engine were improved as well. Integrally cast turbine vane rings were incorporated in both the compressor and power turbine stages.[65] This reduced parts count, improved hot section durability, and reduced the internal leakages in the turbine area, improving efficiency. In earlier engines, nonstructural turning vanes were used at the power turbine exit, but they were subject to cracking. A "ski-jump" inner duct was developed that eliminated the vanes. A dual fuel manifold was introduced to phase in fuel delivery and to better atomize fuel to eliminate hot streaks during starting.[66]

Over time a large variety of engine models evolved in the small PT6A family. A few of the more important models utilized on popular aircraft included the PT6A-25, PT6A-34, PT6A-135, and PT6A-114.

Certificated in May 1976 with a 550-shp take-off rating, the PT6A-25 was an aerobatic derivative of the PT6A-27 with PT6A-20 first-stage reduction gearing. The oil system was modified for inverted flight and other aerobatic maneuvers. The A-25 was used on the Beech T-34C, and the A-25A powered the Pilatus PC-7. Both aircraft were military trainers. Introduced in the 1970s, the Beech T-34C was a standard trainer aircraft for the U.S. Navy and was used by military services around the world. The Pilatus PC-7 was also an important trainer aircraft, and by the 1980s, it had become one the world's leading military trainers.

The PT6A-34, with a 750-shp take-off rating, was also a PT6A-27 derivative. It incorporated air-cooled gas generator turbine vanes which allowed a higher temperature. This engine was certificated in November 1971, and used on a variety of airplanes including the extremely popular, twin-engine Embraer Bandeirante EMB-110 and -111 commuter aircraft models and IAI Arava aircraft.

Also part of the small PT6A turboprop family were the Century (100) Series engines. One of these engines was the PT6A-135, a PT6A-34 derivative with an increased ratio gearbox for 1,900 propeller rpm. The purpose for this was to keep the power turbine speed up at a given condition, thus increasing cruise power at a lower propeller speed. Turbine improvements were also made in this engine. Certificated in January 1978 with a 750-shp take-off rating, the A-135 was used in a variety of aircraft, among them the Beech Model F90 King Air.

---

65. J. P. Beauregard, "Progress Report on a Small Turbine for STOL Aircraft and High Speed Surface Vehicles," *Canadian Aeronautics and Space Journal*, Vol. 14, No. 1 (January 1968): 18.

66. Wrong, "Ten Years of PT6 Power," 17-19.

Seating 7 to 10 occupants, the Model F90 was a sequential step-up in the King Air series.

In the early 1980s, an important new utility aircraft was introduced. Capitalizing on the small-package over-night delivery industry pioneered by Federal Express, Cessna developed the Model 208 Caravan I. A popular, large, single-engine utility aircraft, the Caravan was equipped with a 600-shp PT6A-114 engine. This engine was certificated in January 1984 and went into production in April 1984. It was a derivative A-34/A-36 engine with a single port exhaust and a 1,900-rpm low-speed reduction gearbox.

THE LARGE PT6A TURBOPROP ENGINE FAMILY

Market surveys in the late 1960s showed that larger business aircraft and new commuter aircraft were being planned, so the PT6A-40/50 series was launched. The A-40/50 series was the start of what became known as the large PT6A turboprop engine family, importantly distinguished by a two-stage power turbine. Other series in this family included the PT6A-60, PT6A-65, and PT6A-67. Due to the number of highly specialized engine models in the large PT6A turboprop engine family, only the major models are highlighted.

PT6A-40/50 SERIES

In October 1973, the PT6A-41 became the first in the series of large PT6A engines to receive certification. Flat rated at 850 shp (with thermodynamic capability of 1,120 shp), it featured a completely redesigned compressor with higher axial stage loadings (pressure ratio per stage) and a higher airflow. In redesigning the axial compressor, the outer diameter was held constant, in contrast to earlier PT6 axial compressors, which had a constant inner diameter. Two inter-stage bleed valves helped to achieve the more difficult stage matchings resulting from the higher pressure loadings. A second power turbine was added to the engine along with a redesigned reduction gearbox to handle the higher power. The A-41 was originally built for the Beech Super King Air 200, which proved to be a crowning achievement for Beech Aircraft in the decade of the 1970s as one of the most popular turboprop airplanes in the world.[67] The A-41 was also used in the Piper Cheyenne III aircraft.

Following a 1972 Civil Aeronautics Board rule change, commuter air carriers were allowed to operate aircraft carrying up to 30 passengers. The Shorts 330

67. Phillips, *Beechcraft Pursuit of Perfection*, 65.

De Havilland Canada Dash-7 regional transport aircraft powered by four 1,120 shp P&WC PT6A-50 engines. Photograph courtesy of P&WC.

commuter aircraft was designed to take advantage of this new ruling and was in service with two 1,173 shp PT6A-45A engines by August 1976. The A-45A engine had a new 1,700-rpm, higher power reduction gearbox and was the first PT6 engine to have water injection to augment power output. The A-45A also powered the Mohawk 298, a modified Nord 262, used by Allegheny Airlines. The A-45A was certificated in April 1976.

The 1,120-shp PT6A-50 was designed at the same time as the A-41 and A-45 for use on the de Havilland DHC Dash-7, a four-engine 50-passenger STOL commuter aircraft. The A-50 was the first PT6 engine to be installed in an airplane operated under full airline rules (FAR 25 and FAR 121).[68] Aerodynamically, it was identical to the A-41 and A-45, but with differences in its lower rpm gearbox, mounting system, and structure. These reflected the unique design and mission of the Dash-7 aircraft. A new flying testbed was needed, so a Vickers Viscount was purchased from Air Canada and the A-50 mounted in the nose. The A-50 was certificated in September 1976.

The Dash-7 was intended for operation between city centers: it had to be quiet and have STOL performance for short runways. Slow turning, large-

---

68. Previously, PT6 engines had been installed in aircraft which operated under FAR 23 and CAB Part 298 regulations. Essentially, earlier PT6 engines had been used in commuter aircraft as opposed to aircraft which met airline regulations.

diameter propellers would contribute to low noise levels and STOL performance. Part of the lift for the STOL performance would come from prop wash over a large part of the wing. To lower the propeller speed (to 1,210 rpm), a new higher ratio reduction gearbox was designed. To isolate propeller vibration from the airframe, a new two-plane mounting system was required (normally PT6 engines had a single-plane mount) in which the engine was soft mounted immediately behind the propeller and at the rear of the engine. The engine structure was redesigned to take a different loading pattern, including a strengthening of the struts to resist the large torque. Among other changes, the exhaust case was repositioned to direct exhaust gases over the top of the wing, reducing soot deposition on the wing and nacelle and shielding exhaust noise from the ground.

PT6A-60 SERIES

In the early 1980s, P&WC introduced the PT6A-60 series for higher-performance business aircraft. The A-60 series was a parallel development to the A-65 series (a series of engines designed for larger commuter aircraft and certificated prior to the A-60 series), but had a lower thermodynamic power-to-weight ratio. The A-60 engines were derived from the PT6A-41.[69] Among the engines in this series were the: A-60; A-61; A-62.

The A-60 had a 1,050-shp take-off rating. Among the changes (from the A-41) incorporated in the A-60 were an increased cruise power rating, a new flame tube, a larger A-41 compressor, an A-45 gearbox, and a Woodward hydromechanical fuel control. The A-60 was the first to use P&WC's "jet flap," a device that introduced air swirl into the compressor inlet at low speeds to provide easier starting and low-speed acceleration. The jet flap was used in all subsequent PT6 growth models. The A-60 was certificated in November 1982 and was used on the Beech King Air 300.

The A-61 was certificated in November 1982 and installed on the Piper Cheyenne IIIA aircraft. The A-61 had an 850-shp take-off rating and used an A-41 gearbox. The A-62 (950-shp take-off rating), an A-61 with an aerobatic oil system, was certificated in February 1985 and installed on the Pilatus PC-9 military trainer aircraft.

69. There was an intermediate engine model known as the PT6A-42 which followed the A-41, but preceded the A-60. The A-42 was rated at 850 shp, was first produced in September 1979, and was used on the Super King Air B200, Beech 1300, and Beech C-12F.

PT6A-65 SERIES

Late in 1977, P&WC approved development of the PT6A-65, an important series in the PT6A large engine family, designed in response to market research that anticipated new growth in commuter airlines. Aircraft that could carry more passengers farther and faster than existing 19-seaters were foreseen.[70] To get the required extra power without making costly alterations, P&WC decided to design an all new, four-stage axial compressor, with a new centrifugal stage. Pressure ratio and airflow were increased, yet only 1.98 inches were added to the overall length. A new compressor turbine was designed with increased blade speed using new-technology airfoils. These combined to give the increased work needed, yet at increased efficiency. A new fuel control was added along with an automatic power recovery system that would cut in if an engine failed on takeoff.

The lead customers for the A-65 series were Shorts and Beech. The initial engine in the A-65 series was the A-65R, designed for low-altitude operation, and certificated in August 1982. With a take-off rating of 1,376 shp, the A-65R was installed in the Shorts 360, a 36-passenger commuter aircraft, while a derivative version, the A-65B, was designed for high-altitude operation. The A-65B was similar to the A-65R but had no automatic power recovery system. Rated at 1,100 shp and certificated in August 1982, the A-65B powered the Beech 1900, a 19-passenger commuter airliner designed to meet the increased commuter traffic following airline deregulation.

PT6A-67 SERIES

The next major series of PT6A large turboprop engines was the PT6A-67 series, derived from the PT6A-65B. The engine's compressor was redesigned and delivered 10% more airflow through the use of broadened and lengthened airfoils. The turbine was redesigned as well to accommodate the increased airflow. The A-67, the initial model, was certificated in January 1987 at a 1,100-shp take-off rating. A number of subsequent PT6A-67-series models were developed, including the 1,200-shp A-67B powering the Pilatus PC-12 utility aircraft, the 1,279-shp A-67D for the the Beech 1900D commuter aircraft, and the 1,424-shp A-67R installed in the Shorts 360-300 regional airliner.

Among the first aircraft to fly with the A-67 engines was the Beech Starship, an innovative composite business aircraft that earned considerable publicity in the 1980s. Two PT6A-67A engines were installed in a pusher configuration on

70. Sullivan and Milberry, *The Pratt & Whitney Canada Story*, 189.

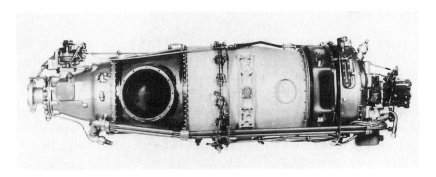

P&WC PT6A-67 turboprop engine. Photograph courtesy of P&WC.

the Starship, but it was not a commercial success. Aircraft performance fell substantially short of predictions, development was lengthy and costly, and sales
were poor. In 1995, Raytheon Aircraft Co. (successor to Beech Aircraft Corp.)
terminated production of the Starship.[71]

The PT6A-68 series was a specialty application developed for two aircraft
candidates in the Joint Primary Aircraft Training System (JPATS) competition
in the mid-1990s. The Beech/Pilatus PC.9 Mk.II and the Northrop/Embraer
Tucano, the only two turboprops in the JPATS competition among a bevy of
turbojet- and turbofan-powered competitors, were powered by the PT6A-68,
which had a 1,250-shp take-off rating. The A-68 series had an A-67 core with a
2,000-rpm gearbox, inverted-flight capability, and a full-authority digital electronic control. In mid-1995, the Air Force announced the Beech/Pilatus PC.9
Mk.II as the winner of the JPATS competition with a plan to purchase as many
as 711 of the aircraft over a 22-year period. The new PT6-powered PC.9 was to
replace the aging fleet of Air Force Cessna T-37B turbojet and the Navy Beech
T-34C turboprop trainers.[72]

The PT6A large turboprop engine family was very successful for P&WC
and gained many important executive and commuter aircraft applications. The
maximum power output of the small PT6A turboprops was basically limited
by their single-stage power turbine. The PT6A large turboprop family's two-
stage power turbine and numerous other changes made for specific airframe

71. Edward H. Phillips, "Poor Sales Kill Beechcraft Starship," *Aviation Week & Space Technology*,
    Vol. 141, No. 24 (December 12/19, 1994): 47-49.

72. Alton K. Marsh and Peter A. Bedell, "Beech/Pilatus Trainer Selected for $7 Billion
    Pentagon Contract," *AOPA PILOT*, Vol. 38, No. 8 (August 1995): 30.

applications kept the PT6 highly competitive in new as well as derivative air-craft markets.

A SPECIAL PT6A TURBOPROP APPLICATION: SPRAYING AND DUSTING

By the mid-1970s, the PT6 was well-established in the business aircraft and commuter aircraft markets and had made some headway into the twin-engine helicopter market. The agricultural market was an area that the company had not yet considered, according to P&WC Director APU Business and Special Projects Derek Emmerson.[73]

One day in 1974, an aerial applications operator came into P&WC's overhaul shop to have an R-1340 crankshaft replaced and was shocked to find that a new one would cost him more than a newly overhauled engine. After discussing this and related engine problems with the operator, Emmerson realized that an opportunity had presented itself. As he recalled, ". . . perhaps we could replace those engines in that market. . .there was a very serious need simply because those engines were up to 45 years old. It was getting difficult to overhaul the engines because we weren't making spare parts for them any longer."[74]

Further investigation confirmed the need and interest on the part of operators. P&WC approached manufacturers of agricultural aircraft, but found that they were not interested because the cost of a turboprop engine was more than that of the entire airplane. Returning to the operators, Emmerson did a detailed survey of large-scale aerial spray operations along the Gulf Coast of Texas and Louisiana. As a result of a visit and an off-the-cuff sales pitch about the benefits of the turbine engine to one receptive operator, a deal was struck whereby P&WC lent a PT6A-34 engine, and the operator provided a Grumman Ag-Cat and the money for the engine conversion.

By spring 1976, this Turbo-Cat was flying and operational data was collected showing that the turbine-powered aircraft had a 60% greater productivity than that of comparable piston-powered aircraft.[75] The reduced weight of the engine meant potentially bigger payloads. The PT6 engine was more powerful, more reliable, and had a longer TBO than the Wasp. The speed and power of the turbine airplane meant that pilots could turn around much faster for another

73. Derek C. Emmerson, interview by Rick Leyes, 6 October 1988, Pratt & Whitney Canada Interviews, transcript, National Air and Space Museum Archives, Washington, D.C.

74. Emmerson to Leyes, 6 October 1988. Actually, P&WC stopped manufacturing spare piston engine parts in 1978.

75. Sullivan and Milberry, *The Pratt & Whitney Canada Story*, 181.

pass. Pilots could fly longer hours because the smooth, relatively quiet turbine greatly reduced fatiguing engine noise and vibration. Dispersal patterns were much better and less overlapping was required because the smooth, small-diameter nose of the aircraft reduced air turbulence behind the engine.

Overhead could be reduced by reducing fleet size, number of pilots, and maintenance costs. An economic analysis of the data collected suggested that two Turbo-Cats could accomplish the same work as three Ag-Cats in less flying time per year.[76] The additional investment required for the turbine-powered aircraft could typically be paid back in two and one-half years.

The benefits of the turbine-powered agricultural aircraft were soon realized by operators. Conversion companies were the first to offer turbine engine installations and later aircraft manufacturers offered them. In the 1970s and 1980s, PT6A-11AG, -15AG, and -34AG engines found their way into a number of ag planes, including among others, the Ayres Turbo Thrush, Frakes Turbo-Cat, Air Tractor AT-502, and Grumman (later Schweizer) Ag-Cats. Because of the rugged operating conditions, the engines had to be made more corrosion resistant and stronger. High capacity, self-cleaning vortex generator filters and replaceable barrier filters were devised to provide cleaner intake air. A deluge-type compressor wash ring and a chromate-coated gas generator and impeller shroud were added to reduce corrosion.

PT6B TURBOSHAFT ENGINES

In the early 1960s, P&WC put significant effort into developing a turboshaft version of the PT6 in anticipation of single-engine helicopter applications described previously. In May 1965, it certificated the PT6B-9 turboshaft engine. The B-9 engine was used in several experimental helicopters, including the Lockheed 286 (military designation XH-51A), but never went into volume production. The company's efforts to find a major customer for a PT6 turboshaft engine were not successful until 1970 when its first certificated TwinPac engines (dual turboshaft engines driving into a combining gearbox) were delivered.

P&WC did not resume a concentrated effort on the PT6B series turboshaft engine until 1976, when the PT6B-34, derived from the B-34, was provisionally certificated at 750 shp. It ran in a prototype Westland Lynx helicopter, but was

76. Derek C. Emmerson, "An Economic and Technical Perspective of the Turboprop Engine in Ag-Aviation," (Paper presented at the Canadian Aeronautics and Space Institute annual general meeting, Quebec City, Quebec, 18 May 1977), 19-20.

not produced. The next attempt was the PT6B-35F, a PT6A-135 with a single port exhaust duct suitable for left- and right-hand installation. It was certificated in February 1982 at 650 shp and ran in the Learavia Learfan prototype, a pusher turboprop aircraft that did not go into production.

The PT6B-36 was a PT6T-3B TwinPac power section with a modified exhaust assembly, new reduction and accessory gearboxes, and a dual-channel full authority digital electronic control (FADEC). The first production date of the B-36 engine was June 1984. It was certificated in September 1984 with a 960-shp take-off rating for the Sikorsky S-76B helicopter.

PT6T TWINPAC TURBOSHAFT ENGINES

In the mid-1960s, P&WC explored various opportunities for PT6 multiple-engine helicopter installations. Important selling points for multiple-engine helicopter use were increased power and safety of flight. In 1967, in response to increased market interest, P&WC began design of a helicopter engine which consisted of two PT6 power sections coupled to a single output gearbox. This engine was known as the PT6T-3 TwinPac.

P&WC's decision to launch the TwinPac was influenced by commercial and military market opportunities. In the 1960s, Bell Helicopters was exploring multiple turbine engine installations in helicopters with various engine companies, including P&WC and Continental Aviation and Engineering (CAE). Bell's first twin-turbine helicopter, a U.S. Army YUH-1D powered by a CAE T67-T-1 twin turboshaft engine, flew in 1965. Notwithstanding CAE's initial success with Bell, P&WC continued to offer Bell a twinned PT6 with a combining gearbox. In negotiations, one competitive advantage that P&WC had was that the PT6 engine was a developed and certificated engine with substantial flight time behind it. The Canadian government had also agreed to support the R&D effort on such an engine. In October 1968, Bell announced its Model 212 (military UH-1N) with twinned PT6 engines.[77]

A military market for twin-turboshaft engines provided further incentive for P&WC. During 1968, the U.S. Navy held a competition for such a power plant, which was won by P&WC. Also in 1968, the Canadian military placed an order for TwinPac-powered versions of the Bell UH-1N.

Winning U.S. military contracts was not easy for P&WC. There was resistance in the U.S. to buying "foreign" engines in spite of the fact that P&WC was wholly owned by UTC. It became clear that the TwinPac would have to be

77. Sullivan and Milberry, *The Pratt & Whitney Canada Story*, 231.

"made in the U.S.A." With guidance from U.S. congressional representatives, P&WC chose to locate a subsidiary in West Virginia. Legal considerations led to Pratt & Whitney West Virginia (P&WWV) becoming a subsidiary of UTC. Engines for the U.S. Navy, Air Force, and Marine Corps were built at P&WWV.

The preliminary design of the TwinPac began in April 1967. From 1968 through 1970, PT6 development work at P&WC concentrated on the TwinPac. The power section was based on the PT6A-27. Increased power was achieved by increasing the turbine inlet temperature. This in turn required the development of a cooled HP (high pressure) turbine vane ring and the use of a new higher-temperature alloy for the HP turbine blades. The PT6T-3 was the first PT6 to incorporate cooled turbine stators and the first P&WC engine to use directionally solidified turbine blades.

The most challenging part of the TwinPac was developing the gearbox and the controls. Strong, lightweight, and highly loaded gears were required to reduce engine speed from 33,000 rpm to 6,230 rpm on a single output shaft; a number of redesigns were required to achieve reliability. Separate oil systems were needed for each power unit and the gearbox so that the failure of any one unit would not contaminate the other. The development of effective rotating oil seals was a problem. Developing a suitable engine control proved to be a major undertaking. A governor, which measured rotor shaft rpm and in turn regulated engine speed through the fuel control, had to be built. In order for the engines to work together, a torque equalizing control was developed that monitored output speeds and adjusted output from one power section to compensate for any loss from the other. The unique configuration of the engine required additional certification tests, which further added to the complexity of the development.

The TwinPac engine first ran in June 1968 and was flown in a Bell 205 test-bed in May 1969. The PT6T-3 was certificated with a 1,800-shp take-off rating in July 1970, and initial deliveries took place the same year. The PT6T-3 was installed in the Bell Model 212, Augusta Bell 212, and Sikorsky S-58T helicopters.

Military derivatives of the PT6T-3 included the T400-CP-400, T400-CP-401, and T400-WV-402. Passing its MQT in March 1970, the -400 was a U.S. Navy, U.S. Air Force, and Canadian armed forces version of the PT6T-3. The -400 incorporated aluminum castings (commercial engines used magnesium castings for weight reduction); it was installed in the Bell UH-1N Iroquois (Huey), Bell AH-1J SeaCobra, and Bell CUH-1N (for the Canadian Armed Forces) helicopters. The -401 was a U.S. Army designation for the -400. The conversion to the -401 configuration took place in June 1972. The -401 was used in the Army VH-1N helicopter. Passing its MQT in February 1975, the -402 was a U.S. Navy designation for an uprated version of the -400 engine. The -402 had hot-end improvements to accommodate increased power and had a power turbine overspeed protection system. The -402 powered the Bell AH-1J and Bell AH-1T attack helicopters.

The commercial demand for higher power led to the introduction of a civil version of the -402, designated the PT6T-6. With a take-off rating of 1,875 shp, this engine incorporated hot-end improvements to accommodate the higher power. The PT6T-6 was certificated in December 1974 and used in the Augusta Bell Model 212 and Sikorsky S-58T helicopters. In August 1979, another version was certificated, designated the PT6T-3B, a basic PT6T-3 with the hot-end improvements used in the PT6T-6. The T-3B was designed to permit higher output with single power section operational. The T-3B was used in the Bell Model 212 and 412 transport helicopters.

Later models included the 1,875-shp PT6T-6B and 1,800-shp PT6T-3BE, both powering the Augusta Bell Model 412 helicopter. In August 1993, the PT6T-3D was certificated at 1,800 shp. The PT6T-3D powered the Bell Model 412HP and 412EP helicopters.

U.S. Huey and SeaCobra helicopters, powered by PT6T TwinPac engines, were used in combat during the Vietnam War. The twin engines provided extra power and margin for survivability. Bell Model 212s, essentially twin-engine civilian adaptations of UH-1 Hueys, were 15-place executive, corporate, and utility transport helicopters. Model 212s were used extensively to transport personnel to and from off-shore oil operations. Again, the TwinPac provided extra reliability for extended overwater operations as well as increased payload. The success of single-engine, single-rotor helicopters converted to dual-engine installation proved to be an important interim measure while the next generation of helicopters, designed from the start with twin engines, were developed.

IMPROVING THE PT6 WITH TECHNICAL INNOVATION

In 1963, the first production engine, the PT6A-6, had a take-off rating of 550 shp and an sfc of 0.65. The PT6A-65AR certificated in February 1984 had a 1,424-shp take-off rating and an sfc of 0.509.[78] In 21 years, the power output of the engine had been increased by 159% and the sfc reduced by 22%. Another measure of turbine engine excellence is power-to-weight ratio. At 550 take-off shp and weighing 270 lb, the PT6A-6 had a 2.04 power-to-weight ratio. In

78. The PT6A-65AR was used on the Shorts 360, Shorts Super Sherpa, and AMI DC-3.

comparison, the PT6A-65AR, rated at 1,424 shp and weighing 486 lb, had a power-to-weight ratio of 2.93 demonstrating a 43.6% improvement in power-to-weight ratio over the PT6A-6. These performance improvements were accomplished within the same 19" PT6 engine diameter. Later developments had even higher take-off power ratings.

In order to achieve higher power-to-weight ratio and lower sfc, the airflow through the engine, the cycle pressure ratio, and the turbine inlet temperature were increased. Along with these steps, increases in efficiency of both the compressor and turbine were required.

One of P&WC's targeted goals was to improve the efficiency of its compressor, which was relatively low in the early 1960s. Early production PT6 engines had a 6.3:1 compressor pressure ratio and a 5.4 lb/sec mass flow.[79] By comparison, 21 years later, the PT6A-65AR compressor had a pressure ratio of 10.8 and an airflow of 9.7 lb/sec.

The pressure ratio of the axial-flow compressor was increased by using improved aerodynamics that allowed more work to be done by each stage, along with increasing the average blade speed (not the rpm).[80] The higher blade speeds required a change from steel to titanium throughout the compressor. Increased airflow was achieved by reducing the hub-to-tip-diameter ratio in the front stages, thereby increasing the cross-sectional area at the compressor inlet. All these steps drove up the Mach number through the centrifugal compressor and required development based on research and more accurate analytical modeling of the flow.

With higher pressure ratio centrifugal compressors came a high degree of swirl, supersonic flow, and compressibility problems in the diffuser at the compressor outlet. Some methods used by P&WC to overcome such problems included backswept blades, zero prewhirl inducer designs, increasing the number of impeller blades, and adding splitters. In addition, John Vrana's invention of the pipe diffuser contributed significantly to reducing diffuser losses. Original PT6 models used a vaned radial diffuser and its losses were relatively high.[81]

79. Smith, "Development of the T-74 Engine," 2.

80. Early PT6 engines had a constant diameter axial compressor rotor with a tapered outer casing. Starting with the PT6A-40, the diameter of the outer casing was held constant and the compressor rotor diameter was increased at the rear stages and reduced at the front stages. This increased the compressor blade diameter of the rear stages, which increased blade tip speeds and resulted in an increased pressure ratio per stage.

81. Colin B. Wrong, "Turbine Engine Design," (AIAA Paper No. AIAA-81-0915 presented at the AIAA Meeting and Technical Display "Frontiers of Achievement," Long Beach, California, 12-14 May 1981), 3.

Summarizing the importance of the pipe diffuser, retired Vice President of Engineering Colin Wrong said:

> We already knew from the experiments that we had done on the rigs that we could get a minimum of 5 or 6 points in efficiency from these new components relative to the old ones, and in due course, it became more like 10 points, really. So, I mean it made a massive difference to the performance of the compressor, and therefore, the performance of the engine.[82]

To reduce cost, weight, and mechanical complexity, highly loaded power turbines were typically used in PT6 development efforts. In turbine engines, power output and fuel efficiency were improved by raising turbine temperature, but temperature was governed by material limitations. Several steps were taken to permit higher turbine temperatures. One was to provide cooling air to the first-stage turbine stator vanes. Another was to design highly loaded turbines in which both the temperature and pressure drop across the turbine rotor were high. Increasing the temperature drop through the turbine permitted the inlet temperature to be raised without raising the average gas temperature through the turbine, thus keeping the temperature of the blades relatively unchanged and eliminating the need to cool them.[83]

Early PT6 engines had only one power turbine stage, which was adequate for the power level required. However, in order to raise the engine power output substantially within the same 19-inch engine diameter, a second power turbine was needed. As power level increased, the single turbine became too highly loaded. The swirl from the single stage was high, increasing exhaust losses, and its efficiency was reduced. The addition of a second turbine allowed the higher power outputs while eliminating these problems. The PT6A-41 was the first engine model so equipped. For P&WC, the PT6A-41 was a benchmark because it marked the transition to the PT6 high-power turboprop engine family.

Another important development in the PT6 turbine section was integrally cast turbine stator vane rings for the compressor and power turbine stages. Early PT6 turbines used individual castings for each vane; this resulted in gas leakage past stator vane platforms. An integrally cast turbine stator ring could have eliminated this leakage problem. However, in the 1960s, such integral rings were considered a development risk because of the possibility of cracks due to ther-

---

82. Colin Wrong, interview by Rick Leyes and William Fleming, 16 January 1991, tape recorded conversation, National Air and Space Museum Archives, Washington, D.C.

83. Hardy to Leyes, 6 October 1988.

mal transients and temperature gradients.[84] In time, the problems were resolved, and P&WC developed successful cast vane rings. The resulting HP turbine efficiency was improved approximately three points. The manufacturing costs of these components were reduced as well.

While P&WC did not do basic materials development, it did benefit from materials development research done at P&WA. One such area was the application of directionally solidified turbine blades first used in the T-402 TwinPac engine. The grain was not randomly oriented but uniform along its length, resulting in a stronger blade, capable of withstanding higher temperatures. In their first experiments, P&WC engineers found that the first-stage turbine blades in the HP rotor failed in creep in spite of the new material because the blades were running too hot. To resolve that problem, the engineers reduced the centrifugal force and thus the stress on the rotor blades by tapering them, thereby reducing their mass. The blades could then withstand the increased temperature with no blade creep and little loss in efficiency. It is interesting to note that the successful tapered blade approach was derived by the "eyeballing and testing it" procedure rather than any analytical work done on a computer.[85]

As a general commentary on PT6 component technology, it is worthwhile mentioning that the PT6 set design precedents used in several subsequent P&WC engine families. Among these were the use of a centrifugal compressor, reverse-flow annular combustor, single-stage uncooled gas generator turbine, free power turbine, and a compressor bleed valve system (instead of a variable geometry compressor). Later P&WC engines also incorporating this component technology included the JT15D turbofan, PW100 series turboprop, PW200 series turboshaft, and PW901A APU.

SUMMARIZING THE PT6 ENGINE

In 1990, the small turboprop engine family, broadly characterized by a single-stage power turbine, ranged in power from 550 to 750 shp. The large turboprop engine family, with a two-stage power turbine, ranged from 850 to more than 1,400 shp. There were also two PT6 turboshaft series: the PT6T TwinPac, comprising two power sections driving a power output shaft through a single gearbox; and the PT6B, a single PT6 power section.

The PT6 also had numerous non-aircraft applications. As early as January 1966, P&WC had formed its Industrial and Marine Division. This division was

84. Peterson to Leyes, 7 October 1988.

85. Peterson to Leyes, 7 October 1988.

responsible for the ST6, a derivative PT6A-27 engine largely intended for industrial and marine applications. Engines of this family incorporated corrosion protection and aluminum castings and were manufactured for a variety of experimental and some production purposes. Marine versions powered or were used in applications including naval vessels, small boats, hovercraft, and hydrofoils. Industrial versions were tried in a wood chipping unit, snowplow, trucks, oil pump, and a train. One version powered race cars at the Indianapolis 500 in 1967 and 1968. The widest application for the non-aircraft engines was in oil and gas fields powering gas pumping compressors.

From 1963 through the end of 1995, 25,234 PT6A turboprops and 7,238 PT6T/B/C turboshafts engines were delivered. By March 1991, there were 122 types of turboprop aircraft in which PT6 engines had been installed.[86] A total of 63 individual PT6A, PT6B, PT6T, and ST6L engine models had been certificated since 1963. As of January 1989, there were 12 certificated or military qualified airframe applications for PT6 turboshaft engines. Aircraft PT6 engines were exceptionally well established in the corporate, regional transport, training, and utility markets.

As a result of the number of engines produced, the number of applications, and the length of production, by the early 1990s, the PT6 was ranked among the most successful small turbine engines ever manufactured in North America. For that reason, it is important to identify the principal reasons for its success and to enumerate the contributions made by the PT6 to the general aviation industry and also to P&WC itself.

The success of the PT6 engine program can be attributed to a number of factors. According to retired P&WC Chairman of the Board Elvie Smith: "It was the right product at the right time. We correctly perceived that there was a real market niche for a small turboprop, so we got it on the market at the right time and made it work. We spent a whole lot of money afterward fixing the problems."[87] Smith felt that the post-certification development work encouraged by P&WA in solving problems and improving the engine were key components in the success of the PT6. P&WC was also able to take advantage of P&WA's worldwide product support network. P&WC trained P&WA technicians to look after routine PT6 problems and also supplied the product support network

86. Pratt & Whitney Canada, *PT6 Status and Information Report*, March 1991. This included certificated aircraft and some non-certificated aircraft, but did not include helicopters. The total number of PT6 aircraft applications was a higher figure. However, the total number of non-certificated, experimental applications in which the PT6 was used could not be easily determined from published information.

87. Smith to Leyes, 6 October 1988.

with its own PT6 technicians. Contributing to this combination of factors was the fact that the engine was reliable and that it worked. According to Smith: ". . . it (the PT6) has an absolutely outstanding reputation for reliability. The average over the whole small PT6 fleet now (in 1988) is about one in-flight shutdown for every 350,000 hours."[88] By 1995, this had improved to one in every 500,000 hours.

Marketing style was another factor that contributed to the success of the PT6. An important aspect of marketing was working closely with airframers on a proposed engine installation. According to Ken Sullivan who did much of the marketing to airframers: "There was no magic formula to selling PT6 engines. A sale was often the result of working with many different airframer engineers and managers for months on technical details and price."[89] In the case of larger PT6 engines, responsiveness to airframers' needs for improved performance engines was an important driver in P&WC's marketing strategy. More powerful and fuel efficient engines were needed, but engine weight, mounting installation, and price were parameters that were limited by airframe or market constraints. The willingness to accommodate such requirements and invest heavily in designing new engine models was considered an important contributing factor to the PT6's success in the marketplace.[90]

While competition for the PT6 market got tougher in later years, up until the late 1960s, the PT6 did not have any real competition in its power class. For P&WC, the closest turbine engine models competing were the 250-shp Allison T63 (civilian Model 250), the Boeing 350-shp T60 (civilian Model 520), and the Turboméca Astazou II rated at 530 shp and Bastan rated at 750 shp.[91] Garrett's 575-shp TPE331, under development at the same time as the PT6, was certificated in February 1965, a little more than a year after the PT6. In time, the TPE331 would make a significant penetration of the general aviation market and directly challenge the PT6.

The PT6 was selected for the King Air, the first small twin-turboprop-powered airplane in the U.S., giving it a significant competitive advantage. The PT6's principal initial competition came from considerably cheaper turbo-

---

88. Smith to Leyes, 6 October 1988.

89. Sullivan to Leyes, 5 October 1988.

90. McLachlan to Leyes, 7 October 1988.

91. The Teledyne CAE 525-shp T72 was originally intended as a turboprop engine for general aviation aircraft, but was converted to a turboshaft engine for the Assault Support Helicopter competition. The development effort on this engine was discontinued following the competition, and therefore, the T72 was not a competitor after this time.

charged reciprocating engines. However, those who could afford to pay more for the speed, reliability, and lower noise and vibration of turbine-powered aircraft did so, one of the reasons for the success of the King Air and its PT6 engine.

In the broader picture, the PT6 was both a beneficiary of and a contributor to the general aviation boom of the mid-1960s. In 1965, with 11,852 aircraft shipments and 15,768 shipments in 1966, general aviation manufacturers were making record sales.[92] Assisted by these boom years, PT6A sales rose from 103 engines in 1964, the first year of production, to 564 engines in 1966. The PT6 also benefitted from the dramatic growth in commuter airlines between 1965 and 1968. The fact that the PT6 was selected for the popular Twin Otter and Beech 99 was a critical element in the PT6's success. Conversely, the fact that the PT6 was available for these new aircraft and that it performed well operationally were factors contributing to their success.

In the early 1970s, the general aviation market went into a sharp decline, and aircraft shipments fell to 7,292 in 1970 and 9,774 in 1972. Following the market, PT6A turboprop sales also fell, declining to 223 in 1970 and 230 in 1971. However, the PT6 TwinPac turboshaft engine picked up the slack for P&WC. Of the 762 PT6A/B/T sales in 1970, 539 (71%) were turboshafts. PT6 turboshaft production jumped again to 1,080 units in 1975 when off-shore oil exploration was booming as a result of the 1973 Arab oil embargo, and operators were looking for reliable twin-engine helicopters.

Executive and commuter aircraft business continued to grow during the 1970s. That growth was accompanied by rising PT6 production as new applications were found for the engine, and growth versions continued to power newer models of already successful aircraft. The effects of the general aviation recession (decline in aircraft shipments), starting in 1979, did not stop the growth of turbine aircraft sales until 1982, partly as a result of U.S. airline deregulation and additional growth in the commuter market. PT6A/B/T engine production peaked in 1981 at 2,815 engines, but, with the economic crash in 1982, fell to 997 engines in 1983. Annual production of PT6A/B/T/C engines from the mid-1980s to the mid-1990s ranged from just over 700 to less than 1,000 engines.

In summary, PT6 sales were a reflection of the condition of the economy and the growth of the commuter and general aviation markets. The engine became the main driver in the growth of P&WC. In the late 1950s, when P&WC made its first steps toward the development of turbine engines, the company had built

---

92. General Aviation Manufacturers Association, *General Aviation Statistical Databook* (Washington, D.C.: General Aviation Manufacturers Association, 1989), 6.

its financial and manufacturing base largely on the aircraft piston engine business. Primarily with the profits generated from P&WC's piston engine business and through Canadian government sponsorship, P&WC was able to underwrite the cost of the PT6 pre-certification development program.

In the 1960s, the piston engine spare parts business continued to support PT6 engine development and manufacturing, but by 1969, the PT6 had begun to contribute to company profits. Gradually, piston engine component production declined, and in 1978, P&WC stopped manufacturing these spare parts altogether.[93] The company's transformation to a turbine engine manufacturer was complete. P&WC had made the transition from a manufacturer of piston engines and parts as a branch-plant licensee to being an integrated manufacturer of gas turbine engines of its own design for worldwide sale. The PT6 engine program was principally responsible for this transformation.

The JT15D

As early as 1958, P&WC had been exploring the possibility of a turbofan engine. However, this project had been set aside as a result of the priority given the PT6 program. In the mid-1960s after the PT6 program was underway, P&WC began to look at other possibilities that would broaden their product line. Marketing analysis showed that a gap existed between business turboprop aircraft such as the 270-mph King Air and the 500+ mph Learjet 23, a pure turbojet aircraft. While existing turbojet corporate aircraft were fast, they were also expensive, burned a lot of fuel, and required highly skilled pilots. An intermediate speed (400-mph) entry-level fanjet offered the alternative possibility of lower cost and lower fuel burn than turbojets and more speed than turboprops.

P&WC proceeded with a design study designated DS-32, an 1,800-lb-thrust fanjet engine. The company also discussed this engine with various airframers. At the time, Cessna was investigating possibilities for a turbine-powered aircraft that would give it access to the market the Beech King Air was beginning to monopolize—the market between the turboprop and turbojet.[94] Subsequently, in early 1966, Dwane Wallace, Chairman of the Board and CEO of Cessna Aircraft called Bill Gwinn, President of United Aircraft Corporation explaining that Cessna wanted to build a small jet, and that they wanted a Pratt & Whitney

93. By 1991, P&WC still supplied some reciprocating engine parts manufactured by subcontractors for P&WC.

94. Sullivan and Milberry, *The Pratt & Whitney Canada Story*, 249.

engine on it. Gwinn then called P&WC President Thor Stephenson, and the next day, P&WC engineers were designing their first fanjet engine.[95] This relatively quick decision provided an interesting contrast to the more measured market analysis and customer survey done on the PT6 program.

The aircraft that Cessna had in mind, the Citation (originally called Fanjet 500), was to have short and unpaved field capability. The aircraft would be an entry-level business jet priced lower than existing business jets. It was to be a simple, easy-to-fly jet to which a multi-engine rated pilot could easily transition without too much instruction. With special certification, it could be flown with one pilot. With fanjet engines, the Citation would be quieter and cheaper to operate than turbojet-powered business aircraft.

JT15D-1

In June 1966, P&WC began detailed design of the engine designated JT15D-1 for the new Cessna aircraft. It was also the company's first attempt to design an engine to cost. Without previous turbine engine manufacturing experience to use as a guideline, the company had been having a difficult time determining and controlling costs on the PT6. A selling price had been picked for the PT6, and as the program developed, the company began to find out what it actually cost to build the engine. Reacting to this experience, a decision was made at the beginning of the JT15D program to select a selling price and and then design individual components within a designated cost. Thus, low cost and simplicity became major drivers on the JT15D program.[96]

The JT15D-1 was a twin spool, turbofan engine with a full-length annular bypass duct. The single-stage LP (low pressure) axial fan was driven by a two-stage LP turbine. A single-stage HP centrifugal compressor was driven by a single-stage HP turbine through concentric shafting. Concentric shafting was chosen to avoid an awkward aerodynamic configuration. P&WC's patented pipe diffuser delivered the air from the centrifugal compressor to the annular reverse-flow combustor. The engine bypass ratio was 3.3:1, and the total mass airflow was 74 lb/sec. Originally planned at 2,000-lb-thrust, in 1969, the thrust was raised to 2,200 lb with an sfc of 0.540. The higher thrust requirement followed refinements in Cessna's design of the airframe.

Because P&WC had not built a fan engine before, it received help from P&WA on the aerodynamic design of the fan stage for the JT15D. Aerody-

95. Smith to Leyes, 6 October 1988.

96. Smith to Leyes, 6 October 1988.

namically, the resultant fan was related to the fan on the Pratt & Whitney JT9D engine, scaled to be suitable for the much smaller JT15D. Mechanically, the fan consisted of solid titanium blades with part-span shrouds and a pressure ratio was 1.5:1.

Some additional points about the JT15D-1 fan design are noteworthy. One of the original design objectives of the engine was low noise. Turbofan engines were less noisy than turbojet engines largely as a result of their lower exhaust jet velocity. Early fan engines had inlet guide vanes, as was usual with compressors. However, they had noise and anti-icing problems, and technology advances allowed their elimination. The JT15D followed this practice. Its fan blade-to-stator vane number ratio and axial spacing were selected to minimize the wake interaction noise. When the engine was later installed on the Citation, the take-off, sideline, and approach noise level was significantly lower than the FAA Part 36 noise standards adopted in 1969. The Citation was the first aircraft to be certificated under the new noise regulations, and, at the time, was said to be the quietest jet aircraft in the world.[97]

Another important problem related to the JT15D fan design was the problem of bird strike and foreign object ingestion testing. The required certification tests were standardized, and no differentiation was made between large and small turbine engines. This meant that the JT15D engine had to undergo an ingestion test of a 4-lb bird. All known earlier certifications, including the largest high-bypass fan engines, had used a combination of smaller birds to meet the 4-lb requirement. The damage resulting from the test required strengthening of the fan bearing support structure and thickening of the engine casing over the centrifugal compressor discharge. Bird ingestion tests did considerable damage to the fan blades, primarily outboard of the mid-span shrouds; a stiffener was incorporated to solve that problem.

When the engine first ran on September 23, 1967, the JT15D had a large diameter single-stage LP turbine driving the fan. In charge of engineering at the time, Elvie Smith felt that the resultant large engine diameter was going to be "too much of a clunker" and initiated a redesign.[98] While compromising on the cost, a more streamlined engine design emerged when two smaller diameter turbines were substituted for the larger single-stage turbine used previously.

Starting on August 22, 1968, the JT15D was flight tested in a nacelle mounted

97. H. C Eatock, J. C. Plucinsky, and J. A. Saintsbury, "Designing Small Gas Turbine Engines for Low Noise and Clean Exhaust," (AIAA Paper No. 73-1154 presented at the CASI/AIAA Aeronautical Meeting, Montreal, Canada, 29-30 October, 1973), 11.

98. Smith to Leyes, 6 October 1988.

Cutaway of P&WC JT15D-1 turbofan engine. Now part of the National Air and Space Museum collection, this was one of two engines that powered the prototype Cessna Citation on its first flight. National Air and Space Museum, Smithsonian Institution (SI Neg. No. 95-1220-2).

under a modified Avro CF-100 testbed aircraft. On September 16 of that year, P&WC received from Cessna its first order consisting of 50 JT15Ds. The prototype engines were delivered to Cessna in August 1969. The prototype Citation flew on September 15, 1969. (One of the two engines used on this flight was donated to the National Air and Space Museum by P&WC in February 1978). Certification testing was completed in March 1971 with an accumulated engine running time of 7,600 hours. Certification followed in May 1971.

Superseding the JT15D-1 was the JT15D-1A, certificated in July 1976. The D-1A model had a higher fan speed limit, offering an increase in power at altitude. The JT15D-1B was certificated in July 1982. This model had yet greater cruise thrust at higher altitudes.

JT15D-4/D-5

The first growth version of the engine was the JT15D-4. This engine first ran on January 28, 1972 and was certificated in September 1973 at 2,500 lb static thrust. The additional thrust was obtained with the insertion of an axial boost stage on the LP shaft between the fan and HP centrifugal compressor. The boost stage increased the core airflow by 21% and the overall compressor pressure ratio by 15%. The bypass ratio was lowered from the JT15D-1's 3.3:1 to 2.6:1.

Prototype D-4 engines were delivered to Aerospatiale for the Corvette aircraft in the second half of 1972; production deliveries commenced in September 1973. The D-4 also powered the Citation II (JT15D-4), Cessna Citation S/II (JT15D-4B), SIAI-Marchetti 211 (JT15D-4C), the Mitsubishi Diamond I (JT15D-4), and the Diamond 1A (JT15D-4D) business jet aircraft. The model D-4C was certificated in September 1982, the D-4B in May 1983, and the D-4D in December 1983. All had changes and improvements geared toward specific airframe applications.

To meet customer requirements for more thrust and improved fuel economy, individual component testing on a new model, the JT15D-5, began in early 1976, and the first run of a complete engine was in late 1977. In order to maintain the same external engine dimensions, core engine pressure ratio and airflow were to be increased and bypass ratio decreased. To minimize the necessary reduction in bypass ratio, a new higher-flow fan was needed. The fan part-span shroud and local stiffeners that penalized JT15D-1 through JT15D-4 aerodynamic performance were eliminated. Fewer, wide-chord blades were used to maintain structural integrity and avoid airfoil resonance.

Other improvements were made in the JT15D-5. The compressor boost stage was aerodynamically redesigned for improved efficiency. A larger diameter centrifugal compressor with exducer tip backsweep was designed and demonstrated a significant performance gain over the D-4 compressor.[99] A limited authority electronic fuel control was chosen primarily because of the advantages it offered in surge avoidance at high altitude. Directionally solidified MAR-M-200 HP turbine blades were used for improved turbine temperature tolerance and creep characteristics. Pipe diffusers were redesigned with cast steel ends and integral vanes in place of the sheet metal configuration of earlier models. This was done

---

99. D. I. Boyd, "Development of a New Technology Small Fan Jet Engine," (ASME Paper No. 85-IGT-139 presented at the 1985 Beijing International Gas Turbine Symposium and Exposition, Beijing, People's Republic of China, 1-7 September 1985), 3. Exducer tip backsweep refers to that portion of the centrifugal compressor blade which intercepts the compressor outer diameter at an angle which is less than 90 degrees as measured from the tangent. Exducer tip backsweep lowers the Mach number of the compressor discharge air. The result is better efficiency and broader range of operation between the limits of surge and choke at the expense of lower total compressor pressure rise.

Cessna Citation V executive aircraft powered by two P&WC JT15D-5A turbofan engines. Photograph courtesy of P&WC.

to eliminate aerodynamically excited vibration and to withstand higher pressures and temperatures.

The D-5 model was rated at 3,190 lb thermodynamic take-off thrust, 28% more than the D-4. The D-5 first flew in a flight test aircraft in April 1978, and prototype engines were delivered to Cessna in December 1982. When certificated in December 1983, the engine was flat rated at 2,900 lb take-off thrust at an atmospheric temperature of 80 degrees F. At this rating, the engine had 25% more altitude cruise thrust and approximately 3% lower specific fuel consumption than the D-4 model. Production began in June 1984.

The D-5 model was used on a variety of jet aircraft. The engine powered the Mitsubishi Diamond II (later the Model 400A Beechjet) and U.S. Navy T-47A jet aircraft. The JT15D-5A (certificated November 1988) was used on the Citation V business jet, and the JT15D-5D (certificated in 1993) entered service in July 1994 on the Cessna V Ultra executive jet. In February 1990, the Beechjet was selected by the U.S. Air Force as the T-1A Jay Hawk TTTS (Tanker Transport Trainer System) aircraft, powered by the JT15D-5B engine (certificated in 1990). This was a significant sale for P&WC. The Jay Hawk's D-5B engine was assembled at P&WC's West Virginia facility. An aerobatic version, the JT15D-5C (certificated November 1991), powered the SIAI-Marchetti/Grumman S.211A and the Rockwell International/Deutsche Aerospace Ranger 2000 jet trainers, both candidates for the U.S. Joint Primary Aircraft Training System competition in the mid-1990s. The D-5F (certificated in 1993) powered the Beech TCX. The models D-5, D-5A, D-5B, and D-5F had a take-off, flat-rated thrust of 2,900 lb. The D-5C and the D-5D had flat-rated, take-off thrusts of 3,190 lb and 3,045 lb respectively.

JT15D SUMMARY

With the installation of the JT15D on the Citation, P&WC had once again found a popular airframe application. By mid-1988, 1,519 Citations had been sold by Cessna, more than any other business jet.[100] The JT15D was used on the Citation models 500, I, II, S/II, and V. Not only a popular aircraft, the Citation also won the 1985 Collier Trophy Award for its unparalleled flight safety record.

It's use on a popular jet made the JT15D a commercially successful engine. By the end of 1995, 5,161 JT15D series turbofan engines had been delivered. At that time, the number of JT15D engines produced was second only to the number of PT6 turboprop/turboshaft engines.

There were two important contributions that the JT15D made to P&WC. Technically, the JT15D gave P&WC the opportunity to broaden its technical expertise, as it gained experience on turbofan engines and on concentric-shaft turbine engines. From a marketing viewpoint, Cessna's and P&WC's gap analysis had proved correct. There was a commercial need for a low cost, entry-level business jet.

The PW100

In the early to mid-1970s, P&WC's advance design group and marketing department began to focus on the possibility of a turboprop engine larger than the PT6. P&WC engineers felt that they had begun to reach a limit to the growth

---

100. Mark R. Twombly, "Honor Roll: Citations are Found at the Head of the Class," *AOPA PILOT*, Vol. 31, No. 10 (October 1988): 39.

that they could get out of the PT6. As Vice President of Marketing Richard McLachlan described it, "We were running out of space to put the air through. We felt that we had to have another generation engine that addressed a new market and incorporated higher technology."[101]

Although it had no request from an airframer for a power plant of this type, P&WC bet on the possibility that it could find an application. Two possibilities for a larger engine were the commuter and the executive aircraft markets. While the commuter aircraft industry was growing and showed some future promise, most commuter airlines were using aircraft seating 19 or fewer passengers. Inasmuch as the movement toward developing larger commuter aircraft appeared to be slow or nonexistent, the company focused on the executive aircraft market.

One possible application was a new turboprop business plane that would outperform existing turboprops and compete directly with corporate jets. If such an aircraft were developed, there would be a future market need for a higher performance turboprop engine.

In addition to the various airframe possibilities that would help determine the power level of a next-generation turboprop engine, there were other considerations. One was that the new engine design not compete directly with the high-power versions of the PT6.[102] A clearly defined growth plan was another.

In 1974, P&WC initiated studies to define an optimum layout. During these studies, alternative engine configurations were examined. To achieve the desired HP ratio of 15:1, two compressor designs were seriously considered. One design was a two-shaft, free turbine engine consisting of an axial compressor with variable geometry (i.e. variable inlet guide vanes and variable stator vanes) followed by a centrifugal compressor, with both compressors on a single shaft. The other design was a three-shaft engine using two centrifugal compressors, each on its own shaft. The studies suggested that the performance, weight, and cost of the two concepts were nearly identical.[103] A final decision had to be made on other grounds.

It was decided to trade the complexity and potential maintenance problems of the variable geometry engine for the potential shaft dynamics and bearing/seal problems that were associated with a three-shaft engine. P&WC engineers felt that the three-shaft engine was a more conservative approach with lower development risk. They believed that the inherent strength and rugged-

---

101. McLachlan to Leyes, 7 October 1988.

102. Hardy, "Development of Gas Turbine Engines," 59.

103. R. E. Morris, "The Pratt & Whitney PW100- Evolution of a Design Concept," (Pratt & Whitney Aircraft of Canada Ltd. Report No. ZZISSN-8-2821-28-3-211, 1982).

ness of the centrifugal compressor would result in lower maintenance costs and higher reliability than an axial-flow version. Also, over the years, P&WC designers had developed considerable expertise working with centrifugal compressors and were confident that they could reach their performance targets with minimum development.[104] By mounting centrifugal compressors on separate shafts, stability problems associated with high pressure ratio single-spool compressors could be avoided.

The pressure ratio selected for the engine sacrificed about 1% sfc in the baseline engine design, but resulted in nearly optimum performance in pre-planned growth versions of the engine. In order to improve the engine specific weight, a turbine inlet temperature was chosen that was high enough to require cooling the first-stage blade, but low enough so as not to require cooling of the second-stage blade. Up through 30% growth, second-stage blade cooling would not be necessary.

THE TDE-1

In view of the novel features of the proposed design, it was considered prudent to build a proof-of-concept engine. Thus, in October 1976, the company started to design the Technology Demonstrator Engine (TDE-1). This was the first demonstrator engine and first three-shaft engine that P&WC had built. The 1,665-shp engine had two centrifugal compressors, each on its own shaft. Variable inlet guide vanes were placed ahead of the LP compressor. A free power turbine with a front power takeoff was mounted on the third engine shaft. The rotors and shafting were designed as if they were for an operational engine, but the static structural elements were more industrial in style (i.e boilerplate).

By the end of 1977, the gas demonstrator ran, and in August 1978, the complete demonstrator engine first ran. Fundamentally, it performed well and seemed viable.[105] The most important finding from the test was that the shaft bending critical speed of the power turbine rotor was somewhat lower than expected due to difficulty in estimating the effective bearing support stiffness.[106] Because the problem was encountered and understood during the demonstrator engine phase of the program, it could be remedied before the start of flight en-

104. Morris, "PW100- Design Concept."

105. Hardy to Leyes, 6 October 1988.

106. David L. Cook, "Development of the PW100 Turboprop Engines," (SAE Paper No. 850909 presented at the General Aviation Aircraft Meeting and Exposition, Wichita, Kansas, 16-19 April 1985), 4.

gine development. It was felt that this finding alone justified development of the demonstrator engine; encountering this low-speed critical vibration in an engine development program would have caused serious delay and extra cost. Compressor performance and stall margin were found to be good without the variable geometry, and it was removed. The cooled first-stage turbine blades, P&WC's first, generally performed well.

CHANGES IN DIRECTION

P&WC approached some of its traditional customers such as Beech about the possibility of designing a business aircraft around such an engine, but found them not ready for such a venture. Also, the development of new business jets cut short the hope of a large turboprop aircraft competing with the corporate market. However, another trend in the airline industry soon reshaped P&WC's marketing strategy and the design of its new engine.

Commuter airlines, which had experienced a rapid growth in the 1970s, continued their accelerated growth following the 1978 Airline Deregulation Act.[107] This act scheduled a four-year phase out of the rate and route authority of the Civil Aeronautics Board, the federal agency that regulated airline passenger service. Entrepreneurial opportunities opened up as access to domestic routes became available to any qualified carrier. Also, carriers were increasingly given freedom to set their own fares. Freedom to add routes and set fares allowed new carriers to enter new markets.

A significant part of this post-1978 growth resulted from new routes acquired by commuter airlines. As trunk carriers began to withdraw from small towns and short routes that were not economical with large aircraft, commuter airlines seized the opportunity and began to provide replacement service with their smaller aircraft. About 23 percent of the growth in passengers carried between 1978 and 1980 could be attributed to deregulation as a result of trunk carrier terminations.[108] As a group, commuter airlines experienced a high rate of growth in both the number of markets served and passengers carried. Between 1976 and 1980, the number of markets served grew at an annual rate of 22.2 percent, while passengers carried increased at 22.8 percent.[109] Another important trend that developed before and after deregulation among commu-

---

107. Bailey, Graham, and Kaplan, *Deregulating the Airlines*, 80.

108. John R. Meyer and Clinton V. Oster, Jr., *Deregulation and the New Airline Entrepreneurs* (Cambridge: The MIT Press, 1984), 144.

109. Meyer and Oster, *Deregulation and the New Airline Entrepreneurs*, 141.

ters was a move toward longer hauls and more passengers carried in these longer hauls.[110]

Traditionally, not only market considerations, but government regulations,
had influenced the size of aircraft used by commuter airlines. In 1952, with the
adoption of Part 298 of its economic regulations, the CAB granted air taxi operators (commuters) an exemption to provide scheduled air service as long as
they operated aircraft that did not have a gross take-off weight exceeding 12,500
lb.[111] This meant that such operators were exempted from CAB route and fare
regulation. The CAB 12,500-lb gross-weight limit was approximately equivalent to a 19-passenger seat aircraft. In 1972, the CAB raised the limit and allowed commuters to operate aircraft seating no more than 30 passengers with a
payload capacity of no more than 7,500 lb. Following deregulation in 1978, the
CAB raised the limit to 60 seats.

The growth in the commuter market and the ability to operate substantially
larger aircraft after deregulation provided increased opportunities for commuter
airliners. Accordingly, P&WC saw a growing opportunity for regional aircraft
seating between 19 and 50 passengers.[112] In addition to this North American
commuter aircraft market, a potentially large market for similar aircraft existed
in other countries.

Airframers such as Embraer and de Havilland, long-standing customers of
P&WC, were also investigating these new market opportunities. Embraer had
been studying a successor to its Bandeirante commuter aircraft since 1974. Deregulation in the U.S. was the catalyst that finally moved Embraer to begin work
on the EMB-120, a 30-passenger commuter aircraft.[113] As a result of Embraer's
and de Havilland's serious interest in developing larger commuter aircraft and
because no executive aircraft market could be found, P&WC redirected its
efforts toward a regional aircraft engine rather than an executive aircraft engine.

The significant operational differences between the business aircraft and the
proposed regional aircraft meant that the original engine design criteria would
have to be changed. Business aircraft typically flew at higher altitudes than did
regional aircraft. Accordingly, engine performance would have to be opti-

110. Meyer and Oster, *Deregulation and the New Airline Entrepreneurs*, 143.

111. R. E. G. Davis, *Airlines of the United States Since 1914* (Washington, D.C.: Smithsonian
Institution Press, 1972), 483.

112. M. D. Stoten and R. A. Harvey, "Regional Airline Turboprop Engine Technology,"
(AIAA Paper No. AIAA-83-1158 presented at the AIAA/SAE/ASME 19th Joint Propulsion
Conference, Seattle, Washington, 27-29 June 1983), 1.

113. Sullivan and Milberry, *The Pratt & Whitney Canada Story*, 265.

mized for the lower altitudes. Because regional aircraft flew more hours per year, they required engines with a longer life. The engine itself would have to be scaled up because the regional airplanes under consideration were bigger than the corporate aircraft originally anticipated. As was required of many engines, but particularly of airliner engines, the new one would have to have good fuel consumption, be easy to service, and be rugged and reliable. First cost would have to be kept as low as possible without compromising operating costs.

To meet these requirements, the preliminary design specifications called for an emphasis on good fuel consumption. The engine would not be at the lowest possible cost because such an engine would have to emphasize mechical simplicity at the expense of fuel consumption. Some increase in complexity and cost was acceptable to achieve low fuel consumption. High reliability and low maintenance costs, of primary importance to regional operators, were also important design considerations. In view of the uncertainty whether the target market would develop as anticipated and in order to ensure wide application, the engine needed to be designed with significant short-term growth capability.[114]

THE PT7A-1

In February 1979, an engineering team was formed, and in June, full-scale development efforts began on a 2,000 shp class engine, designated PT7A-1. The PT7 was a free turbine turboprop engine with three concentric shafts. The engine had two centrifugal compressors mounted on separate shafts, with each compressor driven by its own single-stage turbine. An annular reverse-flow combustor was used to reduce the compressor shaft length and keep the engine compact. A two-stage power turbine drove the propeller through an offset reduction gearbox.

The air entered the engine through an "S"-shaped inlet duct equipped with a bypass duct designed to inertially separate birds and other foreign matter from the airflow. The first-stage LP centrifugal compressor had a 4.8:1 pressure ratio and was machined from forged titanium. This delivered air to a pipe diffuser leading into constant area ducts that blended into a full annulus at the entry to the second-stage, HP compressor. This second-stage compressor operated at just under 3.0:1 pressure ratio at its design point and was also machined from forged titanium. With each compressor stage independently driven by its own turbine,

114. Morris, "PW100- Design Concept."

it was possible to keep each stage operating at its own best speed and efficiency as the power level was varied.[115]

The engine's combustor was constructed of sheet metal for low-cost manufacture and repair. Refined cooling schemes were incorporated to keep wall temperatures low for long life. The principal advantage of the combustor's reverse-flow (folded) configuration was to permit the HP turbine to be very close-coupled mechanically to the HP compressor. Because of the PT7's three-concentric-shaft configuration, short shaft lengths were desirable in order to avoid shaft vibration problems.

The HP turbine had segmented, cooled stator vanes and cooled rotor blades. For reasons of cost, complexity, maintainability, and ease of development a decision was made to cool only the HP turbine for the first models of the engine. The second-stage LP turbine, which drove the LP compressor, had an uncooled vane ring and uncooled rotor for high efficiency.[116] Tip clearance on both stages was closely controlled by cooling of the turbine casings.

Because of the relatively HP ratio of the engine, two stages were required for the power turbine. Shrouded tips were incorporated in each stage. The two-stage power turbine rotated in the same direction as the HP spool and opposite to the LP spool, providing a small advantage in turbine efficiency from less turning of the airflow in the stator row. The offset gearbox that drove the propeller provided installation flexibility in a range of aircraft types and freedom in locating accessories for ease of maintenance.

The power control system consisted of a hydromechanical unit to provide essential fuel control functions for safe engine control and flight in the event of power failure and an electronic control unit to provide a number of power management functions for high quality control and significantly reduced pilot workload. One of the workload reduction functions of the electronic control was to automatically compute target torque for prevailing conditions and provide a signal to drive a target bug on the cockpit torque indicator, relieving the pilot of the task of manually computing torque settings. One of the power management functions was to ensure that engine acceleration and deceleration were controlled to a constant rate of change of speed with time. This resulted in

115. M. D. Stoten, "The PW100 Commuter Powerplant," (AIAA Paper No. AIAA-81-1731 presented at the AIAA Aircraft Systems and Technology Conference, Dayton, Ohio, 11-13 August 1981), 2.

116. An engine designed without turbine cooling, given the same turbine inlet temperature, engine pressure ratio, and basic component efficiencies, is more efficient overall since the use of cooling air always has a performance cost, i.e. pressure losses of cooling flow, cost of its re-introduction into primary flow, etc.

highly repeatable acceleration times regardless of fuel type, accessory horse-power, and airbleed.[117]

THE PW100 SERIES AND ITS DEVELOPMENT PROBLEMS

In 1980, the PT7 was redesignated the PW100, a family of engines. The first of the engines in this family was the PW115. The first run of the PW100 gas generator took place in December 1980, and the first run of the complete engine was conducted in March 1981. The first flight test took place in February 1982 in the nose of P&WC's Viscount testbed aircraft. As with any development program, problems occurred.

Because the PW100 was a three-shaft engine, it had two compartments with three labyrinth seals keeping the oil in by flowing air inward. Nearly identical air pressures were required outside each labyrinth, otherwise the flow in one might reverse, resulting in loss of oil. Great care and some trial and error was required to ensure that the seals did not rub during normal operating conditions and that the proper balance of air pressure was maintained.[118]

Other oil and air systems problems occurred. The system that scavenged oil from the bearing compartments and prevented oil escape from them had to be redesigned when it was discovered that oil was drawn into the gaspath during starting. The overboard deoiler breather also had to be redesigned to separate the flow of oil and air during engine operation.[119]

Some of the castings gave problems. According to Vice President of Operations Roy Abraham, the thin-wall castings presented an engineering and manufacturing challenge.[120] Foundry operators believed that they had developed casting technology to the point where thin-wall structural castings were practical, yielding substantial cost, weight, and parts reduction, and so P&WC moved from sheet metal to thin-wall castings in key areas. Some worked out well, but the nickel-alloy-based castings used around the HP compressor gave serious quality problems, not solved until well into series production. With this practical experience, newer designs moved back to sheet metal in a number of cases.

Other problems occurred as well. During tests in the engine cells, gearbox casings cracked. This turned out to be due to stresses imposed by the propeller

---

117. Stoten, "The PW100 Commuter Powerplant," 3.

118. Cook, "Development of the PW100 Turboprop Engines," 5-6.

119. Cook, "Development of the PW100 Turboprop Engines," 5.

120. Roy C. Abraham, interview by Rick Leyes, 6 October 1988, Pratt & Whitney Canada Interviews, transcript, National Air and Space Museum Archives, Washington, D.C.

in the test cell atmosphere. These problems went away once the engine was installed in the Viscount test aircraft.[121] Containment of separated power turbine
blades also was a problem and several redesigns of the containment hoops were
necessary to find a solution.[122] Another containment problem resulted from the
loss of a LP impeller inducer tip that punctured the LP diffuser pipe. Reinforcing patches on the diffuser pipe solved the puncture problem; removing a
source of vibration in the impeller shroud prevented further impeller tip separa-
tions. A problem with retaining labyrinth seals that had been snapped onto disks
or shafts was resolved by a redesigned seal attachment that was bolted on.

SELLING, CERTIFICATING, AND GROWING THE PW100 SERIES

P&WC's long-standing relationship with Embraer and de Havilland and the
fact that its engine was designed specifically for regional aircraft gave P&WC a
competitive advantage in selling the engine to those companies. Nonetheless,
the PW100 engine had to compete against the General Electric CT7 for the
final selection. The CT7 was a commercial turboprop engine derived from the
General Electric T700 turboshaft engine, originally designed as a military helicopter engine. While the PW100 had some competitive advantages over the
CT7, it also had some disadvantages. The PW100 had been designed with
significant growth capability built into its initial size. Because of this, it was
somewhat heavy and its fuel consumption was a bit high compared to the CT7.

   After weighing the merits of each engine, in September 1979 Embraer chose
the PT7 (PW100) engine for its new EMB-120 Brasilia commuter. Assessing the
competition, P&WC's Chairman of the Board Elvie Smith felt that the higher
power offered by the PW100 engine was an important factor in the airframer's
final decision.[123]

PW115/120 SERIES

The two original PW100 models, both of which were certificated in December
1983, were designated as the PW115 and PW120. Other engines in this series,
which followed later, included the PW118, PW119, and PW121.

   The PW115, selected for the EMB-120, had a take-off rating of 1,600 shp.
Later, the PW118 (certificated in March 1986) and the PW118A (certificated in

---

121. Hardy to Leyes, 6 October 1988.

122. Cook, "Development of the PW100 Turboprop Engines," 6.

123. Smith to Leyes, 6 October 1988.

June 1987), each with an 1,800-shp take-off rating, were used on the EMB-120. Derived from the PW115, the PW118 was mechanically similar, but had increased shaft horsepower and other improvements. Further derivatives included the PW119 and PW119B, built for the Dornier 328 30-seat regional airliner. The PW119B was certificated in June 1993 at 2,180 shp.

In 1980, de Havilland, another important airframer, selected the PT7 over the General Electric CT7 for its new design, the Dash 8-100 commuter aircraft. Again, the power level was important in the selection. The engine selected for the Dash 8-100, with a 2,000-shp take-off rating, was later designated the PW120, a new model with a new reduction gearbox and reserve power rating. The PW120 was certificated in December 1983, and the first production engines were delivered in January 1984.

Minor uprated variants of the PW120 were the PW120A and PW121. The 2,000-shp PW120A (certificated in September 1984) and the 2,150-shp PW121 (certificated in February 1987) were also used on the de Havilland Dash 8-100. The PW120A was derived from and mechanically similar to the PW120, but had increased maximum continuous and cruise power ratings.

In its continuing competition against the General Electric CT7 for the new regional airliner market, P&WC was not selected for the Saab-Fairchild SF-340 and Casa-Nurtanio CN-235. However, in June 1981, the PW120 was chosen for the Aerospatiale/Alenia ATR 42 commuter. A later version of the ATR 42 used the PW121.

PW124 SERIES

While developing the PW100 for the Brasilia, a 30-passenger aircraft; the Dash 8, a 37/40-passenger aircraft; and the ATR 42, a 42/49-passenger aircraft, it became obvious to P&WC that the market was going to grow into progressively larger aircraft.[124] The next wave of regional aircraft were being designed for 50 and more passengers. Also, to leverage their substantial investment in these new aircraft, P&WC perceived that airframers would continue to manufacture them for some period of time. In order to meet the demands of their customers, airframers would probably stretch the aircraft and build faster versions of them.

In July 1982, before the certification of the PW120, a decision was made to design and offer a growth engine, the PW124. This engine had a 2,400-shp take-off rating. The added power was to be achieved by reconfiguring the LP compressor to increase airflow through the engine by 15% and the pressure ratio

124. McLachlan to Leyes, 7 October 1988.

by 23%. Primarily, this was done by substituting a new larger diameter, higher pressure ratio LP compressor. The engine also had a new inlet, intercompressor bleed valve, LP diffuser, cooled LP turbine inlet vanes, power turbine blades, and control system. A new gearbox would be designed for a six-blade propeller.

Essentially, the company then had two-brand new engines, the PW120 and the PW124, under development before the first one had gone into production. This was an unusual approach. Typically, an engine company would develop engines in series to take advantage of the learning on the first engine before starting on the second. However, the decision turned out to be a good one for P&WC.

The first two models of the new growth engine, the PW124 and PW124A were certificated in November 1985. As new airframe applications were won, the PW124 series evolved into a number of derivative engine models including, in order of certification, the PW125, PW123, PW126, and PW127.

Competing against a modernized version of the venerable Rolls-Royce Dart—the world's first operational commercial turboprop engine—the PW124 was selected in 1982 for the British Aerospace ATP, a 64-passenger commuter. Again, competing against the Dart, P&WC was successful in signing a contract with Fokker in May 1983. In this instance, the PW124 would power the Fokker 50, a 50-passenger commuter. In 1985, the PW124 was announced for the Aerospatiale/Aeritalia ATR 72, a 72-passenger aircraft.

A spin-off from the PW124 was the 2,400-shp PW124A, mechanically similar to the PW124, but with an increased maximum continuous rating and no thermal rating change. Certificated in May 1988, the PW124B had PW124A gearbox and turbomachinery, but incorporated the PW123 LP rotor. The ATR 72-200 used the PW124B.

The next engine model in the PW124 series was the PW125. The PW125 was mechanically similar to the PW124, but was re-rated in accordance with certification rules regarding engine contingency ratings. The PW125A was similar to the PW125, but had increased contingency ratings and increased thermal ratings and torque limits. In May 1987, the PW125B was certificated with a 2,500-shp take-off rating. The PW125B was used on the Fokker 50 Series 100.

In June 1987, the PW123, the next model in the PW124 series, was certificated. The PW123 had a 2,380-shp take-off rating; it had a PW120A gearbox module and a PW124 turbomachinery module. The de Havilland Dash 8-200 used the 2,150-shp PW123C/123D, and the Dash 8-300 used the PW123 and the 2,500-shp PW123B (certificated in November 1991). Another variation of this model was the PW123AF (certificated in February 1990), used on the Canadair CL-215T/CL-415 water bomber.

TOP: Pratt & Whitney Canada XPW123 turboprop engine similar to that used in the de Havilland DHC-8 Dash 8 Series 300 regional transport aircraft. This engine is in the National Air and Space Museum's collection. National Air and Space Museum, Smithsonian Institution (SI Neg. No. 95-1220-23).

BOTTOM: The P&WC PW123 turboprop engine powers regional transport aircraft such as this de Havilland DHC-8 Dash 8 Series 300 parked on the ramp of Washington Dulles International Airport in Virginia. The PW123 engine cowling has been opened for inspection. Photograph courtesy of Brian Nicklas.

The PW126 was certificated in May 1987 with a 2,653-shp take-off rating. It was followed in June 1989 with the certification of the 2,662-shp PW126A. The PW126 and 126A powered the Jetstream Aircraft (formerly British Aerospace) ATP commuter.

The PW127 (certificated in February 1992) was an increased power version of the PW123 with a take-off rating of 2,750 shp. The ATR 72-210, developed for operation in hot-and-high conditions, was powered by the PW127. The Fokker 50 Series 300, a high-performance variant of the Fokker 50, was powered by the PW127B rated at 2,750 shp. The PW127D, certificated in January 1994 and also rated at 2,750 shp, powered the Jetstream 61 commuter. The 2,400-shp PW127E was selected for the ATR 42-500 commuter.

In the spring of 1995, P&WC initiated a major redesign designated PW150. It was selected to power the proposed 70-seat high-speed de Havilland Dash 8 Series 400 transport. The PW150 was selected over Allison's AE2100 and AlliedSignal's emerging Lycoming Series 500 engines. The PW150 had a new three-stage axial-flow LP compressor and single-stage centrifugal HP compressor. Prior PW100 series engines used single centrifugal stages for both their LP and HP compressors. The axial/centrifugal design was chosen to increase engine airflow while maintaining essentially the same installation envelope. In addition, the PW150 had more turbine cooling and a new 5,000-shp-class reduction gearbox. With these changes, the PW150 had the thermodynamic capability to generate 6,500 to 7,500 eshp, a significant increase from the 3,200-eshp thermodynamic rating of the PW127.[125]

FIELD EXPERIENCE WITH THE PW100

By 1992, several problems had emerged with the PW115/PW120 series engines.[126] A great deal had been learned by P&WC about electronic controls and their interaction with airframe systems and mechanical fuel controls. The biggest problem was to ensure that the electronic control box, the harness, and the connectors were protected from moisture.

The PW124 series engines shared the electronic control problems and had more problems in general than did the smaller engine series. Two principal problems were turbines that deteriorated prior to TBO and internal engine fires. The turbines were redesigned for additional cooling of the turbine vanes and blades

125. "Bombardier Picks PW150," *Aviation Week & Space Technology*, Vol. 142, No. 18 (May 1, 1995): 36.

126. Hardy, "Development of Gas Turbine Engines," 70.

to prevent premature oxidation. The fires were traced to the failure of the No. 5 bearing, which caused oil to leak into a hot cavity in the engine. This problem was fixed by improving the bearing mounting and running clearances and by venting any leaking oil to an area where the fire could be contained within the engine and by installing a sensor to alert the pilot to shut down the engine.[127]

PW100 SUMMARY

In the mid-1970s, P&WC initiated studies which led to the PW100 program. This speculative effort paid off for P&WC. The three-shaft engine configuration, the first of its type for P&WC, worked well, and the power range and growth capability were properly chosen. As a result of again having the right engine at the right time, P&WC was able to successfully capture a very large segment of the growing regional aircraft market.

The production of PW100 engines, beginning in 1984, augmented PT6 sales to regional aircraft manufacturers. In 1981, 57% percent of P&WC engine sales dollars were for corporate aircraft. However, by 1988, 61% of the sales dollars were for regional airline aircraft. Part of this market shift was due to the depressed general aviation market and part was due to increasing regional aircraft sales. Production timing was excellent, and the PW100 engine was a product that helped P&WC meet this changing market demand. By the end of 1995, 3,791 PW100 series engines had been delivered.

The success of this engine program was influenced to a significant extent by changing government regulations and by a resultant growth in the regional aircraft market. For its part, the PW100 was one of the key turboprop engines that made possible a new generation of larger regional aircraft to serve a growing commuter market.

The PW200

The goal of producing an inexpensive small turbine engine for low-power general aviation applications had interested manufacturers for many years. In the late 1960s, when the development of the TwinPac PT6 was underway, P&WC's advance design group thought of designing a smaller, less expensive turboprop or turboshaft engine. Jokingly, the designers called it the "flivver" engine.[128]

127. Hardy, *Development of Gas Turbine Engines*, 70-71.

128. McLachlan to Leyes, 7 October 1988.

Various very simple turboprop engines were designed for fixed-wing aircraft and cost estimates were made. In almost every case, however, the design could not beat the lower cost of the well-established supercharged or turbocharged reciprocating engine in the 400-shp range. However, a much greater opportunity existed for a low horsepower turboshaft engine. Helicopters were more expensive than fixed-wing aircraft, they were extremely sensitive to weight, and operators would be willing to pay a higher price for an engine with a high power-to-weight ratio.

From P&WC's viewpoint, the biggest problem with entering the low-power turboshaft market was Allison's incumbency in that marketplace.[129] With its Model 250 engine, Allison had captured essentially all of the business in North America and a sizeable share of the market in the rest of the world. Being far down on the learning curve and the cost of their engine, Allison could offer a lower price than P&WC. Also, Allison's incumbency meant that airframers would be reluctant to re-engine their helicopters at great expense unless a there was a clear-cut advantage. With the proper design, cost control, and level of technology, P&WC felt that it could make a more reliable engine and use reliability as its primary lever to enter the field. This potential advantage set the stage for a new engine, the PW200.

An incentive for P&WC entering the field was offered by market studies in the early 1980s suggesting that a major growth in helicopters would take place over a 15-year period.[130] The commercial helicopter industry was booming in the late '70s and early '80s largely as a result of gas and oil exploration.[131]

In response to the market situation and because Canada was the second largest market for commercial helicopters, the Canadian government solicited helicopter manufacturers to build plants in Canada. Bell Helicopter Textron responded to the offer, and in 1983, the Canadian and Quebec governments announced their financial support for a new Bell plant at Mirabel, north of Montreal, to manufacture the Bell Model 400 TwinRanger.[132] The Canadian

---

129. McLachlan to Leyes, 7 October 1988.

130. Emmerson to Leyes, 6 October 1988.

131. Sullivan and Milberry, *The Pratt & Whitney Canada Story*, 299.

132. Since the plant opened in 1984, it took over production of the Bell 206B JetRanger, 206L LongRanger, Model 212, and Model 412. According to Bell, the plant was manufacturing one-third of all the commercial helicopters being made in the world. David Hughes, "Canadian Aerospace Industry Prepares for Rising Competition, Falling Defense Sales," *Aviation Week & Space Technology*, Vol. 134, No. 11 (March 18, 1991): 68.

Cutaway of P&WC PW200 turboshaft engine. Photograph courtesy of P&WC.

government also agreed to support P&WC in developing the PW209T for the TwinRanger. In a separate arrangement, government assistance was also given to Messerschmitt-Boelkow-Blohm (MBB) of Germany to set up a factory in Fort Erie, Ont., to build BO 105 helicopters.

Work on an engine known as the PW209T began in October 1983. Rated at 937 shp, the PW209T consisted of two power plants driving a single output shaft through a combining gearbox. Extensive discussions were held with Bell Helicopter and MBB in the initial stages and throughout the design process to integrate their requirements into the basic engine concept.[133] A single-stage machined titanium centrifugal compressor with an 8:1 pressure ratio was driven by a single-stage axial turbine with uncooled vanes and blades. Compressed air was directed through individual pipe diffusers into a reverse-flow annular combustion chamber. A through-shaft, single-stage free power turbine drove a front-mounted combination reduction and accessory gearbox. Dry engine weight was 468 lb, and engine diameter was only 13 inches.

133. Kevin A. Goom, "Anatomy of a New Helicopter Engine—Pratt & Whitney Canada PW200 Series Turboshafts," *Vertiflite*, Vol. 32, No. 5 (September/October 1986): 20.

The primary design objective was low cost.[134] To achieve it, parts count was reduced until the engine had 45% fewer parts than a typical PT6. Also, thin–wall castings replaced sheet metal parts in many areas of the engine. Engine reliability and durability were fundamental design objectives in view of the increasing importance that operators were placing on low cost and low maintenance.[135] Thus, simplicity and proven technology were emphasized in the design.

Detail design was completed in mid–1985, and the PW209T first ran on November 26, 1985. The PW209T engine underwent development from November 1985 to May 1987 during which time three development engines were run for more than 1,300 hours.

The original PW209T did not meet its required specification levels during the development program.[136] In late 1986, a redesign of all of the engine's aerodynamic components was initiated to incorporate the successful designs of the PW205B and to increase the take-off power of each power section from 470 to 520 shp to accommodate increases in the gross weight of the Bell 400A. Another development problem was the gearbox lubrication system, which proved to be inadequate because of excessive oil churning and insufficient oil tank capacity. A redesign of the gearbox, completed in June 1986, corrected these problems.

The decline of oil prices in the 1980s and the subsequent collapse of the helicopter market prevented P&WC from certificating a PW200 series production engine in the 1980s. The earlier P&WC market analysis had proved overly optimistic. In May 1987, all work on the twinned PW209T was stopped when Bell decided not to produce the Bell 400A TwinRanger. Emphasis then shifted to single power-section PW200 series engines.

In July 1985, design and development of the PW205B, a derivative, single-power section, growth version of the PW209T, had begun. The engine was developed for use in the MBB Bo 105LS-B1 helicopter. Detail design was completed in June 1986, and the first engine ran on February 2, 1987. The PW205B had a single power section, a new gearbox, and an electronic enhanced fuel management system. The power section of the PW205B was a growth version of the PW209T power section and had increased airflow and turbine temperatures for a thermodynamic take-off power of 590 shp. Three engines were used in the PW205B development program. Tests proved successful with performance exceeding specifications. The only significant problem was wear on the

---

134. Hardy to Leyes, 6 October 1988.

135. Goom, "Anatomy of a New Helicopter Engine," 23.

136. Hardy, "Development of Gas Turbine Engines," 79.

gearbox output shaft caused by deflection of the rear gearbox casing; this was fixed by redesigning the casing and by changes in the gearing.

The first two experimental PW205B engines were shipped to MBB Canada Limited in November 1987. The PW205B-powered MBB BO 105LS-B1 helicopter made its first flight on October 12, 1988. After an interruption early in 1989, BO 105 flight testing was completed in June. At that time, MBB terminated the PW205B-powered BO 105 helicopter program.

Early in 1989, during the time that the BO 105 flight testing had been temporarily suspended, P&WC made a proposal to McDonnell Douglas Helicopter to power a new helicopter known as the MDX. The MDX (later designated MD Explorer) used a unique, patented Notar (no tail rotor) anti-torque and directional control system. In 1989, McDonnell Douglas announced that it planned to certificate its twin-turbine MDX with both P&WC PW206A and Turboméca TM 319-2 engines.[137] Thus, McDonnell Douglas became the launch customer for the PW200 program.

The PW206A used in the MD Explorer was an uprated derivative of the PW205B. Development of the PW206A took place between 1990 and 1991 with six engines, which ran 3,800 hours. As a result of earlier experience on the PW209T and PW205B programs, no major problems were discovered. All specified performance, weight, and operability targets were met. The PW206A was originally certificated in December 1991 with a take-off rating of 621 shp. Production deliveries for the PW206A began in January 1992. In November 1993, the PW206A was recertificated and uprated to 640 shp (take-off).

Other airframe applications were found. MBB renewed its interest in the PW206. In January 1991, the company announced its intention to produce a new light twin-turbine helicopter. Known as the BO 108, the aircraft would be available with the customer's choice of a pair of P&WC PW206B or Turboméca TM319-1B turboshaft engines.[138] In 1993, Eurocopter Deutschland (ECD), a joint Aerospatiale/Deutsche Aerospace company, launched a civil multipurpose helicopter known as the EC135. The EC135 was a dual-engine helicopter, and it was offered with either Turboméca TM319-1B1 Arrius or PW&C PW206B turboshaft engines.

The PW206B had a take-off rating of 635 shp. The PW206B was a PW206A power section with a new angled drive gearbox to fit the application. The

---

137. "Turboméca and P&W Engines Selected for MDX Helicopter," *Business & Commercial Aviation*, Vol. 65, No. 2 (August 1989): 28.

138. Seth B. Golbey, "MBB Gives Green Light to Long-Awaited BO 108 Twin," *AOPA PILOT* (March 1991): 28.

PW206B test program began in 1992, and the first engine ran in February of that year. The first two prototypes of the engine were delivered to ECD in November 1993. The PW206B-powered EC135 prototype made its first flight in April 1994. Later, a 640-shp (take-off) PW206C was selected to power the Agusta A109 Power, a high-performance twin-engine helicopter. The PW206C was similar to the PW206A, but configured specifically for the the the Agusta A109 Power. The first production PW206C engines were shipped in January 1996.

While the PW200 engine originally evolved from an attempt to design a small, inexpensive turbine engine for either fixed- or rotorwing aircraft, the first applications for the engine were helicopters. A commercially successful turbine engine that could substantially replace the turbocharged piston engine in the small fixed-wing aircraft market had yet to be built. P&WC was successful in finding important helicopter applications for the PW200.

The PW300

The mid-size or transcontinental corporate jet market was an area that had interested P&WC since the 1970s. In 1972, Garrett had certificated its TFE731, and the first production models were installed on the Dassault Falcon 10. It was rated at 3,500 lb thrust compared to the 2,200-lb-thrust JT15D. The TFE731 significantly improved the usefulness of the first-generation business jets such as the Lockheed JetStar and Learjet by improving their range, performance, and reducing their noise level. In the years that followed, the TFE731 went on to become the most popular engine in its class for the mid-size corporate jet.

In the early 1970s, with an eye toward this growing market, P&WC initiated a design study (ADS509) of a 5,000-7,000-lb-thrust turbofan engine. As the study progressed, it was clear that development of such an engine, later designated JT25D, would be very costly. At P&WA's suggestion, P&WC contacted the Bristol Engine Division of Rolls-Royce to discuss a co-development agreement. Rolls-Royce was well into the design of its RB401, in the same power range as the JT25D. Subsequently, an agreement was reached to pool engineering and financial resources, and the RB401 became the lead project because of its advanced status.[139]

At the time these discussions were taking place, P&WA, Rolls-Royce, Motoren- und Turbinen-Union GmbH (MTU), and Fiat SpA had received per-

139. Sullivan and Milberry, *The Pratt & Whitney Canada Story*, 258.

mission from the U.S. Department of Justice for a partnership arrangement on the JT10D. At a later stage, an unofficial opinion was sought from the U.S. Department of Justice relative to a cooperative venture between P&WC and Rolls-Royce. The advice received was that such a program would not be acceptable, since the cost was such that either company could handle it, while that was not the case with the much more expensive JT10D program. Thus, in 1976, the P&WC/Rolls-Royce partnership was dissolved. It is ironic to note, that within the next decade, joint engine partnerships became standard as engine development costs rose beyond the capability of many individual manufacturers.

MARKET REASSESSMENT AND DESIGN RATIONALE

In the 1980s, P&WC resurrected the JT25D/RB401 concept with a new engine, the PW300, similar to its predecessor in terms of thrust class, weight, and sfc. As a result of the economic recession in 1982, P&WC's JT15D fanjet engine production had dropped radically from a high of 499 engines in 1981 to a low of 86 engines in 1983. Aircraft at the lower end of the price and performance scale, such as the JT15D-powered Citation, were owned by smaller companies, hard hit by the recession. More expensive, medium-size jets owned by larger corporations were less susceptible to economic setbacks. As conditions stabilized, P&WC began to re-examine opportunities in the mid-size turbofan market.

In the process of examining this market, P&WC came to the conclusion that it would be unlikely that new aircraft would be developed.[140] The market was saturated with candidate aircraft, and high development costs meant that entirely new designs were prohibitively expensive. However, to maintain their market share, it seemed likely that airframers might be convinced to upgrade and improve their existing aircraft. Improvements in range, speed, and cabin volume would require a larger engine than the TFE731, then rated between 3,500 to 4,500 lb take-off thrust. Aircraft such as the British Aerospace BAe 125, Falcon 20, Learjet 55, and Citation III were very good airframes and could effectively use more power.[141]

Strategically, it also made sense to avoid a direct challenge to Garrett's strong position in the mid-size corporate jet market. It was better to jump over the TFE731 and develop a larger fanjet engine. A successful engine would strengthen and stabilize P&WC's corporate jet engine sales. The company

---

140. Graham Warwick, "Small Turbines Step Forward," *Flight International*, Vol. 132, No. 4074 (August 8, 1987): 27.

141. Bill Sweetman, "PW300, FJ44 and Dual Pac," *Interavia* (June 1988): 482.

would have the JT15D for entry level corporate jets and the PW300 for mid-size corporate jets. Analyses at P&WC indicated that Garrett's TFE731 was near the upper limit of its growth capability.[142] This left an opening in the thrust class in which P&WC was interested. If the PW300's thrust range were sized properly, higher thrust PW300 engines could be directed primarily to this market. Lower thrust PW300s could be marketed as improved performance retrofit engines for existing mid-size corporate jets, capturing some of the TFE731 market. From the beginning, the PW300 was envisaged as a family of engines with a possible thrust range between -15% and +30% of a baseline engine rated at approximately 5,000 lb take-off thrust.[143]

Weight and performance were to become the big drivers in the PW300 program.[144] The engine could not be a conservative design. It would have to replace a generally rear-mounted engine on a stretched airframe, and the airplane would have to balance properly without major modifications. Therefore, engine weight was critical and could be no more than 950 lb. Significant performance enhancements, specifically power and sfc, were critical as well.

In October 1984, testing of engine components began. Experimental tests of high-risk technology concepts were conducted prior to detailed design in order to determine as early as possible whether the concepts were workable. The objective was to spend development dollars near the start of the new program to reduce long-term risk and cost. Emerging from preliminary design in the middle of 1985, the engine was defined at 4,750 lb take-off thrust and 950 lb weight. The engine's 30% growth potential could be realized either in one step or in 10% increments. Both a regional jet engine and a turboprop core engine variants had been planned during preliminary design in order to accommodate unforseen opportunities.

Because the estimated cost of developing the engine (approximately $500 million Canadian) was very large, a risk-sharing partner was sought. Motoren-und Turbinen-Union München GmbH was interested in picking up 25% of the investment and sharing technical development of the engine. While this would lower P&WC's profits by 25%, it also strengthened European market potential. MTU assumed responsibility for design, development, and manufacturing of the LP turbine, turbine exhaust casing, and shared the test program costs. The Canadian government underwrote 26% of the PW300 development costs. This

---

142. Sullivan and Milberry, *The Pratt & Whitney Canada Story*, 301.

143. R. A. Harvey and D. Karanjia, "Optimising Turbofans for Business Aircraft," (Pratt & Whitney Canada, Inc. Paper No. CASI-870008, May 1987), 1.

144. Hardy to Leyes, 6 October 1988.

P&WC PW305 turbofan engine. Photograph courtesy of P&WC.

federal assistance was in the form of a repayable loan based on future engine sales royalties.

The resulting PW300 was a twin-spool turbofan engine. A four-stage axial, single-stage centrifugal HP compressor was driven by a two-stage HP turbine. The first two axial compressor stages incorporated variable guide vanes. The first-stage stator vanes and rotor blades of the HP turbine were cooled. The engine incorporated P&WC's pipe diffuser design. A straight-through combustor featured air-blast fuel nozzles derived from PW100 technology. The front fan with a bypass ratio of 4.5:1 was driven by a three-stage LP turbine.

The engine also incorporated a full authority digital electronic control with no hydromechanical backup for optimum high-altitude handling. The PW305 was the first general aviation engine to be certificated with a FADEC.[145] With FADEC, the engine thrust levers were not linked physically to the engines, but rather electrically to the FADEC computers. These computers continuously

145. Fred George, "Full Authority Digital Engine Controls," *Business & Commercial Aviation,* Vol. 71, No. 3 (September 1992): 111.

monitored temperature and altitude and adjusted the power so that the engines always operated at their most efficient settings.

The PW300 was P&WC's first high-bypass ratio turbofan engine. It was designed for minimum weight and maximum efficiency at a design point of 1,113 lb thrust at 40,000 ft at Mach 0.8. It was designed to give medium-size business jets transcontinental range in either direction. The fan, above all, conferred the extra range necessary for the transcontinental mission.[146] Further weight saving was made possible with what P&WC termed a "Fan-High" configuration meaning that no LP compressor was incorporated in the original design of the LP shaft. However, P&WC did intend that higher powered versions would have LP compressor boost stages to supercharge the engine core with either the same fan or a larger fan.

The PW300 engine was the highest-technology engine that had yet been developed by P&WC.[147] The greatest risk was achieving the prescribed weight at the specified thrust. Among the components not used previously in P&WC engines were variable vanes in the HP compressor. While similar components were considered during early stages of the PW100 program, ultimately they were not used. To obtain the lightest weight engine, variable vanes were a necessity in the PW300. P&WA assisted in the development of these vanes. A straight-through-flow combustor was used for the first time, principally to reduce engine diameter. Some derivative technology was used in the engine. The fan section was styled after the JT15D, the compressor was similar to that used on some PT6 models, a PW100-style HP turbine section was used, and the LP turbine system was patterned after P&WA's PW3000.[148]

Full design of the PW300 started in September 1985. The test core of the engine was first run on September 28, 1987. The first run of a complete PW300 was on March 15, 1988 at P&WC's plant in Mississauga, Ont. Ten engines were used in a fast-paced two-year test program. The first flight test of the PW305 on P&WC's Boeing 720B testbed aircraft was on May 1, 1989. Certification was obtained in August 1990. The engine met its weight target, and sea-level thrust and sfc were within specification. Additional work was required to meet the specified high-altitude cruise sfc. This was achieved by tightening up compressor and turbine tip clearances and stiffening both turbine and engine structural

---

146. "The Business Jet Power Battle," *Flight International*, Vol. 135, No. 4166 (May 27, 1989): 38.

147. Sullivan and Milberry, *The Pratt & Whitney Canada Story*, 303.

148. Robert L. Parrish, "The Power Behind the BAe 1000," *Business & Commercial Aviation*, Vol. 65, No. 4 (October 1989): 100.

casings as well as more careful use of turbine cooling air and controlling leaks of HP air.[149]

WINNING AIRFRAME APPLICATIONS

In 1989, the PW305, rated at 5,200 lb thrust was selected for the British Aerospace BAe 1000, a stretched and longer-range version of the successful 125 series. The PW305 was selected in preference to the 5,600-lb-thrust General Electric/Garrett CFE738 and the Garrett TFE731-5BR.[150] The PW305 was selected over the CFE738 because the CFE738 was not yet certificated, and its use would have delayed the BAe 1000 program an estimated 18 months. Higher thrust and better sfc were the reasons that the PW305 was selected in preference to the TFE731. The first flight of the PW305-powered BAe 1000 took place on June 16, 1990. The PW305 made the BAe 1000 the first intercontinental mid-size business jet.[151] Production versions of the BAe 1000 were powered by PW305B engines, flat rated at 5,266 lb thrust and certificated in January 1993.

In 1989, the PW305 also competed against the CFE738 for the Falcon 2000, an aircraft similar in concept to the BAe 1000 in that it was conceived as a business jet larger than mid-size, but smaller and less expensive than top-end executive aircraft.[152] However, in 1990, the CFE738 was selected for the Falcon 2000.

In 1989, P&WC's overall marketing strategy for the PW300 engine family was realized in part when Volpar Aircraft Corporation initiated plans to certificate a PW305 engine for retrofitting a Dassault Falcon business jet. The engine was certificated in August 1990, and the first flight of the Volpar PW300-F20 aircraft powered by a PW305 took place in March 1991. As a retrofit engine, the PW305's important selling points were that it had more thrust and burned less fuel than a comparable TFE731 and accomplished this at a similar size and weight.

An application for the PW305A was the Learjet Model 60 business jet, which made its first flight powered by two of these engines, take-off, flat rated at 4,679

149. Hardy, "Development of Gas Turbine Engines," 77.

150. John Fricker, "BAe's 1000: The Ultimate 125?," *Business & Commercial Aviation*, Vol. 65, No. 4 (October 1989): 97.

151. Nigel Moll, "BAe 1000: First Intercontinental Midsize Bizjet," *Flying*, Vol. 118, No. 11 (November 1991): 68.

152. "Dassault 'Medium Large' Falcon 2000," *Aviation Equipment Maintenance*, Vol. 8, No. 12 (December 1989): 48.

lb thrust each, in June 1991. The Learjet 60 had the largest cabin and longest range in its family.[153] In the late 1980s, the PW305A had edged out a Garrett TFE731-5B in the contest for this airframe. The PW305A was certificated in December 1992 and entered service on the Learjet 60 in January 1993.

In its continuing battle with the CFE738, the PW306A won an upset victory when it was selected in 1994 for the Israel Aircraft Industries Galaxy long-range business jet. The CFE738 was then a certificated engine, a competitive advantage, whereas the PW306A was not. However, the PW306A offered the same installed performance, weighed 600 lb less, and had a simplified starting system.[154] In October 1993, P&WC launched design and development of the PW306A.

A derivative growth version of the PW305, the PW306A was take-off flat rated at 5,700 lb thrust. The primary difference between them was the fact that the PW306A had a new fan design that was one inch larger in diameter with a one-quarter inch increase in blade chord and two fewer blades. The fan was also moved forward one inch, and the fan stator behind it was repositioned from a rearward sweep to a near vertical position to improve noise control. The first-stage axial compressor was redesigned to produce more airflow and greater efficiency. The second-stage HP turbine blades were redesigned to incorporate "shower-head"-type cooling and were cast of a single-crystal alloy in order to accommodate higher temperatures. The three stages of LP turbine blades were restaggered, and a forced exhaust mixer was added for better altitude performance and noise control. These changes resulted in a 20% higher flat-rated thrust than the baseline PW305 with little change in external engine dimensions. The first PW306A engine run took place in March 1994, and the first flight was on August 17, 1994. The PW306A was certificated November 22, 1995.

The PW300 series engine successfully achieved its design goals by providing improved performance and transcontinental range for derivative mid-size business jets. It also proved to be a highly successful competitor against the General Electric/Garrett CFE738 and Garrett's TFE731. It was a design departure for P&WC in that it incorporated features such as variable vanes in the HP compressor and straight-through-flow combustor. It was also P&WC's first high-bypass ratio turbofan engine.

153. David M. North, "Learjet 60 Stakes Claim in Corporate Market," *Aviation Week & Space Technology*, Vol. 138, No. 26 (June 28, 1993): 38.

154. "P&WC Scores with its PW306A," *Business & Commercial Aviation*, Vol. 75, No. 2 (August 1994): 84.

P&WC PW500 turbofan engine mock-up. Photograph courtesy of P&WC.

The PW500

With its JT15D and PW300, P&WC had bracketed the low and high end power requirements for the light- to medium-size corporate jet market. Powering the lower end of this market, the JT15D was rated from 2,200 lb thrust (JT15D-1) to 3,190 lb thrust (JT15D-5). The higher end was covered by the PW300, with approximately 4,750 to 5,700 lb thrust. By designing an engine in the 3,000 to 4,500-lb-thrust class, P&WC could fill its gap in the middle power range and have a seamless product offering for the low- to medium-thrust turbofan engine market.

P&WC filled this gap with the PW500 turbofan engine. With the JT15D, PW500, and PW300, P&WC had a set of products that could compete, depending on the power requirement of a given airframe, against the various models of the AlliedSignal (Garrett) TFE731 and AlliedSignal/General Electric CFE738 turbofan engines. Another important justification for the PW500 program was that advances in technology made it possible to build a substantially more efficient engine than the JT15D. Because the JT15D would be replaced eventually in the marketplace, it was in P&WC's best interest to be the company that replaced it.[155] A trend away from turboprop to turbofan powered aircraft in the lighter end of the market made this an important future market, a fact that further justified the decision.

In the mid-1980s, P&WC market surveys indicated a need for an engine to power derivative aircraft with 10 to 15% better sfc and more thrust at altitude compared to existing corporate jet engines. There was an interest in getting airplanes to cruising altitudes faster and to higher altitudes than were being used. Such an engine had to be light, compatible with existing airframe engine mountings, and optimized for two-hour missions.[156]

Originally, two variants were conceptualized to meet this need. The first version was the PW530, rated at 3,000 lb thrust. The second, larger version was given a temporary engineering designation of PW500/10 and was rated at 4,000 lb thrust.

In January 1992, P&WC began design work on a PW500 demonstration engine known as the Turbofan Technology Integrator/Demonstrator (TTID). The purpose of the TTID was to reduce the development time and to assess the development risk. The TTID was built to be close to production configuration and first tested as a complete engine in November 1992. The first test flight of the demonstrator engine was in November 1993 on P&WC's Boeing 720 flying testbed.

Design work on the PW500 program was begun after a launch customer had been secured. This customer was Cessna Aircraft, which had selected the PW530A, rated at 2,750 lb thrust, for its new Citation Bravo medium-size corporate aircraft.[157] The Bravo was designed as a Cessna Citation 2 replacement. The PW530A engines would allow the Bravo to climb to cruising altitude much more quickly than the engines on the Citation 2; the Bravo also cruised faster and burned less fuel.[158] Preliminary design of the 2,750-lb-thrust PW530A be-

155. John Morris, "Pratt & Whitney Canada Aims to Give More Value," *Business & Commercial Aviation Show News*, NBAA 1994 (October 4, 1994): 36.

156. David Esler, "A New Player in the Medium-Thrust League," *Business & Commercial Aviation*, Vol. 75, No. 2 (August 1994): 80.

157. Francine Osborne, "The Pratt & Whitney Canada PW530A Turbofan Selected to Power the New Cessna Citation Bravo," P&WC Press Release, Farnborough, England, 5 September 1994.

158. "Cessna's Last New Plane—The Bravo—Was Launched Only a Month Ago," *Business & Commercial Aviation Show News*, National Business Aircraft Association 1994 (October 4, 1994).

gan in August 1992, and detailed design work began in November. The first test cell run of the PW530 was on October 29, 1993. Initial altitude performance and operability were demonstrated on P&WC's Boeing 720 aircraft on May 27, 1994. The PW530 was certificated on December 22, 1995.

On October 4, 1994, it was announced that Cessna had selected the second model in the PW500 family, the PW545A, rated at 3,640 lb thrust, for its new Citation Excel medium-size corporate aircraft.[159] Detailed design on the PW545A began in April 1994, the first engine run was on December 20, 1994, the first flight on P&WC's Boeing 720 was on May 24, 1995, and the PW545-powered Citation Excel first flew on February 29, 1996.

TECHNICAL ASPECTS

The PW500 was a new or "clean sheet of paper" engine design and a new engine family for P&WC, approximately the same size and weight as P&WC's JT15D, but with higher thrust and improved sfc. The PW530 was flat rated at 2,600 lb thrust with a take-off sfc of 0.468 and a dry weight of 605 lb. The PW545 was flat rated at 3,876 lb thrust with a take-off sfc of 0.436 and a dry weight of 765 lb.

The PW500 series engine was a two-spool, high-bypass turbofan engine. The PW530 and the PW545 shared a common gas generator core consisting of: a two-stage axial, single-stage centrifugal compressor; reverse-flow combustor; single-stage cooled HP turbine vane assembly; and a single-stage HP turbine with non-cooled blades. The axial compressor components were of an integrally bladed rotor (IBR) design. Both engines used a forced mixer in the LP turbine assembly, incorporated to mix the core and bypass flows efficiently for improved high-altitude performance. In the PW530, the LP rotor included a 23-inch IBR fan driven by a two-stage turbine. Among the additional features in the PW545 were: an increased fan diameter to 27.3 inches (utilizing an IBR wide-chord design); an IBR compressor boost stage to increase airflow into the engine; a third, un-cooled LP turbine stage; and single crystal HP turbine blades for higher temperature operating capability.

The IBR technology in the fan and first two compressor stages was chosen to reduce weight, improve low-cycle fatigue characteristics due to reduced blade root stresses, enhance foreign object damage protection due to more robust blade profiles, and improve efficiency by eliminating blade root leakage.[160]

---

159. Francine Osborne, "The Pratt & Whitney Canada PW545A Turbofan Selected to Power the New Cessna Citation Excel," P&WC Press Release, New Orleans, Louisiana, 4 October 1994.

160. Stanley W. Kandebo, "PW500 Flight Tests to Begin by Year-End," *Aviation Week & Space Technology*, Vol. 139, No. 18 (November 1, 1993): 65.

As was the case with the PW300 series engines, the LP turbine module components of the PW500 series engines were designed, developed, and produced by MTU-München. MTU joined the PW500 program as a 15% risk sharing partner, scheduled for increasing participation over time. With the PW500 engine, P&WC filled its gap in the middle power range (between the JT15D and the PW300) and had a seamless product offering for the low- to medium-thrust turbofan engine market.

Non-Production Engines

Every turbine engine company has developed to various degrees engines that did not go into production. The PW400 and the T800 are examples of P&WC engines that did not reach production.

In the mid-1980s, with an eye toward fast commuter aircraft including the Embraer/FAMA CBA-123, Shorts/DHC NA90, and Dornier Do328, PW&C decided to offer a "super" PT6 to be known as the PW400. Announced in 1987, the proposed 2,000-shp engine was to be longer and slightly wider than the PT6 with a similar layout. It was to have been a newer technology engine, weighing 600 to 700 lb, with a free turbine configuration suitable for pusher or tractor installation.[161]

At the time, the biggest PT6 had a capability of about 1,500 shp. P&WC's PW100 series commuter aircraft engine had been conservatively designed for significant growth capability. This meant that the engine was a bit heavy and less competitive in terms of weight at the lower end of its 1,600-to 1,800-shp power range. It was thought that the PW400 could bridge the gap between the PT6 and PW100.[162]

The engine was designed and negotiations were started with Embraer. Competing against Garrett and General Electric, P&WC Canada offered Embraer a relatively high price based on what it felt was a limited volume market. In what P&WC felt was a very aggressive commercial proposal, Garrett offered Embraer a new free turbine engine (the TPF351-20).[163] In the final competition, Embraer told P&WC that it wanted to split the engine market 50-50 between the PW400 and the Garrett TPF351. From P&WC's viewpoint, an already limited volume market was likely to be cut in half, and the company

161. Graham Warwick, "Small Turbines Step Forward," *Flight International*, Vol. 132, No. 4074 (August 8, 1987): 27.

162. McLachlan to Leyes, 7 October 1988.

163. McLachlan to Leyes, 7 October 1988.

withdrew from the competition. P&WC then discontinued work on the PW400.

After the competition, P&WC worked on improving the competitiveness of its existing engines in the 2,000-shp power range. With a weight reduction program, the company sought to make the PW100 more competitive in the 1,800-to 2,100-shp range. It also sought ways to increase the power of the PT6A-67.[164]

Another non-production engine was the T800. In 1982, the U.S. Army invited helicopter manufacturers to submit design concepts for a multi-mission helicopter, the LHX (Light Helicopter Experimental) to be powered by a new turboshaft engine, the T800. In August 1984, Textron Lycoming and P&WA jointly submitted a proposal for the LHX engine. In July 1985, competitive contracts were awarded to two teams: P&WA/Textron Lycoming and Garrett/Allison. P&WC joined the P&WA/Textron Lycoming team and took responsibility for developing the centrifugal compressor and two-stage power turbine. The T800-APW-800 built by P&WC, P&WA, and Textron Lycoming ran on July 7, 1986.[165] Late in 1988, the Army selected the Garrett/Allison engine, at which time further work by P&WA on the T800 was discontinued.[166]

Research and Technology Development

Research and development (R&D) was traditionally a company priority. In the 1980s, an average of 20% of the company's total budget was spent on R&D compared to a 3% average in Canadian industry.[167] Of that, about one-third was invested in improving its various production engine models.

P&WC devoted significant research effort toward improving its centrifugal compressor technology. In the mid-1960s, P&WC developed a viscous pipe analogy design method for rotors incorporating swept blading and a new concept in diffusers known as the pipe diffuser.[168] Backswept blading lowered the

---

164. Stanley W. Kandebo, "Pratt & Whitney Canada Applies Expertise in Small Gas Turbines to New Markets," *Aviation Week & Space Technology*, Vol. 129, No. 3 (January 18, 1988): 42.

165. Details on the development, demonstration, and testing of the T800-APW-800 are in this book's Lycoming chapter.

166. Sullivan and Milberry, *The Pratt & Whitney Canada Story*, 303.

167. Sullivan and Milberry, *The Pratt & Whitney Canada Story*, 308.

168. R. E. Morris and D. P. Kenny, "HP Ratio Centrifugal Compressors for Small Gas Turbine Engines," (A P&WC Paper prepared for the 31st Meeting of the Propulsion and Energetics Panel of AGARD "Helicopter Propulsion Systems," Ottawa, Ontario, 10-14 June 1968), 6-5. The pipe analogy method was used to design impellers by setting up an analogy between the flow in an impeller and the flow in a curved diffusing pipe.

Mach number of compressor discharge air and resulted in improved compressor efficiency. The pipe diffuser, which significantly improved compressor efficiency, became a hallmark of P&WC engines. In parallel with these developments, P&WC initiated a pioneering two-dimensional finite element stress analysis for complex shapes such as swept blades.

Using these concepts, a series of research demonstrations were conducted by P&WC. Pressure ratios from 5 to 14:1 at efficiencies of 80 to 70% respectively, above the then state-of-the-art compressor technology, were demonstrated by P&WC in 1970 and published in the early 1970s.[169] The first uprating of the PT6 was significantly improved by the application of this technology. The JT15D, certificated in 1971, also incorporated such a centrifugal rotor as its core compressor.

Improvements continued to be made in the rotor viscous analysis and the design of the diffuser passages leading to the PW100 turboprop, based on a twin spool, two-stage centrifugal compressor. In this same time frame, 10:1 compressor pressure ratio at an efficiency of 80% was achieved in a single stage for the first time in a P&WC technology demonstration.

As a result of this background in centrifugal compressor technology development, the U.S. Army Aviation Applied Technology Directorate awarded P&WC a demonstration program in 1981. The target objective of the program was 15:1 pressure ratio in a single centrifugal compressor stage at 78% efficiency. In 1985, P&WC met the target objective.[170] In 1988, efficiency was raised to 80%. This later figure represented a gain of 10.0 efficiency points over the original 14:1 compressor pressure ratio demonstration in 1970.[171]

In the early 1990s, P&WC concentrated on optimizing efficiency in the 8 to 10:1 compressor ratio range for single centrifugal stages in 600- to 900-shp engines. Also, the centrifugal stages of the axial-centrifugal compressors used in P&WC's larger turboprop and turbofan engines were the subject of an intense technology program.

Another component technology that P&WC developed extensively for its engines was highly loaded turbines. According to retired Vice President of Engineering Colin Wrong:

169. David Kenny, "P&WC Centrifugal Compressor Highlights," to Rick Leyes, unpublished notes, November 1991, National Air and Space Museum Archives, Washington, D.C.

170. David P. Kenny, "The History and Future of the Centrifugal Compressor in Aviation Gas Turbines," (SAE Paper No. 841635 (SP-602) presented as the First Cliff Garrett Turbomachinery Award Lecture, 16 October 1984), 8.

171. Kenny, "P&WC Centrifugal Compressor Highlights."

What we always did as Pratt Canada, with a limited budget and so on, was specialize in stuff that we needed for our little engines which we couldn't get from the people (P&WA) that were doing the big engines, for instance, single-stage compressor turbines. The PT6 was the first engine designed with that philosophy, and nobody could tell us how to do that. We had to sort of do that ourselves. That was a matter of jacking up the blade speed (and thereby) making a big temperature drop across the turbine. We got good efficiencies and had only one stage whereas everyone else at that time period would have been using two stages. Today any new engine will use much fewer stages than they used to. Pratt Canada was the first one there with these highly loaded turbines.[172]

Overview

THE STATUS OF THE CORPORATION

In the worldwide turbine engine market, P&WC's market niche was principally commercial fixed-wing turboprop and turbofan corporate aircraft and turboprop regional transport aircraft. P&WC also made inroads in military and commercial helicopter markets.

P&WC's head office, principal engineering center, and main manufacturing plants were located in Longueuil, Que. An engineering and assembly and test facility was situated in Mississauga, Ont., and an assembly and test facility in Lethbridge, Alberta. A computer-integrated manufacturing plant was located in Halifax, N.S. P&WC flight testing was conducted from St. Hubert, not far from Longueuil. In 1994, P&WC facilities totaled 2.7 million square feet.

In the late 1980s and continuing into the early 1990s, P&WC was the largest aerospace company in Canada. In 1990, sales reached a high of more than $1.5 billion dollars, and the company employed approximately 9,400 people. Consistent with the rest of the Canadian aerospace industry, the majority of P&WC's sales were destined for export. Between 1983 and 1993, exports constituted approximately 90% of the company's production.[173] Over that same time, P&WC's business shifted from primarily general aviation aircraft engines to regional transport engines.

In the early 1990s, a worldwide recession combined with the influence of the Persian Gulf War in 1991, resulted in a decline in the number of airline passen-

---

172. Wrong to Leyes and Fleming, 16 January 1991.

173. Stanley W. Kandebo, "Pratt-Canada Focuses on Export Markets," *Aviation Week & Space Technology*, Vol. 140, No. 17 (April 25, 1994): 24.

Pratt & Whitney Canada's main facilities in Longueuil, Quebec. Photograph courtesy of P&WC.

gers.[174] With regional transports being P&WC's primary business at that time, its subsequent production of engines and its employment level declined.[175] Engine production declined from approximately 1,800 units in 1990 to 1,200 in 1992, and employment declined from the 1990 peak of 9,400 to about 7,000 people at the start of 1993.[176]

By the end of 1995, P&WC had delivered a total of 43,027 turbine engines since 1963. By far, the largest number of these were PT6A turboprop and PT6T/B/C turboshaft engines, accounting for 32,472 of the total. By 1994, P&WC's star product, the PT6A turboprop engine powered the majority of light turboprop aircraft in the world.[177]

THE COMPANY'S MARKETS AND INFLUENCING FACTORS

Over the years, P&WC faced shifting markets. In 1958, when P&WC conducted its first market surveys, the 450-shp general aviation engine market was a

174. Pratt & Whitney Canada, *1991 Annual Report to Employees* (Longueuil, Quebec: Pratt & Whitney Canada, December 1991), 2.

175. David Hughes, "Slip in Canadian Growth Expected Before Expansion Rate Rebounds," *Aviation Week & Space Technology*, Vol. 136, No. 11 (March 16, 1992): 48.

176. David Hughes, "Difficult Market Conditions Pose Challenges for Canada," *Aviation Week & Space Technology*, Vol. 138, No. 11 (March 15, 1993): 51.

177. Pratt & Whitney Canada, *World Leader in Small Gas Turbine Technology* (Longueuil, Quebec: Pratt & Whitney Canada, January 1994).

promising market that was open for competition and in which P&WC could compete. The U.S. military jet engine market was dominated by two large companies, Pratt & Whitney Aircraft and General Electric. Commercial jet aircraft were also coming on line in the U.S., but only large turbine engine companies were able to compete for this business. P&WC did not have the financial ability, staff, or manufacturing capability to compete for either the military or large commercial jet engine market and undertook development of the PT6 for the general aviation market. The PT6 established a technical and financial foundation for P&WC's turbine engine business and became by far its most important product.

During the 1960s, the general aviation economic boom and regulatory rule changes helped increase the number of turbine-powered executive and commuter aircraft. The Beech King Air launched the PT6 engine program for executive aircraft. The Beech 99 and de Havilland Twin Otter expanded PT6 applications to the regional transport market. From the 1960s through the 1970s, most PT6 sales were for commercial applications, primarily corporate aircraft.

In the late 1960s, the PT6T TwinPac penetrated the U.S. medium-size helicopter market with Canadian government financial help and with a U.S. production facility. In 1970 and 1971, most PT6T (T400) series engines sold were for U.S. military helicopters. In 1975, the petroleum industry's off-shore oil drilling boom drove a demand for TwinPac-powered commercial helicopters.

In the mid-1960s, P&WC's first turbofan engine, the JT15D, was selected for the highly successful Cessna Citation. An important engine for P&WC, the JT15D continued over the years to find a variety of commercial and military jet applications.

In the early to mid-1970s, P&WC began to focus on a turboprop engine larger than the PT6. The resulting PW100 engine was originally intended for use in executive aircraft. However, growth in commuter airlines following the 1978 Airline Deregulation Act had a strong influence on the design of the engine, which was ultimately optimized for regional airline use and sold well in that market. With the PW100, the majority of P&WC sales shifted from corporate to regional aircraft. In 1981, 57% of P&WC engine sales dollars were for corporate aircraft, but by 1988, 61% were for regional aircraft. In 1991, P&WC derived about 60% of its sales from PW100 series engines.[178]

A decline in JT15D turbofan engine sales in the early 1980s for small corporate jets, precipitated by an economic recession, spurred the development of the

178. David Hughes, "Canadian Aerospace Industry Prepares for Rising Competition, Falling Defense Sales," *Aviation Week & Space Technology*, Vol. 134, No. 11 (March 18, 1991): 69.

PW300 turbofan engine. In the late 1980s and early 1990s, the PW300 won its first applications on higher-thrust class, mid-size corporate jets.

The PW100 and the PW300 were part of P&WC's efforts started in the late 1970s to expand and diversify its product line. In the mid-1980s, P&WC developed the PW200 primarily for the low-cost, low-power helicopter turboshaft engine market. In the early 1990s, P&WC began efforts to further consolidate its position in the mid-size corporate jet market with the PW500 turbofan engine. With initial airframe applications in the early 1990s, the PW500 filled the P&WC's power gap between the JT15D and the PW300.

Because engine development time required four to five years and was very costly, accurate market forecasting was critical in order to have the right engine available when a market opened. However, having developed a seamless product line of available engines, the company could respond quickly to new market opportunities and switch its emphasis toward any given model depending on the market situation. Its broad product base helped to offset transitions in market emphasis.

Despite the economic recession which occurred in the early 1990s, the reduction in Cold War tensions led to new business opportunities in former communist and communist countries. P&WC pursued a policy that it described as trading technology for market presence.[179] The objective was to form partnerships that would result in production of P&WC engines in foreign countries at prices that would make the engines affordable within those countries. A key ingredient in this strategy was to have vendors within those countries make as many components and subassemblies as possible. After indigenous sources of production were established, the long-range objective was the codevelopment of new products (versions of existing engines) followed by new products. For such ventures, P&WC had targeted the Commonwealth of Independent States (CIS), formerly the Soviet Union; eastern Europe; and China. One such venture can be illustrated.

In 1993, P&WC and the Klimov Corporation formed a joint venture known as Pratt & Whitney/Klimov Ltd. in St. Petersburg. P&WC owned 51% of the company. The mission of the company was to establish itself as the leading supplier of small gas turbine engines for civil aircraft in the CIS.[180] Klimov was the leading engine design bureau with 90% of the existing helicopter fleet in the former Soviet Union and also a leader in military turbofan engines. For the ini-

179. Kandebo, "Pratt-Canada Focuses on Export Markets," 24.

180. Francine Osborne, "United Technologies and Klimov to Sign Joint Venture Agreement," P&WC Press Release, Le Bourget, France, 15 June 1993.

tial product line, P&WC granted licenses to the joint venture for its PW200 and PT6A-67 engines. This venture gave P&WC access to a potentially very significant market.

CONTRIBUTIONS

The PT6 was one of the most commercially successful small gas turbines ever designed, helping to make possible the first generation of turboprop corporate and commuter aircraft. Starting as an aircraft piston engine sales and service organization, by the 1980s, P&WC had become a worldwide exporter of small gas turbine engines and the largest aerospace company in Canada. A specialist in turbine engines for corporate and regional transport aircraft, P&WC also contributed to the utility, performance, and success of a wide-variety of military and general aviation aircraft and helicopters.

ALLISON | 9

Company Origin and Early Growth

The origin of the Allison Engine Company stemmed from a passion for automobile racing shared by a group of businessmen in the Indianapolis automotive community. Public zest for auto racing was fueled by such events as the festive Indiana State Fair car race on Labor Day in 1905 when Barney Oldfield was cheered across the finish line in his Green Dragon.[1] In the stands that day were Jim Allison and Carl Fisher, owners of the Prest-O-Lite company, Frank Wheeler of Wheeler Schebler Carburetor Company, Arthur Newby, principal owner of the National Motor Company and Vehicle Corporation of Indianapolis, and several other automobile manufacturers. So enthusiastic were these men, that they decided then and there to arrange for a 24-hour race on the same track. The success of that race led to their founding of the Motor Speedway, a two-and-one-half-mile paved and banked oval track. Allison served as Secretary-Treasurer of the Motor Speedway firm and became President in 1923, a position that he held until 1926 when the project was taken over by Edward V. Rickenbacker and his associates.

James Allison became a wealthy businessman involved in a variety of businesses. He was Vice President of the Allison Coupon Company, founded by his

---

1. Paul Sonnenburg and William A. Schoneberger, *Allison Power of Excellence* (Malibu, Cal.: Coastline Publishers, 1990), 15.

father and operated at that time by family members. However, Jim Allison's principal interest lay in his ventures in the automobile industry and automobile racing.

Allison was approached by his friend John Aitken, a speedway driver, who suggested that Allison start a racing team of his own. He enlisted the support of his associates, and on September 14, 1915, they established the Indianapolis Speedway Team Company, operated by Allison and Fisher.

Setting up shop in downtown Indianapolis, they began redesigning and re-building domestic and foreign cars and entering them in the races. Their early efforts were relatively unsuccessful, in part because they had to run the cars three miles between the shop and track for each test run. In late 1916, Allison suggested that a shop be built near the Speedway where it would be more con-venient to make test runs on the track, and that he become sole owner of the company. All agreed, and Allison took over the Indianapolis Speedway Team Corporation, which at that time had a workforce of 20 mechanics and engineers. The new shop was furnished with the latest mechanical precision machinery, and in January 1917, Norman H. Gilman, previously Assistant Superintendent of National Motors, was hired as the Chief Engineer and Superintendent of the Allison Speedway Team Company.[2] Gilman was to become a predominant figure in the company's development and growth.

SUPPORTING THE WAR EFFORT

The company was gaining recognition and success in the auto racing business, when suddenly its course was permanently altered by America's declaration of war against Germany on April 16, 1917. The day after war was declared, Allison came into the shop and announced, "Can't have any more races. Quit work on the cars, but hold the men and keep on paying them." He turned to Gilman and said, "Go out and find how we can get the war orders rolling. Take any jobs you like, especially the ones the other fellow can't do."[3] The final statement of that instruction signaled the emphasis that the company would place on solving unique technical and engineering problems in the years that followed.

Allison's prompt commitment of his shop resources to support the war effort resulted in a flow of challenging design and manufacturing jobs. The Allison shop developed production models of superchargers, trucks, Whippet tanks, and high-speed tractors. Allison also built the tools, fixtures, gauges, and model

2. Sonnenburg and Schoneberger, *Allison Power of Excellence*, 13-21.

3. "World War I Has its Day at Allison," *Allison News* Vol. 1, No. 3 (1 August 1941): 1-2.

parts for the first Liberty aircraft engines built by the Nordyke & Marmon Company.[4] As the war orders flowed in, the 20-man shop grew to 100 engineers and mechanics augmented by a temporary staff of 150 draftsmen. Of all the projects that the company worked on during the war, the most significant for its future was work on the Liberty engine. It led to Allison's entry into the aircraft engine business.[5]

AIRCRAFT ENGINES

After the war ended, Jim Allison returned to auto racing only briefly. He commissioned a winning car for the 1919 Indianapolis 500, then abandoned racing and sold all of his cars. Allison instructed Gilman to redirect the company's work to focus on mechanical and engineering jobs similar to those worked on during the War. In the Fall of 1918, Gilman had been elevated to Vice President, General Manager, and Chief Engineer. By 1920, operation of the plant had been placed in his hands. In the Fall of 1920, the company was renamed the Allison Engineering Company to reflect the nature of its business. By that time, Jim Allison had become actively involved in Florida real estate development with Carl Fisher. Allison's visits to the plant became relatively infrequent.[6]

Throughout the 1920s, Gilman and his staff shaped the company's mastery of high performance technology and the challenge of transforming power into propulsion. The company's solution of unique engineering and technical problems with bearings, gearing, and superchargers during the 1920s made important contributions to the aircraft engine industry.

The Liberty engine was Allison's stepping stone into the aircraft engine business. Although the Liberty engine had been successfully developed and manufactured during the War, it had a grave shortcoming. Its connecting rod bearings routinely failed after only about 50 hours. Through Norman Gilman's insight, the cause of this problem was diagnosed as a distortion of the rod bearing shell at high power levels, leading to failure and disintegration of the bearing. His solution, which consisted of strengthening the bearing shell and employing a unique technique for casting the bearing into the shell, extended the Liberty's service life from tens to hundreds of hours. The company's success with the redesign of the Liberty's bearings led to a series of Army contracts to retrofit new bearings into surplus Liberty engines.

4. "World War I Has its Day at Allison," 1-2.

5. Sonnenburg and Schoneberger, *Allison Power of Excellence*, 27-28.

6. "World War I Has Its Day at Allison," 1-2.

Gilman's steel backed cast bronze bearing design was later used in all U.S. and most foreign made piston-type aircraft engines. Allison was soon supplying bearings to engine makers throughout the world. From 1927, the Gilman bearing became a major portion of Allison's business and helped sustain the company well into the 1930s.[7]

Following World War I, the military had also turned to Allison for special engineering jobs. One important job was the development of high-speed gearing for aircraft engines. For piston engines to produce high power outputs efficiently, it was necessary that they rotate at speeds higher than those at which propellers could operate with good efficiencies. Consequently, a mechanical linkage was needed between the engine turning at high speeds and the propeller rotating at lower speeds. That need translated into high speed reduction gearing.

Gearing of the type needed did not exist in the U.S. at the time. Some engineers questioned the possibility of making gears strong and light enough to withstand operation at rotational speeds required for aircraft engines—speeds half again as fast as engineering handbooks considered feasible. Limits on gear speed had been established through observation that, at high speeds, small irregularities in the gears pounded on each other, leading to gear failure. Allison's solution was to manufacture gearing with very high precision, to use proper alloys, and to use appropriate heat treating.[8]

In addition to its aircraft bearing and gearing work, in the mid-1920s, Allison became involved in manufacturing high-speed, Roots-type superchargers, or "blowers" for aircraft engine manifold pressure boost. These blowers were used on Pratt & Whitney R-1340 Wasp engines, which set a number of altitude records.

Allison built the first aircraft engine of its own design for the Army Air Corps in 1924, an X-type, 24-cylinder, air-cooled engine of more than 1,200 hp. Unfortunately for Allison, the Army Air Corps abandoned interest in large single-engine airplanes before the Allison engine could be used.

During the 1920s, Allison also built a number of experimental marine engines, and a diesel engine for the Navy's airships. Under a Navy contract, work began in 1927 on a six-cylinder, in-line, 765-hp, two-stroke-cycle diesel engine. This engine passed its preliminary tests successfully but was eventually dropped by the Navy.[9]

---

7. Sonnenburg and Schoneberger, *Allison Power of Excellence*, 28–31

8. Sonnenburg and Schoneberger, *Allison Power of Excellence*, 32–33.

9. Robert Schlaifer and S. D. Heron, *Development of Aircraft Engines, Development of Aviation Fuels* (Boston: Graduate School of Business Administration, 1950), 274.

Also, during the 1920s, Jim Allison continued to provide financial support for the company he had founded, personally financing losses when they occurred. Although his time and business activities by then were almost wholly devoted to Florida real estate, he retained pride and interest in the company. Then, in August 1928, the fate of Allison Engineering was cast to the winds with Allison's untimely death at age 55.

GENERAL MOTORS

At the time of Allison's death, Allison Engineering could hardly have been classed as a business in the usual sense of the word. Profits were erratic and losses were persistent because the shop engaged only in experimental work. From a financial perspective, it was a hazardous enterprise. The executors of Allison's estate planned to get rid of it as quickly as possible. On December 31, 1928, the company was purchased by the Fisher brothers, owners of the Fisher Brothers Investment Corporation, a Detroit firm, and Eddie Rickenbacker was installed as President.[10] The sale stipulated that operations must continue in Indianapolis.

Just three months later, General Motors (GM) bought the company from the Fishers. Gilman was made Allison's President and General Manager, retaining the title of Chief Engineer. The company continued to operate under the name Allison Engineering Company.

GM's purchase of Allison was part of an overall strategy to reenter the field of commercial aviation. GM had become involved in aviation during World War I with the manufacture of 2,000 Liberty engines at its Buick and Cadillac plants.[11] In 1919, GM acquired the Dayton-Wright Airplane Company, founded in 1916 by Charles Kettering and five other Dayton businessmen. The company had built the British de Havilland DH-4 aircraft during the War.[12] However, in the post-war period, companies in the aircraft field lost heavily on government contracts, and a number of them folded. Likewise, Dayton-Wright was losing money. At the end of 1922, when Dayton-Wright's post-war losses had reached $400,000, GM Vice-Presidents John J. Raskob and Alfred P. Sloan

10. "Post War Events are Related in Allison History," *Allison News*, Vol. 1, No. 3 (15 August 1941): 1-3.

11. Report of Benedict Crowell, *America's Munitions 1917-1918* (Washington, D.C.: Washington Government Printing Office, 1919), 274.

12. Stuart W. Leslie, *Boss Kettering* (New York, N.Y.: Columbia University Press, 1983), 71-75, 90-97.

decided to withdraw GM from the aircraft field and henceforth liquidated Dayton-Wright. Raskob wrote, "The present market is entirely Army and Navy and so much difficulty has been found in dealing with the Government that it is considered useless to carry on."[13]

However, in the late 1920s, GM reversed its position as it became clear that aviation was likely to be one of the great American growth industries. As an automotive producer, GM was particularly concerned about the development of a market for the "flivver" plane, conceived as a small plane for everyday family use. In Sloan's words, "The development of such a plane would have large, unfavorable consequences for the automobile industry, and we felt that we had to gain some protection by 'declaring ourselves in' the aviation industry."[14] In 1929, GM reentered the aviation field. Two major investments were a 24% interest in Bendix Aviation Corporation and a 40% interest in Fokker Aircraft Corporation of America at a cost of some $23 million. In addition, GM purchased the entire capital stock of the Allison Engineering Company at a cost of only $592,000. Years later, Alfred Sloan said of the Allison purchase:

> By our standards, it was a small operation: the company had fewer than 200 employees in 1929 and its manufacturing facilities occupied only about 50,000 square feet of floor space. We considered it to be only of minor importance in our plans to enter the aviation industry. Yet as events turned out, we were to make Allison our principal link to the industry.[15]

GM's long-range view in making these acquisitions was summarized in its 1929 annual report as follows:

> General Motors, in forming this association (with the aviation industry) felt that, in view of the more or less close relationship in an engineering way between the airplane and the motor car, its operating organization, technical and otherwise, should be placed in a position where it would have an opportunity to (come into) contact with the specific problems involved in transportation by air. What the future of the airplane may be, no one can positively state at this time. Through this association General Motors will be able to evaluate the de-

13. Jacob A. Vander Meulen, *The Politics of Aircraft* (Lawrence, Kansas: University Press of Kansas, 1991), 43.

14. Alfred P. Sloan, Jr., *My Years With General Motors* (New York, N.Y.: Mcfadden-Bartell Corporation, 1963), 362-363.

15. Sloan, *My Years With General Motors*, 369.

velopment of the industry and determine its future policies with a more definite knowledge of the facts.[16]

GM moved quickly to chart the company's future. At its May 14, 1929, meeting the GM Operations Committee appointed a special committee that was charged, ". . . to consider the airplane program and formulate a plan for the Allison Engineering Company to decide what types of engines for aviation we should build."[17] Members of the special committee were Charles Wilson, Ormand Hunt, Charles F. Kettering, and Allison's Norman Gilman. For some time, Gilman had believed that Allison should build a modern engine, more powerful than anything yet attempted, as a replacement for the Liberty engine. Reflecting his views, the plan developed by the special committee was to build a 1,000-hp engine that would begin at a relatively conservative 750 hp. Based on the company's years of experience with the Liberty engine, it was to be liquid cooled. The decision to develop this new engine marked another major transition in the nature of Allison's business, from a technically skilled and well-equipped precision job shop to a developer and manufacturer of aircraft engines.

THE V-1710

On May 7, 1929, Norman Gilman and his design team began to sketch their new engine. It was to be a 12-cylinder, V-type, liquid-cooled engine having a displacement of 1,710 cubic inches, leading to its designation as the V-1710. Liquid cooling was selected to give the engine a slimmer profile than an air-cooled engine. Ethylene glycol coolant, with its boiling temperature of 357°F, would permit operation at higher coolant temperatures than the water-cooled Liberty engine. This resulted in a smaller radiator, which reduced aircraft drag. The initial design of this engine was made at company expense.[18]

The V-1710 underwent a turbulent period of development and marketing that extended through the 1930s. Gilman first took his design to Wright Field, but at that time, the Air Corps believed Curtiss-Wright would continue with liquid-cooled engine development, and the Army was unimpressed with Allison's comparatively small facilities and tiny engineering force. Gilman then

---

16. Sloan, *My Years With General Motors*, 363.

17. Sonnenburg and Schoneberger, *Allison Power of Excellence*, 44.

18. Sonnenburg and Schoneberger, *Allison Power of Excellence*, 48.

turned to the Navy and found that an interest existed in a high-power liquid-cooled engine to power the Navy's rigid dirigibles.[19]

Allison's first contract for the engine was received in 1930, from the Navy, for one V-1710-A engine rated at 650 hp. In 1932, the engine passed its 50-hour acceptance test at 750 hp. In 1933, the Navy gave Allison a contract for three reversible V-1710-B engines for use on its airships. (These engines could reverse from full power in one direction to the same condition in the opposite direction in 8 seconds). Two engines of this type were built and were ready to ship at the time the airship *Macon* was lost. With its loss, the airship program ended, and Allison's Navy contract was terminated.

Meanwhile, during the early 1930s, the Army developed an interest in using a 1,000-hp, turbosupercharged version of Allison's V-1710 as a fighter engine. In 1932, the Army ordered a modified, 1,000-hp version of the Navy's original engine. In March 1937, after extensive redesign, the V-1710-C6 passed its 150-hour type test, becoming the first American engine to qualify at 1,000 hp on a military approval test. In early 1937, the Army began flight testing the V-1710. The breakthrough for the engine came in 1939 when a Curtiss XP-40 powered by the V-1710-C13 won the fighter competition held by the Army to determine the relative superiority of various planes. The success of the XP-40 brought orders for the V-1710, and put an end to nine years of financial losses suffered by Allison in funding its development.[20]

FROM JOB SHOP TO PRODUCTION PLANT

Following the success of the Curtiss XP-40 in winning the fighter competition, there was a surge of interest in the V-1710—not only by the Army Air Corps, but also by the British and French armed forces. However, Allison had a serious problem. Although it had grown to some extent during the 1930s, it was still a small engineering firm with no facilities for quantity production.

The Assistant Secretary of War, Louis Johnson, visited William S. Knudsen, president of GM, to see what could be done about producing Allison engines. There were firm orders for only 836 engines and Johnson offered no assurance that more orders would be following. Although it did not make sense from a business point of view to set up a production plant for such a small quantity of engines, the leaders at GM felt that the V-1710 would probably be in great de-

---

19. Page Shamburger and Joe Christy, *The Curtiss Hawks* (Kalamazoo, Michigan: Wolverine Press, 1972), 227.

20. Schlaifer and Heron, *Development of Aircraft Engines, Development of Aviation Fuels*, 276-280.

mand. The decision was made to establish a new Allison production plant in Indianapolis.[21]

On May 30, 1939, ground was broken for a new Allison factory and office building with floor space of 360,000 square feet. In February 1940, production began in this new plant, and by December 1941, engines were being produced at a rate of 1,100 per month. In developing the plant, the far-flung organizing capability of GM was enlisted to put the Allison Division on a mass production basis. Highly experienced design and production engineers arrived from all points of the compass to help form a nucleus around which Allison ultimately built an organization of more than 10,000 employees.[22]

The V-1710 powered the Lockheed P-38, Bell P-39 and P-63, Curtiss-Wright P-40, and early models of the North American P-51. Production soared. In March 1944, the 50,000th V-1710 rolled off the production line, and when the production run ended in December 1947, 74,125 engines had been shipped. The final V-1710 order was for 750 engines to power the North American F-82 Twin Mustang. With these engines, piston engine development and production at Allison came to an end.[23]

As the power plant used in four frontline fighters during World War II, the V-1710 contributed significantly to the U.S. and Allied war efforts, particularly during the War's early years. Numerically, it was one of the most important engines built in the U.S. during the War, and it was the most important American designed liquid-cooled engine. It was also responsible for building a production base for future Allison aircraft engines.

THE MULTIBANK V-3420

By 1937, high officials in the Army believed that large liquid-cooled engines would soon displace air-cooled engines in weight-carrying aircraft as well as in fighters. Accordingly, Allison was awarded a contract to develop a double engine, the V-3420, basically two V-1710s geared together.[24] Considered the most powerful engine in the world at that time, the V-3420 was rated at 3,000 hp. In 1943, the V-3420 made its first flight in the Fisher P-75 Eagle, a high-speed low-altitude fighter. Late in the war, plans were made for producing the V-3420 in very large quantity for use in the P-75. However, because the P-75 lost a

21. Sloan, *My Years With General Motors*, 370, 371.

22. "Final Installment of History," *Allison News*, Vol. 1, No. 6 (19 September 1941): 2, 7.

23. Sonnenburg and Schoneberger, *Allison Power of Excellence*, 99.

24. Schlaifer and Heron, *Development of Aircraft Engines, Development of Aviation Fuels*, 279-293.

flight competition to the latest model P-47 airplane and because the war was nearly at an end, only about 100 engines were produced before the project was dropped.

Redefining GM's Role in Aviation

Early in World War II, it became apparent that GM's involvement in aviation had grown so large as to raise questions about the company's future in the industry. GM was not only building aircraft engines at Allison, but was also a major producer of military aircraft at its North American Aviation plant and of aircraft accessories at its Bendix plant. Alfred Sloan's recommendations in a 1942 report to GM's Postwar Planning Group became the basis for establishing the company's postwar aviation program. The essence of the report was that GM should develop a complete position in the manufacture of aircraft accessories, meaning basically aircraft engines, but not produce either military or commercial aircraft. He felt that the company could not sell engines effectively to its aircraft customers while competing with them in the aircraft marketplace. Following Sloan's recommendations, GM retained Allison to continue developing and manufacturing aircraft engines, and in 1948, disposed of North American and Bendix.[25]

The Gas Turbine Comes to Allison

In mid-1944, as V-1710s were rolling off the production lines at record rates, the Army Air Forces turned to Allison as a second production source for General Electric's I-40 (J33) turbojet engine. A few months later, the Army turned to Allison again as a second source for production of GE's I-16 (J31) engine. In early 1946, there followed an Army Air Forces decision to transfer almost all production of GE's TG-180 (J35) turbojet to Allison. By late 1947, Allison had been given complete engineering and production responsibility for both the I-40 and TG-180 engines. Introduction of I-40 production at Allison, followed by I-16 and TG-180 production, signaled yet another major transition in the history of the company, its move to become a major developer and manufacturer of aircraft gas turbines. It was through the acquisition of engineering data on these GE-developed engines and their early production and testing that Allison gained its initial turbine engine expertise.

25. Sloan, *My Years With General Motors*, 371-374.

THE I-40

At GE, several turbine engine programs competed for limited company resources. The Army Air Forces had placed high priority on the I-40 engine, the power plant for the Lockheed P-80 Shooting Star, considered by the Air Corps to be the most important of its jet aircraft projects. However, concurrent development of the I-16 and TG-180 turbojets and the TG-100 turboprop at GE absorbed a considerable portion of the company's engineering and manufacturing resources. The resulting I-40 production schedule that was offered by GE proved unsatisfactory to the Army Air Forces.[26]

On April 6, 1944, the Production Engineering Section at Wright Field recommended that a second source be considered for production of the I-40. The Resources Control Section at Wright Field then recommended that Allison be selected for this task. After brief consideration of other possible production sources, Allison was selected. Allison estimated that with a slight decrease in V-1710 and V-3420 production, it could provide adequate floor space to support a production program of 500 I-40 engines per month. By June 26, 1944, Allison had been brought into the I-40 production program with an Army Air Forces order for 2,000 engines, and in February 1945, Allison shipped its first I-40.

Following the termination of hostilities in the Pacific on V-J Day, August 15, 1945, the Army Air Forces determined that only a single source was required to produce the 910 I-40 engines remaining in its production program. Based on a cost quotation lower than GE's, Allison was given a contract on September 24, 1945, to produce the remaining I-40 engines. At that point, the Army terminated the GE I-40 contract and turned over to Allison all engineering responsibility for the I-40. Allison ultimately produced 15,525 J33 engines.

I-16

In October 1944, only four months after receiving the I-40 production contract, Allison was given a contract by the Army Air Forces to produce 720 GE I-16 engines plus spare parts. In February 1945, 200 engines were added to this contract for use by the Navy. Following V-J Day in August 1945, the Navy scaled

26. Historical Office, Wright Field, "Summary of Case History of Turbo-Jet Engine J33 (I-40) Series," (Declassified Secret Report), 1946, Historical Study No. 95, Office of History, Headquarters Air Force Logistics Command, Wright-Patterson Air Force Base, Dayton, Ohio, 1-5.

back its aircraft procurement, and the Army announced that it had no further requirement for the I-16. The production orders were canceled with only 22 I-16s delivered by Allison.[27]

THE TG-180 TRANSITION

In mid-1944, it became apparent to the military that GE's manufacturing capacity needed to be augmented further to satisfy demand for the TG-180 engine. In December 1944, the Army Air Forces selected Chevrolet to provide engineering and development assistance to GE on the TG-180. The manufacture of TG-180 engines was started at a GM facility in Tonawanda, N.Y., where 1,200 of the engines were to be produced. Work proceeded slowly, and in response to a proposal submitted by GM in December 1945, the Army Air Forces decided in April 1946 that Chevrolet would produce only 131 engines at Tonawanda with remaining production moved to Allison in Indianapolis. In September 1947, Allison received a contract giving it responsibility for all further production and engineering development of the J35 engine. Allison redesigned the engine, raising its rated thrust from about 3,700 to 5,000 lb, and developed an afterburner.[28] A total of 14,169 J35s were built.

THE T38 AND T40, ALLISON'S FIRST TURBOPROPS

In 1945, preliminary design work began on the company's first turboprop engines. One was the T38, an axial-flow turboprop rated at 2,925 shp.[29] The other was the 5,500-shp T40 turboprop, which consisted of two T38 power sections driving co-axial contrarotating propellers through a complex reduction gearbox. In June 1946, a contract was signed with the Navy to develop the T40. Although the T38 and T40 engines were not particularly successful—only about 240 T40s were built—their design was the first original creative work by Allison on gas turbines. In addition, by establishing the technology base leading to the

27. Amy C. Fenwick, Historical Office, Air Materiel Command, "Supplement to Case History of the Whittle Engine - Whittle, Types I, I-16 (J31), I-18, and I-20(J39) Turbo-Jet Engines," (Declassified Secret Report), July 1949, Historical Study No. 94, Office of History, Air Force Logistics Command, Wright-Patterson Air Force Base, Dayton, Ohio, 27-47.

28. John Wetzler, telephone interview by Rick Leyes and William Fleming, 24 June 1991, Allison Interviews, transcript, National Air and Space Museum Archives, Washington, D.C.

29. *Jane's All the World's Aircraft 1954-1955* (London: Jane's All the World's Aircraft Publishing Co. Ltd., 1955), 334-335.

highly successful T56, these engines set the company's course for the future. In that regard, Jim Knott, who became General Manager of the Allison Division in 1965, commented:

> The Navy put us into the turboprop business with their request for the T38 power section that became the T40. Yes, we said, of course we'd like to do that. But it sealed our fate, detracted from our interest in the pure jet business. We carved a niche for ourselves, but (the focus on turboprop technology) relegated us to third place in terms of gross sales (behind General Electric and Pratt & Whitney).[30]

The first flight of the T38 was in May 1949 as the nose-mounted fifth engine in Allison's Boeing B-17G flying testbed. Allison's first substantive entry into turbine powered commercial aviation came in 1950 when two T38s were installed in the Convair Turbo-Liner, a converted Convair 240 commercial transport first flown in December 1950. The B-17G and Turbo-Liner flight tests provided vital operating information, used later during flight tests of the Allison T56 in transport-type aircraft. Beginning in 1953, the T38 was also mounted in the nose of a McDonnell XF-88B testbed, used for research on supersonic propellers in a joint program of the Air Force, Navy, and National Advisory Committee for Aeronautics.[31]

The large T40 turboprop was used by the Navy in several of its experimental aircraft. The first of these was the Convair XP5Y-1, a prototype aircraft powered by four 5,850-hp T40 engines and first flown in April 1950. The XP5Y program was ultimately canceled due to high costs and instability of the aircraft in flight. The 7,100-hp T40-A-6 was also selected for use in two Navy competitive prototype VTOL aircraft, the Lockheed XFV-1 VTOL Pogo Stick and the Convair XFY-1 Pogo. The T40 was also tested briefly in the Douglas A-2D Skyshark, a fast, maneuverable attack aircraft.[32]

THE T56

Allison's experience in developing and operating the T38 and T40 engines led to the T56 turboprop, Allison's crown jewel, the engine that established Allison as the U.S. leader in the development and manufacture of large turboprop en-

30. Sonnenburg and Schoneberger, *Allison Power of Excellence*, 104-107.

31. Sonnenburg and Schoneberger, *Allison Power of Excellence*, 114-115, 117.

32. Sonnenburg and Schoneberger, *Allison Power of Excellence*, 106-108, 111-114.

gines. In 1949, Allison proposed to the Navy that it underwrite the development of a new twin-turboprop engine, the T54, with power 128% that of the T40. Soon after the T54 project began, the Air Force recognized the possible utility of an axial-flow turboprop for use in future transport aircraft. In 1951, the Air Force announced a requirement for a medium-sized logistic and tactical transport, from which would come the four-engine YC-130 Hercules. Following establishment of this requirement, the T56 was born. With the support of the Air Force, design of the 3,460-hp T56 began in 1951, and its development quickly surpassed work on the T54. Production of the T56 engine began in 1955.[33]

Jack Wetzler, who had been Project Engineer of the J71 developed earlier, observed that at the time the T56 was developed, it was the most modern turboprop engine in the marketplace. In addition, it happened to fit the power class that the airframers wanted for modern turboprop transport airplanes.[34] Eloy Stevens, Director of Marketing, expressed the view that Allison probably became the turboprop leader because they had the capability of being a low-cost manufacturer.[35]

The development and application of the T56, the military version of the engine, and the 501, the commercial version, proceeded on parallel paths. The first production installation of the T56 was on the Lockheed C-130 Hercules, which made its maiden flight on August 23, 1954. The 501-powered Lockheed Electra first flew in December 1957 with Electra commercial service inaugurated by Eastern Airlines in January 1959. Later, the 501 was used in the Convair 580 commercial transport. With continuing improvements and uprating, the military version of the engine was used extensively on the Lockheed C-130 Hercules transport, the Lockheed P-3 Orion anti-submarine warfare aircraft, the Grumman C-2 Greyhound transport, and the Grumman E-2 Hawkeye early warning aircraft.[36]

By the late 1980s, the engine had undergone four generations of uprating. The maximum power rating of the 4th generation engine, which went into production in 1987, had been increased to 5,250 hp, 40 percent above its initial

33. John Wheatley, D. G. Zimmerman, and R. W. Hicks, "The Allison T56 Turbo-Prop Aircraft Engine," (SAE Paper No. 500 presented at the Society of Automotive Engineers Golden Anniversary Aeronautic Meeting, New York, N.Y., 18-21 April, 1955), 1.

34. Wetzler to Leyes and Fleming, 24 June 1991.

35. Eloy Stevens, interview by Rick Leyes and William Fleming, 6 June 1991, Allison Interviews, transcript, National Air and Space Museum Archives, Washington, D.C.

36. Sonnenburg and Schoneberger, *Allison Power of Excellence*, 123, 137-146.

rating.[37] By the end of 1990, after 35 years of production, nearly 15,000 T56 and 501 engines had been built.[38]

THE J71 TURBOJET

After Allison had redesigned the J35 during the late 1940s, it was decided that a larger engine would be developed. In April 1949, work began on the J71, the first turbojet engine wholly designed by Allison. Design of Allison's new axial-flow turbojet drew on the company technology base developed during its uprating of the J33 and J35 engines and development of the T38 and T40 turboprops.

The J71 was used during the mid-to-late 1950s in the initial version of Northrop's S-62 Snark intercontinental missile, prototype and pre-production versions of Martin's XP6M-1 Seamaster flying boat, the McDonnell F3H-2N Demon carrier based fighter, and the Douglas B-66 bombe. However, the J71 was never able to penetrate the high-performance combat aircraft marketplace dominated by the more technically advanced GE J79 and Pratt & Whitney J57 turbojet engines, which were lighter and had better fuel economy than the J71.[39] Consequently, with the phaseout of the F3H program, failure to attract any further airframe installations, and with no new orders in sight, the program was dropped in the late 1950s after producing 1, 707 engines.[40]

THE ROLLS-ROYCE SPEY/TF41

Allison's initial turbofan engine venture stemmed from a relationship with Rolls-Royce. With a view to capitalizing on Rolls' turbofan engine experience while providing them with a U.S. marketing, engineering, and production base, Allison and Rolls-Royce established a partnership in 1958 to develop a family of turbofan engines. The principal engine that emerged was the adaptation of the Rolls-Royce RB.163 Spey into the 12,170-lb-thrust AR963-6 turbofan for use on the Boeing 727 tri-jet airliner. Following an intense 727 competition with Pratt & Whitney's JT8D turbofan, the Allison-Rolls engine lost.

37. Gordon E. Holbrook, "Workhorse of the Transport Industry: The Allison T56," *Casting About*, (1988): 10-11.

38. Sonnenburg and Schoneberger, *Allison Power of Excellence*, 206.

39. Wetzler to Leyes and Fleming, 24 June 1991.

40. Sonnenburg and Schoneberger, *Allison Power of Excellence*, 167, 196, 206.

There was no further market interest in this engine until the mid-1960s, when the Vietnam War created a demand for both military and civil transport jet engines that strained U.S. production capability. An engine source to supplement GE and Pratt & Whitney was desirable. At the same time, the Department of Defense was planning to update Ling-Temco-Vought's versatile A-7 Corsair II fighter-bomber, which was powered by the Pratt & Whitney TF30 turbofan. The Defense Department subsequently asked Allison to Americanize and refine the Allison-Rolls turbofan design for use in the A-7.

In mid-1966, Allison, with Rolls-Royce as a subcontractor, assembled an integrated Allison-Rolls design team and began work on that engine, which became the 14,250-lb-thrust TF41-A-1. As prime contractor, Allison was manufacturer and tester of the production engines, while the two firms jointly designed and developed the engine through qualification testing. Rolls provided approximately half of the final parts, and Allison the balance. Making its maiden flight in September 1968, the TF41-powered LTV A-7D was used during the latter years of the Vietnam War. A total of 1,414 TF41 engines had been manufactured for use in the A-7 by the time the engine went out of production in 1984.[41]

RESEARCH AND DEVELOPMENT FACILITIES

Engine and component test facilities played an important role in the evolution of the company's turbine technology base. During the late 1940s, several V-1710 test cells were modified to test jet engines and component test rigs were built. However, on July 12, 1951, a calamitous explosion caused by a fuel leak seriously damaged a number of the company's engine test facilities.[42]

In 1953, Allison began construction of a new research and development complex, marking the beginning of an extensive program to build and progressively expand a 214-acre research and engineering plant and laboratory adjacent to its engineering offices and manufacturing plants. It evolved over nearly four decades.[43] By 1990, the research and engineering plant totaled 920,000 square feet. This complex supported five functional activities vital to the design, development, and testing of new products: research, materials development, information processing, experimental manufacturing, and full-scale product testing.

41. Sonnenburg and Schoneberger, *Allison Power of Excellence*, 130-136, 206.

42. Sonnenburg and Schoneberger, *Allison Power of Excellence*, 115-117.

43. Richard B. Fisher, telephone interview by William Fleming, 3 February 1993, Allison Interviews, National Air and Space Museum Archives, Washington, D.C.

The research facilities consisted of 12 well-equipped laboratories used to develop and test specialized components such as small-scale fans, turbines, and combustors. Research also was conducted in such areas as heat transfer, combustion mechanics, component aerodynamics, rotor dynamics, fuel nozzles, and seals.

The materials laboratory supported the company's development and production of both aircraft and industrial turbine engines. Work in this facility covered materials qualification, failure analysis, and the research, development, and improvement of materials and processes.

Information processing was performed using general purpose and specialized equipment located at Allison as well as at facilities throughout the Midwest. Allison's equipment included a computer-aided design and manufacturing (CAD/CAM) center, a secure computer center for classified work, and a real-time simulator used by the electronics and controls group. A large information processing center located in Indianapolis was used to support the various Allison equipment. In addition, Allison had access to a CRAY/XMP vector computer at General Motors Research Laboratories in Warren, Mich.

Allison maintained an on-site experimental manufacturing facility to do developmental fabrication. Equipped with more than 360 conventional and computer-controlled machines, this facility supported the development of fabrication processes for new products and hardware configuration changes for old products.

Testing activities were conducted in a variety of full-scale engine and component test facilities that were supported by a large air supply and conditioning plant. Component testing was performed in compressor, turbine, combustor, and gearbox test facilities. Full-scale engine test facilities included 16 turboshaft dynamometer test stands of varying capacity, a variable attitude test stand, four turboprop test stands, and three turbofan and turbojet test stands. Conditioned air could be supplied to both the component and full-scale engine test stands for altitude simulation. New test facilities continued to be added as they were required to support development testing of new products.[44]

Allison by the Late 1950s

By the late 1950s, when Allison was about to enter the small gas turbine engine business, the company had become one of the big three in the large gas turbine

44. Allison Gas Turbine Division General Motors Corporation, *Allison Facilities/Capabilities* (Indianapolis, In.: Hilltop Press, undated), 1-48.

aircraft engine community, alongside General Electric and Pratt & Whitney. With its T56, Allison was becoming a major manufacturer of large turboprop engines in the U.S. As a result of its concentration on turboprop engines, it had also become the country's recognized leader in high-speed reduction gearing.

During more than a decade of engineering and development work on turbojets and turboprops, Allison's turbine technology base and expertise had evolved to a more mature level. As the 1950s drew to a close, the development of a small gas turbine engine was about to offer Allison a new challenge.

The Model 250 (T63)—Allison's First Small Turbine Engine

ORIGIN OF THE MODEL 250

John Wetzler, T63 Chief Project Engineer from 1959 to 1966, recounted that his first exposure to the possible need for a small gas turbine engine came around 1956-57 when he was a member of the NACA Subcommittee on Compressors and Turbines. At a meeting of the subcommittee, Don Weidhuner, Chief of the Engine Research and Development Office for the Army, said that what the Army really wanted was a 250-shp gas turbine. Wetzler recalled that, "Everyone in that room had been working on bigger and better engines, higher and faster and everything else, and we all laughed."[45]

Weidhuner's comment was based on the results of studies by the Army and evaluation of its future aircraft requirements. In 1956, the Army decided that to improve the capabilities of its future aircraft, the development of new, lightweight power plants was required.[46] It was believed that gas turbine engines would, on the basis of their low-specific-weight characteristics, offer greater potential than reciprocating engines. The probable power requirements for Army aircraft of the 1960s indicated that a 250-shp engine would be the size to develop. Subsequently, design study contracts were awarded to Lycoming, Garrett, Teledyne CAE, and the TurboMotor Division of Curtiss-Wright (a short-lived division established by Curtiss-Wright for the purpose of entering the small gas turbine business) to conduct parametric design studies of an engine that had the following characteristics. It was to be a free-shaft, turbine-type engine and have a 30-minute rating of 250 shp under standard sea-level condi-

---

45. Wetzler to Leyes and Fleming, 24 June 1991.

46. S. H. Spooner, "Summary of Design Studies of a 250-HP Gas Turbine Engine," (ASME Paper No. 59-GTP-7 presented at the Gas Turbine Power Conference and Exhibit, Cincinnati, Ohio, 8-11 March 1959), 1.

tions. It needed to have the capability for operation in single- or multiple-engine installations and be adaptable to helicopters, fixed-wing aircraft, and possibly STOL/VTOL aircraft. Weight, fuel consumption, and overall dimensions were to be minimized, but dimensions could not compromise engine efficiencies. These studies, completed in early 1957, showed that it was possible to design an engine of the type that the Army required. They also provided a basis on which to establish engine specifications for use in the following design competition.[47]

PRE-COMPETITION PRELIMINARY DESIGN

For many months prior to the crystallization of firm requirements for such an engine, Allison was engaged in activities directed toward evaluation of design and market possibilities for an engine in the 250 shp class.[48] Heading preliminary design was Russ Hall. After a period of activity devoted to extensive design studies and their evaluation, a promising configuration was defined. Installation studies and a market survey were carried out. The results of the market survey were very encouraging and suggested a wide range of application possibilities for an engine in the 250 shp class. A full-scale mock-up of the engine was made to study its maintainability and serviceability features, and it was presented to airframe customers and the military for familiarization and review.

The result of this program was the original Allison Model 250 that was presented to the military in a formal proposal on March 10, 1958.[49] Allison proposed two versions: the 250-B1 turboprop engine and the 250-C1 turboshaft. The proposed engines had a seven-stage axial and single-stage centrifugal compressor, single combustor, and a single-stage gas producer turbine and two-stage power turbine. The engine was rated at 250 shp and had a 0.70 lb/hp/hr sfc.

WINNING THE COMPETITION AND THE CONCEPT STUDY CONFIGURATION

In June 1958, Allison was awarded the contract by the Air Force (who managed it for the Army) for the design and development of the Model 250 (military desig-

---

47. Spooner, "Summary of Design Studies of a 250-HP Gas Turbine Engine," 5.

48. R. S. Hall, "Design Approach to the Allison Model 250 Engine," (SAE Paper No. 49R presented at the SAE National Aeronautic Meeting, New York, N.Y., 31 March–3 April 1959), 1.

49. Allison Division of General Motors Corporation, "Allison Model 250 Gas Turbine Engines - Proposal for Purchase Request 07607, Volume II - Technical Description," 10 March 1958, National Air and Space Museum Archives, Washington, D.C., 1-3.

nation T63) engine. The engine was to produce 250 shp at an sfc of 0.71 for a weight of 110 lb as a turboprop and 95 lb as a turboshaft engine. It was to be a simple, low-cost engine designed for a 1,000-hour life and capable of operation over a wide range of attitudes for a wide variety of airframe and other applications.[50]

Factors believed to have contributed to winning the contract were the following: General Motors was a dependable contractor and had the resources to complete the contract; the proposal promised 250 shp, 0.71 sfc, and 95 lb weight; engine price was estimated at $4,000 each; Allison accepted a fixed-price, cost-sharing contract; the engine was to be FAA certificated at the same time that it was military qualified.[51] Additionally, advanced technical features incorporated in the proposed design may have contributed to the decision.

Contractually, Allison and the Army each paid $3.2 million, a total cost of $6.4 million; all costs above $6.4 million were to be paid by Allison. The original design studies had estimated that the cost of the development program would be $16 to 20 million dollars.[52] However, to provide a cost advantage in the competition, Allison's management had drastically reduced the cost estimate in the proposal. As it turned out, the actual development cost exceeded the original design study estimate.

The design and development team was headed by Bill Castle, Chief Project Engineer, who also worked on the compressor mechanical design.[53] Beuford Hall did the turbine mechanical design work, and Bob Larkin designed the gearbox. The combustor was designed by Joe Barney, who was also in charge of all turbine and compressor aerodynamic design. The control work was performed by Bob Wente, and Russ Hall provided the compressor and turbine aerodynamic design data. Organizationally, this design team was under the direction of Charles McDowall, Chief of Preliminary Design, and his assistant John Wheatley.

50. John M. Wetzler and William S. Castle, "Development of the T63 Engine," (ASME Paper No. 61-AV-63 presented at the Aviation Conference, Los Angeles, Cal., 12-16 March 1961), 1.

51. John Wetzler, Boca Raton, Fl., to Rick Leyes, unpublished notes, 12 May 1991, National Air and Space Museum Archives, Washington, D.C.

52. Spooner, "Summary of Design Studies of a 250-HP Gas Turbine Engine, 5.

53. William Castle, interview by Rick Leyes and William Fleming, 4 June 1991, Allison Interviews, transcript, National Air and Space Museum Archives, Washington, D.C. For this interview, Castle prepared a set of written notes and additional, unpublished data. Unless otherwise indicated, all subsequent references to Mr. Castle are from his transcript, notes, or data.

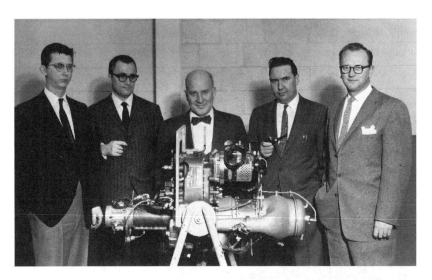

Some members of the Allison Model 250 design team with the first Model 250 engine just prior to its first test run in March 1959. Left to right: Wylie Johnson, Dick Galune, John Wheatley, Bill Castle, and Bob Wente. Photograph courtesy of Allison.

The group proceeded immediately with detail design and fabrication of parts for the first development engine, later referred to as the concept study configuration. Because it was an extremely difficult job to meet the specifications, the concept study configuration included many aerodynamic and mechanical features that proved later to have gone beyond the then state of the art.

Between November 18-20, 1958, a military-run mock-up review was conducted to establish production configurations for the engine. Responding to airframers in attendance, some suggested changes were made in the engine.

Following an extensive series of mechanical arrangement studies, the designers selected a unique configuration that was simple and compact. The gear case was the engine's primary structure and was located in the middle of the engine. The compressor assembly was bolted to the forward face of the gearbox, and the turbine combustor assembly, including the exhaust hood, was attached to the rear face. With this arrangement, air was brought directly into the axial-centrifugal compressor located at the front of the engine. The compressed air was delivered through a vaneless diffuser to a collector scroll from which it flowed through two external tubes to the single combustor mounted on the rear

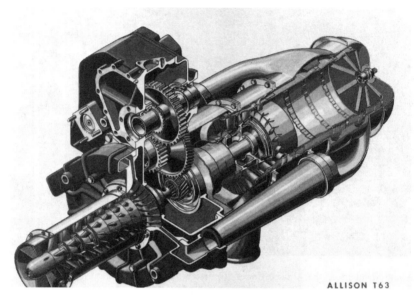

ALLISON T63

Cutaway drawing of the Allison Model 250 (military designation T63) turboshaft engine illustrating the mechanical arrangement selected by the design team in 1958. Photograph courtesy of Allison.

of the engine. Combustion gases then flowed forward from the combustor through the single-stage core and two-stage power turbines into an exhaust hood in the middle of the engine where they were collected and exhausted downward.[54]

The gear case, located in the middle of the engine, contained a spur gear train that engaged a small gear on the power turbine shaft and transmitted the output power vertically upward to a horizontal shaft parallel with and directly above the compressor turbine centerline. The output shaft was accessible from both the front and rear face of the gearbox. A planetary reduction gearbox, including provisions for mounting a propeller, was attached to the forward portion of the gearbox on turboprop models of the engine.

The mechanical arrangement of the engine offered several advantages. The gearbox casting was the main structural element of the engine and carried the mounting pads. With the complete compressor assembly and the turbine combustor assembly bolted to the gearbox much in the same manner as were the ac-

54. Hall, "Design Approach to the Allison Model 250 Engine," 23-24.

cessories, both assemblies were readily accessible for maintenance and servicing. Location of the single combustion chamber on the rear of the engine also made it easily removable.

There were two principal reasons for selecting the center gearbox arrangement over the alternative of locating the power drive at the front of the compressor. Because there was to be both a turboprop and turboshaft version of the engine, the front power drive arrangement would have required the propeller gearbox for the turboprop to be located ahead of the compressor, which was considered undesirable for good inlet flow aerodynamics. In addition, the front power arrangement would have required the power shaft to be run co-axially through the rotating compressor shaft. The right-angle power takeoff of the center gearbox arrangement was considered a simpler design than the concentric-shaft arrangement.

An axial-centrifugal compressor arrangement was selected for the engine during the original Army design studies as a practical means of achieving the desired compression ratio with acceptable efficiency and minimum complexity. A design compressor pressure ratio of about 6.2:1 was selected to achieve the desired 0.70 sfc. A seven-stage axial, single-stage centrifugal compressor was chosen by Russ Hall.[55] The axial portion of the compressor was chosen because of the high efficiencies obtainable with it and because the 6.2 pressure ratio was beyond the capability of a single-stage centrifugal. The centrifugal compressor, which followed the axial compressor, was better suited to handle the well-developed boundry layer and minimized the space required to deliver the air to the collector scroll before being passed to the combustor. The compressor rotor was planned to be constructed of sheet-metal disks and blades which were inset into a high-strength, fiberglass-impregnated plastic. To drive the compressor, a single-stage turbine was considered satisfactory at the time, although it was very highly loaded.

Quenching effects occurring in extremely small combustors dictated the use of not more than one or two combustion chambers to maintain practical combustor dimensions. The arrangement selected was a single combustion chamber that approximated the T56 engine combustor with which there had been a great deal of experience. However, this combustor had an unwieldy volume relative to the size of the engine, an example of an important problem facing small gas turbine engine designers, i.e., major elements of small engines were out of proportion to their size relative to what they were in larger turbine engines.[56]

55. Castle to Leyes and Fleming, 4 June 1991.

56. Hall, "Design Approach to the Allison Model 250 Engine," 22.

ORIGINAL RUNNING-ENGINE CONFIGURATION

In reducing the concept engine configuration to detail drawings and actual hardware (i.e., the original running-engine configuration), numerous changes were made to improve its mechanical design and simplify its fabrication. The sheet-metal-and-plastic rotor assemblies were replaced with cast wheel and blade assemblies, and the case design was modified. The original design was not used because of the weight of the compressor case, the blind spots in the diffuser casting, and the lack of engineering confidence in the ability to produce good rotor blade airfoils. The design of the turbine wheel and blade castings was refined, and the turbine stator blade assemblies were changed from fabricated structures to castings. In the concept study configuration, the core and power turbines were each overhung. In the running-engine configuration, the gas generator turbine bearing was moved aft of its turbine to permit better control of oil flow to and from the bearing.[57] Recognizing the potential cost advantage of using investment castings, a great deal of effort was made in both the compressor section and the turbine section to design an acceptable end product balance of machineability, surface finish, material selection, castability, endurance life, and configuration. Some performance compromise and scheduling delays were experienced to achieve this balance.[58]

Director of Engineering Dimitrius Gerdan, had instructed Castle to have the engine running in nine months following the June 1958 contract award.[59] A turboshaft version of this engine, fully instrumented, first ran on March 26, 1959. However, it immediately ran into serious difficulties, both aerodynamic and mechanical. Excessive turbine inlet temperatures prevented operation at full rotational speed. There was also distortion in the turbine section which terminated many engine runs because of turbine rotor seizure. Aerodynamically, the compressor was 3 1/2% low in airflow at the design point and 7% low in efficiency. The single-stage core turbine was 14% low in efficiency; when this turbine was pressed to deliver the level of power needed to drive the compressor, its efficiency became unsatisfactorily low. Mechanically, distortion of the structure supporting the power turbine bearings resulted in severe blade tip rubbing and seal interference of the two-stage power turbine.

---

57. Wetzler and Castle, "Development of the T63 Engine," 2-7.

58. Bill Castle, Speedway, Indiana, to Rick Leyes, fax, 3 March 1999, National Air and Space Museum Archives, Washington, D.C.

59. Castle to Leyes and Fleming, 4 June 1991.

In mid-1959, after attempting to raise performance by refinements and tune-up and by making such mechanical improvements as beefing up structure and opening clearances, the project faced a critical decision. Should an attempt be made to continue development of the engine on the present course in the face of the major problems that existed or should a new design be started that would avoid those problems? The decision made was to scrap the original aerodynamic design of the compressor and turbine as well as the mechanical arrangement of the turbine. All engine running was stopped and full effort was focused on the new design.[60]

REVISED CONFIGURATION

In September 1959, about the time the redesign of the T63 began, John Wetzler was named Chief Project Engineer of the T63, a position he held until 1966.[61]

In redesigning the T63, several changes were made to improve the compressor performance. The aerodynamics of the axial stages were completely revised using a different set of aerodynamic parameters. Also, the diameter of the compressor was enlarged by 1/4 inch to increase the airflow. Initial tests of the redesigned compressor indicated good efficiencies for the axial portion, but a low centrifugal-stage efficiency. In addition, the centrifugal stage had not been operating at its design pressure ratio. This deficiency was corrected by removing the seventh axial stage, thereby driving the centrifugal stage pressure ratio upward to its design level. An intensive aerodynamic redesign effort was begun to improve the centrifugal stage efficiency.

The turbine section was mechanically and aerodynamically redesigned to increase the gas generator turbine efficiency, to reduce exhaust duct losses, and to avoid the distortions that had caused blade tip rubbing of the gas turbine and power turbines. The aerodynamic design of the turbines was completely revised, and a second stage was added to the core turbine. To minimize turbine tip clearance, rings of an improved material were mounted around each turbine. The support of both the core and power turbines was changed from being overhung to being straddle-mounted between bearings located forward and aft of each turbine. In addition, the turbine vane assemblies and clearance rings were keyed to the turbine case structure in a manner that permitted radial growth while maintaining concentricity and avoiding distortions from case

60. Wetzler and Castle, "Development of the T63 Engine," 2-3.

61. Wetzler to Leyes and Fleming, 24 June 1991.

loading. In tests of the modified turbine assembly, each turbine stage exceeded its design efficiency.[62]

The gases leaving the turbine were collected, deswirled, and turned at high velocities, leading to excessive pressure loss in the original single exhaust duct. To decrease this loss, a dual outlet exhaust duct was designed that increased the flow area and reduced the gas velocity in the exhaust duct.

The center gearbox incorporating the output gearing was only changed to accommodate the slightly reduced speed of the power turbine. Thus, the arrangement for operation as a turboprop and turboshaft engine was retained. The redesigned engine configuration was first run as a turboshaft in February 1960 and as a turboprop in March 1960. In April 1960, Allison ran and completed an unofficial in-house 50-hour PFRT.

As could be expected, engine performance and endurance development problems continued. Ever present during the entire program was the continual battle of performance improvement, while at the same time maintaining essentially zero clearance between the rotating and stationary turbine elements. The balance was particularly difficult considering the small diameter and high rotational speed of the compressor and turbine units. Compressor and turbine rotational vibration problems were corrected by using spring mass isolators at selected bearing locations. The centrifugal section of the compressor design was modified several times to match acceptable tip running clearances, diffuser configuration, and diffuser vane settings. In the combustor section, the fuel nozzle and ignitor location went through a number of modifications to provide satisfactory cold weather starting, the use of a variety of fuels, and an acceptable burner outlet radial temperature profile. Main shaft bearings subjected to the high rotational speed and high thrust loads were modified several times before satisfactory operation was achieved.[63]

In March 1961, the first official YT63-A-3 PFRT was run satisfactorily, but the failure of a third-stage turbine wheel on a development engine caused Allison and the Air Force to mutually agree to hold engine delivery until the problem was solved. This failure was attributed to vibration of the relatively long turbine blades. To make the turbine less susceptible to blade failures resulting from excitation by the exhaust duct vanes, both power turbine stages were

62. The decision, at this time, to use two gas producer turbine stages and two power turbine stages, each straddle mounted, proved fortunate. It provided the basic design for the extraordinary power growth capability attainable in future models of the engine.

63. Castle to Leyes, 3 March 1999.

shrouded.[64] The official YT63-A-3 PFRT was then run in July 1961. While the test was passed, an impeller inducer vane failed on a development engine in August 1961; for this reason, Allison did not request PFRT approval. A first-stage turbine wheel disk failed in September 1961 during yet another PFRT test due to wide temperature disparities in its mass. This series of failures, while typical of engine development programs, caused the Army to reconsider its position. Allison had not been meeting the established developments dates, and the Army had a lot at stake.

RECOMPETITION FOR AN ALTERNATE ENGINE

As a result of the September 1961 turbine failure, in November, the Army decided to request proposals from other engine manufacturers for development of an alternate engine.[65] The Army was concerned that Allison would not be able to meet the engine delivery schedule for the Light Observation Helicopter (LOH) prototypes.[66]

In March 1960, the Army had announced a design competition for a new Light Observation Helicopter to replace the Hiller H-23 Raven, Bell H-13 Sioux, and Cessna L-19 Bird Dog. Missions specified for the new LOH were visual observation and target acquisition, reconnaissance, command control, and utility tasks at the company level. Twelve firms submitted 22 proposals for this competition. On May 19, 1961, three LOH finalists, Bell, Fairchild-Hiller, and Hughes, were announced by a joint Army-Navy selection team. The selected helicopters were designated the Bell OH-4A, Fairchild-Hiller OH-5A (Hiller became Fairchild-Hiller in 1964), and Hughes OH-6A. In June 1961, the Army selected the Allison T63-A-5 turboshaft engine to power each of these helicopters.

In response to the Army's request for alternate engine proposals, CAE and Boeing competed for the alternate engine contract. CAE won, and in late 1961, began developing the T65 engine. The T65 was a potential replacement for the T63 should problems continue.

Meanwhile, at Allison, Wetzler and Castle were aware of action by the military to acquire an alternate engine. However, they knew little about the CAE

64. Wetzler and Castle, "Development of the T63 Engine," 8-9.

65. "T63 Engine Tests," *Aviation Week & Space Technology*, Vol. 75, No. 22 (November 27, 1961): 24.

66. Larry Booda, "Redesigned T63 Engine Ends 50-hour. Test," *Aviation Week & Space Technology*, Vol. 75, No. 26 (December 25, 1961): 46.

and Boeing proposals or the engine that CAE was developing.[67] In Castle's words, "I guess we at engineering didn't know much about it while the competition was going on. We weren't invited, obviously. The military hadn't said much to us." When questioned as to what impact CAE's development of the T65 had on work at Allison, Castle responded, "I don't think it excited us. We couldn't have worked harder. We were working full bore, 24 hours a day, seven days a week the way it was, or at least a few of us were."[68] At the time, Castle felt that most of their problems were behind them and that they were making good progress.

Their confidence proved well founded. October and November 1961 marked the turning point of the T63 program. A turbine wheel change incorporated strengthening, cooling, and lower operating speed. Subsequently, in November 1961, the YT63-A-3 turboshaft model passed its official 50-hour PFRT, and the corresponding YT63-A-1 turboprop model passed its 50-hour PFRT in December 1961.

EARLY FLIGHT TESTS

After Allison's in house (unofficial) 50-hour PFRT had been completed (in April 1960), a YT63-A-3 turboshaft engine was delivered, in October 1960, to the Bell Helicopter Corp., Fort Worth, Texas, for flight testing. It was installed in an HUL-1M (a converted Bell UH-13R purchased by the Navy specifically for adaptation to gas turbine power) helicopter. In December 1960, the first runs of the engine were made in the HUL-1M; in February 1961, the first lift-off was made. The helicopter could not be flown until after the official military PFRT, and this was completed in November 1961. Allison's Chief Test Pilot Jack Schweibold and copilot E. J. Smith made the first test flight of the YT63-powered HUL-1M helicopter in February 1962 at the Bell Helicopter Company.[69] In May 1962, the Bell HUL-1M was delivered to Allison for flight testing at the factory. In June, flight testing of the YT63-A-3-powered HUL-1M began at Allison, which produced some unexpected and critical problems.

A major problem encountered at the beginning of flight tests involved the control of engine power. It was desirable to avoid the need for the pilot to con-

67. Wetzler to Leyes and Fleming, 24 June 1991.

68. Castle to Leyes and Fleming, 4 June 1991.

69. Jack Schweibold, interview by Rick Leyes and Bill Fleming, 4 June 1991, Allison
    Interviews, tape recording, National Air and Space Museum Archives, Washington, D.C.

trol engine power and helicopter flight attitude independently. Therefore, the engine control had been designed to rely on the turbine governor to control engine power in response to the pilot's input to the helicopter flight control lever for more or less lift from the rotor. The pilot could change engine power at will without needing to monitor engine parameters to avoid exceeding engine limits. This system required that the engine response and the torsional characteristics of the rotor/transmission system be properly matched within the control, a problem that turned out to be considerably more difficult than anticipated.

Allison's previous experience in matching engine and aircraft dynamics had been with turboprop installations in fixed-wing aircraft. However, as the helicopter flight tests began, it was found that the matching problem was considerably different and much more complex. In fixed-wing installations, the propeller itself had little flexibility and there was a short, hard coupling between the engine and propeller. In contrast, the helicopter had a large floppy rotor, often with lead-lag hinges for the rotor blades that provided a very flexible coupling. In addition, the drive shaft was as much as three times as long as any propeller shaft. Coupled to the drive shaft was the input shaft between the transmission and the engine. There was also a long shaft from the engine back to the tail rotor. Consequently, there was a great deal of torsional flexibility between the helicopter and the engine. The result was extremely complex torsional effects in which the torsional frequencies of the various elements overlapped.

Initial steps taken to attenuate these forces and bring them into balance included changing spring weights and fuel control metering valve spring forces in the mechanical systems, installing piston displacement cylinders in hydraulic portions of the system, and adjusting the pneumatic cushioning between the compressor discharge pressure and the pressure chambers in the fuel control system.

After these adjustments were made, the first test flight was made at the Bell Helicopter Company. Immediately after liftoff, a torsional divergence occurred in hover that resulted in an uncontrolled increase in power requiring engine shutdown and an autorotation landing. The engine fuel control was so sensitive that it recognized the torsional inputs of each rotor blade pass and was so fast in responding to them that the engine power increased uncontrollably. Proper matching was achieved by putting more volume in the engine control pneumatic damping system and changing spring weights.

This experience made it clear that helicopter rotor response was going to be very sensitive to rates of engine power change regulated by its control system. Consequently, considerable attention would be called for in matching the engine control to the rotor dynamics. In fact, it was later discovered during flight tests in Hughes and Hiller prototype helicopters that the rotor's torsional dynamics were different for each helicopter. Therefore, it was necessary to match engine control dynamics individually with the rotor torsional dynamics of each helicopter. This requirement was a universal one shared by all turboshaft engine and helicopter systems.

Schweibold stated that the matching of the helicopter's torsional acceleration and deceleration equation with the engine control was probably the single biggest problem encountered in the T63's flight testing. In this regard, he observed, "In fact, almost 95% of all of our flight test work on aircraft (that) we have put through flight test here in Indianapolis has been put into the fuel control match to the helicopter."[70]

Following early flight test operations with the YT63-A-3 in the Bell HUL-1M helicopter, both Bell and the Army decided that the engine should be installed with the exhaust directed upward instead of downward. The primary reason that the exhaust had been directed downward originally was that the Army's intended initial application for the engine was in a fixed-wing airplane similar to the Cessna Bird Dog light reconnaissance and observation aircraft. In early 1962, Allison began modifying the engine to direct the exhaust upward, basically turning the engine upside down with the gearbox on the bottom and the exhaust on the top. This became the YT63-A-5.

As mentioned, in June 1962, flight testing of the YT63-A-3-powered HUL-1M began at Allison, which produced some unexpected and critical problems. According to Schweibold, who conducted the flight testing, the first problem encountered in flight testing had nothing to do with the engine.[71] Rather, it entailed miniaturizing instrumentation and providing means for remote data recording. Most of Allison's previous flight testing experience had been in C-130s, Convairs, and other large transport aircraft. In those aircraft, large instrument boards and manometer panels could be installed, and engineers were carried aloft to read and record the engine data. NASA's work on instrument miniaturization had not yet impacted engine flight testing. Consequently, engineers were still in the era of manually reading instruments. Nevertheless, Allison had some experience in flight testing engines in fighter aircraft. Drawing on that experience, miniaturized instruments and recording devices were developed that permitted recording of T63 engine data in flight without the presence of an engineer.

---

70. Schweibold to Leyes and Fleming, 4 June 1991.

71. Schweibold to Leyes and Fleming, 4 June 1991.

Later, as the A-3 and A-5 models of the T63 were being flight tested, another problem was encountered, that of slow engine acceleration. To solve the earlier torsional divergence problem, the engine controls had been desensitized. But during subsequent flight tests, it was found that, in an aborted landing during descent, the engine was slow in responding, allowing rotor speed to drop to 85 to 90% of rated speed with a loss of lift and tail rotor effectiveness. Although engine acceleration was found to be within specifications, the response was too slow to meet the requirement for a rapid power increase during an aborted landing, especially at high ambient temperatures.

The modification made to provide faster response of the T63 was to incorporate a compressor bleed valve. Upon demand for a rapid power increase, the valve was opened at low engine power and rpm, thereby increasing the compressor stall margin and permitting the engine to be accelerated more rapidly. The valve was then closed at about 85% of rated speed. Operating with the bleed valve, the engine had a two to two and one-half second response, which was as fast as desired.

Necessary modifications and improvements in the engine control system were introduced during the years that followed as the engine was progressively modified and uprated. Then, in the early 1990s, the engine's control system was taken one step further with the introduction of a Full Authority Digital Electronic Control (FADEC), permitting elimination of the various pneumatic and hydromechanical spring-force types of elements used to accommodate the torsional interaction between the helicopter and the engine. With the FADEC, all of the desensitizing was done with filters, resisters, and so on within the electronic control. In addition, the FADEC system made it possible to further reduce the acceleration time of the engine.[72]

The first upward-exhaust model, the YT63-A-5 turboshaft engine, passed its 50-hour PFRT in March 1962. In May 1962, the delivery of the first flight-qualified YT63-A-5 engines were made for the Army's LOH program. Over several months they were delivered to Bell, Hiller, and Hughes for installation and flight testing in their prototype candidate LOH helicopters. In September 1962, the official 150-hour qualification test was passed. In December 1962, the T63-A-5 received its military approval, and its commercial equivalent, the Model 250-C10, received its FAA certification. Production of the T63-A-5 began the same month. Powered by the T63-A-5, the Bell YOH-4A first flew on December 8, 1962; the Hiller YOH-5A on January 26, 1963, and the Hughes YOH-6A on February 27, 1963.

72. Schweibold to Leyes and Fleming, 4 June 1991.

SAND AND DUST DAMAGE

Recognizing the potential problem of sand and dust ingestion in field operations, flight tests were begun in the early 1960s to assess and alleviate such damage. This program was partly funded by GM and partly by military contracts. The program's objectives were to define: (1) those operational environments where significant quantities of sand and dust were encountered, (2) the problems resulting from operation in these environments, and (3) practical solutions to those problems.

Field tests were performed on the T63-A-5A engine installed in the Bell OH-4A, Hiller OH-5A, and Hughes OH-6A helicopters at Fort Rucker, Fort Benning, the Yuma Proving Grounds, and Edwards Air Force Base. Early observations in both field and laboratory tests revealed that two factors contributed to performance degradation: erosion within the compressor and turbine and accumulations within the engine. The effect was manifested as loss in compressor and turbine efficiency and a loss of compressor stall margin. Steps were taken both to increase the tolerance of the engine to sand and dust ingestion and to minimize the amount of contaminant ingested by the engine through the use of inlet particle separators.[73]

The results of these tests were supplemented by early field experience during the Vietnam War. Because light observation helicopters in Vietnam were operated from unprepared sites, the rotor kicked up large amounts of dirt and debris ingested by the engine during liftoff, landing, and hover near the ground. Due to the resulting performance degration, the time between removals of T63 engines in Vietnam was only 100 to 300 hours.[74]

During these early operations in Vietnam, some unexplained losses in engine performance were also encountered. These losses were traced to an accumulation of a glass-like substance on the first-stage turbine guide vanes.[75] Similar accumulations were also found during the flight tests at the Yuma Proving Grounds. The material found on the guide vanes was an agglomeration of contaminant particles which had been heated to a semi-molten state in the combustor and then formed a clinker-like coating. These accumulations occurred during prolonged constant power operations at turbine inlet temperatures above

73. George V. Bianchini, "T63 Sand and Dust Tolerance Development and Field Experience," (SAE Paper No. 670334 presented at the SAE National Aeronautic Meeting, New York, N.Y., 24-27 April 1967), 1-8.

74. George Bianchini, interview by Rick Leyes and William Fleming, 5 June 1991, Allison Interviews, transcript, National Air and Space Museum Archives, Washington, D.C.

75. Bianchini to Leyes and Fleming, 5 June 1991.

1,550°F while continuously ingesting contaminants. Thus, hovering over sand or dust for prolonged periods of time at fixed high-power settings resulted in rapid performance degradation.[76]

It was found that the accumulations could be cleaned out of the turbine by intermittently reducing power settings or by an engine shutdown and restart. As a result, a recommendation was made to pilots in the field to minimize their time hovering in sand and dust if possible, but if such operation was necessary, they were instructed to change power settings periodically in order to shed the accumulations.[77]

Some difficulty was also experienced during the field tests with contamination of the engine control system. The control system used compressor discharge pressure as a control parameter. Although the pressure sensing port was small, some contaminant found its way into it and was carried into the control system. This problem was solved by relocating the port to an area of the compressor discharge that was free of contaminant.[78]

During the flight tests, both inertial particle separator and barrier-type filters were evaluated on the T63 in a Hughes OH-6 helicopter. Much of this work was done with filters manufactured by the Donaldson Company; one was a Donaldson Strata-Panel self-cleaning inertial particle separator filter. Tests with that filter on the T63 engine provided filtration efficiencies ranging from 81 to 93%. No accumulations occurred in the engine, and significant erosion was encountered only by the plastic compressor tip seal coating. Filtration efficiencies of two Donaldson Duralife barrier filters tested on the engine averaged 99%. However, these filters required cleaning after each test run, and the pressure drop across them increased progressively even after they were cleaned. These factors led to the conclusion that use of the barrier type filter on military aircraft would pose major maintenance and logistics problems, and the work to develop a satisfactory filter for the engine was focused on inertial separator type filters.[79] Field tests demonstrated that use of this type of filter when operating in a contaminated environment increased the life of the engine by about tenfold.[80]

76. Bianchini, "T63 Engine Sand and Dust Tolerance Development and Field Experience," 5-10.

77. Bianchini to Leyes and Fleming, 5 June 1991.

78. Bianchini, "T63 Engine Sand and Dust Tolerance Development and Field Experience," 5-10.

79. Bianchini, "T63 Engine Sand and Dust Tolerance Development and Field Experience," 10-21.

80. Bianchini to Leyes and Fleming, 5 June 1991.

MODEL 250/T63 ENGINE EVOLUTION AND APPLICATION

As Allison began selling the Model 250 and T63 engines, a pricing policy was adopted that was designed to control the market for 250-hp class engines. Allison kept the price of the engine low enough to discourage competition. Allison's pricing policy was made possible by the fact that, during the 1960s, GM was not pressing the division to become a big money maker. Eloy Stevens recalled that, during a meeting he attended in 1960 with Allison's Executive Vice President and several GM executive officers, the GM position as expressed by one of the executive officers was:

> We do not intend to make a lot of money with Allison as a division of General Motors. What we intend to do, as good citizens of the United States, is to plan to have that division as a stand-ready division. If the country needs us to mobilize and do things, we're going to be there and have the capability to do it. On the other hand, we don't intend to lose a lot of money with the division either.[81]

The Model 250/T63 became an extremely successful engine and enjoyed a rapidly growing popularity during the 1960s. The key to its initial success lay in winning the Army's 250-shp turboshaft engine competition in 1958 and the selection of the engine in 1961 to power each of the Army's LOH candidate helicopters. Between January and June 1964, LOH tests were conducted by the Army. On May 26, 1965, the Hughes OH-6A Cayuse, priced at 32% below its nearest competitor, was declared the winner of the LOH competition and received a production contract. Production delivery began in 1966. In 1967, the Army decided to reopen the LOH competition as a result of its displeasure with substantial increases in the OH-6A's unit price and lagging production schedule. On March 8, 1968, the Bell Model 206A was named the winner of this LOH competition, and subsequently designated OH-58A Kiowa. Both the OH-6 and OH-58 were used extensively during the Vietnam War and contributed importantly to the Army's airmobility operations.

On the commercial side, a completely redesigned derivative of Bell's OH-4A, known as the Bell Model 206A JetRanger, was certificated in October 1966. Similarly, Hughes and Fairchild-Hiller developed commercial versions of their military helicopters, the Hughes Model 500 and the Fairchild-Hiller FH-1100. All of these helicopters were powered by the Model 250 engine. These commercial helicopters were followed by several other Model

81. Stevens to Leyes and Fleming, 6 June 1991.

From left to right: Bell OH-4A, Hiller OH-5A, and Hughes OH-6A helicopters. Powered by Allison T63 turboshaft engines, these helicopters were competitors in 1964 for the Army's Light Observation Helicopter program. Photograph courtesy of John Wetzler.

250-powered U.S. and European light helicopters. A large number of helicopters using the Model 250 engine were built and became widely used as a result of the newly found and expanding commercial market for turbine-powered utility helicopters. By 1975, the Model 250 also powered several larger twin-engine helicopters, among which were the MBB BO105, Nurtanio NBO105, and Aerospatiale AS355 Ecureuil 2/TwinStar. By the mid-1970s, the light helicopter had become extremely popular worldwide for both military and civilian uses, and Allison's engine had captured most of the market. In fact, Allison's total sales of T63 and Model 250 engines up to the late 1980s accounted for more than 80% of the light turbine helicopter engine market in non-communist nations.[82]

In addition to low price, three other factors were considered important to capturing the major market share. Technically, the Model 250/T63 engine's

82. Rick Butler, Allison Division of General Motors, "Model 250 Engine History," unpublished and undated (ca. mid-1980s) paper, National Air and Space Museum Archives, Washington, D.C., 4.

high power-to-weight ratio combined with continuing product improvement was important. Operationally, the engine established a favorable in-service record for good performance and reliability. Finally, the support provided to the end user through Allison's worldwide distribution and servicing network was important to the engine's success.[83]

As new and larger helicopter models were introduced, the Model 250/T63 kept pace. Allison developed four series of T63 and Model 250 engines with progressively higher power outputs to accommodate these increasing performance and payload requirements.[84] The development history of these four series of engines follows.

SERIES I ENGINES

The Series I turboshaft was the initial Army and commercial offering of the engine. Ranging in power from 250 to 317 shp, this series included the T63-A-3, A-5, A-5A, and A-700 military turboshafts and the Model 250-C10 and 250-C18/18A commercial versions. A total of 6,410 Series I turboshafts were manufactured between 1962 and 1975 when production of this engine series was terminated. The principal Series I turboprop engine was the 317-shp 250-B15G, 92 of which were produced between 1969 and 1974. The main use of this engine was in conversions of Cessna and Beech aircraft.[85]

The 250-shp T63-A-5 powered the three original LOH condidates, the Bell OH-4A, Fairchild-Hiller OH-5A, and the Hughes OH-6A. The production version of the winning OH-6A was powered by the 317-shp T63-A-5A.[86] On March 24, 1966, a YOH-6A powered by the T63-A-5 set a helicopter world speed record of 171.85 mph.[87]

The commercial version of the A-5 series, the 317-shp Model 250-C18/ C18A, powered the Bell Model 206A, Hughes Model 500, and the Fairchild-

---

83. Castle to Leyes and Fleming, 4 June 1991.

84. E. C. Stevens and F. F. Conn, "Evolution of the Allison Model 250 Engine," (SAE Paper No. 800603 presented at the Society of Automotive Engineers Turbine Powered Executive Aircraft Meeting, Phoenix, Arizona, 9-11 April 1980), 1-6.

85. Castle to Leyes and Fleming, 4 June 1991.

86. Norman Polmar and Floyd D. Kennedy, Jr., *Military Helicopters of the World* (Annapolis, Maryland: Naval Institute Press, 1981), 176, 211, 227.

87. *World and United States Aviation and Space Records and Annual Report* (Arlington, Virginia: National Aeronautics Association of the USA, 1992), 161.

Hiller FH-1100. The Bell 206A JetRanger and its successors became one of the world's most popular light turbine helicopters. The Hughes Model 500 and its derivatives were also very successful in both commercial and military export/license-built roles. By comparison, relatively few FH-1100 helicopers were built for civil and military users.[88]

The 317-shp Model T63-A-700, an improved version of the Model T63-A-5A, powered the winner of the reopened LOH competition, the Bell OH-58A Kiowa. The OH-58A was first delivered to the Army on May 23, 1968.[89] In November, the OH-58A joined the Hughes OH-6A in Vietnam, performing observation missions and making life-saving flights to rescue wounded troops.[90] Operating in this environment, the T63 proved to be reliable and rugged and the engine received high praise from helicopter pilots.[91] The primary problems encountered were compressor erosion due to sand and dust ingestion, and distress of the turbine components due to over-temperature operation, both of which reduced engine performance and life.[92]

Soon after the 250-C18 and C18A turboshaft engines were certificated and became operational in 1965, they began encountering a number of in-service difficulties. Customers began expressing distress over the engine's reliability, the number and frequency of failures and removal demands, and the repair costs. Among the problems encountered, the three most serious were in the compressor, fuel control, and bearings. Compressor problems were traced to resonance at certain operating conditions. The fix was to rework the third- and sixth-stage compressor blades and the first- and second-stage guide vanes. Fuel control problems, which were considerable, were traced to improperly set controllers, which Bendix corrected. Bearing problems were almost wholly from wear believed to stem from heat and were corrected by replacing the bearings with ones made of M50 steel, using one-piece retainers.[93]

88. Polmar and Kennedy, Jr., *Military Helicopters of the World*, 176, 211, 227.

89. Polmar and Kennedy, Jr., *Military Helicopters of the World*, 176-177.

90. Giorgio Aspostolo, *The Illustrated Encyclopedia of Helicopters* (New York: Bonanza Books, 1984), 49-50.

91. Jerry Flanders and George Mayo, interview by Rick Leyes and William Fleming, 4 June 1991, Allison Interviews, transcript, National Air and Space Museum Archives, Washington D. C.

92. Castle to Leyes and Fleming, 4 June 1991.

93. "Tempest in a Turbine," *Business & Commercial Aviation*, Vol. 25, No. 1 (July 1969): 53-55.

As a result of the number of in-flight failures, Allison set up a "Blue Ribbon" program to correct the difficulties. In the words of George Mayo, Product Line Director of the Model 250 engine:

> The Blue Ribbon activity was essentially one that said: "Let's take all the engineering expertise and the service experience and look at the thing and say, What is it that we can do, should do, now that we know what we (should) do to make these (engines) better?"[94]

The program was ultimately successful in improving the service life and reliability of the engine.

SERIES II ENGINES

In response to the demand by military and civilian users for more power, higher reliability, and longer engine life, development of the Series II engines began in 1966 with the Model 250-C20, which was certificated in May 1970. That engine, rated at 400 shp, was a derivative of the 317-shp C18. Higher power output was obtained by increasing the diameter of the compressor to raise the airflow by 13%, and by increasing the turbine temperature. This temperature increase was made possible by advances in turbine materials. Higher reliability and longer engine life were obtained by applying engineering design improvements derived from the "Blue Ribbon" program. Series II turboshafts included the Model 250-C20, C20B, and C20R civilian models and the T63-A-720, the military version of the C20B. The turboprop engines in this series were the 420-shp 250-B17B/C/D/F models, the first of which was certificated in April 1971.[95] Between 1971, when production of Series II engines began, through 1995, 15,301 Series II turboshaft engines and 958 Series II (250-B17/All) turboprops were produced.[96]

A major growth pattern in the Model 250 commercial market began to evolve in the mid-1960s. Enhanced versions of Bell's Model 206 JetRanger and Long-Ranger were major users of the Series II Model 250 engines. In addition, as im-

94. Flanders and Mayo to Leyes and Fleming, 4 June 1991.

95. Castle to Leyes and Fleming, 4 June 1991.

96. Tommy H. Thomason, Indianapolis, Indiana, to Rick Leyes, letter and production data, 17 April 1996, National Air and Space Museum Archives, Washington, D.C.

proved versions of the Hughes Model 500 and the Fairchild-Hiller FH-1100 helicopters were introduced, they too were equipped with the more powerful Series II C20 engines. Several foreign airframers also adapted Series II engines to their helicopters. These applications included the French Aerospatiale AS355, German MBB BO105, and Italian Agusta A109, twin-engine utility helicopters used for a variety of civil and military applications.

Most of these helicopters were operated by fleet owners who were experienced in the way the helicopters and their engines should be operated, and thus handled them properly. However, there were also a number of users who operated only one or two helicopters, and due to their inexperience, subjected the engines to conditions for which they were not designed. As a result problems occurred. In one such instance, a logging firm in the Grand Canyon area that was using a helicopter to lift logs reported engine problems. The helicopter would not lift as many logs as the operators desired. Their solution was to put the logs on a truck, hover the helicopter over the truck, and then drive the truck as fast as it would go. The truck would then stop near the edge of the canyon at which point the helicopter lifted the logs and flew out over the canyon where it lost some altitude, then gained enough forward speed to lift the load. Their engine problems resulted from the fact that, to perform this procedure, it was necessary to operate the engine beyond its redline.[97]

The Model 250-B17 turboprop was used in a number of small fixed-wing aircraft which were converted from piston to turboprop engines. One example was the Beech Bonanza A36, a 6-seat utility aircraft. At 420 shp and 195 lb, the 250-B17 installed in the A36 produced about 50% more power and weighed less than half that of the Continental piston engine it replaced. This power boost produced a spectacular improvement in the Bonanza's rate of climb. Although the 250-B17 burned 24 gallons per hour (gph) of fuel compared to 16 gph for the Continental engine, the use of tip tanks raised the aircraft's fuel capacity sufficiently to bring its endurance into line with the piston powered model, i.e. about five hours. In that length of flight time, the turboprop powered A36 could travel farther than its piston-powered counterpart due to its 30-knot advantage in airspeed.[98] However, in order to realize this higher airspeed advantage, it was necessary to fly the turbine conversions at altitudes of 15,000 to 17,000 feet where oxygen was needed in unpressurized aircraft like the A36. To

97. Stevens to Leyes and Fleming, 6 June 1991.

98. Nigel Moll, "Turbine Torque Show: The Bonanza A36 Joins Allison Wonderland," *Flying*, Vol. 115, No. 5 (May 1988): 42-46.

Beechcraft Bonanza Model A36 aircraft powered by Allison Model 250-B17C turbo-prop engine. Photograph courtesy of Allison.

fly at lower altitudes where oxygen was not needed meant operating at reduced engine power levels because of FAA restrictions on maximum indicated air-speed. At or near full power, the engine could drive the aircraft above its struc-turally limited never-exceed speed.[99] By 1991, about 25 A36s had been con-verted to the Model 250 engine.[100]

In addition to the A36, the 250-B17 turboprop powered a variety of small commuter aircraft used in the U.S., Europe, and Australia and the South Pacific basin. These aircraft included the 16-passenger, twin-engine Australian N-22B Nomad made by the Commonwealth Aircraft Corporation Ltd., the Italian nine-passenger, twin-engine SF.600TP Canguro (Kangaroo) made by Siai-Marchetti SpA, and the United Kingdom twin-engine PBN Islander made by Pilatus Britten-Norman Ltd. that was outfitted for a variety of purposes.[101] The

99. Mark R. Twombly, "Power Tool," *AOPA PILOT*, Vol. 35, No. 12 (December 1992): 84-91.

100. Flanders and Mayo to Leyes and Fleming, 4 June 1991.

101. *Jane's All the World's Aircraft 1984-85* (New York: Jane's Publishing Inc., 1984), 7, 149. Also see
  Jane's All the World's Aircraft 1987-88 (London: Jane's Publishing Co. Ltd., 1987), 320-321.

250-B17 engine was also used in three two-place, single-engine primary trainer aircraft, Italy's SF260TP made by Siai-Marchetti, Japan's KM-2Kai made by Fuji Heavy Industries Ltd., and India's HAL HTT-34 made by Hindustan Aeronautics Ltd.[102]

SERIES II LICENSING VENTURE

In 1984-85, Allison investigated an approach to produce lower cost versions of the basic Model 250 turboshaft and turboprop engines through cooperative ventures. This approach was based on fabricating selected portions of the engines in various Asian countries. Using these parts, the engines would then be assembled at Allison and sold back to the countries for use in their proposed aircraft projects and worldwide to compete with general aviation piston engines. Thus, the company introduced the Model 225 engine, based on the Series II Model 250-C20B, and identified it as the Asian engine.

The venture began as a partner program and later became a pure vendor program. The partners were China Aero-Technology Import and Export Corp. in China, Samsung United Aerospace Co. in South Korea, and Hindustan Aeronautics in India. Each partner was to make a portion of the engine and ship the parts to Indianapolis for assembly.[103] The approach was never fully developed due to a slowdown of the world aviation markets. Although the engines reached certification status, the required production volume was less than required for full program launch and the partners became Allison suppliers.

Although based on the C20B, the Model 225 was to cost less and be rated at a lower power level than the C20 Series II engines. The power rating was to be 350 shp as compared to the 420-shp rating of the C20B, and the unit price was anticipated to be $55,000 to $60,000 as compared to $85,000 for the Allison built Model 250.[104]

Allison's new "low-cost" Model 225-C10 turboshaft/turboprop engine was certificated in March 1987.[105] However, soon after it was certificated, Allison backed away from the program. Although low in cost, the engine was still very

102. *Jane's All the World's Aircraft 1984-85*, 148-149, 157. Also see *Jane's All The World's Aircraft 1986-87* (New York: Jane's Publishing Inc., 1987), 99.

103. Flanders and Mayo to Leyes and Fleming, 4 June 1991.

104. "Allison Model 225 Turboshaft/Prop Attains March Certification," *Turbine Intelligence*, Vol. 11, No. 50 (March 30, 1987).

105. "Allison Model 225 Turboshaft/Prop Attains March Certification."

sophisticated to make. Consequently, it appeared that to carry off the whole program as planned, Allison would need to subsidize it far more than the company could afford. Although the Model 225 program was stopped, the three Asian partners were kept as vendors to fabricate those parts of Model 250 engines for which they were qualified. In this role, they were used in the same way as stateside parts suppliers.[106]

SERIES III ENGINE

In the 1972-73 period, development of the Model 250-C28 turboshaft was begun in response to the requirement of commercial helicopter operators for still more power and speed. That engine, a derivative of the C20 rated at 500 shp, was certificated in May 1976. The compressor underwent a major redesign. The four axial stages of the Series II C20 engine were eliminated. Instead, a single-stage, front-entry centrifugal compressor produced a higher pressure ratio and handled considerably more airflow than the axial-centrifugal compressor it replaced.

Pete Tramm, a compressor design expert, and his aerodynamic technology group made a substantial advance in compressor technology at Allison in the early 1970s. As a result, the new single centrifugal-compressor stage of the Series III engines produced a pressure ratio of 8:1 compared to the 6:1 pressure ratio of the six-stage axial and single-stage centrifugal compressor of the Series I engines and did so with a higher efficiency. Reflecting on Pete Tramm's contributions to the Model 250 compressor system and those later on the T406 engine, Mike Hudson, Allison's Director of Engineering, observed, "I think he'll probably go down as one of the great compressor designers in our industry in the sixties, seventies and eighties."[107]

The C28 compressor also incorporated a unique patented device, called a slotted inducer, used later in the T800 turboshaft engine. That device was a slot system in the shroud covering the inducer that acted as a "smart bleed." It allowed air to be bled off through the inducer shroud at low to intermediate rotating speeds when the compression system was choked at the discharge. At high speeds it allowed air to inflow through that section. The patent was held by Dennis Chapman and David Sayer. In recognition of their work on this device,

106. Flanders and Mayo to Leyes and Fleming, 4 June 1991.

107. Mike Hudson, interview by Rick Leyes and William Fleming, 6 June 1991, Allison Interviews, transcript, National Aeronautics and Space Museum Archives, Washington, D.C.

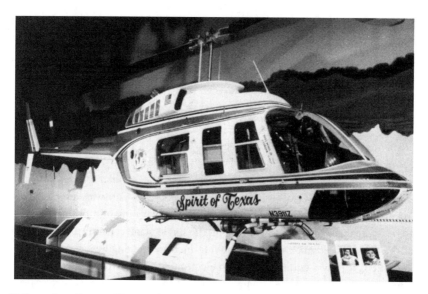

Bell 206 LongRanger "Spirit of Texas" helicopter powered by an Allison Model 250-C28B engine, set an around-the-world record for helicopters in 1962. This helicopter is in the National Air and Space Museum collection. National Air and Space Museum Photo by Mark Avino, Smithsonian Institution (SI Neg. No. 91-15701).

they jointly received the General Motors Kettering Award, the corporation's highest award for technical achievement.[108]

The turbine aerodynamics were modified to match the higher airflow and pressure ratio of the compressor and to provide improved efficiency. The C28 also had a redesigned gearbox to handle its higher power output, and an inlet particle separator was installed. Put into production in 1976, 869 Model 250-C28B/C engines were produced through 1995.

The 250-C28 was used in two commercial utility helicopters that were growth versions of models powered by Series II engines, the Bell 206 Long-Ranger and the MBB BO105-LS.[109] Piloted by H. Ross Perot, Jr., a Bell 206 LongRanger powered by a 250-C28 engine set an around-the-world speed record for helicopters in September 1982 of 29 days and 3 hours.[110] Perot's

108. Hudson to Leyes and Fleming, 6 June 1991.

109. Castle to Leyes and Fleming, 4 June 1991.

110. *World and United States Aviation and Space Records and Annual Report*, 160.

LongRanger and a model of the C28 engine that powered it are in the National Air and Space Museum's collection. The Bell LongRanger and the MBB BO105-LS were used principally for air ambulance service, executive transport, and pilot training.

SERIES IV ENGINES

Demands for even further increases in power for light helicopters led to development of the 650-shp Series IV turboshaft engines beginning in the mid-1970s. This series included the 250-C30, the civilian model, and the T703-AD-700 (250-C30R) military version. The 250-C30 was certificated in March 1978, and the T703-AD-700 passed its military qualification test in July 1981. To achieve a higher power output, the single-stage centrifugal compressor of the Series IV engines was redesigned to increase the engine airflow. In addition, the C30R was equipped with an electronic fuel control.[111] With production beginning in 1978, 1,855 Model 250-C30/All and 514 T703-A-703 Series IV turboshaft engines were produced through 1995.

Early in its commercial service, a few C30 engines experienced turbine bursts caused by overspeeding of the turbine wheel following failure of the shaft between the turbine and compressor or failure of the spline adapter on the shaft. After extensive spin pit work, and at considerable cost, an energy absorbing ring internal to the engine was developed. This ring could stop a burst turbine wheel and keep it contained within the engine.[112]

With the development of the Series IV engines, the power output of the Model 250 had been increased by a factor of 2.6 above that of the 250-shp T63-A-3 and A-5 models with little growth in engine size (i.e., an increase in length from 40.5 to 41 inches, in width from 19 to 21.9 inches, and in height from 22.5 to 25.1 inches. Although weight grew from 136 to 240 lb, the power-to-weight ratio had increased from 1.84 to 2.71).[113]

Key technical advances from the Series I to the Series IV engines that produced the engine's increase in power output were threefold. First, advances in compressor technology permitted redesign of the compressor from an axial-centrifugal configuration to a single centrifugal stage that produced both increased pressure ratio and airflow. Secondly, improved turbine aerodynamics

---

111. Castle to Leyes and Fleming, 4 June 1991.

112. Flanders and Mayo to Leyes and Fleming, 4 June 1991.

113. Castle to Leyes and Fleming, 4 June 1991.

technology produced higher turbine efficiencies and better compressor-turbine matching. Thirdly, material improvements allowed turbine operating temperatures to be raised.

The Series IV engines were used in the Bell 206 LongRanger, Hughes Model 530, Aerospatiale AD350G Ecureuil/AStar, Sikorsky S-76 Mark II, and Bell OH-58D Kiowa. The C30-powered LongRanger was a growth version of the C28-powered LongRanger. The Hughes Model 530 was an improved military variant of the Hughes Model 500 and was used primarily for point-attack and anti-armor missions. The twin-engine Sikorsky S-76 Mark II, a 12-passenger commercial helicopter, was used mainly for offshore oil rig support, corporate executive transport, air ambulance service, and general utility operations.[114] Between February 4 and 9, 1982, a 250-C30 powered production version of the S-76 Mark II, without special modifications, set 12 world records including a 500-km closed-circuit speed record of 214.833 mph and an altitude record of 26,050 feet.[115] Bell's OH-58D had been selected by the Army in September 1981 as the winner of the Army Helicopter Improvement Program (AHIP), with the task of developing a near-term scout helicopter. Later, Bell also developed a weapon-carrying version of the OH-58D. In 1991, this armed version flew combat missions in Operation Desert Storm.[116]

FIELD SERVICE ORGANIZATION

One of the key factors in the Model 250's worldwide acceptance by commercial operators was the field service provided by Allison. Early in the Model 250 program, Allison took a hard look at the nature of the civil market. The company realized that there were going to be many individual owners of small fixed-wing aircraft and helicopters, much like owners of automobiles and diesel equipment. As Jerry Flanders, Senior Engineer on the Model 250, observed, "There were a lot of similarities in what we perceived for the Model 250 engine in fixed-wing applications as well as helicopters, to the then current General Motors dealership and distributorship arrangements in the 1960s."[117]

114. *Jane's All the World's Aircraft 1989-90* (Coulson, England: Jane's Information Group Limited, 1989), 462, 514.

115. *Jane's All the World's Aircraft 1987-88*, 517.

116. *Jane's All the World's Aircraft 1989-90*, 362-363.

117. Flanders and Mayo to Leyes and Fleming, 4 June 1991.

Allison went to 66 companies around the world and examined their capabilities. Nine were selected as distributors to support the Model 250 engine worldwide, providing all of the after-market sales, service, and support for the engine. The distributors were responsible for providing spare parts, technical assistance and publications, warranty administration, overhaul, training, and individual service and support. By 1969, all overhaul and repair of the Model 250 engine in Indianapolis had been terminated. Distributors were also responsible for the establishment of subdealers, which were Authorized Maintenance and Overhaul Centers (AMOC). By the end of 1991, 28 of these AMOCs had been established by the nine distributors.

The distributors and their AMOCs were staffed by 500 field service technical representatives. The effectiveness of this organization was such that in 1988 the Army began using Allison's commercial distributor system to overhaul and repair its engines in lieu of the overhaul base that it had established in Corpus Christi, Tex. Bids were accepted from the distributors for the overhaul of engines numbering in the hundreds and contracts were awarded by the Army to the low bidder.[118]

MODEL 250/T63 ENGINE SUMMARY

As the only American turboshaft engine in its power class for many years, the Model 250 engine was largely responsible for the creation of the light turbine helicopter industry and the widespread commercial use of light turbine helicopters. Whereas the piston engine-powered helicopter had limited payload and speed capability, the small gas turbine with its great power-to-weight ratio overcame these limitations and made the light turbine helicopter a very versatile and productive product.

Beginning in the early 1960s, the turbine-powered light helicopter had a major effect on military operations, business operations, the aircraft industry, and society in general. Light turbine military helicopters were used in combat for aerial observation, rescuing wounded troops, and weapon platforms. They helped transform the Army's warfare techniques and air support of its ground troops. While turbine engines became the power plant for high-performance military combat helicopters and many civilian helicopters, a substantial number of military trainer and civilian helicopters continued to be powered by reciprocating engines. Nevertheless, light turbine helicopters were used widely for such

---

118. Flanders and Mayo to Leyes and Fleming, 4 June 1991.

purposes as transporting personnel and equipment to off-shore oil drilling plat-forms, oil exploration, logging operations, air ambulance service, police work, fire fighting, and executive transport. Finally, the growth in demand for the light turbine helicopter produced a growing and healthy light helicopter indus-try in America.

From 1962 through the end of 1995, Allison had produced a total of 24,949 turboshaft engines (18,447 commercial and 6,502 military) for use in light heli-copters. The engine dominated the light turbine helicopter engine market throughout the world for many years. During the early 1990s, Allison continued to add new applications to the Model 250 family, such as the Enstrom TH-28, the Schweizer 300, the Kamov 226, and the Bell 430, which, combined with the existing customer base, offered the potential of extending production of the en-gine well into the future. In view of the tremendous number of Model 250 en-gines in the field and the continuing market for the engine, Allison could also look forward to the manufacture of spare and replacement parts as a significant portion of its business for many years.

In contrast to the many thousands of Model 250 turboshaft engines manufac-tured for light turbine helicopters, from 1962 through the end of 1995 Allison built only 1,059 turboprop engines for small fixed-wing aircraft. Fixed-wing aircraft performance benefits offered by the turboprop engine's high power-to-weight ratio, compared to that of the piston engine, most often could not be justified because of the significantly higher initial cost of the turboprop. For the helicopter, the performance gains offered by the gas turbine far outweighed its higher initial cost. Nevertheless, the reliability, smoothness, low noise character-istics, long operating life, and performance of the small turboprop had attracted a degree of interest in the general aviation community.

For specialized fixed-wing aircraft operations, such as primary flight trainers for military pilots or the need to operate in remote areas, price was not the driv-ing factor. Student military pilots found the transition to jet aircraft was smoother when they trained in aircraft powered by gas turbines rather than pis-ton engines. Model 250-powered military training aircraft were used by Italy, Japan, and India. An example of a specialized civil operation favoring use of the turboprop was the Allison Model 250 powered N-22B Nomad built by the Commonwealth Aircraft Corporation Ltd. in Australia and used by flying doc-tors in the South Pacific who visited remote areas with unimproved runways. In those operations, the take-off distance of the turboprop-powered Nomad was half of that required by comparable piston engine-powered aircraft.

In the course of its design and development, the Model 250 engine faced some difficult design challenges involving miniaturization of engine components. For example, it was necessary to maintain very tight clearances between spinning and stationary parts to minimize gas path leakage, which could compromise engine performance. Another challenge was to manufacture the tiny compressor and turbine blades with shapes and finishes that would yield high aerodynamic efficiencies. The development of in-house casting technology for making single-piece turbine wheels and axial-compressor rotor stages was a milestone technology. The major technical contribution of the Model 250 engine was to overcome these challenges by stimulating the development of techniques and technology that led to a small, reasonably priced, reliable engine with good fuel efficiency and a high power-to-weight ratio.

Although the Model 250/T63 had a long history, it had not become an "old" engine in terms of the design technology that it embodied. For example, as turbine engine technology advanced over the years, the engine design was progressively upgraded to incorporate those advances. By the early 1990s, its technology was as advanced as the new competing engines that had emerged. It also still enjoyed confidence and popularity among its users.

In addition to progressive technical improvements, an important factor enabling the Model 250/T63 to capture its major market share and become the predominant American engine in its power class was Allison's deliberate pricing at a point low enough to discourage competitors. As a result, the engine faced little competition in the light turbine helicopter marketplace until 1975 when Lycoming introduced its 600- to 700-shp LTS101 turboshaft engine models, and no serious competition until a decade later. By the late 1980s, Pratt & Whitney Canada had introduced its new PW200 series turboshaft engines and Turboméca had entered the U.S. light turbine helicopter market with its TM319 series engines. The TM319 had been delivered to the French military, which was a protected market, and both the P&WC and Turboméca engines had been selected for the Eurocopter 108 and the McDonnell Douglas Explorer.

By the early 1990s, the existence of competing engines had given airframers a choice of power plants for new and conversion helicopters and aircraft. Nevertheless, the Model 250, with its known reliability and operating cost, was selected for use in a number of new derivative helicopters and for conversions of piston-powered fixed-wing aircraft. Bell introduced the TwinRanger, and Tridair Helicopters, Inc., introduced the Gemini ST, a twin-engine conversion of the Bell LongRanger. Both of these helicopters were powered by a pair of Model 250-C20R engines. In the same timeframe, Agusta embarked on its flight test activity with the Allison Model C20R/9 in the upgraded A 109D helicopter in preparation for certification and production release. In addition, the fixed-wing market had begun to expand with the O & N Aircraft Modifications, Inc., conversion program for the piston-powered Cessna P-210. That, with the installation of the Model 250 in trainers, such as the Valmet Aviation Industries, Inc., L-90 TP Redigo in Finland and the Enaer T-35 DT Turbo Pillan in Chile, gave rise to increased production of the turboprop version of the engine.

Technology Demonstration and the GMA 500 Program

Beginning in 1965, Allison had been awarded several contracts as a participant in the military's programs to demonstrate advanced gas turbine engine technology. The first of these contracts involved gas generators for a series of turbojet and turbofan lift engines. These Air Force contracts were included in the Advanced Turbine Engine Gas Generator (ATEGG) program. Later, under the ATEGG program, Allison developed the GMA 100 gas generator to demonstrate intermediate turbine temperature technology. That work was followed by another Air Force ATEGG contract, extending from 1972 to 1977, under which Allison demonstrated high turbine temperature technology in the GMA 200 gas generator. This work was complemented by Navy and Air Force Joint Technology Demonstrator Engine (JTDE) contracts, extending from 1972 to 1979, to demonstrate advanced engine components. The new technology that evolved from the work under these contracts became part of Allison's technology data base used in designing the GMA 500 demonstrator engine.

Knowing that the Army was planning a competition for development of an Advanced Technology Demonstrator Engine (ATDE) in the 800-shp class, in 1975, Allison began assembling a baseline of component technologies for design of the engine. The Request for Proposals for this engine was released by the Army in 1976. The objective of the contract was to demonstrate the achievable level of performance attainable by turboshaft engines in the 800-shp class and to perform a limited demonstration of its durability.[119] Specific objectives were to demonstrate improvements in specific fuel consumption of 17 to 20% and in power-to-weight ratio of 25 to 35% compared to then current turboshaft engines in the 800-shp class.[120]

119. Ray Funkhouser, interview by Rick Leyes and William Fleming, 5 June 1991, Allison Interviews, transcript, National Air and Space Museum Archives, Washington, D.C.

120. Ron Alto and Walt DeRoo, "Can the Propulsion Industry Truly Share R&D Technology to the Customer's Benefit? - The LHTEC T800 Story," *Vertiflite*, Vol. 34, No. 3 (May/June 1988): 121, 122.

Five companies submitted proposals: Allison, Garrett, Lycoming, General Electric, and Pratt & Whitney's Florida Division. On February 1, 1977, two competing, four-year, fixed-price contracts were awarded, one to Allison for development of the GMA 500 and the other to Lycoming for its PLT34. An Allison news release reporting the award read, "An $11,300,000 contract has been awarded Detroit Diesel Allison Division of General Motors for the design and demonstration of an advanced technology gas turbine engine with potential application as a turboshaft power plant for future military helicopters."[121] The award was made by the U.S. Army Fort Eustis Directorate Air Mobility Research and Development Laboratory at Fort Eustis, Virginia.[122]

ENGINE DESIGN AND DEVELOPMENT

Bill Castle was responsible for the preliminary design work on the GMA 500. His initial decision in designing the engine was the selection of the compressor configuration. Two basic arrangements were considered. One was a two-stage centrifugal compressor and the other was an axial-centrifugal arrangement. After considering these two alternatives, the two-stage centrifugal compressor approach was selected. Allison had confidence in this approach, having used the high pressure ratio, high-efficiency centrifugal compressor in the Model 250 Series III and IV engines.

Two other factors influenced the selection of the two-stage centrifugal compressor arrangement. According to Castle, one factor was cost.[123] Castle believed that a centrifugal stage would be less expensive to manufacture than an axial design. The other factor was that the Army had specified that the engine have concentric straight-through shafting with the power takeoff at the front of the engine. The centrifugal front-stage arrangement was favored because the hole that could be run through it was considerably larger than that through an equivalent axial compressor. The larger-diameter hole through the centrifugal stage permitted use of a larger-diameter, stiffer shaft, reducing shaft vibration problems.

Cast turbine blades were chosen for the engine as a result of the company's considerable experience with that technology. The blades of each stage were

121. Castle to Leyes and Fleming, 4 June 1991. Allison news release, 18 March 1977. Believing that defense business would decline with the end of the Vietnam War, in September 1970, GM combined the Allison Division with the Detroit Diesel Division to form the Detroit Diesel Allison Division.

122. Castle to Leyes and Fleming, 4 June 1991.

123. Castle to Leyes and Fleming, 4 June 1991.

cast onto a ring which in turn was pressed into and diffusion bonded to a hub made of a higher-strength material. This arrangement was called a hybrid turbine wheel. The gas generator turbine had two cooled stages and the power turbine had two uncooled stages.

To satisfy the Army's requirement for compactness, the gas turbine compressor and turbine were close coupled with a reverse-flow type of combustor wrapped around the gas generator turbine. To promote cooling, the combustor liner was fabricated of Lamilloy® (a registered trademark of General Motors Corporation), a porous material employing a concept developed by Allison. Lamilloy® was a quasi-transpiration cooling system comprised of a multi-layered wall made by a combination of investment casting, electrochemical etching, and isostatic bonding that produced a pattern of flow paths through the material. Lamilloy® could be formed or cast into complex shapes, and its porosity could be tailored to the needs of the particular hot-part environment in which it was used.

An integral inlet particle separator was located ahead of the compressor, and the accessory gearbox was located on top of the engine. The gearbox was driven off the gas generator turbine shaft by a gear drive located between the two compressors.[124]

Chief Project Engineer of the GMA 500 was Ray Funkhouser. Under his direction, testing of GMA 500 components began in the spring of 1977, followed by the start of gas generator and complete engine development and demonstration testing in early 1979. Demonstration testing proceeded until late 1980 when a rubbing problem emerged. The design of the turbine was such that the entire turbine assembly was overhung from a bearing located ahead of the gas generator and turbine. There was no support structure between the power and gas generator turbines.

At that time, there was also a study of supercritical shafting, which involved designing the bearing system to permit acceleration of the engine through its critical speed range. It was decided to redesign the turbine support structure and bearing system to correct the turbine rubbing problem and also permit operation of the engine at speeds above its critical speed. However, insufficient time was left to modify the turbine support structure and complete the demonstration program before the contract expired a few months later. In addition to the turbine support redesign, there were certain other components that both Allison and the Army felt should be improved. Simultaneously, development of the competing Lycoming PLT34 engine was faced with a delay. Additional time

124. Castle to Leyes and Fleming, 4 June 1991. See also Funkhouser to Leyes and Fleming, 5 June 1991.

was needed to complete some of the component improvement work on the PLT34 engine. As a result, the Army decided to release a solicitation confined to Allison and Lycoming for a two-year follow-on contract. Called the Critical Component Improvement Program (CCIP), the two contracts were awarded in late 1981 and extended through 1983.

Under the CCIP contract, Allison redesigned the turbine bearing support and successfully completed its demonstration of the engine. In redesigning the turbine assembly, an additional bearing was located between the gas generator and power turbines to provide further support. Allison gave the redesigned engine the designation of Model 280-C1 to conform to the company's then standard system of engine model designation. By early 1984, the sfc, power output, and other objectives of the program had been successfully demonstrated. In fact, the engine demonstrated an output of over 1,000 shp, which substantially exceeded the contract requirement of 850 shp.

Following completion of the CCIP contract, Allison continued testing the engine and working on improvements using its Independent Research and Development (IR&D) funds. By 1985, when engine testing came to an end, there had been a total of 560 hours of testing on the demonstrator engine. In addition, helicopter flight tests had been made on a Full Authority Digital Electronic Control and an Engine Monitoring System (EMS) for the engine.[125] The EMS provided life monitoring parameters and fault detection for line-replaceable components.

During the early 1980s, Allison had begun marketing the Model 280-C1 engine and envisioned its use for a variety of civil and military helicopter and fixed-wing applications. One of the most promising of these applications was the Army's planned Advanced Scout Helicopter (ASH). Unfortunately, that program was canceled, and other applications for a new 1,000-shp-class turboshaft engine failed to materialize. By the time development work on the engine ended in 1985, the Army had developed a requirement for a new more powerful turboshaft engine, the T800.

GMA 500 SUMMARY

The technology that Allison derived from the GMA 500 program served a useful purpose. Technical features developed and demonstrated during the GMA 500 program were incorporated in the T800. Those features included the GMA 500's very efficient inlet particle separator, the highly efficient small power tur-

125. Funkhouser to Leyes and Fleming, 5 June 1991.

bine, the newly developed FADEC and EMS, and the gearbox and accessory drive system which was smaller than its contemporaries.

The T800

Beginning in 1984, the T800 turboshaft engine was developed jointly by Allison and Garrett for use in the new Light Helicopter Experimental (LHX) to be developed by the Army. Development of the engine from Garrett's perspective was described in the Garrett chapter along with a detailed description of the engine, the procurement cycle, engine development and performance, and potential military and civil applications. The T800 discussion in this chapter was written to complement the discussion of that engine in the Garrett chapter by presenting additional program highlights as seen from Allison's viewpoint.

During the time that Allison had been developing the GMA 500 engine, Garrett was developing a 1,500-lb-thrust F109 turbofan engine for the Air Force T-46 Next Generation Trainer. In January 1983, following the Army's announcement of the LHX program, Garrett decided to make a turboshaft out of the F109 engine. That engine, rated at 1,000 shp, first ran in August 1984. As a result, Allison with its GMA 500 and Garrett with its TSE109 were both well situated technically to enter the race for the T800 contract.

In early 1984, as Allison and Garrett were positioning themselves for the T800 engine competition, the Army let it be known that a teaming arrangement would be preferred and might even be required to win the contract. The purpose of this arrangement was to provide dual and competing T800 production sources. The form of the teaming arrangement was left to the bidders.

According to Blake Wallace, then General Manager of the Allison Gas Turbine Division (later President and CEO of the Allison Engine Company):

> We felt that they (the Army) were going to do that. We felt they were serious, and that turned out to be correct. We said: "If that's really going to happen, its best to be the first one to choose your partner, because otherwise you get whoever is left over." So we had done a study and decided that Garrett was a pretty good candidate.[126]

Wallace did not recall who made the initial contact, but he remembered that he and Bob Choulet, Allied Signal Aerospace Company, Engine Group

126. Dr. F. Blake Wallace, interview by Rick Leyes and William Fleming, 15 August 1991, Allison Interviews, transcript, National Air and Space Museum Archives, Washington, D.C.

President, had dinner together one evening in Indianapolis during which they discussed the subject of teaming and agreed that it should be explored further.

Two days later, following discussions between Blake Wallace and Frank Roberts, Executive Vice President of Garrett, a memorandum of understanding had been developed that described how the two companies would work together. Wallace observed:

> We made an agreement that everything would be on top of the table, no hidden agendas, nobody was going to try to position himself for more of the spares or something like that. Everything on the table, let's split it right down the middle, and that's what we did. It has worked very well for us.[127]

It was on this basis that their partnership known as the Light Helicopter Turbine Engine Company (LHTEC) was formed. Blake Wallace believed that there were two important reasons why Allison and Garrett would fit together. One reason was that their engine concepts were identical. Both companies used the two-stage centrifugal compressor approach, and the other parts of their engines were similar. He stated:

> If you took the Garrett engine concept and our engine concept and you laid them out on a piece of paper, unless you were a real aficionado, you wouldn't have been able to tell the difference between them. So that made it easy to integrate.[128]

His second reason for believing they would be a good fit was that the two companies were similar in size and capability. In his view, it was difficult to have companies that were different in these respects in a teaming arrangement, thus:

> If you have one big company with lots of technical capability and another little company that can't gain the respect of the big company, you're not going to have a very good 50-50 teaming arrangement. We and Garrett are about the same size, about the same size engineering organizations, so we made a good match.[129]

In May and June 1984, as business arrangements were still being settled, teams including Joe Byrd, LHTEC Program Manager from 1985 to 1988, and Norm

---

127. Wallace to Leyes and Fleming, 15 August 1991.

128. Wallace to Leyes and Fleming, 15 August 1991.

129. Wallace to Leyes and Fleming, 15 August 1991.

Egbert, who became the T800 Chief Engineer and was placed in charge of engineering management in 1989, went to Garrett for a series of meetings to compare notes and decide how to proceed.[130] Mike Hudson, who led the Allison team at the original meeting, acknowledged that neither company knew exactly how to do it, because in those days the industry was extremely guarded with respect to its technology data base and other proprietary information. Although Allison had teamed with Rolls-Royce earlier on the TF41 engine, they had never teamed with an American company. As Mike Hudson recounted:

> So, we agonized over it and decided: "Well we've got to show them the engine (the GMA 500) if we're ever going to team." So, I stood up and showed a cross-section of the engine and kind of talked our way through it. We sat down and said: "Okay, you guys talk." Frank (Roberts, Garrett's Executive Vice President) said: "Gee, we didn't bring any viewgraphs, but if I could use your cross-section it looks just like ours (the TSE109)." So that was probably the first omen that we could fit these two companies together. Our configuration was almost exactly the same. It was the twin-centrifugal-compressor, two-stage turbine driving it, foldback annular combustor, and two-stage power turbine.[131]

From that point on, the barriers of secrecy tumbled and the two companies began sharing technology in pursuit of their joint objective.

HOW THE PARTNERSHIP FUNCTIONED

Soon after the May-June meetings, the Army indicated that the power level of the T800 was to be 1,200 shp, which meant that the engine would be larger than either Allison's GMA 500 or Garrett's TSE109. As the Allison and Garrett engineers proceeded with the preliminary design of their T800, it was agreed that the TSE109 core should be used as it was larger than that of the GMA 500 and more nearly met the power requirements. Keeping in mind the agreed upon 50-50 split of work, Allison was given responsibility for the power turbine, inlet particle separator, electronic engine control, and gearbox and accessory drive. Assignment of responsibility for these components to Allison was logical as the advanced technology embodied in each had been well demonstrated during the GMA 500 program.

130. Norm Egbert, interview by Rick Leyes and William Fleming, 6 June 1991, Allison Interviews, transcript, National Air and Space Museum, Washington, D.C.

131. Hudson to Leyes and Fleming, 6 June 1991.

Although each partner was assigned prime responsibilities, the partnership
was structured in such a way that both partners were also obligated to bring forward requisite technology to solve any shortcoming. The result was that both
partners made extensive contributions to the other partner's components. To
ensure that both companies contributed, a partnership sign-off was required on
every drawing. Norm Egbert, the T800 Chief Engineer, observed:

> That is probably one of the strengths of the partnership. In certain areas I think
> all of us feel that the partnership is capable of producing a superior engine be-
> cause of the fact we bring two design methodologies, two data bases, and two
> potential solutions to any problem."[132]

LHTEC was established as a joint venture company, the staffing and management shared by Allison and Garrett. Its function was to provide the engineering
and project support functions for the T800 program. Once production of the
engine began, production contracts would go to Allison and Garrett, not to
LHTEC. Engines shipped from Allison would be T800-AD-800s and those
from Garrett would be T800-GA-800s. The two companies would compete for
military production contracts. It was expected that production for commercial
applications would be on a co-production basis, shared by the two companies.
LHTEC would continue as an independent entity to provide the engine program such services as product support, warranty administration, configuration
management, and continuing support for future engine growth.[133]

ENGINE DEVELOPMENT

With the power requirement of 1,200 shp set by the Army, LHTEC proceeded
to develop a T800 technology prototype engine, also known as the ATE109,
that could produce that level of power output. The turbine temperature of the
engine's TSE109 core was increased about 200°F, and the engine incorporated
Allison technology in the gas generator turbine and other engine components.
The ATE109 engine first ran in October 1984 and was flown on a Bell UH-1B
on March 7, 1985, only two days after LHTEC had submitted its proposal to the
Army for full-scale development of the T800.

In addition to LHTEC, two other teams submitted proposals. One was a
Lycoming/Pratt & Whitney joint venture called APW. The other was a GE/
Williams International leader-follower arrangement. On July 19, 1985, LHTEC

132. Egbert to Leyes and Fleming, 6 June 1991.

133. Egbert to Leyes and Fleming, 6 June 1991.

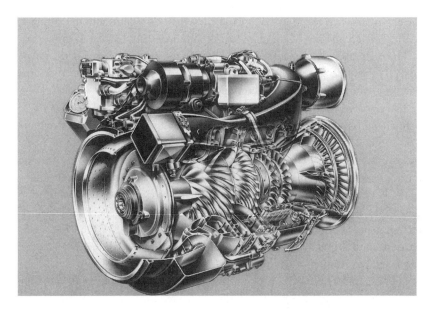

Cutaway LHTEC T800 turboshaft engine developed jointly by Allison and Garrett. Photograph courtesy of Allison.

and APW were selected as winners to enter a Full-Scale Development (FSD) competition. Upon award of its contract, LHTEC embarked on the full-scale design and development of its engine, designated the T800-LHT-800.

Three months into the program, the Army held an engine mock-up review that was attended by the competing LHX airframe teams and Army user personnel. Approximately 200 concerns were registered by the attendees. Norm Egbert observed, "As we sat back after they left and looked at these concerns, we said, 'We can't address these concerns without making some significant changes.' Management got together and we decided to bite off a big chunk."[134]

In spite of the fact that they were three months into a contract in which they had agreed to have an engine running in 15 months, an effort was undertaken to completely redesign the external arrangement of the engine. The redesign effort was an example of concurrent engineering concentrating on maintainability and survivability. Tradeoff studies were made on every aspect of the configuration with inputs of both the maintainability people and the survivability people factored into the design. In spite of the time required for this work,

134. Egbert to Leyes and Fleming, 6 June 1991.

the engine was first tested in August 1986, less than 14 months into the contract, beating the competing Lycoming-Pratt & Whitney team by one month.

The preliminary flight rating (PFR) phase of the program ended in the fall of 1988. The engine was rated at 1,382 shp and had a sfc of 0.46. The T800 first flew in October 1988 on an Agusta A129 helicopter.

The government proceeded with its original plan to "downselect" to a single partnership for completion of the T800 qualification and production program. The downselect was based on the demonstrated achievements in the PFR phase and on the value to the government of the competitors' rebid for remaining portions of the program. William Schnell replaced Joe Byrd as Program Manager. LHTEC was awarded a sole-source contract to complete the Qualification Test (QT) and other program elements on October 28, 1988. Resolution of technical and administrative issues extended Army issuance of the PFR qualifications until September 1989. The PFR qualification cleared the T800 engine for use in government-sanctioned flight tests.

LHTEC initiated the QT program at the time of the October 1988 contract award. On February 22, 1991, the Army contracted for 32 T800 engines for use during the demonstration and validation phase of the program.[135] By the following December a total of over 13,400 hours of testing had been done on the 18 QT and PFR engines, 1,200 in QT alone.[136] On September 1, 1992, LHTEC responded to a request for proposal for an Army funded T800 growth program that would increase the power output by 15% to about 1,500 shp. At that time, LHTEC expected to complete all contractual responsibilities for the T800 development program by the end of 1992. Consequently, the growth program would provide continued Army funding support for the engine.[137] The boost in power output was to be obtained by lengthening the compressor blades to increase the engine airflow and by incorporating hot section changes that would permit a minor increase in turbine temperature. The additional power was required to accommodate increases in the weight of the RAH-66 Comanche, the helicopter that was to use the T800.[138]

---

135. William Schnell, LHTEC Manager, to William Fleming, fax, 9 September 1992, National Air and Space Museum Archives, Washington, D.C.

136. Ray Horn Jr., "Airbreathing Propulsion," *Aerospace America*, Vol. 29, No. 12 (December 1991): 58.

137. Schnell to Fleming, 9 September 1992.

138. Stanley W. Kandebo and David F. Bond, "Allison/Garrett to Boost T800 Power 12% With Higher Turbine Temperatures, Airflow," *Aviation Week & Space Technology*, Vol. 134, No. 3 (January 20, 1992): 62.

ENGINE TESTING

The engine testing throughout the PFR and QT phases was divided evenly between the two partners. Each partner tested the components for which it was
responsible. Engine testing was similarly divided. If the prime purpose of a test
was to demonstrate a certain component, an attempt was made to run that test
in the house that had design responsibility for the component. For example, inlet particle separator and control system tests were performed at Allison. Endurance testing was shared by the two partners.

Several more rigorous types of testing were introduced into the program by
the Army. These included gyroscopic testing, loss-of-oil testing, and electromagnetic interference (EMI) testing. The gyroscopic tests were made to demonstrate
mechanical integrity of the engine in simulated flight maneuvers. For example, it
was required that the engine accelerate from idle to maximum power and return
to idle while being rotated on a test rig at various rates. There was also a requirement for the engine to run steady state while being rotated at a fixed rate. The
T800 successfully demonstrated its mechanical integrity during these tests.

Loss of oil testing became a qualification requirement to address combat survivability in a wartime environment. The engine was required to operate at
power for six minutes with no oil provided to any oil wetted components.
Fulfillment of this requirement demonstrated the capability of the engine to
operate in the event the oil tank was drained due to ballistic or shrapnel damage.
After extensive development, which included the design of an internal sump
emergency oil system, the engine successfully demonstrated the capability to
meet this requirement.

EMI testing was emphasized in the program because the military had experienced a large number of instances in which aircraft incorporating electronics
had encountered problems when operating in the proximity of radio towers, radars, and other electromagnetic signal generating devices. Spurious electromagnetic signals generated by such devices were found to create serious problems in
aircraft electronic systems, and in some cases, caused crashes. The EMI tests revealed the need for modifications to a number of electronic components to
meet the test requirements.

Although most of the testing was done in Allison and Garrett facilities, some
testing that required specialized facility capabilities was performed at government test installations. Inlet pressure and temperature distortion tests that simulated armament gas ingestion were performed at NASA's Lewis Research Center. In those tests, inlet temperature transients of up to 1,000°C per second for
1.5 seconds were imposed on a 60-degree sector of the engine inlet. Complete

compressor inlet flow distortion tests were also performed at NASA Lewis. Corrosion tests and anti-icing tests were run at the Naval Air Warfare Center (NAWC) in Trenton, N.J. The gyroscopic testing was done at the NAWC facility in Lakehurst, N.J. Armament gas ingestion tests with live rockets that had been scheduled to be conducted at the Aberdeen Proving Ground were not completed.[139]

LHTEC completed the battery of QT requisites in early 1993, and the Army issued full T800 qualification on July 29, 1993. FAA certification of the civil T800 variant, the CTS 800, was awarded September 10, 1993.

T800 APPLICATIONS

The prime application for the T800 was the RAH-66 Comanche. The contract for the Comanche had been awarded to the Boeing-Sikorsky team in April 1991 following the downselect competition with the Bell-McDonnell Douglas team. By 1993, the Comanche had been approved and funded through the demonstration/validation phase. Although the Comanche had a high priority within the Army at that time, initiation of its production, with a resulting requirement for engines, was yet to be approved due to budget uncertainties. Nevertheless, there was the potential that the Boeing-Sikorsky team might exploit the foreign market for its Comanche, which would be reflected in orders for the T800 engine.

By mid-1991, LHTEC had also begun pursuing a number of other possible applications for the T800. A potential application promoted by Global Helicopter Technology was use of the T800 to replace the Lycoming T53 in the Army's several thousand Bell UH-1H Huey utility helicopters. Advantages offered by Global in its promotion of the T800 were improved fuel economy, a significant increase in payload and range capabilities, and modern avionics. It was claimed that the T800 had a 40% lower sfc than the T53 and at least a 20% advantage over any other engine being considered for retrofit in the Huey. In addition, a FADEC was used to control the engine, improving operating characteristics and power response over those of the T53. In 1992, Global demonstrated its Huey 800 to Army and National Guard officials.[140]

Among other possible applications was the Agusta A-129, a T800-powered scout attack aircraft with attributes similar to those of the Comanche. There had also been work with the U.K.'s Westland, including flight evaluations of a

---

139. Egbert to Leyes and Fleming, 6 June 1991.

140. David A. Brown, "T800 Engine Boosts UH-1H's Performance," *Aviation Week & Space Technology*, Vol. 138, No. 1 (January 4, 1993): 48–51.

T800-powered Lynx in late 1991. Although the T800 had been successfully demonstrated by mid-1991 as a replacement for Lycoming's LTS101 in the Coast Guard's Aerospatiale HH-65, the Coast Guard's position was that it did not plan to proceed beyond demonstration flight testing with the T800. LHTEC was also working with American Eurocopter (formerly a joint LTV/Aerospatiale venture) on the Panther 800 program, using a T800 in a Dauphine helicopter. Extensive flight demonstrations were conducted on the Panther 800 in late 1992 and early 1993. One other different type of application explored with the military was use of the T800 to drive an electrical generation unit.[141]

T800 SUMMARY

Formation of the Allison-Garrett partnership resulted in an alliance of two competitors in which they effectively shared their technology to successfully develop, demonstrate, and qualify the T800 engine. LHTEC provided them with the understanding of how to make such a partnership work, recognizing that future partnerships might become necessary to satisfy customer requirements or to effectively cope with competitive pressures within the industry in a limited-market environment.

The T800 was an advanced technology engine with performance and capabilities superior to those that could have been achieved in an engine developed by either partner alone. By establishing the teaming requirement, the Army not only achieved its objective of providing dual production capabilities for the engine, but obtained a better engine.

Although the T800 engine had successfully met its design objectives and had been accepted by the Army for use in its new Light Helicopter, the Comanche, by the end of 1991, the engine faced an uncertain future. The Army's plans for further development of the Comanche were in flux, which meant that the requirement for engines remained undefined. Although other possible applications for the T800 were being pursued, no firm commitments had been made.

The T406 Engine Family

According to Nick Blaskoski, T406 Program Manager, the T406 engine family came about as a result of a progression of several interrelated events.[142] In 1982,

---

141. Egbert to Leyes and Fleming, 6 June 1991.

142. Nick Blaskoski, interview by Rick Leyes and William Fleming, 6 June 1991, Allison Interviews, transcript, National Air and Space Museum Archives, Washington, D.C.

Allison had proposed a new engine design for the Army's Modern Technology Demonstrator Engines (MTDE) program competition. Allison lost this competition in 1983 to P&W and GE, which were selected to develop competing demonstrator engines. It was generally expected that one of the two competing engines developed for the MTDE program was destined for use in the Navy's planned experimental tilt-rotor aircraft, the JVX (Joint-Services VTOL Experimental), later the Bell-Boeing V-22 Osprey. The V-22 was a twin-engine aircraft with VTOL and STOL capabilities. When the V-22's two engine nacelles, located at the wing tips, were pointed upward, the aircraft could operate as a helicopter. Tilting the nacelles forward in flight to the horizontal position permitted operation with the speed and efficiency of a turboprop aircraft.

In 1983, shortly before the Navy released the Request for Proposals for the V-22 engine, there had been a major reorganization at Allison. In September 1983, Allison was separated from Detroit Diesel, and the following April, it was established as a separate division of General Motors. In March 1983, Blake Wallace had been brought aboard as Allison's Director of Gas Turbine Sales and Engineering. When Allison was separated from Detroit Diesel the following September, Wallace was named General Manager.[143] Under Wallace's direction an aggressive effort to expand the company's product base began. Up to that time, Allison had not planned to compete for the V-22 engine contract because it was thought that GE and P&W, which were developing demonstrator engines under the MTDE program, had the inside track. However, after Wallace came on board, he altered that attitude. With its T56 engine, Allison had become the industry leader in the arena of large turboprops and had cornered the market for 5,000- to 6,000-shp-class turboprop engines. Wallace wanted to protect that market segment for Allison by assembling a serious bid for the V-22 engine, which was in the same power class as the T56 and could potentially be its next generation successor.[144]

In 1984, Blake Wallace had instructed Mike Hudson, his newly appointed Director of Engineering, to develop a strategic business plan for use in helping to shape the company's future. The essence of that plan was to protect Allison's traditional business while moving into the transport engine marketplace. The tactical plan to support that move was to concentrate competitive activity on the company considered the prime competitor, GE. That move was pursued

143. Sonnenburg and Schoneberger, *Allison Power of Excellence*, 199.

144. Blaskoski to Leyes and Fleming, 6 June 1991.

Allison T406-powered Bell-Boeing V-22 Osprey tilt-rotor aircraft in flight. Photograph courtesy of Allison.

with vigor by entering the V-22 engine competition. Mike Hudson and John Arvin, then Chief of Preliminary Design, developed a technically conservative all-axial-flow engine design while Frank Verkamp, the Large Engine Product Line Director, shaped the business strategy and led the proposal effort.[145]

At the time of the V-22 engine competition, Allison believed that the engine developed under the V-22 engine contract could also be used as a turboprop. They felt that the turboprop version could satisfy the 5,000-shp-class needs of the Air Force, Navy, and Army in replacements or upgrades of the Lockheed C-130 cargo aircraft and the Lockheed P-3 surveillance aircraft, thereby protecting the market dominated by Allison's T56 turboprop. It was also believed that the turboprop would be used in advanced cargo aircraft, that the turboshaft version would be used in Boeing Vertol CH-47 helicopter upgrades, and that, at some point in the future, a turbofan version might find a place in the commercial transport and executive aircraft market. With the prospect of a gradual phase-out of the 30-year old T56, Allison feared that if the T406 engine family

145. Hudson to Leyes and Fleming, 6 June 1991.

lost out in being chosen for those potential future applications, the company would revert to being a builder of small engines such as the Model 250 and T800.

THE V-22 ENGINE COMPETITION

The strategy used by Allison in the V-22 engine competition was to propose an engine based on a conservative design approach. Pressure ratio and turbine temperature were selected at levels that would permit future engine power increases without changing the physical dimensions and configuration of the engine. In contrast, Allison's competitors proposed high pressure ratios and turbine temperatures to obtain maximum efficiency and low engine weight. Allison's design approach produced an engine that had more power margin for hot day operation and with a single engine out than did its competitors. Although this design approach resulted in an engine with a slightly lower efficiency and higher weight than the competing engines, each of those differences amounted to less than 5%.

The rationale behind Allison's strategy was that the design would be very competitive with those offered by its competitors, and in addition, would have an advantage in terms of low technical risk. Another factor influencing the competition was that John Lehman, who was then Secretary of the Navy, believed that firm fixed-price contracts would reduce the government's risk and minimize its cost. Thus, the V-22 engine contract was to be a firm fixed-price contract.[146]

Secretary Lehman made the final selection and, on December 24, 1985, announced his decision to select Allison. Allison's engine was picked because of its lower risk, established reliability and support system, and lower research and development costs.[147] The final contract was signed on May 2, 1986.

In addition to being a fixed-price contract, there were price guarantees for the first four lots of production up to 625 engines and a guaranteed second source that the bidder was required to qualify. Following announcement of the contract award, the Navy informed Allison that P&W, GE, Garrett, and Lycoming were acceptable second-source candidates. GE did not bid, and from among the other three, Allison selected P&W because they agreed to the business terms that the

---

146. Blaskoski to Leyes and Fleming, 6 June 1991.

147. "Allison Selected to Develop Engine for VTOL V-22 Osprey," *Aviation Week & Space Technology*, Vol. 124, No. 1 (January 6, 1986): 23.

Navy was looking for, which was critical, and they brought with them technical input that Allison felt was important.[148] Later, because of the reduced production requirement for the V-22 and T406, P&W withdrew from the second-source position with the general agreement of all parties involved and with a request from the Navy.

Selection of the T406 was an important win for Allison. In addition to protecting its large turboprop market as a replacement for the T56, it became a building block for two closely related derivative engines that later followed, the GMA 2100 turboprop and GMA 3700, both of which opened new markets for Allison.

T406 ENGINE DESIGN

The T406 was a 6,000-shp-class turboshaft, front-drive, free turbine engine. The engine had all axial-flow components. It combined the best design philosophies embodied in the T56 turboprop and the XT701, a large experimental turboshaft engine that Allison had developed for the heavy lift helicopter before that program was canceled by the Army. The compressor and gas generator turbine design evolved from the T56 and the power turbine and shafting from the XT701. Unique features to the T406 that were not part of those two engines were the four-bearing shaft design, a removable accessory gearbox to enhance maintainability, and a patented scavenge system that was part of the engine lubrication system.[149]

The compressor had 14 axial-flow stages with the inlet guide vane, and the first five stator stages variable to enhance starting and part-speed performance. The early stages were lightly loaded to enhance tolerance to inlet flow distortion. The diffuser/combustor was an integrated aerodynamic flow system extending from the compressor outlet to the turbine inlet. The diffuser/combustor casing was a major structural member of the engine that contained a full annular combustor liner. The gas generator turbine had air-cooled first- and second-stage guide vanes with air-cooled, single-crystal, first-stage rotor blades and solid second-stage blades. The power turbine had two uncooled stages, and it co-rotated with the gas generator turbine.

Because the engine was required to operate in a range of positions from horizontal to vertical, a lubrication system was designed with a unique, patented

148. Blaskoski to Leyes and Fleming, 6 June 1991.
149. Blaskoski to Leyes and Fleming, 6 June 1991.

Allison T406-AD-400 turboshaft engine. Photograph courtesy of Allison.

scavenging system that used centrifugal force generated by scavenge pump impellers in the sumps at each of the rotor bearings. The benefit of this system was that oil scavenging was unaffected by engine attitude because the centrifugal head developed by the impellers rather than gravity was the major force used to scavenge the oil.[150]

An important aspect of the engine's mechanical design was the necessity that it satisfy the maintainability requirements. The engine specifications required that any flight-line engine repair task should not require more than 45 minutes. This requirement covered anything from replacing a module to removing and replacing the engine. As a result, detailed attention was given to such factors as the type and number of bolts used, where they were located, and the special tools needed to service the engine. In discussing this subject, Blaskoski re-

150. John R. Arvin and Mark E. Bowman, "T406 Engine Development Program," (ASME Paper No. 90-GT-245 presented at the American Society of Mechanical Engineers Gas Turbine and Aerospace Congress and Exposition, Brussels, Belgium, 11-14 June 1990), 1-6.

marked, "We've incorporated over 100 design changes strictly for maintainability reasons."[151]

The T406 was capable of producing nearly 7,000 shp at sea level static standard day conditions operating at a turbine inlet temperature of 2,208°F. However, the engine was flat rated at 6,150 shp out to 103°F. At its thermodynamic power output and weighing only 975 lb, the engine had a power-to-weight ratio over 7 to 1, the highest of any engine in its class.[152]

T406 DEVELOPMENT AND TESTING

Allison recognized that development of this engine would require additional test facilities with some unique capabilities. Accordingly, an old test cell was modified to accommodate T406 development and altitude testing. In addition, a new facility was constructed for attitude testing of the engine.[153]

The performance test cell was designed to permit engine operation in a horizontal position at simulated altitudes from sea level to 20,000 feet and at inlet temperatures from -60 to 120°F. This facility was used primarily for engine performance evaluation and for instrumented component diagnostic evaluation.[154]

The attitude test facility was used to operate the engine at various attitudes in an ambient environment. It had the capability to simultaneously pitch an operating engine from -90 to +115 degrees from horizontal while rolling it from +55 to -55 degrees. Development of this facility proved to be a major task. According to Dick Fisher, Department Head, Test Services, "We had lots of trouble."[155] During its early operation, problems were encountered with the test stand bearings and with the lubrication system. These problems led to a costly and time-consuming redesign effort before the test stand functioned properly.

The attitude capabilities of this facility were used to evaluate the engine lubrication system and sump operation, vertical-mode start and shut down, engine oil consumption and static oil leakage, and vertical heat rejection when the engine exhaust was pointed downward. The T406 underwent extensive testing

151. Blaskoski to Leyes and Fleming, 6 June 1991.

152. Arvin and Bowman, "T406 Engine Development Program," 4.

153. Fisher to Fleming, 3 February 1993.

154. Howard A. Chambers, "Turboshaft Engine Development for Commercial Tiltrotor Aircraft," (SAE Paper No. 911017 presented at the SAE General, Corporate, & Regional Aviation Meeting and Exposition, Wichita, Kansas, 9-11 April 1991), 8-11.

155. Fisher to Fleming, 3 February 1993.

in this facility at transient attitude conditions and with prolonged operation at various attitudes. The test results showed that the engine was insensitive to attitude conditions.[156]

A complete engine was first run in December 1986, just 12 months after announcement of the contract award. Although numerous minor problems customarily associated with development of a new engine were encountered, there were no major development problems. A number of minor interface problems between the aircraft and engine were encountered, mostly because the aircraft and engine were each developed under different sets of government specifications. For example, it was necessary to shorten the engine slightly to satisfy engine nacelle dimensions and to increase the allowable back pressure on the engine to accommodate the Bell-Boeing infrared suppressor.

A major interface challenge was computer-to-computer mating of the aircraft and engine controls. The aircraft had a Flight Control System (FCS) that communicated with and was highly integrated with each engine's control system. Each engine had a Full Authority Digital Electronic Control. During operation of the aircraft, the FCS gave the FADEC a power demand signal, and it also controlled rotor speed. The FADEC then adjusted fuel flow to satisfy the FCS while protecting the engine from overtemperature and overspeed. The development challenge was getting these two controls to work together.[157]

V-22 PROGRAM UNCERTAINTIES

In 1989, the V-22 Osprey faced an uncertain future. Reflecting Defense Department's recommendations, the President's budget submission to Congress in early 1989 contained no funds for the V-22. In January 1989, Allison had been put under contract to start production of the T406. However, that contract was terminated by the Navy the following December. Meanwhile, in response to strong lobbying, Congress restored funds in the Defense budget for the V-22. This tug-of-war between the administration and Congress continued through 1992 with zero V-22 funds being requested in the administration's budget each year followed by the restoration of V-22 funds by the Congress.

Meanwhile, the T406 engine began flight test operations in the V-22 on March 19, 1989. By mid-1991, four V-22 aircraft were flying, eighteen T406 engines had been delivered to support the flight testing, and those engines had ac-

156. Chambers, "Turboshaft Engine Development for Commercial Tiltrotor Aircraft," 8-11.

157. Blaskoski to Leyes and Fleming, 6 June 1991.

cumulated over 2,000 hours of ground and flight testing on those aircraft. At that time, Allison was continuing work on the engine under its Full-Scale Development contract, scheduled to be completed in May 1992.[158]

COLLIER TROPHY AWARD

On May 17, 1991, the National Aeronautic Association awarded the Collier Trophy to the Bell-Boeing V-22 Osprey tilt-rotor development team. That trophy was awarded annually for the greatest achievement in aeronautics or astronautics in America demonstrated by actual use in the previous year. Allison had also been one of the recipients of the 1987 Collier Trophy for its role in the development of the 578-DX propfan propulsion system. Although Allison was not named directly as a recipient of the Collier Trophy in 1991, in their acceptance speeches, the presidents of both Bell and Boeing acknowledged and recognized Allison for making the V-22 possible.

GMA 2100 DESIGN AND DEVELOPMENT

Beginning in 1984, a market forecast model was developed annually under the direction of Mike Hudson to support the formation of Allison's marketing strategy. As a result of these forecast models, 1988 became a positioning year for the company. By that time, the market forecasts had solidified the thinking within the company that Allison needed to have a greater presence in the commercial engine business. That thinking translated into development of the GMA engine family, which stemmed from the T406 engine.[159]

Work began on the GMA 2100 turboprop engine, designed and developed to accommodate the rigorous requirements of the regional airlines. These included high-use factors and multiple engine starts, changes in power setting, and numerous shutdowns per day. The engine consisted of a T406 power section, a modern reduction gearbox embodying the experience and lessons learned with the T56/501D turboprop, and a FADEC.

Based on the highly successful T56/501 two-stage planetary system, the main reduction gearbox was designed with emphasis on reliability and maintainability. Reliability was achieved through reduction in parts count and by giving at-

158. Blaskoski to Leyes and Fleming, 6 June 1991.

159. Jerry Wouters, interview by Rick Leyes and William Fleming, 5 June 1991, Allison Interviews, transcript, National Air and Space Museum Archives, Washington, D.C.

tention to lessons learned in the field. Maintainability features included a line replaceable accessory gear case, a modular propeller control, replaceable lubrication pumps, and borescope ports for visual internal inspection. In addition, maintainability features of the power section included design for use of standard hand tools; use of couplings, fasteners, and connectors that permitted ease of engine removal and installation; a line-replaceable accessory drive; and an engine monitoring system.

The GMA 2100 electronic engine control was based on the T406 FADEC design. Engine and propeller functions were coordinated by a single power lever. A power management system controlled propeller tip speed to improve cabin noise levels and flight-segment power levels.[160]

Development testing of the engine, described by Jerry Wouters, GMA Marketing Manager, included both ground and flight tests.[161] Ground tests were performed on the engine alone and on the complete engine nacelle system including the propeller. These tests were conducted in the Allison indoor turboprop test facility. Tests were also conducted in Allison's gearbox test facility on the newly designed reduction gearbox. Flight tests of the prototype engine began on the Lockheed P-3 flying testbed aircraft in 1990. The purpose of these single-engine flight tests was to demonstrate engine and propeller control system compatibility; assure acceptable propeller blade stresses, propeller shaft moments, and vibration levels; perform air starts; and demonstrate reverse thrust operation.

GMA 2100 TURBOPROP APPLICATIONS

Following Blake Wallace's initiative to preserve the 6,000-shp-class turboprop market for Allison, a decision had been made to have a turboprop version of the T406 for military and commercial applications. Consequently, even before winning the T406 contract, Allison began marketing a turboprop version of the engine. That engine, later designated the GMA 2100, was offered as a replacement engine for the Lockheed C-130 and P-3 and the Grumman E-2. The engine was also proposed to Lockheed for its new Long Range Air Anti-submarine

---

160. R. E. Riffel, T. F. Piercy, T. F. McKain, and E. T. Lewis, "The GMA 2100 and GMA 3007 Engines for Regional Aircraft," (AIAA Paper No. AIAA-90-2523 presented at the AIAA/SAE/ASME/ASEE 26th Joint Propulsion Conference, Orlando, Florida, 16-18 July 1990), 2-3.

161. Wouters to Leyes and Fleming, 5 June 1991.

Capable Aircraft (LRAACA), but lost to GE's GE38 turboprop. As it turned out, that loss was inconsequential as the LRAACA program was canceled later by the Navy.[162]

By 1988, a commercial market for the GMA 2100 was beginning to emerge led by Saab, interested in growing from its 35-passenger Model 340 regional commuter to the 50-passenger size. In addition, de Havilland saw a need for a high-speed, turboprop-powered regional commuter aircraft in the 70-passenger size. Also, the Indonesian manufacturer, Industri Pesawat Terbang Nusantara (IPTN), was considering a 50-passenger aircraft that appeared to be the right size for the kind of island-to-island commuter traffic that it was endeavoring to support.[163]

Allison made its initial offering of the GMA 2100 to Saab for use in the high speed 50-passenger Saab 2000 regional transport in competition with the GE/Lycoming GLC38.[164] On July 11, 1989, Saab-Scania and General Motors announced the selection of the GMA 2100 engine for the Saab 2000.[165] The engine was to be certificated at 4,550 shp for the Saab 2000. However, according to Allen Novick, Allison's Large Aircraft Engine Product Line Director, the GMA 2100 had the capability to grow easily to more than 6,000-shp with essentially no changes.[166]

IPTN was next to enter the 50-passenger commuter aircraft market with its N-250 twin-engine turboprop transport. For this application, the GMA 2100 was offered in competition with the P&WC PW130 and the GE/Lycoming GLC38.[167] On July 12, 1990, IPTN announced that the GMA 2100 was the winner of the N-250 engine competition. Commenting on the announcement, IPTN President Dr. Ing B. J. Habibie said, "The GMA 2100 was selected for its modern technology, high-growth capability, and low-risk development plan."[168]

---

162. Blaskoski to Leyes and Fleming, 6 June 1991.

163. Blaskoski to Leyes and Fleming, 6 June 1991.

164. "Allison Powers Saab 2000," *Flight International*, Vol. 4174, No. 136 (July 22, 1989): 18.

165. Allison Gas Turbine Division, "Saab Selects Allison Engine for the Saab 2000," press release, 11 July 1989.

166. "Allison Engine Chosen for Saab 2000," *Aviation Week & Space Technology*, Vol. 131, No. 3 (July 17, 1989): 29.

167. "Allison Wins Regional Powerplant Battle," *Flight International*, Vol. 4225, No. 138 (July 18-24, 1990): 18.

168. Allison Gas Turbine Division, "IPTN Selects Allison GMA 2100 To Power N-250 Regional Aircraft," press release, 12 July 1990.

Saab 2000 commuter aircraft. Photograph courtesy of Allison.

This announcement was followed on June 16, 1991, by the signing of an order for the first 46 GMA 2100 engines for the N-250.[169] Expanding on Dr. Habibie's reference to the GMA 2100's high growth capability, Jerry Wouters, GMA 2100 Marketing Manager, observed that both Allison and IPTN felt that there were likely to be growth versions of the N-250 up to the 80-passenger size. He affirmed that the power margin of the GMA 2100 could easily handle such growth.[170]

Wouters enumerated the points that Allison stressed in selling its GMA 2100 engine.[171] These included the company's commitment to the customer, its experience as this country's leading turboprop manufacturer, product assurance, engine reliability and maintainability, economies of scale by virtue of its high volume production, and attention to airframer and customer concerns about engine noise and exhaust pollution.

169. Allison Gas Turbine Division, "IPTN Places GMA 2100 Engine Order," press release, 16 June 1991.

170. Wouters to Leyes and Fleming, 5 June 1991.

171. Wouters to Leyes and Fleming, 5 June 1991.

One selling point that was particularly attractive to Saab was the offer to furnish the complete engine nacelle system and to integrate the propeller system with the engine. Saab was interested in purchasing the complete propulsion system package from the engine supplier, who would be responsible for not only supplying the unit but also supporting it with the airlines. Allison used five major subcontractors in the design and manufacture of its nacelle system to provide the nacelle skin, inlet and air/oil coolers, engine mounts and vibration isolators, electric power generator, and air turbine starter. To provide customer service for their GMA engines, Allison established a worldwide service system with on-site field managers, overhaul and repair centers, and parts supply centers.[172]

The GMA 2100 enjoyed a substantial power margin when installed in the Saab 2000 and N-250. The T406 had a thermodynamic power rating of nearly 7,000 shp and, for the V-22, had been flat rated at 6,150 shp to 103°F. To provide an engine for commercial applications with their demands for high reliability and long life, much of the GMA 2100 development testing was conducted at 5,700 shp to 90°F. The power requirements for the Saab 2000 and N-250 applications were even lower, resulting in ample turbine temperature margins. For the Saab 2000, the engine was flat rated at 4,150 shp to 100°F. For the N-250, it was flat rated at 3,600 shp to 115°F. These power ratings resulted in turbine temperature margins of over 235°F.

As a result of flat rating the engine at relatively low power levels, the aircraft were carrying somewhat heavier engines than they needed. However, the temperature derating of the GMA 2100 meant extended life for the engine's hot section, an important factor for regional airline operations. In addition, there remained sufficient power margin to accommodate later growth versions of the aircraft by merely increasing the turbine operating temperature.[173]

The Saab 2000 first flew on March 26, 1992. By January 1993, it was more than half way through its certification flights, and Saab had received orders for 46 Saab 2000s with initial delivery scheduled for late 1993.[174] The first flight of the IPTN N-250 was scheduled to follow in 1995.[175]

---

172. Wouters to Leyes and Fleming, 5 June 1991.

173. Wouters to Leyes and Fleming, 5 June 1991.

174. David M. North, "Speed, Range Boost Saab 2000's Appeal," *Aviation Week & Space Technology*, Vol. 138, No. 5 (February 1, 1993): 44-48.

175. "Allison GMA 2100 Selected," *Aviation Week & Space Technology*, Vol. 133, No. 5 (July 30, 1990): 11.

Allison GMA 3007 turbofan engine. Photograph courtesy of Allison.

THE GMA 3007 TURBOFAN

Although there had been a commitment for a turboprop version of the T406 engine at the time its development began in 1985, there had not been a similar commitment for a 7,000-lb-thrust-class turbofan version. It was not until around 1988 that a commercial market for that size turbofan engine began to evolve. At that time, Short Brothers was evaluating a 50-passenger regional jet called the RJX, requiring an engine in that thrust class. Later, the Brazilian aircraft company Embraer began designing the twin-engine EMB-145, a 50-passenger commuter aircraft for which an engine in the 7,000-lb-thrust class was suitable. Cessna was also beginning work on an upgraded 8- to 12-passenger very-high-speed executive jet, the Citation X, which required a 6,000-lb-thrust engine.

In response to the evolving commercial transport market, Allison began discussions with Rolls-Royce concerning the possible joint development of a turbofan, to be designated the RB580. Allison would provide its T406 core and Rolls-Royce would use its technology to develop the fan, power turbine, and bypass ducting. As explained by Gary Williams, GMA 3007 Program Manager, such an arrangement would combine Allison's experience and recognition as the developer of the T406 core with Rolls-Royce's turbofan technology and its recognition and experience in the commuter/regional airline marketplace. The proposed arrangement was recognized by both parties as one that effectively used the respective strengths of each. However, before there had been sufficient time to formalize the business relationship, Embraer began to press for a program launch decision on engine development. Rolls-Royce was also in the process of heavily committing itself to its large engine programs, including the Trent. In the face of these factors, both parties mutually agreed to end their discussions. Thus, in 1989, Allison bit the bullet to go alone on a multimillion-dollar investment to develop the GMA 3007 turbofan engine.[176]

GMA 3007 TURBOFAN DESIGN AND DEVELOPMENT

The GMA 3007 was a modern-technology, high-bypass ratio turbofan engine designed for the 50-passenger-class regional jet market. As a result, primary design objectives were to provide competitive sfc, long engine life, high reliability, low noise and emissions levels, and minimum engine weight. Williams pointed out that particular emphasis was placed on assuring that the engine could satisfy airline expectations for engine life and reliability. Basic design criteria included a 20,000-cycle life on rotating components and ample turbine temperature margin to obviate the need for engine removals due to performance degradation.[177]

The GMA 3007 had a bypass ratio of 5:1, a maximum thrust rating of 7,200 lb, an sfc of 0.39, and an engine weight of 1,581 lb. Using the T406 and GMA 2100 gas generator core, the engine had a single-stage, direct-drive fan with wide chord blades providing excellent performance and foreign object resistance. The fan was driven by a three-stage low pressure turbine. The engine was enclosed within an acoustically-treated bypass duct, which was a structural member of the engine. The arrangement offered several advantages, including improved bypass duct aerodynamics and reduced deflection of the engine core. The bypass and core streams exhausted into a co-annular mixer, producing a low noise signature. The T406/GMA 2100 core was basically unchanged in order to achieve the objective of maximum commonality within the T406/GMA engine family, an important factor in achieving low engine costs. The engine also had a FADEC based on the T406 and GMA 2100 systems.

Mechanical design emphasis was placed on maintainability, provided in part by the modular construction of the engine. Fan blades and vanes were on-wing

---

176. Gary Williams, interview by Rick Leyes and William Fleming, 5 June 1991, Allison Interviews, transcript, National Air and Space Museum Archives, Washington, D.C.

177. Williams to Leyes and Fleming, 5 June 1991.

replaceable items and access to the engine core was obtained through large, strategically-placed access panels. All gearbox-mounted accessories were readily accessible line-replaceable units.

With its high-bypass ratio, the attention given to minimizing airfoil interaction in the turbine and fan assemblies and the exhaust duct surface treatment, the noise signature of the engine was competitive with the quietest power plants in service during the early 1990s. In fact, the engine achieved FAA Stage III noise requirements with margins in excess of 20 EPNdB.[178]

Following two years of engine design and component testing, full-scale testing began in July 1991. Several hundred hours of earlier fan tests had been relatively trouble free. Fan efficiency and airflow had been within 1% of the predicted values.[179] During early sea-level testing, the engine developed about 4% more thrust than its planned 7,200-lb-thrust certification rating, and the sfc was 2 to 3% better than predicted.[180] Gary Williams recounted that altitude testing at the Navy's Trenton facility produced similar results. On August 21, 1992, the GMA 3007 made its maiden flight mounted on the right side of the twin-engine Citation VII.[181] The Citation X, a medium-size executive jet with transcontinental range and the ability to fly at Mach 0.9 at 41,000 feet, made its first flight powered by two GMA 3007 engines on December 21, 1993.[182]

GMA 3007 APPLICATIONS

The GMA 3007 was offered to Embraer in competition with the GE/Garrett CFE738 and Lycoming's ALF 507.[183] On March 5, 1990, Embraer announced the selection of the GMA 3007 to power its 50-passenger EMB-145 on the basis

178. Riffel, Piercy, McKain, and Lewis, "The GMA 2100 and GMA 3007 Engines for Regional Aircraft," 4-6.

179. Williams to Leyes and Fleming, 5 June 1991.

180. Stanley W. Kandebo, "Allison GMA 3007 Turbofan Generates More Power, Uses Less Fuel Than Predicted," *Aviation Week & Space Technology*, Vol. 135, No. 14 (October 7, 1991): 70.

181. "GMA 3007 Turbofan Completes First Flight on Citation 7," *Aviation Week & Space Technology*, Vol. 137, No. 10 (September 7, 1992): 51.

182. "First Citation 10 in Flight Test," *Aviation Week & Space Technology*, Vol. 140, No. 2 (January 10, 1994): 44.

183. Hudson to Leyes and Fleming, 6 June 1991.

of better fuel economy and higher projected reliability than its two competitors.[184] Later, the GMA 3007 was offered in competition with the GE/Garrett
CFE738 and P&WC PW305 to power Cessna's Citation X. At the unveiling of
plans for the Citation X in late 1990, Cessna's Chairman Russ Meyer announced
selection of the GMA 3007 as its power plant. He attributed the ability of the engine to grow with the airplane as a major factor behind the choice.[185]

Although the GMA 3007, sized for 50-passenger-class regional jets, had a
maximum thrust rating of 7,200 lb, it was downrated to 7,000 lb for the
EMB-145 and 6,000 lb for the Citation X.[186] The downrating for the Citation
X lowered the turbine temperature by about 250°F below the design maximum
temperature. Cessna expected this reduction in turbine temperature to allow
6,000 hours between overhauls with no midpoint hot-section inspection.[187]
Further, by raising the turbine temperature of the engine in either airplane, the
engine's thrust output could be increased to accommodate aircraft growth.

Although development of the GMA 3007 engine was proceeding smoothly,
in late 1991, Embraer announced that the EMB-145 program had been put on
hold. This action was reportedly taken to permit company officials to reexamine
the market, the economy, and the aircraft design. Other factors involved in taking this action were that the EMB-145 was behind schedule and that Embraer
was embedded in Brazil's unpredictable economy.[188] Then, in mid-1992, following a reevaluation of the GMA 3007 versus GE's CF34, Embraer recon-
firmed its choice of the GMA 3007, and the first flight of the EMB-145 was
made in mid-1995.

It was in the 1994-1995 time period that the GMA 3007 was selected to
power the Teledyne Ryan long endurance Tier 2+ unmanned reconnaissance
aircraft being developed for the Air Force. Powered by the GMA 3007 engine,
the Tier 2+ was designed to fly for a distance of 14,000 miles or to loiter over a
battlefield for more than 42 hours at a time. Much of its flight time was to be
spent at altitudes over 60,000 ft with a ceiling of approximately 70,000 ft. Ac-

184. "Another Major Victory for Allison," *The Power Letter*, Issue 95 (21 March 1990): 1.

185. Nigel Moll, "Citation X is a 10," *Flying*, Vol. 118, No. 6 (June 1991): 108-110.

186. "Allison Begins Ground Tests of GMA 3007 Commercial Turbofan Engine," *Aviation Week
& Space Technology*, Vol. 137, No. 10 (August 5, 1991): 64.

187. Nigel Moll, "Citation X is a 10," 108.

188. Sao Jose Dos Campos, "Embraer Slows Program Development While Reviewing
Commuter Market," *Aviation Week & Space Technology*, Vol. 135, No. 19 (November 11,
1991): 53.

cordingly the GMA 3007 was to be requalified to 70,000 ft in wind tunnel tests.[189]

T406 SUMMARY

Development of the T406/GMA engine family proved to be a technically successful venture for Allison. Each engine progressed through its development with relatively few problems. Much of that success was due to Allison's design approach, which incorporated a mix of evolved technology with lessons that were learned in the development and operation of its earlier engines. Built-in design margins for engine growth were also an important part of Allison's design philosophy.

The T406/GMA family also placed Allison in a position to maintain and expand its market segment. The GMA 2100 allowed Allison to maintain a strong competitive position in the 5,000- to 6,000-shp turboprop market that had been long dominated by its T56. In addition, with the successful marketing of the GMA 2100 and GMA 3007 for new regional transport and executive jet aircraft, Allison was able to enter new marketplaces.

SMALL EXPENDABLE ENGINES

Allison's heritage in missile propulsion dated back to the 1950s when its J33 turbojet was used in Chance Vought's Regulus, Martin's Matador and Mace, and the J71 in Northrop's Snark missiles. As those missiles were phased out of the Air Force and Navy inventory, Allison left the missile propulsion market to concentrate its efforts on turboshaft engines, the T56 (Model 501) and Model 250 (T63).

In 1987, about 30 years later, Allison decided to make a detailed assessment of business opportunities for small limited-life gas turbine engines. From that assessment came projections of a significant growth in the use of gas turbine-powered tactical and strategic missile weapon systems. It was discovered that within the military there was a growing interest in the use of such systems stemming from the success of the Tomahawk cruise missile and the Harpoon sea-launched missile and from the increasing acceptance of them as viable weapons.

Allison knew that the key to succeeding in a growing market for limited-life gas turbines was the ability to develop and sell small, compact, low-cost, efficient

189. David Fulghum, "Tier 2+ Tricks Enemy Missiles," *Aviation Week & Space Technology*, Vol. 143, No. 2 (July 10, 1995): 49.

engines. According to Harold Stocker, Chief Project Engineer for Unmanned Subsonic Military Engines, Allison believed that it could compete successfully in this market.[190] That belief was based on the company's depth of small engine technology demonstrated by the dominance of its Model 250 in the worldwide light turbine helicopter market, and its access to General Motors advanced manufacturing technology for low-cost production of small engines that had been established during GM's development of a small automotive gas turbine.

In 1987, Allison took steps to reenter the expendable gas turbine business, concentrating its efforts on three types of engines. The first was the tactical missile turbojet engine, normally thought of as being used for missions in which the engine life was a matter of minutes and engine cost was of prime importance. The second was the propfan cruise missile engine, which could provide the high efficiency needed to achieve very long range. The third was a supersonic expendable turbojet aimed at very high altitude operation at Mach 3.5 or higher.

TACTICAL MISSILE ENGINES

During its 1987 assessment of the limited-life gas turbine business, Allison had observed an explosion of interest in the tactical arena for smart weapons, or standoff weapons as they were later called. The Army's Tacit Rainbow, formerly a "black program," that had just been made public, appeared to offer an initial opportunity to enter the expendable gas turbine market. Allison proposed that it would complete qualification and produce a small turbojet engine for the Tacit Rainbow under license from Noel Penny Turbines Ltd. of Coventry, England. However, the ground-launched Tacit Rainbow program was canceled before the Army had selected an engine for it.[191]

In 1989, Allison began development of a 500-lb-thrust short-life turbojet, the Model 150, under contract to the Navy for use in the Fiber Optic Guided (FOG) Skipper missile. The FOG Skipper was a turbojet-powered derivative of the rocket-powered AGM-123, providing it with increased range. With a bifurcated inlet and exhaust, the engine was buried within an outer casing that served as a structural member of the missile. The engine was first run in August 1990, and by mid-1991 the Navy had ordered six engines, one of which had been shipped in June 1991 for use in a FOG Skipper demonstrator standoff missile. In

190. Harold Stocker, interview by Rick Leyes and William Fleming, 5 June 1991, Allison Interviews, transcript, National Air and Space Museum Archives, Washington, D.C.

191. Stocker to Leyes and Fleming, 5 June 1991.

1993, the Model 150 engine, under the direction of Emanuel Papandreas, was also being considered for use in the Hughes Longhorn Maverick air-to-ground missile. This missile was a follow-on to the rocket-powered Maverick, more than 5,000 of which were launched from A-10 and other aircraft against tanks and other armored vehicles during the Persian Gulf War.

In 1990, Allison also received a contract from the Army to develop and demonstrate the 200- to 300-lb-thrust Model 120 high-performance turbojet engine, later rated at 274 lb thrust. The Model 120, while based on the Model 150's design, contained several aerodynamic and production-oriented improvements. The 9-inch-diameter, 16.1-lb engine had a centrifugal compressor, axial turbine, and reverse-flow combustor. It was developed as a propulsion candidate for the Army's advanced tactical Long-Range Fiber Optic Missile (LONGFOG). The Model 120 engine was to serve as the sustainer in an integrated boost/sustainer propulsion system.[192] Under the direction of W. James Dickerson, the Model 120 successfully completed initial tests in July 1993.[193]

Through these efforts, by the early 1990s, Allison had positioned itself to enter the tactical weapon marketplace. Among its U.S. competitors were Williams International and Teledyne CAE with their miniature short-life turbojets developed for tactical weapons.

PROPFAN CRUISE MISSILE ENGINE

At the time that Allison began pursuing the tactical missile market in 1987, the Air Force announced the Expendable Turbine Engine Concepts (ETEC) program. Under that program, multiple contracts were to be awarded for demonstration of engines for various type missions. One of the goals of that program was to increase the fuel economy of cruise missile engines by as much as 30%. Traditionally, the two-spool turbofan, such as the Williams F107, had been used to power cruise missiles. However, an analysis made by Allison indicated that the fuel economy goal of the ETEC program for long-range missiles could not be achieved by a turbofan without high technical risk. In considering alternatives that could achieve the fuel economy goal, Allison came up with the concept of scaling down its commercial propfan. The combination of propfan

---

192. Dr. J. S. Lilley, "The Development of Low-Cost Expendable Turbojet Engines at MICOM," (U.S. Army Missile Command, Redstone Arsenal, Alabama, 1992), 4, 5, 13.

193. Stocker to Leyes and Fleming, 5 June 1991.

blades that could be folded out after launch and a clever gearbox arrangement produced a very efficient, compact engine concept.[194]

Propfan propulsion systems for commercial transport aircraft had been evaluated by Allison and others during the 1970s and 1980s. Such systems offered the high propulsive efficiency of the turboprop combined with the flight speed capability of turbojets and turbofans. This work was a part of NASA's Aircraft Engine Efficient Program, which was a response to the fuel shortages and increased fuel costs resulting from the Middle East oil embargo in 1973.[195] The thin, highly-swept, scimitar-shaped airfoils used in the multi-blade propfans minimized the adverse compressibility effects experienced by contemporary propellers at high flight speeds. The use of the propfan in commercial or military aircraft could improve the fuel economy by as much as 30% over that of conventional turbofans.

Allison participated in design studies and experiments with NASA, Boeing, McDonnell Douglas, Lockheed, Hamilton Standard, Pratt & Whitney, and the military during the 1970s and into the 1980s. Those studies and experiments were followed by flight tests of both single and contrarotating propfan systems. Technology was verified during 1987 in flight tests of a 9-foot-diameter, 8-bladed propfan on the wing of a Gulfstream II. That demonstration was part of NASA's Propfan Test Assessment (PTA) program. The fan was developed by Hamilton Standard, and it was driven by the Allison Model 501-M78 engine, a Model 570 industrial gas turbine driving a modified T56 gearbox.

Meanwhile, in December 1986, Allison joined with P&W to develop and demonstrate a contrarotating propfan propulsion system designated the 578-DX. That system consisted of Hamilton Standard 11.6-foot-diameter, 6-bladed contrarotating propfans powered by an Allison 10,500-hp power section with a contrarotating gearbox system. Flight tests of the system were successfully performed on a McDonnell Douglas MD-80 testbed aircraft in April and May of 1989. In recognition of the success and technical contributions of this program, the propfan development team, which included Allison, was selected as the recipient of the Robert J. Collier Trophy for 1987.[196] It was technology developed

---

194. Stocker to Leyes and Fleming, 5 June 1991.

195. Roy D. Hager and Deborah Vrabel, *Advanced Turboprop Project* (Washington, D.C.: National Aeronautics and Space Administration, 1988), v-vi.

196. David H. Quick, Indianapolis, Indiana, to Rick Leyes, unpublished comments prepared by Dan Riffel in attachment to letter, 16 May 1994, National Air and Space Museum Archives, Washington, D.C.

in this program that Allison planned to integrate into a small propfan cruise missile engine.

The Air Force's request for proposals under the ETEC program permitted bidders to submit a prime candidate engine and an alternate. However, a bidder could win a contract for only a single engine. Because Allison was developing a "stoichiometric" hot section for a supersonic missile under contract to the Air Force Propulsion Laboratory, bids were submitted for both a supersonic turbojet and a propfan propulsion system. As Stocker said:

> The Government was true to their word. They liked them both, but they were only going to let us have one, and so at that time, the supersonic looked more attractive to us for a lot of reasons. It was using a lot of high technology which we developed here, and we saw a good place to use it. There were a lot of questions about the propfan, and it had a lot of doubters shooting at it.[197]

Allison accepted a contract for its supersonic engine. In addition, contracts jointly funded by the Air Force and Navy were awarded to Garrett for a second supersonic turbojet, to Williams for a subsonic turbofan, and Teledyne CAE for a subsonic turbojet.

After winning the supersonic engine contract, Mike Hudson directed Stocker to find a home for the propfan. Hudson believed that it had a fuel economy unmatched by anything else, and that there were certainly needs for it. In fact, it was projected to exceed the 30% goal of ETEC.

At that time, both the Air Force and Navy had developed a renewed interest in long-range missiles. The Air Force's Engine Model Derivative Program (EMDP) office developed a special interest in the propfan. In July 1987, the Air Force awarded a four-month contract to Boeing to study the use of a propfan in the Boeing Air-Launched Cruise Missile. In addition, Allison, Williams, Teledyne CAE, and Garrett were each awarded contracts by Boeing to support the Boeing propfan study. That effort was followed by further propfan design and tradeoff studies by the four companies for EMDP. However, EMDP's funds were limited. It soon became evident to Allison that, with four contractors being kept in the program, there would be a need for each of them to invest a fair amount of company money to augment the limited military funds.

The outcome was an in-house strategy meeting in which Allison management decided that the best course of action was to share the financial risk of developing a new engine by joining with another company in a partnership similar to that of LHTEC. Allison contacted Williams in late 1988. Soon afterward, Blake Wallace and Sam Williams met and agreed that a partnership would be a good thing for both companies. In the spring of 1989, the Williams-Allison Propfan Engine Company (WAPEC) was formed. The arrangement was for a 50-50 split between the two companies with Williams being given the lead in the program because of its extensive experience in long-range strategic cruise missiles. Williams was responsible for the gas generator portion of the engine, and Allison had responsibility for the power turbine, gearbox, and propfan.

In September 1989, the Naval Air Systems Command (NAVAIR) awarded contracts to five airframe companies for concept definition studies of the Long-Range Conventional Standoff Weapon (LRCSW). The LRCSW was a joint Navy/Air Force program intended to provide the services with relatively inexpensive and highly accurate standoff weapons that could deliver conventional payloads against fixed landbased targets over extended distances. The LRCSW was to employ propfan propulsion, giving it 50% greater range than was possible with a turbofan.[198]

In late 1988, a joint Defense Department/NASA program was begun to demonstrate the potential of the highly fuel-efficient propfan engine to extend the range of next-generation cruise missiles. With support provided by researchers at the Naval Weapons Center, China Lake, and NASA's Ames and Lewis Research Centers, tests of a scale model Tomahawk 12.1 inches in diameter powered by a propfan were conducted in the 14-foot transonic wind tunnel at Ames. These tests were successfully completed in June 1991.

In mid-1990, proposals had been submitted by Williams/Allison WAPEC and a Teledyne CAE/General Electric team to demonstrate a propfan engine that could power the LRCSW. These proposals had been evaluated and officials were ready to issue a contract when, in January 1991, work was canceled. The LRCSW program was dropped at that time when the Navy and Air Force failed to justify requirements for the weapon or the money to back it.[199]

Thus, in early 1991, the WAPEC propfan, designated WA100, was placed on hold. The Williams/Allison partnership remained in place although WAPEC

---

198. Guy de Bakker and Mark Hewish, "The US Navy League Exhibition; Five Contractors Study LRCSW," *International Defence Review*, Vol. 23, No. 6 (June 1, 1990), 715.

199. Breck W. Henderson, "Propfan Engine May Be Suitable For Next-Generation Cruise Missile," *Aviation Week & Space Technology*, Vol. 136, No. 1 (January 6, 1992), 62-63.

was unstaffed. Meanwhile, both companies continued to search for potential applications for their small propfan engine.[200]

SUPERSONIC TURBOJET

Having won a contract in 1988 under the ETEC program to demonstrate a limited-life Mach 3.5 high-altitude turbojet, Allison proceeded with development of the engine to provide the Department of Defense a weapon capability that it did not then have. Although the program was still classified at that time, some information was available.

Allison's supersonic engine, the GMA 900, designated the J102-AD-100 by the Air Force, was a high-technology engine that operated at very high temperatures. The use of Allison's Lamilloy®, a semi-porous laminated material, was the key to the engine's high operating temperature capability. According to an interview with Mike Hudson reported in *Aviation Week & Space Technology* magazine, the 274-lb power plant, 45-inches long and slightly more than one foot in diameter, was equipped with a single-stage Lamilloy® turbine rotor, a Lamilloy® combustor, and a Lamilloy® turbine guide vane assembly.[201] Use of a heat exchanger, which used fuel as a heat sink, reduced the temperature and the amount of compressor discharge air required to cool the Lamilloy® turbine rotor blades. Other technologies used in this power plant included hybrid ceramic roller and ball bearings that could be lubricated and cooled by fuel, thus eliminating the need for an oil lubrication system.

The hot section of the engine was first demonstrated in early 1990. In engine tests that followed in early 1991, the engine ran satisfactorily at conditions simulating Mach 3.5 and altitudes up to 90,000 ft. Engine inlet temperatures, which were believed by Allison to be record-setting, exceeded 900°F. In addition, the classified combustor and turbine temperatures were also believed to have set records, according to Allison officials.[202]

Stocker described the program as being both high risk and high payoff. Successful demonstration of the engine in early 1991 stimulated a growth of interest within the government and the airframe community in terms of using the engine for both lethal and non-lethal supersonic missions. Although potential missions were classified or proprietary, an illustrative advantage offered

---

200. Stocker to Leyes and Fleming, 5 June 1991.

201. "Allison GMA 900 Turbine Tested at 90,000 Ft., Mach 3.5 Conditions," *Aviation Week & Space Technology*, Vol. 134, No. 21 (May 27, 1991): 91.

202. "Allison GMA 900 Turbine Tested at 90,000 Ft., Mach 3.5 Conditions," 91.

by a Mach 3.5 vehicle was its ability to go quickly to a key location such as an enemy's mobile launcher site that had been pinpointed during a ground-to-ground missile attack. It would thus be possible to reach and destroy the launchers before the enemy would have time to move them. As a result of the J102-A-100 engine's successful demonstration, by mid-1991, Allison had stepped up its work with airframe companies on the possible use of the engine.[203]

In 1993, the Allison J102, under the direction of William S. Kennedy, was being considered for the Navy's Supersonic Sea Skimming Target (SSST) program and for ARPA's Relocatable Target Detection and Destruction (Warbreaker) program. The engine was capable of missions requiring operation at Mach 0.8 to 2.5 at sea level, up to Mach 3.5+ at 90,000+ feet, as well as at Mach numbers and altitudes intermediate between these conditions.

SMALL EXPENDABLE ENGINES SUMMARY

Between 1987 and 1993, Allison had entered and positioned itself well in the small limited-life gas turbine engine arena. This success came largely as a result of three factors: its small engine technology and manufacturing experience developed over a period of 30 years with the Model 250 engine; its access to General Motors' advanced manufacturing technology for low-cost production of small engines developed in its automotive gas turbine program; and, particularly important in its supersonic engine development, its high-temperature technology, a key element of which was Lamilloy®.[204]

Allison's Years With GM

Over the years, Allison had an interesting roller-coaster relationship with General Motors with a surprise ending. In 1929, the company was bought by GM as the engine development portion of a corporate entree into commerical aviation. Following GM's purchase, Allison was focused on the development of the V-1710. GM supported development of that engine through the 1930s, during which time Allison remained a small engineering shop, growing slowly and developing a modest manufacturing capability. In 1939, the V-1710 suddenly found itself in great demand by the Army Air Corps as the power plant for a

---

203. Stocker to Leyes and Fleming, 5 June 1991.

204. Stocker to Leyes and Fleming, 5 June 1991.

new pursuit aircraft. GM poured its resources into transforming Allison into a major manufacturing plant for production of the V-1710.

In the years following World War II, GM viewed Allison as the corporation's "good citizen contribution" by maintaining it as a standby resource for mass production of aircraft engines should the country need it to be mobilized in the future. The corporation's philosophy from the perspective of Eloy Stevens was: "This was an organization, that as long as it didn't lose a lot of money, could kind of hang in there and be ready for the country, if needed."[205] Allison proceeded in that mode until 1973, at which time it was was merged with the Detroit Diesel Division to form the Detroit Diesel Allison Division. Allison operated within that alliance for more than a decade during which time it received little attention from the GM board room.

In 1980, Garrett expressed an interest in acquiring Allison, following which GM announced that Allison was for sale. However, the offer ultimately made by Garrett was too low to suit GM. With the unexpected interest expressed by others in acquiring Allison, GM decided in 1982, that rather than selling Allison, a better business strategy would be to keep the company and invest in its growth and further development. Representing only about six-tenths of one percent of GM's gross sales at that time, Allison was neither a major fiscal nor management concern for the corporation.

To implement its decision, GM separated Allison from Detroit Diesel in 1983 and brought Blake Wallace, who had broad gas turbine management experience, from GE as General Manager. Under his direction, and with the support of GM, he developed and implemented an eight-step action plan to strengthen the company. In implementing his plan, Wallace strengthened the company management, enhanced its technology and manufacturing capabilities, expanded its engine portfolio, and entered new markets. Wallace's efforts were supported by major GM investments in the company. In fact, by 1991, Allison was operating with five times the reinvestment dollars that it had received in the late 1970s and early 1980s.[206]

In January 1992, following the end of the Cold War, with the country in recession and GM facing major business losses, it once more announced that Allison was for sale. In September 1993, General Motors announced that it had sold the division. The new owners were a group of Allison executives and the New York investment banking firm of Clayton, Dubilier & Rice, Inc. Blake Wallace, formerly General Manager of the Allison Gas Turbine Division, be-

---

205. Stevens to Leyes and Fleming, 6 June 1991.

206. Stevens to Leyes and Fleming, 6 June 1991.

came President and CEO of the new company, named the Allison Engine Company.[207] The reported purchase price was $310 million.[208]

Although GM had been largely supportive of Allison during its long ownership of the compay, the relationship was indeed a roller-coaster affair. During periods when Allison's sales and GM's corporate profits were high and the national economy was expanding, as was the case during World War II and during the 1980s, there was considerable corporate interest in and support for Allison. However, during leaner times, corporate interest diminished. This lack of interest reached bottom in 1992 when GM was losing money and the country found itself in a recession; GM then sold Allison.

Then, on November 21, 1994, only 14 months after GM's sale of Allison, Rolls-Royce Inc. announced that it had agreed to purchase the Allison Engine Company. With this purchase, Rolls-Royce anticipated establishing a portfolio of engines covering the spectrum of power levels from Allison's little 500-shp class Model 250 turboshaft engine to Rolls-Royce's high-bypass turbofans with a potential of 100,000 plus lb thrust. Further, the acquisition of Allison would give Rolls-Royce a significant manufacturing presence in the U.S., considered the world's most important center of aerospace activity.[209] The purchase of Allison was finalized in early 1995 at a price of $535 million, resulting in a $215 million profit for the owners in little over a year's time. Allison's former Executive Vice President of General Engineering, Mike Hudson, became the new President and Chief Operating Officer of Allison, which operated as a unit of the Rolls-Royce Aerospace Group based in Derby, England.[210]

Overview

STATUS OF THE CORPORATION

During more than three decades from 1958 through 1992, the Model 250 was the only small turbine engine that Allison produced. Further, it was not until the mid-1980s that Allison began development of any other small gas turbine en-

207. Anthony L. Velocci, Jr. and Stanley W. Kandebo, "Allison Buyout Spurs Consolidation," *Aviation Week & Space Technology*, Vol. 139, No. 12 (September 30, 1993): 35.

208. "Pilot Briefing," *AOPA PILOT*, Vol. 37, No. 4 (April 1994), 38.

209. Stanley W. Kandebo and Carole A. Shifrin, "Rolls to Buy Allison in $525-Million Deal," *Aviation Week & Space Technology*, Vol. 141, No. 22 (November 28, 1994), 22-24.

210. Arnold Lewis, "Allison Engine Company: New Emphasis on Quality," *Business & Commercial Aviation*, Vol. 77, No. 4 (October 1995): C14.

Allison Plant No. 5 manufacturing facility (top) and Plant No. 8 development facility (bottom) in Indianapolis, Indiana. Photographs courtesy of Allison.

gines. As a result, until the mid-1990s, the company's small engine sales were confined to military and civil users of light turbine helicopters and small fixed-wing aircraft. Having no real U.S. competition in its power class until the late 1970s, the engine captured a huge market share, becoming the primary power plant for U.S. military light turbine helicopters through the 1980s and dominating the worldwide civil light turbine helicopter market. From 1962 through the end of 1995, 26,008 Model 250/T63 turboshaft and turboprop engines had been built.

Located in Indianapolis, Ind., Allison began production of its World War II V-1710 engine in a newly-constructed 390,000 square foot factory and office building in 1941. Subsequent additions were made to that facility during the 1940s as Allison began developing and manufacturing large turbojet and turboprop engines. In 1953, development began on its Research and Engineering Plant on 214 acres of land adjacent to the company's manufacturing facilities. With progressive additions, by 1990, that plant totaled 920,000 square feet.

With the large T56 turboprop and the small Model 250 turboshaft engines as its leading income producers, Allison's revenues grew over the years reaching nearly $1 billion in 1990. Just under half of those revenues came from selling spare parts and engineering services. However, Allison's business experienced a downturn along with the rest of the aerospace industry in 1991 with revenues dropping to between $500 and $750 million. As a result of the reduced revenues and expenses related to development of the new GMA 2100 and GMA 3007 engines, Allison reportedly posted a loss of $500 to $700 million in 1991. With this reduction in revenues, the employment level at Allison dropped from 7,012 in 1991 to 4,276 in November 1994.[211] Nevertheless, following its sale by GM, Allison posted a 1994 pre-tax profit of $6 million, following several years of losses.[212] In 1996, Allison had approximately 4,400 employees and an annual revenue of some $622 million.

THE COMPANY'S MARKETS AND INflUENCING FACTORS

The 1950s were formative years in reshaping Allison's niche in the aircraft gas turbine marketplace. During the early 1950s, the company began losing its segment of the market in the area of large military turbojets and fell behind its competitors, GE and P&W. Meanwhile, Allison had been putting a good deal of effort on turboprop engines and high speed gearboxes for them. Out of this effort had come development of the T56 turboprop for large military and civil transport aircraft. Thus, as GE and P&W forged ahead with their development of large turbojets and turbofans for military combat and commercial transport aircraft, they essentially left the large turboprop market to Allison. With its suc-

211. Anthony L. Velocci, Jr., "Push For Profits May Drive GM From Aerospace Sector," *Aviation Week & Space Technology*, Vol. 137, No. 22 (November 30, 1992): 25.

212. Lewis, "Allison Engine Company: New Emphasis on Quality," C15.

cessful T56, Allison became essentially the unchallenged producer of large tur-boprops for the U.S. military.

Allison's entry into the small gas turbine aircraft engines market came in 1958 when it won the competition for the development of the Army's T63 turbo-shaft engine. Without any significant competition in its power class for many years, Allison dominated the U.S. military and civilian light turbine helicopter market; by the early 1980s, the Model 250 also powered the majority of the free world's light turbine helicopters. The Model 250/T63 became the second tall pole in the company's business tent. Thus, for more than two decades, Allison was essentially a two engine company.

Production of the Model 250/T63 engine grew during the late 1960s and early 1970s with the deployment of the light turbine helicopter in the Vietnam War and its use in civil and commercial applications. A major stimulus to the commercial production rate of the engine was the oil crisis in the mid-1970s. With the sudden rise in the price of oil and the projection of continued in-creases, there was a flurry of activity in expanding the amount of off-shore oil drilling. To support this activity, light helicopters were needed to transport crews and lightweight equipment to and from the oil rigs. As a result, the sale of Model 250 engines soared during the late 1970s and early 1980s. Average year production rates remained at a lower, relatively stable rate through 1995.[213]

With the Model 250 engine being used extensively throughout the world, Allison enjoyed a continuing spare parts business in addition to its sale of new engines. In servicing the more than 26,000 engines that had been produced, by the end of 1995, a significant part of Allison's sales revenues came from supply-ing spare parts for them.

A broadening of Allison's engine portfolio began following General Motors' 1982 decision to make some significant investments in Allison for new product development and facility enhancement and Blake Wallace's 1983 arrival as the new general manager. Implementing GM's investment strategy, Wallace spear-headed the development of several new engines with the objective of entering new marketplaces, growing business volume, and becoming one of the "big boys" in the aircraft gas turbine industry. The expanded portfolio included the T800, the T406 family, and several short-life engines.

Allison's philosophy of joint ventures, as expressed by Blake Wallace, was that the company strongly believed in program-by-program joint venturing. He ob-served that there seemed to be a new trend in international joint ventures, led by such companies as MTU and Daimler-Benz of Germany, toward omnibus

---

213. Thomason to Leyes and Fleming, 17, April 1996.

company-to-company relationships that covered all products. He did not believe that to be a wise policy. He felt the more prudent approach was the pro-
gram-to-program arrangement in which the partner was chosen because a tan-
gible benefit or expertise was brought by that partner to a specific program.
Wallace observed that international joint ventures would be a continuing way
in which the company would do its business. In addition, he anticipated occasional domestic joint ventures such as the ones that it had formed with Garrett
and Williams. However, he emphasized that all of Allison's existing domestic
joint ventures had been driven by U.S. military programs in one way or another, not by commercial programs.[214]

CONTRIBUTIONS

Allison's most notable contribution in the small gas turbine aircraft engine business was the development and manufacture of the Model 250 turboshaft engine
for use in light helicopters. Blake Wallace, Allison's Former General Manager,
called the Model 250 a milestone engine that made the gas turbine-powered
light helicopter an economically viable and practical machine.[215]

The Model 250 was largely responsible for the creation of the light turbine
helicopter industry in North America and the widespread commercial use of
light turbine helicopters throughout the world. These helicopters were used to
transport personnel and light equipment for a variety of purposes. The T63 also
contributed importantly to the Army's light helicopters. They were used extensively during the Vietnam War and later conflicts as part of Army airmobility
operations.

214. Wallace to Leyes and Fleming, 15 August, 1991.
215. Wallace to Leyes and Fleming, 15 August 1991.

GARRETT (ALLIEDSIGNAL ENGINES) | 10

Pre-Turbine Company History

With the financial backing and encouragement of top Douglas Aircraft Company executives, corporate officers from Douglas's Northrop division, and others, Cliff Garrett formed his own company, Aircraft Tool and Supply Company, in 1936 in Los Angeles, Cal.[1] The purpose of the company was to sell, distribute, and act as a sales representative for the East Coast manufacturers of aircraft tools, equipment, and supplies.[2] Some of Garrett's first customers were Consolidated Aircraft, Douglas, Lockheed, Northrop, Vultee, North American, Ryan, and others. In addition to the Southern California aviation industry, sales were also made to the oil and railroad industry by the late 1930s.

Before starting his own company, Cliff Garrett had studied engineering at UCLA, but was unable to finish because of the Great Depression. Nonetheless,

1. The company was incorporated on May 23, 1936 as the Aircraft Tool and Supply Company in Los Angeles, California. On July 2, 1936, the company was renamed Garrett Supply Company. The name was again changed August 18, 1938 to the Garrett Corporation. By the end of 1939, three divisions existed, one of which was the Airesearch Manufacturing Company. In May, 1942, the spelling of Airesearch was changed to AiResearch with a capital "R". Also in 1942, the AiResearch Manufacturing Company of Arizona was formed, a wholly owned subsidiary located in Phoenix. Over the years, many other divisions were established by Garrett. To prevent an unfriendly takeover by Curtiss-Wright following the death in 1963 of Cliff Garrett, Garrett merged with Signal Oil and Gas Company on

611

he was interested in anything technical and wanted his company to become more involved in technology development. He was also keenly interested in aviation. As an employer, he insisted on and got the most from the people who worked for him.

Collectively reminiscing about Cliff Garrett, early turbine engine engineers Roy Schinnerer and Wilton Parker commented on his character:

> Cliff was a very, very powerful man in terms of his dynamics. He was just on the go all the time . . . He ran the operation, he ran the show . . . He was a very dynamic individual, but he was loved by everybody in the company . . . He'd chew you out, all right, if you weren't doing things right, but at the same time he'd give you a pat on the back about the things you were doing right, and encourage you to solve whatever problems you had . . . The important thing with Cliff, he knew every nut and bolt that was going on, and everybody under him had to know it, too. Cliff would frequently come down the hall and drop in your office, unannounced, and chat about things. He was constantly roaming the shop. Cliff was not sitting on his behind on the office chair all the time. He was out knowing what was going on in the laboratory.[3]

Cliff Garrett was also an outstanding salesman, and his emphasis on sales continued through the history of the corporation to be a driving force at Garrett.

January 20, 1964. Together they became The Signal Companies. Late in 1985, the Signal Companies merged with Allied Corporation to become Allied-Signal, Inc.

In 1979, the Garrett Turbine Engine Company was created from the AiResearch Manufacturing Company of Arizona, encompassing both APU and propulsion engine product lines. In 1988, these two product lines were split into the Garrett Engine Division, for propulsion engines, and the Garrett Auxiliary Power Division of Allied-Signal Aerospace Company. In 1993, the two companies were rejoined and named AlliedSignal Engines.

In 1993, the identification of "Garrett" aircraft engines was changed to "AlliedSignal" engines. This reflected a transition of original product names to a unified AlliedSignal corporate identity. However, for purposes of historical simplification, the engines and original company which produced them will be referred to as "Garrett" in this chapter. In addition, AlliedSignal acquired Lycoming Turbine Engine Division from Textron in 1994. That division's history will also be identified as "Lycoming" in this book. The combined "Garrett" and "Lycoming" enterprises, now known as AlliedSignal Engines, located in Phoenix, Arizona, are discussed at the end of this chapter in the Overview section.

2. William A. Schoneberger and Robert R. H. Scholl, *Out of Thin Air: Garrett's First 50 Years* (Los Angeles, Cal.: The Garrett Corporation, 1985), 29.

3. Roy Schinnerer and Wilton Parker, interview by Rick Leyes and Bill Fleming, 8 March 1990, Garrett Interviews, transcript, National Air and Space Museum Archives, Washington, D.C.

According to Roy Ekrom, President and Chief Executive Officer of Allied-Signal Aerospace Company in 1990:

> Historically, the differentiating feature in this company was the way we sold. There are a lot of companies who are good technically, and we're an outstanding company technically, but we started as a selling company. Mr. Garrett set up a sales company first, and the design, development, and manufacturing capabilities came in later. So, we always had a very powerful central sales organization and that set us up so that our divisions worked very well together, because the central sales thing was kind of a forcing mechanism.[4]

In the late 1930s, Cliff Garrett's interest in having his company develop and sell high-technology aviation products began to coalesce. In 1938, a decision was made to expand the company by manufacturing and selling patented products. With a personal interest in high-altitude flight and a clear need for aircraft pressurization equipment to make it possible, Cliff Garrett formed a division in 1939, the Airesearch Manufacturing Co., to design, develop, and manufacture such equipment. Under the direction of Walt Ramsaur, the first product designed and built by the company was an all-aluminum aircraft intercooler, first incorporated into the turbosupercharger system of Boeing B-17 bombers to give them high-altitude capability and more speed. Aircraft engine oil coolers, a successful companion product, were used on many World War II combat aircraft.

With business expanding rapidly as a result of the war, a larger factory was built in 1941 on the open fields at the northwest corner of Century and Sepulveda Blvd. in Los Angeles, now the entrance to Los Angeles International Airport. With the escalating war in the Pacific, and the possibility of a west coast invasion of the continental U.S., and the potential for sabotage, another operation was established in a newly built Defense Plant Corporation factory in Phoenix, Ariz. in 1942. A Garrett subsidiary, the new company was called AiResearch Manufacturing Company of Arizona. It manufactured copper and aluminum heat exchanger extrusions, aluminum intercoolers, and electrical actuators.

Other wartime activities included building a high-altitude flight research chamber at its California plant to test cabin pressurization valves and other equipment. By the end of the war, in addition to its intercoolers, aircoolers, and cabin pressure regulators, Garrett had begun work on aircraft air conditioning

4. Roy H. Ekrom, interview by Rick Leyes and Bill Fleming, 8 March 1990, Garrett Interviews, transcript, National Air and Space Museum Archives, Washington, D.C.

equipment, starter assemblies, and shaft-driven secondary power units. In planning for the post-war era, Cliff Garrett was interested in what he perceived would be an expanding need for transport aircraft pressure regulation and heat transfer equipment.

In the spring of 1945, with World War II nearing its conclusion, Garrett was already investigating the possibility of using gas turbine engines and variable speed drives to power cabin pressurization and air conditioning systems.[5] With board approval for research and development funding up to one million dollars, the financial foundation was set in place for the work on what would eventually become Garrett's gas turbine auxiliary power units. Thus, the preliminary research and development work leading to Garrett entering the gas turbine engine era had started in the early 1940s.

Garrett Enters the Turbine Era

PROJECT A AND THE XP-80 REFRIGERATION TURBINE

In early 1943, a proprietary technology demonstration project was launched at Garrett that would have an important and lasting impact on the company. Northrop had a need for a pressurization compressor for the XB-35 "Flying Wing," a high-altitude, long-range bomber then under development. Because he felt that it would complement the cabin pressure valve equipment the company was building, and because he wanted to be in the business of pressurizing post-war commercial aircraft, Cliff Garrett was interested in providing the compressor for Northrop.[6] However, Garrett had no experienced turbomachinery designers. Jack Northrop, a close friend of Garrett, was co-owner with him of the Northill Company, which produced water pumps, among other things. Northrop lent Garrett Waldemar F. Mayer, an experienced Northill pump designer.

As a student, Mayer had studied under the eminent European professor Aurel Stodola, who conducted the first systematic experiments on supersonic nozzle flows for steam turbine engines. After emigrating from Hungary, Mayer worked for Byron Jackson and other companies designing pumping machinery. Mayer was convinced that good compressible flow turbomachinery could be built

5. Schoneberger and Scholl, *Out of Thin Air*, 80.

6. Roy Schinnerer, Long Beach, California, to Neil Cleere, letter, 1 March 1990, National Air and Space Museum Archives, Washington, D.C.

around incompressible (water pump), backward curved impeller designs to achieve broad operating ranges and high efficiencies.[7]

In the spring of 1943, Mayer started to design and develop a two-stage compressor for a cabin air compressor. This company-classified technology development program was known as "Project A." The cast aluminum compressor wheels were backward curved and shrouded, very similar to those of a water pump.[8] The unit had a double volute crossover duct (interstage diffuser and turning duct) also similar to that of a water pump. It was sized for 45 lb/min (Std. Atm.) at a pressure ratio of 1.75:1. With 7.25-inch-diameter, 8-vane, 30-degree backward curvature (measured from the tangent), and shrouded cast aluminum impellers, it demonstrated an adiabatic temperature rise efficiency of 78% at the design point.[9]

Although the unit never went beyond being a laboratory development tool, Project A was important because it demonstrated early on that high efficiencies over broad operating ranges were characteristices of the Mayer backward curved water pump impeller designs as applied to centrifugal air compressors. This knowledge and experience later became an important consideration for aircraft cabin air conditioning equipment. It also was the foundation for the

7. Schinnerer to Cleere, 1 March 1990. A backward curved centrifugal compressor (impeller) contains a set of blades that intercept the outer diameter at an angle significantly less than 90 degrees measured from the tangent, the sweep effect being counter to the direction of rotation and along the entire blade length. The purpose of backward curved blades is to improve the compressor surge margin by allowing the compressor to operate over a wider range of airflows at any given speed. Pressure ratio per stage and rotor discharge air velocity of a backward curved compresssor are always lower than for a 90 degree compressor of comparable rotor tip speed and flow path efficiency. When matched with a proper diffuser, the lower discharge velocity assisted the diffusion process. Proper exploitation of these factors resulted in a compressor with high efficiency and a broad operating characteristic (higher stall margin) at the expense of lower total compressor pressure rise.

8. A shrouded centrifugal compressor consists of a set of blades, which for structural and aerodynamic reasons, are attached to a rear disk. On the forward side of the blades (moving forward in an axial direction toward the air inlet) is a shroud which follows the contour of the blades and is an integral part of the compressor. Shrouds on centrifugal water pumps and compressors prevent blade-to-blade leakage and also reduce rotor thrusts. Furthermore, it is usually true that shrouding is more economical with small cast rotors and large fabricated rotors in air compressors up to a stage pressure ratio of 2. Above this level of pressure ratio, the higher speeds required resulted in unacceptably high shroud stresses.

9. Roy Schinnerer, Long Beach, California, to Montgomerie Steele, letter, 20 December 1991, National Air and Space Museum Archives, Washington, D.C.

compressor design on the first Garrett small gas turbine engine, called the "Black Box."

At the same time that he was working on the centrifugal air compressors, Mayer became involved in a project to provide an air refrigeration turbine for the Lockheed XP-80A.[10] This aircraft made its first flight on June 10, 1944 from the Muroc Flight Test Base in California's Mojave Desert. However, on the third test flight the cockpit temperature became so high that test pilot Tony LeVier found it difficult to touch anything in the cockpit, including the throttle, and he developed heat blisters on his left forearm.[11] On the basis of Garrett's experience building cabin pressure regulating valves for the Boeing B-29 bomber and other aircraft environmental control valves, Lockheed turned to Garrett for help. Cliff Garrett assigned the project to Mayer who came up with several feasibility studies that summer.

The concept was to take jet engine bleed air, put it through a pre-cooler (a heat exchanger that partially cooled the bleed air), further cool it with a miniature refrigeration turbine and direct the refrigerated air to the cockpit. Working from Mayer's sketches, Roy Schinnerer, a mechanical engineer who had joined Garrett in 1942, designed the prototype. In August, the prototype machine was built, using a fan as a loading device. The 1.5-inch-diameter turbine wheel was sized to give a cabin airflow of about 5 lb per minute; it was a radial-inflow, full-entry, cantilever bucket type with a rotating speed of 60,000 to 70,000 rpm.[12] First testing revealed a remarkable 65 to 70% adiabatic turbine efficiency.[13]

10. Refrigeration turbines are used in a variety of thermodynamic cycles. Terminology is not standardized in referring to them and includes cycle descriptions such as air cycle, bootstrap, refrigeration, and expansion turbines. Regardless of the type, the principle of operation is the same. A source of compressed air is provided to the turbine, and the compressed air is expanded through the turbine to produce a refrigeration effect. Cycles vary in terms of how the power output from the turbine is used. It is this variation in turbine power use that leads to the variations in terminology. Because the common factor between each system or cycle is to provide refrigeration by expanding the air supply through a turbine, the generic term "refrigeration turbine" is used to apply to the early Garrett expansion turbine cycles.

11. Tony LeVier to Rick Leyes, telephone conversation, 29 August 1990. The high temperatures resulted when LeVier attempted a speed run at low altitude. The cockpit was pressurized with compressor bleed air. A bleed air valve was stuck wide open and ram air compressible heating along with the desert temperature made the cabin unbearably hot. Even when the defective valve was replaced, the cockpit controls were still too hot to hold, and LeVier felt that the aircraft was unsafe to fly.

12. There are several types of radial-inflow turbines. In the case of the XP-80 refrigeration turbine, the inlet air flowed through peripheral nozzles and entered the turbine blade passages in an inward radial (centripetal) direction. The nozzles created a strong

Excited by the results, Lockheed engineers wanted the breadboard (proto-type) machine installed in the aircraft. The turbine unit was modified slightly, using an enlarged diameter wheel for a small increase in airflow, and a better fan was provided for absorbing turbine power. The unit was mounted on a quickly-fabricated pre-cooler (air-to-air heat exchanger) and was installed in the airplane. The turbine could drop the 175 degree F inlet air temperature from the heat exchanger to 40°F. By the end of 1944, the prototype unit had been flown on an XP-80. For the first test airplane, ram air provided the cooling airflow through the heat exchanger. Production units used the suction air pumped by the loading fan. The XP-80 expansion turbine became the first re-frigeration turbine to be flown on an aircraft, and the first Garrett turbomachine to be built and shipped to a customer. It put Garrett into the turbomachinery business, and led to the development of many aircraft air conditioning sys-tems.[14]

To keep pace with this rapidly expanding business, Garrett began adding ex-pert turbomachinery designers, especially those with experience in aircraft en-gines and superchargers. One of the key figures was Homer Wood, who had come to work for Garrett as a development engineer in 1943 and was later made responsible for turning the prototype refrigeration turbine into production hardware.

At the end of the war, aircraft manufacturers had many new designs in prog-ress, especially new commercial transports. Garrett was scrambling to carve out its future in environmental control systems and was using design groups led by both Wood and Mayer to respond to manufacturers. In 1946, Wood's group won the design competition to supply nearly all the environmental system components for the Lockheed 649/749 Constellation. At the heart of the sys-tem (a variable-speed, engine-driven, cabin air compressor, integrated with a

radial-inflow vortex (a miniature tornado) with high angular momentum, which was converted into shaft torque by the turbine blading. Following the exhaust duct, the discharge airflow left the refrigeration turbine in an axial direction.

A cantilevered turbine bucket type refers to the set of turbine blades which are located on the periphery of the turbine disk and which are parallel to the axis of rotation. Because the blades "stick out" from the back of the disk without support (except where they are attached to the disk), they are called cantilevered.

13. According to Schinnerer, the only comparable high rpm turbine wheels existing in the U.S. at the time were of the tool grinder type. Such wheels demonstrated relatively low efficiencies of about 20 to 30%.

14. Roy Schinnerer, Long Beach, California, to Harry H. Wetzel, letter, 24 July 1990, National Air and Space Museum Archives, Washington, D.C.

The prototype Garrett refrigeration turbine for the Lockheed XP-80A aircraft. Photograph courtesy of Garrett.

three-wheel bootstrap[15] air cycle refrigeration turbine) lay an improved compressor design.[16] During that same period, Douglas Aircraft chose Garrett to supply much of the environmental control system for the DC-6, including a simple air cycle refrigeration turbine also designed and developed by Wood's group.

15. The term bootstrap refers to a system in which the power from the refrigeration turbine is fed to a compressor, which takes the output of the original compressed air source (an engine driven cabin air compressor in this case) and boosts it to a higher pressure level before it passes through the refrigeration turbine. A three-wheel machine adds a fan rotor (also driven by the turbine) which draws cooling air across an interstage heat exchanger as part of the system.

16. Wood had designed a two-stage, backward curved, 2:1 pressure ratio machine in 1945 for pressurizing the Lockheed Constellation. It's blading geometry differed from Meyer's design primarily to reduce stress levels. The crossover flow path was a continuous annulus, also common in water pumps, instead of a double volute transfer. Homer Wood, Rancho Murieta, California, to Rick Leyes, letter, 20 December 1991, National Air and Space Museum Archives, Washington, D.C.

THE BLACK BOX: GARRETT'S FIRST TURBINE ENGINE

Unlike Lockheed and Douglas, which specified individual units for different environmental control functions, Boeing conceived an integrated auxiliary airborne power plant for its Model 377 Stratocruiser. Boeing wanted a lightweight, compact, self-powered unit that could furnish AC and DC current for the plane's electrical systems; provide airflow for cabin pressurization and cabin air conditioning; and supply hot air for the wing anti-icing system. Boeing wanted the complete unit in 18 months. Cliff Garrett accepted the job and turned the responsibility over to Mayer.

To get the job done, a decision was made to develop a small gas turbine engine.[17] Preliminary work was started in the spring of 1945. In August, at Cliff Garrett's request, Mayer and Schinnerer went on a "brain picking" tour to the top turbine engineers in the country—in the Navy, at Westinghouse, Wright Field, and other organizations and companies—that were working on large gas turbine engines. While the pair were well received by companies and had some good interviews, at least one senior engineer said to them, "Well, fellows, I wish you all the luck in the world, but I don't think that you're going to make it."[18] The problems of building a small turbine engine were considered formidable by experts in the field. Nonetheless, the project proceeded.

Mayer did the theoretical work on the unit. Schinnerer was responsible for development and testing. In September, Wilton Parker was hired as the Chief Mechanical Designer of the engine. As the design progressed, the unit was nicknamed the "Black Box" due in part to the secrecy of the project and the fact that because it had so many gadgets in it, it was like a magical black box.[19]

The gas generator portion of the Black Box consisted of a three-stage, shrouded, backward curved compressor, a combustor, and a single-stage axial turbine, which was principally an impulse turbine. A geared power take-off shaft ran a blower and a generator/alternator. Compressor bleed air was routed through a cooling turbine; it generated power (about 40 hp) back into the main shaft. A portion of the gas generator exhaust gases were ducted to the wing

---

17. At this time, it is no longer clear whether the decision to develop a small gas turbine engine as an integrated auxiliary airborne powerplant was a Boeing, Garrett, or joint company decision.

18. Schinnerer to Leyes and Fleming, 8 March 1990.

19. Wilton Parker, interview by Rick Leyes, 22 August 1990, Garrett Interviews, tape recorded telephone conversation, National Air and Space Museum Archives, Washington, D.C.

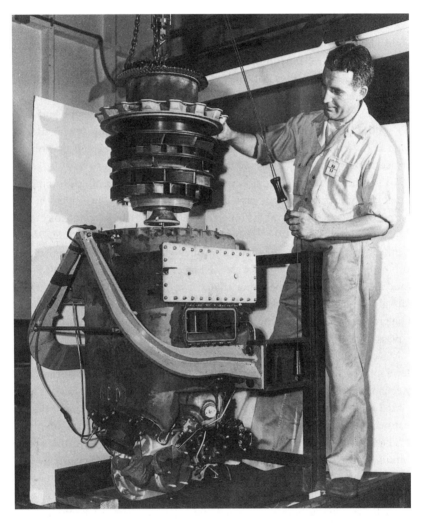

The "Black Box," Garrett's first turbine engine. Photograph courtesy of Garrett.

anti-icing system. Also in the Black Box package were primary and secondary heat exchangers, automatic controls, regulators, and air ducts. As Schinnerer later described the complexity of the unit, "It did everything but fry potatoes!"[20]

20. Roy Schinnerer, interview by Rick Leyes, 22 August 1990, Garrett Interviews, tape recorded telephone conversation, National Air and Space Museum Archives, Washington, D.C.

Component testing began by mid-1946, as soon as hardware was cut. The tests showed excellent overall compressor efficiency in the neighborhood of 81 to 82%. The high compressor efficiency was not surprising as the technology flowed directly from Project A. The pressure ratio had been increased to 3:1 from Project A's 1.75:1, primarily by adding a third stage. The combustor, the design of which had been directed by engineering team member Bill Parrish, also performed well in tests. This was an important accomplishment for Garrett as the company had never before built a combustor.

The turbine wheel could not be tested as a component because Garrett did not have an available test rig with the capacity to absorb its power. Turbine wheel development couldn't be done until the complete machine was ready to run. Assembled late in the fall of 1946, the Black Box was driven by an external power source, but it could not generate sufficient power to run by itself. The team tried for a month to get the Black Box to run. The untested turbine component was discovered to be the problem. With an efficiency of less than 70%, the turbine engine was on the borderline of being self-supporting. By the time it was discovered that the turbine efficiency was too low, there was insufficient time left to redesign the turbine and meet the contract deadline. Subsequently, the Black Box program was canceled at the end of December 1946.

In the end, the complexity of the unit, the low turbine wheel efficiency, and the tight development schedule killed the Black Box. As Parker later put it, "I think the objective was a little ahead of our technology at that time. The compressor portion of it and the general mechanical parts of it worked fine. The turbine was its Achilles' Heel. By today's standards, it would be classed as extremely unsophisticated in turbine design. I think the machine probably would have performed nicely, if it had an adequate turbine."[21]

Despite the cost of the program to Garrett and the problems that it caused with Boeing, there were some important lessons learned, particularly what not to do. Further work on axial turbines was discontinued at Garrett in favor of the radial-inflow turbine. The highly successful backward curved centrifugal compressors were continued in future Garrett projects. The knowledge gained from building a successful combustor was part of the technology base gained from the program, and it was carried through on future Garrett engines. Finally, a young engineering team had gained further turbomachinery experience, and the company had developed additional turbine test facilities.

21. Parker to Leyes and Fleming, 8 March 1990.

GARRETT'S FIRST SUCCESSFUL GAS TURBINE ENGINE: THE GTC43/44

About the same time that the Black Box program was under development, the Navy was looking for a lightweight source of airborne auxiliary power to replace reciprocating engine APUs. The Navy Bureau of Aeronautics (BuAer) awarded Garrett a contract in 1947 to develop a 50-shp gas powered turbine (GTP50).[22] BuAer was also interested in finding a means to start a new generation of turbine engines. The first generation of jet engines had been started with electric starters, but as the size of the engines grew, it was clear that starters would become unacceptably cumbersome and heavy. Navy Aeronautical Engineering Duty Officer Bill Pattison was given the responsibility of finding an alternative starting system for BuAer.

After looking at a high pressure pneumatic control system on the Convair B-46 in San Diego, Pattison decided to tour Garrett in Los Angeles. At Garrett, he saw the XP-80 expansion turbine and projected that its roughly 10-shp output could be easily scaled to 35 shp, the amount of power needed to start the 5,525-shp Allison XT40 turboprop engines in the Navy-sponsored Convair XP5Y-1 flying boat. Returning to BuAer in Washington, D.C., Pattison drew up sketches for an air turbine starter with a Bendix-type starting engaging mechanism. BuAer then approached Garrett about building the starter. Agreeing to develop such a unit, Garrett was given a contract in early 1947 by BuAer for the 35-shp ATS35 air turbine starter.[23]

The ATS35 turbine needed a source of high presure air to drive it. When the Black Box project had been terminated, Garrett's gas turbine staff was merged with the environmental systems staff. Both engineering groups were then under the direction of Homer Wood, who became Assistant Chief Engineer. At the same time that BuAer and Garrett were discussing the air turbine starter, Wood suggested to Pattison that the compresssor for the GTP50 turbine engine (not yet under design or development) could be enlarged.[24] Air could be bled from the compressor section of the GTP50 and be used to drive the air turbine starter. The GTP50 contract was later modified. The modified contract covered a compressed air supply machine which ultimately became the GTC43/44, a gas turbine compressor.

22. W. R. Ramsaur and Douglas J. Ingells, "Twisting the Big Jet's Tail: The Story of the Small Gas Turbine," *New Frontiers* (Fall 1959): 4.

23. Bill Pattison, interview by Rick Leyes, 18 April 1990, Garrett Interviews, transcript, National Air and Space Museum Archives, Washington, D.C.

24. Pattison to Leyes, 18 April 1990.

The GTC43/44 had its genesis as a design in scaled sketches that Wood made.[25] The design work was done between March and April of 1947. The GTC43/44 contained a two-stage backward curved, shrouded compressor with an overall pressure ratio of 3:1. This was basically an uprating of the Lockheed Constellation cabin air compressor developed two years earlier by Wood, with a higher pressure ratio and higher flow rates. From the compressor, two outlets connected with elbows led to combustion chambers and two outlets were designed to connect with the ATS35 air turbine starter. The GTC43/44 had two independent tubular steel, straight-through-flow combustors. The turbine was a single-stage radial-inflow turbine with radial blades. The compressor unit delivered 43 lb of air at 44 lb per square inch absolute pressure, hence its name, GTC43/44.

Wood's compressor design was influenced by his respect for Mayer's water pump compressor technology demonstrations on Project A and the Black Box. Like Mayer, Wood had a strong interest in centrifugal water pump technology which he could trace back to his Caltech education. In designing the pressurization compressor for the Lockheed Constellation, Wood used a two-stage backward curved compressor based on Mayer's work. While Wood's GTC43/44 compressor was an important step up from the Black Box and Lockheed compressor (it had a 54% increase in adiabatic head compared to the Black Box), he departed from Mayer only in more agressively modifying the pump geometry to improve the structural characteristics of the impellers. As for his reason for continuing with backward curved impellers, Wood said: ". . . it had to pass from full bleed to zero bleed and not surge and not become ridiculously inefficient at full bleed."[26] The shrouded compressor configuration was chosen to raise the efficiency.

The turbine designed by Wood was a radial-inflow turbine of the radial element type.[27] On the basis of stress analysis and spin pit testing, Wood had determined that a cantilever radial-inflow turbine of the type used in the XP-80 expansion turbine was not strong enough to withstand the hot combustor discharge air.

As a result of his own engineering studies and his knowledge of the experimental work done by Prof. Dr. Ing. W. T. von der Nuell, Wood knew that a

25. Homer Wood, interview by Rick Leyes and Bill Fleming, 9 March 1990, Garrett Interviews, transcript, National Air and Space Museum Archives, Washington, D.C.

26. Wood to Leyes and Fleming, 9 March 1990.

27. A radial element means that the intersection of a blade with a plane normal to the rotor center line is (to a good approximation) a straight line passing through the center if extended through the hub.

centrifugal compressor could be run backward with high efficiency. As Chief Director of the DVL Institute for Turbomachinery in Germany, von der Nuell had conducted tests on Daimler-Benz aircraft superchargers.[28] In 1939, centrifugal compressor tests exceeded 80% efficiency. When run backward as radial-inflow turbines, these same centrifugal compressors showed an efficiency exceeding 90%. The higher efficiency was due to the fact that the airflow was accelerated when the compressor was used as a turbine in contrast to the deceleration of the airflow when it was used as a compressor.[29] Von der Nuell's earlier studies influenced Wood's decision to use a radial-inflow turbine.

Wood also chose a radial turbine because he did not believe that axial turbines could compete in production cost and efficiency with radials at the time.[30] In small rotor sizes, it was expensive to maintain the profile accuracies of axial rotor blading with high efficiencies.[31] Radial-inflow turbine blades were less sensitive to surface flow deviations as long as the basic flow area relationships were maintained. Fewer blades were required as well. The radial-inflow turbine also had broad operating range, an important consideration in a bleed air compressor unit. Finally, within the design parameters of the engine, the radial turbine had high efficiencies which were equal to those of axial turbines.

In fact, the choice of a radial turbine proved very effective. On July 1, 1947, cold air tests of an aluminum model of the GTC43/44 hot turbine wheel showed 82–84% efficiency. Working within a tight schedule, Garrett engineers conducted the first self-sustaining test run of the GTC43/44 on August 23. In October, Wood conferred with von der Nuell and Dr. Hans von Ohain, both of whom were German "Operation Paperclip" engineers employed at Wright Field. In fact, in 1937 Von Ohain had designed and successfully tested a prototype of the first German aircraft gas turbine engine, which incorporated a radial-inflow turbine. Late in 1947, the Air Force lent von der Nuell to Garrett to

28. W. T. von der Nuell, "Single-Stage Radial Turbines for Gaseous Substances with High Rotative and Low Specific Speed," *Trans. ASME* Vol. 74, No. 4 (1952): 499–500.

29. W. T. von der Nuell, interview by Rick Leyes, 23 October 1990, Garrett Interviews, tape recorded telephone conversation, National Air and Space Museum Archives, Washington, D.C. Dr. von der Nuell's definitive paper on radial turbines is: "The Radial Turbine," T-2 Report No. F-TR-2149-ND, Headquarters Air Materiel Command, Wright Field, Dayton, Ohio (January 1948). It was also published as: "The Radial Turbine," *Technical Data Digest*, United States Air Force Headquarters, Air Materiel Command (September 1947).

30. Homer J. Wood, Sherman Oaks, California, to Neil Cleere, letter, 21 February 1990, National Air and Space Museum Archives, Washington, D.C.

31. Homer J. Wood, "Current Technology of Radial-Inflow Turbines for Compressible Fluids," *Journal of Engineering for Power* (January 1963): 83.

contribute to the project. The following year, von der Nuell officially joined Garrett as Senior Development Engineer and became an important part of the team that improved the GTC43/44.

On June 2, 1948, the GTC43/44 passed its 200-hour Navy endurance test, and it is believed that it was the first small gas turbine engine to pass a 200-hour endurance test.[32] The GTP70, a derivative GTC43/44 that produced shaft power to drive a 40 kVA alternator, passed a similar test in June 1949. Garrett's production of the GTC43/44 began in 1948. Field use of the GTC43/44 started in 1950.

The first flight service of a GTC43/44 was on April 18, 1950 in the Convair XP5Y-1 flying boat. Two GTC43/44s provided compressed air for starting the main engines and for driving alternators that powered the electrical systems on the P5Y. In a broad push for further applications, a mobile ground power version was also used to start the North American A2J. In one application, the GTC43/44 was put into a streamlined external store pod with retractable wheels that could be air-ferried to aircraft carriers and other military bases to start aircraft. Other units were placed in ground support trailers and also used for starting aircraft. The first GTC43/44 production contract was for a derivative unit known as the GH4, a ground heater used to preheat Convair B-36 aircraft deployed in arctic operations. The first commercial use was in a ground vehicle for starting the Lockheed Electra.

As the GTC43/44 was deployed, the inevitable field service problems developed. While the choice of a radial turbine had been logical, problems occurred that proved difficult to solve. The first turbine wheels developed radial cracks in the turbine rims. In 1951, a team headed by von der Nuell traced the problem to low-cycle fatigue due to thermal compression and tension stresses during startup. Von der Nuell's team discovered that deep gashes in the rim relieved the stresses. When the new wheel was backed with a stator shroud (to fill the gaps in the rim), the aerodynamic flow was maintained. This redesigned turbine became a production standard in 1952.

The GTC43/44 was packaged in a self-contained unit that was fire and weather-proof, had automatic controls, and contained its own fuel and oil supply. Automatic controls, which had been designed to provide fully automatic starting and overload protection, proved unreliable in service. The twin-combustor design also proved to be a problem. The twin combustors made the turbine end of the machine extremely hot, which made it difficult to find a suitable fire-proof

32. Homer J. Wood, "Has the Teapot Tempest Come of Age?," (Presented at the SAE
    National Aeronautic Meeting, Los Angeles, California, 29 September-3 October 1953, but
    not published), National Air and Space Museum Archives, Washington, D.C.

enclosure. Due to the size contraints of the package, the GTC43/44 did not have cross-firing tubes; thus, imbalanced lightoffs could and did occur. Considerable engineering effort was devoted to solving such field service, packaging, and design problems.

Despite such problems, the GTC43/44 was a commercial success and more than 500 GTC43/44 units were manufactured between 1949 and the early 1950s for a variety of applications. The GTC43/44 was Garrett's first successful turbine engine. It was the start of a major new product line, the gas turbine auxiliary power unit (APU), with which Garrett dominated the world markets through the 1990s. The engine also provided a technology base for future Garrett prime propulsion engines. For their pioneering paper on the use of pneumatic auxiliary power systems for aircraft (based on the GTC43/44 and GTP70 experience), Homer Wood and Fred Dallenbach (a Garrett thermodynamic systems engineer) received the Society of Automotive Engineer's Wright Brothers Medal for 1949.[33]

GARRETT BUILDS AN INDUSTRIAL BASE ON THE GTC85

In May 1948, the Equipment Laboratory of the Air Materiel Command became interested in gas turbine compressor-type units as ground and air transportable support equipment for aircraft.[34] Helmut Schelp, another German Operation Paperclip scientist who was with the Equipment Laboratory, had worked with Garrett on the GH4 ground heater, which employed six GTC43/44 engines. The GH4 heater met the interim needs of the Air Force, but a higher performance engine was sought. Schelp wrote a specification for a unit consisting of a two-stage radial compressor and single-stage radial turbine.[35] Garrett bid on the RFP and won a $49,000 contract.[36] This contract was the start of an APU that would eventually be sold in the tens of thousands and would be an important factor in building an industrial base for Garrett's turbine engine business.

Designed by Homer Wood, the winning gas turbine compressor was Garrett's

33. H. J. Wood and F. Dallenbach, "Auxiliary Gas Turbines for Pneumatic Power in Aircraft Applications," *SAE Quarterly Transactions* 4 (April 1950): 197.

34. Ivan E. Speer, "Design and Development of a Broad-Range, High-Efficiency Centrifugal Compressor for a Small Gas Turbine Compressor Unit," (ASME Paper No. 52-SA-14 presented at the ASME Semi-Annual Meeting, Cincinnati, Ohio, 15-19 June 1952), 395.

35. Helmut Schelp, interview by Rick Leyes and Bill Fleming, 9 March 1990, Garrett Interviews, transcript, National Air and Space Museum Archives, Washington, D.C.

36. Harry Wetzel, interview by Rick Leyes, 9 April 1990, Garrett Interviews, transcript, National Air and Space Museum Archives, Washington, D.C.

GTC85 (Air Force Type F2). It consisted of double-entry first-stage and single-entry second-stage centrifugal compressors, a single tangential combustor, and a radial-inflow turbine. The compressor was designed for a 2.5-lb-per-second through-flow and a 3.36:1 pressure ratio. The double-entry compressor was chosen to increase the through-flow without increasing the diametral size of the basic machine.[37] Its blading was copied from the 1945 Constellation compressor. An important improvement over the GTC43/44 was the enclosure of the turbine and nozzlebox in a patented compressed air plenum chamber. This layer of insulation improved fire safety and eliminated the design problem of shielding the turbine hot section from the aircraft structure. The single combustion chamber eliminated the problems associated with the dual combustion chambers on the GTC43/44, as well as the extra parts required to solve the problems.

The process of developing the compressor fell to Ivan Speer, a Garrett Assistant Project Engineer. As in its predecessor, the first and second stages of the compressor were backward curved and shrouded.[38] First compressor component tests were run in December 1949. When Speer completed the final compressor calibration in April 1951, the design point efficiency was measured at 81.2%. At the time, it was believed to be the best obtained for that size class of equipment.[39]

By early 1952, a prototype version of the GTC85 was running. However, by mid-1952 an event occurred which caused a major modification to the original design. The Lockheed C-130 transport aircraft, then under development, needed an APU to make it independent of ground power systems. Needing more compressed air than could be delivered by the GTC85-3, Lockheed engineers had installed a mock-up of the CAE Turboméca Palouste in the aircraft. To compete for Lockheed's business, Wood and his group of engineers, literally working over a weekend, doubled the compressor through-flow from 2.5 lb/second to 5.0 lb/second. Fortunately, the original design was very conservative, and the flow path could be enlarged. The new version, known as the GTC85-10, succeeded in winning the competition against the Palouste.

The redesigned GTC85 also included a significant change in the turbine. The "gashed disk" geometry of the GTC43/44 turbine was replaced by a "star" configuration design developed by Dr. von der Nuell and his engineering team.

37. Wood, "Has the Teapot Tempest Come of Age?," 11.
38. Beginning with the higher flow version GTC85-10, the second-stage shroud was eliminated. An adjacent bearing allowed the close face clearances to be maintained without the shroud. However, as a conservative move, the first-stage shroud was kept.
39. Speer, "Design and Development of a High-Efficiency Centrifugal Compressor," 403.

The turbine exducer length and solidity were also substantially increased. Tests on GTC85-class radial turbines revealed a 93 to 94% total-to-total turbine efficiency rating. Subsequently, Garrett became the first company to exploit radial turbines close to their full potential in terms of efficiency, tip speed, and temperature.[40]

While the GTC85 was still under development, several external events occurred that had an immediate impact on the corporation and eventually effected the entire GTC85 program. With the advent of the Korean War in 1950, the Defense Department wanted dual production sources for critical manufacturers. The success of the GTC43/44 auxiliary gas turbine power units had given Garrett a leading position in this market. The GTC85 had the promise of becoming a standard turbine engine starting system for both the Navy and Air Force. The Defense Department wanted Garrett, a key defense company located in Los Angeles, to license their product to a company that was located at least 400 miles inland from the coast. In order to protect his company's market position and still comply with the requirements, Cliff Garrett negotiated an arrangement to "dual tool" his company's gas turbine production rather than license the product. During World War II, the company had leased a plant in Phoenix from the Defense Plant Corporation and operated it as a wholly owned subsidiary of Garrett. According to Ivan Speer, Cliff had always liked Phoenix and so, "We very scientifically picked Phoenix as a place to build the plant."[41]

The other major event was a $36 million Navy contract awarded in October 1951 for auxiliary gas turbine engines, air turbine starters, and control valves for seven types of turboprops and turbojets manufactured by five aircraft companies.[42] It was the largest production order yet awarded for small gas turbine engines. As a result of the record Navy order and increasing business, the Phoenix operation was expanded. Between 1952 and 1954, GTC43/44 and GTC85 production was transferred from Los Angeles to Phoenix. Economy of scale required that the units be built in one place, but the Los Angeles plant was kept tooled for production as a potential backup source in accordance with the agreement.

Prototype GTC85-10 engines were built in 1953 and production started in 1954. The MA-1A trailer unit, a portable GTC85 ground APU, was accepted by the Air Force in June 1956. In addition to military sales, many variations of the GTC85 were sold to the air transport industry. One important variant was the

40. Wood to Cleere, 21 February 1990, 4.

41. Speer to Leyes and Fleming, 8 March 1990.

42. Schoneberger and Scholl, *Out of Thin Air*, 100.

GTCP85, which produced both shaft power for driving an alternator and compressed air for engine starting and air conditioning. Eastern Airlines was the first company to buy the GTCP85s from Garrett. Eastern had Garrett install them in panel trucks to provide ground support for their Lockheed Electras and early jets. Later, Eastern worked closely with Garrett to convince Boeing that it was feasible and desirable to install these units inside early Boeing 727s.[43] Eastern wanted to reduce the APU cart clutter on the ground and wanted their aircraft to have self-contained APUs. The installation of 85 series APUs on Boeing 727s, and later Douglas DC-9s, became an extremely important part of Garrett's business.

With burgeoning military and commercial sales, the 85 series APU was the machine that put Garrett into the gas turbine business. Also, the 85 series would eventually become the most successful APU ever sold. Between 1952 and 1996, more than 30,000 were produced. Popular applications included the Lockheed C-130, Boeing 727 and 737, Douglas DC-9, and Air Force MA-1A ground support carts. The GTC85-20 used in the start cart serviced the Boeing B-52 and Douglas B-66 bombers and the entire "Century" series of fighter aircraft. Some 2,200 model GTCP85-180 APUs were installed aboard Boeing 727s. Another 4,700 model GTCP85-98 were ordered by the military and used in M32A-60 ground carts. The sales of this popular APU built an industrial base for Garrett's propulsion turbine engine business.

PIONEERING APU WORK BUILDS A FOUNDATION FOR PRIME MOVERS

The pioneering APU work done by Garrett engineers in the 1940s and early 1950s built a technical, financial, and manufacturing foundation for the company's movement into prime propulsion. It also assisted Garrett in moving into additional diversified turbomachinery applications.

The two key component technologies that were developed during this time were the two-stage backward curved centrifugal compressor and the radial-inflow turbine. Historically, designers of lower speed industrial compressors had used backward curved blades for many years. However, the pressure ratio of these units was too low for a practical gas turbine. Higher pressure ratios de-

43. Speer to Leyes and Fleming, 8 March 1990. Also, Pattison to Leyes, 18 April 1990. According to Speer and Pattison, Boeing engineers originally objected to the additional onboard weight of the APUs. A mock-up of two-thirds the fuselage length of a Boeing 727 was built by Garrett and an APU installed. Engineering tests were conducted to convince Boeing that the installation was feasible. Eastern convinced Boeing that it would not buy 727s without APUs installed.

manded much higher tip speeds from the rotor, but the resulting centrifugal forces would induce bending stresses on the backward curved blades that were believed to be intolerable. Demonstrating designs that refuted this belief was an important contribution by Garrett. No other organization had really confronted this problem, and the success of the total effort was an important breakthrough in centrifugal compresssor technology.[44]

The compressor introduced by Mayer would become a hallmark of Garrett's aircraft turbine engines. The radial-inflow turbine was not used in the company's aircraft engines; it was used in APUs and turbochargers. Garrett was the first company to exploit the radial-inflow turbine close to its full potential.

Garrett made other technical advances as a result of its development work on APUs, and these were later incorporated into gas turbine aircraft engines. Bearings and gears that could operate reliably at very high rotational speeds were developed. Turbine engine controls and combustor technology were also advanced. In-house engineering work on controls led to a type of corporate vertical integration which meant that the company could provide some of its own engine accessories. Considerable effort was applied to the development of manufacturing procedures aimed at improving aerodynamic efficiencies of the compressor and turbine. Important improvements in efficiencies came from the ability to achieve very small compressor and turbine blade tip clearances and profiles.

In a broader technical perspective, Garrett's efforts to improve its manufacturing procedures for compressor and turbine blading touched on one of the difficult problems historically associated with small gas turbine engines. The required tip clearances, blade profiles, smoothness, and so forth, necessary for aerodynamic efficiency approached the limits of manufacturing technique. For this reason, most designers of large gas turbine aircraft engines of the time felt that building an efficient, small gas turbine engine in a cost effective manner seemed unlikely.

Other engineers, such as Homer Wood, felt that the effects of small size on performance had been exaggerated. In his opinion, skeptics in big turbine engine companies had confused the failure to use exact scaling when employing scale model testing with genuine Reynolds number influences, and had lumped the problems together as a size problem. To solve the problem of reducing tolerences with part size, he would say to the shop machinists, "If people can build Ford V-8s with crank pins that are held to less than 1,000ths and wrist pins

44. Robert O. Bullock, Phoenix, Arizona, to Bill Fleming, letter, August 1990, National Air and Space Museum Archives, Washington, D.C.

that are held in 10,000ths, you can jolly well hold my compressor or my turbine blade profiles within errors of a couple of thousandths." The resulting close tolerance parts produced the desired efficiencies. Reflecting on this early work, Wood felt that ". . . the real clue is not in the gas dynamics, but in the manufacturing method and the recognition that little parts have to be made the way you make watches, and you can still do it for a price."[45] Cliff Garrett had a talented manufacturing team oriented to finding new fabrication methods in parallel with new product development. His conscious goal was to market machinery which competition could not produce at matching performance and prices."[46]

Low cost manufacturing techniques developed for APUs were later incorporated into aircraft engine production. The APU business also built a solid financial base for the company. Garrett's pioneering work in air turbine starters and APUs eventually led to its dominant position in the field. By the mid-1960s, Garrett was said to have produced about 80% of the total number of gas turbines with power ratings from 30 to 850 hp built in America and Europe.[47] Early competitors, including Boeing, Continental and Solar, eventually either got out of the business or took a secondary position to Garrett.

Garrett APUs, particularly the 85 series, made important contributions to the aviation industry. These APUs made it possible to free aircraft from their dependency on ground support systems and provided additional in-flight safety.

Moving into Prime Propulsion

THE PRESSURE JET HELICOPTER

The first use of a Garrett engine actually used as a prime mover occurred in October 1957. The McDonnell Aircraft Corporation used three GTC85 compressor turbines mounted in parallel to power its Model 120 pressure jet helicopter. The engines supplied compressed air to pressure jet burners, into which fuel was injected, at the tips of three rotor blades. The exhaust gases were vented against the rudders for directional control. This project was discontinued due to low propulsive efficiency, high noise, and high visibility because of the "halo" effect of tip-burning rotors, but it did help stimulate Garrett's interest in future propulsion engines.

45. Wood to Leyes and Fleming, 9 March 1990.

46. Wood to Cleere, 19 April 1996.

47. *Jane's All the World's Aircraft 1966-67*, ed. John W. R. Taylor (New York: McGraw-Hill Book Co., 1967), 515.

The TPE331

The fact that Garrett was well-positioned in the APU business led it to seek new markets for its turbine engine technology. The development of a new engine, the TPE331, proved very successful. It would become one of the most important fixed-wing, general aviation engines, and it would put Garrett in the prime propulsion business.

In late 1959, a project group under Division Chief Engineer Helmut Schelp carried out some preliminary design exercises on a hypothetical 200-300-shp engine concept called the Model 231. One of the research efforts under this project was a program funded by Wright Field to design and test several mixed-flow compressors to improve compressor technology and thereby improve APU fuel consumption. Other aspects of this program were company-sponsored efforts to scale GTC85 radial turbines and to identify state-of-the-art aerodynamic components. One aspect of this program may also have been to study a turboshaft engine for the upcoming 250-shp Light Observation Helicopter (LOH) engine competition. The results of the various studies were used to determine the trade-offs of various design versions of the 231.[48]

Following these studies of the Model 231, Schelp embarked on a personal initiative to define in general the specifications and layout of a small 330-shp turbine engine called the Model 331.[49] Schelp felt that Garrett had taken a leading position in the APU business and that Garrett should enlarge its business by entering the small turboprop field. He was also spurred by the fact that in the early 1950s Boeing had flight tested its Model 502 turbine engine and was interested in the same field.

About the same time, Ivan Speer, then Assistant Division Manager at Garrett's Phoenix division, was interested in a second generation APU or "gen set" (generator set). Many of Garrett's APUs had been used as military generator sets, or when installed on aircraft, as combination generator set and air-conditioning supply. These 43/44 and 85 series APUs were low pressure ratio engines. Speer felt that in order to get further generator set business, it would be necessary to have much better fuel consumption than the existing APUs had. A new generator set design would have to have a much higher pressure ratio and be more fuel efficient and compact than the existing APUs. According to Speer, developing a new generator set, " . . . would not be a dramatic deci-

---

48. Robert O. Bullock, Scottsdale, Arizona, to Neil Cleere, letter, 16 August 1990, National Air and Space Museum Archives, Washington, D.C.

49. Schelp to Leyes and Fleming, 9 March 1990.

sion for the company to make, rather a logical progression of the existing product line."[50]

Thus, Schelp was interested in pursuing prime propulsion and Speer was interested in improving the existing APU product line. The deciding vote was cast by Cliff Garrett, who was not interested in prime propulsion. According to Speer:

> The thing that he (Cliff Garrett) told me was (that) he was reluctant to put the company on the line for something that was a prime safety-of-flight item, flight propulsion. We had always been in the important things in an airplane, but there were always alternatives. Systems were such that you could fall back on something. You weren't really a prime safety-of-flight item. Even if you lose your pressurization, you can always drop the mask and get down, you see.[51]

Cliff Garrett's decision was conservative and perhaps prudent from a business perspective. At the time, the market for small gas turbine aircraft engines was not clearly defined in terms of application and size. Established large gas turbine engine manufacturers seemingly would have been formidable competitors. Also, as a manufacturer of small components for large engine companies, Garrett may have been concerned about his company competing with its important customers. Another factor could have been the greater product liability associated with a prime propulsion engine. What can be said with certainty is that developing an improved APU rather than a prime mover engine was a less costly and less risky decision.[52]

THE DESIGN OBJECTIVES OF THE MODEL 331

In September 1959, design work was initiated on a generator set known as the GTP331. The objective of this company-sponsored effort was to design and develop an engine in the 400-shp class with growth potential to at least 700 shp. By December 1959, working layouts were underway to produce an engine that would meet the initial design goals of 400 shp, 200 lb weight, 0.66 lb/hp/hr sfc,

50. Speer to Leyes and Fleming, 8 March 1990.

51. Speer to Leyes and Fleming, 8 March 1990.

52. From a historical perspective, Cliff Garrett's reluctance to allow the company to get into the prime propulsion business cost Garrett the opportunity to have its turbine engine available to the general aviation market ahead of Pratt & Whitney Canada. Pratt & Whitney Canada's PT6 was certificated in December 1963, and Garrett's TPE331 was not certificated until February 1965. With a head start, the PT6 eventually became the dominant (in terms of production numbers) general aviation and small transport engine.

and a 1,000 hour Time Between Overhaul (TBO). The basic configuration consisted of a single-shaft engine with a two-stage centrifugal compressor, reverse-flow combustor, and three-stage axial turbine.

The design philosophy of the basic Model 331 engine was to design a low-cost, rugged, lightweight aircraft engine with good specific fuel consumption without undue complication in the mechanical design of the control system.[53] A high pressure ratio, two-stage centrifugal compressor was chosen. To keep the cost down, a three-stage axial-flow turbine with integrally cast blades was selected. A reverse-flow annular combustor was selected to minimize engine length, weight, and combustor pressure loss. The rotors had the same hub diameter and blade profile for all three stages, with the blade tips cut down for the first-and second-stage rotors.

The fact that the original design of the Model 331 called for a straight radial compressor rather than a backward curved compressor was a departure from Garrett's previous practice. According to Model 331 Project Engineer Curt Bradley:

> . . . the motivation to go to radial blades was derived from experience in the prior research work done on a variety of mixed-flow compressors which were not especially good overall designs. The mixed-flow designs suffered from a high hub weight and in most cases did not develop the pressure ratio expected. Further, the use of shrouded compressors on other older engines had fabrication costs which were much higher than desired. My direction from Helmut Schelp was that the 331 had to be a 'light, light, lightweight and low-cost design' and would use the radial compressor to eliminate the machining costs and high stresses found in older backward curved designs.[54]

In the process of selecting a turbine for the Model 331, Garrett engineers considered both a radial-inflow turbine and an axial-flow turbine. Factors favoring the radial turbine included Garrett's experience with this type of turbine, and fewer (by comparison) aerodynamic problems associated with it. On the other hand, the axial turbine had the advantage of being lighter and more compact. Furthermore, the cost of the axial design could be reduced by using the same design and casting for all three stages, and by grinding down to the diameters required in the first and second stages, the same tooling could be used for

53. A. A. Aymar, "Engineering Report: Development History of the AiResearch Model TPE331 Aircraft Turboprop Engine," (Engineering Report No. GT-7182-R, AiResearch Manufacturing Co., Phoenix, Arizona, 25 October 1963), 4.

54. Bradley to Leyes, 12 June 1991.

each stage. A final consideration favoring an axial turbine was the possibility of making provisions for a free turbine, dual-shaft engine design. In the final analysis, Garrett chose an axial turbine design[55].

The engine was designed as a single-shaft engine principally to meet the requirements for its original use, i.e. a generator set. Although consideration was given to making the engine a free turbine by disconnecting the third stage axial rotor and using it as a power turbine, a free turbine was not suitable as a generator set. The single-shaft turbine had the inertia to run a generator without a significant loss of speed when subject to shock load conditions.

The original design team of the Model 331 was headed by Project Engineer Curt E. Bradley under the general supervision of Bill Spatz, Chief, Propulsion Engines. Under the direction of Chief of Aerodynamics Bob Bullock, Aerodynamicist Jules Dussard spearheaded the aerodynamic design effort and directly supervised most of the work on the compressor design. Aerodynamicist Ray Davis was responsible for the compressors, and Aerodynamicist Jack Rebeske was responsible for the turbine design. This design group worked closely with Bob Van Nimwegen, Supervisor of Stress and Vibration, and Montgomerie C. Steele, Stress and Vibrations Engineer.

EXPLORING MARKETS FOR THE MODEL 331 DESIGN

With the design of the engine underway, various possible applications appeared that provided incentives to spur the development of a marketable engine. As an industrial turbine, such a unit could be a total energy package for shopping centers and apartment complexes. It could be used as a power source to provide electricity, heating, and air-conditioning. Other anticipated uses were ground, marine, and airborne generator drives as well as vehicular drives for military and off-highway equipment. An expected program by BuShips for marine propulsion in this range of power was also a possibility.

In 1959, the first contract for the development of the Model 331 as an APU, called the GTP331, came from the Army's Engineering Research and Development Laboratory (ERDL). Two APUs were developed for this program, and the program contributed some development money and technology to the Model 331.

In December 1960, the first complete power section of the Model 331 was coupled with a reduction gear from another model engine and testing began. On the first day of operation, the engine was run to full speed. Immediately

55. Wood to Leyes and Fleming, 9 March, 1990.

thereafter, the power section went into a full-scale development program which included performance and endurance testing. In the early fall of 1961, several events occurred which stimulated Garrett's interest in a turboshaft version of the engine.

The first was the Navy's announced need for a 400- to 500-shp turbine engine for an Assault Support Helicopter (ASH). Garrett responded by proposing a turboshaft version of its Model 331 known as the TSE331-7. This was basically a GTP331 APU, rated at 440 shp, with a new gearbox with higher rpm operation. Garrett installed a TSE331-7 engine in a Republic Lark helicopter (license-built Sud Aviation Alouette II) and flight tests began on October 12, 1961. Flight tests were successfully completed with no engine malfunctions.

The second event that stimulated Garrett's interest in turboshafts was an Army request for proposal in November 1961 for an alternate Light Observation Helicopter engine. The engine originally chosen for the LOH, the 250-shp Allison T63, had encountered a series of problems. Garrett was one of several companies invited to submit proposals and compete for this potentially very large helicopter engine acquisition.

In August 1962, a TSE331 was shipped to Piasecki Aircraft Corporation for installation and flight testing in their Airjeep I. Although the flight tests of the Republic Lark and the Piasecki Airjeep I were satisfactory, the fixed-shaft TSE331 was not as ideally suited for helicopter operation as free turbine engines were. In order to start a helicopter with a fixed-shaft engine, a heavy clutch was required. To accelerate rapidly from idle, a heavy torque was required to bring the main and tail rotors up to the desired rotating speed. The fixed-shaft engine did not have this torque available at idle speeds. On the other hand, a free turbine had maximum torque available under idle conditions, and the helicopter rotor blades could be accelerated rapidly. Thus, while the TSE331 would work as a helicopter engine, it was at a competitive disadvantage with respect to free turbine engines on the market. In fact, Garrett had considered a free turbine engine known as the Model 331-50.[56] However, by 1964 Garrett's attempt to market a free turbine engine ended unsuccessfully.[57]

The original ASH was never procured by the Navy, Allison provided the engine for the LOH, and no orders developed from Piasecki. The Model 331 had been converted from a generator set APU to a turboshaft engine as a result of

56. William S. Reed, "AiResearch's 331 Engine Tested on Lark," *Aviation Week & Space Technology*, Vol. 75, No. 25 (December 18, 1961): 70.

57. C. M. Plattner, "Garrett Uprates T76 Engine to 660 shp," *Aviation Week & Space Technology*, Vol. 81, No. 8 (August 24, 1964): 97.

what appeared to be a developing helicopter market. However, the helicopter market did not develop. It became evident to Garrett, about the same time as Pratt & Whitney Canada reached a similar conclusion, that the first market was going to be for a general aviation turboprop engine.[58] There had been no new development in general aviation piston engines for many years. In fact, the horizontally-opposed engines had less performance than the generation of radial engines preceding them, resulting in a negative performance plateau. It was at this point that Ivan Speer began to try to convince Cliff Garrett to let the company get into the propulsion business and to make a turboprop out of the Model 331.

There were other factors that affected Speer's attempt to change Cliff Garrett's thinking. As the gen set began to reach the stage in its development when it was about ready for sale, not much of a market existed for it. Garrett already had nearly all of the APU market and the market was not expected to grow that much. With a saturated APU market, Speer had begun to cast about for other markets. It was clear to him that the company would have to go into propulsion engines in order to expand. With respect to the potential problem of competing against the big engine companies, Speer felt that Garrett would have some natural advantages. It was easier to build up than to scale down. It would be difficult for large engine manufacturers to be cost competitive due to overhead, and certain aspects of large engine technology were not suitable for small engines. Garrett had been producing turbine engines since the 1940s, which gave it as much experience in the business as most large engine companies. Also, Garrett's building-block approach toward evolving a product and getting markets for it a piece at a time would enable it to get into prime propulsion under a condition that it could afford.[59]

About a year before Cliff Garrett's death in 1963, Speer was allowed to proceed with a turboprop 331 engine. According to Speer:

> I finally got a tacit approval out of Cliff to say that he just gave me license to bootleg it. He says, 'Well, okay, go ahead. Don't let me hear about it being any problems. I don't even want to know about it.' So that was the way we started. He died about a year later, and then of course, Harry (Wetzel) took over. Harry Wetzel was always gung-ho: 'Sure, let's go into the damn propulsion business.' From then on, it was not a bootleg project, but an honest-to-goodness 'let's go.'[60]

58. Speer to Leyes and Fleming, 8 March 1990.

59. Speer to Leyes and Fleming, 9 March 1990.

60. Speer to Leyes and Fleming, 9 March 1990.

In fact, the top echelon of Garrett's managers were aviation oriented, and they had a natural inclination to go into the prime propulsion business. Harry Wetzel was an Air Force pilot during World War II, and Bill Pattison flew PBYs for the Navy. Ivan Speer was interested in flying, and later became a pilot. All of these top managers, and others in the corporation, clearly saw the advantages of being in the propulsion business.

THE T76

The decision to proceed with a turboprop version of the TSE331 was stimulated by another important event. By the fall of 1962, the requirements for a counterinsurgency (COIN) aircraft sponsored by the Navy had begun to jell. The COIN aircraft required a 500-shp engine. Encouraged by the successful operation of two previous helicopter installations of the turboshaft engine, Garrett decided to proceed with the design of a turboprop engine for the COIN aircraft.[61] The power requirements were soon raised to 600 shp. Garrett needed to uprate the power section and develop a turboprop reduction gearbox. This turboprop engine was called the TPE331-25, and its military designation was the YT76-G-2 (clockwise rotation) and -G-4 (counterclockwise rotation).

While the basic engine retained the TSE331 gas generator, key components for the turboprop engine needed to be selected and developed. For the engine air inlet, the scoop type was found to be the most attractive configuration among the alternatives examined. It provided an acceptable inlet channel height (to reduce inlet losses under all conditions and distortion at high angles of attack), minimized aircraft ducting requirements, and included a controlled area reduction from the engine inlet to the compressor that was built into the gearbox housing.[62] Demonstrating the validity of the configuration, tests conducted later showed total pressure recoveries varying between 0.995 and 1.00 with low distortion. With the ram effect of the propeller, pressure recoveries were found to be as high as 1.007.

The original design allowed the offset gearbox to be placed in two positions. This meant that the scoop could be located either up or down which gave the airframer the flexibility of locating the engine nacelle above or below the wing.

61. Aymer, "Development History of the AiResearch Model TPE331," 3.

62. J-P. Frignac and E. J. Privoznik, "The Growth and Evolution of the TPE331," (ASME Paper No. 79-GT-164 presented at the Gas Turbine Conference and Exhibit and Solar Energy Conference, San Diego, Cal., 12-15 March, 1979), 3.

In the case of the COIN aircraft, the location of the inlet scoop on top placed it a greater distance from the ground which offered better protection from foreign object damage (FOD) on unprepared landing strips. An additional advantage of the inlet selection was that a wall of the gearbox formed a portion of the inlet. Thus, heat from the gearbox oil contributed significantly to anti-icing the inlet and reduced the amount of bleed air required for anti-icing. It also eliminated the need for a special inlet separator for ice and super-cooled water like that developed for the Pratt & Whitney Canada PT6. The added weight and complexity of such an airframe inlet was avoided.

A two-stage gear reduction was required to achieve the desired ratio of turbine shaft-to-propeller speed. A combination spur gear and single-stage planetary gear set was selected, the principal advantage of such an arrangement being that the offset gearbox resulting from this design was complementary to the scoop inlet.[63] Also, the offset provided a convenient area on the back face of the gearbox to mount the accessories, minimizing frontal area. The engine mounts could be located on the back face of the gearbox, which also minimized the engine nacelle cross-section.

The basic gas generator section of the YT76-G-2 retained the same configuration as the 440-shp TSE331, but the engine was uprated to 575 shp. This power increase was achieved by redesigning the compressor to increase the airflow and pressure ratio and by increasing the turbine inlet temperature.

The reverse-flow annular combustor had been chosen for several reasons. It could use the space surrounding the turbine assembly and still fit within the outer diameter of the diffuser for the second-stage compressor.[64] It kept the engine compact, and it permitted removal of the entire combustion system during maintenance without disturbing the rotating assembly. The cross-sectional area available for combustion was sufficiently large to allow relatively low velocities and thereby low pressure losses in spite of the reverse-flow gas path.

The earlier choice of a fixed-shaft rather than a free turbine configuration proved to have special advantages in fixed-wing aircraft. The single-shaft power response to throttle movement resulted in virtually instantaneous power recovery in the event of a go-around and maximum reverse thrust for an aborted takeoff.[65] Also, rapid, steep descents could be scheduled by varying the amount of propeller drag. Conversely, to limit the amount of drag of a windmilling propeller in the event of a sudden power loss, the TPE331 had a negative torque

63. Frignac and Privoznik, "The Growth and Evolution of the TPE331," 3.

64. Frignac and Privoznik, "The Growth and Evolution of the TPE331," 4.

65. Frignac and Privoznik, "The Growth and Evolution of the TPE331," 5.

sensing (NTS) system that automatically moved the propeller blade toward the feathered position. The fixed-shaft design itself was mechanically simple and required only a simple two-bearing support for the rotating group.

Testing of the first Garrett turboprop, the 575-shp YT76-G-2 (TPE331-25), was initiated on a propeller test stand on July 31, 1963. In August 1964, the Navy selected the North American NA-300 (later designated the OV-10A by the military) as the winning design for the COIN competition. At first, North American said that either the Pratt & Whitney Canada T74 (military designation for the PT6) or the Garrett T76 would be suitable for its aircraft. However, the Navy insisted that North American specify only a single engine, and North American selected Garrett.[66] The contract for the prototype YT76 was issued on October 26 by the Navy.

Bill Pattison, retired Vice President of Sales for the Garrett Corporation, explained the technical and political advantages of the T76 relative to the T74 which made it an attractive powerplant for North American to use on its COIN aircraft:

> There are technical differences between the two engines. One is a free spool engine, and the other is a fixed-shaft engine. One of the things that we felt was very important for a military airplane was quick response, particularly for an aircraft of the COIN type. In the case of the free turbine engine, you have to spool up the free turbine propeller, particularly if you are in an idle descent or you're loitering or something and you want to get out of there in a hurry. There's an appreciable length of time between pushing the throttle forward and getting full power out of the engine. In the case of the 331, the response is almost instantaneous. With the 331, you could have a beta (pilot) controlled propeller, and it was a very, very effective dive brake. You could go from that kind of a steep descent to an almost instantaneous application of take-off power. Those are some of the reasons. Moreover, the PT6 was a Canadian-built engine, and this was a Navy airplane. The 331 was a U.S.-built engine.[67]

Despite losing the original design competition, Convair continued to push the case for its COIN candidate aircraft, the Charger, and Pratt & Whitney Canada pressed for further consideration of its T74 engine. As a result of company and political efforts, further evaluations and fly-off competitions were held in 1965 and 1966. The Convair Charger was dropped from the competition fol-

66. George C. Wilson, "Large-Scale COIN Buy Remains in Doubt," *Aviation Week & Space Technology*, Vol. 81, No. 8 (August 24, 1964): 24.

67. Pattison to Leyes, 18 April 1990.

The test run of the first Garrett turboprop engine, the 575 shp TPE331, was witnessed in August 1963 by (from left) Division Chief Engineer Dr. Helmut Schelp, Garrett Corporation Chairman of the Board Eddie Bellande, Corporate Vice President of Engineering Walt Ramsaur, and Assistant Division Manager Claude Kirk. Photograph courtesy of Garrett.

lowing loss of the airplane when it crashed in October 1965. Pratt & Whitney Canada's T74 was evaluated further in the North American OV-10 during 1966. However, these evaluations failed to dislodge Garrett from its installation on the OV-10. For Garrett, the continuing pressure of the highly competitive atmosphere following the original selection resulted in the loss of previously promised military development and qualification money and Garrett had to assume these costs.[68]

The 50-hour preliminary flight rating test for Garrett's prototype YT76-G-2 engine was completed in the spring of 1964. Flight testing was done with a YT76-G-2 engine installed in the nose of Garrett's flight testbed, a World War II-era Douglas A-26 bomber.

T76-G-6/-8 AND -10/-12

Because of the military's need for additional power and because of problems that had occurred with early versions of the engine, the YT76-G-2 was redesigned. By 1964, component testing for the the first uprated version (660 shp), the YT76-G-6/-8 (clockwise and counter-clockwise rotation), had been accomplished. Compressor splitter vanes were added and the turbine was redesigned for increased efficiency. The turbine efficiency was raised by insetting the turbine blades by machining a circumferential recess in the static turbine tip shroud and increasing the length of the turbine blades to block the passage of air around the blade tips. On July 16, 1965, the prototype YOV-10A Bronco aircraft flew with the YT76-G-6/-8 engines installed.

Design work on a more powerful engine, the T76-G-10/-12 model, was initiated in 1966. This engine was rated at 715 shp. In the YT76-G-2, target compressor performance had not quite been met. A characteristic of the YT76 radial compressor was that peak efficiency occurred very close to the surge line. In order to provide an adequate surge margin, it was necessary to lower the operating line of the compressor. The result was a reduction in both pressure ratio and efficiency and a loss in performance. To solve this problem, a completely new backward curved centrifugal compressor, the B2, was designed.[69]

---

68. Pattison to Leyes, 18 April 1990.

69. The name B2 was derived from the fact that the new compressor had a backward (B) curved compressor in the first stage and in the second stage. Thus, it had two backward curved compressors or two Bs.

The new compressor used an annular 13-vaned crossover duct instead of the 8-vaned duct used previously. Efficiency improved significantly.[70] Garrett returned to the backward curved compressor blades successfully developed much earlier in its APUs.[71] While the backward curved blade compressor was more expensive to manufacture than one with straight radial blades, it was also more efficient, had a greater surge margin, and improved engine performance. Over the years, Garrett's stress engineering group had made significant improvements in backward curved blade technology which permitted higher rotational speeds, greater stress capability, and compressor pressure ratios. Garrett aerodynamicists had developed new computer programs to predict highly accurate data on blade performance for the shapes actually being generated by production. In addition, new instrumentation techniques for measuring blade thermal growth distortions caused by centrifugal force and pressure gradients permitted uniform blade clearances which improved efficiency and reliability.

The turbine also underwent a major redesign in the T76. In the original Model 331, engine performance had been traded for manufacturing and maintenance economy. However, the YT76 had experienced some hot tears in the casting. That problem, coupled with the military requirement for more power, led to a new turbine design.

In contrast to the single blade profile for all three stages, each stage of the redesigned turbine had different blade contours to optimize performance. Additional turbine blades were added to the second and third stages. At the same time that these improvements were being made to the T76, a research program was undertaken to study the effects of acceleration across the turbine nozzle and the impact of end-wall contouring on the stator. It was determined that stage performance was very sensitive to the amount of acceleration across the nozzle. The more acceleration across the nozzle, the better the performance. Also, the contouring had a significant effect on efficiency. This research was used on later model TPE331 engines to improve their efficiency.

In 1966, both the first engine run and the Preliminary Flight Rating Test (PFRT) were completed on the T76-G-10/-12. In 1967, the T76 Model Qualification Test (MQT) was passed. The improved T76 was flown on the first

70. Frignac and Privoznik, "The Growth and Evolution of the TPE331," 5.

71. Actually, the first reintroduction of the backward curved compressor was on the Model 331 APU designed for the Army Engineering Research and Development Lab (ERDL) project. A backward curved compressor was used on the second stage to solve a thrust bearing problem and to change the operating point to a region of higher efficiency.

production OV-10A on August 6, 1967. Eventually, some 1,078 engines were delivered for the OV-10 between 1967 and 1992. Heavily used in the Vietnam War, the T76 engines on the OV-10 proved reliable in service. The aircraft often operated from unimproved fields. A number of engines operating in this environment were examined upon disassembly, and were found to have ingested a considerable amount of sand and dust.[72]

The titanium first-stage compressor impeller, which might at first be expected to suffer the worst effects, showed little damage. The reason was that the backward curved blade design of the impeller caused the engine to reject stones and other large particles on the landing strip rather than ingest them. This was also found to be the case during laboratory testing when 1/4-inch bolts, ice balls, and wiping rags were fed into the engine inlet. From 1967 to 1970, the in-service engine removal rate actually decreased from 1.2/1,000 hrs to .15/1,000 hrs.

One field service problem of note was manifested during the war. Subject to heavy particle and salt ingestion, the T76 engines were washed daily to reduce this contamination. Trapped in the lower engine section where the magnesium gearbox housing joined the steel compressor housing, puddles of contaminated wash water contributed to heavy corrosion. The splitter blades were found to be relatively intact, and this permitted the engine to function adequately although power and efficiency were reduced. Drain holes were drilled in the casing and protective coatings were used as a stop gap measure, but eventually the solution was a new gearcase made of aluminum.

The TPE331's selection for the COIN aircraft was one of the most important turning points in the history of the engine program. Not only did it give Garrett access to an important military market, it was a major factor in the TPE331 becoming a production engine.

TPE331 SERIES I AND II

In 1962, at the same time that the COIN competition had spurred work on the the Model 331, Garrett was working on a commercial version of the engine, known as the TPE331 Series I and Series II (TPE331-25/-43). This was the same as the military YT76-G-2/-4, first run in July 1963. The first engine in this series received its type certification from the FAA on February 25, 1965, at a rat-

---

72. Howard Buchner, Jr. and John W. Peach, "Turboprop Engine Operational Experience in STOL Aircraft Operating from Rough Fields," (SAE Paper No. 680228 presented at the Business Aircraft Meeting, Wichita, Kansas, 3-5 April 1968).

ing of 575 shp. The TPE331 Series I and II engines were actually in commercial service before the YT76 engine had been qualified for military service.

Lou Kell, a mechanical engineer with Garrett, was responsible for working with the FAA to get the first TPE331 certificated. This was a new experience for Garrett engineers, who had never certificated an aircraft engine. Kell recalled that Garrett's primary test facility consisted of an outside test and prop stand.[73] Hay bales had been piled around the test stand in an attempt to pacify neighbors who complained about the noisy engine, which had been running at take-off power for long periods. With these basic facilities, ingestion tests were run on the TPE331 including sand and dust, birds, rags, bolts, ice, and water. The rugged centrifugal compressor showed relatively minor damage after these tests.[74] Altitude tests were conducted at Garrett's altitude facility to verify such parameters as performance and surge. Considerable testing with strain gauges and internal instrumentation was performed as well.

Even before the TPE331 was certificated, Garrett had prospects for selling the engine commercially. The first commercial versions of the TPE331 were installed on the Volpar Super Turbo 18 commuter aircraft, Mitsubishi MU-2B and D, and Aero Commander Turbo Commander 680 (later Gulfstream Commander 680). Other applications included the Air Asia Porter (Pilatus Turbo-Porter), Carstedt Jet Liner 600A, and Fairchild Heliporter.

The first TPE331 was shipped to Volpar, an aircraft conversion company located in Van Nuys, California, in April 1963. The engine was installed on a Beech C-45 (military version of the civilian Beech Model 18). This aircraft was flown on April 1964, making the first flight of the TPE331 on a fixed-wing aircraft. The TPE331 significantly improved the performance of the airplane. Volpar continued with its turbine conversion work. The first of its turbine-powered aircraft, known as Volpar Super Turbo 18s, were put into service in February 1966 with 575-shp TPE331-47 engines. Another Volpar conversion, the Turbo Liner, was certificated in March 1968 as a 15-passenger commuter aircraft. Volpar conversions were also used by Air America in the Vietnam War.

The launch customer for the TPE331 was Mitsubishi who selected it in July 1964 for the MU-2, a twin turboprop STOL utility transport. The MU-2 was originally designed around the Turboméca Astazou, but difficulties with the en-

73. Lou Kell, interview by Rick Leyes and Bill Fleming, 5 March 1990, Garrett Interviews, transcript, National Air and Space Museum Archives, Washington, D.C.

74. Garrett Corporation, "TPE 331/T-76 Turboprop Engine: Operationally Proven and Supported," (Publication No. MS-2964-0, undated).

Garrett TPE331 turboprop engine. Photograph courtesy of Garrett.

gine and a desire to sell the aircraft in the U.S. led Mitsubishi to seek another powerplant.[75] Again, working a deal with Mitsubishi, Garrett agreed to assemble the first three models of the MU-2 at its AiResearch Aviation Service Company in Los Angeles and install the TPE331 on the aircraft. The first TPE331 was delivered in 1964 for installation on a Mitsubishi aircraft. (Production deliveries of the TPE331 began in 1965). Mitsubishi then had a demonstrator aircraft, and Garrett had an engine installed on the fastest turboprop aircraft in the business. FAA certification for the TPE331-25A powered Mitsubishi MU-2B was received in November 1965.

The first key commercial sale of the TPE331 was to Aero Commander in 1964.[76] Garrett was particularly interested in getting the TPE331 installed on Aero Commander's piston-powered Grand Commander. Unlike the Beech Model 18, it was a high-performance airplane that was built for high speed and pressurized for high altitudes. It also was a high-wing aircraft which gave it good

75. Speer to Leyes and Fleming, 8 March 1990.

76. Speer to Leyes and Fleming, 8 March 1990.

The first key commercial sale of Garrett's TPE331 was to Aero Commander. Shown is an Aero Commander 840 executive aircraft. Photograph courtesy of Garrett.

ground clearance for the propellers. Such an application would make an excellent testbed aircraft, and the full capability of the turboprop engine could be used.

Making a deal with Aero Commander, Garrett supplied the engine and de-signed and tested the engine inlets. Aero Commander did the engine installation. Garrett then did the flight testing to gather data for itself and for Aero Com-mander. The prototype aircraft flew for the first time on December 31, 1964. By replacing the 380-hp reciprocating engines with 605-eshp turboprops, a 12% in-crease in cruising speed and a 55% increase in the rate of climb was demon-strated. Also, the range was increased by 10%.[77] Once Garrett got on the Com-mander, it could point to something flying and give demonstration rides to potential customers.[78] Deliveries of the TPE331-43-powered Aero Commander Turbo Commander, an executive transport aircraft, began in June 1965.

Thus, Garrett was off to a good start with the TPE331. With Volpar, Aero Commander, Mitsubishi, and other applications, the TPE331 had penetrated the commercial market. With sales to both military (North American OV-10) and commercial markets, Garrett had established itself as an aircraft propulsion company.

77. Howard A. Buckner, Jr. and Robert R. Van Nimwegen, "Growth and Operational Capabilities of the Turboprop Engine," (SAE Paper 670236 presented at the Business Aircraft Conference, Wichita, Kansas, 5-7 April, 1967), 3.

78. Speer to Leyes and Fleming, 8 March 1990.

THE CENTURY SERIES (THE TPE331-1 AND -2)

In 1965, at the same time that the design work was initiated on the T76, Garrett made an arrangement with the FAA to follow the T76 MQT process and to accept most of the documentation for civil certification of the engine. Thus, in December 1967 shortly after the T76 had been qualified, three higher horsepower versions of the TPE331 were certificated as civil versions of the T76-G-10/-12. They were the the TPE331-1 and -150, both flat rated at 665 shp, and the TPE331-2 rated at 715 shp. These engines, known as the Century Series (starting with the TPE331-1-101), used the same compressor and turbine as the T76 and were available in inlet-up or inlet-down configuration.

The 715-shp engine was just what Ed Swearingen, President of Swearingen Aircraft, was looking for.[79] In 1966, Swearingen Aircraft had certificated an eight-passenger, twin-turboprop executive transport known as the Merlin IIA. It was powered by 550-shp Pratt & Whitney Canada PT6 engines. Ed Swearingen had a reputation in the industry for speed, and in order to get more performance out of the Merlin, he needed more powerful engines.

The military-derivative Garrett engine not only had the power, but its compact, straight-through configuration had inherent drag reducing and performance enhancing features. It had a straight-in air intake which gave it good ram recovery and a straight-out, in-line exhaust pipe which added jet thrust. Such features were important to Ed Swearingen. Given an alternative when the Garrett engine became available, Swearingen was said to have felt that the PT6 engine had too much drag for his objective.[80] The PT6 had front, side-mounted exhaust stacks (so called "steer horns"), less ram recovery (by comparison to the TPE331), and had a de-icing door (inertial air separator) mounted below the engine. These were features that Swearingen did not want, given a choice.[81] With TPE331-1-151G engines installed, the new aircraft was certificated in 1968 as the Merlin IIB, and Ed Swearingen had a 300-mph airplane.[82]

The case of the Garrett engine installation on the Swearingen Merlin is use-

---

79. Speer to Leyes and Fleming, 8 March 1990.

80. Speer to Leyes and Fleming, 8 March 1990.

81. Speer to Leyes and Fleming, 8 March 1990.

82. It is possible that another factor may have contributed to Swearingen's decision to make the expensive change-over from the PT6 installation on the Merlin. Said to have been caught in a cash-flow squeeze, Swearingen could not put money up front for PT6 engines, but could get TPE331 engines on consignment from Garrett. See Kenneth H. Sullivan and Larry Milberry, *Power: The Pratt & Whitney Canada Story* (Toronto: CANAV Books, 1989), 168.

ful for illustrating several points. The PT6 and the TPE331 had become direct competitors for the same military and general aviation markets. Competing against the PT6, the key sales points for early Garrett commercial TPE331s were performance and fuel economy.[83] Contributing factors to performance and fuel economy were low-weight and low-drag installation, straight-in ram air recovery, exhaust jet thrust, and a high pressure ratio compressor. The single-shaft design of the TPE331, which gave it a very limited speed range, was also a contributing factor. For example, at aircraft cruising speed, the engine rpm would be within 5 to 10% of its take-off rpm. This "single-point" operation made it possible for designers to optimize the engine efficiency, which resulted in good fuel economy.

Aside from the technical selling points, the human dimension and persistence were considered important aspects of the early success of the TPE331. Reflecting on the company's relationship with Swearingen, retired Executive Vice President, Sales and Service Bill Pattison said:

> A number of us knew Ed Swearingen quite well when he was involved (with his original company), again, with Fairchild later on. We also helped to market that airplane. But I think it's just a relationship that we had with Ed, and we liked him and he liked us. I think people sometimes look for spectacular reasons why somebody gets the business, but 90% of the time it's just pure hard work and dogged persistence and trying to satisfy the customer and trying to get him to see your way of thinking and believe in the merits of what you're trying to sell.[84]

Good customer rapport, persistence, and good technical features were an important part of the early TPE331 marketing strategy. In fact, they were compensating factors for other disadvantages for Garrett. The facts that the TPE331 had entered the market after the PT6 and that Garrett had been recognized within the industry as an APU company, not a propulsion company, had created some marketing difficulties for the TPE331. According to retired CEO and president Harry Wetzel, "What we didn't have (at the time) was the Pratt & Whitney name and their very, very competent selling ability. Even though it was a Canadian company, the name Pratt & Whitney was devastating to us, and it cost us an awful lot of money just to get a ticket at the poker table."[85]

In fact, the TPE331 did provide Garrett with a ticket at the poker table. The

---

83. Speer to Leyes and Fleming, 8 March 1990.

84. Pattison to Leyes, 18 April 1990.

85. Wetzel to Leyes, 9 April 1990.

Century Series TPE331 powered uprated models of aircraft previously powered by the Series I and II such as the Volpar conversions, Mitsubishi MU-2 (Models DP, F, and G), and Aero Commander. It also powered the Short Brothers Skyvan 3, CASA 212-100, and other miscellaneous aircraft. As the TPE331 penetrated the commercial market, significant sales were made for executive aircraft, and sales for regional aircraft began to grow. In 1965, the first year of TPE331 production, 153 engines were shipped for executive aircraft. By the end of 1969, 1,098 such engines had been shipped. In 1966, the first 25 TPE331 engines for regional airlines were shipped; by the end of 1969, 255 engines were delivered.

THE TPE331 CONTINUES TO GROW IN THE LATE 1960S AND 1970S

In response to the needs of their customers, Garrett continued to increase the performance of the TPE331 in the late 1960s and the decade of the 1970s. The first group of more powerful engines, the TPE331-3, -5, and -6, were collectively known as the 840 Series (840 shp). Following the 840 Series was the the 1,040-shp T76-G-420, an engine with 45% more power. The commercial derivatives of the 1,040-shp T76-G-420 were the TPE331-8/-9 and -10/-11 engines rated at 865 and 1,000 shp respectively. During this period, the majority of TPE331 sales were for executive aircraft. TPE331 regional aircraft annual sales typically constituted slightly less than one third the number of executive aircraft sales. The first significant number of sales for regional aircraft TPE331 engines did not come until the late 1960s as a result of an explosive growth in commuter airlines.

THE 840 SERIES (THE TPE331-3, -5, AND -6)

One airframer that wanted to penetrate this growing commuter market was Swearingen. In a joint venture with the Fairchild-Hiller Corporation, Swearingen started construction in 1968 of a 19/20-passenger airliner to be known as the Model SA-226TC Metro (later Metro I and II). The Metro was considerably larger than the Merlin and required more take-off capability. Subsequently, it needed more power. In 1967, in response to the need for more power, Garrett had started design of the TPE331-3.

To increase the power in the TPE331-3, Garrett initially considered raising the turbine inlet temperature. Concerned about the risk of going to significantly higher temperatures without adequate turbine heat transfer test data, Chief of Aerodynamics Bob Bullock proposed increasing the airflow as a less risky alternative to raise the power output.[86] Using the principles of transonic axial-flow compressor technology that he had demonstrated while at NACA's Lewis Research Center, Bullock and his engineering team were able to push the inlet airflow relative Mach number up to 1.3. To minimize shock losses within the compressor, the airfoils were designed to constrain supersonic over-velocities on the suction surface of the inducer.

This new "high-flow" design raised the pressure ratio as well as the airflow. For Garrett, this solution bought time until a high-temperature research program and test facilities could be developed. In recognition for his earlier part in the development of transonic compressor technology while working at NACA, Bullock and two of his colleagues received the prestigious Goddard Award in 1967.

The TPE331-3 was certificated in March 1969 at 840 shp and installed on the Metro, certificated in June of 1970. Eventually, the Metro would become a successful commuter aircraft. In contrast to other early turbine-powered commuter aircraft, such as the unpressurized and relatively slow Twin Otter, the Metro was a pressurized, high-speed, high-performance aircraft. The advantage of such an aircraft was that it could climb to altitude quickly where it would burn less fuel and cruise faster; it could then let down fast. With longer routes and mountainous terrains in the western U.S., such aircraft were especially important. In hub-and-spoke operations, commuters often operated in a narrow time window between the main airport and remote locations. In such cases, high performance and high speed were important factors. Another important aspect was the cost savings that resulted from burning less fuel at altitude.

For Garrett, the importance of getting its engine on the Metro was the entree that it provided to the commuter market. The TPE331-3 was also used on the Merlin III and IIIa, and the Century Aviation Jetstream (Handley Page C-10A). The Jetstream was later put into production in the 1980s as the BAe Jetstream 31, and like the Metro, it would become an important commuter aircraft.

One off-shoot conversion of the TPE331-3 was the TSE331-3U, a turboshaft engine. Minor modifications were made in such areas as the torque sensing system and fuel control. The engine was rated at 800 shp. The engine was used by Aviation Specialities, Inc., to convert the Sikorsky S-55 helicopter to turbine power. Removing the S-55's heavy radial engine resulted in a weight savings of about 900 lb. The S-55T was certificated for production in 1971. Due to a limited market, only 48 TSE331-3U engines were sold.

---

86. Bob Bullock, interview by Rick Leyes, 19 December 1990, Garrett Interviews, tape recorded telephone conversation, National Air and Space Museum Archives, Washington, D.C.

Other airframers saw the advantage of the 840-shp Garrett engine and wanted the same power at cruise. The TPE331-5 and -6 followed, both identical to the TPE331-3, but flat rated. These designs started in 1969 and were first run and certificated in 1970. Both engines had an 840-shp thermodynamic rating for better high-altitude, hot-day performance, and were flat rated by lower-speed gearing to 776 shp (TPE331-5) and 715 shp (TPE331-6) at takeoff. With full rated power at cruise, the airframers had faster flying airplanes. The TPE331-5 was used on aircraft such as the Rockwell Aero Commander Turbo Commander 690 (later Gulfstream Commander 690), CASA 212-100, Mitsubishi MU-2N, and Dornier 228. The TPE331-6 was used on aircraft such as the Mitsubishi MU-2S/K/L/M and Beech B-100.

There were problems with the 840 Series engines, however. The engine had been stretched to the limit. Getting the engine through performance acceptance testing was made difficult due to the lack of margin. Subsequently, a lot of tolerance tightening and hand polishing of parts was required to get the engines through the tests. Complicating the problem was the relatively new process of using integrally cast blades on turbine wheels. On longer blades, every blade had to be measured and bent or straightened to achieve the areas and blade settings desired.

That every bit of power had been squeezed out of the TPE331-3 for its installation on the original Metro was a problem that followed the engine into service on that aircraft. With no performance margin, the installation losses and normal engine performance degradation due to use resulted in significant power loss. Pilot reports of the inability to obtain flight manual take-off power led to an FAA investigation in 1980. As a result of this investigation, a revision to the flight manual allowing reduced power takeoffs was issued in June 1982. This problem was eventually solved by giving operators an opportunity to upgrade their Metro engines with a more powerful engine in the Garrett TPE331 series.[87] This upgraded engine was a derivative of an improved T76 engine.

T76-G-420/-421 AND TPE331-8/-9 AND -10/-11

In response to a need for more power for the North American OV-10A Bronco, Garrett boosted the power of the 715-shp T76 to 1,040 shp. The new

---

87. The 840-shp TPE331-3U-303 engine was installed on the original Swearingen Model SA-226TC Metro. The TPE331-10UA-511G was later installed on the Fairchild Metro II-10. The improved engine had a 1,000-shp thermodynamic rating and a 840-shp gearbox rating. This gave the aircraft considerable power margin. It also improved the potential for less unscheduled maintenance because the engine was not being pushed to its limit.

model was designated the T76-G-420/-421 and was installed the Rockwell International OV-10D. The design was initiated in 1974, and the first engine run was in 1975. It passed its MQT in 1977.

This 45% increase in power was accomplished within the same engine frame size. The compressor from the 840-shp civil engine was used, although some changes were made. The main emphasis of the redesign was to increase the turbine efficiency and turbine inlet temperature.[88] To withstand the increased temperature, the first stage of turbine stator vanes and rotor blades were air cooled. The first-stage stator incorporated segmented vanes with trailing edge discharge slots for cooling air. Instead of the uncooled, integrally cast turbine rotors previously used in past TPE331 engines, the first stage had air-cooled cast blades with a fir tree attachment to a forged disk. To avoid increasing the engine diameter and to prevent diametral interference from the larger turbine, the combustor was shortened. The shorter combustor reduced the overall length of the engine.

When combined with existing commercial engine gearboxes, new civil engines resulted. They were the TPE331-8/-9 and -10/-11. The design for the the TPE331-8 was initiated in 1974, and both the -8 and -9 were certificated in 1976. The TPE331-8 was flat rated at 865 shp, and it powered the Cessna Conquest 441, a twin-turboprop executive transport. The TPE331-8 retained the -3 compressor, but used a different turbine to handle more efficiently the higher compressor flow at altitude cruise conditions. Although the TPE331-8/-9 were derived from the 1,040-shp T76 engine, they operated at a lower temperature. Therefore, they used an uncooled first-stage turbine. Despite the substantial power increase over the TPE331-3, -5, and -6, the TPE331-8/-9 were more fuel efficient engines. For example, at an altitude of 30,000 ft and a 300 KTAS (Knots True Air Speed), the -8/-9 had a 12.5% improvement in cruise fuel consumption.

The TPE331-10/-11 were direct civilian versions of the the T76-G-420/-421. The design of both engines was initiated in 1975. Certification for the -10 was received in 1978 and in 1979 for the -11. The TPE331-10 was rated at 1,000 shp, and the -11, which had a higher gearbox limit, was rated at 1,100 shp. By 1990, 31 different versions of the TPE331-10/-11 powered a variety of aircraft including the Gulfstream Commander (980 and 1000), CASA212-300, BAe Scottish Jetstream 31, Mitsubishi Solitaire and Marquise, Fairchild Merlin (IIIB and IIIC) and Metro (II, III and IV).

During the late 1960s and 1970s, the TPE331 had grown considerably in power without an increase in frame size. Much of this power was attributable to

---

88. Frignac and Privoznik, "The Growth and Evolution of the TPE331," 6.

a higher-flow compressor and a cooled first-stage turbine. Most of the sales were for executive aircraft, but a transition to commuter applications would soon begin.

THE TPE331 IN THE 1980S: COMMUTER AIRCRAFT ENGINES (TPE331-14/15, TPE331-12, AND TPE331-14GR/HR)

Until the early 1980s, the majority of TPE331 engines had been sold for use on executive aircraft. From 1965 (the first year of TPE331 production) through 1981, 1,738 engines had been shipped for installation on regional airlines and 5,562 for executive aircraft. The first significant surge of TPE331 engines shipped to regional airlines began in the late 1960s in response to the first boom in the commuter aircraft industry. Again, in the late 1970s, as a result of the growth in the general aviation economy and the 1978 Airline Deregulation Act, the number of engines shipped to regional airlines increased again.

After reaching a record number of sales (731 engines shipped) in 1981, the number of TPE331 engines for executive aircraft declined significantly in 1982 (250 shipments) and continued to drop through the 1980s. This reflected a downturn in the national economy in 1982 as well as a decade of declining general aviation aircraft sales. In 1982, TPE331 regional airline airline sales (253) first passed TPE331 executive aircraft sales and continued to remain strong in the 1980s.

In the late 1970s, as an increasing number of hours were put on the TPE331 as a regional aircraft engine, it became evident to Garrett that technical changes would have to be made. According to Chief Engineer of Turboprop Engines, R. D. Miller:

> When we basically started the turboprop business, it (the TPE331) was designed for executive aircraft . . . Then we entered the commuter marketplace. That's where all of a sudden we found out what you really have to do to make an engine. The commuter marketplace is a much different animal than the executive. Whereas an executive puts 300 hours a year on, maybe, a commuter will put up to 2,000 hours a year on, and the engine has to be much more rugged, more durable, can't break as often . . . When the commuter quits, 19 people suddenly have to be rerouted, and they're upset and unhappy. So we learned an awful lot about how you need to ruggedize and improve the reliability of the turboprop engine.[89]

89. R. D. (Dee) Miller, interview by Rick Leyes and Bill Fleming, 6 March 1990, Garrett Interviews, transcript, National Air and Space Museum Archives, Washington, D.C.

For example, the TPE331-10 engine had a carbon erosion problem with the combustor. While not a significant problem for an executive aircraft running several hundred hours per year, commuter operators which were flying several thousand hours each year found their expensive turbine stators and rotors eroding. Undertaking a major program, Garrett engineers were able to develop a non-carbon-generating combustor which could be retrofitted to existing engines.

Another important upgrade focused on the gearbox. Wider gears as well as heavier and higher capacity bearings helped to make the TPE331-10 and -11 more rugged for commuter operation. It was discovered that it was really necessary to overdesign in the gearbox area, particularly in non-flat-rated engines where the power was being pushed up to the design limits of the gearbox.

In 1979, the design was started for the TPE331-14/-15. These engines were originally sized for 1,645 thermodynamic shp, approximately 60% more power than their immediate predecessors. In order to achieve this, the engine was scaled up 20%. This was the first time that the basic TPE331 frame had been enlarged. The TPE331-14 was designed specifically as a commuter engine. It was also the first TPE331 engine designed to be truly modular, a response to an airline need for improved maintainability. The first run of the engine was in 1981, and it was certificated in 1984 at 1,250 shp. The TPE331-15 was also first run in 1981. It was updated in 1987 and certificated in November 1988 at 1,645 shp.

Unfortunately, the TPE331-14 lost an important competition in 1980 to the GE CT7 on the Saab 340A, a 35-passenger commuter aircraft. Following the 1978 Airline Deregulation Act and the growth in the commuter market, new regional aircraft seating 19 to 50 passengers were developed. The TPE331-14 was really too small for such applications and was bypassed in an important market.

Competing engines such as the P&WC PW100 and the GE CT7 were new generation designs that had the higher power ratings required by commuter aircraft that were growing in size. They could be flat rated over a wider range to suit a variety of power needs and aircraft growth. In contrast, it is ironic to note that the design philosphy of the TPE331 was traditionally that of a compact, lightweight, high-performance engine. Such a tradition was fine as long the engine had the power to meet what the market demanded, but in this case it was a handicap against engines with more room for growth. An alternative application was sought and the first found for the TPE331-14 was the Piper Cheyenne 400 LS, an eight-seat business aircraft.

The last new TPE331 engine model developed in the 1980s was the TPE331-12, derived from the TPE331-10. This engine was originally designed for the Gulfstream Commander 1200, but the aircraft did not materialize. The turbine was based on the TPE331-10/-11, and its compressor was scaled down from the TPE331-14. It was certificated in December 1984 at 1,070 shp (flat-rated). The bulk of the shipments went to power the BAe Jetstream Super 31 and the Metro V regional airliners. The TPE331-12B version was selected for the Shorts Tucano trainer aircraft in the late 1980s, and in 1996, was selected by CASA for its new version of the 212 regional airliner, the 212-400.

The TPE331-14GR/HR was a very successful, more powerful derivative of the TPE331-14. Rated at 1,759 shp, this engine had a higher rotating speed, higher temperature operation, improved materials, and an improved stator. It had an automatic power reserve rating of 1,960 shp thermodynamic. The engine was chosen for the BAe Jetstream 41, a 24-27-passenger airliner in April 1989. Certification was achieved in mid-1992, and deliveries to BAe began in 1993.

In 1992, the TPE331-14GR/HR was also selected for a new version of the Russian Antonov AN-38, 27-passenger transport/utility aircraft, when the division announced a cooperative agreement with Antonov Design Bureau of Kiev, Russia and the Novosibirsk Aircraft Production Associations to provide the entire propulsion package. The initial flight took place in June 1994. A $4.6 million contract for first production deliveries was signed in the summer of 1996.

By the end of the 1980s, the number of new production TPE331 engines destined for the commuter market exceeded those for executive aircraft. In 1988 and 1989, no executive aircraft engines were shipped, a fact which paralleled the dismal state of general aviation aircraft sales industry-wide. Concurrent with this market shift, the TPE331 models developed during this decade were specifically designed for commuter aircraft.

Despite its setback in the larger regional aircraft market in the 1980s, Garrett did well in the 19-passenger commuter aircraft market. By 1990, Garrett claimed to have 70% of this market, in which the primary contenders were the Beech 1900, BAe Jetstream 31 and Super 31, and Fairchild Metro series, which were top sellers in that category of aircraft.[90]

The 19-passenger commuter aircraft market was an important counterbalance to the absence of executive aircraft sales in the 1980s and the consequent decline in the total number of TPE331 engine sales. The high-usage commuter aircraft engines accounted for a strong aftermarket in parts and maintenance that more than compensated for the loss of revenue on initial shipments.

Although TPE331 sales during the 1980s were primarily for commuter aircraft, the widest application of the engine during its lifetime had been in execu-

---

90. Miller to Leyes and Fleming, 6 March 1990.

Garrett TPE331-powered British Aerospace BAe Jetstream 31 commuter aircraft used by Piedmont. Photograph courtesy of Garrett.

tive aircraft. By 1996, more than 12,700 TPE engines were shipped. About 69% were for executive aircraft compared to about 20% for regional airlines. During this time the military aircraft market accounted for just about 9% of shipments.

The TPE331 had an important impact on Garrett, and it made a significant contribution to the aircraft industry. It established Garrett as a prime propulsion company. It was one of two gas turbine engines (along with the P&WC PT6) that altered the performance of general aviation executive and commuter aircraft, greatly increasing speed, rate of climb, ceiling and payload.

NO ENGINE IS PERFECT

Like all engines, the TPE331 had its share of development problems and growing pains. Because such problems were spread out over time, it is useful to group the most significant problem areas without regard to specific models. The company also had problems as a result of its major maintenance policy on early models of the TPE331.

As a result of field adjustment problems encountered in APUs, Garrett decided to locate the three-speed start switch inside the TPE331 gearbox. The switch was preset at the factory so that it could not be improperly adjusted in the field. However, once in service, this "Murphy-proof" fix proved to be a nuisance. If the switch drifted or malfunctioned, the engine had to be taken apart in order to access and reset the speed switch. Later, the switch was made an electronic device and mounted outside the engine.

An early development problem was lean burner blowout during cold starts. The ten Simplex fuel atomizers in the first TPE331 engine exhibited start difficulties. The problem was remedied by adding five radial atomizers for starting and normal operation, while the ten axial atomizers were actuated after engine start.

A propeller reduction gear problem was traced to the high-speed pinion bearing which did not have adequate durability. The problem was solved with extremely precise machined and matched gear and bearing sets. Also, the torque sensor reliability was another problem which was improved by going to closer tolerances and by matching metering piston sets. A single-shaft engine required a more complex fuel schedule, and a closer synchronization with the propeller governing system. The entire propulsion system operated at a single speed, due to the coupling of the propeller, compressor, gas generator turbine, and power turbine. To avoid compressor surge, provide timely starting, and system acceleration and deceleration, fuel scheduling had to be precise.

In addition, the interaction between the propeller speed operation and the gas generator governor (all coupled together by the single shaft) was critical to prevent interference and engine instability. This presented a severe challenge both to Garrett and the control supplier, Woodward Governor Company. Successful resolution was achieved using a complex hydromechanical fuel control employing 2-D and 3-D cams for accurate scheduling, as well a fuel underspeed and overspeed governor phased with a conventional propeller governor.

A relatively high number of FAA Airworthiness Directives were issued with respect to the TPE331. Typically, such directives were requested by Garrett in order to correct undesirable conditions. Analyzing the directives, Garrett grouped the problems into four major areas: low-cycle fatigue of components in the turbine section, due primarily to start-stop thermal stresses; high-cycle fatigue of static components throughout the engine due to vibration; fretting, wear, and/or loosening of components and fasteners resulting from vibration; and wear and eventual failure of couplings or splines due to lack of adequate or proper lubrication.

Low-cycle fatigue affected turbine rotors as well as hot-section static components such as turbine nozzles and seals. Originally, such components were life-limited on the basis of operating hours, but in the late 1970s, the limiting factor was changed to the number of cycles when it was discovered that this method was more accurate. However, further experience showed that the number chosen for cycle life exceeded the actual component life, and the number of cycles had to be readjusted. Ultimately, the turbine rotors were redesigned to prevent premature cracking and failure. Garrett worked closely with the FAA to develop inspection methods allowing in-service engines to continue operating until new turbine wheels were available.

High-cycle fatigue of static components, fretting, wear, and/or loosening of fasteners were successfully resolved by redesigning each affected component so that it would tolerate its specific operating conditions better. Wear and eventual failure of couplings or splines, especially in the fuel control drive, were the result of improper servicing or the lack of proper alignment. A new oil-lubricated spline for this drive all but eliminated the wear problems.

Another maintenance related problem stemmed from Garrett's original TPE331 policy requiring that all major maintenance be done at Garrett for engines in the U.S. This policy enabled Garrett to monitor what was happening to its engines in the field and to have better quality control over what maintenance was being performed. Garrett was concerned that the TPE331's early reputation could be damaged by sub-standard maintenance.

This policy, while advantageous to Garrett, began to create customer relations problems. Many operators apparently felt that they were captive to the engine manufacturer and resented it.[91] The need for an alternative policy began to become more acute as the engine was increasingly used for commuter aircraft. Unlike executive aircraft, commuter aircraft needed rapid maintenance turnarounds. Commuter aircraft operators, many with limited capability for doing their own maintenance, wanted service centers more conveniently located. In the 1980s, responding to this need, Garrett started to establish service centers across the U.S.

The TPE331 continued to improve in both reliability and serviceability. Initial TBO's of 2,000 hrs grew to 6,900 hrs on a fixed time basis, and to over 9,000 hrs on a continued airworthiness program on later models. Reliability improved from .13/1,000 hrs Mean Time Between Failures (MTBF) and .36/1,000 hrs In-Flight Shut Down (IFSD) in 1980 to .03/1,000 hrs MTBF and .10/1,000 hrs IFSD in 1996. TPE331 engines had acculmulated more than 66 million operating hours by 1996.

TPE331 SUMMARY

Between 1965 and 1996, 12,774 TPE331 engines were shipped, powering more than 70 airframe applications, including many variations of basic aircraft models. The number of engines delivered, the number of applications, and the length of

91. Robert L. Parrish, "The Ultimate Power Struggle Part I: The Garrett TPE331," *Business & Commercial Aviation*, (September 1985): 134.

production put the TPE331 in an exclusive club. The TPE331, along with the Pratt & Whitney Canada PT6, was one of the two most important turbine engines used for fixed-wing general aviation aircraft in North America.

The TPE331 was the engine that put Garrett into the prime propulsion business. The TPE331 was principally important as an executive aircraft engine and as a 19-passenger regional aircraft engine. The TPE331 had only limited military service. However, the selection of the YT76 for the COIN aircraft resulted in an important contract for the T76/TPE331, which enabled it to become a production engine.

From a technical aspect, the engine performance improved considerably from the first model. The TPE331 grew from 575 shp (Series I and II) to a flat-rated 1,070 shp (TPE331-12) in the same frame size. The scaled-up TPE331-14 and -15 versions achieved 1,645 shp, and the derivative TPE331-14HR/GR output grew to 1,759 shp. In the TPE331-1 to -12 series, the shaft horsepower almost doubled, the specific fuel consumption improved by about 18%, and the power-to-weight ratio improved by about 58%. The improvements attained in the same frame size were due primarily to increased airflow and pressure ratio in the two-stage centrifugal compressor and to increased turbine inlet temperature.[92]

The hallmark component of the TPE331, which was used in many of Garrett's other engines, was the two-stage centrifugal compressor. The transonic inducer and the highly backward curved blades were characteristics of most TPE331 compressors. Such technology contributed to the improved compressor airflow and pressure ratio while maintaining the same or better efficiencies and acceptable surge margins.[93] Another hallmark of the TPE331 was its high-power density (power per installed engine volume).[94] In fact, the TPE331 was noteworthy for its short and compact configuration, low installed weight and drag, good specific fuel consumption, aerodynamic efficiency, and performance.

The TPE331 was one of the key engines that significantly improved the performance and the utility of general aviation executive and commuter aircraft. In the case of early piston engine aircraft conversions, significant gains were made

---

92. Montgomerie C. Steele, "Some Issues in the Growth of Small Gas Turbine Aircraft Propulsion Engines," (AIAA Paper No. 21-7354 presented at the International Symposium on Air Breathing Engines, 9th, Athens, Greece, 3-8 September, 1989), 3.

93. For example, the airflow on the TPE331-25/61 and -25/71 engines (redesignation of TPE331 Series I and II) was 5.78 lb/sec and the pressure ratio was 8.0. With the scaled-up compressor on the TPE331 840 series, the airflow on the TPE331-6-251 increased to 7.75 lb/sec, and the pressure ratio increased to 10.37.

94. Don G. Caldwell, interview by Bill Fleming, 7 March 1990, Garrett Interviews, transcript, National Air and Space Museum Archives, Washington, D.C.

in aircraft performance due to lower installed weight, less frontal area (and consequently less parasitic drag), and more power.

Turboshaft Developments in the 1960s and 1970s

Like other small turbine engine companies, Garrett wanted to build engines for helicopters as well as for fixed-wing aircraft. As discussed earlier, the first use of a Garrett engine as a prime mover was in 1957 when three GTC85s APUs powered a McDonnell Aircraft Model 120 "pressure jet" helicopter. In October 1961, a TSE331-7 engine was flown in a Republic Lark. In August 1962, a TSE331 was shipped for testing in a Piasecki Aircraft Airjeep I. However, it was not until the early 1970s that Garrett found a limited production application for its TSE331-3U engine in the Sikorsky S-55.

During the late 1960s and early 1970s, Garrett took two further steps to penetrate the helicopter market. One was the development of the TSE36-1 which became an FAA type certificated engine. It was a designed as a low cost replacement for 150-250-shp helicopter piston engines. The other, the TSE231-1, was a new free turbine design intended for use in a high-performance executive helicopter.

TSE36-1

In mid-1968, Garrett announced its entry into the light helicopter engine market with the start of the TSE36-1 engine program.[95] The TSE36-1 was a direct descendant of the model GTCP36-4 APU which had been developed in 1963 for use as an APU on commercial and executive transport aircraft. Approximately one year before the announcement, flight tests were conducted on a Hughes 269A helicopter powered by a modified GTCP36-17 to which a reduction gearbox and a modified fuel control were added.

The performance of the helicopter was considerably improved. The installed weight of the 220-shp flight test engine was 184 lb less than the 180-shp piston engine that it replaced. The higher power, lower weight turbine engine lifted the helicopter hover ceiling from 6,000 ft to over 14,000 ft. The useful load, true airspeed, and rate of climb were improved considerably as well. While the fuel

95. Jere G. Castor, "The Garrett TSE36-1 Propulsion Engine: Design and Operational Considerations," (SAE Paper No. 690310 presented at the National Business Aircraft Meeting, Wichita, Kansas, 26-28 March, 1969), 1.

consumption was approximately double that of the piston engine, the weight advantage of the turbine made it possible to carry enough additional fuel to exceed the range of the piston-powered helicopter.

The production engine definition was completed in May 1968. The program objectives were to provide a simple, low cost (initial cost and cost of ownership), highly reliable, and easily maintainable 240-shp turboshaft engine.[96] Low cost was predicated on a large APU production base and a simple engine design. Company expenditure was expected to be minimal because the TSE36 was based on an existing APU for which the major design and tooling were already in place. The engine was FAA certificated in April 1970.

The TSE36-1 was a single-shaft turboshaft engine with integral wet-sump gearbox. It consisted of a single-stage centrifugal compressor, single-can reverse-flow combustor with torus transition chamber, single-stage radial-inflow turbine, and flanged exhaust duct. It was rated at 240 shp with a 0.83 lb/hr/lb sfc. The power-to-weight ratio was 1.35:1.

The TSE36-1 was originally designed for the Enstrom T-28 two/three-seat general aviation helicopter. This was to be a turbine version of the Enstrom F-28. Garrett had been approached to supply a turbine engine for the helicopter. The prototype Enstrom T-28 flew in May 1968 with a TSE36-1 engine. However, the Enstrom T-28 project died when Enstrom operations were shut down in February 1970. Efforts to market the engine to Hughes were not successful. When its intended application did not materialize and other marketing efforts did not bear fruit, Garrett shelved the program in 1970.

TSE231-1

The TSE231-1 was designed for use in the Gates Learjet Corporation's Twinjet helicopter. The Twinjet was to be a high-performance executive helicopter with production scheduled for 1972. Engines available at the time, such as the Allison Model 250, did not have the power required for this application. Lear's Aircraft Division Executive Vice-President and General Manager Malcolm Harned worked with Garrett Group Vice President Ivan Speer to define an engine for the Twinjet.

The result was a new turbine engine design for Garrett. It was also the company's first free turbine turboshaft engine. The engine had a two-stage centrifugal compressor and single-stage axial-flow gas generator turbine. The compressor was a high pressure ratio, scaled-down version of the T76-G-10 compressor.

96. Castor, "The Garrett TSE36-1 Propulsion Engine," 4.

The engine had an annular reverse-flow combustor. A front-end drive was linked through a two-stage reduction gear to a single-stage axial-flow power turbine. It had a sea-level rated power of 474 shp, a 0.605 lb/hr/lb sfc, and a 2.77:1 power-to-weight ratio.

The TSE231 was designed to be a direct competitor to the Allison Model 250 turboshaft engine. It was laid out to fit in Allison-powered airframes. Designed to sit on the same mounts, the engine's shaft centerline would correspond to that of the Model 250. The objective was to have an engine of approximately the same volume but twice the power.

The development of the TSE231-1 turboshaft engine began in September 1969. The first full engine run was in September 1970, and the first run in a Bell JetRanger testbed helicopter was in December 1970. With the onset of a general aviation recession in the early 1970s, Lear ran into financial problems and had to make a decision whether to continue with the Twinjet helicopter development or to pursue a parallel development on its Learjet Model 35. The company decided to drop the helicopter in favor of the business jet. Production orders which had been placed by Lear for the TSE231 were canceled.

Garrett continued its efforts on the TSE231-1 and tried to sell it to Bell Helicopter Company because Bell was the place to sell helicopter engines at the time. However, Garrett could not penetrate this important market. Discussing their efforts with Bell, Ivan Speer recalled:

> I found out later they were really using us as a way to get Allison to fix their troubles (with the Model 250). But they were seriously thinking about making an engine change if they didn't get a better deal out of Allison. So we did do a pretty extensive flight test with the JetRanger . . . It would have been a fine power plant, but they obviously had a lot of money invested in various models around the Allison engine, and then Allison did improve their engine and got out of the woods. So we just faded away. Besides that, we probably had about all the business we could handle, anyway, with the turboprop at that time.[97]

Two additional attempts were made to find a home for the TSE231. Garrett tried to interest the Army in using the TSE231 as a replacement engine for the Allison T63 (military designation for the Model 250 engine)-powered light observation helicopters used in the Vietnam War. Though encouraged by the Army to continue development, no funds were forthcoming. The TSE231 received a short reprieve when it was selected and successfully demonstrated in the Marine Corps's Small Tactical Aerial Mobility Platform (STAMP). How-

97. Speer to Leyes and Fleming, 8 March 1990.

ever, this program was concluded and no further applications were found for the TSE231.

Despite the fact that no production applications were found for the TSE231, the program led to very important technical developments that influenced future Garrett turbine engine design. To drive the high pressure ratio compressor, a high-work single-stage axial-flow gas turbine was used. Bill Waterman, Chief of Aero-thermo Component Design, summed up the important technical lesson learned:

> We learned a lot from the TSE231 program, particularly in the turbine area, with regard to the design of high expansion ratio nozzles. We employed the end-wall contouring that we had learned from the TPE331/T76 program, but we missed the flow capacity of that turbine nozzle by a substantial amount. We were assigning the losses through the nozzle incorrectly, and had been doing so all along. The 231, because of the large pressure drop across the nozzle, finally opened our eyes, and we altered the method by which we assigned the loss distribution. Subsequently, on all our turbines, we have never missed the flow by more than a percent. The TSE231 gas generator turbine was the first at Garrett that really utilized what we call non-free vortex design, which permits setting the blade angles from hub to tip to control flow distribution. This technology was certainly responsible for why the turbine, which had nearly doubled the work capacity of previous Garrett axial designs, was reasonably successful.[98]

Garrett's first successful helicopter engine development contract did not come until the late 1980s.[99] While the TSE36 and TSE231 were an interesting sidebar to Garrett's turbine engine development, the main focus of the company at that time was on fixed-wing aircraft engines with continued improvement of the TPE331 and development of a new engine, the ATF3 turbofan.

The ATF3

In the 1960s, Garrett President and Chief Operating Officer Harry Wetzel had a strong interest in the fan engine field.[100] Wetzel saw an opportunity to re-engine

98. Bill Waterman, interview by Bill Fleming, 7 March 1990, Garrett Interviews, transcript, National Air and Space Museum Archives, Washington, D.C.

99. In 1988, Garrett in a joint venture with Allison, known as LHTEC, won an important helicopter engine development contract. This was the T800 engine program.

100. Wetzel to Leyes, 9 April 1990.

first generation business jets. Aircraft such as the North American Sabreliner and Lockheed JetStar were both powered by the Pratt & Whitney JT12D and the Learjet by the GE CJ610. These turbojet engines had a high fuel consumption, and consequently, the aircraft had limited range.

When Garrett's Director of Product Planning, Anthony DuPont, proposed a 4,000-lb-thrust fan engine, Wetzel was ready to move on the idea. In 1966, the ATF3-1 (AiResearch Turbofan No. 3) design effort began at Garrett's AiResearch Manufacturing Division in Torrance, Cal., where DuPont was located. Accordingly, the ATF3 was set up as a project that proceeded separate and apart from Garrett's gas turbine engineering team in Phoenix, and having little or no engineering interface with the Phoenix group.

THE ATF3 DESIGN RATIONALE AND CONFIGURATION

The performance objective of the ATF3 was to decrease cruise thrust specific fuel consumption (tsfc) by 30 to 40% compared to small engines then in production.[101] Through its low specific fuel consumption (approximately 0.40 at takeoff), the engine was intended to give transcontinental U.S. range to 10/12-seat business aircraft.[102] Garrett also planned the engine for application to VTOL and STOL aircraft in the 1970s. In particular, it was intended that by contributing to a high aircraft thrust/weight ratio, the new turbofan could be used to re-engine smaller business jets, providing them with STOL capability and increased range.[103]

Because it was a medium-bypass ratio turbofan design, the engine had inherently better sfc than the existing lower-bypass turbofan and turbojet engines of the time. Important contributing factors to the sfc were the engine's three-spool design, a high compressor pressure ratio, and a high turbine inlet temperature.

The ATF3 was a three-shaft axial-flow turbofan engine having a bypass ratio of 3.19:1. It was said to be the first three-shaft design in the United States.[104] A single-stage fan was originally driven by a two-stage axial-flow intermediate

101. Robert R. Van Nimwegen, "Design Features of the Garrett ATF3 Turbofan Engine," (SAE Paper No. 710776 presented at the National Aeronautic and Space Engineering and Manufacturing Meeting, Los Angeles, Cal., 28-30 September 1971), 1.

102. *Jane's All the World's Aircraft 1969-1970*, ed. John W. R. Taylor (London: Haymarket Publishing Group, 1970), 721.

103. *Jane's All the World's Aircraft 1970-1971*, ed. John W. R. Taylor (London: Haymarket Publish Group, 1971), 758.

104. Nimwegen, "Design Features of the Garrett ATF3 Turbofan Engine," 11.

Garrett ATF3 turbofan engine. Photograph courtesy of Garrett.

pressure (IP) turbine located between the high pressure (HP) and low pressure (LP) compressor turbines.[105] A five-stage axial-flow LP compressor was driven by a two-stage axial turbine. The fan and LP compressor shafts were concentric. A single-stage HP centrifugal compressor was driven by a single-stage axial turbine. The HP compressor and turbine assembly was opposed to the other two spools and was located at the rear of the engine.

The engine had a complicated airflow path. A portion of the air that passed through the fan entered the LP compressor. After leaving the LP compressor, the airflow was split among eight ducts within which it flowed to the rear of the engine. The air was then turned 180 degrees to enter the HP compressor. From the HP compressor, the air entered a reverse-flow annular combustor. The hot gases continued forward sequentially through the HP compressor turbine, the IP fan turbines, and the LP compressor turbines. The LP turbine discharge air was bifurcated into eight streams and turned 180 degrees through a cascade of vanes into the fan discharge stream completing the cycle.

The unusual order of flowing gases through the HP compressor turbine, the IP fan turbine, and then the LP compressor turbine tended to give a better cycle and turbine match at low power than the more conventional order (i.e. the first turbine drives the HP compressor, the second drives the LP compressor, and the third drives the fan). This arrangement resulted in a better partial power sfc.

---

105. A third stage fan turbine was added in the ATF3-3 (military designation YF104-GA-100).

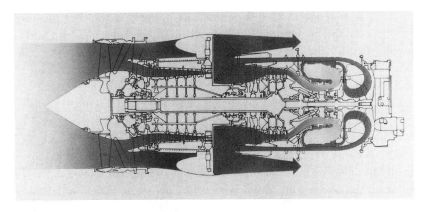

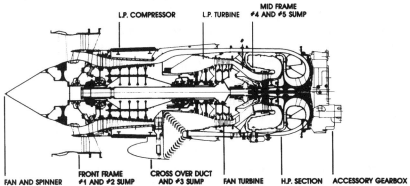

Cross sections showing complicated airflow through Garrett ATF3 turbofan engine. Photographs courtesy of Garrett.

From a practical viewpoint, it would have been difficult at the time to make a small engine with three concentric shafts. The inner shaft would have been very thin, leading to serious vibration problems. The rotor arrangement in the ATF3 was considered mechanically simpler than conventional designs. Each spool was straddle-mounted on two bearings. From a maintenance point of view, putting the high-temperature, high pressure component at the back of the engine made it more accessible. The opposed-shaft configuration also contributed to a lower engine frontal area and thus lower drag, gaining the thermodynamic advantage of a three-spool engine without the mechanical disadvantage of three concentric shafts. The design also took advantage of Garrett's experience with high-speed centrifugal compressors and reverse-flow combustors.[106]

106. Nimwegen, "Design Features of the Garrett ATF3 Turbofan Engine," 3.

The ATF3 was one of the first small engines to capitalize on the benefits of the "turbofan thermodynamic cycle" already in use on large engines. Gains in overall efficiency and fuel economy were achieved by optimizing bypass ratio, compresssor pressure ratio, and turbine inlet temperature. At takeoff, the bypass ratio was 3:19. Contributing greatly to its fuel efficiency was its high pressure ratio. The overall pressure ratio of 25:1 was unusually high for a small engine.[107] By comparison the Rolls-Royce RB211, believed to be the only other working three spool engine at the time, had an overall compressor ratio of 16:1. The Pratt & Whitney JT9D had a 24:1 compression ratio. Also contributing significantly to fuel economy was a high turbine inlet temperature of 2,200°F. To withstand this temperature, the first-stage turbine nozzle vanes and HP rotor blades were cooled. This was Garrett's first cooled turbine of very small size, a step forward for the company.

The ATF3 had other advantages. To reduce maintenance, it was designed so that major sub-assemblies could be removed as modular components. The engine was inherently quiet. The absence of fan inlet guide vanes, the mixed-flow exhaust, and the internal airflow reversals reduced noise. The mixed-flow exhaust reduced infared signatures for application in various "black programs." The engine also had low emissions.

However, there were disadvantages inherent in the design. One critical disadvantage was that the forward flow through the IP fan and LP compressor turbines produced a thrust in the same direction as the fan and LP compressor thrust. The result was heavy thrust loads on the bearings and high air leakage through the seals. These pressure forces were additive on the ATF3 and an excessive forward thrust was thus generated. In order to live with this situation, pressure differences were artifically created across pneumatic pistons attached to the shaft. The diameter of these rotating pistons had to be large, and very low seal clearance was necessary to prevent the waste of high pressure air through the seals. Excessive leakage around the seals could not be maintained without expensive manufacturing and inspection techniques. This was a major source of the cost and performance problems.

Another disadvantage was increased cost, weight, and complexity of the crossover ductwork. Also, with four airflow reversals (including the reverse-flow combustor) and extended ducting from the LP to HP compressor, pressure losses were relatively high.

107. C. M. Plattner, "Garrett Tests Corporate Turbofan Engine," *Aviation Week & Space Technology*, Vol. 89, No. 15 (October 7, 1968): 77.

Despite the ATF3's technical promise, the disadvantages inherent in the design plagued the development program. Eventually, they contributed significantly to the engine's limited application.

DEVELOPING, MARKETING, AND CERTIFICATING THE ATF3

As designer of the engine, DuPont was given his own organization in Los Angeles to design and test the engine. A $15 million dollar development testing facility was built there expressly for the ATF3 program. The design proceeded as a confidential project separate from Garrett's engine program and engineering team in Phoenix. As work on the engine proceeded, the uniqueness of its design dealt the first serious setback to the program. When the prototype engine was tested for the first time, it would not self-sustain. The problem proved difficult to find and to solve. An engineering team from Phoenix was brought in to assist.

To resist the combined forward LP compressor and turbine thrust and the IP fan and turbine thrust on the bearings, balance seals were installed. At the outer periphery of the engine frame, rotating disks (seals) were installed and high pressure air was injected on one side and low pressure on the other side. However, because of the large outside diameter of the rotating disks, it was difficult to maintain the required close tolerance in the manufacturing and alignment process. Leakage of compressor air through the disks kept the engine from self sustaining. The problem proved difficult to solve and delayed development of the engine.

Development testing of the ATF3 began in May 1968. In October 1968, North American Rockwell awarded Garrett a $59 million contract to supply ATF3s, nacelles, and reversers to replace the Sabreliner Series 60 Pratt & Whitney JT12A-8 turbojet engines. The ATF3 would have given the Sabreliner nonstop transcontinental range. Deliveries were planned for 1970. However, as a result of delays in the development program due to its technical problems, Garrett was not able to certificate and deliver the ATF3 when North American needed it. In 1971, North American canceled a 300-engine order. For contract nonperformance, North American brought suit against Garrett.[108] An out-of-court settlement was made. Despite this problem, Garrett's long-term relations with Rockwell were not damaged, and Rockwell continued to use Garrett's TPE331 engine on executive and military aircraft.[109]

108. Schoneberger and Scholl, *Out of Thin Air: Garrett's First 50 Years*, 175.

109. Rockwell also used the Garrett TFE731-3R turbofan engine on the Sabreliner Model 65.

Garrett's loss of an important commercial contract was a significant blow to the ATF3 program. Other early marketing efforts and design studies also did not result in contracts. For instance, a 4,000-shp turboshaft variant was studied for possible application in the U.S. tri-service Heavy Lift Helicopter (HLH) project. Another scaled-down version of approximately 2,000 shp was considered for the Army's Utility Tactical Transport Aircraft System (UTTAS). A number of other design studies for various applications were also performed, but nothing came of them.

The engine was also considered for re-engining the Lockheed JetStar and the Dassault Falcon, but no orders were forthcoming at the time. When the engine was part way through its development, it was recognized that the ATF3 was a larger engine than the airframe companies really needed. For example, it was too big for the Learjet, one of the original applications sought. The ATF3 was more appropriate for medium-weight business jets (25,000-35,000 lb).

The ATF3-1 was rescued when it was chosen to power fly-off models of the Ryan Compass Cope RPV. In the late 1960's, Garrett was given a $6 million contract by the Air Force to continue development of the engine. Under this contract, a third stage was added to the fan turbine. The engine was given a military designation YF104-GA-100 (civilian model ATF3-3). It successfully completed preliminary flight rating tests in 1972 at 4,050 lb thrust. The engine was installed on the Teledyne Ryan YQM-98A Compass Cope R prototype high-altitude, long-endurance strategic RPV. In 1974, this RPV set an unofficial endurance record for RPVs by remaining aloft for well over 24 hours and reached an altitude above 55,000 ft. This record demonstrated the efficiency of the ATF3. With the expiration of this contract, the ATF3 was put on the back burner.

The ATF3 program was brought to life again in the spring of 1976 when Dassault-Breguet selected the engine to power its Falcon business jet. In 1977, the Coast Guard awarded a contract to Falcon Jet Corporation, a distributor and support center for Falcons in the U.S. The contract was for a medium-range surveillance aircraft, a modified Falcon 20G (later given the military designation HU-25A). Garrett made a contract with Falcon Jet to certificate the ATF3 for its Falcon aircraft sales. With the start of development work on the Coast Guard application, all development work on the ATF3 was moved from Torrance, Cal., to Phoenix. However, the test cells that had been built in Torrance continued to be used for the development program.

A major redesign known as the ATF3-6 was undertaken for this application. Rated at 5,050 lb take-off thrust, the engine had to be rematched and optimized to run at higher operating temperatures. Garrett had an extraordinarily difficult time certificating the ATF3 for the Falcon application. While Garrett originally predicted certificating the engine for the Coast Guard in October 1978, this did not happen until 1981. While various problems contributed to this delay, it was the FAA bird ingestion test that was the biggest obstacle. Under a set of stringent regulations, the ATF3-6 was required to withstand the ingestion of two 1.5-lb birds at 250 knots and still maintain a minimum of 75% thrust.

The engine failed to pass this test in September 1979 because of performance degradation. Garrett redesigned the fan blades, incorporating improved midspan dampers and ribs on the underside of the blade and in a chordwise direction. It also reprogrammed the electronics computer in the fuel control system. Despite these changes, the ATF3 failed a second bird ingestion test in May 1980.[110] In all fairness to the ATF3 program, it is important to note that the stringent bird ingestion criterion developed by the FAA for large turbofan engines were very challenging for smaller size engines.

Additional delays were encountered during the 1980 FAA 150-hour endurance tests that resulted from a failure of a fan blade mid-span damper, fuel nozzle malfunction, and a main shaft bearing failure. The problems were resolved. The bird ingestion test was passed in January 1981. The model ATF3-6A was certificated at 5,440 lb static thrust on December 24, 1981.

In 1982, the ATF3-6A began service with the Coast Guard on the HU-25A Guardian and in 1983 on the Falcon 200 business jet. The ATF3-6A's excellent specific fuel consumption gave the HU-25A Guardian the capability of spending four hours on station at very low altitudes, perfect for its role of search and rescue, surveillance, and drug interdiction. However, the Coast Guard experienced a higher maintenance activity and spare engine usage than anticipated. The HU-25's low-altitude flights over salt water, and attendant spray ingestion, resulted in corrosion of the abradable shrouds in the LP compressor, degrading performance. To resolve the corrosion problem, a corrosion resistant bronze-based material was selected to replace the aluminum-based abradable shrouds. Also, an integral water wash system was developed for the engine to remove corrosive residue. A primary problem in the Falcon 200 business jet was the failure of bearing sump carbon oil seals due to excessive heat. Garrett solved the problem by redesigning the seal rotors, changing to a different grade of carbon, and reducing seal face loading.

110. "Bird Ingestion Trips ATF3-6," *Aviation Week & Space Technology*, Vol. 112, No. 22 (June 2, 1980): 21.

ATF3 SUMMARY

By 1990, only 204 ATF3 engines were in service in approximately 100 airplanes. The distribution between the military (HU-25A) and civilian (Falcon 200) was about equal. After years of effort developing and certificating the engine, the engine finally became available for the general aviation market just before the country entered the 1982 recession. By the time the marketplace had recovered, the engine had been overtaken by improved technology. Other Garrett engines, such as the TFE731-5B and the CFE738, bracketed the ATF3's thrust class and had better specific fuel consumption and thrust-to-weight ratios. It was also eclipsed by competitor engines such as the Pratt & Whitney Canada PW305. By 1985, Garrett had discontinued its marketing efforts on the ATF3.

Reflecting on the program, Al Stimac, then Director of Turbofan Programs and Earl Cummings, then Turbofan Program Manager, felt that the unique technology that gave the engine its good fuel consumption also made the engine difficult to manufacture and to support.[111] The engine had a very high parts count when compared with other Garrett engines. Numerous welded sheet metal components such as the crossover duct, HP case, and fan turbine entry duct were labor intensive and the maintenance of required tolerances during fabrication was challenging. The resulting high cost of the engine was a factor that limited its market acceptance.

The ATF3 was not a successful commercial venture for Garrett. Supporting such a small engine population was very expensive for the corporation. Nonetheless, providing support for the engine throughout its service life was a virtual necessity for Garrett. Dassault had been one of Garrett's most important customers for many years, and the Coast Guard and the various commercial operators were potential future customers.

The ATF3 introduced innovative technology for its time. However, the very technology which gave it excellent fuel consumption created a complex, costly-to-manufacture engine with difficult-to-solve technical problems. From a historical perspective, it was the first U.S. three-shaft turbine engine.

The TFE731

Development of the TFE731 turofan was spurred by the ATF3. In contrast to the ATF3 program, however, the TFE731 was a highly successful engine pro-

111. Al Stimac and W. E. Cummings, interview by Rick Leyes and Bill Fleming, 7 March 1990, Garrett Interviews, transcript, National Air and Space Museum Archives, Washington, D.C.

gram. The right engine at the right time, the TFE731 succeeded in replacing a generation of turbojet business jet engines and within a decade captured most of the market in its thrust class. It contributed to making the executive jet a really practical business tool by giving it transcontinental range for the first time.

Reflecting on the TFE731, retired Chairman, Chief Executive Officer, and President Harry Wetzel credited the ATF3 program as an important stimulus for the TFE731: "The ATF3 was a big disappointment. . . . But it did stir the organization to the point that they came up with the 731, which has been an enormously successful engine."[112] In the 1960s, Wetzel wanted to get Garrett into the fan engine field for business jets because it was an obvious niche that was not being filled. The dominant business jet engines then existing, the P&W JT12D, the GE CJ610, and the Rolls-Royce Viper, were fuel guzzlers; the aircraft that they powered had limited range.[113]

Wetzel's problem was finding someone within the organization to tackle the problem. The ATF3 was the first proposal forthcoming. The fact that Wetzel approved having the ATF3 development done at Garrett's division in Los Angeles fostered a highly competitive atmosphere at Garrett's Phoenix division, which had always been responsible for prime propulsion engines. As Wetzel put it, "That turned the guys in Phoenix wild, and pretty soon we had the 731 proposal for something they could do in Phoenix."[114]

The Phoenix proposal came from Ivan Speer. According to Speer, the conceptual TFE731 engine started at the 1968 National Business Aircraft Association show in Houston, Texas.[115] In a discussion with his business associates and friends, a consensus was reached that the derivative military jet engines on display at the show were old technology, used too much fuel, and were too noisy. A change was in order for a next-generation engine. A high-bypass ratio fan in the order of perhaps 3:1 would be about the right design for small business jets.

---

112. Wetzel to Leyes, 9 April 1990.

113. Other business jet turbofan engines existed or were under development at the time, but most of the business jet market was powered by these three turbojet engines. The GE CF700 turbofan, a derivative CJ610 turbojet engine, had been certificated in 1964 and was in service on business jet aircraft such as the Falcon 20. The Turboméca-Snecma Larzac turbofan engine program was launched in 1968, and the development program took place about the same time as the TFE731 program. Design of the Pratt & Whitney Canada JT15D turbofan engine began in 1966, and it was certificated for the Citation in 1971. The 11,400 lb thrust Rolls-Royce Spey turbofan engine was in service at the time on large Grumman Gulfstream II executive transport aircraft.

114. Wetzel to Leyes, 9 April 1990.

115. Speer to Leyes and Fleming, 8 March 1990.

In the course of the conversation, Speer recalled that the company had recently completed development of two new-generation APUs. One was the TSCP700, a twin-spool APU for the McDonnell Douglas DC-10, and the other was the single-spool GTCP660 APU. Speer felt that, "Somewhere in that box of goodies maybe we can come up with a design of an engine that would satisfy this second generation requirement."[116]

Speer began to study the possibilities. In selecting the rating of the engine, Speer decided on an engine that everywhere within the flight envelope would stay on top of the GE CJ610 turbojet. Working with Learjet's Aircraft Division Executive Vice-President and General Manager Malcolm Harned, Speer began running computer simulations to create aircraft-engine matching for the Learjet. Speer also worked with Ed Swearingen, President of Swearingen Aircraft, to do preliminary engineering studies for a small delta-wing, transonic business jet known as the SA-28T.

The first design was a paper engine designated TFE731-1. It was a 2,710 lb thrust two-spool, geared turbofan engine.[117] On April 4, 1969, Speer proposed the engine to Garrett's Board of Directors.[118] A strong part of Speer's argument was that the DC-10 APU core would be the building block for the TFE731; many TFE731 parts would be common with the APU parts. This would materially reduce the development risk and the manufacturing cost of the engine.[119] The Board gave tentative approval, but wanted customer commitments for the engine before proceeding. On April 11, Garrett officially announced plans for its new engine.[120] In short order, Speer was able to get Swearingen Aircraft to place an engine order for its SA-28T.[121]

In his ongoing discussions with airframe manufacturers, it became evident to Speer that the original TSCP700 compressor would not give the needed performance, particularly for the Learjet aircraft. However, if the higher flow GTCP660 LP compressor from the Boeing 747 APU were substituted for

116. Speer to Leyes and Fleming, 8 March 1990.

117. C. M. Plattner, "Garrett Developing New Turbofan," *Aviation Week & Space Technology*, Vol. 90, No. 17 (April 28, 1969): 92.

118. George C. Larson, "Reflections," *Business & Commercial Aviation*, Vol. 47, No. 1 (July 1980): 40-42.

119. While this was a logical assumption at the time, by the time the TFE731 was certificated, not one part was interchangeable with the DC-10 APU.

120. Gordon Fletcher, "Garrett-AiResearch Develops New 2,700 lb thrust Turbofan," Garrett Corporation, Los Angeles, California, News Release, 11 April 1969.

121. This order was eventually dropped when Swearingen did not produce the SA-28T aircraft.

TSCP700 LP compressor, and other changes were made in the engine, air-framers would have the performance that they needed. In May 1969, these changes were made, resulting in an improved, 3,500-lb-thrust turbofan engine, the TFE731-2.[122] One month later, Learjet placed an order.

As part of its marketing effort, Garrett had given a press conference on the TFE731 at the 1969 Paris Air Show. Afterward, Executive Vice President of Sales and Service Bill Pattison was summoned to meet the principals of Avions Marcel Dassault, including founder Marcel Dassault. With the assistance of a TFE731 design engineer, Pattison negotiated a TFE731 specification and contract for a model to power the Falcon 10.[123] This was the first major order for the TFE731. This historic meeting also set the precedent for Garrett's future close association with Dassault. With the Dassault order in hand, Garrett had three commitments for use of the engine, and Harry Wetzel approved the TFE731 project.

TFE731 DESIGN RATIONALE AND DEVELOPMENT PROGRAM

The 3,500-lb-thrust TFE731-2 engine was designed originally to give coast-to-coast range to U.S. business jet aircraft in the 12,500- to 15,000-lb category. This two-spool geared front-fan engine had a bypass ratio of 2.52:1 at the design point of 40,000 ft and Mach 0.8.

The LP spool consisted of a three-stage turbine driving a four-stage ax-ial-flow compressor and the fan. The LP spool drove the the fan through a plan-etary gear reduction system. The HP spool was made up of a single-stage tur-bine driving a single-stage, backward curved centrifugal compressor. The spools were concentric with the LP shaft housed within the HP spool. A reverse-flow combustor was used to reduce engine length.

The TFE731 was derived essentially from Garrett APU and TPE331 technol-ogy. The TFE731 was an amalgamation of the four-stage axial compressor from the GTCP660-4 APU and the two-spool, axial-centrifugal configuration of the TSCP700 APU. The TFE731 compressor differed from the GTCP660 com-pressor in that inserted titantium blades were adopted for the rotor and the stator to minimize blade vibration, reduce maintenance cost, and guarantee a closer control of the aerodynamic contours.[124] Also, the first-stage aerodynamic

---

122. Lad Kuzela, "Ivan Speer's Little Engine that Did," *Industry Week* (November 12, 1979): 109

123. Pattison to Leyes, 18 April 1990.

124. M. C. Steele and F. L. Roberts, "Highlights of the Design and Development of a Modern Geared-Fan Jet Engine," (SAE Paper No. 720351 presented at the National Business Aircraft Meeting, Wichita, Kansas, 15-17 March 1972), 5.

Dassault Falcon 10 aircraft powered by Garrett TFE731 turbofan engines. Photograph courtesy of Garrett.

components were redesigned to minimize the transition length from the fan discharge and to improve low-speed surge characteristics. The centrifugal compressor and the reverse-flow combustor designs were also derivative designs and were based on development work on the TPE331 series engine.

To use this existing technology, particularly the LP compressor, which was developed and working well, the geared-fan concept was chosen. The geared fan, the first fan designed by Garrett Phoenix Division engineers and unique to the industry, was required to avoid excessive fan tip speeds. It was scaled down from a NASA design. Garrett's experience with turboprop gear trains provided the confidence to proceed with this configuration even though more than 3,000 horsepower was transmitted within a compact gear train envelope only 10.2 inches in diameter and 3.5 inches long.[125] Speed reduction from the LP spool was in the ratio of 0.55:1.

Another key technology transferred from the APUs was the analog fuel control used on the TFE731-2, which was the first TFE731 production model. Because APUs had to operate unattended and sustain large shock loads (as during main engine starts) without surging or being unstable, Garrett had developed extensive fuel control expertise. Garrett fuel control specialists had quickly fo-

Garrett TFE731-3R turbofan engine. This engine is in the National Air and Space Museum collection. National Air and Space Museum photo by Eric Long, Smithsonian Institution (SI Neg. No. 97-16503).

cused their efforts on the development of full authority analog electronic fuel controls. These were already in production for Garrett's largest APUs when the 731 was conceived, and AiResearch Manufacturing of California was the established supplier of the necessary "black boxes" (i.e. fuel controls).

The TFE731-2 fuel control consisted of a single-channel, full authority analog electronic control with a limited-capability hydromechanical backup. The electronic primary control provided all of the control functions and protections in normal operation. When the electronic control became inoperative, the hydromechanical back up provided unlimited power lever movement, achieved at least 80% thrust at takeoff and full HP spool overspeed protection. Phoenix engineering was responsible for the overall control system design integration and for the hydromechanical design while other Garrett divisions manufactured the hydromechanical unit and detail designed and manufactured the electronic unit. The TFE731-2 was the first successful production propulsion engine, of any size, to rely on such an electronic control. This control provided the engine with precise power setting (N_2 control) with T_5 topping (inter-turbine temper-

ature), overspeed protection, surge valve operation, a nearly automatic start sequence, and compatibility with a custom-made synchronizer to minimize cabin
noise.

On September 15, 1970, the first TFE731-2 engine run to partial thrust was
accomplished, and two days later it achieved rated take-off thrust. Extensive rig
testing of individual components, begun as early as March 1969, contributed to
the success of the essentially trouble-free first runs.

The first major setback to the fast-moving TFE731 program was the FAA bird
ingestion test. During 1971, the FAA had become increasingly concerned about
engine bird ingestion and required turbine engines to pass a four-pound bird ingestion test at 250 knots. The TFE731 had not been designed to handle such a
large bird and came apart completely on its first test on September 17, 1971. Significant effort had to be devoted to redesign the fan and part of the engine structure to withstand the impact. Thick, blunt blades were needed, but they created
an apparent choking problem that had to be resolved. Beefing up the fan blades
and the structure also added weight to the engine. However, the redesign efforts
were successful, and the four-pound bird test was passed uneventfully in 1972.

Two problems were also encountered with the fan reduction gearbox. One
was the need for improved restraint of sun gear movement. The other was that
the heat rejection to the gearbox lubrication system was 50% greater than predicted. Both problems were solved with gearbox modifications which also resulted in a smoother, quieter running gear train.[126]

Two of the TFE731-2 engine design objectives were to minimize noise and
emission levels. To reduce noise, fan rotor-to-stator spacing was increased, and
the number of fan blades and stator vanes was optimized. In addition, an acous-
tically treated fan inlet housing was incorporated. To reduce exhaust emissions,
the reverse-flow combustor had a longer flow path, resulting in a more homo-
geneous fuel-air mixture. The development program tests showed significantly
lower noise levels than comparable turbojets and smoke characteristics of the
combustor were found to be below the level of visibility at all power settings.[127]

In January 1971, the first altitude performance evaluation was run at the Air
Force Marquardt Altitude Test Facility located in Van Nuys, Cal. Altitude tests
continued at NASA Lewis in September as a result of NASA's interest in testing
the first geared-fan engine. It is interesting to note that in tests to induce surge
of the fan, LP compressor, and HP compressor, NASA could not get the HP

---

126. Steele and Roberts, "Highlights of the Design and Development of a Modern Geared-Fan
    Jet Engine," 10-11.

127. M. C. Steele and F. L. Roberts, "Highlights of the Design and Development of a Modern
    Geared-Fan Jet Engine," 12-13.

centrifugal compressor to surge, a testimony to Garrett's hallmark backward curved centrifugal compressor design. According to Ivan Speer:

> . . . they had surged that LP compressor and the fan literally hundreds of times. But no matter how hard they tried, they could not surge that high spool. That old backward curved centrifugal wouldn't do it . . . So that, of course, is one of the reasons why the 731 was so damn stable, because it wouldn't react to the two spools (i.e. the disturbances upstream). It would just eat whatever it had eaten and calmed right down.[128]

The first test TFE731-2 flight was made on May 19, 1971 with the engine installed in the right nacelle of a Learjet Model 25 aircraft. On October 15, 1971, the Dassault Falcon 10 made its flight fully powered by TFE731 engines. In August 1972, FAA certification of the TFE731-2 and first production deliveries to Dassault for the Falcon 10 took place. The TFE731-2 entered service on the Gates Learjet Models 35 and 36 in 1974. Variations of the TFE731-2 later included applications on the Spanish Air Force CASA C-101 Aviojet trainer/attack aircraft, Republic of China (Taiwan) AT-TC-3 trainer, and Argentina's FMA IA.63 Pampa trainer.

THE BUSINESS CLIMATE IN THE EARLY 1970S AND THE TFE731-3

When the TFE731-2 went into production in 1972, it held a unique position in that it had been sized just right for the mainstream business jet marketplace. There was essentially no serious competition from other engine manufacturers in the TFE731's thrust class. Large engine manufacturers were concentrating on producing turbofan engines for large transport aircraft, and Garrett was given a grace period in which to enter the market.

About the same time, a fuel crisis, public concern about noise, and a healthy economy had a significant impact on the program. The excellent fuel consumption of the TFE731 promised to significantly improve the usefulness of the business jet to its operators by extending its range. While the cost of jet fuel was not a key issue for operators at the time that the TFE731 design was underway, an Arab oil embargo in 1973, resulting in greatly increased prices, contributed to the early success of the engine. Also, the TFE731 arrived at a time when greater focus on environmental issues, such as noise, were becoming a concern. Communties were beginning to restrict access of noisy turbojet aircraft to airports by imposing curfews. The general aviation market was healthy, and business jet

128. Speer to Leyes and Fleming, 8 March 1990.

shipments increased during the 1970s. All of these factors contributed to the early success of the TFE731 program.

In the early 1970s, several new medium-size business jet aircraft and aircraft re-engining programs called for an engine with increased thrust and improved altitude performance. Garrett sought to beat the performance of the Pratt & Whitney JT12, used on several important business aircraft, including the Lockheed JetStar. This requirement led to the development of the TFE731-3 which had a 3,700-lb-thrust take-off rating. Using the basic power section of the TFE731-2, a new cooled HP turbine rotor capable of a higher inlet temperature was added. Cruise thrust was increased 11% at a comparable tsfc.[129]

Derivative technology and in-service experience was factored into the new model. The cooled turbine rotor was Garrett Phoenix Division's first, and was derived from development work on a cooled HP turbine for the ATF3 in Los Angeles. In-service experience on the TFE731-2 was also factored into the design of the TFE731-3. To ease maintenance and inspection procedures, components such as the outer transition liner, the outer fan duct, and the rear turbine support case were redesigned.

The TFE731-3 was chosen for the Lockheed JetStar II executive transport in October 1972. It first ran in September 1973, received its FAA certification in September 1974, and engine deliveries began in April 1974. The prototype conversion version JetStar first flew in July 1974, and the first production aircraft was delivered in 1976. The JetStar conversion was indicative of what the TFE731 did for the first-generation business jet aircraft. According to aviation writer and editor George Larson:

> The most celebrated installation of the TFE (for turbofan engine) 731 is in the Lockheed JetStar II, for which the new engine was the equivalent of a heart transplant. The great Lockheed executive barge, the first business jet built in the U.S., had suffered from the passage of time, from a range that wouldn't stretch from one coast of the U.S. to the other and from engines that burned lots of fuel and made a monstrous thunder. To some, it appeared that the JetStar, though an admirable exercise in design at the time it was conceived, had now become a machine that simply turned expensive fuel into noise and smoke, with travel being only incidental. It had been out of production since 1973.
>
> The simple conversion to the modern 731-3 rejuvenated the JetStar so that it now spans the U.S. with aplomb, climbs faster and, most important, is so much quieter that it sounds as if takeoffs are being made at partial-thrust settings. The

---

129. Karl R. Fledderjohn, "The TFE731-5: Evolution of a Decade of Business Jet Service," (SAE Paper No. 830756 presented at the Business Aircraft Meeting and Exposition, Wichita, Kansas, 12–15 April 1983), 1.

specific fuel consumption of the 731 at full static thrust is about 50% that of the JetStar's Pratt & Whitney JT12s at take-off power.[130]

The TFE731-3 also replaced the Rolls-Royce Viper on the Hawker-Siddeley HS 125 and the Pratt & Whitney JT12 on the Rockwell Sabreliner Model 65 business jets. Various versions of this model 731 were eventually used on the JetStar II, Dassault Falcon 50, CASA C-101, and Israel Aircraft Industries Model 1124 Westwind.

In the late 1970s, two variations of the TFE731-3, the -3A and the -3B, were developed in response to new mid-sized aircraft requiring higher altitude and increased range. Using the same basic TFE731-2 and -3 power section, a new higher-pressure-ratio fan with improved altitude performance and several turbine improvements were incorporated to provide a 51,000-ft-altitude engine with a significant reduction in cruise tsfc.[131] The -3A entered service on the Learjet 55 and the -3B on the Cessna Citation III. The -3A was also used on the Israel Aircraft Industries Model 1125.

The mid-span dampers and the blunt airfoils developed to solve the bird ingestion problem on the first TFE731 were later discovered to have contributed to a downstream shockwave choking problem. As a result of new three-dimensional computational methods (viscous flow analysis) developed by Garrett, fan transonic flow problems were resolved by a new fan design that met the ingestion strength requirements without sacrificing performance. The -3A and the -3B both incorporated the new fan.

The -3B used uncooled directionally solidified HP turbine blades to achieve its performance. The HP turbine blade was developed during a NASA-Garrett research program in the 1970s called the Materials for Advanced Turbine Engines (MATE) program. The objective was to develop a casting process and material properties data for directionally solidified blades. The new blade and its stator demonstrated a 1.4% increase in aerodynamic stage efficiency. This resulted in an improvement in engine tsfc by permitting a 75% decrease in the HP compressor bleed airflow which was used for turbine cooling.[132]

THE TFE731-4

In October 1979, Garrett announced a 4,000-lb-thrust TFE731-4, based on the TFE731-3 core and fan with a five-stage LP compressor. A possible application

130. George C. Larson, "Falcon 10 Power Up," *Flying*, Vol. 96, No. 6 (June 1975): 43.

131. Fledderjohn, "The TFE731-5: Evolution of a Decade of Business Jet Service," 2.

132. Fledderjohn, "The TFE731-5: Evolution of a Decade of Business Jet Service," 2.

was a version of the Hawker-Siddeley HS 125 for the Middle East market. The program was re-evaluated when it was determined that sfc estimated for the design had risen more sharply than the power. However, by 1990, another evaluation was underway to consider using it as an upgrade engine to enhance the performance of aircraft using the TFE731-3. In February 1991, the Cessna Citation VII, a higher powered version of the Citation III, made its first flight powered by the TFE731-4. This version of the TFE731-4 was rated at 4,080 lb thrust and used the TFE731-3A/3B fan coupled with the TFE731-5 core.

A summary of the principal problems of the Garrett TFE731-2 and -3 models, included the LP compressor, the HP turbine blades, bearings, and HP compressor impeller. LP compressor problems included a rotation problem of the first-stage through third-stage compressor stator vane assemblies and fatigue cracks in the compressor disk dovetail. These problems were respectively resolved with anti-rotation pins in the stators and a large root-radius dovetail and disk modifications. HP turbine blade failures were solved with a new blade design and casting process. Main shaft bearing and planet bearing problems were tackled with improved inspection techniques for better quality control of new main bearings and a redesigned planet bearing assembly. HP compressor impellers tip failures were resolved with modifications to both the impellers and diffusers.[133]

In 1991, an emergency airworthiness directive was issued by the FAA on selected TFE731-2 and -3 models requiring inspections for cracks in the LP fan disks of an estimated 3,000 engines.[134] The focus was on fan blade dovetail slots that had not received a lance shotpeening process following machining.

By 1976, Garrett's TFE731 powered new-production aircraft from all the leading general aviation jet manufacturers.[135] Between 1973 and the end of 1977, the TFE731 succeeded in capturing more than 45% of the small-to-medium business jet market for use on 10 different types, either as original equipment or retrofit.[136] Much of Garrett's success by then was at the expense of GE, whose engines losing market share to the TFE731 were the GE CJ610 on the Gates Learjet 23/24 and IAI Westwind 1121 and 1123 as well as the CF700 on the Dassault Falcon 20 and Rockwell Sabreliner 75. By the late 1970s, Garrett was well positioned for its next major model in the TFE731 series, the TFE731-5.

133. Fledderjohn, "The TFE731-5: Evolution of a Decade of Business Jet Service," 5–8.

134. "FAA Issues AD on TFE731," *Aviation Week & Space Technology*, Vol. 134, No. 16 (April 22, 1991): 17.

135. Ken Fulton, "TFE731 Inherits General Aviation Fan Market," *Flight International*, Vol. 11, No. 3557 (May 14, 1977): 1315.

136. J. Philip Geddes, "Giants battle in US Small Turbine Market," *Interavia*, Vol. 33 (March 1978): 180.

THE TFE731-5

Between 1975 and 1977, Garrett participated with NASA in a joint program known as the Quiet, Clean, General Aviation Turbofan (QCGAT) engine. Using the core components of a TFE731 engine, advanced acoustic features were introduced into new fan and LP turbine components, as well as the nacelle and exhaust nozzle.[137] Exhaust jet noise was decreased by increasing turbine power extraction and thus reducing core jet velocity. A forced mixing of the core and fan streams further decreased peak jet velocity. Turbomachinery noise was decreased by the selection of fan spacings and airfoil counts as well as the use of multiple layer acoustic absorbers in the inlet and discharge ducts. This technology was incorporated into the TFE731-5 engine.

The 4,305-lb-thrust TFE731-5 was a move by Garrett to close the thrust gap that existed between the top of the TFE731 series and the bottom of the ATF3 engine. The principal improvements to the engine included a new and larger fan, an improved fan gearbox, and a high-work LP turbine to match the fan and provide increased efficiency.

The new, higher-flow fan was a .97 scale of the ATF3-6A fan that was developed in the QCGAT program. The larger fan had a bypass ratio of 3.33:1, which, combined with improved efficiency of the fan and LP turbine, and an increase in turbine inlet temperature, increased thrust and reduced sfc at all conditions.[138]

To provide optimum fan aerodynamic performance, the gearbox reduction ratio was lowered from 0.555 to 0.496. The gearbox was once again redesigned, not only to accommodate the new reduction ratio, but to improve its operation. Emphasis was given to such improvements as increasing the gear contact ratio to smooth out vibration and reducing the planet gear rotational speed and increasing its bearing design load capacity.

Other important changes were made to the LP turbine, the engine control system, and engine acoustics. Derived from the QCGAT program, the new LP turbine was designed for improved efficiency, longer life, and 14% more work capacity. For improved performance and reliability, a digital electronic control system computer replaced the analog computer in earlier models. The new computer also had a Built-In Test Equipment (BITE) feature which monitored selected parameters, identified faults, and took corrective action to permit continued engine operation when faults occurred. With a careful redesign of the

137. F. B. Wallace, "Business Aircraft Turbofan Technology Progress in the 1970's, Challenges in the 1980's," (SAE Paper No. 800601 presented at the Turbine Powered Executive Aircraft Meeting, Phoenix, Arizona, 9-11 March 1980), 3.

138. Fledderjohn, "The TFE731-5: Evolution of a Decade of Business Jet Service," 8.

fan and stator, no noise increase over earlier, lower-powered TFE731s was measured.

The first flight of the TFE731-5 engine was conducted on October 4, 1982 on a Garrett Falcon 20 test aircraft. It was certificated in November 1983. Its applications were in retrofitting Dassault-Breguet Falcon 20s, on the British Aerospace BAe 125-800 business jet, and the CASA C-101CC Aviojet.

In March 1985, the TFE731-5A was certificated at 4,500 lb thrust. It powered the Dassault-Breguet Falcon 900. A compound mixer exhaust nozzle was added to the engine. The technology for the nozzle was derived from the QCGAT program. The nozzle used a fluted-core exhaust duct to mix the hot-core exhaust gas with the bypass air before discharging the mixed flow through a convergent-divergent nozzle.[139] Mixing the flows prior to expanding them to ambient conditions improved thrust by more efficiently using the thermal energy in the hot-core flow. It also lowered noise levels and smoke emissions. Previous TFE731 engines had co-annular exhaust systems wherein the fan stream and core stream were expanded to ambient conditions separately.

In February 1991, an uprated version called the -5B was certificated at 4,750 lb thrust. The 5.5% thrust increase was achieved through an improvement in gear ratios and an increase in turbine operating temperture. It quickly found application on the Dassault Falcon 900B and on Falcon 20B retrofit programs; it was also subsequently adopted on the Hawker 800XP.

The use of a fan gearbox provided the TFE731 family of engines with great flexibility in matching different fans to a given core and vice versa, or in re-matching the fan to the same core to achieve different optimizations. The family included the TFE731-3 and -3A/-3B, the TFE731-5/-5B, the TFE731-4 and -5, and the next series, the TFE731-20/-40/-60 engines.

THE TFE731-20/-40/-60 SERIES

In 1992, full-scale design and development was launched on a new family of TFE731 engines, the TFE731-20/-40/-60. These engines, by capitalizing on a common core based on the proven TFE731-5/-5B core enhanced with 1990s technologies, covered and exeeded the thrust range of the previous TFE731-2 through -5B engines, with major improvements in fuel consumption, cost of ownership, and durability.

The performance improvement was obtained by increasing the fan airflow and bypass ratio, which in turn necessitated an increase in overall pressure ratio

139. Steele, "Some Issues in the Growth of Small Gas Turbine Aircraft Propulsion Engines," 7.

and turbine inlet temperature. Suitable material changes were also made to achieve the durability and reliability objectives. Major improvements were made in the fan, HP spool, LP turbine, and fuel control.

The -20 and -40 were equipped with a 28.1-inch-diameter fan, a scaled-down version of the -5/-5B fan, embodying the latest technology with respect to mid-span-damper-equipped fans. The TFE731-60 used a 30.7-inch-diameter, wide-chord, damperless fan derived from aeromechanical advancements developed on the TFE1042 program. Both types of fans provided their respective engines with 15-16% increased airflow, compared to previous engine models, for the same frame size. On the -60, for example, this resulted in a 27% bypass ratio increase over the -5B at cruise conditions and translated to lower sfc and increased thrust.

All three engines used a higher performance HP spool incorporating an improved geometry and higher pressure ratio (33%) centrifugal compressor, contributing to a 22% increase in overall engine pressure ratio. The HP spool also incorporated an improved HP turbine aeromechanical design to match the increased pressure ratio and higher turbine inlet temperature.

The LP turbine also underwent modifications. Improvements were made, some taking advantage of the CFE738 design, to reduce tip clearances and to match the HP turbine exit conditions while improving durability and reliability.

This family of engines used new, second-generation Digital Electronic Engine Controls (DEEC) which took advantage of lessons learned over 20 years with TFE731 electronic controls. The first generation DEEC was introduced on the TFE731-5 in the early 1980s. Improvements in the new DEEC included, among other things, an automatic setting of flight manual N1s (fan speed, the FAA approved thrust setting parameter) and "trim-free" capability (i.e. there was no need to run the engine to set up a new control).

These major component improvements, as well as others, provided significant advantages to engine owners and operators including: substantial temperature margins to "red line"; climb power setting not limited by mechanical limits; reduced pilot workload; improved reliability and durability; improved maintainability; and significantly reduced cost of ownership.

The TFE731-20, rated at 3,500 lb take-off thrust, provided 15% more cruise thrust and 9% lower sfc than the TFE731-2. The TFE731-40, rated at 4,250 lb take-off thrust, provided 24% more cruise thrust and 7% lower sfc than the TFE731-3. The TFE731-60, rated at 5,000 lb take-off thrust, provided 17% more cruise thrust and 12% lower sfc than the TFE731-5.

The -40 and -60 were certified by the FAA in 1995, and the -20 was certified in 1996. They entered production in 1996. The -20 was selected by

Bombardier to power the Learjet 45, and the -40 powered the Dassault Falcon 50EX and IAI Astra SPX. The -60 was selected for the Dassault Falcon 900EX.

TFE731 SUMMARY

TFE731 performance improved significantly since the introduction of the first model. Performance ranged from 3,500 lb take-off thrust with 0.504 sfc on the TFE731-2, the first production engine, to 5,000 lb take-off thrust with 0.424 sfc (at standard day conditions) on the TFE731-60.

Assessing Garrett's most important technical achievements in the TFE731 engine, Chief of Aero-thermo Component Design Bill Waterman stated, "Probably the most significant thing in the 731 line were the fan aeromechanical developments using advanced analytical techniques both in the aerodynamic and mechanical area to solve this bird ingestion problem, and the incorporation of cooled turbine aerodynamics into the Garrett Phoenix Division repertoire."[140] The computational methods that made possible a 3-D transonic aerodynamic analysis of the fan were largely developed by Garrett engineers. The cooled turbine technology was principally derived from Garrett-sponsored film and impingement cooling research at Arizona State University.

It is interesting to note that Smithsonian Institution Zoologist Roxie Laybourne helped Garrett improve the TFE731's resistance to bird strikes, improving flight safety. From the remnants of birds ingested by TFE731 engines, Laybourne was able to identify bird species and sizes. This information was studied in combination with the knowledge of flight speed, altitude, and condition with an analysis of the engine damage resulting from bird strikes. These analyses enabled Garrett to make improvements that minimized bird strike damage.

In reviewing the mechanical achievements in the engine, Vice-President of Engineering J.-P. Frignac described the fan gearbox as one significant development, "We were the first people to introduce a geared fan and to develop it to the very successful, reliable configuration which it is today . . . It was a challenge, not only from the gear design–high-speed, high-power density gear system—but it's a challenge from the lubrication standpoint, because you have to get the oil in and out of that gearbox without creating what we call churning losses in the gearbox."[141] Furthermore, according to Frignac, Garrett designed all of the high-

---

140. Waterman to Fleming, 7 March 1990.

141. J.-P. Frignac, interview by Rick Leyes and Bill Fleming, 5 March 1990, Garrett Interviews, transcript, National Air and Space Museum Archives, Washington, D.C.

speed (50,000 rpm) gearbox bearings. This experience extended the state of the art in high-speed bearings and was useful in later Garrett turboprop engines.

Obviously, good performance and solid engineering achievements were factors that contributed to excellent production figures. Between 1971 and the end of 1996, 7,900 TFE731 business jet engines were shipped. Peak production years were 1980, 1981, and 1982 with respective shipments of 704, 752, and 559 engines. There were 19 airframe applications, principally general aviation business aircraft and a few military applications. By the end of the 1980s, the TFE731 had the distinction of powering the majority of business jet aircraft types. As a result of such impressive statistics, the TFE731 earned its right as one of the key small gas turbine aircraft engines.

In assessing the success and the significance of the program, Senior Chief Engineer Montgomerie C. Steele who was in charge of propulsion at the time the TFE731 program was started said:

> Historically, I think the most important thing for the success of the engine was the thrust class that was chosen. It was designed with great thought towards reliability in the field, and that's proven out to be the case over the years. But the right thrust class was selected, the right mechanical designs for the reliability that was projected as needed to be for this particular marketplace. So if the 731 had to be remembered for any two things, I would say reliability and selection of the thrust class. You might add to that, incidentally, low noise, because obviously we were just at the beginning of the time when all the engine companies were talking about being good neighbors and, of course, there were significantly lower noise levels than in a pure jet.[142]

Supplementing Steele's comments, J.-P. Frignac added:

> It was the right thrust class and the right fuel consumption level, and at the right time. I think the timing was perfect. It was good for this company, but I think the engine allowed the business jet market to expand the way it has expanded in the subsequent years. It would not have expanded if the engine had not been there, and the engine was successful because the market was expanded.[143]

As for the future of the TFE731 engine, Garrett intended to continue to improve the engine and increase its capability, and to develop derivative engines to

142. Montgomerie C. Steele, interview by Rick Leyes and Bill Fleming, 5 March 1990, transcript, National Air and Space Museum Archives, Washington, D.C.

143. Frignac to Leyes and Fleming, 5 March 1990.

upgrade existing airframes. According to Chief Engineer of Commercial Turbo-fan Engines Elmer Wheeler, Garrett intended to push the TFE731 to the baseline of its new commercial turbofan engine, the CFE738: "We . . . as part of the over-all scheme of things, have set the 738 at a point above the market. We intend to grow the 731s closer to the 738."[144] In fact, Garrett had set a 5,600-lb-thrust ceiling on the growth of the TFE731 so as not to compete with the CFE738.[145]

The other continuing and related trend was the retrofit market. Garrett marketed the TFE731-5 for Falcon 20 retrofits. By replacing the GE CF700 engine, Garrett claimed that the Falcon 20's range could be increased by 50%, climb performance improved, fuel consumption reduced by 42%, and noise footprint reduced.[146] In fact, such claims were verified, and the TFE731 had once again improved the utility of the business jet.

The TFE731 filled a unique gap in the market place for a fuel efficient, low-noise turbofan engine family in the 3,500-5,000-lb-thrust class to power medium- (15,000 lb) to heavy- (40,000 lb) business jets with transcontinental range capability.[147] On the lower end of the scale, the TFE731 ran into competition with the P&WC's JT15D turbofan engine, and on the upper end, with P&WC's PW300 series turbofan. However, the fact that so many business jet models fell within the exact thrust range of the TFE731 gave it a premier position within the business jet market. Technically, the TFE731 was notable because it was the first geared turbofan engine.

New Directions: The Need to Diversify

In the early years of the small gas turbine engine industry, the participants had staked out their own individual marketplaces. They found niches within the market and generally established incumbent positions within them. Garrett and Pratt & Whitney Canada shared the general aviation market. Williams International and Teledyne CAE had established themselves in the expendable engine

---

144. Elmer L. Wheeler, interview by Rick Leyes and Bill Fleming, 6 March 1990, Garrett Interviews, transcript, National Air and Space Museum Archives, Washington, D.C. The CFE738 was a joint Garrett/GE turbofan engine, development of which started in 1988.

145. Robert L. Parrish, "Garrett Engines: Still Growing with Technology," *Business & Commercial Aviation*, Vol. 68, No. 6 (June 1991): 70.

146. Richard N. Aarons, "Falcon 731 Retrofit Program," *Business & Commercial Aviation*, Vol. 64, No. 6 (June 1989): 81. See also Robert Bailey, "Garrett puts Pep in Falcon 20 with TFE731s," *Professional Pilot*, Vol. 24, No. 9 (September 1990): 62-65.

147. Three or four TFE731 engines were used to power 40,000 lb class jets.

business. Allison and Lycoming were incumbents in the helicopter business. GE had a stake in both the helicopter and small jet business. Although established in their market niches, it was generally recognized within the industry that each company would need to broaden its product line to expand sales and remain competitive. It was in this environment at the end of the 1970s that Garrett began to diversify into new markets.

Reviewing this trend, then Vice President, Sales and Marketing Edmund Dobak said:

> Each one of the companies in this business has seen the great advantage of trying to broaden its product scope and its marketplace—the marketplaces that it services—because the technology is so expensive to gain that unless you can take that technology and take advantage of it over a wider product base and wider markets, you could be put at a very difficult competitive position sometime into the future. So all of the companies participating in this market have been looking for ways to broaden their product and market base. There has been a different degree of success in different companies as they've tried to do this, but it has reshaped the industry considerably . . . Those who are successful at this broadening will have a much stronger position than those who continue to have a rather narrow focus within the business.[148]

By the late 1970s, Garrett was essentially a two-engine-program company. The TPE331 and the TFE731 had become self-sustaining programs and had given Garrett a strong position in the general aviation executive aircraft market and an entree into the commuter market. Other programs, including helicopter turboshaft and expendable turbojet engines had been started, but no applications had been found. This left Garrett highly dependent on the general aviation economy. When this economy began to decline at the end of the 1970s, Garrett was vulnerable to market erosion. As President of Garrett Turbine Engine Company Malcolm Craig explained in 1990:

> And then came the crash, and the market started to disappear in the early 1980s. We were sitting there not probably in as bad a shape as some people, but still highly dependent upon the general aviation marketplace with only a very modest amount of military business. We said at that time that we really needed to go about the process of broadening the business to be successful. Even though the 731 and 331 were successful in terms of quantities produced and

148. Edmund Dobak, interview by Bill Fleming, 7 March 1990, Garrett Interviews, transcript, National Air and Space Museum Archives, Washington, D.C.

still looked like they still had long-term outlooks, in order to make a true viable business, it was necessary to have a much larger production base and greater number of markets with a greater number of engines.[149]

One of the key drivers behind the strategy to implement the needed diversification effort was Allied-Signal Aerospace Company Engine Group President Robert Choulet. Commenting on the shift in company direction, Choulet said:

> At that point in time (late 1970s and early 1980s), we set about a strategic plan to get into different marketplaces . . . markets other than the general aviation and executives, such as the helicopter marketplace, both civil and military, such as the fighter engine markeplace with the 1042 (TFE1042), and always pretty much in a horsepower range which matched the capability of our factory in terms of what size hardware we could manufacture.[150]

Getting more military business was one obvious channel that the diversification strategy could pursue. Garrett had made some inroads in this area with the T76 program and with its participation in government technology demonstration programs in the 1970s. The focus of technology demonstration programs was component research and development. Having established its technical credibility with the military, Garrett was in a position to compete for additional, full engine programs. Winning an important engine competition would not only help Garrett diversify and balance its product base, but would further its technical capability. According to Choulet:

> . . . we wanted to be in the military business because we felt that over the long haul, we needed much more help from the government, if you will, to maintain our technology base. A good balance of military and commercial business over the long haul was going to be better than a very narrow market focus in either general aviation at that time or commuters. We set about a strategy by stair-stepping from derivative engines . . .[151]

In the 1980s, this stair stepping led to five additional major engine programs. These included the TFE1042, F109, T800, CFE738, and the TPF351 programs.

149. Malcolm Craig, interview by Rick Leyes and Bill Fleming, 7 March 1990, Garrett Interviews, transcript, National Air and Space Museum Archives, Washington, D.C.

150. Robert Choulet, interview by Rick Leyes and Bill Fleming, 8 March 1990, Garrett Interviews, transcript, National Air and Space Museum Archives, Washington, D.C.

151. Choulet to Leyes and Fleming, 8 March 1990.

The TFE1042 turbofan engine was originally based on TFE731 technology and was a movement into the light fighter aircraft field. The F109 turbofan engine used derivative TPE331 technology and branched out into the military trainer aircraft field. The T800 turboshaft engine was derived in part from F109 technology and was an opportunity to get into the light military helicopter market. The CFE738 turbofan incorporated TFE731 technology and was an attempt to penetrate the medium-large business jet market. The TPF351 turboprop was based on TPE331 technology and sought to gain further access to the commuter market. By the end of the 1980s, its diversification program had moved Garrett toward becoming a seven-engine company.

The TFE1042

Executive Vice President, Sales and Service, Bill Pattison along with Garrett marketing people wanted to move the company in the direction of the low-cost military fighter and trainer engine business. Pattison had developed a business friendship with Volvo Flygmotor AB in Trollhättan, Sweden. Volvo Flygmotor had made Garrett engine components. It was looking for a U.S. engine partner for an aircraft which later would become the Saab-Scania IAS 39 Gripen. Subsequently, the two companies struck an agreement for a joint engine development effort.

On March 2, 1978, Garrett signed an agreement with Volvo Flygmotor to develop a light fighter aircraft engine, the TFE1042. Volvo Flygmotor had made RM8 supersonic turbofan engines for the Saab 37 Viggen. The RM8 was a Swedish military version of the P&W JT8D to which had been added a multi-stage fan and an afterburner. Volvo Flygmotor became responsible for the fan, gearbox, and the afterburner on the new engine. For its part, Garrett contributed the engine core, a TFE731 and the LP turbine. Both companies intended to market the engine.

The prototype TFE1042-5 was a low-bypass ratio derivative of the TFE731 engine—a two-shaft turbofan with geared front fan. The engine incorporated a wide-chord, two-stage fan with high surge margin. The engine used the TFE731-3A core.

The TFE1042 went from the drawing board to test in 16 months with the prototype, non-afterburning engine running in August 1979. It was tested in Sweden and achieved predicted performance with a rated thrust of 4,100 lb. Additional variations of this non-production engine were the TFE1042-6 and -7. The TFE1042-6 was an improved non-afterburning engine rated at 4,840 lb thrust. The TFE1042-7, the afterburning version of the 1042-6, had a dry rating of 4,820 lb thrust and an afterburning rating of 8,340 lb thrust.

In the early 1980s, the TFE1042 was entered in the Jakt/Attack/Spanning (JAS) competition for the next generation fighter aircraft, the Gripen. However, the TFE1042 lost this competition to GE's F404 engine, known as the Volvo Flygmotor RM12. The Gripen, orginally designed as a twin-engine fighter, became a single-engine fighter. The GE F404 was an existing engine with high thrust and considerable growth potential. In contrast, the TFE1042 was a prototype engine, without afterburner at that time, and without sufficient thrust for single-engine application.

Despite the loss of the competition, Garrett continued looking for an application. At the 1980 Farnborough Airshow, Garrett was approached by Dr. Mike Hua, a pilot, a general in the Republic of China (Taiwan) Air Force, and the Deputy Director of Taiwan's Aero Industry Development Center (AIDC). Dr. Hua was given an opportunity to see the demonstrator engine run in Sweden. The TFE1042 was approximately the right size for his country's proposed Indigenous Defensive Fighter (IDF). An engine partnership with Garrett would offer design experience and a production opportunity for his country. For its part, Garrett was familiar with AIDC as a result of supplying its TFE731-2 for the AT-TC-3 trainer aircraft.

A three-partner arrangement was discussed with Volvo Flygmotor, which declined the invitation to participate due to Sweden's existing diplomatic relationship with the People's Republic of China. The original Volvo Flygmotor-Garrett partnership was then discontinued by mutual agreement. After obtaining an export license in 1982, Garrett entered into a joint development partnership with AIDC. Each partner was to have equal share in the development cost. Both partners would also share manpower and facilities. The partnership would focus on the development of an engine for the IDF, a next generation, lightweight strike/fighter aircraft. The joint venture was called the International Turbine Engine Corporation (ITEC).

Garrett engineers took a second look at the TFE1042 as a result of the time that had elapsed since the start of the program. The engineers decided that they could improve the engine performance if they started with a clean sheet of paper and developed a new centerline engine specifically for the IDF. The result was the XTFE1042-70. Cycle studies were begun in 1982, and at the same time, further development of the original TFE1042 was discontinued. Full-scale development work on the engine began in April 1983.

The TFE1042-70 was a two-spool turbofan with variable compressor inlet geometry, in-line annular combustor, afterburner, and variable exhaust nozzle.

The LP spool consisted of a three-stage fan driven by a single-stage cooled turbine. The HP spool consisted of four-stage axial compressor followed by a single-stage centrifugal compressor driven by a single-stage cooled turbine. With a bypass ratio of .36:1, the rated sea-level afterburning thrust of the engine was 8,350 lb. This was Garrett's first production engine with an axial-centrifugal compressor on a single shaft. Prior axial-centrifugal compressors, such as those on the ATF3 and TFE731, had been incorporated in dual-shaft designs.

Rig testing of components was started in 1984 at Marquardt in Van Nuys, Cal. The first engine to test (FETT) took place in April 1985, and the first run to full power maximum afterburning in August.[152] Captive flight tests to determine how the afterburner would operate at altitude were conducted on a Falcon 20 in 1987.

While designing the IDF, AIDC decided that additional power would be required to improve aircraft acceleration. The fan and LP turbine were redesigned. The fan had a higher flow capacity and higher pressure ratio; the turbine inlet temperature was increased. External engine accessories were rearranged to make the engine more compact to accomodate a narrow aircraft fuselage design. Engine thrust was upgraded.

Two basic versions of the engine were made available. One, the TFE1042-70 used in the IDF, was flat rated at 5,000 lb thrust dry and 8,350 lb thrust with afterburning. The other version was designated by the Air Force as the F124-GA-100 (non-afterburning), at 6,300 lb thrust, and the afterburning F125-GA-100 (9,250 lb thrust A/B). The first TFE1042-70 engines were delivered in October 1988. Initial Flight Release was completed in February 1989, and the twin-engine IDF made its first flight in May.

The TFE1042 was Garrett's first afterburning turbofan design and was the company's first venture into the military fighter business. Senior Project Engineer, Military Fan and Jet Engines Tom Bruce was responsible for the afterburner development. According to Bruce, "The afterburner itself did present a challenge for us, because again, we'd never done it before, and it's not a forgiving technology."[153] In order to develop the afterburner technology quickly, Bruce and his associates worked with consultants who were experts in the field.

---

152. A. R. Finkelstein, "The TFE1042- A New Low Bypass Ratio Engine," (ASME Paper No. 90-GT-385 presented at the Gas Turbine and Aeroengine Congress and Exposition, Brussels, Belgium, 11-14 June 1990), 4.

153. Thomas W. Bruce, interview by Rick Leyes and Bill Fleming, 6 March 1990, Garrett Interviews, transcript, National Air and Space Museum Archives, Washington, D.C.

With respect to the fact that the TFE1042 was Garrett's first fighter aircraft engine Bruce commented:

> We're breaking new ground . . . It's getting into different types of markets than we'd ever had before. So it's not only getting to know new technologies, it's getting to know new customers and new philosophies. The maintenance philosophies and the technological requirements for military products are vastly different.[154]

Although the ITEC TFE1042 was a new engine design and Garrett's first engine with an afterburner, the development program proceeded normally. During the development, there were only three major technical problems. LP turbine blade fatigue was resolved by increasing the natural frequency of the blade, using a cooled airfoil instead of a solid uncooled configuration. HP compressor fatigue was addressed by redesigning the blade attachment to reduce dovetail contact stress. A problem with high-altitude compressor surge was solved by optimizing both the control system schedules and compressor inlet guide vane and schedules.[155]

The TFE1042 began production in 1992, with the engines being assembled by AIDC in Taiwan. The engines achieved Initial Operational Capability in 1994 and went into service on the Indigenous Defensive Fighter for the Republic of China Air Force. In 1990, Al Finkelstein, then Program Manager for the TFE1042 and also Vice-President and General Manager for ITEC, commented on the importance of the TFE1042 to AIDC and Taiwan, ". . . it's an important program two ways. One is for the defense (of the country). The other is for national pride and utilization of their people with a (high) technology background."[156]

In 1994, the non-afterburning F124-GA-100 version was chosen by Czech Republic Aero Vodochody for its new L-159 light attack/advanced trainer aircraft. The first customer was the Czech Air Force which committed to an initial order of 72 aircraft. In addition, McDonnell Douglas Aerospace selected the F124 in early 1996 as the baseline power plant for their T-45A Lead-in-Fighter Trainer candidate being offered to the Royal Australian Air Force (RAAF). Although not selected as the trainer, the F124-powered BAe Hawk

154. Bruce to Leyes and Fleming, 6 March 1990.

155. Garrett, unpublished program analysis provided by Garrett for 1990 Garrett interviews by Rick Leyes and Bill Fleming, National Air and Space Museum Archives, Washington, D.C.

156. Al Finkelstein, interview by Rick Leyes and Bill Fleming, 5 March 1990, Garrett Interviews, transcript, National Air and Space Museum Archives, Washington, D.C.

Garrett TFE1042 turbofan engine. Photograph courtesy of Garrett.

also was a candidate for the RAAF Lead-in-Fighter aircraft program. The initial flight test engine was delivered in June 1996, and the first flight followed in October.

By 1996, a growth version of the F124/125 had been defined. Also under development was an industrial version, designated AS1042. For Garrett, the TFE1042 was a major step into a new market area.

The TFE76/F109/TFE109

In the late 1970s, an opportunity arose that coincided with Garrett's strategy to diversify and acquire additional military business. The Air Force had a need to replace its Cessna T-37 primary trainer aircraft. Embodying 1950s' technology, this aircraft was nearing the end of its original design service life. The aircraft was hampered by such things as limited performance, lack of pressurization, and limited range. The engines were not fuel efficient by existing standards, and they were noisy and emitted excessive pollutants. To address such problems, the Air Force specified requirements for a Next Generation Trainer (T-46) to replace the T-37. This led to a competition among airframers, who then needed to select engines to power their aircraft.

In preparation for this competition, Garrett took core engine technology from the TPE331 engine and fan technology from the TFE731 and ATF3 engines and integrated them into a 1,500-lb-thrust-class demonstrator engine called the TFE76. The designation 76 meant that the core engine was based on Garrett's T76 (TPE331-10). The thrust level selected for the TFE76 came from

T-37-derived requirements. The development of the Garrett-sponsored demonstrator engine started in 1979. Component rig tests were conducted between 1979 and 1981, and the testing of the demonstrator engine began in 1981.

Garrett supported three airframe companies in the proposal competition. They were Fairchild, Rockwell, and Ensign Aircraft. In July 1982, the Fairchild candidate was selected as the winner and the Air Force awarded a contract to Garrett to develop the engine, designated the F109-GA-100. The initial contract was valued at $121 million and covered the development of the F109 and delivery of 29 engines, with an option for a further 119 engines. This was Garrett's first sale to the Air Force.

In the process of designing the T-46A, Fairchild felt that it would not need the 1,500 lb thrust of the TFE76 demonstrator engine. They calculated that two 1,200 lb thrust engines would be adequate. However, the Air Force felt that the airplane would inevitably grow, so they added 10% to the thrust requirement. Thus, the TFE109 was developed at 1,330 lb thrust. At this power level, the F109 was expected to give the T-46A much higher flight performance with considerably lower fuel consumption than the Continental J69-powered T-37 that it would replace.

The F109 was a two-shaft contrarotating high-bypass turbofan engine. The LP spool consisted of a single-stage fan directly coupled to a two-stage axial turbine. The HP spool included a two-stage centrifugal compressor driven by a two-stage axial turbine. The engine had a reverse-flow annular combustor. The bypass ratio was 5.0:1, and the maximum take-off sfc was an incredibly low 0.392 lb/h/lb.

To get the low sfc, Garrett used a high pressure ratio compressor design with accompanying high stage efficiency and surge free operation. In 1979, Garrett had initiated an R&D program to develop a high pressure ratio (13.5:1) two-stage centrifugal compressor beyond the 10:1 ratio of the TPE331 program. Because the pressure ratio of the TPE331-10 compressor used in the TFE76 engine was too low for good sfc in a turbofan engine, a scaled compressor based on the 13.5:1 R&D compressor was integrated into the F109 engine. During the course of the F109 program, the two-stage compressor pressure ratio was raised to 14:1. The efficiency objective was met and part-speed surge margin was obtained with the use of an inducer recirculation port called a ported shroud.

The F109 used high pressure turbine technology derived from Garrett's Low Aspect Ratio Turbine (LART) program with the Air Force, which had been undertaken in the mid-1970s. In this program, Garrett used 3-D viscous analysis to design and test turbomachinery blading, which resulted in significantly reduced nozzle losses and increased rotor efficiency. The HP turbine

Garrett F109 turbofan engine. Photograph courtesy of Garrett.

was cooled in the first stage and uncooled in the second stage. The LP turbine had a higher work coefficient than previous Garrett LP turbines. The technology of the LP turbine that drove the fan was derived directly from scaled-down TFE731-5.

The F109 was the first engine to meet the new Air Force Engine Structural Integrity Program (ENSIP) requirement (Mil Spec 1783). The goal of the ENSIP program was to provide engines with higher reliability, improved durability, and lower life cycle costs. It had two major elements. The first, design verification, began with assessing the thermal, vibrational, and stress environments of critical engine components all through evaluation and service. The second element, fracture-free performance, was concerned with the structural integrity of critical engine parts. It required that the engine design provide fracture-free performance of critical parts in the presence of material, manufacturing, and process defects (flaws) for a period not less than the scheduled inspection interval for the engine. For example, under ENSIP, if a small crack were not detected during an inspection, the engine was supposed to be able to tolerate the defect and continue to run without failure until the next inspection interval.

Thus, the F109 was designed to stringent crack propagation criteria, and qualification testing was conducted to verify analytically determined inspection intervals. Garrett and the Air Force jointly developed the methods and procedures for implementing ENSIP. This was done without delay to the program. The technology and methodology developed under this program was the first of its kind to be introduced to the turbine engine industry and set a precedent for future military engines.

The engine was also the first to be developed under a new Air Force qualification process. Previously the Air Force had a Preliminary Flight Rating Test and a Model Qualification Test qualification process. Under the new plan, the F109 completed an Initial Flight Release (IFR), Full Flight Release (FFR), and an Initial Service Release (ISR). In this new process, the pass/fail criteria were based on endurance tests that included the actual mission duty cycle, accelerated damage accumulation, reliability/maintainability/safety, and life cycle cost.

Other technical highlights of the program included a Built-in Test Equipment feature, low noise signature, and reduced emissions. The F109 incorporated a Full Authority Digital Electronic Control (FADEC) with BITE capability. The BITE feature monitored selected parameters, identified faults, and took corrective action to permit engine operation when faults occurred; it also monitored engine performance trends and provided diagnostics. The F109 had a low noise signature that was the result of fan and compound exhaust nozzle technology adapted from the TFE731. Normal conversation could be carried on eight feet from the F109 at ground idle compared with 80 feet from the Continental powered T-37B. The visible exhaust level was minimized by incorporating a reverse-flow annular combustor and piloted airblast fuel nozzles.

The F109 first ran in December 1983. Altitude testing was conducted at the Arnold Engineering Development Center in Tullahoma, Tenn. The F109 also underwent testing in the Naval Air Propulsion Center's gyroscopic test facility in New Jersey. By July 1985, 5,000 hours had been logged on 11 test engines. First flight of the F109 on the Fairchild T-46A was October 15, 1985. Unfortunately, as a result of problems encountered in the development of the Fairchild T-46A, the Air Force canceled the program in March 1986. The Air Force decided to extend the life of the T-37 fleet through a Structural Life Extension Program (SLEP). This was considered to be a stopgap measure until a new, cost-effective trainer became available.

Despite the cancellation of the T-46 program, the Air Force continued with the development, qualification, and Lot 1 production of the F109 engine. In 1987, 25 engines were completed in the first production lot, delivered to the Air Force, and qualified.

Although the F109 did not go into full-scale production, the engine program was considered a technical success. Undoubtedly, this was a factor explaining why the F109 development program was continued after the T-46 was canceled. Reviewing the program, Garrett Engine Division President Malcom Craig said, "In the case of the F109, the Air Force has, I think, constantly given us very, very good marks publicly and privately. That has been one of the best run engine programs that they've had the pleasure to be involved in, and we're quite pleased with the fact that they were happy with it."[157] Garrett had dedicated considerable effort to the program to make it a success. According to Senior Project Engineer, Military Fan and Jet Engines Tom Bruce, who assumed responsibility for the F109 program in 1988, " . . .we dedicated very diligently the resources to do a good job on the program, because it was a showcase program for us. We had a good team working on it."[158]

After the completion of the F109 development program, Garrett continued to explore alternative markets. In 1985, a civilian derivative of the F109 known as the TFE109-1 was selected for the Promavia Jet Squalus, a single-engine, side-by-side, low-cost primary trainer. Potential markets for the Squalus included civilian commercial pilot training and military primary training programs.

While it was a genuine disappointment that that the F109 engine did not go into production as a result of the cancellation of the T-46 program, the engine program achieved recognition in several important ways. It was Garrett's first sale to the Air Force. As an exemplary military development program, it opened the door to future military contracts. For Garrett, it was an important stepping stone toward achieving a military market position. Technically, it was the first engine developed under the ENSIP program and, in fact, helped develop that technology for future military engine programs. Finally, of great importance to Garrett was that they were able to parlay the core technology of the F109 program into yet another military engine program, the T800.

The LHTEC T800

In 1982, the Army invited manufacturers to submit design concepts for its Light Helicopter Experimental (LHX) program. The purpose of the program was to replace Vietnam War-era light scout/attack helicopters such as the AH-1 Huey-Cobra, UH-1 Iroquois, the OH-58 Kiowa, and the OH-6 Cayuse. With a po-

157. Craig to Leyes and Fleming, 7 March 1990.

158. Bruce to Leyes and Fleming, 6 March 1990.

tential military requirement of 10,000 engines, the program presented Garrett with an opportunity to broaden its military market and to enter the helicopter engine business. The LHX engine program represented a substantial production base, both civilian and military, which would contribute to the company's becoming a well-balanced business.

Responding to this opportunity, Garrett decided to make a turboshaft out of its F109 engine. Known as the TSE109, this was a free turbine engine with a two-stage centrifugal-flow compressor, reverse-flow annular combustor, and a two-stage power turbine. The TSE109 was rated at 1,000 shp. The core differed from the F109 principally as a result of a redesigned turbine capable of higher temperature operation.

Garrett started work on the TSE109 demonstrator engine in January 1983. It first ran in 1984. As work was proceeding on the TSE109 in 1984, Garrett learned about the proposed Army acquisition strategy on the LHX engine. In order to minimize life cycle costs, the Army was interested in a Request for Proposal (RFP) that would provide continuous competition throughout the engine development and production phases.[159] To meet this competitive acquistion strategy, Garrett believed it should team with another engine company.

In considering a partner, Garrett was attracted to the Allison Gas Turbine Division of General Motors. In part, the reason for this was that in 1975, Allison had received partial funding from the Army Applied Technology Directorate at Ft. Eustis, Va. for an exploratory development component program.[160] This program became the Advanced Technology Demonstrator Engines (ATDE) program. In 1977, Allison was awarded a four-year cost-sharing ATDE contract to develop an advanced technology 800-shp-class turboshaft engine for future Army and Department of Defense applications. Given the Allison designation GMA 500, the resulting 850-shp engine had a two-stage centrifugal compressor, reverse-flow annular combustor, and a two-stage axial uncooled power turbine. After meeting or exceeding all of the Army specifications, Allison received a follow-on contract for the engine under the Army's Critical Component Improvement Program.

Garrett had competed for the ATDE program, but the ADTE contracts had been awarded to Allison and Lycoming. Suspecting that the Army would lean

159. Lt. Colonel Arnold E. (Sandy) Weand, Jr., "T800 Engine Program: A Break from Tradition," *Army Aviation*, Vol. 38, Nos. 8 & 9 (August-September, 1989): 32.

160. Ron Alto and Walt DeRoo, "Can the Propulsion Industry Truly Share R&D Technology to the Customer's Benefit? -The LHTEC T800 Story," *Vertiflight*, Vol. 34, No. 3 (May/June, 1988): 121-122.

toward awarding a contract to one of the ATDE participants, Garrett studied the Allison and Lycoming engines. The Lycoming engine had been built around a multi-stage axial-centrifugal core and the Allison engine had a two-stage centrifugal compressor. Although it was a lower power engine, the configuration of the Allison GMA 500 was nearly identical to Garrett's TSE109. Allison was also a logical choice for Garrett because Allison's principal market was helicopter engines, and Garrett had not had significant success in this market. For these reasons and the following, Allied-Signal Aerospace Company Engine Group President Robert Choulet summed up Garrett's teaming rationale by saying, ". . . we concluded that Allison would be the best partner because we thought alike and had a lot of Allison people in the company, and had traded people. Allison's General Manager Blake Wallace was with us 23 years before he went there, and he was running Allison. So that was a natural partnership activity."[161]

In June 1984, discussions were initiated with Allison. On October 12, an agreement was signed between the two companies and a joint partnership known as the Light Helicopter Turbine Engine Company (LHTEC) was formed. On December 5, 1984, the Army released an RFP for a 1,200-shp engine known as the T800. LHTEC responded with a joint proposal in March 1985. Also responding were the teams of Avco-Lycoming/Pratt & Whitney, which was a joint venture called APW, and GE and Williams International with a leader-follower arrangement. On July 19, 1985, LHTEC and APW were selected as winners to enter a Full-Scale Development (FSD) competition.

A unique aspect of the proposal and full-scale development competition was the evaluation criteria. The T800 selection was the first in which maintenance and initial cost were given equal weight along with more traditional technical issues.[162] The Army had developed a maintenance criteria called Reliability, Availability, and Maintainability/Integrated Logistic Support (RAM/ILS). What the Army was looking for in RAM was the type of quality that ensured that an engine would rarely fail in flight or fail to start and that could be inspected and repaired easily. With respect to ILS, the Army viewed the engine as a set of spares; it was to be designed in a manner such that the necessary logistics support would be achieved. In the course of the development competition, another maintenance requirement, Manpower Personnel Integration (MANPRINT), was included in the RAM/ILS criteria. It required that all new Army equipment had to be compatible with existing levels of skill and training.

---

161. Choulet to Leyes and Fleming, 8 March 1990.

162. Charles Gilson, Ramon Lopez, and Bill Sweetman, "Turboshafts on Tenterhooks," *Interavia Aerospace Review*, (November 1989): 1117.

The cost of the engine was as important to the Army as its maintainability. An unprecedented agreement was obtained from the two competing teams by the Army.[163] The teams were required to guarantee contractually that they would not exceed the average design-to-cost of $245,000 for the intended 10,000-engine buy and to guarantee that the engines would cost less than $120 per hour to operate and support.

Although all of the evaluation criteria were equally weighted, the Army allowed ranges of performance and weight in its specifications to allow the contracting teams to make trade-off studies to define the optimum engine at the lowest cost. LHTEC eventually made 400 such studies evaluating weight, cost, reliability, maintainability, etc. For example, a trade study would be done to determine how much weight would have to be given up to make the engine more maintainable.

With respect to the turbine engine industry as a whole, another unique aspect of the T800 full-scale development program was that it was the first government-sponsored partnership competition. It was also the first such engine program to emphasize continuous competition from RFP throughout development and production phases. The purpose of this arrangement was to get a better blend of technologies as a result of the partnership arrangement and to avoid problems with single-source procurement.

The LHTEC T800 technology prototype engine was originally 1,000-shp-class (later 1,200-shp) turboshaft engine designed for more than 50% growth. It consisted of a two-stage high pressure ratio centrifugal-flow compressor, annular reverse-flow combustor, a two-stage axial-flow air-cooled gas generator turbine, and a two-stage axial-flow free power turbine. Garrett provided the core technology, which was based on the TSE109/F109. This core had the proper flow size. The core had high performance components, it was a proven design with many hours running time, and it had considerable damage tolerance capability having been designed under ENSIP. Allison took responsibility for the Inlet Particle Separator (IPS), the gearbox, the control system, and the LP turbine. The IPS had excellent efficiency, the gearbox and accessory drive train were of considerably reduced size over earlier generation engines, the full-authority, adaptive digital electronic engine control had already been flight tested, and the power turbine technology was developed. The T800 technology prototype engine (also known as the ATE109), which incorporated both Allison and Garrett

163. Lt. Colonel Arnold E. (Sandy) Weand, Jr., "T800 Engine Program: Smart Contracting," *Army Aviation*, Vol. 38, No. 11 (November 30, 1989): 54.

LHTEC T800 turboshaft engine. Photograph courtesy of Garrett.

technology, was first run in October 1984 and first flown on a Bell UH-1B on March 7, 1985. Testing on this engine continued through 1986.

Work on the full-scale development engine, known as the T800-LHT-800, started in July 1985. In selecting the cycle temperature and pressure ratio, specific power, sfc, and life cycle cost were considered the significant drivers.[164] Basic engine design was completed by November 1985 and was followed by component rig tests. First engine operation was in August 1986. Tests showed that part power specific fuel consumption levels were approximately 6% higher than required specification levels. A performance diagnostic program resulted in studies that focused on lower than expected compressor and gas generator turbine performance and a review of inlet pressure loss. A secondary effort was conducted to advance goals of those areas that were satisfying their requirements. In particular,

164. Andy Cosner and Glyn Rutledge, "Successful Performance Development Program for the T800-LHT-800 Turboshaft Engine," (SAE Paper No. 891048 presented at the General Aviation Aircraft Meeting and Exposition, Wichita, Kansas, 11-13 April 1989), 3.

effort was placed on improving power turbine efficiency, reducing IPS blower power extraction, and a reduction of exhaust diffuser losses.

An evaluation of the gas generator turbine showed that dramatic improvements in sfc could be obtained with better flow path sealing. The power turbine also showed improvements with better sealing. Compressor tests cleared the primary compressor elements, but led engineers to focus on the engine inlet, static structure/clearances, secondary flows, and the deswirl cascade. Tests showed that impeller shroud clearances could be reduced, that secondary flow leakages could be reduced, and that compressor efficiency could be improved by removing the inducer vent flow. Following the results of these evaluations, numerous trial modifications were made throughout the engine during the remainder of the development program. The net result was that engine performance significantly bettered specification requirements and nearly achieved full production margin goals at the 600-shp point (of 1,200 shp). All other performance rating points (power as well as sfc points) more than surpassed the goal production margins.[165]

The T800-LHT-800 had a maximum rated power (5 minute take-off rating) of 1,334 shp and a contingency (2 minute) rating of 1,399 shp. At maximum rated power, the sfc was 0.45 lb/hr/lb. Its base weight was only 310 lb.

The first flight of the T800 was accomplished in October 1988 on an Agusta A-129 helicopter. On October 28, it was announced that LHTEC had won the full-scale development competition over the competing APW team.[166] LHTEC was awarded a $75.4-million contract to continue development of the engine through full military qualification. LHTEC was selected by the Army "because of their aggressive and successful test program during the Preliminary Flight Rating (PFR) phase, and their extensive commitments and guarantees for the remaining portion of the T800 program."[167]

Both Garrett and Allison were scheduled to manufacture jointly the first two production lots. Beginning with the third lot, the two companies would begin to compete with one another for the majority of production engines. The losing bidder would have a guaranteed minority percentage of the production en-

---

165. Cosner and Rutledge, "Successful Performance Development Program for the T800-LHT-800 Turboshaft Engine," 6-10.

166. "Army Selects LHTEC's T800 Engine for Future LHX Fleets," *Aviation Week & Space Technology*, Vol. 129, No. 19 (November 7, 1988): 22.

167. Lt. Colonel Arnold E. (Sandy) Weand, Jr., "The Aviator's and Mechanic's Engine," *Army Aviation*, Vol. 39, No. 2 (February 28, 1990): 50.

gines for that lot. Further competitions would be scheduled to ensure future competitive bidding. In 1988, the original 10,000-helicopter buy for the LHX program was reduced as a result of military budget reductions. This left a smaller Army engine market than either participating company had originally counted on, and it drove LHTEC to seek a variety of other military and commercial applications to broaden the production base.

The Coast Guard, with the Army, contracted with LHTEC on February 15, 1990, to conduct "proof-of-concept" flight tests in a Coast Guard HH-65 helicopter. The Coast Guard had considered the T800 as a possible replacement engine for the Textron Lycoming LTS101 engine installed in the HH-65, but in 1991 decided to stay with the Lycoming engine. Foreign military flight test evaluation programs were also conducted on the Westland Lynx, LTV/Aerospatiale Panther 800, and Augusta A129 Mangusta helicopters.

The T800-LHT-800 was military qualified in July 1993, and 32 production engine hardware sets were shipped that year to the Army for assembly and testing. Some of these engines were used to support the Boeing Sikorsky Comanche RAH-66 flight test program. Rollout of the Comanche took place in May 1995, and the first flight of the prototype number one Comanche took place in January 1996. With the Army doing the work, five UH-1H Hueys were converted to T800-LHT-800 engines and delivered to the U.S. Justice Department's Border Patrol for a flight evaluation program that began in June 1995.

In December 1992, the Army awarded LHTEC a contract to increase the power of the T800. The need for more power followed a weight increase in the RAH-66 Comanche. Designated T800-LHT-801, this first growth engine was rated at 1,550 shp for 10 minutes. The engine airflow size was increased 17% (over the T800-LHT-800) maintaining the same frame size, customer inlet interface, and configuration. The turbine temperature was slightly increased (+30 F). The power increase was accomplished with new, low-risk compressor technology. Testing of the -801 began in March 1994, and Critical Design Review was completed in February 1996.

Summarizing the T800 program, director, Military Programs Carl Baerst felt that the unique thing about it, " . . .was the fact that for us it was the first big partnership effort and represented learning how to do business in that kind of environment . . ."[168] In order to reconcile differences on technical points between the engineers from each company, Baerst felt that the primary objec-

---

168. Carl F. Baerst, interview by Rick Leyes and Bill Fleming, 7 March 1990, Garrett
Interviews, transcript, National Air and Space Museum Archives, Washington, D.C.

tive was "just a matter of good communication and trying to avoid the NIH (not invented here) syndrome." Baerst further felt that it was just as important not to "sell the product short and put in something that's not the best in order to compromise." The formal tradeoff studies done by LHTEC were one way to resolve differences of engineering opinions. By formally weighting and scoring different approaches, the facts had a way of determining the best way to go.

Commenting on the unique design requirements for the T800 engine, Senior Chief Engineer for Advanced and Support Projects Chuck Corrigan said:

> The guidelines were pretty clear from the Army that a user-friendly engine was as important to them in their evaluation and procurement cycle as was a high performance engine. Retaining that performance over a hostile environment was also important. So we had to balance all those. It's truly probably the most balanced engine design that I have worked on in the 30 years I've been in the business. It was not just driven by performance, thrust-to-weight, or horse-power-per-lb of engine weight; it was . . . driven by all of those requirements from the Army.[169]

Corrigan felt that a point of emphasis on the T800 program was, "The fact that the T800 essentially opens up a new market for Garrett, and that is the helicopter market. Up until the T800, we were not successful in entering the helicopter market, so the T800 has offered us that product for the helicopter market. That is a real, real plus for both ourselves and our partner."[170]

The T800 improved Garrett's access to both the military and helicopter markets. In a larger perspective, the T800 program was the first government-sponsored engine program that emphasized continuous competition from RFP throughout development and production phases. The purpose of the program was to reduce life cycle costs, improve the final product, and to have dual-source production capability. From the Army's point of view, the T800 program was a new way of doing business, in which it was able to obtain unprecedented contract guarantees in the areas of engine performance, reliability, maintainability, produceability, cost, and other considerations.[171] The design of the T800 was also unique in that it emphasized maintenance and cost as much as the more traditional performance requirements.

169. Chuck Corrigan, interview by Rick Leyes and Bill Fleming, 6 March 1990, Garrett Interviews, transcript, National Air and Space Museum Archives, Washington, D.C.

170. Corrigan to Leyes and Fleming, 6 March 1990.

171. Weand, "T800 Engine Program: Smart Contracting," 47.

CTS800 AND CTP800

Commercial derivatives of the T800 were also developed. In January 1989, LHTEC unveiled the CTS800, a civil version of the engine.[172] The CTS800 was offered with and without a speed reduction gearbox giving the engine broader application in civilian markets.[173] The FAA certification program for the CTS800 ran concurrently with the military T800 program, and the CTS800-0A was certificated in September 1993. The CTS800-0A was a civilian version of the military T800-LHT-800.

The first commercial production engine, designated CTS800-4N, was derived from the CTS800-0A and was similar to it in power, fuel economy, weight, life, reliability/maintainability, and operational characteristics. To reduce cost, T800 military requirements were eliminated. The CTS800-4N included a speed reduction gearbox for commercial helicopter applications. It was designed to produce excellent fuel economy over a range of 700 to 1,300 shp for either single- (6,500-9,500 lb) or twin- (9,500-13,000 lb) engine helicopter applications. For applications requiring more power than the CTS800-4N, the CTS800-50 was planned as a civilian derivative of the T800-LHT-801.

The CTS800 civil version was selected in 1995 by Hindustan Aeronautics Limited for a new Advanced Light Helicopter (ALH) being developed. The first flight aboard a naval version of the ALH took place in January 1996.

In the fall of 1996, a contract award by Federal Express to Ayers Corporation for a new-concept, small, containerized cargo aircraft led to the first selection of a turboprop version of the CTS800-4N, called the CTP800-4T. Two engines, connected by a gearbox to a driveshaft, would power a single propeller, thus providing twin turboprop redundancy without the drag of a second propeller. The combined power of the two engines was 2,400 shp. Designated Ayers Loadmaster LM200, the aircraft was designed as a new feeder aircraft for the Federal Express system.

The TPF351

Between 1986 and 1987, pursuing its interest in additional commuter aircraft applications, Garrett discussed with Embraer their requirements for an engine to

172. "LHTEC Develops Commercial Version of T800 Engine," *Aviation Week & Space Technology*, Vol. 130, No. 4 (January 23, 1989): 19.

173. Ron L. Alto and L. "Skip" Scipioni, "Next-Generation Civil Rotorcraft," *Vertiflite*, Vol. 35, No. 4 (May/June 1989): 54.

power a new aircraft. The proposed commuter and executive transport aircraft was called the Embraer/FAMA CBA-123 and was intended to replace aircraft such as the EMB-110 Bandeirante. The CBA-123 was to be a high-speed (400 mph), twin-engine, pusher-type aircraft. The CBA-123 had originally been designed around a Pratt & Whitney PT6 engine, but Embraer sought other candidate engines after it became clear that the PT6 did not have enough power when changes were made in the aircraft design.

For this aircraft, Garrett originally proposed a pusher version of the TPE331-14. Although the existing TPE331 design could be adapted for pusher application, the losses associated with this installation caused the company to focus on a rear-drive gearbox configuration.[174] Also, Garrett soon found that Embraer needed more power than the TPE331-14 could provide. Garrett then proposed a paper engine known as the TPE331-16, a free turbine, rear-drive version of the TPE331-14. However, the CBA-123 continued to grow in size and weight and even more power was needed. Thus, a new free turbine, rear-drive engine principally based on TPE331-14/15 and T800 technology was proposed. This 2,100-shp engine was known as the TPF351-20 Turboprop Fan.

During this time, P&WC proposed a more powerful engine known as the PW400. It was against the PW400 that the TPF351 competed. In September 1987, Embraer selected the TPF351 for the CBA-123. This was a new market for Garrett as P&WC had traditionally been Embraer's engine supplier. Among Garrett's selling points were a very advanced full authority digital control made by Bendix, an offer to design the engine nacelle, and an agreement to build a new propfan engine specifically for Embraer's requirements. Subsequently, on October 1, Garrett kicked-off the TPF351 program.

The TPF351-20 free turbine engine consisted of a two-stage centrifugal compressor, two-stage axial-flow gas generator turbine, and three-stage axial-flow power turbine. In order to get the required airflow, Garrett's latest technology T800/F109 compressor was geometrically scaled upward by 39%. The two gas generator turbine stages of the TPF351-20 were based on the first two turbine stages of the TPE331-14/15. The three-stage power turbine was new, but similar in design to the TFE731 and F109 fan turbine. A single, annular reverse-flow combustor, derived from the TPE331-14/15 was used. In order to minimize development time as well as reduce risk and cost, Garrett took this derivative approach to the TPF351.

---

174. R. D. Miller, "Garrett TPF351-20 Turboprop Fan Engine Development," (SAE Paper No. 891047 presented at the SAE General Aviation Meeting and Exposition, Wichita, Kansas, 11-13 April 1989), 1.

Garrett TPF351 turboprop engine. Photograph courtesy of Garrett.

The TPF351 was Garrett's first free turbine turboprop engine. Having built free turbine turbofan and turboshaft engines, Garrett had the technology to proceed with the design. For the power range required, the free turbine design provided better cycle efficiency and was less difficult to start than a fixed-shaft engine.

The free turbine design was an in-line, opposed-shaft configuration. The inlet and the accessory gearbox were on the front of the engine followed by the gas generator shaft. Facing rearward was the power turbine shaft, which drove the propeller shaft through an in-line reduction gearbox. Garrett found this configuration to be the only viable option available to meet the free turbine, pusher requirement. While the configuration was a virtual necessity, it offered important future advantages for the company. Howard Krasnow, Chief Project Engineer on the TPF351 program, assessed the layout by saying:

> . . . there really was not another way to lay out the engine without making it a free turbine, because we still have the inlet and the accessory gearbox on the front of the engine, and we had to drive the propeller from the back. That was a customer requirement. There really was not a way to do that with one shaft. So we were actually sort of forced into making the design, but we're happy with it, because now what that gives us is almost a launch pad to go

head-to-head with some of the competition and make a traditional front driver—as we call them, tractor-drive engines—with the two shafts. It's something we have been wanting to do for many years as a company, and this program gave us that opportunity . . . I think we're happy that it turned out this way, because this does give us a whole new family of engines that we really will be in a better position in the market now to compete with the Pratts (Pratt & Whitney Canada) and the other people that have this style and size class of engine.[175]

In a tractor configuration, the TPE331 engine could only be done as a single-shaft engine because there was insufficient room in the hub of the engine to carry another shaft through the engine core. Thus, when Garrett layed out the design of the TPF351, sufficient room was left for a concentric configuration which would permit a tractor configuration.

The design criteria for the engine was largely driven by the fact that the CBA-123 was a commuter aircraft and its engines were rear mounted, making the aircraft center of gravity a special concern. For Garrett, this meant that engine weight reduction and a high power-to-weight ratio were important drivers. The result was a 2,100-shp engine weighing 763 lb which gave it a 2.87 power-to-weight ratio. The 2,100-shp thermodynamic capability was also necessary to provide the 400-mph speed that the customer wanted. Because the CBA-123 was a commuter aircraft with high utilization rates, it was particularly important to have a rugged, durable engine. This resulted in a calculated low-cycle fatigue design life of approximately 100,000 cycles for the compressor and 45,000 for both sets of turbine wheels. The engine would be run about 60°F hotter than the TPE331-14, which meant that enhanced materials would be required for higher temperature capability and strength in hot section components. For instance, the first-stage HP turbine nozzle used directionally solidified cast vanes and the second-stage HP turbine blade used single-crystal technology

Some interesting technical features in the design of the engine included a bi-cast second-stage HP turbine stator assembly, a new style gear train called a compound star helical gear system, and an improved fuel nozzle/manifold system. The bi-cast assembly consisted of machined vanes of high temperature MA754 material cast into inner and outer bands of MAR-M-247 with slip joints. This yielded an integral ring nozzle with minimal leak paths and very high temperature capability. The new low-profile gear system was chosen in preference to a TPE331-type offset gearbox to reduce deflections and vibrations. This

---

175. Howard Krasnow, interview by Rick Leyes and Bill Fleming, 6 March 1990, Garrett
    Interviews, transcript, National Air and Space Museum Archives, Washington, D.C.

in-line gearbox also had an internal prop control system. The whole unit fitted into a streamlined, smooth-contour nacelle. Because the new gear teeth required very precise timing as a result of the high load level, special grinders had to be imported from Germany to fabricate the parts. To reduce weight, Garrett, for the first time, used a precision investment cast gearbox. A new fuel nozzle and manifold system that allowed individual replacement of fuel nozzles without removing the manifold was added to improve engine maintainability.

The first engine was run between May 19-23, 1989, only 19 and a half months after the start of the design. The engine was also run to full take-off power at this time, testimony to the fact that the engine was, for the most part, a known-technology, derivative design. Four engines were used in the development program and additional engines were to be added for the FAA certification program. The TPF351 was first flown on July 9, 1990 on Garrett's Boeing 720 flying testbed. On July 18, 1990, the Embraer CBA-123 made its maiden flight powered by two TPF351-20s.

By 1990, various options were being considered for the TPF351 while the certification program was underway. The engine had been designed to provide for growth up to 30% in the same frame size and to 50% within the same basic mounting envelope. Thus, Garrett was looking at growth versions up to 3,000 shp. Tractor versions of the engine were also under study. One such version, a paper engine known as the TPE341-21, had been proposed for the Dornier 328 regional airliner, but lost to the P&WC PW119 in 1988. Helicopter applications were also being explored. Another development of the TPF351 program was announced in 1990.[176] In an agreement between Volvo Flygmotor AB and Garrett, Volvo became a 5.6% risk- and revenue-sharing partner in the engine.

However, in 1992, Embraer and Allied-Signal mutually agreed to suspend furthur development of CBA123 and the TPF351. The engine was nine months from certification and had behaved very well in both development and flight tests at Embraer. Disappearing demand for 19-passenger aircraft, as well as rising costs on the development of the CBA123, dictated the cancellation.

The CFE738

In the late 1970s, Garrett was doing conceptual design work on a next generation TFE731 engine (TFE731-X) in the 5,000-lb-thrust class. The nature of this work was largely focused on the engine's axial-centrifugal compressor. This was

176. "Business Briefs," *Aviation Equipment Maintenance*, Vol. 9, No. 1 (January 1990): 36.

a four-stage axial, single-stage centrifugal compressor, which, for the sake of flexibility, would allow an axial stage to be added or deleted to vary engine performance.

Garrett's studies on the next generation TFE731 in the late 1970s and early 1980s coincided with the start of its diversification efforts to broaden both its product base and market share. Building on the success of its TFE731 program in the medium-size business jet market, a more powerful engine could expand this market to include heavier business jets.

Market studies conducted by Garrett in the early 1980s revealed a significant gap between medium-size jets and heavy-, top-of-the-line jets. These studies suggested that a plane designed to fit between medium-size jets such as Falcon 20s and BAe 125s and larger jets including the Falcon 900 and Gulfstream IV could find wide market acceptance. However, a new engine would be required for such an airplane. Garrett was not alone in perceiving this market opportunity. GE had come to a similar conclusion.

In March 1983, GE had received a Department of Defense contract to design and build an advanced turboshaft demonstrator engine in the 5,000-shp class. This demonstrator engine was called the GE27, and it was developed for the Army's Modern Technology Demonstrator Engines (MTDE) program. The potential application for this engine was the Bell-Boeing V-22 Osprey. The HP spool of the GE27 engine developed under the MDTE program was very similar to the HP spool that Garrett was studying for its next generation TFE731.

Garrett became aware that GE also had its eyes on the same business jet market, and that the GE27 core was a means to enter it. In addition, Garrett saw the similarity in the size and configuration of the Garrett next generation TFE731 and the GE27 engine. It became apparent that the best course of action was to approach that marketplace in partnership with GE. It would have been difficult for either company individually to justify the type of investment required to go it alone. Thus, starting in November 1984, Garrett and GE began to study the possibility of jointly developing a turbofan engine in the 5,600-lb-thrust class.

In June 1987, the CFE Company was formed by Garrett and GE to manage all phases of the development, manufacture, and marketing of an engine known as the CFE738 turbofan. Commenting on teaming arrangements and the CFE738 team in particular, Roy Ekrom, then Chief Executive Officer of Allied-Signal Aerospace Company said:

> There are three things that drive you on any teaming arrangement. One is financial; second is technical, and the third is market. And it depends on what

you're looking for at that particular time. For example, on the 738, we wanted to have an engine that came on top of our 731 engine. GE had a core. So it was availability of the core that was a driver for us. When you talk about investing, you talk about investing two things; one is money and the other is your critical technical talent. We had a full table for our technical talent, and we also didn't want to expend the money to develop that core. So that was a partnership that enabled us to conserve our technical asset and our financial resources and get an engine that was well positioned, we thought.[177]

The CFE Company was a 50-50 joint venture with Garrett essentially responsible for the low spool and GE for the high spool. Garrett provided the fan, front frame, accessory gearbox, LP turbine, rear frame, and lubrication system. GE provided the compressor, combustion system, HP turbine, control system, and fuel system. Both companies would be responsible for selected development and certification tests and endurance tests. GE would be responsible for core tests and Garrett for flight tests. Complete cores would be shipped to Garrett for final assembly, acceptance tests, and shipping.

The CFE738 was a high-bypass ratio, two-shaft turbofan engine. The thermodynamic cycle was optimized around the high-altitude cruise performance of medium-large business jets. The HP core contained a five-stage axial, one-stage centrifugal compressor (with the first three axial stages variable), an in-line annular combustor, and a cooled two-stage HP turbine. The LP spool consisted of a single-stage fan driven by a three-stage LP turbine. The LP spool and fan were based on the Garrett TFE731-5. The LP turbine was aerodynamically derived from Garrett's participation in the NASA QCGAT program and Army high-work turbine program. The engine was rated at 5,725 lb take-off thrust with a .371 lb/hr/lb sfc. The bypass ratio was 5.3:1 and the thrust-to-weight ratio was 4.32:1.

The launch of the full-scale engine development program was in May 1988. The core for the prototype CFE738 was shipped to Garrett in March 1990 where it was mated to the rest of the engine. Testing of the complete engine began in May 1990. Ground tests were temporarily suspended in the summer of 1990 when a vibration problem in the engine's HP turbine led to a turbine blade separation.[178] The CFE738-1 was FAA certificated December 17, 1993 and Joint Aviation Authority (JAA) validated in August 1994. In 1996, CFE was evaluating derivative engine concepts in the 7,000-lb-thrust category.

177. Ekrom to Leyes and Fleming, 8 March 1990.

178. "Turbofan Tests Suspended," *Aviation Week & Space Technology*, Vol. 133, No. 6 (August 6, 1990): 11.

The CFE738 turbofan engine was jointly developed by Garrett and General Electric. Photograph courtesy of Garrett.

An interesting aspect of the CFE738 development program was the technology interface between the two companies. The division of responsibilities was generally determined by the company that had the better or already developed technology. In the strictest sense, the venture was not a technology sharing program. For example, fundamental design, stress, and aerodynamic codes were not transferred between companies. However, there were examples where the best of the two companies' technologies were incorporated in the engine. In the case of the mixer nozzle, when it was demonstrated that GE had a better technology, Garrett used that technology to design the nozzle. On the other hand, a strong suit of Garrett's was shaft dynamic techniques used on hydraulically–dampened bearings that were incorporated in the engine. In the case of some material properties, which were particularly expensive to develop, some technical exchanges were made between the companies.

Another important part of the program was its management. While higher level management agreed on the principles of the cooperative venture, it was up to the program participants to make it work. To resolve potential problems and set up an open door policy, both companies placed resident representatives in each other's plants. These representatives participated in the program as though they were employees assigned to the program. In the case where a difference of engineering opinion existed between the companies, the person responsible for that particular element would make the final decision. If the other company disagreed, a provision was made for a Technical Advisory Committee, composed of senior specialists in the area, to resolve the problem.

To make joint venture programs work, participants had to do more than meet their contractual requirements. According to Jerry McCann, Program Engineer in charge of the CFE738:

> One of the men at GE who has some other experience on some of these joint venture programs—and this particular program is supposedly a 50-50 joint venture—but he said it as though it's another one of these 60-60 joint ventures. That is probably something that is really true. We're seeing you can't just do half. You have to feel as though you're doing more than half and just live with it, because we oversee, or feel an obligation, to look at something that they have a responsibility for, and vice versa . . . In order to feel comfortable, we just have to do a little bit more. We overlook what they're doing a little bit, and they do the same. We do joint design reviews of each other's components, and we have to respond to each other's concerns.[179]

The competition for airframe applications proceeded in parallel with the CFE738 development program. In 1989, in its first competition for the medium-large business jet market, the CFE738 lost to P&WC's PW305 for the British Aerospace BAe 1000. However, success did come in 1990 when the CFE738 engine was selected for the Dassault Falcon 2000. While the CFE738 was originally designed for medium-large business jets, a secondary potential application was the relatively new regional jet market. In competition against Allison's GMA3007 and Textron Lycoming's LF507 for the Embraer EMB-145 regional jet, the CFE738 lost to the GMA3007 in March 1990.

The CFE738-1 made its first flight on the Falcon 2000 on March 4, 1993. The first production CFE738 was delivered in September 1993, and the engine was certificated in December 1993. By the end of 1996, 104 CFE738 engines had been produced.

179. Jerry McCann, interview by Rick Leyes and Bill Fleming, 6 March 1990, Garrett
   Interviews, transcript, National Air and Space Museum Archives, Washington, D.C.

In 1996, the CFE company was evaluating derivative engine concepts with thrust ratings exceeding 7,000 lb. The concepts included use of a higher efficiency fan to produce the higher thrust and improved fuel efficiency.

Garrett felt that one of the important aspects of the program was the good fuel efficiency of the engine. Program Engineer Jerry McCann stated, "One of the things that we believe this engine is going to bring to the business jet area—it's really an engine that was designed for a business jet—is that it brings the levels of fuel economy down to the levels of some of the big turbofans that are in commercial service right now. Our specific fuel consumptions are down in the ranges of 0.644 (at maximum cruise, M=0.8 at 40,000 ft), and those are in line with some of the modern large turbofan commercial engines."[180]

Expendable Turbine Engines

Another aspect of Garrett's diversification effort, but somewhat closely held by the company, was its continuing work, beginning in the early 1970's, on low cost, expendable turbine engines. The ETJ131 (Expendable Turbojet) engine in the mid-1970s was an attempt to turn an existing automotive turbocharger product into an ultra-low-cost turbojet engine to power small missiles and decoys. The ETJ331 was an engine designed for Garrett's attempt to secure the Harpoon missile business in the early 1970s. Another early 1970s program was the ETJ341. While these programs did not result in production applications, a renewed effort in the early 1980s resulted in some important technology developments. This was the ETJ1081 program, followed by the ETJ1091 program.

ETJ331, ETJ341, AND ETJ1000

In late 1971, Garrett started development of a low-cost, limited-life turbojet engine called the ETJ331 for use in RPVs, drones, and missiles. On March 20, 1972, Garrett submitted a proposal for the Navy Harpoon missile engine development competition. Garrett's entry was its ETJ331; this engine was given the military designation YJ401-GA-400.

The YJ401 compressor was based on the GTCP660 APU four-stage axial compressor. It had an in-line annular combustor. The single-stage axial-flow turbine was a new-design, one-piece casting. It was a straight-through-flow de-

180. McCann to Leyes and Fleming, 6 March 1990.

sign with a 500-1,000-lb-thrust rating. Additional details on the engine performance were not available.

Late in entering the Harpoon missile engine competition, the YJ401 ran into problems getting ready for the qualification tests. The principal obstacles were the fuel control and starter. A fluidic fuel control (using air-driven computer logic) was chosen because it was believed to be low cost, and impervious to electrical interference. However, buried inside the engine, it proved to be very temperature sensitive and would not maintain its setting. Failure of planetary gears in a small cartridge starter resulted in uncontained failures of the starter during laboratory tests. Because of the fuel control and starter problems, the YJ401 only made two out of five required starts during captive flight tests.

Shortly after those unsuccessful tests, Teledyne CAE won the Harpoon missile engine competition with its JA402-CA-400 engine. However, the Garrett YJ401 Flight Readiness Qualification engine was completed in 1973 under the Navy contract and delivery "YT" engines were used successfully in a modified prototype "Condor" standoff tactical missile. Powered by a YJ401, the prototype flew a 100-mile mission (three times the range of the rocket-powered Condor) and hit its target, a fast moving patrol boat, in the pilot house.

The Harpoon missile engine program was followed by the ETJ341. The ETJ341 was proposed by Garrett in May 1974 as an engine for the Tomahawk missile. This was a 1,000-lb-thrust-class turbojet engine with a reusable rating for possible use in medium-range, recoverable RPVs. Simplicity and low production cost were design goals. The engine configuration was the same as the ETJ331/YJ401, with a zero-stage fan added to the compressor. The engine was successfully tested under Navy contract in the Garrett high-altitude test facility.

In June 1974, Garrett completed component testing on the ETJ1000 for another Navy missile application. The engine was similar in configuration to the Harpoon missile in that it was a three-stage axial-flow compressor engine. However, this was a higher thrust application for up to Mach 2.5 application. No application was found.

ETJ131

In 1976, perceiving a market for small radar-killer missiles, RPVs, and target drones, Garrett initiated a short-term project to build an ultra-low-cost turbojet engine to power such missiles. Senior Project Engineer, Military Fan and Jet Engines, Bill Norgren was a participant in this project, known as the ETJ131. Norgren recalled, "The way we felt that it was possible to get a truly low cost

engine was to take an existing component. So in Los Angeles, they started doing some work on that. They took the biggest compressor they had and the biggest turbine that they had out of the turbocharger, and started building a pure
jet engine out of it."[181]

A production Model T-18A turbocharger from the AiResearch Industrial
Division, which was used on truck engines, was converted to a turbojet engine.
In order to use an unmodified turbocharger, a combustor was fabricated and
placed in a "sore thumb" (perpendicular to) configuration with respect to the
radial compressor and radial turbine elements of the turbocharger. A rudimentary fuel control was developed. Various lubrication systems were tried with a
fuel mist/oil mist found to be the most desirable for a production engine.

The design criteria specified a 100-lb-thrust engine with a one-start, one-
hour life at maximum thrust.[182] The operating envelope was sea level to 20,000
ft and up to Mach 0.9. The engine could burn gasoline, JP-4, or JP-5. By September 1976, when the project was completed, the engine had demonstrated a
zero-speed, zero-altitude thrust of 95 lb with a sfc of 1.7 lb/hr/lb.

A follow-on, augmented version of the engine was built and demonstrated
between 1976 and mid-1977. This engine, known as the ETJ131 Model 1030,
was partly funded by the Air Force Aero Propulsion Laboratory (AFAPL). To
the basic ETJ131, a ramjet-type sudden-expansion burner (dump combustor)
was added. The engine demonstrated a net thrust of 200 lb and sfc of 3.09
lb/hr/lb at sea level Mach 0.7.

The original ETJ131 was designed to compete against two-stroke-cycle reciprocating engines powering existing missiles. This was part of the reason why
a very low per-unit cost engine was necessary. However, no application was
found for either version of the engine despite its demonstrated performance and
promise of low cost. According to Bill Norgren, finding a sponsor to underwrite the development cost was the problem:

> We always felt that it was the kind of thing that had great promise if some-
> body would put up the up-front money to get the development under way, get
> into some sort of a qualification program. At that time, nobody felt that they
> could afford the couple of hundred thousand dollars that it would take to get
> the engine to a point where it would be qualifiable or part way through some

181. William M. Norgren, interview by Rick Leyes and Bill Fleming, 6 March 1990, Garrett
Interviews, transcript, National Air and Space Museum Archives, Washington, D.C.

182. C. F. Baerst and W. M. Norgren, "Development of an Ultra-Low-Cost Gas Turbine,"
(NARPV Paper, Garrett No. 21-2575, presented at the NARPV Symposium, Washington,
D.C., 6-8 June, 1977).

sort of a preliminary qual (qualification) program. So it was dropped since it wasn't in the mainstream of the Garrett family at that time.[183]

ETJ1050, ETJ1081, AND ETJ1091

After a number of attempts to enter the expendable engine business in the 1970s, no production applications were found. In the early 1980s, Garrett decided to make a renewed, dedicated effort to get into this business when new opportunities arose in the expendable engine business. This effort coincided with its major diversification drive and relied on Garrett's superior technology base and low-cost turbocharger manufacturing capability. The rationale for this effort was outlined by Robert Choulet, Allied-Signal Aerospace Company President, Engine Group:

> . . . there were a couple of major "black" competitions coming up for roughly 1,000, 1,200 lb thrust engines, which needed a new technology engine that didn't exist either at Williams or at Continental or in the foreign competitor. So it looked like an ideal opportunity with the marketplace shifting towards unmanned vehicles, big quantities, and a gnawing threat in the market sense, that if you let this guy alone long enough, that eventually he'd make enough money to come after your man-rated business, which he's doing. Let's take him on.[184]

Garrett was also prompted to re-enter this business by the military, concerned that existing expendable turbojet engines were both too expensive and too inefficient. The challenge for Garrett was to produce a turbojet engine in the 1,000-lb-thrust class at half the cost and 25% better fuel consumption than first generation engines such as the Teledyne CAE J402 and Microturbo TRI-60. The funding for the program came from both Garrett and the military.

A paper engine proposed by Garrett in August 1981 was the ETJ1050, intended as an engine for an advanced cruise missile. It offered potentially better performance and improved range for the missile. It was a two-spool, low-bypass ratio fan engine. Some components and the core engine were built and rig tested. The core engine had a high pressure ratio centrifugal compressor and a laminated, cooled turbine wheel incorporating low aspect ratio turbine blades. However, the program was canceled by the potential customer.

In 1984, Garrett started development of the ETJ1081 engine. The ETJ1081 was a high-technology single-spool turbojet engine, using a first-stage mixed-

183. Norgren to Leyes and Fleming, 6 March 1990.

184. Choulet to Leyes and Fleming, 8 March 1990.

flow and a second-stage centrifugal compressor, an air-blast atomizing combustor, a high-work, axial-flow single-stage turbine and fuel-lubricated ceramic bearings. The engine used a cartridge-type tip impingement starter for the turbine. Because it made extensive use of castings, the engine had a very low parts count. The engine was designed to cost substantially less than $50,000 per copy based on the number of engines ordered. It was only a little over 13 inches in diameter, 24 inches in length, and weighed 131 lb.

The mixed-flow compressor was developed by Garrett in Air Force and Navy technology programs in the late 1970s and early 1980s. The mixed-flow compressor was described by Chief of Aero-thermo Component Design Bill Waterman:

> The mixed-flow compressor has high pressure ratio in a single stage by virtue of the radial mixed-flow flow path, where at least some of the pressure rise is achieved by centrifuging the flow. It requires relatively high exit tip speeds, but it does not have the high frontal area of the centrifugal compressor because the diffusion system is flowing in the axial direction rather than radial. So it has a relatively high flow per unit of frontal area, not as high as an all-axial machine, but much higher pressure ratio capability than an all-axial machine by virtue of the higher exit wheel speeds of the rotor that can be achieved.[185]

The single-stage, mixed-flow compressor generated a pressure ratio rise in excess of 3:1. It had other advantages as well. It was very tolerant to distortion, an important characteristic for unmanned engines which use flush inlets to conserve space, but which induce high distortion. The mixed-flow compressor tolerated such distortion due to its large surge margin.

Other interesting aspects of the engine were the high-work turbine rotor, ceramic ball bearings, and combustor. The turbine rotor incorporated Low Aspect Ratio Turbine technology. A Garrett patented ceramic ball bearing was used for the thrust bearing. This allowed the use of fuel as the lubricant as well as the bearing coolant, eliminating an oil subsystem and providing ease of long-term storage. A small, Garrett-patented integral fuel pump provided high pressure fuel to pure airblast nozzles in the combustor. The integral design provided high speed with low volume to improve packaging.

The early configuration ETJ1081 suffered from a mismatch between the first- and second-stage compressor. Also, the turbine exit guide vane was not optimized for tested rotor exit conditions. In early 1987, these components were modified, and predicted component efficiencies were achieved and exceeded in

185. Waterman to Fleming, 7 March 1990.

some cases. Early versions of the engine achieved 1,000 lb thrust in 1985 with the non-optimized hardware when the turbine temperature was pushed up and the combustor outlet temperature pattern factor improved. Later, in the corrected configuration, the engine demonstrated 1,325 lb thrust.

At the end of 1987, Garrett, in partnership with MTU and Fiat, proposed the ETJ1081 for a NATO-initiated, five-nation Modular Standoff Weapon (MSOW) program. The reason for joining with MTU and Fiat was a requirement that the major subsystem had to be produced by a consortium which allowed company and country sharing of work and production. MTU had responsibility for the compressor testing and fabrication, and Fiat had responsibility for the turbine. Both Rockwell and Convair represented the U.S. in the two competing MSOW airframe consortia. Garrett was selected by General Dynamics, which lost in 1989 to the Alliance Defense Corporation headed by Rockwell. Shortly afterward, the MSOW was canceled.

As of 1990, Garrett continued to investigate domestic and foreign possibilites of using the engine for RPV target vehicles, an anti-submarine missile, and other standoff weapons. Primary competitor engines for the ETJ1081 were the Teledyne CAE Model 382-10 and Williams WR8300. To a lesser extent, the Microturbo TRI-60 was also a competitor.

The ETJ1091 was a supersonic version of the ETJ1081. Development started in 1985, and by 1990, the ETJ1091 had been run as a demonstrator engine. Development proceeded under a classified Air Force program called Expendable Turbine Engine Concepts (ETEC). Like the ETJ1081, it used a mixed-flow compressor, but the second-stage compressor was axial-flow to reduce volume. The engine was designed to be a very high-performance, high-temperature, high thrust-to-weight, small-diameter engine using advanced materials technology.

Despite two decades of trying to enter the expendable missile engine business, Garrett had yet to find a production application by the mid-1990s. However, its missile programs advanced both the company's and DOD's technology base for these types of engines. The ETJ1081 and the ETJ1091 provided the company with a potential expanded product base.

Research and Technology Development

There were many technical highlights associated with Garrett. Among these, the twin backward curved centrifugal compressor developed by W. F. Mayer in his successful 1940s technology demonstration program known as Project A was a

The enduring influence of W. F. Mayer's backward curved centrifugal compressor rotor design can be seen by comparing the Garrett GTC43/44 first-stage rotor (the smaller rotor) to the Garrett T800 first-stage rotor (the larger rotor). Photograph courtesy of Garrett.

keystone in the genesis of the company's compressor technology. Mayer's legacy became Garrett's hallmark compressor for many of its engine designs. Analyzing the lasting effect of Mayer's work, Homer Wood observed, "I do not know anyone associated with the T800, but I do see in it a lingering influence of W. F. Mayer."[186]

It is interesting to note that it was the ruggedness and the compactness of the centrifugal compressor design, making it ideal for small gas turbine engines (at least up to about 2,000–3,000 shp), that made Garrett continually invest in improving both its design technology and its manufacturing technology. The initial technology developed by W. F. Mayer for small pressure ratio compressors was based on incompressible water pump technology. In order to achieve pressure ratios of 4–5:1 for APUs and 8–18:1 for aircraft engines, Garrett had to develop the viscous flow analysis tools required to design compressors with supersonic relative inlet Mach number and compact diffusers optimized to obtain high stage efficiencies. Commenting on the significance of Garrett's compres-

186. Homer J. Wood, Sherman Oaks, California, to Neil Cleere, letter, 21 March 1990,
    National Air and Space Museum Archives, Washington, D.C.

sors, Chief of Aero-thermo Component Design Bill Waterman said, " . . . we remain, I believe, the leader in the U.S., if not the world, in two-stage centrifugal compressor technology in terms of pressure ratio, efficiency, and surge margin."[187]

An important factor in bringing Garrett's compressor technology to today's level was the development of the transonic inducer. Chief of Aerodynamics Bob Bullock and his engineering team were able to improve significantly the airflow and pressure ratio of the TPE331, and thereby engine power, by raising the inlet airflow relative Mach number to 1.3. This contribution was carried on in subsequent Garrett compressor designs.

Garrett continuously invested in manufacturing equipment required to carve very precise and complex aerodynamic shapes out of titanium forgings, economically. Very sophisticated 4-spindle, 5-axis milling machines were custom developed by Garrett for this purpose. They were initially cam operated and subsequently numerically controlled programmable. The accuracy, precision, and geometric complexity of the typical backward curved centrifugal impeller mandated a very close collaboration between the aerodynamicist, the design engineer, the manufacturing engineer and the machine operator—"concurrent engineering" as it became known. A key decision made early on in this team approach was the adoption of the straight element technique to generate airfoils, which permitted the use of a simple conical cutter. While this imposed an inherent constraint on the aerodynamicist, it allowed high performance designs to be manufactured economically.

Resolving the bird ingestion problem by finding a proper balance between the conflicting mechanical and aerodynamic requirements was a difficult task in both the TFE731 and ATF3 programs. On one hand, blade integrity during bird impact necessitated the use of thick leading edges, while on the other hand, thinner leading edges were required for good aerodynamics. This was not only a difficult problem for the ATF3 and the TFE731 programs, but for small engines in general because the size of a medium to large bird was large in proportion to the engine fan. The solution to this extremely difficult technical problem was achieved by supplementing outstanding inter-discipline teamwork with the advanced 3-D computational methods developed by Garrett for handling transonic flow with relatively blunt airfoils.

From an environmental perspective, the technology jointly developed by Garrett and NASA-Lewis in the Quiet, Clean, General Aviation Turbofan

187. Waterman to Fleming, 7 March 1990.

program was important for reducing engine noise and emissions in later versions of the TFE731. Of special interest in the QCGAT program was the development of the compound mixer nozzle, which pre-mixed fan and core airflow prior to discharge to ambient conditions. This had the double benefit of reducing noise and improving thrust.

Garrett's participation in the Air Force Low Aspect Ratio Turbine program, as well as earlier work on the TSE231, yielded very important applied technology in the high-work turbine design area. According to Bill Waterman, Garrett was ". . . the first company in the country to utilize three-dimensional (3-D) viscous analysis methods for the design of turbomachinery blading. Many people had been working in 3-D analysis, but nobody at that time (mid-1970s) had ventured to use it in design."[188] The 3-D analysis, developed by Garrett aerodynamicist Paul Dodge and his associates, resulted in reduced nozzle losses and improved rotor efficiency. This technology continued to be used in all Garrett high pressure ratio turbines. It was also especially useful as an analytical tool to predict engine performance for small engines that were difficult to instrument for testing.

Several aspects of the F109 turbofan and T800 turboshaft programs were noteworthy. Garrett's F109 engine was not only the first engine designed to meet the Air Force Engine Structural Integrity Program requirements, but helped develop the methods and procedures for it. The result was a new standard for higher reliability, improved durability, and low life cycle costs for future military engines. The LHTEC T800 engine developed under the Reliability, Availability, and Maintainability/Integrated Logistic Support criteria resulted in a more "user friendly" engine in terms of maintenance and also reduced life cycle cost.

Other Garrett technology included the first production application of a full authority analog electronic control, which was used first on the TFE731. It was upgraded, in the early 1980s on the TFE731-5, to a digital electronic engine control to take advantage of that technology in terms of programming flexibility, self-test, fault annunciation, and trim-free capabilities. Garrett developed high-speed, highly-loaded, precision tolerance bearings which exceeded the state of the art for its engines. While generally not specializing in materials development, Garrett did leading research on next-generation composite, metal-matrix, and ceramic turbine engine components. Examples included ceramic combustors, silicon-carbide-fiber-reinforced disks, titanium-aluminide compressors, and carbon-carbon-type turbine wheels.

---

188. Waterman to Fleming, 7 March 1990.

Overview

THE STATUS OF THE CORPORATION

In 1996, the AlliedSignal Engines Division of AlliedSignal Aerospace Company, was a leading supplier of propulsion engines for business aviation and regional airlines, and had emerging business across a broad front of military and commercial engines. With the acquisition of the Lycoming Turbine Engine Division of Textron in 1994, it also had greatly strengthened its presence in military and civil helicopters, and in higher-thrust business jet and regional jet engine applications. The company was the leading world supplier of APUs for commercial and military aircraft, and it produced gas turbine engines for industrial and marine applications.

AlliedSignal Engines was headquartered in Phoenix, Arizona at Sky Harbor Airport. At this location in 1996, the division had 6,300 employees and a plant size of 2.5 million sq ft. Starting in 1994, the Lycoming operation in Stratford, Conn. was gradually transitioned to Phoenix. AlliedSignal completed its operations at Stratford and turned the facility over to the Army in September 1998. In addition, a manufacturing site in Greer, S.C. had some 450 employees working in 200,000 sq ft of facilities. The 1996 annual sales for the combined "Garrett" and "Lycoming" engines was $2.1 billion.

Historically, the majority of Garrett's aircraft propulsion engine business came from two engine types: the TPE331 turboprop and the TFE731 turbofan. The majority of TPE331 engines shipped were for executive aircraft. However, as a result of a shifting market in the 1980s, most of the subsequent TPE331s produced were for regional airline aircraft, a market that Garrett had aggressively pursued; in fact, in 1989, the majority of the world's 19-passenger regional airline aircraft were powered by the TPE331. Sixty percent of all models of business jet aircraft were powered by TFE731 engines.

THE COMPANY'S MARKETS AND INFLUENCING FACTORS

On the basis of an extremely successful gas turbine APU business, Garrett was able to build a financial, technical, and production base for the development of major gas turbine aircraft engines. As a result of the success of the TPE331 and TFE731, until the early 1990s, the principal markets for Garrett engines were general aviation and small transport aircraft. Pratt & Whitney Canada was Garrett's primary competitor. Garrett's TPE331 and P&WC's PT6 were historic

AlliedSignal Engines (formerly Garrett) main manufacturing facility (foreground), hangars, flight test facility, and flying testbed aircraft border the Sky Harbor Airport runway. The company is located in Phoenix, Arizona. Photograph courtesy of AlliedSignal Engines.

competitors, and to a lesser extent, Garrett's TFE731 turbofan and P&WC's JT15D also competed with one another.

To broaden its product base, as early as the 1970s, Garrett pursued engine diversification through joint ventures with other engine manufacturers. As a result, development began on five new engine types during the 1980s. By the early 1990s, the majority of these engines had been developed for possible applications or committed for production. These included the: TFE1042, produced in partnership with Taiwan's AIDC for their IDF lightweight military fighter; the Air Force F109 turbofan for military trainers; the Army T800 turboshaft engine produced by the Garrett/Allison joint venture Light Helicopter Turbine Engine Company for the Comanche helicopter; the TPF351 for regional airliners; and the CFE738, produced by Garrett/GE joint venture CFE company for medium-heavy business jets.

In 1994, Garrett, which had become AlliedSignal Engines, further diversified its product base with the acquisition of the Lycoming line of gas turbine en-

gines. The ALF502/507 line of regional aircraft turbofan engines nicely reinforced AlliedSignal's presence in this market. The LTS101, T53, and T55
turboshaft engines provided AlliedSignal with increased presence in the helicopter engine market. The TF40 established a position in marine and industrial
applications. Further, AlliedSignal acquired additional engineering expertise
and manufacturing capability.

By the mid-1990s, the company's original diversification strategy had achieved
mixed results. While Garrett's TFE1042, T800, and CFE738 had found applications, the F109 and TPF351 had been less successful. The emphasis of the original
diversification strategy was on military engine developments to remove the company from its dependence on the general aviation and regional airline markets and
to develop advanced technology with financial support from the military. The
emphasis on military programs paralleled a decade of expanding defense budgets
associated with President Ronald Reagan. With shrinking defense budgets in the
1990s, the success of military programs such as the T800 hinged on the production of airframe applications. AlliedSignal's strategy for the 1990s included selling
its developed products, looking for additional applications, and continued diversification of its product lines, which it accomplished in large part by acquiring
Lycoming. For its military engines, the company continued to look for retrofit
opportunities on existing airframes, civilian applications, and foreign military applications.

In a larger perspective, the paths taken by Garrett's various engine programs
mirrored other major trends which occurred in the aerospace industry. The
partnerships entered into on the TFE1042, T800, and CFE738 engine programs
reflected the high costs of development, limited and high risk markets for the
products, new government procurement strategies, and foreign market access
trends occurring internationally. Of the engines developed in the 1980s, only
the TFE1042 and ETJ1081 were new centerline engines, due to the high development costs. That the TFE1042 could be developed as a wholly new engine
was an opportunity made possible by Garrett's partner, the Republic of China's
Aero Industry Development Center, which brought with it an application and
joint financing. The CFE738 program, which brought together Garrett and GE,
required partners who could contribute largely derivative engine components
for a shot at a potential, but high-risk, market for medium-heavy business jets.

The direction taken by the T800 program responded to a shift in military
procurement strategies from engines emphasizing maximum performance toward more maintainable engines with lower life cycle costs. The F109 program
demonstrated another shift in military procurement strategy. To avoid the risk

and sizeable investment required for new engines and airframes, the military was looking for already developed "off-the-shelf" technology. While Garrett benefited by the government paying for most of the development costs of both the F109 and T800 programs, the contractual guarantees extracted from the T800 program illustrated the increased cost of doing business with the government. More development costs were shared by participating companies.

In the commercial area, at least one market driver remained generally constant over time. The need for higher, faster, and more fuel efficient aircraft continued to drive the TPE331 and TFE731 engine programs. Operating costs, also a constant driver, were addressed in particular in the 1980s with a Garrett Maintenance Service Plan. This was essentially an insurance program for the TFE731 and some models of the TPE331 that allowed operators to have predictable engine maintenance costs and continuous coverage of engine service and repair. The changing trends in the commercial area were the increased demand for small transport engines, such as the TPE331, and the increasing globalization of the market as more foreign sales became a major portion of Garrett's business.

CONTRIBUTIONS

Historically, Garrett specialized in aircraft APUs and was a leading supplier of propulsion engines for business aviation and regional aircraft. In 1948, Garrett manufactured its first gas turbine APU. While other companies produced small gas turbine aircraft engines before then, Garrett remained the oldest company in the U.S. to continuously manufacture small gas turbine engines. The TPE331 engine contributed to making possible the first generation of high performance turboprop executive and commuter aircraft. The TFE731 engine, which powered most models of medium-size business jets, helped transform the utility of the business jet by giving it transcontinental range and expanded corporate market areas.

MINIATURE AND MODEL TURBINE ENGINES 11

Miniature Turbojets

In the mid-1980s, a need emerged for miniature turbojet engines to propel small tactical missiles. Earlier, the pacing technology in small tactical missiles had been on-board electronics. Because of the limited capability of such electronics to perform complex tactical missions, the solid rocket with its low cost, high reliability, and simplicity of operation was the choice for sustainer engines in most small tactical missiles. But, by the mid-1980s, significant advances had been made in the areas of microprocessors, seeker/sensor technology, and navigation systems. As a result, a more fuel efficient sustainer propulsion system than the solid rocket was needed to exploit the potential of enhanced electronic system capabilities now available to small tactical missile systems.

The turbojet engine was viewed as the answer because of its substantial fuel economy advantage over that of the solid rocket. The turbojet offered an order of magnitude improvement in fuel economy, which translated into significant increases in range. In addition, its operating time was limited only by the amount of onboard fuel capacity, and it had a minimal visual and infrared signature compared to the rocket. Although the turbojet was the engine of choice, no turbojets that were small enough to be suitable for small tactical missiles existed in the mid-1980s. The only operational turbine-powered tactical missiles at that time were the turbojet-powered Harpoon and the turbofan-powered Tomahawk, both relatively large and expensive missiles powered by correspondingly large (i.e. for small tactical missiles) and expensive engines.

For the turbojet to replace the solid rocket in small tactical missiles, it was necessary that its performance advantages be accompanied by production costs, reliability, maintenance requirements, and service life that were comparable to those of the solid rocket. Low cost was considered the design driver. It was evident that if such turbojet propulsion systems could be developed, the capabilities of the next generation of tactical missiles could be greatly enhanced.

It was also clear that development of miniature turbojets satisfying the requirements of small tactical missiles would require a radical departure from traditional man-rated engine design practices. In 1985, the Propulsion Directorate of the U.S. Army Missile Command (MICOM) initiated a program to develop what became a series of small turbojet engines exclusively intended for small tactical missile applications. By 1992, 19 such engines had been developed or were under development by six companies in support of MICOM's program. These engines ranged from four to eight and one-half inches in diameter with static thrust ratings from 32.5 to 274 lb. The six companies that developed these engines and the number of engines developed by each were as follows:

Allison Gas Turbine Division	1
Sunstrand Power Systems	5
Technical Directions, Inc.	2
Teledyne CAE	7
Turbine Technologies Ltd.	1
Williams International	3

Three of these companies, Allison, Teledyne CAE, and Williams International were major manufacturers of small gas turbine aircraft engines, and their history and work is discussed in detail in their respective chapters in this book. Sunstrand Aerospace, headquartered in San Diego, Cal., had been developing and manufacturing small gas turbine auxiliary power units since the mid-1950s. Following its acquisition of Solar's Turbomach Division from the Caterpillar Tractor Company in 1985, Sunstrand began developing a series of miniature turbojet engines for MICOM.

Turbine Technologies, Ltd. was a small firm in Chetek, Wis. involved in advancing turbine technology. The miniature turbojet engine that the company had developed and subsequently offered to MICOM was the 33-lb-thrust SR-30. That engine was also the central element of a miniature propulsion laboratory, the Mini Lab, which Turbine Technologies had developed and was marketing as a turbine engine teaching aid for engineering schools. In response to interest expressed in 1993 by homebuilt aircraft enthusiasts, Turbine Technologies developed an uprated version that produced up to 80 lb thrust. Two of those engines were installed on a Bede Aircraft BD-5 homebuilt airplane for display at the 1994 Experimental Aviation Association Convention in Oshkosh, Wis.[1]

Technical Directions, Inc. was a small company located in Ortonville, Mich., 50 miles north of Detroit. The company's specialty was small low-cost gas turbine engines. In the design of its engines, the company attempted to use existing production components wherever possible. For example, automotive and truck turbosuperchargers provided a variety of components of the size and type that could be used in small gas turbine engines.

The 19 engines developed by these six companies were designed to be low cost, reliable, and expendable. All of the engines were relatively simple single-spool turbojets with annular combustors and single-stage turbines. Beyond that, their design features varied.

The Allison (Model 120) and Turbine Technologies (Model SR30) engines had single-stage centrifugal compressors, axial turbines, and reverse-flow combustors. All five of Sunstrand's engines (Models TJ-20/-70/-90/-50/-50C) used a back-to-back compressor and turbine arrangement in a single component known as a monorotor. The monorotor design feature, which had been developed by Solar in 1947, was passed on to Sunstrand in its acquisition of Solar's Turbomach Group in 1985. The radial-flow turbine was characteristic of the monorotor design. Three of Sundstrand's engines (TJ-20/-70/-90) used single-stage centrifugal compressors, and the other two (TJ-50/-50C) had single-stage mixed-flow compressors. The two Technical Directions engines had single-stage centrifugal compressors with one engine (Model TD4) using a mixed-flow and the other (Model TD7) a radial turbine. All seven of Teledyne CAE's engines used single-stage compressors, three (Models 305-4, 305-7, 318) of which were centrifugal and four (Models 304, 304C, 305-10, 305-10C) of which were mixed-flow. Of these engines, two (Models 304, 304C) had mixed-flow turbines, and the other five used single-stage axial turbines. Two (WJ-119, P8910) of the Williams engines used six-stage axial-flow compressors, and the other (P9005) had a single-stage mixed-flow compressor. All three Williams engines had single-stage axial turbines.

The stimulus for the development of these engines came from the Fiber Optic Guided Missile (FOG-M) program, a technology demonstration program that had been conducted in-house by the Research Engineering and Development Center of MICOM during the early to mid-1980s. The FOG-M system used a missile with a television camera seeker and a two-way fiber optic guidance link. Several versions of the FOG-M, using solid rocket propulsion, were built and

1. *Sport Aviation*, Vol. 43, No. 6 (June 1994): 13.

flight tested. The most mature variant, the Initial Operational Evaluation (IOE) System, was approximately six inches in diameter, 66 inches long, weighed 79 lb, and had a range of 6 miles at a flight speed of 225 miles per hour. During the course of testing this system, it became evident that the solid rocket sustainer severely limited the performance and range of the missile, leading to MICOM's program for the development of a more efficient replacement sustainer engine.[2]

LOCAAPS PROGRAM

Beginning in 1985 and continuing through 1992, MICOM conducted six programs for the development of miniaturized expendable turbojet engines. The first of these was the Low Cost Air Augmented Propulsion System (LOCAAPS) program, which extended through 1988. The purpose of that program was to develop high efficiency sustainer technology for long-range versions of FOG-M. During the course of the program, three engines were developed and demonstrated.[3] Contracts for engine development, fabrication, and testing were awarded to Sunstrand for its 43-lb-thrust TJ-20 and to Teledyne CAE for its 40-lb-thrust Model 305-4.[4] Both engines had been designed using corporate Independent Research and Development (IRAD) funds. At the same time, Turbine Technologies Ltd. had developed the 33-lb-thrust SR30 using company funds. The design of all three engines was basically conservative with emphasis on low production cost at the sacrifice of fuel economy and thrust-to-weight ratio. Both the TJ-20 and 305-4 were successfully demonstrated in a one-fifth-scale MiG 27 radio controlled target vehicle, and MICOM procured two SR30 engines for experimental evaluation. Other than the miniature engines developed earlier by the Dreher Engineering Company that were described in Chapter 3, these three six-inch-diameter-class engines were the smallest U.S. turbojet engines that were known to have been built up to 1988.

NLOS PROGRAM

Meanwhile, based on the success of the FOG-M program, a decision was made to initiate full-scale development of a turbojet powered FOG-M-type weapon sys-

2. Dr. J. S. Lilley, "The Development of Low-Cost Expendable Turbojet Engines at MICOM," (U.S. Army Missile Command, Redstone Arsenal, Alabama, 1992), 1-13.

3. Lilley, "The Development of Low-Cost Expendable Turbojet Engines at MICOM," 1-13.

4. A. Jones, H. Weber, and E. Fort, "GEMJET—A Small, Low Cost, Expendable Turbojet," (AIAA Paper No. AIAA-87-2140 presented at the AIAA/SAE/ASME/ASEE 23rd Joint Propulsion Conference, San Diego, Cal., 29 June - 2 July 1987), 1-12.

tem. During 1987 and 1988, performance requirements were drawn up for that system, the Non-Line-Of-Sight (NLOS) missile. In 1988, a contract was awarded to the Boeing Company for the NLOS, a Fog-M based anti-helicopter weapon system. The NLOS was designed to have a significant increase in flight speed and range over the IOE FOG-M solid rocket system, mandating that a turbojet be used as its sustainer engine. In addition, the NLOS was significantly larger and heavier than the IOE FOG-M. As a result, the NLOS required a higher thrust turbojet sustainer engine than those that had been developed and demonstrated in the LOCAAPS program. Key requirements of the NLOS sustainer system were that it have a low unit cost and be a complete self contained "wooden round" propulsion system that could be structurally integrated into the missile airframe.

Three companies competed for the contract to provide a new larger turbojet sustainer engine for the NLOS. In anticipation of a contract award for the NLOS, Sunstrand, Williams, and Teledyne CAE each conducted aggressive IRAD programs to develop an engine that would satisfy NLOS requirements. Sunstrand began design of the 97-lb-thrust TJ-90, an uprated version of the TJ-20. Teledyne CAE initiated work on the 90-lb-thrust Model 305-7E, an upgrade of its 305-4. Williams began development of the 92-lb-thrust WJ119-2, based heavily on the gas generator of its F121 turbofan. Following an intensive competition, in 1989, the Williams WJ119 was selected as the NLOS sustainer engine. Development of the engine proceeded until 1991 at which time the program was terminated.

At the time of the development of the Williams WJ119-2 NLOS sustainer engine, Technical Directions undertook development of its 101-lb-thrust TD7, designed to satisfy NLOS performance requirements. The objective of that engine was to reduce cost by 50% through the use of off-the-shelf low-cost automotive components. The key design feature was use of an automotive turbocharger compressor and turbine. Design and component testing of the engine was performed under a contract from the U.S. Army Laboratory Command. In 1991, following component rig tests, MICOM awarded a contract to Technical Directions for development and demonstration of the complete engine.

NLOS TRR PROGRAM

In parallel with the NLOS program, the NLOS Technical Risk Reduction (TRR) program was initiated during 1989. The purpose of that program was to reduce the technical risk of the NLOS program by developing and demonstrating an alternate technology turbojet sustainer for the NLOS missile system. Two engine development contracts were awarded under the NLOS TRR program.

One was to Sunstrand to develop an alternate sustainer technology module based on the TJ-90, which the company had proposed in the NLOS competition. Under this contract, Sunstrand was to incorporate the TJ-90 design into a complete sustainer module satisfying NLOS performance requirements. As part of the TJ-90 development effort, Sunstrand developed an interim engine, the TJ-70, an uprated version of the TJ-20. The TJ-70 demonstrated a 68% increase in thrust within the same engine size and using nearly all the same components as the TJ-20.[5] The TJ-90 underwent uninstalled static tests and both static and wind tunnel tests installed in the NLOS TRR sustainer module. Performance of the engine met or exceeded the design requirements under all conditions in which it was tested.[6]

The second NLOS TRR sustainer engine contract was awarded to Williams for development of the 144-lb-thrust P8910, essentially a growth version of the WJ119-2. The P8910 achieved a 42% increase in thrust over the WJ119 while maintaining the same engine envelope. Increased compressor airflow and higher turbine temperatures yielded the performance increase.

LONGFOG PROGRAM

Between 1988 and 1991, concurrent with the NLOS and NLOS TRR programs, advanced tactical missile system concepts evolved into what became the Long Range Fiber Optic Guided Missile (LONGFOG). The range of LONGFOG was to be 24 miles with a goal of 60 miles and the desired fly-out Mach number was 0.7, thus requiring a significantly larger sustainer engine than the NLOS missile.

Development of the Williams P8910, which satisfied the missile's sustainer thrust requirement, was continued under the LONGFOG program following termination of the NLOS TRR program in 1991. Meanwhile, using its IRAD funds, Teledyne CAE had been developing the 178-lb-thrust Model 305-10, which also satisfied LONGFOG sustainer requirements. In 1991, a contract was awarded to Teledyne CAE to develop and demonstrate a ceramic turbine stage for the 305-10, the upgraded version of which was designated the 305-10C and had a rated thrust of 218 lb.

As part of the LONGFOG program, a LONGFOG boost/sustainer effort was initiated to develop an integrated propulsion system that could perform both the

5. Lilley, "The Development of Low-Cost Expendable Turbojet Engines at MICOM," 3-4.

6. J. S. Lilley and S. L. Pengally, "Experimental Evaluation of the NLOS TRR Turbojet Sustainer Module," (U.S. Army Missile Command, Redstone Arsenal, Alabama, January 1992), 5-14.

boost and sustainer functions. The core of that system was to be a high performance turbojet. In 1990, a contract was awarded to Allison for development of its 274-lb-thrust Model 120 engine, specifically designed to be integrated with an augmentation system to perform both the boost and sustainer functions.

SENGAP PROGRAM

The success and steady progress of these small turbojet engine programs that had been inspired by FOG-M led to the establishment of the Small Engine Applications Program (SENGAP) by the Defense Advanced Research Projects Agency (DARPA) in 1990. The objective of the SENGAP program was to develop and demonstrate very small-diameter (4 inches), low-cost, expendable turbojet engines. Such engines were expected to make possible a new class of miniature, low-cost, high-performance flight systems including tactical missiles, submunitions, decoys, remotely piloted vehicles, and targets. Due principally to MICOM's success with its previous small engine programs, it was selected as the contracting agent and technical manager for the SENGAP program.

Under the SENGAP program, engine development contracts were awarded in 1990 to Sunstrand, Technical Directions, Teledyne CAE, and Williams. Six engines were developed under these contracts. Five of the engines were in the 50 to 60-lb-thrust class. Sunstrand developed the TJ-50, which had ceramic rotating parts, turbine nozzle, and exhaust duct. The original version of the engine was designed to operate at metal compatible turbine temperatures. A modified version, the TJ-50C, was designed to operate at elevated turbine temperatures. Technical Directions developed a turbocharger based engine, the TD4, based on its off-the-shelf automotive parts design philosophy used in its TD7 engine. Teledyne CAE developed the Model 304, a scaled-down version of its 305-10. In addition to the 304, which was an all metallic engine, a non-metallic version of the engine, the 304C, was also designed. The 304C, designed to operate at elevated turbine temperatures with a rated thrust of 81 lb, included a composite inlet housing and compressor rotor and a ceramic turbine rotor. The Williams engine, the P9005, was an all new design. These six engines, all of which were in the 4-inch diameter class, were the smallest U.S. military sponsored turbojets built up to 1992.

GLTR MISSILE PROGRAM

The sixth program directed by MICOM for development of small, low-cost expendable turbojet engines entailed development of the sustainer engine for the Ground-Launched Tacit Rainbow (GLTR) Missile Program. In 1989, Raytheon Corporation and McDonnell Douglas Astronautics Company were announced as the contractors for a ground-launched version of the Tacit Rainbow for the Army. They were also qualified as second-source contractors for the AGM-136A air-launched version of the Tacit Rainbow, developed earlier by Northrop for the Air Force and Navy. The AGM-136A was powered by the Williams F121-WR-100 turbojet, and discussion of it is included in the Williams chapter.

The GLTR used a solid rocket booster and a turbojet sustainer. The thrust and fuel efficiency requirements were such as to permit extensive use of LOCAAPS technology, minimizing engine development cost. In 1989, the Teledyne CAE 177-lb-thrust Model 318 turbojet, developed using corporate funds, was selected as the GLTR sustainer engine. That engine was a scaled-up version of the Model 305-4 developed for the LOCAAPS program and incorporated its low cost design features. Development of the Model 318 engine had been nearly completed, when in 1991, the GLTR program was terminated by the military.

TECHNOLOGY BASE

These six programs directed by MICOM established a broad spectrum of engine technology, which could be drawn upon in the future development of engines for various types of expendable flight vehicles. At one end of the technology spectrum were low-cost engines such as the TD-4, TJ-20, 305-4, SR30, TD7, and 318. Those engines used easily producible or commercially available components, thereby achieving minimal production cost while providing acceptable performance. In the mid-range of technology were engines such as the 304, 305-10, and 120, which used advanced aerodynamic rotating components to achieve significant performance improvements—accompanied by increased production costs. At the upper end of the technology spectrum were the very high-performance engine models such as the TJ-50C, 304C, and 305-10C. Those engines used ceramic components that further increased production costs. In 1992, MICOM was continuing its work to extend the technology of miniature turbojets.

The development of this family of engines made possible a new class of small tactical missiles. With its performance advantage over the solid rocket, the miniature turbojet enabled small missile systems to accomplish missions that had not been feasible previously using solid rockets for sustainer propulsion. Further, the broad spectrum of technology embodied in the turbojet engines developed under MICOM's direction provided the missile designer with the latitude to trade engine performance, and thus engine technology level, against engine production cost in selecting the engine for a given missile system.[7]

Model Turbine Engines

As the miniaturization of turbojet engines became feasible, a market also emerged for tiny engines to power model aircraft. These engines were similar in some respects to the miniature engines developed for tactical missiles. Although gas turbine engines for model aircraft are outside the scope of this history, some examples of them are briefly mentioned here for the purpose of being comprehensive.

Full working turbojet engines small enough to propel a standard model aircraft became available in the late 1980s. The first such engine to go into production was the Turborec T240 turbojet. It produced a thrust of 10.1 lb with a 1.57 sfc, had an exhaust gas temperature of about 1,200 degrees F, and operated at a rotational speed of 122,000 rpm.

Kurt Schreckling was the first modeler to successfully construct very small lightweight jet engines using amateur means. His efforts resulted in production versions in kit form of his FD3 series of engines. Thanks to his drawings, modelers in many parts of the world were able to produce their own versions of his engines.[8] As a further example of model aircraft engines, the characteristics of two Schreckling engines are as follows:[9]

	FD3/64	FD3/67
Thrust	5.4 lb	6.7 lb
Weight		2.42 lb
Rotating speed	75,000 rpm	85,000 rpm

These engines are representative of engines also developed by other modelers in various areas of the world. The development of such engines has been reviewed in other publications.

7. Lilley, "The Development of Low-Cost Expendable Turbojet Engines at MICOM," 4-13.

8. Thomas Kamps, *Model Jet Engines* (Worchestershire, United Kingdom: Traplet Publications Limited, 1995), 17, 24, 26.

9. Kurt Schreckling, *Gas Turbine Engines for Model Aircraft* (Worchestershire, United Kingdom: Traplet Publications Limited, 1996), 97, 103. Schreckling engine kits were manufactured by the Schneider-Sanchez Company based in St. Lambrecht, Austria.

THE HISTORICAL EVOLUTION OF THE INDUSTRY: A Comparative Summary, Analysis, and Conclusion

12

Introduction

In previous chapters, the histories of individual corporations and their products have been described. In part, the purpose of this chapter is to look comparatively and critically at these corporations in the context of the airframe markets in which they competed. Another objective is to describe and analyze the industry's major development periods. The external influences and internal considerations which shaped the industry are examined. Technology highlights are noted. Finally, the important contributions made by the industry, its companies, and their engines are reviewed.

By the end of the 20th century, the small gas turbine aircraft engine industry in North America had evolved through four major periods of development. These could be characterized broadly as:

1. Entry into the industry (1940s and 1950s).
2. Market expansion, establishment of niche markets, and corporate specialization within niche markets (1960s through mid-1970s).
3. Maturing of the industry (late 1970s through 1980s).
4. Restructuring and consolidation of the industry (1990s).

During each of these periods, there were major external influences that shaped the direction and the development of the engine industry. Among these were the: prevailing national, international, and aerospace economic conditions;

government policies, legislation, and budget; social and political events; and war or the threat of war.

Historically, an anticipated or existing airframe market had an especially important influence on companies considering the development of new aircraft engines. However, for pioneer companies in this industry, there existed a potential rather than a clearly defined airframe market for small gas turbine engines. What market did exist was for military aircraft. As the industry matured, the clear-cut performance advantages offered by small turbine engines helped create major military and commercial airframe markets.

While influenced by these external factors, companies had additional internal considerations to evaluate when making a decision to enter the small gas turbine engine business or to expand their product lines once they were in it. Such considerations included existing corporate policies; available financial, technical, and human resources; manufacturing capability; market studies; and competition.

Thus, the development of the small turbine engine industry was shaped by external influences and internal considerations. Conversely, this industry had a significant influence on and made important contributions to military and commercial operations, the aerospace industry, and society as a whole.

1. Entry into the Industry (1940s and 1950s)

THE HISTORICAL CONTEXT FOR THE FORMATION OF THE INDUSTRY (1940S)

In the 1940s, the small gas turbine aircraft engine industry began to emerge within a wartime environment. The jet aircraft engine was one of the new technologies pushed forward by the war effort. The small gas turbine engine was a relatively minor aspect of this new technology, but a promising one. Development of jet engines, large and small, was also heavily influenced by the existing policies of military engine procurement. These policies and procedures were described in Chapter 3.

The procurement of early gas turbine engines conformed to the procedures established for the development of military piston engines. In general, engines were developed independent of the aircraft in which they were to be used and were provided to airframe companies as government furnished equipment that the airframer was responsible for integrating into its aircraft. With the exception of specific circumstances designated by military authority, corporations developing engines were prohibited from cooperating with one another by anti-trust laws.

During the 1940s, funding for research and development was relatively plentiful. As a result, a number of new gas turbine engines as well as a variety of propulsion systems were under development by 1945. Such diverse developments were undertaken in order to establish the relative feasibility and usefulness of the many competing concepts. However, by the end of the decade, engine development efforts were largely focused on turbojet engines for fighter and bomber airplane applications.

INFLUENCES ON CORPORATE ENTRY

The genesis of the small gas turbine aircraft engine industry took place during the 1940s and 1950s. The seven major North American companies that developed small gas turbine aircraft engines entered at this time as did other companies that eventually phased out of the business. The industry's chronological development during this time period was traced in Chapter 3. This section looks at the factors that influenced corporate entry.

The obvious common internal motivation for these companies to enter this business was to increase their revenues by expanding their portfolio of products for existing or future markets. However, factors influencing the decision of each company to enter the business varied. Certain external factors were especially important influences including the drive for new technology during World War II, the availability of military research and development funding, the opportunities associated with the begining of the "Jet Age," and the needs made evident by the Korean and Cold Wars.

During World War II, military interest in jet aircraft engines was the primary driving force in the U.S. that pushed forward rapid gas turbine engine technology development and corporate entry into the new field. GE, Westinghouse, and Allison were the principal military-sponsored corporations. In 1941, GE and Westinghouse, each with turbomachinery experience and an interest in gas turbine engines prior to the War, were given military contracts for the development and production of gas turbine aircraft engines. Westinghouse began work on its 9.5A small turbojet engine in late 1942. GE's first experiments with small turbine engines were Operation Bootstrap in late 1943 and the B-1 turbojet engine in the spring of 1944. In 1944, Allison was designated by the Army Air Forces as a second production source for selected GE-designed jet engines.

Other corporations also entered the business as a result of war-related needs. In each case, the focus of these corporations was directed toward smaller turbine engine or component design. In 1943, during Project A, Garrett began design and development of a two-stage compressor for an aircraft cabin air

compressor. Garrett's subsequent turbomachinery work, largely military sponsored, became the basis for the company's ability to later enter the small gas
turbine aircraft engine business. Also in 1943, in an independent initiative,
Boeing inaugurated jet propulsion studies in response to military interest in
large jet aircraft. With military contracts, Fredric Flader founded his company
in 1944 to develop small turbine engines.

After World War II, some corporations explored new markets, including
small turbine engines, to offset their loss of business as a result of canceled military contracts. In 1946, Fairchild began preliminary design studies of a small, expendable turbojet engine, the J44. A post-war development, Continental Aviation and Engineering's first gas turbine unit was the TR125 APU. In 1946, Solar
began work on a turbine-driven APU.

In the late 1940s and early 1950s, there was widespread interest in the new
"Jet Age" as a result of the performance of jet aircraft. The military services and
corporations were highly motivated to exploit this technology. As a relatively
new invention, the gas turbine engine promised aircraft performance advantages
so significant that it rapidly created its own demand and market. This market
was almost exclusively controlled by the military. The military placed priority
on the development of large turbine engines to power jet fighter and bomber
aircraft. Lesser importance was placed on the development of drones, trainer
aircraft, and missiles, and correspondingly, the small turbojet engine to power
them. Even so, the many potential airframe applications for jet engines was
sufficient motivation for the military to sponsor jet engine development, large
and small, and for companies to respond with proposals or to initiate their own
market studies on jet engines. Important turbine-powered helicopter and light
aircraft demonstration projects in the early 1950s suggested the possibility for
even more airframe applications, which solidified interest in turboshaft engines
and stimulated interest in turboprop engines.

In the early 1950s, the Korean War and the Cold War pushed military-
sponsored large jet engine development and production into high gear. Major
companies in the turbine engine business, such as GE, Allison, and Pratt &
Whitney, produced large quantities of jet engines for military aircraft. These
events had an impact on the development of the small turbine engine. During
the Korean War, the helicopter proved its worth by providing rapid transport
for wounded troops, rescue, observation, command and control, airmobile
troop landings, and resupply.[1] The Korean War also accelerated the growth of

1. Lynn Montross, *Calvalry of the Sky: The Story of U.S. Marine Combat Helicopters* (New York:
Harper & Brothers, 1954), 109-200.

Army aviation.[2] The piston-powered helicopter had demonstrated its capability, and the need for even higher-performance combat helicopters was made clear. In the 1950s, the Army became a sponsor for gas turbine engine development, and later, the primary user of turbine-powered combat helicopters. In the environment of the Korean War and the Cold War, the need for jet aircraft of many types, including trainers and target aircraft, was pushed forward. Small jet trainer aircraft provided an economical and safe alternative to training in high-performance jet combat aircraft. The need for realistic, cost-effective, jet-powered drones for target practice was also evident.

In the 1950s, corporations continued to enter the small gas turbine engine business. While companies like Lycoming, Williams International, Pratt & Whitney Canada (P&WC), and Wright Aeronautical Corporation were undoubtably influenced by external events, their decision to enter the business was more directly related to specific internal considerations. Under the direction of a new president, in 1950, Wright entered the gas turbine aircraft engine business and obtained licenses from Bristol and Armstrong-Siddeley to manufacture and sell their engines in the U.S. Lycoming's move into the business in 1951 was instigated by Anselm Franz who successfully convinced the company president that an unfulfilled market niche existed for medium-power small gas turbine engines. On the basis of his experience and continuing interest in gas turbine engines, in 1954, Sam Williams formed his own company to develop small gas turbine engines for diversified uses including aviation. Seeing a potential market for small gas turbine engines for light aircraft and knowing that its piston engine spare parts business would eventually decline, in 1956, P&WC began hiring a design team of gas turbine specialists.

Other corporations, that had entered the turbine engine business earlier, expanded or formalized their small gas turbine operations in the 1950s. Continental Aviation and Engineering (CAE) negotiated with Turboméca a licensing agreement for eight small turbine engines in 1951, and thus quickly launched itself into the business. In 1953, in response to favorable market studies and other internal corporate considerations, GE formed its Small Aircraft Engine Department. In 1958, Allison received a military contract for its Model 250 turboshaft engine, the first small turbine engine developed by the company. Garrett started design work on a generator set in 1959; after Cliff Garrett's death in 1963, the company officially entered the aircraft propulsion business, and the generator set became the Model 331 turboprop engine.

2. Frederic A. Bergerson, *The Army Gets an Air Force: Tactics of Insurgent Bureaucratic Politics* (Baltimore: The Johns Hopkins University Press, 1980), 71.

Some generalized observations can be made about corporate entry into the small turbine engine business in the 1940s and 1950s. A common characteristic of almost all of the corporations entering the small engine business was that they had manufactured some type of aviation product and thus had an interest in continuing in the aviation business. Of these companies, GE, Westinghouse, CAE, Lycoming, Allison, P&WC, Wright Aeronautical, and Fairchild, were actually in the business of manufacturing aircraft engines (i.e. large turbine or piston engines) prior to building small gas turbine aircraft engines. Another basic trait common to all companies was a belief that there was a future for small turbine engines, and each company wanted to expand its business to take advantage of that market and potential profits.

Previous turbine engine experience was beneficial for corporations entering the small gas turbine aircraft engine business, but not a prerequisite. About half of the major corporations entering the business had some prior gas turbine engine or component experience; among these were GE, Westinghouse, Allison, Garrett, and Williams. Those that did not, such as Lycoming and P&WC, generally had available expertise to work successfully with gas turbine engines. In the case of CAE and Wright, licensing agreements provided a technology transfer and immediate access to the business. Allison also originally benefited from the transfer of technology when it manufactured GE-designed turbojet engines during and after World War II.

GE, Westinghouse, and Allison entered the large jet engine business as a direct result of the military need for jet engines during World War II. Relatively few corporations entered the business as Williams did, in a strictly self-financed, independent entrepreneurial initiative, and none did so without the prospect for acquiring some military or commercial contract. For subsidiaries of major diversified corporations, such as GE, P&WC, Allison, and Lycoming, the technical or financial support of the parent corporation generally played a key role in assisting their entry into this new business.

MARKET DEVELOPMENTS

Starting in the late 1940s and early 1950s, the first significant military market was for small turbojet engines to power target drones, missiles, and small trainer aircraft. Following successful flight demonstrations in the early 1950s, another market rapidly began to emerge for turboshaft engines to power military helicopters. The key event that led to this development was the flight of the first turbine helicopter, the Kaman K-225, on December 11, 1951.[3] Flight demonstrations were also conducted with light aircraft equipped with turboprop engines, but no significant market emerged for them in the 1950s. A strong market developed for turbine engine APUs, starters, and environmental systems. Toward the end of the 1950s, a market began to develop for small military utility cargo jets and utility trainer jets which shortly thereafter evolved into a booming business jet aircraft market. Also at that time, a market for the first commercial helicopters began to emerge. In general, a potential, rather than an actual commercial market existed in the 1950s for small gas turbine aircraft engines.

Among the major turbine engines which first emerged in the 1950s were the CAE J69 turbojet engine, Lycoming T53 and T55 turboshaft engines, and GE T58 turboshaft and J85 turbojet engines. All of these engines found pivotal airframe applications, were produced in substantial quantities, had long production runs, and became major products for their companies. Development of these engines began in the 1950s, and by the late 1950s and 1960s, significant engine production runs were underway. Produced in smaller quantities, but also successful, were the P&WC-designed and Pratt & Whitney Aircraft (P&WA)-manufactured JT12 (military designation J60) turbojet engine and the GE T64 turboshaft/turboprop engine. The Fairchild J44 turbojet engine was an important engine, although, unlike the other engines, its development started in the late 1940s, and its production run ended in the late 1950s. The critical link in the success of all of these engines was winning either a major military engine competition or being selected to power an important airframe which won a major military contract.

A number of small turbojet engine competitions and airframe applications materialized, some of which resulted in significant production. While the Westinghouse 9.5 was the first small turbojet engine to be successfully developed and manufactured in the U.S., the Fairchild J44 was the first small turbojet engine to find a significant production application. The J44 was first used in the Navy's Fairchild Petrel missile and subsequently in the Ryan Q-2 Firebee target drone, a joint Army, Navy, and Air Force project. In 1952, CAE's Marboré II (military designation J69) was chosen to power some versions of Air Force and Navy Firebee drones; in 1953, the Air Force selected Cessna's Marboré II-powered design for its T-37 jet trainer aircraft. GE's J85 won an Air Force contract in 1954. In 1956, the J85 was selected for the Air Force GAM-72 Green Quail mis-

3. Charles H. Kaman, *Kaman: Our Early Years* (Indianapolis, Indiana: The Curtis Publishing Company, 1985), 85-91.

sile built by McDonnell; the same year, a J85-powered Northrop design was selected for the Air Force T-38A Talon supersonic trainer. Additional Air Force
requirements in 1956 for a UTX (Utility Trainer Experimental) aircraft and a
UCX (Utility Cargo-Transport Experimental) aircraft later led to the North
American T-39 and Lockheed C-140, both powered by P&WA J60 turbojet
engines. Each of these contracts contributed to the success of these first-
generation small turbojet engines.

In the decade of the 1950s, major military competitions were also held for
turboshaft and turboprop engines. In 1952, the Army (through the Air Force) issued an RFP for a turboprop/turboshaft engine, and Lycoming won the contract for the T53 engine. By the mid-1950s, the T53 was selected for the Army's
first turbine-powered aircraft (fixed-wing or rotary), the Bell UH-1 Iroquois
(Huey). The Bell UH-1 was procured by the Army in a larger quantity than any
other helicopter. In 1954, Lycoming received another important contract from
the Army/Air Force for the T55 turboprop engine. In 1959, the Army selected
the T55-powered Boeing Vertol helicopter, the CH-47A Chinook. In 1953,
GE's T58 won a Navy contract. Important Navy helicopter contracts awarded
in the 1950s for T58-powered helicopters, included the Sikorsky SH-3 Sea King
and the Kaman SH-2 Seasprite. In 1958, Allison's Model 250 (military designation T63) turboprop/turboshaft engine was awarded an Army/Air Force contract. In 1961, the Army selected the T63 for its finalist competitors in the Light
Observation Helicopter (LOH) competition, which included the Bell OH-4A,
Fairchild-Hiller OH-5A, and Hughes OH-6A.

The Army and Navy contracts in the 1950s were a major turning point for
the small turboshaft engine. They led to a rapidly expanding turbine-powered
helicopter market in the years that followed with the turboshaft engine completely dominating the intermediate to heavy military and commercial helicopter markets, and capturing, to a notable extent, the light military and commercial helicopter markets. The Army, in particular, became a key sponsor for small
turbine engine development and the largest single market for turbine-powered
helicopters, largely as a result of military policies that carefully defined and limited its aviation function vis-à-vis that of the Air Force. These contracts assured
the successful entry of Lycoming, GE, and Allison into the small turboshaft engine business.

In the 1950s, development of turboprop engines was carried out in parallel
with their turboshaft counterparts. In an Army demonstration program, CAE's
XT51 and Boeing's XT50 engines were installed in Cessna L-19 light observation aircraft where fantastic performance advantages were realized, but draw-
backs such as high sfc and cost precluded any competitive advantage over exist-

ing reciprocating engines. The first operational Army turboprop aircraft was the Lycoming T53-powered Grumman OV-1 battlefield surveillance aircraft. In 1958, P&WC began market studies on small general aviation aircraft, and in 1959, with Canadian government support, began design of its PT6 turboprop/turboshaft engine. While there was a strong interest and a number of prototype applications for small turboprop engines, this market lagged significantly behind the market for small turbojet and turboshaft engines.

A substantial market, primarily military sponsored, developed for gas turbine APUs, starters, and environmental systems in the late 1940s and 1950s. This market was eventually dominated by Garrett, whose APU business built a technical, financial, and manufacturing foundation for the company's movement into prime propulsion in the early 1960s. Other corporations which had a significant stake in this business included Boeing, CAE, and Solar. In some instances, these corporations also had small gas turbine products for non-aircraft-related uses. Williams had small industrial, marine, and jet turbine engines under development in the mid- to late 1950s, but had not yet secured important applications for its products as had the other companies.

By the end of the 1950s, the primary market for small engines was military turbojet and turboshaft engines. Selected small turbine engines, which had been under development in the 1950s, would become major production engines in the 1960s. Ultimately, each successful company that entered this business had won a major military engine competition, powered an important airframe which won a major military contract, or had strong future prospects for civilian applications. Conversely, those companies that did not win an important application and did not sustain their success did not remain in the business. By 1960, three corporations that had entered the small turbine engine business had left it. Westinghouse and Fairchild were not able to maintain competitive positions within the industry. Flader left due to its inability to develop a suitable engine. In 1965, Wright Aeronautical decided to stop development of gas turbine aircraft engines because of an executive conviction that the military services would not support another major engine company. While Wright had manufactured under license a successful large turbine engine, it had not found production applications for its small turbine engines.

TECHNOLOGY CHALLENGES

The design and manufacturing of early small turbine engines presented problems that were sufficiently difficult as to bring into question whether the technology itself was technically and economically viable. These problems contributed to the technology of small gas turbines engines lagging behind that of large engines. The optimum configuration of small turbine engines was an open-ended question to which a variety of approaches appeared possible; these made preliminary design studies both an opportunity and a challenge. Another challenge was the highly developed and relatively inexpensive piston engine, which was the primary competitor to the small turbine engine, especially for light aircraft and early helicopter applications.

A detailed explanation of small turbine engine design and manufacturing complications was presented in Chapter 2. These included problems associated with size effects on aerodynamic performance, manufacturing limitations, combustor design constraints, turbine blade cooling and thermal disk stress, mechanical design, and high cost.

Limited small gas turbine design, development, and manufacturing experience was another difficulty. While most corporations had large turbine engine design experience, they did not always have experience with the unique design problems associated with small gas turbine engines. The lack of facilities, test equipment, and instrumentation sized appropriately for small turbine engines were an obstacle. In some cases, the design objectives far exceeded the possible.

Engines are currently designed with the expectation that they will achieve on their first run performance and efficiency approximating calculated projections. Pioneer small turbine engine engineers weren't always certain that their engines would run at all, and sometimes they didn't. In 1945, Garrett's engineers were tasked to design an aircraft APU (The Black Box) that would provide an unrealistic number of functions. With a tight schedule, no test rigs, and marginal turbine efficiency, the engine would not self-sustain; inexperience and overly optimistic design objectives were contributing factors. When GE's XT58 turboshaft engine first went to test in 1955, it would self-sustain only as a gas generator, performance was low, and considerable aerodynamic and mechanical difficulties were experienced with the compressor. A weight-saving compressor design expected too much performance from too few stages, test instrumentation was not available for such a small compressor design, and a tight development schedule and budget constraints precluded component testing.

Both a challenge and an opportunity for pioneer turbine engine designers was the variety of design approaches that were possible for component technology and engine configuration. For example, it was possible to select from a larger variety of compressor designs in small engines than large engines. In the early to mid-1940s, some designers of large turbine engines had selected axial-flow compressors while others opted for centrifugal-flow designs. However, by the late 1940s, the axial compressor had become the choice of large engine designers in the U.S. due to its smaller frontal area and its inherently higher efficiency. In contrast, small turbine engines lent themselves to a variety of mechanical arrangements including either axial or centrifugal-flow compressors or combinations of both.

There was no best way to design a small engine compressor and different approaches were taken by different companies in the small turbine engine business. After internal engineering debate, companies found solutions that were satisfactory for their proposed designs. Often the configuration that companies adopted for their first small turbine engines set a precedent for their future engine designs. To illustrate the variety of approaches in turbojet engines, for instance, the Westinghouse 9.5 had an axial-flow compressor, the Fairchild J44 used a mixed-flow compressor, and the CAE J69 incorporated a centrifugal compressor. Among the turboshaft/turboprop engines, the Lycoming T53 and T55, the Allison T63, and the P&WC PT6 used an axial-centrifugal approach, whereas the Garrett TPE331 used dual centrifugal compressor stages. GE used all-axial stage compressors on its J85 turbojet, T58 turboshaft, and T64 turboshaft/turboprop engines.

For its part, GE introduced several noteworthy technologies for compressors. The variable stator compressor resulted in higher compressor compression ratios while preventing the rear compressor stages from stalling at part-engine speed by reducing compressor airflow. Variable stator compressors later became a mainstream turbine engine technology. GE also developed a smooth spool compressor which was noted for dimensional stability, improved aerodynamic efficiency, and longer life. The T58 was the first small engine to use a variable stator compressor and the first GE engine to incorporate a smooth spool compressor.

Another component for which alternative approaches was taken was the combustor. Reverse-flow annular combustors were used in the Garrett TPE331, P&WC PT6, Lycoming T53 and T55; a reverse-flow single-can combustor was used in the Allison T63. In contrast, the Westinghouse 9.5, Fairchild J44, CAE J69, and GE T58, T64, and J85 utilized straight-through-flow combustors. The P&WA JT12 incorporated a straight-through-flow cannular combustor. The type of combustor selected for an engine was the one that best suited the overall mechanical arrangement of that engine's components.

The overall configuration of a small turbine engine was an especially important consideration. Initially, for turboshaft and turboprop engines, it was necessary to decide whether a single-shaft or a free turbine arrangement was preferable. A strong case could be built for either design, with the free turbine arrangement offering more operational flexibility and the single shaft a simpler design. An important additional advantage of the free turbine was that it was inherently suitable as a multiple-function engine (i.e. turboshaft or turboprop), and its gas generator could be adapted to a wide variety of aircraft types. The multiple-function turbine engine was itself an innovation in the early 1950s, earlier turbine engines having been designed for a single class of aircraft. In the case of the free turbine engine, an additional choice existed between the use of opposed or concentric shafts. At the time, the opposed shaft design was considered a significantly less difficult design problem for small turbine engines, but the concentric shaft was more readily adaptable to front-drive airframe applications.

The selection of an engine's mechanical arrangement was influenced both by engine type and airframe application. The small turbojet for fixed-wing aircraft employed a single-shaft configuration. Turboprop aircraft successfully used both free turbine and single-shaft engines. Early helicopters were also powered by free turbine and single-shaft engines. However, the free turbine arrangement was particularly well suited for helicopter applications and quickly became the configuration of choice. This was because it eliminated the need for a clutch in the drive train and because it had high torque at low output shaft speed.

The free turbine arrangement was, with one exception, universally selected for U.S. turboshaft and/or turboprop engines. Among the early free turbine designs were the P&WC PT6, Allison T63, Lycoming T53 and T55, GE T58 and T64, and Boeing T50. Garrett's TPE331 was the only major U.S. turboprop/turboshaft single-shaft design.

Concentric- and opposed-shaft arrangements were used with equal success in free turbine engines. Those engines that used the opposed-shaft design included the P&WC PT6, GE T58, and Boeing T50. Concentric-shaft engines included the Lycoming T53 and T55 and the GE T64 engines. The drive position also varied from engine to engine with most using a front power take-off drive. Front-drive engines included the Lycoming T53 and T55, P&WC PT6, Garrett TPE331, GE T64, and Boeing T50. The Allison T63 had a parallel drive for front- or rear-drive applications with a mid-engine power takeoff, and GE's T58 had a rear-drive arrangement. The front-drive, concentric-shaft arrangement of the T53 set a design precedent for turboshaft engines widely accepted throughout the industry in the U.S.

As its primary competitor, the piston engine was both a technology and a marketing challenge to the small turbine engine. This was especially true for light fixed-wing aircraft and helicopter applications. To be successful, the small turboshaft/turboprop engine had to displace the well-established piston engine in these aircraft. Advantages of turbine engines included their high power output, low weight, and low vibration. In contrast, reciprocating engine strong points included low sfc and low cost. For some airframes, gas turbine engines could compete effectively against piston engines. For instance, cost was of less importance than performance to the military services. The higher cost of the turbine engine was less significant than the fact that the turbine engine could deliver greater power per pound than the piston engine.

In general, for small fixed-wing aircraft applications in the early 1950s in both military and civilian markets, the performance advantages of the small turboprop engine were outweighed by its disadvantages. However, the small turboshaft engine literally transformed helicopters, giving them significantly greater payload, speed, and altitude capability.

Operating experience with helicopters revealed demands for which reciprocating engines were inherently unsuited. Extended periods of operation at high power resulted in unsatisfactory engine durability. Furthermore, intrinsic piston engine vibration was accentuated in helicopter operations. The advantages of the turboshaft engine compared to the piston engine in the helicopter included its significantly lower weight, higher power, markedly better durability when operated at high power, lower vibration, and its ability to be located externally to the fuselage. Aircraft piston engines had been developed for decades to meet the requirements of fixed-wing aircraft, which generally included good sfc and durability at cruise power (used for most of the flight). However, during high-power operations, piston engine sfc was high. Because helicopters operated at high power for extended periods and because helicopter missions were relatively short, the sfc advantage of piston engines was either eliminated or greatly reduced.

Despite the various technology challenges facing the small turbine engine in the 1940s and 1950s, steady progress was made in all areas of turbine engine design, development, and application. In one sense, inexperience proved to be an asset for turbine engine pioneers. While many of the technology challenges appeared formidable, these engineers did not see why such problems could not be overcome. In time, they were, and in part, this was due to improved tools of the trade. Reflecting on changes over time, GE engineer Frank Pickering said:

> If you look back over 40 years, I think it's almost hard to make an assessment that is significant... dramatic enough of what has happened. When you think about when we were designing and developing engines back in the fifties with slide rules, adding machines, data acquisition equipment (like) manometers and gauges, then look at the capability today and what's happened with the advent

of digital computers, the ability not only to design, but to acquire, analyze, and iterate data (in) real time, do computations with 3-D software, the Cray computer capability that we have, it's mind-boggling to think of the difference.[4]

2. Market Expansion, Establishment of Niche Markets, and Corporate Specialization within Niche Markets (1960s through mid-1970s)

This time period, a "golden age" for small gas turbine engines, was broadly characterized by the rapid expansion of the market for these engines, the establishment of niche markets within this expanded market, corporate specialization within niche markets, and significant advances in technology.

As the result of the favorable combination of factors, from its modest beginnings in the 1940s and 1950s, the market for small gas turbine engines expanded significantly. An important factor was the technological advantage of the small gas turbine aircraft engine with its ability, compared to reciprocating and rocket engines, to enhance the flight performance of aircraft and missiles, respectively. No longer a tentative technology, the small gas turbine was overcoming its initial technical challenges and moving toward a state of refinement. World events and government policies generated a strong military and commercial market demand. This demand motivated companies and individuals to develop and produce innovative new engine designs. Military organizations with plentiful funding and the need to acquire large numbers of these engines were a key factor in the general expansion of the small gas turbine aircraft engine market.

Within this growing market, new markets for small turbine engines emerged. Earlier, the primary market for small engines had been limited primarily to military turbojet and turboshaft airframe applications. In the 1960s through the mid-1970s, a broader spectrum of airframe and commercial possibilities for small engines opened up.

As a result of competitions or major airframe applications that they had won in the 1950s, most corporations were not only well on their way toward establishing a firm foothold in the business, but, as it later became evident, in specialty market niches as well. Often, the first successful small engine developed by a company became one of its most successful products and gave the company its original preeminence in a turbine market niche.

4. Frank Pickering, interview by Rick Leyes and William Fleming, 20 August 1991, GE Interviews, transcript, National Air and Space Museum Archives, Washington, D.C.

At the time, and generally speaking, corporations sought the broadest possible applications for their turbine engines rather than emerging niche markets. Corporations did, nonetheless, exploit advantages that they gained in niche markets. Over time, as their products sold especially well in these narrower markets, corporations essentially found themselves specializing in them.

These market niches were broadly split among the corporations in the following manner. Garrett and P&WC shared the turboprop-powered general aviation and regional transport market, and both developed turbofan engines for the business jet market. Williams International and CAE established themselves in the expendable engine business, including targets, RPVs, and missiles. CAE also had a significant share of the subsonic jet trainer market. Allison and Lycoming became incumbents in different power classes of the helicopter engine business. GE had an important stake in nearly all of the market niches, including the helicopter, jet trainer, small military aircraft, missile, and executive jet engine business.

WORLD EVENTS AND GOVERNMENT POLICIES INFLUENCING THE SMALL ENGINE MARKET

World events during this time period were a major determinant of the military markets that evolved. The Cold War with Russia continued unabated. In the early 1960s, the loss of U.S. spy planes in several highly-publicized and embarrassing incidents triggered the development of reconnaissance drone aircraft. In Vietnam, U.S. involvement changed from an advisory role in the early 1960s to full-scale war from 1964 to 1973, precipitating a huge demand for helicopters of all types as well as small special purpose aircraft and reconnaissance drones. The negotiation of nuclear disarmament treaties beginnning in the early 1970s contributed to the requirement for a strategic cruise missile. The military services responded by sponsoring the development and production of a wide range of aircraft, rotorcraft, and missiles, many of which required small gas turbines.

A significant commercial market for small turbine engines developed during this time period. This market was influenced by various externally-oriented factors, including among others, the Jet Age and changes in federal policies. In the 1960s, world air transportation had increased dramatically largely because of the widespread use of jet transport aircraft, which increased the convenience of air travel while reducing the costs. Clearly, a parallel opportunity existed for smaller aircraft to become part of the Jet Age as well. Some small engine companies put money and resources into commercial engine development for this potentially lucrative market. Their aspirations were realized in the mid- to late 1960s as a market emerged for both executive and commuter aircraft. Intangible commodities associated with the Jet Age, including glamour and prestige, proved to be important selling features that stimulated a rapidly swelling demand for turbine-powered executive aircraft. An influential factor in the growth of commuter airlines was a series of policy changes promulgated by the Civil Aeronautics Board (CAB), a federal agency that regulated airline passenger service. Equipped with new, efficient turbine-powered aircraft, commuter airlines could profit in short-haul, low-density markets that were unprofitable for larger carriers.

An important world event and increased environmental concern had an impact on small turbine engines, especially those targeted for the commercial market. In 1973, the Organization of Petroleum Exporting Countries (OPEC) began a series of actions that drastically affected the availability and price of petroleum products, including aircraft gas turbine fuel. A series of oil price increases and an oil embargo on the U.S. stimulated the development and purchase of helicopters used in commercial off-shore oil drilling operations. The oil shortage resulted in an increased market for turbine engines with significantly lower fuel consumption. In the early 1970s, public concern about the environment increased and engine companies were faced with the need to reduce engine noise and exhaust pollution.

Other external factors affecting the design and development of military gas turbine engines were changes in government policies and procedures. Before 1960, the engine programs of individual military services were integrated primarily by mutual consent among the services with little detailed direction from the centralized Department of Defense (DoD). In 1960, centralized control of military weapons by the DoD, including gas turbine aircraft engines, began to increase, exercised through the centralization of authority over budgeting and allocation of appropriations and through more formalized management procedures. In 1965, a procedure was put into place that required that a specific, approved aircraft program be in existence before any new engine program that was estimated to cost more than $25 million for development or $100 million for production could be initiated for such an aircraft.[5]

Historically, it had taken one to two years longer to develop a new engine than to develop a new aircraft.[6] Before 1965, military engines had been developed to fill a power class rather than being associated with a specific aircraft. En-

---

5. Department of Defense Directive 3200.9, *Initiation of Engineering and Operational System Development*, 1 July 1965.

6. A. Stuart Atkinson and Charles C. McClelland, "Rationale for a Military Aircraft Engine Development Program," *Astronautics & Aeronautics*, Vol. 7, No. 9 (September 1969): 35.

gine development normally was well along before the design of a new aircraft that might use the engine would be initiated. The new DoD procedures prohibited this practice. In response, the three major military services adopted the practice of designating initial engine development efforts as "advanced developments," "demonstrator," or "gas generator" programs to justify the initiation of an engine development in advance of an approved airplane program.

This practice also provided test vehicles for focusing military technology programs. Previously, engine component research and development sponsored by the military had been somewhat random. Now, research and development programs evolved that were focused on contributing to improving the performance first of gas generators and later of full-scale demonstrator engines. In the 1960s, the Air Force led this practice and initiated the Light Weight Gas Generator (LWGG) program.[7] LWGG evolved into the the Air Force's Advanced Turbine Engine Gas Generator (ATEGG) program. ATEGG concentrated on core technologies using a building block concept. In 1975, ATEGG core technologies were incorporated into the Joint Technology Demonstrator Engine (JTDE) program, a Navy and Air Force program. In JTDE, ATEGG technologies were integrated into complete thrust and turboshaft demonstrator engines. Small turbine engine technology development was included in these various programs. The Army focused on small gas turbine demonstrator programs, the first of which was initiated in 1967 as the 1,500-shp Demonstrator Engine. This was followed by the the Small Turbine Advanced Gas Generator (STAGG) program in 1972 and a sequence of others later.[8] Many of the small turbine engine companies participated in these military R&D programs.

As a result of operational problems experienced, the government began to impose additional design and test requirements on both military and commercial turbine engines. For example, helicopter operations during the Vietnam War revealed the need for major improvements including resistance to small arms fire and foreign object damage, ease of maintenance, and extended range. As a result, the Army developed a set of design requirements for next generation engines that emphasized significantly improved reliability, maintainability, survivability, and fuel consumption in addition to performance. All services began

7. Marvin A. Stibich, Chief, Fan/Compressor Branch, Turbine Engine Division, Aero Propulsion & Power Directorate, Department of the Air Force, Wright Laboratory (AFMC), Wright-Patterson Air Force Base, Ohio, to Rick Leyes, letter, 13 January 1995.

8. Mr. Eric Clay Ames, "Power Systems Division, Aviation Applied Technology Directorate's Pursuit of Advanced Turboshaft Engines and Their Benefits to Army Rotorcraft," *U.S. Army Aviation Digest*, Professional Bulletin 1-94-6 (November/December 1994): 32-33.

to put more emphasis on reducing the sum of the development, production and maintenance costs (the life cycle cost) of engines.

NICHE MARKET DEVELOPMENTS

The 1960s through the mid-1970s was characterized by the expansion of all of the markets for the small gas turbine aircraft engine. As the result of both military sponsorship and entrepreneurial initiative within the industry, many new small engines with good performance and reliability became available. The interaction of the possibilities provided by these new engines with the military and commercial demand for aircraft that could exploit the possibilities offered by them resulted in the expansion of the markets previously mentioned. Engines were also developed for commercial markets independent of the military services.

Corporations that had established footholds in emerging niche markets in the 1950s further consolidated their gains in the 1960s through mid-1970s and began to specialize. A key element in corporations dominating or specializing in a niche market was having the first successful engine in a power class or winning an important competition or airframe application. The military niche markets were: military helicopters; military jet trainers; selected small military aircraft; decoy, tactical, and strategic missiles; and target drones and reconnaissance RPVs. The commercial niche markets were: general aviation business jet and turboprop aircraft; general aviation helicopters; and regional transport aircraft.

MILITARY HELICOPTERS

During the Korean War and later military exercises, the helicopter had established itself as an effective vehicle in a number of important military missions. The availability of gas turbines in the late 1950s and early 1960s, combined with their ability to markedly improve the performance of these helicopters, resulted in their being used in a number of important applications, especially by the Army, Navy, and Marine Corps. The turboshaft engines that had won major engine competitions or been selected for important airframe applications in the 1950s began to be produced in quantity in the 1960s as a number of turbine-powered helicopters became operational. To a significant extent, the huge production of helicopters in the 1960s and 1970s, and the subsequent demand for small turboshaft engines, was driven by the Vietnam War.

As the war in Vietnam intensified, the helicopter became an increasingly important weapon. During the war, the Army acquired, by far, more helicopters than any other military service. Army aviation expansion was not only driven by the war, but by evolutionary changes in military policy that had occurred since 1947 when the Air Force became a separate military service. Army aviation units were originally responsible for artillery spotting and liaison work. In the 1950s and 1960s, the Army developed airmobility tactics that included the tactical transport of troops and supplies to the battlefield by helicopter. In the 1960s, the Army was authorized to develop armed, including attack, helicopters for the close air support of its troops in combat.

While piston-powered helicopters were used throughout the war, the defining role was held by turbine engine helicopters. John Tolson, Commanding General, 1st Cavalry Division (Airmobile), Vietnam commented on their role by saying: "The turbine engine helicopter with its great power, its reliability, and its smaller requirement for maintenance, was the technological turning point as far as airmobility is concerned."[9] These powerful, turbine engine helicopters made possible to a great extent the airmobility and close air support combat tactics that subsequently became a cornerstone of conventional Army warfare.

Initially, most turbine-powered helicopters were acquired primarily for utility and transport missions. During the Vietnam War, when airmobility tactics were used extensively, the Army built up a large fleet of Lycoming T53-powered Bell UH-1 Iroquois and AH-1 HueyCobra and Lycoming T55-powered Boeing Vertol CH-47 Chinook helicopters. The result was that, in terms of quantity, Lycoming completely dominated the Army intermediate and medium-heavy turboshaft engine market.

The versatility of the helicopter resulted in military services expanding their use to include search and rescue, medical evacuation, reconnaissance, aircraft recovery and heavy lift operations, close air support and assault, anti-submarine warfare, minesweeping, and many other special missions. In the early 1960s, the Army's need for light observation helicopters formed a niche that was filled by the Hughes OH-6 Cayuse followed by the Bell OH-58 Kiowa, both powered by the Allison T63 engine. Because of the large number of these light helicopters purchased by the Army, the Allison T63 dominated the military light helicopter engine market.

The T58 program gave GE an entree with the Navy which had traditionally bought many of its engines from P&WA. GE's standing with the Navy was further consolidated when it won the T64 turboshaft engine competition. Thus, GE won important Navy helicopter engine programs in the intermediate

---

9. Lieutenant General John J. Tolson, *Airmobility in Vietnam: Helicopter Warfare in Southeast Asia* (New York: Arno Press, 1981), 3.

through heavy power range. P&WA found an application for its T73 (civilian JFTD12) turboshaft engine in the Army's Sikorsky CH-54 Tarhe, a heavy lift helicopter. P&WC's West Virginia-manufactured T400 twin-engine turboshaft provided extra power and reliability for UH-1N and AH-1J helicopters which were primarily used by the Navy and Marine Corps.

Four market sub-niches for turbine-powered helicopters can be identified and are listed below. Included are representative major models of operational combat helicopters developed during this time period. During the Vietnam War (1964 through 1973), the turboshaft engines which were produced in the largest quantities for these helicopters were the Lycoming T53 and T55, Allison T63, and GE T58 engines. The service responsible for the development of the helicopter or the primary service using that helicopter is shown. It should also be noted that classification of helicopters, especially heavy and medium-heavy lift helicopters, changed overtime as a result of the introduction of new helicopter models or engine upgrades. Helicopter designations changed as well.

Niche	Helicopters	Engines	Service
Heavy	Sikorsky CH-54	P&WA T73	Army
	Sikorsky CH-53	GE T64	Navy/Marine
Medium-Heavy	Boeing Vertol CH-47	Lycoming T55	Army
	Sikorsky CH-3	GE T58	Navy
	Boeing Vertol CH-46	GE T58	Navy/Marine
Intermediate	Sikorsky H-52	GE T58	Coast Guard
	Bell UH-1	Lycoming T53	Army
	Bell UH-1N	P&WC T400	Navy/Marine
	Bell AH-1	Lycoming T53	Army
	Bell AH-1J	P&WC T400	Navy/Marine
	Kaman H-2	GE T58	Navy/Marine
Light	Bell OH-58	Allison T63	Army
	Hughes OH-6	Allison T63	Army

Although the investment required to enter a small engine market was within the capabilities of the smaller companies and entrepreneurs, generally the size of a particular market would not support more than one primary company. The helicopter market was the exception. Because of the demand for helicopters by all branches of the military services, there was a relatively large market in all four sub-niches of the military helicopter market.

Both the Army and the Navy sponsored the development of engines for heavy and medium-heavy helicopters, and the Army sponsored the development of an engine for light helicopters. Those companies that had been successful in winning competitions for the design and production of military helicopter engines prior to the mid-1960s were in a very strong position to retain the dominance of their particular niche. After the mid-1960s, not only were there sparse military funds available to develop new competitive engine designs, but also, any company that used its own funds to develop a new engine for the military market would have to overcome the unstated loyalty that each service felt for "their" sponsored engine and its company. Although each service, in order to minimize logistic support needs, tended to use its "own" engine for the helicopters it procured, the threat of competition between engine companies probably assured each service of a better product than would have otherwise resulted.

In the early 1970s, two additional important helicopter engine competitions took place that later radically altered the composition of players in this market. They were the Utility Tactical Transport Aircraft System helicopter (UTTAS) and the Advanced Attack Helicopter (AAH). UTTAS was the first engine competition that the Army had sponsored without going through the Air Force. In 1972, the Army awarded GE contracts for its T700 engine to power both of these helicopter programs. These were GE's first significant Army engine contracts. Later, when the T700 was produced in quantity for these and other helicopter programs, GE began to displace Lycoming as the Army's primary supplier for medium-power turboshaft engines.

MILITARY JET TRAINERS

As essentially all of the military front-line combat airplanes were powered by jet engines, the need for jet trainers became evident. However, trainer developments were normally a low acquisition priority, and engines specifically for trainers were not developed. The dominant engine used in subsonic trainers, the CAE J69, powered the twin-engine Cessna T-37, procured by the Air Force as a trainer for primary flight instruction. The Navy's subsonic trainer, the North American Rockwell T-2 Buckeye, originally equipped with a single Westinghouse J34 engine, was re-engined twice more, first with twin P&WA J60 engines, and then twin GE J85 engines. The first supersonic trainer, the Northrop T-38A Talon, was powered by twin GE J85 engines. GE's J85 engine was the sole U.S. supersonic trainer aircraft power plant, and it had a sizeable portion of the U.S. subsonic jet trainer engine market as well.

SELECTED SMALL MILITARY AIRCRAFT

The military also exploited the capabilities of the small turbine engine in various small aircraft as they became operational in the 1960s. Fixed-wing turboprop observation aircraft included the Army Grumman OV-1 Mohawk, powered by the Lycoming T53, and the North American Rockwell OV-10 Bronco, equipped with Garrett T76 engines, and used by the Marine Corps, Navy, and Air Force. The Vietnam War drove the production of both of these aircraft. The Army played a critical role in the development of the Beech Model 90 King Air turbo-prop aircraft powered by P&WC PT6 engines; for the Army, this became the U-21A Ute utility aircraft. In 1961, the Air Force received initial deliveries of P&WA J60 turbojet-powered aircraft, the North American T-39 and the Lockheed C-140. In each case, these airframe applications were important factors in the initial success of the first generation small turbine engines that powered them.

By the mid-1960s, the demonstrated capability of the large, high-bypass turbofan engine encouraged the development and production of small high-bypass engines. The first of these to go into production was the GE TF34 turbofan engine, developed for the Lockheed S-3A Viking ASW airplane; this engine was subsequently used in the Fairchild A-10 Thunderbolt II close air support aircraft. Both aircraft became operational in the mid-1970s.

In early 1960s, the Department of Defense selected an aircraft for favored nations under the Military Assistance Program. This aircraft became the Northrop F-5A Freedom Fighter, a new less complex and less costly aircraft suitable for the defense requirements of allies and friendly nations outside the U.S. In the early 1970s, a low-cost air superiority fighter designed to counter the threat of late model Soviet MiG-21 fighters became operational. It was the Northrop F-5E Tiger II, a derivative of the original F-5A. Both aircraft were powered by various models of GE's J85 engine, and both were procured in very sizeable quantities. Along with the T-38 trainer, the F-5 generated a substantial production run for GE's J85 turbojet engine. The J85 became the standard Air Force supersonic trainer and light international fighter engine.

DECOY, TACTICAL, AND STRATEGIC MISSILES

Toward the end of the 1960s, the Air Force had an interest in replacing the GE J85-powered McDonnell GAM-72 missile. Air Force-initiated studies of a Subsonic Cruise Armed Decoy (SCAD) validated the feasibility of an air-launched unarmed decoy missile to assist bomber penetration and to provide additional flexibility to bombers as armed decoy or armed attack (standoff) missiles. In an independent effort in 1963, Williams began development of its WR19 turbofan engine. The fuel efficient turbofan engine was a newly developing technology that held promise for significantly extending missile endurance for decoy, tactical, and strategic missions. Thus, coincidentally, the Williams engine was available at the time that the Air Force was interested in a small turbofan engine for a new generation of air-launched missiles.

In 1973, a derivative WR19 engine known as the Williams F107 won the Air Force AGM-86A SCAD missile engine competition. In 1972, the signing of the Strategic Arms Limitation Treaty (SALT) spurred significant additional funding for a Navy Sea-Launched Cruise Missile (SLCM) program. Continuing debates between the Air Force and the Secretary of Defense regarding the role of the SCAD missile killed the Air Force SCAD program in 1973, but SALT negotiation requirements were instrumental in immediately reviving it as the Air-Launched Cruise Missile (ALCM) program. In 1976, the Navy also selected the Williams F107 engine for its cruise missile. Williams was on its way to locking in a very sizeable cruise missile turbofan engine market.

During the late 1960s, as the desired range of tactical military missiles extended beyond the capability of rocket-powered designs, the services looked at turbine-powered tactical missiles. Motivated in part by the sinking of the Israeli destroyer Elath in 1967 by a Egyptian Soviet-made Styx missile, the Navy became interested in an anti-ship missile, and in 1969, the Harpoon program was established.[10] In 1971, McDonnell Douglas won a contract for the Harpoon (AGM-84A) air-launched anti-ship missile. In 1972, Teledyne CAE received a Navy contract for the development of the J402, a low-cost, expendable turbojet engine, to power the Harpoon. Teledyne CAE cornered what would become a substantial portion of the expendable tactical missile engine business.

TARGET DRONES AND RECONNAISSANCE RPVS

From the early 1960s to the mid-1970s, a considerable demand existed for target drones. Just as the extensive use of military turbojet airplanes generated the need for jet trainers, it also generated the need for turbojet-powered targets. Small jet-powered target drones provided an economical alternative to obsolescent fighters and bombers converted into target drones. Highly maneuverable and elusive, they also resulted in more realistic training.

A smaller but important demand developed for reconnaissance RPVs. Historically, because unmanned reconnaissance RPVs had posed a potential programmatic and financial threat to manned reconnaissance aircraft programs and because of some operational limitations, there had been resistance to their development by the military services. The loss of two U.S. reconnaissance aircraft

10. Kenneth P. Werrell, *The Evolution of the Cruise Missile* (Maxwell Air Force Base: Air University Press, 1985), 150.

in 1960 during the continuing Cold War with Russia and the loss of another reconnaissance aircraft during the Cuban Missile Crisis in 1962 were the initial incidents that changed this situation and spurred the development of unmanned
RPVs. The prelude to U.S. involvement in the Vietnam War, the 1964 Tonkin
Gulf incident, led to the operational deployment of these RPVs. During the
Vietnam War, extensive use of unmanned reconnaissance RPVs was made as a
result of the increasing effectiveness of enemy missiles used to destroy high-
flying aircraft. The importance of the reconnaissance RPV was that it eliminated the need to risk the lives of pilots.

CAE was directly impacted by these events. The need for jet targets and unmanned RPVs contributed importantly to the success CAE turbojet engines, in
particular the J69. The Ryan Firebee Q-2 (later BQM-34) target drone, powered by the J69, was produced in the largest quantity at this time, and it was used
by the U.S. Navy, Air Force, and Army as well as Canada. Starting in 1962,
Q-2C target drones were modified for photo reconnaissance as Model 147A/B
RPVs and were used extensively during the Vietnam War. CAE's J69 also powered the BQM-34E Firebee II supersonic target drone, and CAE's J100 powered
a high-altitude version of the Firebee RPV. During the early 1970s, CAE's J402
turbojet engine was selected to power the Beech MQM-107 Streaker Variable-Speed Training Target (VSTT) Vehicle that was to be developed for the
U.S. Army Missile Command. As the power plant supplier for major airframe
applications, CAE dominated both the target drone and reconnaissance RPV
engine market.

Williams also did well in this same market. The Northrop Chukar I
MQM-74A target drone, powered by the Williams WR24 turbojet engine, was
produced in quantity for the Navy and various international customers. The
WR24 was the first Williams high-volume production engine. The Canadair
CL-89 reconnaissance drone was powered by a Williams WR2 turbojet engine.
A smaller market penetration was made by Boeing, whose T50 turboshaft engine
powered the Gyrodyne DSN-3 (military designation QH-50C/D) which was
produced for the Navy's Drone Anti-Submarine Helicopter (DASH) program.
GE's J85 powered the Air Force-sponsored Radioplane Q-4B, a supersonic tar-
get drone which did not go into production.

GENERAL AVIATION BUSINESS JET AND TURBOPROP AIRCRAFT

The small gas turbine engine did not begin to penetrate the U.S. general aviation
aircraft market until the mid-1960s. The fact that the general aviation market was
more price sensitive than the military market and that the piston engine was

well-suited for low-performance, fixed-wing aircraft initially made it difficult for most engine companies to justify investment in developing gas turbine engines specifically for this market.

However, by the early 1960s, a clear need for higher performance business aircraft had emerged. Business executives, who had been using aircraft including World War II-era conversions, Douglas DC-3s, and Beech D-18 aircraft, saw the clear advantage offered by turbine transports, which airlines were rapidly putting into service. Turbine-powered aircraft were faster, which meant less time in the air and more time for business. They were more comfortable (less vibration), were more reliable (less down-time for maintenance), had high-altitude (over-the-weather) capability, and better resale value. Turbine powered aircraft were "more airplane for more money." An additional selling point was the glamour and prestige associated with ownership of turbine-powered aircraft. Within general aviation, the first market to evolve was for business jet and turboprop aircraft. The turbine-powered executive aircraft market expanded rapidly in the 1960s, coinciding with, and to a certain extent, both contributing to and benefitting from the mid-1960s general aviation record sales boom.

The first U.S. business jet aircraft had their roots in 1950's military aircraft requirements. The North American Sabreliner evolved from the Air Force UTX specifications and the Lockheed JetStar from the UCX specifications. In the early 1960s, these jets, powered by P&WA JT12 turbojet engines, were certificated and went into production as successful first-generation business jet aircraft. They were followed shortly by jets developed specifically for business use, including among others, the Aero Commander Jet Commander and Learjet, both of which were powered by derivative GE J85 engines known as CJ610s. Thus, first generation business jets were turbojet powered, and the GE CJ610 and the P&WA JT12 dominated the U.S. business jet aircraft market. Neither engine was specifically designed for the business jet market, but their thrust-class was ideally suited for that application when it emerged.

Small turbofan engines for business jets followed shortly. In 1964, GE's CF700, a derivative J85, became the first small turbofan engine certificated in the U.S. The CF700 was introduced on the Dassault Falcon 20. While it was the first turbofan in this market, the CF700 had only limited commercial success (other applications included the Rockwell International Sabreliner 75A and 80). By the early 1970s, two very successful small commercial turbofan engines were in production, the P&WC JT15D and Garrett TFE731. The JT15D was originally developed for the popular Cessna Citation, a small, entry-level business jet. Within a decade of its introduction in 1972, the TFE731 captured most of the medium-size business jet market and succeeded, through retrofitting, in replacing many turbojet engines on first-generation business aircraft. Small turbofan engines owed their success over turbojet engines to improved fuel economy, generally higher thrust-to-weight ratio, and reduced noise. Better sfc was the key factor that extended the range and inreased the payload of business jets, which greatly improved their utility.

Turboprop business aircraft also emerged in the 1960s. Canada's first turboprop engine, the P&WC PT6, was certificated in both Canada and the U.S. The first PT6 production application was the Beech King Air 90, the first U.S. small twin-turboprop aircraft; it became the most popular turboprop-powered executive transport in the world. Garrett's TPE331 turboprop engine also won commercial applications on executive aircraft, including the Aero Commander Turbo Commander and Mitsubishi MU-2.

GENERAL AVIATION HELICOPTERS

Helicopters using turboshaft engines, following their initial military developments, were modified for, and sold in, the commercial market. In the 1960s, a significant market began to develop for turbine-powered commercial helicopters. Some of the more prominent examples are mentioned.

In 1958, the GE CT58, a derivative T58, became the first gas turbine engine to be FAA type certificated for commercial helicopter use. In the early 1960s, it was introduced into commercial helicopter airline service powering Sikorsky S-62, Sikorsky S-61, and Boeing Vertol Model 107 commercial helicopters. These helicopters were utilized in a very small niche market so relatively few CT58 engines were produced for them.

By a substantial margin, the Allison Model 250, a derivative of the T63, came to dominate the commercial light turbine helicopter market. The Model 250 powered U.S. helicopters, including the Bell 206 JetRanger, Hughes Model 500 series, and Fairchild-Hiller FH-1100, as well as many foreign light helicopters. Starting in the mid-1960s, many thousands of Model 250 engines were produced for this market.

The Lycoming T53 had a stake in the commercial utility and transport helicopter market. In 1962, Bell introduced the T53-powered Model 204B, a derivative of the UH-1B Huey helicopter. The Model 204B was later produced under license by Agusta-Bell in Italy and Fugi-Bell in Japan for civil and military applications. In 1967, Bell introduced a civil derivative of the UH-1D known as the Model 205A which was also powered by the T53. The Model 205 was a stretched and more powerful version of the Model 204 and was also produced under license by Augusta and Fugi.

P&WC's PT6T TwinPac was a pioneering effort in bringing twin-turbine engine safety and power to helicopters. Certificated in 1970, the first PT6T model was installed in the Bell Model 212, Augusta Bell 212, and Sikorsky S-58T helicopters. The Model 212 was a corporate and utility transport and was used extensively in off-shore oil operations where extra reliability for extended overwater flights was critical. In the late 1970s, an improved version of the PT6T was installed in the Bell Model 412 transport helicopter, a four-rotor-blade version of the Model 212.

REGIONAL TRANSPORT AIRCRAFT

From 1964 to 1968, the number of scheduled air taxi operators (commuter companies) grew dramatically, and in the following years, the commuter market continued to grow. The growth in the regional transport market was stimulated in part by important Civil Aeronautics Board regulatory changes. Certificated carriers, which could not generate sufficient passenger boardings in small communities, were allowed to abandon service to these areas; for the first time, commuter airline companies were allowed to fly on routes previously flown only by certificated airlines. This growth was also stimulated by the appearance of new, more efficient turbine-powered commuter aircraft seating 19 or less passengers. The turboprop engine market for these aircraft was split by P&WC and Garrett. Those aircraft powered by P&WC's PT6 included the de Havilland DHC-6 Twin Otter, Beech 99, and Embraer Bandeirante EMB-110. Those equipped with Garrett TPE331s included the Shorts Skyvan, Swearingen (Fairchild) Metro, and CASA C-212. In 1972, further CAB changes led to larger commuter aircraft seating up to 30 passengers. An important new aircraft in this category was the PT6-powered Shorts SD3-30 which entered service in the mid-1970s.

A popular aircraft, which was certificated to airline standards, but used extensively in commuter operations was the 56-passenger de Havilland DHC-7. The DHC-7 entered service in 1978 and was powered by four P&WC PT6 turboprop engines.

Largely as a result of their small turbine engines, these aircraft were able to operate at lower costs and carry more passengers than earlier generation piston-powered commuter aircraft. Turboprop engines delivered the speed, short-field performance, passenger payload, and reliability needed for commuter operations.

The small gas turbine engine was a critical factor in the development, growth, and success of the U.S. commuter industry.

CORPORATE SPECIALIZATION WITHIN NICHE MARKETS

As corporations competed fiercely for military and commercial contracts, a clear pattern of corporate specialization emerged. Often, there were at least two major competitors for each niche market; one competitor typically emerged as the dominant engine supplier. In some cases, niche markets were less-competitively divided by several engine manufacturers. The divisions were by airframe applications, power ratings, or customer preference.

Lycoming, Allison, and GE split the helicopter engine market with Lycoming holding the leading market share. Between the 1960s and mid-1970s, both Lycoming and Allison's small engine efforts were focused on helicopter engines, while GE's small engine division was more diversified. Lycoming was the turbine engine supplier for the Army's first generation intermediate and medium lift helicopters, GE was the Navy's purveyor of engines for intermediate through heavy lift helicopters, and Allison was the Army's manufacturer for light helicopter engines. The Vietnam War drove helicopter engine sales for all three companies, especially Lycoming.

The Vietnam War had been responsible for the rapid growth of Lycoming's T53 and T55 turbine engine business, but it constrained the company's ability to develop new engine products. By the late 1960s, Lycoming had begun efforts which led to new products in the early 1970s. Near the end of the Vietnam War, from 1971 to 1973, Lycoming's combined T53 and T55 helicopter engine sales plummeted as a result of the Army's helicopter procurement reduction. An unsuccessful bid for the Army UTTAS helicopter engine competition and a loss in the Air Force A-X (experimental attack) aircraft competition further exacerbated Lycoming's position. Until 1970, most of Lycoming's turbine sales were for military helicopters and aircraft. Thereafter, incremental improvements were made in both commercial and international sales.

Allison, which had won the all-important Army 250 shp turboshaft/turboprop engine competition in 1958, did well from the mid-1960s through the mid-1970s in both the military and civilian markets with its T63/Model 250. Accounting for the vast majority of light turbine engine helicopter sales, the Model 250 contributed to the rapid, worldwide expansion of the light turbine helicopter. Especially popular Allison-powered helicopters were the Bell Model 206 JetRanger (military OH-58) and Hughes Model 500 (military OH-6). Other companies sought to enter this market with competitive engine designs, including the Lycoming LTS101, the Garrett TSE36 and TSE231, Continental T65, and P&WC "flivver" engine. An important factor in Allison's success in thwarting its competition was a low pricing policy, which was endorsed by its parent corporation, General Motors.

Because they were in the same power class, from the 1960s to the mid-1970s, P&WC and Garrett were head-to-head competitors in the turboprop-powered general aviation, regional transport, and selected military aircraft markets. Both companies found the general aviation market attractive as this market had not yet been penetrated by other turbine engine companies and because the engines required by this market were of the size that the companies could afford to develop. Both companies were additionally motivated by possible military turboprop or turboshaft applications, and their engines were military qualified as the P&WC T74 and Garrett T76.

In this competition, P&WC's assets included the name recognition and worldwide product support network of its parent company, P&WA, as well as Canadian government engine development support. However, despite concerted efforts, P&WC was able to win only a relatively small T74 production run. Although it was U.S.-owned, P&WC was a foreign company, and thus, faced some resistance when competing for U.S. military contracts. While P&WC eventually penetrated the U.S. military market by building a factory in West Virginia, the difficulty that it faced contributed to the corporation's strong focus on the general aviation and regional transport market.

Garrett was successful in the military market. Its T76 was selected for the North American NA-300 which won the Navy COIN (counter-insurgency) competition and became the OV-10 aircraft. However, as a result of company policy, Garrett, delayed entering the prime propulsion field, a fact that cost the company the opportunity to have its TPE331 on the market before the PT6. While the TPE331 and PT6 proved to be successful products, it was the PT6 that eventually powered the majority of light turboprop aircraft worldwide.

There were three principal U.S. companies in the executive jet market, P&WC, P&WA, and Garrett. In the 1970s, while P&WC and Garrett were both marketing business jet turbofan engines, the two companies were less direct competitors than they were in the turboprop market. This was because the P&WC JT15D and Garrett TFE731 were in different thrust classes, and therefore powered different size business jets. GE and P&WA competed directly against each other in the 1960s as GE's CJ610 vied with P&WA's JT12 turbojet engine for business jet applications.

Williams International and CAE had expertise in expendable turbofan engines for strategic missiles and were the primary competitors in this business during the 1960s to the mid-1970s. In the late 1960s, Williams was the first company to have a very small turbofan engine under development and in the process of flight testing. Military interest in a SCAD missile in the late 1960s stimulated CAE's interest in very small turbofan engines. The Williams F107 and CAE F106 were direct competitors for the Air Force SCAD missile engine. In a later rematch, the F107 and a derivative F106 again competed in the Navy cruise missile engine competition. Because the F107 prevailed in both cases, Williams became the principal cruise missile engine manufacturer. In the late 1970s, CAE won a competition to license-build the Williams F107 engine, thus sharing a secondary role in this market. Williams also had a place in the expendable turbojet engine business with its WR2 and WR-24.

CAE's speciality in the expendable engine business proved to be turbojet-powered tactical, drone, and RPV missiles. With its J402 powering the Harpoon and Streaker missiles and its J69 and J100 in various models of Firebee missiles, CAE found an important niche. Over time, expendable turbojet engines accounted for the majority of CAE's engine production. CAE made a considerable effort to enter the turboshaft engine business, but none of those engines ever reached production due to lack of applications. Thus, near the end of the 1960s, CAE focused its primary efforts on turbojet engines.

From the mid-1960s through the late 1980s, CAE also achieved considerable success winning contracts in support of Air Force and Navy gas turbine technology demonstration programs. During that time, much of the corporation's R&D funds were used in support of these military demonstration programs. While important technical achievements were made, they did not result in significant new business opportunities for CAE.

Among all of the small engine manufacturers at the time, GE was the most diversified with important shares in nearly all of the market niches, including the helicopter, jet trainer, small military aircraft, decoy missile, and executive jet engine business. GE also had every manner of small turbine engine under development or in production including turbojet, turboshaft, turboprop, and turbofan engines. Unlike other corporations that tended to specialize in a given market or type of engine, GE had almost all of the bases covered. The airframe markets in which GE did not have production applications at that time included: turboprop-powered regional transport or executive aircraft; tactical and strategic missiles; target drones and reconnaissance RPVs; and light helicopters.

By company policy, GE sequentially: let the military pay for most of an engine's development; secured a military production contract; got production tooling started; and then sold the engine commercially. The "military before commercial policy" was not only done for sound financial reasons, but was used as an important sales tool for GE commercial efforts which followed. With the exception of its very strong presence in the business jet market and an inroad in commercial helicopters, GE was largely focused on military engine products as a result of this company policy.

TECHNOLOGY HIGHLIGHTS

From the 1960s to the mid-1970s, there were significant technological advances in small gas turbine engines and in their components. There was a movement toward refining solutions for or circumventions to the technology challenges of the 1940s and 1950s. Successful small turbofan engines were developed. Significant development activity was focused on engines for VTOL applications. Important technical achievements were made in small turbine engine components.

A major technological advance was the introduction of small turbofan engines. Large, low-bypass turbofans had been demonstrated, and they stimulated the development and use of small turbofan engines. Most of the small engine companies had a turbofan engine under development. It was a new category of small turbine engines, and the configuration was an open question. A number were tried, including front and rear fans as well as one-, two-, and three-shaft engines. Both low- and high-bypass ratio turbofan engines were developed. Of these variations, the high-bypass front fan, two-concentric-shaft configuration eventually became the most widely adopted type. The widespread adaptation of turbofan engines in the 1970s was not only a result of their performance advantages, especially their low sfc, but also the result of environmental and oil shortage concerns.

Certificated in 1964, GE's CF700 was the first entry in the small turbofan market. This was accomplished by the expedient solution of adding an aft fan and making other modifications to the single-shaft J85 turbojet engine. Slightly ahead of its time, Lycoming ran the first North American high-bypass ratio turbofan engine, the PLF1-A2 in 1964, but was unsuccessful in finding sufficient market interest to justify certification of the engine. Other turbofan engines, which began development in the 1960s or early 1970s and which were military qualified or civilian certificated, included the Williams WR19 (later designated F107), P&WC JT15D, GE TF34, Garrett ATF3 and TFE731, and Lycoming ALF502.

During the 1960s, there was continuing military and government research interest in vertical take-off and landing (VTOL) aircraft. Research was devoted to high thrust-to-weight ratio small jet engines and to adapting shaft turbine engines to such aircraft. CAE, in particular, was noted for its XLJ95 lift jet engine which demonstrated a 22:1 thrust-to-weight ratio, higher than any other engine. Other small turbine engines that were used in various VTOL or boost applications included the GE J85, designed as a high thrust-to-weight ratio turbojet engine, the GE T58 and T64 turboshaft engines, Lycoming T53 and T55, GE CF700 turbofan engine, P&WA JT12, and the CAE J69 turbojet engine.

In addition to the specialized work on specific types of small turbine engines, important improvements were made in individual engine components. For example, centrifugal and axial-flow compressor research produced higher compressor efficiencies, pressure ratios, and airflow rates.

A few selected examples of compressor improvements and protection devices were especially noteworthy. For its TPE331, Garrett used transonic axial-flow technology to raise the inlet airflow Mach number to 1.3. Garrett used this axial-flow transonic inducer in combination with its tradition of highly backward curved centrifugal-flow compressor designs to improve compressor airflow and pressure ratio while maintaining the same or better efficiencies and acceptable surge margins. Lycoming introduced a transonic axial-flow compressor on its T53. Originally developed for the PT6, P&WC patented a pipe diffuser that significantly reduced losses and improved compressor efficiency. In its Model 250-C28, Allison replaced a six-stage axial, single-stage centrifugal compressor with a higher pressure ratio and higher efficiency single-stage centrifugal compressor. The -C28 compressor also incorporated a slotted inducer that acted as a "smart bleed," allowing air to bleed off through the inducer shroud at lower speeds (to prevent choking), but permitting air to be flowed in at higher speeds.

An important combustor development was the GE12 low-pressure, vaporizing combustor, which established a design precedent for future GE engines. Among other benefits, it resulted in a shorter and lighter combustor, better pattern factor, reduced smoke, high combustor efficiency over a wide range of operating conditions, and higher temperatures.

Considerable progress was made in the development of high temperature turbines in small turbine engines. To capitalize on the increased pressure ratios available, it was necessary to increase the temperatures at which the turbines could operate. Technology proceeded by improving high temperature materials for turbines, enhancing turbine blade and stator cooling, and increasing turbine loading.

Illustrations of high temperature turbine development efforts included the following. P&WA was on the forefront in developing directionally solidified materials, and its subsidiary P&WC benefitted from this technology by introducing it in its PT6T-3 TwinPac turbine blades. A difficult technology to achieve in small turbine engines, small blades with miniature internal cooling passages and holes were developed to enhance convection cooling. GE's T64 was the company's first small engine to use air-cooled turbine blades. Very high temperature, air-cooled turbine nozzles and blades became a GE signature technology. Garrett's ATF3 incorporated similar air-cooled turbine technology.

Significantly higher turbine loading, which resulted in high temperature drop and cooler blade temperatures, was also achieved. Garrett's TSE231 turboshaft engine used high expansion ratio turbine nozzles and was the first Garrett engine to use a turbine that was designed with non-free vortex design concepts which permitted setting the blade angles from hub to tip to control flow distribution and thereby nearly doubled the work capacity done by previous company axial turbines. Other examples of high-work turbines included GE's T64 and P&WC's PT6.

Major non-performance improvements were made in small turbine engines, especially in the areas of reliability, maintainability, durability, survivability, resistance to foreign object damage (FOD), and reduction in parts count. To protect engines from FOD, barrier filters and inertial inlet particle separators were developed. Garrett made significant advances toward minimizing large bird ingestion damage in its TFE731 turbofan engine through the use of blunt airfoils and mid-span dampers. To reduce parts count on the GE12 (later T700) turboshaft engine, GE introduced the blisk concept with the axial compressor rotor and blades forged integrally in a single rotor. The GE12 also introduced numerous precedent setting techical features which significantly reduced engine maintenance costs. CAE's J402 expendable turbojet engine was designed to be maintenance free with a shelf life of five years. Known as the "wooden round" engine, it powered the fully-fueled, ready-to-go Harpoon missile.

There were also a number of additional miscellaneous technology developments. Notable among these was an afterburner for GE's J85 turbojet, the first U.S. small turbine engine to be so equipped. Williams, in its WR2 engine, successfully developed accessories that were driven at engine rotor shaft speeds up to 60,000 rpm, thereby eliminating the conventional reduction gearbox. Expanded use was made of computers to enhance engine analysis and component design, and Garrett was one of the leaders in this area. In materials development, component casting technology was improved which allowed single-piece castings to replace multiple-piece components. Allison's Model 250 was an example of an engine that successfully incorporated this technology. Finally, Air Force-sponsored R&D programs, including LWGG and ATEGG, as well as Army-sponsored programs, including the 1,500 shp Demonstrator Engine and STAGG, were responsible for a variety of cutting edge technologies which often were integrated into production engines.

3. Maturing of the Industry (Late 1970s–1980s)

A general observation of this period suggests, that while the total number of products continued to expand in the small gas turbine industry, the industry itself had begun to mature. The number of corporations in the business had stabilized. Their specialties in niche markets were well defined. Broadly speaking, the market had begun to move toward a point of equilibrium. Nonetheless, important new military and commercial programs were initiated which had a vital impact on the demand for small engines. World events and government policies continued to influence the small engine market, and the market for small engines became increasingly international.

Two especially profound corporate changes took place. First, a major trend toward corporate diversification was evident as companies sought to expand their product base. An important driver was not only the desire to improve revenue, but also to reduce sensitivity toward market vicissitudes and economic cycles. Secondly, there was a complex movement toward corporate partnerships, both domestic and international, in response to changes in military procurement policy, the cost of technology development, and the desire to penetrate international markets.

Small engine technology was well established and incremental progress continued to be made in performance and efficiency. Increasing emphasis was placed on design-to-cost strategies, improved maintainability and reliability, and designs which addressed environmental concerns. Corporations developed a new generation of engines derived in part from successful original products.

INFLUENTIAL WORLD EVENTS AND GOVERNMENT AND MILITARY POLICIES

A ripple effect from earlier world events continued to have an impact on certain aspects of the small engine business. The oil embargo of the early 1970s and increased environmental concerns continued to influence engine design in a positive manner and to affect the demand for helicopters used in off-shore oil exploration and production. The end of U.S. involvement in the Vietnam War, which had drastically cut the military demand for various small turbine engines, caused some corporations to seek or emphasize new products. Continuing nuclear disarmament treaty discussions did not prevent the operational deployment of cruise missiles in the early 1980s. However, the deployment of ground-launched cruise missiles generated significant controversy, ultimately resulting in an agreement to eliminate specific types of nuclear cruise missiles in 1987. Airline deregulation in 1978 had a broad-reaching influence on both the design and production of small turbine engines. The production of small engines followed closely economic cycles.

Military procurement policies evolved that influenced small engine design. One policy emphasized or required joint service management of engine development. Another policy involved second-sourcing production through a leader-follower approach. One engine program introduced continuous competition from RFP to production as well as government-sponsored partnership competition. Continued emphasis was placed on more maintainable and durable engines with lower life cycle costs. More realistic testing procedures were developed. Affordability of all aspects of engine development and life cycle costs became an increasingly important concern in military engine acquisition programs.

Two procurement policies associated with the cruise missile engine program were joint service management of engine development and leader-follower production sourcing. In 1977, a Joint Cruise Missile Project Office was formed to ensure commonality between the Air Force and Navy cruise missile programs. The air-launched cruise missile program called for a leader-follower approach to manufacturing to second source all components that went into the cruise missile. Ultimately, Williams licensed another manufacturer to build its F107 engine as a result of this requirement.

In 1983, the Army began its Modern Technology Demonstrator Engines (MTDE) program. The objective of this program was to develop a demonstrator turboshaft engine in the 5,000 shp class. A noteworthy aspect of the MTDE program was the fact that additional major design requirements were given equal priority to performance. They included reliability/durability, overhaul/support cost, acquisition cost, weight, and survivability as well as performance.

In 1984, the Army released an RFP for the T800 engine. An important milestone established in this program was a shift from emphasizing maximum performance toward more maintainable engines with lower life cycle costs. Following the precedent established by the T700, the Army continued to emphasize engine reliability, availability, maintainability, and survivability. The T800 program required contractual guarantees for purchase and operating cost. The T800 full-scale development program was the first government-sponsored competition requiring company partnerships. It was also the first such engine program to emphasize continuous competition from RFP throughout development and production phases. The Army's objective was to get a better blend of technologies to improve the final product as a result of the partnership arrangement, to avoid problems with single-source procurement by having a dual-source production capability, and to reduce life cycle costs.

Fixed-price contracts, such as those under which the LHTEC T800 and Allison T406 engines were developed, became a standard way of doing business in the 1980s. The T800 contract also included performance and operating cost guarantees, and the manufacturer was required to underwrite all facilitization and tooling at no cost to the government. While good for the government, such contracts increased the manufacturer's risk and cost of doing business with the government. Toward the late 1980s, as the defense budget began to decline, such contracts became more problematic as original production commitments were delayed and/or diminished. Corporations became increasingly concerned about investing in fixed-price, military development programs due to increased cost share and uncertain production opportunities. Doing business with the government had changed from earlier times when the military had funded the full cost of developing an engine.

By the end of the 1970s, a dramatic change in traditional development test requirements had been made that emphasized accelerated mission-matched endurance testing.[11] Increased low-cycle fatigue testing, "maturity programs," and accelerated simulated mission endurance tests were examples of such tests that attempted to replicate more precisely and realistically actual aircraft mission profiles, including the number of full thermal cycles, time at maximum power, time at temperature, and other relevant parameters. They resulted in significantly improved engines which achieved maturity much earlier in their lives than the previous generation of engines.

MARKET DEVELOPMENTS

The small turbine engine gained important airframe applications in the late 1970s and 1980s. In the military market, helicopter and cruise missile programs, in particular, resulted in significant turbine engine production runs. In the commercial market, deregulation had a major impact on the development of new commuter aircraft and engines for them. The decade of the 1970s was an especially important era in the growth of a variety of applications for small turbine engines.

THE MILITARY MARKET

Many new military helicopter procurement programs were undertaken, largely as replacement for the first generation of turbine-powered helicopters. They had an important impact on the development and production of selected small

---

11. J.-P. Frignac, Frank E. Pickering, William W. Wagner, and J. W. Witherspoon, "Turbine Engines in the 80's," *Astronautics and Aeronautics*, (June 1980): 30.

turboshaft engines. Various small fixed-wing aircraft programs were started. Cruise missile engine production finally got underway after many delays due to competing technologies and politics. A miniature turbojet engine development program was started by the Army. Small turbine engines powering existing airframes were continuously uprated or improved.

A variety of vertical flight programs were in process. Two especially important military helicopter competitions were the Army's Utility Tactical Transport Aircraft System and Advanced Attack Helicopter programs. Sikorsky's YUH-60A had been selected as the UTTAS helicopter and the Hughes YAH-64A as the AAH. In 1977, the Navy selected the Sikorsky SH-60B Seahawk helicopter for its Light Airborne Multi-Purpose System (LAMPS) Mk III program. These procurement programs led to substantial GE T700 engine production, which started in 1978.

In 1981, the Bell OH-58D, powered by the Allison T703, was selected by the Army as the winner of the Army Helicopter Improvement Program (AHIP), which was intended to provide a near-term scout helicopter as an interim measure pending the development of a new design (Light Helicopter Experimental). In 1981, the Joint Services VTOL Experimental (JVX) program was launched with the objective of developing a multiple-service-use medium lift VTOL aircraft. A Bell Helicopter and Boeing Vertol tilt rotor design was chosen in 1983 as the winner of the JVX airframe competition. Allison was selected to develop the T406 engine for this aircraft, the V-22 Osprey. In 1982, the Army invited airframe manufacturers to submit design concepts for its Light Helicopter Experimental (LHX) program, intended to replace Vietnam War-era light scout/attack helicopters such as the AH-1 Cobra, UH-1 Iroquois, OH-58 Kiowa, and OH-6 Cayuse. In 1988, the Light Helicopter Turbine Engine Company (LHTEC) comprised of Garrett and Allison won the T800 turboshaft engine competition for the LHX. In 1979, Aerospatiale won the Coast Guard competition for a helicopter to perform Short Range Recovery (SRR) duties. Designated HH-65A Dolphin, it was powered by Lycoming T702 (civilian designation LTS101) engines. Except for the JVX and LHX, all of these vertical flight programs went into production by the end of the 1980s.

An assortment of small, fixed-wing, turbine-powered aircraft initiatives took place. In the late 1970s, the Air Force had a requirement for a Next Generation Trainer to replace the Cessna T-37. In 1982, Fairchild was selected for this new trainer, the T-46A. However, the T-46A program was canceled in 1986 due to aircraft production delays and escalating costs; this terminated Garrett's promising F109 turbofan engine which powered it. In the early 1970s, Beech re-engined its piston engine T-34 trainer with a P&WC PT6. Designated T-34C, production for Navy and foreign military services began in the mid-1970s. In the 1970s, the Army, Air Force, Navy, and Marine Corps began off-the-shelf acquisition of standard Beech Super King Air 200s (military designation C-12 Huron), powered by P&WC PT6 engines, as staff transport aircraft.

A number of additional Air Force fixed-wing acquistion programs took place in the 1980s, but the number of aircraft procured in each program was small. The following aircraft, all derived from existing commercial products, were among them. In 1983, the Air Force selected the Gates Learjet Model 35A (military designation C-21A) as a replacement for its North American CT-39 jet aircraft; the C-21As were powered by Garrett TFE731 engines. In 1984, the Air Force ordered the P&WC PT6-powered Shorts 330 (military designation C-23) for its European Distribution System Aircraft (EDSA). In 1988, the Air Force chose the British Aerospace BAe 125-800, equipped with Garrett TFE731 engines, as the C-29A for its Combat Flight Inspection and Navigation (C-FIN) requirement to replace its T-39s and C-140s used in a similar role. In 1988, the Air Force placed a contract with Fairchild for its Garrett TPE331-powered Metro 3 (military designation C-26A) regional transport for use as operational support aircraft by the Air National Guard.

Important missile programs began production and development. In 1977, the B-1 bomber program, which had been the top Air Force priority, was canceled, and subsequently, the Air Force canceled its short-range AGM-86A cruise missile program and pushed forward its long-range AGM-86B cruise missile program. In 1977, Williams was contracted to complete development of its F107 turbofan engine as the propulsion system for both the Air Force AGM-86B ALCM and the Navy BGM-109 SLCM programs. Starting in the early 1980s, Williams and its licensee, CAE, began a major engine production run for Air Force and Navy cruise missile engines. The AGM-129A Advanced Cruise Missile entered development in 1983 and was intended as an eventual replacement for the AGM-86B ALCM. Again, Williams was selected to provide the power plant with its F112 turbofan engine, an uprated F107 engine.

In 1985, the Propulsion Directorate of the U.S. Army Missile Command (MICOM) initiated a program to develop a series of miniature turbojet engines for small tactical missile applications. Companies that developed engines for this program in the late 1980s and early 1990s included Allison, Sunstrand Power Systems, Technical Directions, CAE, Turbine Technologies, and Williams. There was a need by the Army for small tactical missiles having longer range and endurance than was achievable with rocket-powered missiles. An important and challenging objective was to minimize the cost of the miniature turbojet engines that powered them.

Small turbine engines selected for earlier military programs or powering existing military aircraft were continuously uprated or improved as required from the mid-1970s to the late 1980s. The yearly production of newer engine models generally increased and older engines tended to decrease, but not in all cases. Williams continued a substantial production run with uprated versions of its WR24 (J400) series turbojet engines for Chukar II and III target drones. CAE began a very important production run of its J402 for the Harpoon missile in 1975 and also continued production, at a reduced level, of its J69 for various models of the Ryan Firebee drone. The yearly production of Lycoming T53 and T55 turboshaft engines generally declined from 1975 to 1990, although uprated versions of each engine continued in production at a relatively low, but constant level. P&WC continued to upgrade its PT6T TwinPac engine and experienced strong sales, both military and civilian, for this engine in the mid-1970s through early 1980s. For GE, by the late 1970s, J85 annual production began a gradual decline as Northrop F-5E/F production tapered off. There was a slight rise in the annual production of GE's T58 in late 1970s and early 1980s and then a decline. Both the J85 and T58 were out of production by the late 1980s. GE's TF34 production peaked from the mid-1970s to early 1980s coinciding with Lockheed S-3A and Fairchild A-10A aircraft programs. Among all of GE's first generation small turbine engines, the T64 soldiered on with continuous performance upgrades and a low, but relatively stable annual production rate primarily for the various models of the Sikorsky CH-53. Allison continuously upgraded its T63/Model 250; Model 250 commercial sales were especially strong in the 1970s and early 1980s relative to T63 military sales.

THE COMMERCIAL MARKET: COMMUTER AIRCRAFT

An important market trend at this time was the rapid proliferation of commuter aircraft as a result of the 1978 Airline Deregulation Act. As the rate and route authority of the federal Civil Aeronautics Board was phased out as a result of deregulation, commuter aircraft were able to acquire a significantly larger market share as trunk carriers withdrew from small towns and short routes. Following deregulation, the CAB again raised the limit on the size of passenger aircraft. Airframers responded to this, and they held a number of engine competitions from 1979 through the early 1990s for larger commuter aircraft.

In the 30-40 passenger commuter aircraft engine competitions, the P&WC PW100 series and GE CT-7 turboprop engines were the primary contenders. Most of these competitions were won by PW100 series which found applications on the Embraer EMB-120 Brasilia, de Havilland Dash 8-100, Aerospatiale/

Aeritalia ATR 42, and Dornier 328. The CT-7 won the Saab-Fairchild SF-340, Casa-Nurtanio CN-235, and Czech Republic LET L-610G commuter aircraft competitions. Garrett's TPE331 was really too small an engine for this market, though a version of the engine specifically designed for commuter applications was selected for the British Aerospace Jetstream 41, a 29 passenger aircraft.

P&WC continued with a string of victories in even larger (50 or more passenger) commuter aircraft engine competitions. The PW100 series won applications on the British Aerospace ATP, Aerospatiale/Aeritalia ATR 72, and Fokker 50. Other new 50-passenger commuter aircraft that selected Allison GMA 2100 turboprop engines were the Saab 2000 and IPTN N-250.

In the traditional 19-passenger commuter aircraft market, Garrett did exceptionally well with its TPE331 turboprop engine. The TPE331 was selected for the British Aerospace Jetstream 31 and Jetstream Super 31. The TPE331 was also selected for various versions of the Fairchild Metro and Dornier Do 228 commuter aircraft. Garrett's new TPF351 turboprop was chosen for the Embraer CBA-123 commuter aircraft. A more powerful version of P&WC's PT6 was picked for a popular commuter aircraft in the 19-passenger market, the Beech 1900.

A new commuter market, the regional jet, began to develop in the 1980s. Introduced in service in 1983, the 82-passenger British Aerospace BAe 146, powered by Lycoming ALF502 turbofan engines, became the first regional jet aircraft. In 1988, GE's CF34 turbofan engine was selected for the 50-passenger Canadair Challenger CL-601 Regional Jet. In 1990, Embraer selected Allison's GMA 3007 turbofan to power its 50-passenger EMB-145 regional jet.

THE COMMERCIAL MARKET: GENERAL AVIATION

The decade of the 1970s was an especially significant growth period for small turbine engines in the general aviation market. By the late 1970s and early 1980s, small turbine engines had substantially penetrated the general aviation market and a broad number of turbine-powered aircraft were developed and manufactured. For example, the number of turbine-powered executive aircraft grew steadily. After a drop in sales in 1971, the number of general aviation turboprop and jet aircraft shipments generally increased each year through 1981.[12] An economic slump led to a drop in sales in the early 1980s after which time the number of general aviation turbine aircraft shipments essentially plateaued

---

12. *General Aviation Aircraft Shipments and Factory Net Billings 1962-1989* (Washington, D.C.: General Aviation Manufacturers Association, 1989).

through the remainder of the decade. More turboprop business aircraft were shipped in the 1970s and 1980s than business jet aircraft. The number of turbine-powered general aviation helicopters increased steadily as well. Numerous uses were found for turbine-powered commercial helicopters which was an important factor contributing to their success. Important turboprop agricultural and utility aircraft were also developed.

Continuing to lead the sales for turboprop business aircraft was the the PT6-powered Beechcraft King Air, which was produced in a variety of progressively more sophisticated models. Other competitor aircraft powered by PT6 engines included the Piper Cheyenne I and II and Cessna Conquest I. The years 1975 through 1985 were peak production years for P&WC's PT6A turboprop engine, and executive aircraft applications were an important component of this production. Garrett's TPE331 also continued to sell well for executive applications including the Cessna Conquest II, Rockwell (Gulfstream) Commander series, Mitsubishi MU-2 series, and Swearingen Merlin series. Garrett's TPE331 sales for executive aircraft increased steadily from 1976 to 1981. Less successful entries in the general aviation turboprop engine market included Lycoming's LTP101, which had a relatively small production run due to a serious initial reliability problem, and Teledyne Continental Motors Aircraft Products Division's TP-500, which did not go into production due to competing company priorities and a limited retrofit market.

Starting in early 1970s, a transition from turbojet to turbofan-powered business jet aircraft took place. This change, which provided significant improvements in fuel economy, and thus range, was essentially complete by the late 1970s and early 1980s. For the most part, new production aircraft were powered by turbofan engines. Retrofitting turbofan engines on first generation turbojet executive aircraft was also an important business. A rapidly expanding market resulted in the introduction of substantially improved and new business jet models. Another trend was the initial development of a second generation of new, improved turbofan engines for this market, including the Garrett/GE CFE738 turbofan and the P&WC PW300 turbofan.

Between 1973 and 1977, Garrett's TFE731 turbofan succeeded in acquiring the leading position in the small to medium-size business jet market, and it steadily accumulated additional applications beyond that time. Among the numerous business jets that it powered as original equipment or as retrofits were: the Dassault Falcon 10/100, Falcon 50, and Falcon 900; Learjet Models 31, 35, 36, and 55; Lockheed JetStar II; British Aerospace BAe HS 125-700 and BAe 800; Israel Aircraft Industries IAI Models 1124 and 1125; Cessna Citation III; and North American Rockwell Sabreliner Model 65. From 1971, when the first engines were shipped, the number of TFE731 engine shipments increased each year peaking in 1981.

P&WC's JT15D turbofan continued to be popular on the Cessna Citation series, including the Citation I, II, and S/II, and an uprated model of the JT15D was later used on the Citation V Ultra. The JT15D was additionally used on the Aerospatiale Corvette and Mitsubishi Diamond I and IA in the 1970s and the Diamond II in the 1980s. Lycoming powered one business jet with its ALF502 turbofan engine, the Canadair CL-600 Challenger. This jet was superceded by the Canadair Challenger CL-601, and in 1979, GE's CF34 turbofan engine was selected for it.

Steady progress was also made in the commercial turbine helicopter market. An important contributing factor was the incredible number of useful duties that such helicopters could perform efficiently. Depending on the specific model, they were used for executive transport, commercial passenger transport, law enforcement and border patrol, offshore oil and natural gas rig support and service, remote resources exploration and extraction, logging, construction, rescue, geological survey, crop spraying, pesticide applications, air ambulance work, fire fighting, pilot training, sight-seeing tours, and general utility.

A number of commercial helicopters were introduced in the late 1970s through the 1980s, and a variety of engine manufacturers supplied turboshaft engines for them. Improved and uprated models of the Bell JetRanger (Model 206B) and LongRanger (Model 206L) were introduced, both types powered by Allison Model 250 turboshaft engines. New Bell helicopters, including the Models 412, 214ST, and 222B/UT, powered respectively by P&WC PT6T, GE CT7, and Lycoming LTS101 turboshaft engines, were certificated. Other commercial Bell helicopters produced during this time included the Model 205A and 214B, which were equipped respectively with commerical derivatives of Lycoming T53 and T55 engines. Such applications assisted Lycoming in increasing its commercial engine sales as its military engine sales declined subsequent to the Vietnam War. New models of Allison Model 250-powered McDonnell Douglas (Hughes) Model MD 500 helicopter series were introduced. A variant of the CH-47 Chinook helicopter, the Boeing Vertol 234, powered by derivative Lycoming T55 engines (designated AL5512), was developed for commercial medium-heavy lift and transport applications. In order for Sikorsky to get a bigger share of the commercial market, the company certificated in 1978 its Model S-76 helicopter powered by two Model 250 engines. Later, Sikorsky introduced its Model S-70C helicopter powered by GE CT7 turboshaft engines.

An important market for North American small turbine engines was foreign-manufactured helicopters. Among them was the high speed, high performance Italian Augusta A 109A powered by twin Model 250 engines. One version of the French Aerospatiale helicopter, the AS 350D AStar, powered by the Lycoming LTS101, was designed to U.S. hospital standards and had an air ambulance role in the U.S. A twin-engine version was the Aerospatiale AS 355 TwinStar powered by Model 250 engines. The German-built and Allison Model 250-powered MBB BO 105 helicopter was upgraded as new models were progressively introduced. An international project, the MBB/Kawasaki BK 117 equipped with LTS101 engines, was introduced in the early 1980s.

To meet these commercial helicopter market needs, North American turbine engine manufacturers continuously upgraded and specially adapted existing engines such as the Allison Model 250, P&WC PT6T, and Lycoming T53 and T55. A GE derivative T700 engine, the CT7, was certificated for this market. A new low-cost, low-power turboshaft engine, the Lycoming LTS101 went into production in 1975, and development of another turboshaft engine, the P&WC PW200, was begun in 1983.

An important development in the mid-1970s through early 1980s was the introduction of turbine-powered agricultural aircraft and a new generation of utility aircraft. P&WC was a major contributor to this market. Among the agricultural aircraft were the Ayres S2R Turbo Thrush, Frakes Turbo-Cat, Air Tractor AT-400, and Grumman (later Schweizer) Ag-Cat G-614, all P&WC PT6 powered. The benefits over comparable piston-powered aircraft included substantially increased productivity, performance, reliability, and lengthened TBOs as well as quieter operation and multiple-fuel capability. Lycoming had a share of this market with its LTP101 powering the Air Tractor AT-302 and several other conversions. A very successful utility aircraft, the PT6-powered Cessna Model 208 Caravan I was certificated in 1984. It could fly fast with a heavy payload, operate from unprepared airstrips, and offered the economy and reliability of single turbine engine operation. Finally, there were numerous experimental and conversion general aviation aircraft applications for small turbine engines in the late 1970s and 1980s.

CHANGES IN THE INDUSTRY: DIVERSIFICATION, LICENSING AGREEMENTS, AND PARTNERSHIPS

Through the late 1970s and 1980s, there was a broad trend among companies within the small gas turbine engine industry toward company product diversification and corporate partnerships. Licensing agreements, which had been the traditional means by which corporations expanded foreign sales, continued to be made, but assumed a diminishing role relative to international corporate partnerships.

DIVERSIFICATION

Strength in a niche market was also a company's Achilles' heel. If market erosion took place, a company highly dependent on only one or two markets was vulnerable. Thus, starting in the late 1970s and continuing through the 1980s, most companies in the small gas turbine engine business began the process of diversifying their product lines. For instance, companies that sold engines to the civilian engine market began to study engine development possibilities for the military engine market, and vice versa. The principal objectives of diversification were to increase corporate financial stability and to reduce engine development costs.

To ensure long-term stability and to diversify their portfolio, corporations invested heavily in new engine development, often for markets for which they did not have a product. As corporations were keenly aware, even with continuous improvements, in time, the sales of their successful engine models would most likely taper off due to market saturation or competition. Complicating their diversification efforts was the time required to develop a new engine, often as long as five years (and sometimes longer). Because engine development was costly, accurate market forecasting was critical. With a good forecast, the right engine would be available when a new or revitalized market opened; if inaccurate, the corporation would incur a significant loss.

One means of reducing risk was to develop a "full-service" product line offering, for example, both civilian and military engines. A company could stabilize its economic base by shifting its emphasis toward any given engine model depending on the market situation. The company could more readily respond to changes in the market and ride out inaccurate market forecasts. A broader product base was a method of dealing with external market forces that were beyond company control. While very few companies actually developed a full-service line for all of the small turbine engine markets, almost all attempted to expand their product line to some degree.

The need to reduce the enormous cost of engine development was another diversification driver. Engine technology was so expensive to gain, that unless a company could use its technology over a broader product base and in a wider number of markets, it could be put in a very difficult competitive position sometime in the future. One example of broadening a product base would be deriving a civilian engine from a military-developed engine. This was a relatively simple and inexpensive operation. An example of increasing access to multiple markets would be the development of a multiple-function engine, a gas generator that could be used either in a turboprop or turbofan engine for fixed-wing aircraft or in a turboshaft engine for helicopters and non-aircraft applications. In either case, the development cost of an engine model could be spread over a larger, and perhaps longer, production run.

Different companies experienced varying degrees of success as they tried to diversify their product base. These efforts reshaped the industry considerably. In general, those corporations that successfully undertook this effort were in a potentially stronger position than if they had continued with a narrower focus.

Lycoming was among the earliest of the corporations to begin diversification efforts. While largely focused on developing and producing medium power turboshaft engines for the Army, by 1963, Lycoming had a variation of its T53 in the Bell Model 204B commercial helicopter. In 1964, Lycoming's parent corporation, Avco, began an overall diversification effort, to acquire through conglomeration, commercial businesses in order to balance its defense-oriented companies.[13] While Lycoming had been interested in new commercial and military projects in the 1960s, the company's resources were largely committed to the development and production of the T53 and T55 engines for the Army. As demand dropped in the early 1970s, Lycoming began a dedicated effort to develop new engine products. However, these efforts really did not begin to have a significant effect until the late 1970s and early 1980s.

Among the new Lycoming products were a low-power LT101 series, which included the LTS101 turboshaft and LTP101 turboprop engines. The turboshaft engines in this series won a few helicopter applications, but their success in the market was limited due to initial in-flight reliability problems. Derived from earlier development projects and military engines, a commercial turbofan engine, the ALF502, began development in 1970, and it captured an executive jet and a regional jet application in the late 1970s and early 1980s. In 1976, Lycoming's AGT1500 tank engine (which had originally been developed in the 1960s) was selected for a new Army tank, later designated M1 Abrams. In the late 1970s, the AGT1500 went into volume production and contributed significantly to Lycoming's revenue in the early 1980s.

In 1989, Lycoming's sales were nearly equally split between the commercial market and the AGT1500 tank engine, which then accounted for most of the

---

13. Avco Corporation, *The First Fifty Years* (Greenwich, Connecticut: Avco Corporation, 1979), 52-73.

company's military sales. While Lycoming's overall military helicopter and air-craft engine sales had declined since the Vietnam War, new commercial prod-ucts and the AGT1500 had offset the loss of business. Thus, the company's efforts to diversify had been successful. By the late 1980s, the company made a decision to focus its component technology efforts on one size class, namely 10-lb-per-second of airflow. That decision resulted from the fact that its most successful engine products, namely the T53 and AGT1500, were in that class, and its test facilities and manufacturing capabilities were well suited for engines moderately larger or smaller than these engines.

In the late 1970s and early 1980s, Garrett initiated a strategic plan to be-come less dependent on its primary general aviation aircraft engine business. In part, Garrett's decision to diversify was motivated by a precipitous decline in general aviation aircraft production which began in the late 1970s. Getting more military business was one obvious channel that such a diversification strategy could pursue. Garrett had made some inroads in the military market with its T76 turboprop program and with its participation in government technology demonstration programs in the 1970s. Winning additional military engine competitions would not only help Garrett diversify, balance, and ex-pand its product base, but would further the company's technical capability as a result of government-assisted development. Starting in 1982, as a result of a shift in the market, Garrett began shipping more TPE331s for regional airline aircraft than executive aircraft. While this was not a company-initiated effort nor was it a new market for Garrett, the company quickly capitalized on this opportunity and made important changes in the TPE331 to better adapt it to regional airline use.

Essentially a two-engine (TPE331 and TFE731) company, Garrett added five major engine programs from the late 1970s through the 1980s as part of its diversification. These included: the TFE1042 turbofan for light fighter aircraft; the F109 turbofan for military trainer aircraft; the T800 turboshaft for light mili-tary helicopters; the CFE738 turbofan for medium-large business jet aircraft; and the TPF351 turboprop for a high-speed commuter aircraft. Garrett also re-newed its continuing development efforts with small expendable turbojet en-gine programs. These engine programs remained within the power range that matched Garrett's production capability.

Defense spending had risen dramatically in the 1980s and had provided Garrett with an incentive to expand in that area. However, by the end of the decade, defense spending declined as a result of a reduction in Cold War ten-sions. Economic recession further dampened the airframe market. By the early 1990s, neither the F109 nor the TPF351 had acquired a successful airframe application. The Boeing Sikorsky RAH-66 Comanche helicopter program, for which the T800 had originally been developed, continued to be delayed as a result of defense cutbacks. The CFE738 and TFE1042 had fared better by winning production applications. Thus, by the mid-1990s, Garrett's diversification strategy had been partly successful.

In the case of P&WC, its small turbine engine market niche was principally fixed-wing turboprop and turbofan corporate and regional transport aircraft. Its principal products were the PT6 turboprop and JT15D turbofan engines. From the mid-1970s through the mid-1980s, P&WC launched new engine programs to diversify its product line and prepare for changing markets, including the emerging larger-passenger, regional transport aircraft market, and the higher-priced, medium-size executive jet market. The PW100 turboprop and PW300 turbofan capitalized on P&WC's existing strength in regional transport and executive jet markets, but sought new niches within those markets. A market that P&WC had been interested in for some time was the low-cost commercial helicopter engine, and the PW200 turboshaft was developed for it. Historically, the vast majority of P&WC's sales were exports, and they continued to remain so during its diversification efforts.

Allison also underwent significant changes. In the 1970s, as a result of a decision by General Motors, considerable corporate resources had been devoted to the development of land vehicle and industrial turbine engines as the development of aircraft turbine engines had slowed. In 1983, under new leadership, Allison was reinvigorated and a major effort to expand its aircraft turbine product base was undertaken.

Like other engine companies, Allison was essentially a two-engine (T56 and T63/Model 250) company. Allison's T56 engine had made the company an industry leader in the 5,000-6,000-shp-class military turboprop market. Allison wanted to protect that market and expand into the transport marketplace. To do so, Allison competed in the JVX tilt rotor aircraft (later Bell-Boeing V-22 Osprey) competition and won it in 1985 with its T406 engine. Expanding on its success, in the late 1980s, Allison successfully launched new commercial engine derivatives from its T406, including the GMA 2100 turboprop for regional transport aircraft and GMA 3007 turbofan for regional and executive jets. This was an excellent example of a corporation adapting a common power section for multiple applications. Allison also developed and demonstrated small expendable engines for tactical and cruise missiles in the late 1980s and early 1990s although no immediate production applications were found. Joining with Garrett in 1984, Allison began development of the T800 turboshaft engine for the Army's LHX (later Boeing Sikorsky RAH-66 Comanche) helicopter. The dominant company in the light helicopter turboshaft engine market, Allison further sought to consolidate its position in that market with this new generation engine.

Unfortunately, in the early 1990s, Allison reported operating losses. Lower sales due to the depressed commercial market and a $1.8 billion investment program in four new engines were cited as the main causes.[14] Seeking to cut its losses and concentrate on its core automotive business, in 1993, General Motors sold Allison to an investment company. The following year, Allison was sold to Rolls-Royce. While Allison's diversification efforts were critical to the long-term viability of the company, the timing proved inauspicious with respect to an unforeseeable economic recession and a reduction in Cold War tensions which resulted in reduced military spending. Allison's 64 year relationship with General Motors had been variously beneficial and problematic. New ownership presented Allison with new opportunities. Also, the investment in new engine programs showed promise as these engines won successive airframe applications.

By the late 1970s, GE already had a more diversified product mix than other corporations in the small engine business. While GE had engine products for most of the major airframe markets, it was largely focused on military engine products, primarily as a result of corporate policy. Although GE's CJ610, CF700, and CT58 achieved important gains in the 1960s, GE's commercial engine revenue remained a relatively small percentage of the company's total engine business through the 1960s. In the 1970s, however, GE's commercial business began to increase. An exception to this was GE's share of the business jet engine market, which it deliberately let decline in the 1970s. However, by the late 1970s, it decided to re-enter this market with the CF34, in part to gain a larger marketshare for the TF34. Influenced by market changes following airline deregulation in 1978, GE entered the regional transport market with its CT7 in the early 1980s. The CT7 made headway in the commercial helicopter market as well. GE continued to make some commercial sales with its CT64 and CT58 engines.

By the early 1980s, GE was not only diversified with products for every major airframe market, but had begun to achieve a much greater presence in the commercial market than it had previously. In 1989, GE's total commercial engine revenue exceeded its military engine revenue for the first time. GE also continued to emphasize its international business, and by the early 1990s, nearly half its total aircraft engine revenue was international business.

---

14. R. Randall Padfield, "Allison Aims at Growth with Rolls-Royce's Help," *Aviation International News*, Vol. 28, No. 11 (June 1, 1996): 40.

Williams was started with the intention of exploring a diversified market of gas turbine products. A mix of commercial and government contracts resulted in the development of industrial, marine, aircraft APUs, and automotive turbine engines, but no major quantity production application was found for these products. In part, the reason for this was that conventional gasoline and diesel engines were well suited for their applications. Except in selected aviation markets, small turbine engines were not commercially competitive against conventional engines. Williams's first high volume product, the WR24, was a drone engine that went into production in 1968. The F107 cruise missile engine was produced in very large quantities, starting in the early 1980s. Thus, the company became largely focused on the expendable small gas turbine engine market for the military. The company specialized in genuinely small turbojets, turbofan, and turboshaft engines.

In 1971, the company started development of the WR44, which was selected for two proposed business jets that never materialized. In the late 1970s, the company renewed its efforts to develop and market a derivative business jet engine, the FJ44, for new generation, entry-level business jet aircraft. In 1989, Rolls-Royce joined Williams on this project. By the early 1990s, the FJ44 had found production applications, and Williams had succeeded in moving into the civilian aircraft engine market. Entering the man-rated general aviation business was an important turning point for the company. Unlike other corporations, and as a matter of choice, the company elected not to pursue civilian turboprop possibilities.

CAE's speciality in the expendable engine business was turbojet-powered tactical, drone, and RPV missiles, and in the man-rated engine market, turbojet-powered trainer aircraft. During the 1950s and early 1960s, CAE had developed turboprop and turboshaft engines in addition to its J69, J100, and J402 turbojet engines, but these efforts did not result in production applications. In the late 1960s, the company made a decision to focus primarily on unmanned, expendable applications. In the 1970s, several concerted development efforts on expendable turbofan engines did not result in production applications, although CAE was licensed to produce Williams F107 engines in 1978. CAE also continued to commit substantial resources to military technology demonstration engine programs. In the mid-1980s, CAE redirected its resources from technology demonstration programs toward new product development. In particular, CAE responded to the Army Missile Command's interest in the development of miniature turbojet engines for small tactical missiles. This was the start of CAE's Model 300 series turbojet engines, and several of the engines in this series found military applications. While CAE had made periodic attempts to diversify its product mix up to the mid-1980s, it remained primarily a niche company specializing in small military turbojet engines, which in part, was a company decision.

LICENSING AGREEMENTS

Licensing agreements were part of the small turbine engine business from the beginning and continued to be so over time. In addition to direct sales, international license agreements for manufacturing turbine engines were an important method of doing business. Without additional investment on their part, firms could collect license fees which contributed to corporate profit. License agreements were popular until the late 1970s, when domestic and international business partnerships began their ascendancy. Selected examples of traditional corporate licensing agreements illustrate how they contributed to the success of important small turbine engines.

More than just adding to its profits, Turboméca's 1951 licensing agreement gave CAE its entree to the aircraft gas turbine engine business. Turboméca's Marboré II was transformed into CAE's popular J69, which was the genesis of all of CAE's successful engines through the early 1980s.

GE's first small engine, the T58, was licensed in 1958 to the de Havilland Engine Co. Ltd. in Great Britain, where it was known as the Gnome engine. The T58 was later licensed to Ishikawajima-Harima Heavy Industries (IHI) in Japan and Klöckner-Humboldt-Deutz (KHD) in Germany. Selected versions of GE's J85 turbojet engine were also manufactured under license by the Orenda Engine Division in Canada and by Alfa-Romeo in Italy. In the 1960s, GE negotiated a T64 manufacturing agreement with IHI in Japan. One T64 model was manufactured under license by MTU in Germany. A partial license was granted to Fiat in Italy to produce yet another T64 model. In each case, these licensing agreements contributed to the foreign sales and longevity of these first generation small GE turbine engines.

Parts for Lycoming's T53 were made by KHD in Germany, Piaggio in Italy, Kawasaki in Japan, and the government of Taiwan. Typically, engines manufactured by these corporations were utilized in various commercial and military versions of Bell Model 204 or 205 helicopters manufactured by or utilized in foreign countries. Spare T53 parts were also manufactured by these companies. Italy and Japan were further licensed to assemble complete T53 engines. Not only contributing to Lycoming's bottom line, the parts manufacturing capability of these companies allowed Lycoming to have a second manufacturing source when the demand for parts exceeded its own capability.

PARTNERSHIPS

Traditionally, U.S. engine companies had for years regarded all technical and marketing information on their engines as being proprietary and such information was carefully guarded from disclosure. Reinforcing these proprietary considerations were U.S. government regulations and laws that prevented non-competitive practices. In both domestic and foreign commercial markets, corporations were prevented from engaging in proposed mergers and joint product ventures that were deemed anti-competitive. In the case of military engine competitions, corporations competed against one another for "winner takes all" development and production contracts. The military did not permit corporations to share technical information. From a set of design specifications established by the military, corporations were typically required to compete on the basis of engine price and/or the capability of a proposed engine to meet those specifications.

By the late 1970s and early 1980s, significant changes within this traditional means of doing business began to have an impact on the small gas turbine aircraft engine industry. Among these changes was a movement toward business partnerships, both domestic and international. Instead of competing individually, companies joined together on the basis of specific products to compete as partners for selected airframe competitions or markets. Depending on the particular case, such agreements were variously called partnerships, joint ventures, co-development and co-production agreements, teaming arrangements, or government required or sponsored competitive partnerships. The major factors driving the formation of partnerships, other than those required by government contract, were financial, technical, and market considerations. Specific factors included an increasingly flat market for turbine engines and the related need for new markets; the changing nature of the global aerospace industry; the high cost of engine development; and changing government regulations, procurement policies, and needs. Partnerships required by government contract were driven by the desirability of dual-source, competitive procurement.

INFLUENCES ON THE FORMATION OF PARTNERSHIPS

Several key events established precedents for commercial engine partnerships and for dual source military engine procurement which led to partnerships.[15] Among these was the 50-50 partnership between GE (U.S.) and Snecma (France) for the development of the CFM56 turbofan for large commercial and

---

15. Earlier successful partnerships between engine companies had been established, but these were less influential than those cited here.

military transport aircraft. This landmark partnership was formed in 1974 and called CFM International, Inc. Another event was the leader-follower program established by the military to provide a second source for all cruise missile components which resulted in a 1978 Williams-CAE license agreement. In 1984, the Secretary of Defense announced a split procurement contract for GE F110 turbofan engines for the General Dynamics F-16 fighter and P&WA F100 turbofans for the McDonnell Douglas F-15 fighter.[16] This watershed event established competitive, dual-source procurement contracts. Finally, in 1984, the Army issued an RFP for what became the T800 turboshaft engine program. The T800 program was also the first government-sponsored competition requiring company partnerships. While such events established precedents for partnerships, market forces played a key role in the transformation of the industry.

An important driver for partnerships was an increasingly flat market for small turbine engines and the related need to find new markets. From the mid-1950s through the late 1970s, the number of airframe applications for small gas turbine engines, especially high production ones, had flourished. However, by the 1980s, the market for small turbine engines began to stabilize as the effects of market saturation set in. There were many competing engine companies for fewer major airframe opportunities. It was clearly better to join forces than compete head-to-head in selected aircraft engine competitions. The increasingly saturated domestic markets led North American engine companies to look with even greater interest at developing markets around the world, and most of these markets required partnerships and offsets to penetrate.

The aerospace industry had begun to change worldwide, and the market for small turbine engines had become increasingly international. Airframe opportunities were available in a number of countries, and in some instances, airframes were jointly developed by corporations from more than one country. Access to compete in foreign airframe markets often required a quid pro quo. For engine competitions sponsored by foreign airframe manufacturers, a business partnership with appropriate foreign engine manufacturers could be a distinct advantage for a North American engine manufacturer, and in some instances, a necessity. Such business partnerships typically shared a percentage of costs, production, and profit. For engine developments or competitions sponsored by foreign government agencies, especially those in developing countries, there was a strong interest in a transfer of North American engine technology, manufacturing expertise, and investment in or partnerships with foreign engine companies. For developing countries, such offsets were often important considerations in the final selec-

---

16. Robert W. Drewes, *The Air Force and the Great Engine War* (Washington, D.C.: National Defense University Press, 1987), 126.

tion process because they could yield highly skilled jobs, critically needed technology, and improved competitive or defensive capabilities.

The high cost of new engine development was another major factor driving partnerships. Fully funding the development of a new engine had gone beyond the reach of some engine companies or put at substantial risk their future should the venture fail. Splitting in varying percentages the costs, production, and profit with another company reduced risk for each partner and made new developments more affordable. This was especially important when the market for an engine was considered to be relatively small and high engine development and production tooling costs needed to be amortized over a smaller number of engines. Development costs also could be driven down by integrating proven components, and to some extent technology, developed by each partner rather than each company developing the entire engine itself. This, of course, was contingent on prospective partners having similar or compatible engine designs. Not only did partnerships reduce development costs, but they conserved critical technical talent. Already spread thin on engineering talent, corporations could share their engineers in joint ventures and use the specialty expertise that each company had developed over time.

Government regulations, procurement policies, and needs had an influence on the formation of partnerships. An important factor that opened the door to partnerships was the relaxing of U.S. government restrictions against what had been earlier considered non-competitive practices. Military dual-procurement policies requiring competitive teaming arrangements led to some U.S. partnerships in the gas turbine engine industry.

From the late 1970s, there were a considerable number of partnerships formed by small engine companies, primarily on the basis of competing for specific military contracts or developments for selected commercial airframe markets. A few examples have been chosen to illustrate U.S. domestic and international partnerships of particular interest and which resulted in military qualified or civilian certified engines. It should be noted that many partnerships on specific products proved unsuccessful when military competitions were lost or canceled and commercial markets did not develop as anticipated. Generally, unsuccessful partnerships were subsequently dissolved, although the partnering corporations, in some cases, found other products that could be developed with their original partners.

U.S. DOMESTIC PARTNERSHIPS

The first domestic partnerships in the U.S. small gas turbine engine industry were driven by changes in military procurement policy. These were later followed by commercial partnerships influenced by precedents set by new military procurement policies and by the commercial factors previously cited.

The initial event in breaking the ice on proprietary restrictions between U.S. companies competing for military engine contracts came when the military established the requirement for a second manufacturing source for all components used in cruise missiles. The resulting leader-follower program mandated that the winning missile engine contractor competitively select a subcontractor to manufacture the same engine. The two contractors were later required to compete on a cost basis for the number of engines that each company would produce. As the winning contractor for the F107 engine, Williams selected CAE as its subcontractor. For these corporations, this was an entirely new business practice, and the transition required considerable effort on the part of each.

In February 1984, the Air Force announced a split GE-P&WA engine award for two fighter aircraft. In contrast to traditional sole source contract awards, this established an important precedent for competitive, dual-source military engine procurement. Unlike a leader-follower program, both manufacturers had equal contracting status with the military. This was not a domestic partnership.

Following this precedent, in December 1984, the Army issued an RFP for the T800, the engine for its LHX program. Among its objectives, the Army sought a better product and dual, competing production sources. In 1988, LHTEC, a 50-50 Garrett-Allison team, was awarded the T800 full-scale development contract.

While Allison had previous partnership arrangements, LHTEC was Garrett's first major partnership. This team fit well together for several reasons. First, the configuration of Garrett's TSE109 and Allison's GMA 500 turboshaft demonstrator engines were nearly identical. Another reason was that the two companies were similar in size and capability, and over time, various personnel had worked for both companies which resulted in a familiarity with each company. Garrett was additionally motived by the fact that Allison's principal market was helicopter engines, a market that Garrett wanted to penetrate. Also, the fact that the GMA 500 had been developed through the Army's Advanced Technology Demonstrator Engine program made Allison a strong candidate. For both companies, the partnership was capable of producing a superior engine because it could bring two design methodologies, two data bases, and two potential solutions to any problem. In addition to getting a better product, the Army anticipated better prices by requiring competitive bidding between Garrett and Allison during production procurement.

Subsequent to the easing of U.S. regulations regarding non-competitive practices, domestic commercial partnerships soon followed the example set by military partnerships. One example was the CFE738 turbofan engine, developed jointly by GE and Garrett for the medium-large business jet market. This program was an excellent illustration of how corporations conserved technical assets and financial resources in expensive, high-risk, limited market programs. In 1987, GE and Garrett formed a 50-50 joint venture known as the CFE Company. This was GE's first domestic partnership, and it was based on the successful CFM International agreement between GE and SNECMA for the development of the CFM56 turbofan engine. In this case, one driver for the formation of this partnership was the strong technical similarity between the GE27 HP spool and the HP spool of Garrett's next generation TFE731. The GE27 HP spool was an already developed core, which saved Garrett considerable development money and time. GE, in turn, benefitted from Garrett's dominant position in the business jet market. Both corporations were interested in the same business jet market and could more effectively compete together than separately against P&WC for applications in this market.

INTERNATIONAL PARTNERSHIPS

Before U.S. small turbine engine companies began forming partnerships with one another, some engine companies had become involved in international partnerships and joint ventures to help open markets for their products. The most influential of these was the 1974 CFM International partnership between GE and SNECMA.

That arrangement was followed by an array of other international partnerships. In order of their formation, several important examples of international partnerships were Garrett's TFE1042, P&WC's PW300, GE's CT7, and Williams's WR44 turbine engines. Garrett's TFE1042 was developed through two successive partnerships, first in 1978 with Sweden's Volvo Flygmotor AB, and subsequently in 1982, as an entirely new engine design, with Taiwan's Aero Industry Development Center. The later partnership was a 50-50 joint venture, which, for Taiwan, was considered important for developing the country's aerospace industry, and for Garrett for sharing costs and for penetration of a new market. In the mid-1980s, with Germany's Motoren und Turbinen-Union as a 25% investment and design/development partner on the PW300, P&WC's risk in developing the engine became manageable, and it strengthened the European market potential of the engine. Starting in 1985, a specific GE CT7 model was co-developed by Alfa Romeo Avio (25%), Fiat Aviazione (25%), and GE (50%). GE had long-standing relations with both Italian companies, and this partnership proved beneficial when a military version of the CT7 was selected for the

Italian Navy's EH 101 helicopter in 1990. In 1989, Williams and Rolls-Royce formed a separate company to produce, market, and service the FJ44. Rolls-Royce, which had a 15% share, participated in off-shore marketing and engine support activities through its existing worldwide network and also contributed to selected component design.

During the 1990s, the number of new international partnerships continued to grow rapidly, particularly in response to new market opportunities in Eastern Europe, the Commonwealth of Independent States, and the Pacific Rim. While partnerships had become a mainstream way of doing international business, some cautionary concerns were commonly expressed within the aerospace industry. In the long run, partnerships had a tendency to level technology between companies, removing some of the traditional competitiveness between them. Also, a standing concern was technology transfer, especially to foreign companies. Essentially, corporations were trading technology for market presence. Another concern was the loss of domestic jobs as component parts were outsourced internationally. Strong competition and technology leadership were the ingredients that had established North America as the world leader in aircraft turbine engines. A final concern was that foreign aerospace companies, originally assisted by North American partnerships, would eventually competitively challenge this leadership position.

TECHNOLOGY HIGHLIGHTS

In the late 1970s through the 1980s, there were continuing noteworthy technology advances in all aspects of turbine engine development. These achievements occurred in turbofan engines, engine components and controls, engine design, and military engine demonstration programs.

The number of new turbofan engines under development proliferated as this type of engine was widely adapted for commercial and military applications. Such engines included Garrett's TFE1042 and F109, P&WC's PW300, Garrett and GE's CFE738, and the Williams FJ44. The Williams F107 turbofan engine received particular distinction by receiving the 1979 Collier Trophy and Federation Aeronautique Internationale awards. The F107 was then the world's smallest, high-efficiency fanjet engine. Another technical advancement in a small turbofan engine was Garrett's TFE1042, which was equipped with an afterburner.

Interesting developments in compressor technology were exemplified by selected Garrett and P&WC programs. Garrett worked with Navy and Air Force technology programs in the late 1970s and early 1980s to develop a mixed-flow compressor which had a relatively high airflow per unit of frontal area and a high pressure ratio in a single stage. Important additional features of mixed-flow compressors were lower frontal area (than a centrifugal compressor) and high inlet airflow distortion tolerance due to large surge margins. While the concept of mixed-flow compressors was not new, Garrett successfully advanced the state of the art in this component. Subsequent to a 1981 Army demonstration program award, P&WC demonstrated a precedent-setting 15:1 pressure ratio in a single centrifugal compressor stage at 78% efficiency. In the early 1980s under an Army Modern Technology Demonstrator Engine contract, the GE27 successfully incorporated a 22-23:1 overall pressure ratio in axial-centrifugal compressor design, contributing to the record-setting performance of that engine.

In combustors, Allison's Lamilloy® was an innovative development. Under a 1977 Army Advanced Technology Demonstrator Engines (ATDE) contract, Allison's GMA 500 combustor was fabricated from Lamilloy,® a quasi-transpiration cooled material comprised of a multi-layered wall made by a combination of investment casting, electrochemical etching, and isostatic bonding that produced a pattern of flow paths through the material. In the early 1990s, a Lamilloy® combustor, turbine rotor, and turbine guide vane assembly contributed to record setting temperatures in Allison's GMA 900 (military designation J102) supersonic turbojet engine.

Turbine technology continued to evolve. Allison developed a hybrid turbine wheel in its GMA 500. The blades of each turbine stage were cast onto a ring which in turn was pressed into and diffusion bonded to a hub made of a higher strength material. Garrett's F109 used HP turbine technology derived from its LART (Low Aspect Ratio Turbine) program with the Air Force in the mid-1970s. In this program, Garrett utilized 3-D viscous analysis to design and test turbomachinery blading which resulted in increased turbine efficiency. Across the board, cooling systems for air-cooled turbine blades became more sophisticated, and in combination with material improvements, higher turbine operating temperatures resulted.

Further advances were made in control technology. An interesting technology that was incorporated on Allison's T406 engine, built for the Bell-Boeing V-22 Osprey, was a computer-to-computer mating of the aircraft and engine controls. The aircraft's flight control system communicated with and was highly integrated with each engine's Full Authority Digital Electronic Control (FADEC) system. Garrett equipped its F109 and TFE731-5 turbofan engines with FADEC controls that had Built-In Test Equipment (BITE) capability. The BITE feature monitored selected parameters, identified faults, and took corrective action to permit engine operation when faults occurred. P&WC's PW300 turbofan was the first general aviation engine to be certificated with a FADEC system. Progress toward fully electronic controls and away from hydromechanical controls was made in small turbine engines as costs came down.[17] While FADEC had been developed and used earlier on large turbine engines, the adaptation of FADEC to small turbine engines continued to make progress.

Several environmental and engine design improvements were especially noteworthy. In the mid-1970s, engine companies participating in NASA's Quiet, Clean, General Aviation Turbofan (QCGAT) program were able to develop improved technologies for reducing engine noise and exhaust emissions as well as increasing thrust. Garrett's TFE731 and Lycoming's ALF101 turbofan engines incorporated QCGAT technology. Garrett's F109 turbofan engine, the development of which started in 1979, was the first engine to meet the Air Force's Engine Structural Integrity Program (ENSIP) requirement. A critical element of ENSIP was that the engine design provide fracture free performance of critical parts in the presence of material, manufacturing, and process defects (flaws) for a period not less than the scheduled inspection interval for the engine. The ENSIP was first presented to the industry by the Air Force in 1975.[18] The technology and methodology developed under this program, which continued to evolve over time, was the first of its kind to be introduced to the turbine engine industry and set a precedent for future military engines. These design standards contributed to the improved safety, durability, and life cycle costs of aircraft gas turbine engines.

Important advances were made in military-sponsored engine demonstrator programs. Development of Lycoming's PLT34A and Allison's GMA 500 proceeded in the Army's Advanced Technology Demonstrator Engine program, which began in 1975. Technology emerging from this ATDE program included single-crystal, cooled turbine blades, FADEC controls, and a high pressure twin-centrifugal compressor. The accomplishments of ATDE and its predecessor, the Army's Small Turbine Advanced Gas Generator, provided the technical base for the Army's T800 program. In 1983, an Army and Navy sponsored program, the Modern Technology Demonstrator Engine, focused on the need to improve the capability of engines in the 4,000- to 6,000-shp class and Pratt & Whitney's PW3005 and GE's GE27 were selected to participate. The U.S. Army Missile Command miniature turbine engine programs, started in 1985, successfully resulted in the development and demonstration of turbojet engines as small as 4-8.5" in diameter.

17. David Hughes, "FADECs Win Role in Smaller Engines," *Aviation Week & Space Technology*, Vol. 137, No. 20 (November 16, 1992): 58.

18. William D. Cowie, "Turbine Engine Structural Integrity Program," *Journal of Aircraft*, Vol. 12, No. 4 (April 1975): 366-369.

The Air Force initiated a series of important demonstrator programs in the 1980s including: High Performance Turbine Engine Technologies (HPTET); Expendable Turbine Engine Concepts (ETEC); and Joint Turbine Advanced Gas Generator (JTAGG). An ambitious effort which originated in 1982, HPTET's goal was no less than doubling turbopropulsion capability by the year 2000.[19] By 1988, this program had expanded greatly as Integrated High Performance Turbine Engine Technology (IHPTET). The HPTET and IHPTET programs were for all turbine engines, large and small. Started in 1989, JTAGG was an Army, Navy, and Air Force program focused on man-rated turboshaft/prop engines. Announced in 1987, ETEC was an Air Force program dedicated to small expendable turbine engines.

4. Restructuring and Consolidation of the Industry (1990s)

Starting in the late 1980s and continuing through the early 1990s, the U.S. aerospace industry underwent drastic restructuring as a result of economic recession, a reduction in Cold War tension, reduced military spending, oversupply, overcapacity, and the pursuit of substantially increased profits. Corporations downsized by permanently laying off thousands of employees. There was a reduction in the number of major aerospace corporations as a result of mergers and acquisitions. The aircraft propulsion industry was hit hard as the demand for gas turbine engines dropped across the board. A major restructuring of the propulsion industry took place when Rolls-Royce purchased Allison and AlliedSignal purchased Lycoming. Military policies focused on centralizing R&D efforts and the use of off-the-shelf engines and airframes, where possible. Later in the decade, the economic cycle reversed itself, and the aerospace industry business improved significantly.

INFLUENTIAL EVENTS AND GOVERNMENT AND MILITARY POLICIES

The end of the Cold War and a shrinking defense budget (beginning in the late 1980s) resulted in cancellations, procurement reductions, and delays in various military airframe and engine programs. Starting in 1987, political and economic reforms and restructuring were initiated in the Soviet Union which ultimately led to the breakup of the country in 1991. Other Soviet-dominated European

19. James S. Petty and Robert E. Henderson, "The Coming Revolution in Turbine Engine Technology," (Turbine Engine Division, Air Force Wright Aeronautical Laboratories (AFWAL/POT), Wright-Patterson Air Force Base, Ohio, 1987), 1-1.

countries underwent similar reforms and achieved independence from communist rule. As a result, U.S. defense spending, which had risen during President Ronald Reagan's tenure in the 1980s, declined. Budget cuts directly impacted engine sales. In 1989, U.S. military engine sales began to decline.[20]

Economic recession, occuring at the same time, also greatly affected the aerospace industry. U.S. aerospace sales began falling in real terms in 1991.[21] From 1989 to the beginning of 1995, employment in the aerospace industry declined an incredible 36% during the recession. Economic recession and severe overcapacity in the air transport industry had a direct negative influence on the propulsion industry as airframe orders were canceled or deferred (orders for transport aircraft began declining in 1989). General aviation was also affected by the recession, although it had been in a persistent slump since the late 1970s. Exacerbating the economic problem for engine manufacturers were the relatively saturated civilian and military small turbine engine markets.

While there was an overall decline of sales throughout the propulsion industry, political events led to some new markets. For example, perceived market opportunities in former communist countries spurred North American engine manufacturers to forge partnerships with engine companies in these countries. Among the objectives of such partnerships were joint product development, indigenous production, licensed sales, service, and parts manufacturing of North American engines. The Persian Gulf War in 1991 also had an important influence on U.S. military weapon exports. The technological success of military weapons systems, including aircraft and rotorcraft, led to increased arms export sales. The U.S. benefitted by the decline in military exports from the former Soviet Union.

Throughout the U.S. aerospace industry, major restructuring and consolidation took place. Corporations were sold off completely or in part. Mergers and consolidation of major aerospace companies were a dominant trend. The world's largest defense contractor was formed in 1994 when Lockheed and Martin Marietta merged.[22] Other examples which took place in the early 1990s included the acquisition of Grumman by Northrop, the sale of General Dynamics's aircraft manufacturing business to Lockheed, and the acquisition of

20. Stanley W. Kandebo, "Market Forces Recast Propulsion Industry," *Aviation Week & Space Technology*, Vol. 140, No. 11 (March 14, 1994): 71.

21. David Vadas, "U.S. Aerospace Industry's Troubles Continue," *Aerospace America*, (January 1995): 14.

22. Anthony L. Velocci, Jr., "Megamerger Points to Industry's Future," *Aviation Week & Space Technology*, Vol. 141, No. 10 (September 5, 1994): 36.

Vought Aircraft by Northrop Grumman.[23] Benefits of such consolidations included improved efficiency by reduced spending on duplicate research and development programs, better ability of corporations to compete domestically and internationally due to price advantages as a result of economies of scale, broader corporate market diversification, much greater financial resources, and substantially improved short-term corporate profitability. The down side included massive layoffs and a reduced defense contractor base resulting in potentially oligopolistic practices. For its part, the Defense Department supported consolidations because it's post-Cold War yearly budgets were no longer able to sustain the number of defense contractors that existed at the peak of President Reagan's defense buildup in the 1980s.

The North American propulsion industry also consolidated. In 1994, AlliedSignal, parent corporation of AlliedSignal Engines (Garrett), acquired the Lycoming Turbine Engine Division of Textron. This acquisition enabled Garrett to expand its engine business to include the larger turbofan-powered regional aircraft market, light, medium, and heavy helicopters, and other commercial and military turbine engine applications. In part, Textron sold Lycoming for debt reduction purposes, the repurchase of common shares, and the financing of acquisitions. Later in 1994, Rolls-Royce purchased the Allison Engine Company from the investment company Clayton, Dubilier & Rice. The acquisition gave Rolls-Royce manufacturing presence in the U.S. and filled gaps in its product line in turboprops, small turbofans, and turboshaft engines. Acquisitions such as these enabled the purchasers to quickly expand and diversify their product line and improve their competitive position in the overall turbine engine market.

In response to declining sales, cost-cutting and shareholder value (profitability) became major drivers for the propulsion industry in the early 1990s. For example, between 1991 and 1994 Pratt & Whitney Aircraft cut spending by nearly 50%, cut inventory by 50%, and slashed work in progress by 75%.[24] Despite the fact that archrival GE's sales dropped from $8 billion in 1991 to $5.7 billion in 1994 and its engine shipments dropped more than 60% between 1990 and 1994, the company had a high 16.4% return on sales in 1994.[25] This perfor-

23. Robert F. Dorr, "The Urge to Merge Reemerges," *Aerospace America*, (November 1994): 10-11.

24. Anthony L. Velocci, Jr., "Cultural Shift Key to Vendor Survival," *Aviation Week & Space Technology*, Vol. 141, No. 22 (November 28, 1994): 45-46.

25. Anthony L. Velocci Jr., "U.S. Quick-Change Act Offers New Lessons," *Aviation Week & Space Technology*, Vol. 142, No. 17 (April 24, 1995): 52.

mance was achieved through productivity improvements that included, among other cost and time cutting measures, reducing one-half of its 1992 workforce by the year 1995. As a result of such efforts, selected propulsion companies achieved record operating profits despite low production rates in a declining or flat market. Ironically, by 1995, record profits in the aerospace/defense industry posed challenges for corporations to find useful outlets for their wealth.[26]

In addition to these important political and economic events that swayed the propulsion industry, government and military policies continued to exert a strong influence. By the late 1980s, most of the engine companies participated in various aspects of the Integrated High Performance Turbine Engine Technology program and also shared a portion of the cost of technology and demonstrator engine development. Some procurement programs were also affected by changes in military policy. One such example was the Joint Primary Aircraft Training System (JPATS), which was introduced by the Department of Defense in 1989. In order to avoid most of the technical risk and sizeable investment required for an entirely new airframe and engine, the military required that candidates be non-developmental, "off-the-shelf" trainer aircraft. Off-the-shelf meant that the airframe and engine met the military qualification or civil certification requirements of the country of origin. Because most military trainer aircraft were manufactured by foreign companies, foreign candidates, teamed with U.S. airframe partners, were allowed to compete. This also represented a trend away from "Buy American."

By the late 1980s there was a noticeable reduction in the development of "new centerline" engines for small military aircraft. Tight military budgets necessitated the continued use of off-the-shelf or derivative engines for many near-term manned aircraft requirements. Increasingly, the focus was placed on upgrading already qualified military engines.

MARKET DEVELOPMENTS

Production authorization was given to a few important military airframe programs in the early to mid-1990s, but more programs were canceled. Despite economic recession and overcapacity in the civilian market, a seemingly contradictory trend occurred as numerous new or derivative airframe programs were introduced. While small turbine engine production declined overall and development programs slowed, new engine models were successfully marketed.

26. Anthony L. Velocci, Jr., "Profit Surge Poses Prickly Dilemma," *Aviation Week & Space Technology*, Vol. 144, No. 7 (February 12, 1996): 20.

THE MILITARY MARKET

Reductions in defense spending impacted military development programs. Some were canceled, others stretched out, and a few received production commitments. One important program that got the green light was JPATS.

The Air Force canceled the Fairchild T-46A trainer aircraft development program in 1986. Because a replacement trainer was still needed for the Cessna T-37B, in 1988, the Air Force called for a new Primary Aircraft Training System (PATS). The Navy also needed a replacement for its Beech T-34C Mentor. To consolidate those needs, the joint Air Force and Navy development program, JPATS, was initiated in 1989, and it superseded the Air Force program. With the exception of an entry from Cessna, the JPATS candidate aircraft were originally designed in either Europe or Latin America.[27] For the competition, most U.S. airframe manufacturers partnered with these foreign aicraft companies. On June 22, 1995, the Raytheon Beech Mk.II trainer aircraft won the JPATS competition. This aircraft was a modified Pilatus PC-9, powered by a P&WC PT6A, and it was to be licensed-built by the Raytheon Aircraft Co. It is interesting to note that P&WC played a strong hand in the JPATS competition as it powered four of the seven candidate JPATS aircraft.

The candidates in the JPATS competition reflected the fact that the military trainer market was dominated by foreign companies. Because of their relatively low development cost and low complexity, trainer aircraft had been developed by many small countries as a way of sustaining their indigenous aircraft industries. It was ironic that the U.S., with the largest market for trainer aircraft in the world, had only one indigenous design in the JPATS competition. The reverse was true for North American small turbine engine manufacturers whose engines powered all but one candidate. The military trainer aircraft market was especially good business for a majority of North American engine manufacturers, and their products were used in numerous military trainers, worldwide.

In the 1980s and 1990s, a bewildering number of North American-engine-

27. The JPATS candidate aircraft included the: Raytheon/Pilatus (Switzerland) PC-9 Mk.II with a P&WC PT6A-68 turboprop engine; Cessna (U.S.) 526 powered by two Williams F129-WR-100 turbofan engines; Northrop Grumman/Augusta (Italy) S.211A with a P&WC JT15D-5C turbofan engine; Lockheed/Aermacchi (Aeronautica Macchi—Italy) MB.339 T-Bird II with a Rolls-Royce Viper RB.582-01 turbojet engine; Northrop Grumman/Embraer (Brazil) EMB-312HJ Super Tucano 2 powered by a P&WC PT6-68/1 turboprop; Rockwell/Deutsche Aerospace (Germany) Fan Ranger powered by a JT15D-4C turbofan engine (later renamed Rockwell Ranger 2000 and re-engined with a P&WC JT15D-5C/3); Vought/FMA (Fabrica Militar de Aviones—Argentina) IA-63 Pampa 2000 with a TFE731-2-2B turbofan engine.

powered trainer aircraft were being marketed, in development, or in service. The P&WC-powered trainer aircraft included the Augusta/Siai-Marchetti S.211 (JT15D), the Embraer EMB-312 Tucano and EMB-312H Super Tucano (PT6), and Pilatus PC-7 and PC-9 (PT6). Allison was also represented, and its Model 250 powered the Augusta/Siai-Marchetti SF.260TP, Enaer T-35DT Turbo-Pillan prototype, Fugi T-5, Hindustan Aeronautics Ltd. HTT-34 (prototype), and Valmet L-90 TP aircraft. Garrett engines equipped the Aero Vodochody L-159 (F124), Aero Industry Development Center AT-3 (TFE731), Construcciones Aeronauticas SA C-101 (TFE731), Fabrica Militar de Aviones (FMA) IA.63 Pampa (TFE731), Nanchang K-8 (TFE731), Promavia Jet Squalus (F109), and the license-built Shorts S312 Tucano (TPE331). A newcomer to the field, Williams powered the re-engined Saab SK 60 (FJ44), the Cessna 526 JPATS candidate (F129), and was offered as an option on the Promavia Jet Squalus F1300 (FJ44). South Korea's Daewoo KTX-1 and India's Hindustan Aeronautics HAS HTT-35 trainers were offered with either a TPE331 or PT6.

Small turbine engines also powered a variety of other military fixed-wing aircraft in the 1990s, and the newer programs during that time included the following aircraft and engine companies. In the early 1990s, International Turbine Engine Company F124 and F125 engines (originally designed by Garrett) were in production for Taiwan's Aero Industry Development Center (AIDC) Ching-Kuo Indigenous Defensive Fighter (IDF) aircraft. Garrett's TPE331 was also being used on the Fairchild C-26A/B transport aircraft. In 1990, the Air Force selected the Beechcraft 400A Beechjet as the T-1A Jayhawk for the Tanker Transport Trainer System (TTTS) competition. This resulted in an important production contract for P&WC whose JT15D powered the Jayhawk. In 1990, the Air Force ordered the GE T64 powered-Alenia G.222 as the C-27A Spartan intra-theater airlift aircraft. The Allison AE 2100 (derivative T406) powered the upgraded Lockheed C-130J transport aircraft. While the number of airframes acquired in these programs was small relative to earlier military procurement programs, they provided an important continuing business for new or upgraded engine products.

The market for military helicopters continued to shrink from the late 1980s through the mid-1990s. Most existing, major military helicopter programs were winding down or continued at relatively low production levels for U.S. military services. Such programs included the: Bell OH-58D Kiowa, which had won the Army Helicopter Improvement Program (AHIP) in 1981; Boeing CH-47D/MH-47E conversion program; Sikorsky UH-60/SH-60 and H-53E; and McDonnell Douglas AH-64. Export sales promised to extend the production of some of these models. On a gradual basis, the Army began to remove from its inventory older turbine-powered military helicopters including the Hughes OH-6, Bell OH-58, and Bell UH-1. In Europe, as a result of the reduced demand for helicopters, one major merger took place. In 1992, the helicopter units of Aerospatiale and MBB/Deutsche Aerospace merged to become Eurocopter, then the largest helicopter manufacturer in the world.

By 1996, two important military vertical flight programs were approved for production and one was not. After years of debate, the Bell-Boeing V-22 Osprey, powered by Allison T406 turboprop engines, received procurement funding in Fiscal Year 1996. The Bell TH-67 Creek, powered by the Allison Model T63 turboshaft engine, won the Army New Training Helicopter (NTH) program in 1993. Flight testing of the Boeing Sikorsky RAH-66 Comanche, powered by LHTEC T800 engines, began in 1996, but no production was authorized by that time.

For their part, engine manufacturers continued to upgrade their turboshaft engines for improved military helicopter performance. Higher performance GE T700 engines were integrated into the Sikorsky UH-60/SH-60 helicopters. The T700 continued to dominate the medium military turboshaft market. A derivative T700 engine was selected in 1990 for the Italian Navy's EH Industries EH 101 helicopter. Other improved or upgraded models of existing turboshaft engines were either marketed or found applications. Among these were the: Lycoming T53 for selected models of the Bell AH-1; Lycoming T55 for the Boeing MH-47E; GE T64 for the Sikorsky MH-53E; and Allison T703 for the Bell OH-58D.

In the mid- to late 1980s, there was a flurry of activity in expendable missiles and small turbine engines for them as a result of the increasing effectiveness of such missiles in combat. However, by the mid-1990s, some important missile development programs had been canceled which resulted in lost opportunities for selected turbine engine manufacturers. The Modular Standoff Weapon (MSOW), a multi-national effort to develop a variety of standoff missiles, some of which were powered by turbine engines, was terminated in 1989.[28] The AGM-136A Air-Launched Tacit Rainbow anti-radiation missile, built by Northrop and powered by a Williams WR36 (F121) turbojet engine, was canceled by the Air Force in 1991 due to a declining defense budget. The Ground-Launched Tacit Rainbow (GLTR) missile (RGM-136), whose airframe developer was Raytheon/McDonnell Douglas and power plant was a CAE Model 318 turbojet, was terminated in 1991 as well. Non-Line-of-Sight (NLOS), a Boeing-

---

28. "Missile Forecast: Modular Standoff Weapon," *Forecast International/DMS Market Intelligence Report*, (March 1991): 1-4.

developed missile which was powered by a Williams WJ119 turbojet, was canceled in 1991 because of fast rising program costs. The Tri-Service Standoff Attack Missile (TSSAM) was ended in 1995 by Defense Secretary William J. Perry due to "significant development problems" and production costs that were "unacceptably high."[29] The AGM-137 TSSAM airframe was developed by Northrop, and a Williams P8300 turbojet was reportedly the power plant. The Long-Range Conventional Standoff Weapon (LRCSW), a joint Navy and Air Force weapon program for an extremely long-range cruise missile, was canceled in 1991. LRCSW was to be powered with a propfan engine. Propfan contender teams were CAE/GE with its Model 235, Williams/Allison Propfan Engine Co. (WAPEC) with its WA100, and Pratt & Whitney/Hamilton Standard.

Important operational turbine-powered missile programs included the McDonnell Douglas AGM/RGM/UGM-84 Harpoon, which dominated the U.S. antiship missile market and McDonnell Douglas AGM-84E Standoff Land Attack Missile (SLAM), which entered service in 1990; both missiles were powered by the CAE J402 engine. Other operational cruise missiles, powered by the Williams F107 turbofan engine, included the Boeing AGM-86B/ALCM (nuclear payload), Boeing AGM-86C/CALCM (conventional payload), and Hughes BGM-109 Tomahawk. Also operational was the Hughes AGM-129A Advanced Cruise Missile, powered by a Williams F112, an air-launched strategic nuclear missile with low observable characteristics. Other development programs continued including supersonic missile engines and miniature turbojet engines for a variety of small military missiles.

By the 1990s, there was renewed interest in Unmanned Aerial Vehicles (UAVs) as they became increasingly effective platforms for real-time, tactical intelligence gathering. A lesson learned from the 1991 Persian Gulf War was the requirement for new, more capable UAVs to support the operational needs of the tactical commander.[30] New UAV programs included the: Teledyne Ryan Tier 2+ reconnaissance UAV equipped with an Allison AE3007 turbofan engine and Lockheed Martin/Boeing Tier 3- DarkStar low observable reconnaissance UAV powered by the Williams FJ44 turbofan engine. A variety of older, turbine-powered target drones and surveillance UAVs continued in operational use or remained under development.

29. Bradley Graham, "Missile Project Became a $3.9 Billion Misfire," *The Washington Post*, April 3, 1995, A1.

30. Office of the Under Secretary of Defense, Acquisition & Technology, Defense Airborne Reconnaissance Office, *Annual Report Unmanned Aerial Vehicles (UAVs)* (Washington, D.C.: August 1995), 1.

THE COMMERCIAL MARKET

The total number of civil aircraft shipments, already low during the 1980s (as compared to the 1970s), declined even further from 1990 to 1994.[31] This included commercial transport aircraft, helicopters, and general aviation aircraft. Despite this, new or derivative airframes were introduced in both the commuter and general aviation markets. Furthermore, new or improved small turbine engines were also introduced in these markets. By the end of the 1990s, business had improved significantly as the economy rebounded.

COMMUTER AIRCRAFT

In the late 1980s, various engine companies had begun development of new engines for the commuter industry. A new market for Allison, the company began working on its GMA 2100 engine for commuter turboprop aircraft and GMA 3007 turbofan for regional jet airlines. In 1988, Lycoming announced plans for an LF500 family of turbofan engines, which was based on the ALF502R. The first engine in this series, the LF507, was selected for the British Aerospace RJ70 Avroliner regional transport, a successor to the BAe 146. P&WC continued development work on significantly uprated PW100 series turboprop engines. Seeking to expand its share of the commuter aircraft market, Garrett began development of the TPF351 for the Embraer CBA-123; however, the CBA-123 program was canceled in 1992 and development work on the TPF351 suspended. Garrett's TPE331 and GE's CT7 turboprop engines continued to remain competitive. In 1992, a program between AlliedSignal, Antonov, and Novosibirsk was launched, and TPE331 engines were selected for the Antonov AN38 transport aircraft.

In the early 1990s, the commuter aircraft market was, for the most part, saturated. Along with the economic recession, there were significantly fewer orders and new aircraft development programs. In 1996, a merger between British Aerospace and ATR, known as Aero International Regional (AIR), resulted in the largest regional aircraft manufacturer, but also the cancellation of the Jetstream 51, 61, and 71 commuter aircraft development programs. Also in 1996, Fokker declared bankruptcy, which impacted the Fokker 50 commuter aircraft.

Irrespective of the temporary economic setback, two important trends were evident in the regional aircraft market. First, regional aircraft continued to grow

31. "Civil Aircraft Shipments—Calendar Years 1972-1995," *World Aviation Directory* (Washington, D.C.: McGraw-Hill, Summer 1995), X-41.

in seating capacity since deregulation in 1978.[32] Second, sales of regional jets were increasing. Both trends were favorable overall for the small turbine engine industry.

GENERAL AVIATION

As with most of the other aviation markets, in general aviation in the early 1990s, too much capacity with too many players and a lot of used aircraft on the market meant that total factory shipments remained low as they had since the early 1980s.[33] Despite this, there were a number of new or derivative business jet developments underway in the late 1980s and early 1990s. The availability of new or significantly upgraded engines made possible both the diversity and the improved performance of this next generation of jets. A strong market interest in light business jets was largely attributable to the Williams FJ44 turbofan engine. Developments in the turbofan-powered, medium-size jet market were focused on aircraft powered by P&WC's PW300 and PW500 turbofans and uprated and improved Garrett TFE731 turbofan models. By 1995, the first signs of a general aviation market revitalization were evident as aircraft deliveries began to increase.

Under development as early as 1971, the William-Rolls FJ44 turbofan engine was designed for a new category of small business jets. Such jets would combine the best characteristic of contemporary executive jets, high speed, with the desirable attributes of business turboprop aircraft, including lower costs, excellent fuel consumption, and short field capability. This new class of jets was directly targeted at replacing the existing market of contemporary small-sized turbojet and turboprop aircraft, which carried the majority of business passengers. In the early 1990s, Williams achieved its goal when the FJ44 was selected for the Cessna CitationJet, Sino-Swearingen SJ-30, and Raytheon PD374 Premier I business jet aircraft.

In the late 1980s and early 1990s, P&WC's PW300 and PW500 turbofan engines were selected for a variety of medium-size business jet aircraft. Designed with high-bypass ratios, they exhibited significantly improved performance and fuel efficiency over the previous generation of turbofan engines. The PW300 was originally designed at a thrust rating higher than Garrett's TFE731 for derivative models of existing business jets, which required increased range, performance, and capacity, and for the retrofit market. The PW300 series found applications on

---

32. Carole A. Shifrin, "New Aircraft Orders Lag Brisk Pace of Traffic Growth," *Aviation Week & Space Technology*, Vol. 144, No. 21 (May 20, 1996): 56–58.

33. *1995 General Aviation Statistical Databook* (Washington, D.C.: General Aviation Manufacturers Association, 1995), 5.

larger, medium aircraft, including the British Aerospace BAe 1000 (later renamed Raytheon Hawker 1000), Learjet Model 60, and IAI Galaxy (a derivative of the Astra SP). P&WC targeted the PW500 series to fill a thrust class gap between the company's lower-powered JT15D and higher-thrust PW300. A new engine, the PW500 was designed for smaller aircraft in the medium-size jet market and won competitions for the Cessna Bravo (a Citation II replacement) and Cessna Excel (a Citation VI replacement). It was anticipated by P&WC that the PW500 would eventually replace the company's JT15D turbofan engine, which continued to have important airframe applications in the 1990s.

Other turbofan engine competitors in the medium-size business jet market included the Garrett TFE731, GE CFE738, GE CF34, and Allison AE 3007. Engines in this new series were selected for the Learjet 45, IAI Astra SPX, and Dassault Falcon 50 EX and 900EX. Earlier TFE731 engine models won applications on the Cessna Citation VI and VII aircraft and the Dassault Falcon 900B and Falcon 20 (retrofit program). The CFE738 was selected for the Dassault Falcon 2000 and the CF34 for the Canadair Challenger 604. The AE 3007 won the competition for the Cessna Citation X very high speed executive jet.

In comparison to business jets, there were relatively few new executive turboprop aircraft introduced in the early 1990s. P&WC's PT6 continued to dominate the turboprop aircraft market through the mid-1990s, and Garrett's TPE331 remained in the number two position. At this time, newer executive turboprop aircraft programs included the Socata TBM 700, Beech 350 Super King Air, and Pilatus PC-12, all powered by PT6 engines.

In the late 1980s through mid-1990s, new models were introduced in the civil rotary wing market despite the recession. Bell announced its Models 206LT TwinRanger, Bell 206L-4 LongRanger IV, 230, 407 and 407T, 430, and 442. All of these Bell helicopters were powered by Allison Model 250 engines except the Model 442 which offered LHTEC CTS800 or MTR 390 turboshaft engines. Additional new Allison Model 250 applications included the Schweizer 330, Enstrom 480 (military designation TH-28), Kamov Ka-226, and Tridair Helicopters, Inc. Gemini ST. Two new McDonnell Douglas helicopters powered by Allison Model 250 engines were the NOTAR (no-tail-rotor) MD 520N and the MD 600N. McDonnell Douglas also introduced its first civil twin-engine helicopter, the MD 900 Explorer; power plant options included the P&WC PW206B and also the Turboméca TM 319 Arrius. The PW206 and Arrius were also offered as alternative options on the Eurocopter EC 135 (formerly known as the BO 108) light twin helicopter. An interesting development, the Kaman Aerospace K-1200 K-Max, known as an "aerial truck" for its external sling load capability, was powered by a Lycoming T53. Sikorsky announced its S-92 Helibus powered by GE CT7 turboshaft engines.

By the mid-1990s, Allison's Model 250 continued to dominate the light civil helicopter market. However, it faced increased competition from P&WC's PW200 series and Turboméca Arrius. To improve the international marketing prospects for a given helicopter model, airframe manufacturers were increasingly offering customers several engine options. Another market development was increased customer demand for twin-engine helicopters. To some extent, surplus U.S. military helicopters offset the demand for new civilian helicopters in certain helicopter markets.

TECHNOLOGY HIGHLIGHTS

Starting in 1988, virtually all government-sponsored R&D efforts devoted to advancing aircraft and missile gas turbine engines for military and commercial use were integrated under the IHPTET program and were working toward unified objectives with the overall goal being a 100% increase in propulsion capability by the year 2000. IHPTET comprised demonstrator programs, component development, and fundamental research. IHPTET participants included the Air Force, Army, Navy, DARPA, NASA, and industry.[34] Industry participants were Allison, Williams, GE, Pratt & Whitney Aircraft, Lycoming, and Teledyne CAE. The scope of IHPTET included turbofan/jet, turboshaft/prop, and expendable turbine engines, large and small.

IHPTET demonstrator programs included Advanced Turbine Engine Gas Generator, Joint Technology Demonstrator Engine, Joint Turbine Advanced Gas Generator, and Joint Expendable Turbine Engine Concepts (JETEC). ATEGG and JTDE technologies were applied to turbojets and turbofans. JTAGG technologies were developed for turboshaft and turboprop applications and JETEC for expendable missile engines. A portion of the funding of these programs was provided by participants from the turbine engine industry.[35] A complete description of these programs is provided in the accompanying two Technology Programs Summary tables.

34. Dick Kazmar, "AIAA Position Paper on the Integrated High Performance Turbine Engine Technology (IHPTET) Initiative," (AIAA Paper, 5 March 1991). An attachment to this paper was entitled "Background Paper on the Integrated High Performance Turbine Engine Technology Initiative."

35. The information on IHPTET programs was taken from generally undated brochures published by the Turbine Engine Division at Wright-Patterson AFB, Ohio. Titles included: *IHPTET Initiative*; *IHPTET Today*; *IHPTET Technology Teams in Action*; *IHPTET Initiative Advanced Technology in Transition* (1993).

Technology Programs Summary

Program	Engine Classification (Large/Small/Etc.)	Program Start Date	Original Sponsor	Program Objectives	Participating Companies	Engine Designations	Major Technology Achievements
Light Weight Gas Generator (LWGG)	Large gas generators	Early 1960s	Air Force	Adv. gas generator development and demonstrators	TCAE Allison P&WA	ATEGG-365 440	Technology transitioned into P&W F100 and GE TF-39
Advanced Turbine Engine Gas Generator Demonstrator (ATEGG)	Man-rated gas generators	Mid-1960s	Air Force	Focus on gas generator technology as building blocks	P&WC, GE TCAE	ATEGG-555 ATEGG-585 ATEGG-689 GMA 100 GMA 200	(Program continuing as part of IHPTET)
1,500 shp Demonstrator Engine	1,500 shp turboshaft demonstrator engine	1967	Army	Combination of greatly improved performance and maintainability (Led to T700)	Lyc.(800 shp) PWA (550 shp) Gar.(250 shp) Wil.(200 shp)	ST9, GE12	Technology transitioned into commercial GE CT700 and P&WC PW100
Small Turbine Advanced Gas Generator (STAGG)	200-800 shp class gas generators	1972	Army	Intended to lead to an engine for light helos. and APU's	Garrett Allison, TCAE Williams	PTL34	Validated 20% reduction in sfc for 200-800 shp class engines
Joint Technology Demonstrator Engine (JTDE)	Man-rated engines	1975	Air Force/ Navy	Full engine demonstration utilizing ATEGG cores	Lycoming Allison TCAE	JTDE-455 GMA 500 JTDE-455	Technology transitioned into engine upgrades
Advanced Technology Demonstrator Engines (ATDE)	Demonstrator using STAGG technology turbine engines	1976	Army	Full-scale demonstrator (Led to T800)		PLT34A, GMA 500	First turboshaft engine to use full authority digital eng. controls

Technology Programs Summary

Program	Engine Classification (Large/Small/Etc.)	Program Start Date	Original Sponsor	Program Objectives	Participating Companies	Engine Designations	Major Technology Achievements
High Performance Turbine Engine Technologies (HPTET)	Air Force turbine engines	1982	Air Force	Dev.tech.reqd.to doubl.turb. capability by yr. 2000	Allison, GE P&WA, TCAE Williams Garrett		(In 1988 became part of DoD/NASA IHPTET prog.)
Modern Technology Demonstrator Engines (MTDE)	4,000–6,000 shp class turboshaft/ turboprop	1983	Army (Later Navy)	Improved engine in the 4,000–6,000 shp class (Led to the T407)	P&WC, GE	PW3005, GE27	Technology incorp. into GE/Allied-Signal CFE738
Expendable Turbine Engine Concepts (ETEC)	Air Force expendable engines	1987	Air Force	Develop tech. appli.to short-life engines	Garrett Allison, TCAE Williams	ETJ1091 GMA 900 (J102-AD-100)	(With Navy participation subsequently changed to JETEC)
Integrated High Performance Turbine Engine Tech.(IHPTET)	All aircraft gas turbines	1988	Army, Navy, Air Force, DARPA, NASA, Industry	Double propul. sys. capability by year 2000	Allison, GE, P&WA, TCAE Williams Lycoming		
Joint Turbine Advanced Gas Generator (TAGG)	Man-rated turboshaft and turboprop eng.	1989	Army, Navy Air Force	Validate critical gas generator core technology	AlliedSignal (Garrett) Lycoming Allison	JTAGG I PLT210	(Part of IHPTET)
Joint Expendable Turbine Engine Concepts (JETEC)	Expendable engine demonstrators	1990	Navy, Air Force (JETEC is part of IHPTET)	Demonstrate advance IHPTET tech. for air breathing missiles	Garrett, TCAE Williams		(Part of IHPTET)

Final Conclusions

THE HISTORY OF THE INDUSTRY

Small gas turbine engines aircraft engines emerged in the 1940s and the 1950s as a distinct class of turbine engines. The early advancement of this technology was largely the result of military development programs carried out by corporations. In the following decades, seven major corporations, which produced these engines in great quantities, became the nucleus of the small turbine engine industry in North America.

By the close of the twentieth century, the North American small gas turbine aircraft engine industry had evolved through four major development periods. In each of these periods, the industry, its corporations, and their engines were influenced by important world events and government and military policies. Internal corporate considerations and policies also helped influence the direction of the industry.

ENGINES OF DISTINCTION

Of the many small gas turbine engines that have been developed in North America, there are a few engine families that are especially noteworthy of distinction. These include the following:

Allison T63/Model 250
Garrett TPE331
Garrett TFE731
General Electric T58
General Electric J85/CJ610/CF700
General Electric T700/CT7
Lycoming T53
Pratt & Whitney Canada PT6
Teledyne CAE J69
Teledyne CAE J402
Williams F107

Four basic criteria were used to select these engines:

1. The engines satisfied basic military qualification tests or FAA certification requirements.
2. They adapted to meet evolving airframe requirements.

3. They were characterized by attributes beyond those common to all engines.

4. They had a unique capability in power and performance that provided a critical match between engine and airframe at a particular time when it was needed.

To examine these criteria in detail, it is important to ask: "What makes an engine successful?" Some very important requirements were common to all small gas turbine aircraft engines. Before any engine could be produced, it had to demonstrate performance, integrity, and flight safety of its design by successfully satisfying rigorous military qualification tests and/or FAA certification requirements. Some 6,000–10,000 development test hours over four to five years were required to develop a new man-rated engine to the point of successful qualification or certification for production.[36] Also, several years of product improvement effort at a level close to that spent on annual development funding was required for correction of service revealed deficiencies and life improvement. Lesser program support than this would not result in a successful engine. With few exceptions, all production engines represented excellent engine designs that were the result of extensive testing and product improvement efforts.

In addition, each engine required the support of the company's sales, marketing, and product support organizations; these functions were critical for initial sales and subsequent aftermarket support to ensure customer satisfaction. Each company also responded to the needs of its customers by adapting the engine to meet evolving airframe designs. These adaptations often involved major design changes and almost always included "growing" the engine by increasing its power to improve aircraft performance and capability and to improve engine durability, reliability, and maintainability to reduce direct operating costs and increase product life.

To achieve enduring recognition, engines of distinction were characterized by attributes beyond those common to all engines. One critical measure was the number of engines produced, a clear indicator of the engine's relative importance, success, and overall contribution. A related but less important factor was the length of time that an engine remained in production. Meeting customer needs for a long period indicated that the engine was providing useful service and contributing to society in more important ways than were those engines with a short production life. The number of airframe applications in which an engine was used was an additional yardstick reflecting both the utility of the en-

---

36. Donald D. Weidhuner, Fort Lauderdale, Florida, to Mark G. Hirsch, letter, 18 May 1997.

gine and its service contribution. Technological contributions were yet another measure of distinction. While competition between similar engines tended to result in comparable performance, some engines achieved significant performance improvements over their competitors or introduced innovations in design that set a standard for engines that followed. Finally, some engines, through their airframe applications, made important historical contributions.

Irrespective of the merits of the design of any engine, none could succeed unless it was used in one or more vehicles that provided a capability for which there was substantial demand. Conversely, the capability of any vehicle was determined, to a large degree, by the merits of its engine(s). In many cases, the engines identified in this section provided a unique capability in power and performance at the particular time in which it was needed. Success was critically dependent upon a technical match between a given engine and airframe design at a time in history during which the combination satisfied an important need.

A factor that most of the successful engines had in common was being the first to be matched with a successful aircraft in a particular market niche. Most of the engines that were first in a market entered the market as the result of winning an important military or civilian competition. With some exceptions, an important factor in being first in a market niche was that design began before there was any customer commitment that the engine would be used in a specific application. This usually meant that the engines would go into production with little or no competition in their power class.

Undertaking the design of an engine in advance of an airframe commitment required significant risk and initiative by individuals in either the company or the government or both. During the early study and preliminary design phases of a new engine, there were usually numerous discussions and exchange of ideas between the engine designer and aircraft manufacturers as to what new designs were feasible, and the potential operator (military, commercial, or both) as to what new designs might be needed. Even in those cases in which the initiation of an actual development was in response to military competition, these competitions were often stimulated by experimental tests, informal studies, or proposals by an engine company. For this reason, in most cases it was difficult to isolate the specific people or organization that could be given credit for "starting" a successful engine design. In a few cases, however, individuals responsible for the overall success of an engine stood out and were identified in the histories of their corporations.

Most significant in the success of an engine were unique and fortuitous market conditions that created a large demand for the aircraft in which the engine was used. Engines that were produced in the greatest quantities and for the longest periods of time resulted from such market conditions. There was no evidence that any of the companies foresaw, or could have foreseen, the magnitude of the number of production aircraft and engines that these unique market conditions would eventually demand. Despite the best planning efforts, it would appear then that the extraordinary success of these engines was often the result of blind luck.

CONTRIBUTIONS

The gas turbine aircraft engine was one of the great inventions of the 20th century. The small gas turbine aircraft engine was an important refinement of this technology. From a tentative technology that appeared difficult to achieve, small gas turbine aircraft engines became mature power plants with improved performance, efficiency, and integrity.

The powerful, lightweight, and highly reliable small gas turbine aircraft engine contributed to major improvements in a broad spectrum of aircraft. These engines literally transformed helicopters by vastly improving their performance and utility. These attributes greatly expanded the use of helicopters in both commercial and military operations. Similarly, small turbine engines led to a major expansion of the general aviation executive and utility fleet by making these aircraft highly productive business assets. The regional transportation industry was made possible by efficient turbine-powered aircraft that could carry more passengers, more quickly, and at lower costs than piston-powered aircraft. Target drones, RPVs, UAVs, missiles, small fighter aircraft, trainer aircraft, helicopters, and a wide variety of other small aircraft contributed significantly to military operations and national defense.

From modest beginnings, this technology expanded to become a major North American industry with a worldwide social, economic, and military impact. For pilots, passengers, and the general public, the net result has been better aircraft, improved in-flight safety and comfort, faster transportation, and an improved environment.

APPENDICES

APPENDIX A

Rationale for the Engine Data Tables and Genealogy Charts

The following Engine Data tables and Engine Development Genealogy charts present a summary of information on the principal engine models developed by the engine manufacturers. These tables and charts include engine data in the text and other data obtained primarily from the engine manufacturers, although some data were obtained from other sources.

The tables and charts include only the principal models of each engine type.[1] Principal models are defined as the initial model of an engine type, models within a type that represented the most significant design improvements, or those models that had a number of important aircraft applications. Most companies developed a number of engine models within a given engine type. Many of those models were basically similar in design and performance, and a number of them either never reached production or had only limited aircraft application. Mechanical differences often amounted to only minor rearrangements of accessories to suit specific aircraft installations or minor internal modifications to improve the engines. Data on the engine models included are representative of the performance, configuration, and principal applications for the

---

1. This terminology follows the terminology specified for military engines. An engine type is generally an engine of original design. An engine model is a specific configuration of an engine type, each of which represents a significant design modification within an engine type.

mechanically similar engines of that model. However, some prototype, non-production, or non-application engines were included in instances where they were historically important or were a key link in the development of progressive engine models.

Data in the Significant Dates tables present the time frames for design start, progression, and production of the principal engine models. In a number of cases, specific dates and production figures were unavailable for a given engine model. Nevertheless, production figures in the tables include, either separately by engine model or in total, all engines of each engine type that were manufactured. Dates indicated for engine certification/qualification represent the FAA Certification date for civilian engines and the Military Qualification date for military engines.

The Engine Development Genealogy charts trace the progression of development and improvement of each engine type and major model. These charts were designed to illustrate the relationship of each engine development step to engine models that preceded it.

The Engine Data tables include significant performance parameters of the principal engine models, their design characteristics, and the aircraft in which they were used. Engine configuration data contained in these charts can be interrelated with similar data in the engine genealogy charts.

Following assembly of the data tables and genealogy charts by the authors, the tables and charts were submitted to the appropriate company to be reviewed for accuracy. However, even with these reviews there were some cases where conflicting information existed in company records and other information sources. In such cases the authors attempted to resolve the conflict by selecting what they believed was the most reliable source.

The final dates for data entry in the tables vary from the early to the mid-1990s among the various companies. The dates reflect the approximate time at which the authors concluded their discussions with the respective companies.

APPENDIX B

Data Tables and Genealogy Charts

Pioneer Small Turbine Engine Companies (Significant Dates)

Company Model No.	Military Model No.	Design/ Devel. Start	Engine First Run	Engine First Flight	Engine Cert./Qual.	First Prod. Engine Delivered	Last Prod. Engine Delivered	Total Engines Produced
Westinghouse								
9.5A	XJ32-WE-2	Late1942		Aug 1945	Jan 1945	July 1944		24
9.5B		1945				1946		20
PD-29	XJ81-WE-3							24 built by Rolls-Royce
Fredric Flader								
J55-FF-1		Jun 1947	Jun 1948					
Boeing								
Model 500	YT50	1945	1947					
Model 502-2E	T50-BO-1	1945	1947	Dec 1951		1949		
Model 502-8				Nov 1952				
Model 502-10C								
Model 502-10VC	T50-BO-8A							
Model 502-10F								
Model 520-2	T60-BO-2	1954	1955		1960		1965	1,500 (All 502 Models)
Model 550-1	T50-BO-12	Early 1960s				Mid-1960s	Apr 1968	
Fairchild								
	J44-R-2	June 1947	Aug 1948	1950		1949		
	J44-R-20			1954				
FT-101E	J44-R-3	1954		1954	1955		1959	
	XJ83	Dec 1954	Early 1957	June 1957	1958			18
West								
XJ38-WS-2		1946						
Wright Aeronautical								
TJ34								None produced by Wright
TJ37						Circa 1957 built by Bristol		

Subsequent Small Turbine Engine Companies (Significant Dates)

Company Model No.	Military Model No.	Design/ Devel. Start	Engine First Run	Engine First Flight	Engine Cert./Qual.	First Prod. Engine Delivered	Last Prod. Engine Delivered	Total Engines Produced
Dreher								
TJD-76A		circa 1969						
TJD-76C								
TJD-76D/76E		1972						
TJD-79A		circa 1970						
Solar								
T62-2	YT62-S-2	1957	Mar 1958	Aug 1959	1959 (40 Hr. PFRT)			12
T66-2	YT66-S-2	1957	1960					

Pioneer Small Turbine Engine Companies

Company Model No.	Military Model No.	S.L. Rated Thrust, Lb	SFC @ S.L. Rated Thrust	Engine Wt. Lb	Thrust-to-Wt. Ratio	Configuration	Applications
Westinghouse							
9.5A	XJ32-WE-2	260	1.60	140	1.86	Turbojet with 6-stage axial-flow compressor, through-flow annular combustor, and single-stage axial turbine.	Navy-built Gorgon II-B and III-B air-to-air missiles and wind tunnel models
9.5B		275–300	1.55	145	1.90 to 2.07	Same as 9.5A with diameter of combustor increased to raise operating altitude.	Martin TD2N-1 drone
	XJ81-WE-3	1,740	1.26	304	5.72	Rolls-Royce RB93/2 Soar engine licensed by Westinghouse. Lightweight turbojet with seven-stage axial-flow compressor, annular combustor, and single-stage axial turbine.	Northrop XQ-4 air-launched target drone (interim engine)
Fredric Flader							
	J55-FF-1	700	1.65	300	2.33	Turbojet with single-stage supersonic compressor, through-flow annular combustor, and single-stage axial turbine.	Designed for Ryan XQ-2 Firebee target drone
Boeing							
Model 500		150 lb	1.3	85	1.76	Turbojet with single-entry centrifugal compressor, two can-type combustors, and single-stage axial turbine.	
Model 502-2E	YT50	175 shp	1.30	230	0.76	Turboshaft using Model 500 as gas generator with opposed single-stage free power turbine plus gearbox.	Kaman K-225 helicopter Kaman HTK-1 twin-engine helicopter

Model 502-8	XT50	210 shp	1.30	267	0.79	Turboprop version of Model 502-2.	Cessna XL-19B Bird Dog (502-8B)
Model 502-10C		240 shp	1.02	320	0.75	Improved turboshaft version of Model 502-2E.	Gyrodyne DSN-3 drone helicopter
Model 502-10VC	T50-BO-8A	300 shp	0.98	334	0.90	Improved turboshaft version of Model 502-10C.	Gyrodyne QH-50C drone helicopter
Model 502-10F		300 shp	0.95	300	1.00	Turboprop version of Model 502-10C.	Radioplane RP-77D radio controlled target
Model 520-2	T60-BO-2	430 shp	0.72	325	1.32	Turboshaft with double entry centrifugal compressor, two reverse-flow combustors, single-stage radial gas generator turbine, opposed single-stage axial free power turbine, and gearbox.	Planned for Navy applications that failed to materialize
Model 550-1	T50-BO-12	365 shp	0.80	250	1.46	Turboshaft derivative of Model 502 engines with single axial stage ahead of centrifugal stage, a radial-type gas generator turbine, and a one-stage free power turbine.	Gyrodyne QH-50D drone helicopter
Fairchild							
	J44-R-2	1,000 lb class		280	3.57	Turbojet with single-stage mixed-flow compressor, through-flow annular combustor, single-stage axial turbine, and structural external cowling.	Fairchild Petrel air-to-underwater missile; Ryan XQ-2 Firebee target drone
	J44-R-20	1,000	1.50	335	2.98	Improved version of J44-R-2.	Same as above
FT-101E	J44-R-3	1,000	1.55	370	2.70	Man-rated version of J44-R-20 structurally altered to increase engine life and reliability.	Bell Model 65 VTOL ATV; Fairchild C-82 and C-123B for thrust augmentation
	XJ83	2,450	0.94	363	6.7	Turbojet with 7-stage transonic compressor, through-flow annular combustor, and 2-stage axial turbine.	Fairchild XSM-73 Bull Goose decoy missile

Pioneer Small Turbine Engine Companies (*Continued*)

Company Model No.	Military Model No.	S.L. Rated Thrust, Lb	SFC @ S.L. Rated Thrust	Engine Wt. Lb	Thrust-to-Wt. Ratio	Configuration	Applications
West							
	XJ38-WS-2	214	1.70	180	1.19	Type B supercharger modified by adding combustion chamber, fuel injection system, intake and exhaust ducts, and redesigned turbine.	Designed for target drones
Wright Aeronautical							
TJ34		2,000	1.01	570	3.51	Armstrong Siddeley "Long Life" Viper ASV.10 turbojet licensed by WAC. Seven-stage axial-flow compressor, annular combustor, and single-stage axial turbine.	Unsuccessfully marketed by WAC for: Ryan Firebee target drone Fairchild Goose decoy missile Boost engine on Martin P5M aircraft
TJ37		4,850	1.06	825	5.88	British Orpheus BOr.3 turbojet licensed by WAC. Seven-stage axial-flow compressor, annular combustor, and single-stage turbine.	Lockheed JetStar CL-329 prototype aircraft

Subsequent Small Turbine Engine Companies

Company Model No.	Military Model No.	S.L. Rated Thrust, Lb	SFC @ S.L. Rated Thrust	Engine Wt. Lb	Power/Thrust-to-Wt. Ratio	Configuration	Applications
Dreher							
TJD-76A		55 lb	1.50	17	3.2	Miniature turbojet with single-stage centrifugal compressor, straight-through-flow combustor, and single-stage axial turbine.	Test flown on Prue 215A sailplane
TJD-76C		55 lb	1.50	14.1	3.9	Lighter version of TJD-76A.	Research tool
TJD-76D/76E		55 lb	1.50	9.9	5.6	Low-cost, short-life versions of TJD-76C.	Designed for small drones and expendable vehicles
TJD-79A		120 lb	1.30	36	3.3	Enlarged version of TJD-76 with same configuration.	Designed for use in sailplanes
Solar							
T62-2	YT62-S-2	83 shp + 14 lb thrust	1.10	52	1.6	Single-shaft turboshaft with back-to-back centrifugal compressor and radial turbine, reverse-flow combustor, and front gearbox.	Gyrodyne YRON-1 one-man rotorcycle helicopter
T66-2	YT66-S-2	83 shp + 14 lb thrust	1.16	53	1.6	Free turbine version of T62-2. Single-stage radial-inflow compressor turbine and single-stage exducer power turbine.	Designed for helicopters and flying platforms

Teledyne CAE (Significant Dates)

Company Model No.	Military Model No.	Design/Devel. Start	Engine First Run	Engine First Flight	Engine Cert./Qual.	First Prod. Engine Delivered	Last Prod. Engine Delivered	Total Engines Produced
Turbojet Engines								
352-4	J69-T-2	1951				1957	1958	23
	J69-T-6					1967	1968	18
352-2	J69-T-9	1951		Nov 1954	mid-1955	1955	1959	1,189
	J69-T-17	1955				1955	1958	27
352-5A/-5B	J69-T-25,-25A			Nov 1959	1958	1959	1976	1,804
354-1	J69-T-19					1954	1959	634
356-7A	J69-T-29					1959	1988	4,924
356-29/-29A	J69-T-41,-41A					1965	1995	1,098
356-34A	J69-T-406					1968	1976	271
356-28A	J100-CA-100		May 1968			1968	1972	96
370-1C	J402-CA-400	1970	1971	Nov 1973	Aug 1974	Apr 1975		6,068 to Mar 1991
372-2A	J402-CA-700					1975	1984	522
373-8B	J402-CA-702		1976			1986		454 to Mar 1991
304		1990	1992					
304C		1990	1992					
305-4		Jan 1986	Jul 1986					
305-7E		Mid-1986	Late 1986	May 1987				
305-10		1990	1991					
305-10C		1991	1992					
312-1		Late 1986	Early 1987					
312-5	J700-CA-400	1989	1991	1994	1993	1992		25 thru 1993
318		1989	1991					
320-1		Late 1986	Early 1987					
320-2		1987	1989					

Turboprop/Turboshaft Engines

Model	Designation							
140/141	MA-1/-1A	1951				1952		1,076
210	XT51-T-1	1951		1954				
220-2	XT51-T-3	1951		1954				
217-5	T72-T-2	1958						
227-4A	T65-T-1	1961			1964			
217-A/-2A	T67-T-1	1963		1965	1964			

Turboprop Developed by **Teledyne Continental Motors**

Model	Designation							
TP500		1978	1980	Sep 1982	Oct 1988			

Turbofan Engines

Model	Designation							
472-5	F106-CA-100	1967						
471-11DX		1973	May 1975		1976 PFRT			
490-4		1973	Mar 1973					
	F107-WR-101	1978 prod. tooling	1981			Nov 1982	Feb 1986	261
	F107-WR-400	1978 prod. tooling	1981			Oct 1982	1991	460
	F107-WR-402						Dec 1991	140
382-10	F408-CA-400	1985	1987	1991	1992			

Demonstrator Engines

Model	Designation							
ATEGG-365	XLJ95-T-1	1962	1964					
ATEGG-440		1967	Jan 1970					
ATEGG-555			Feb 1974					
JTDE-455			Oct 1977					
ATEGG-585			Dec 1982					
ATEGG-589			May 1986					

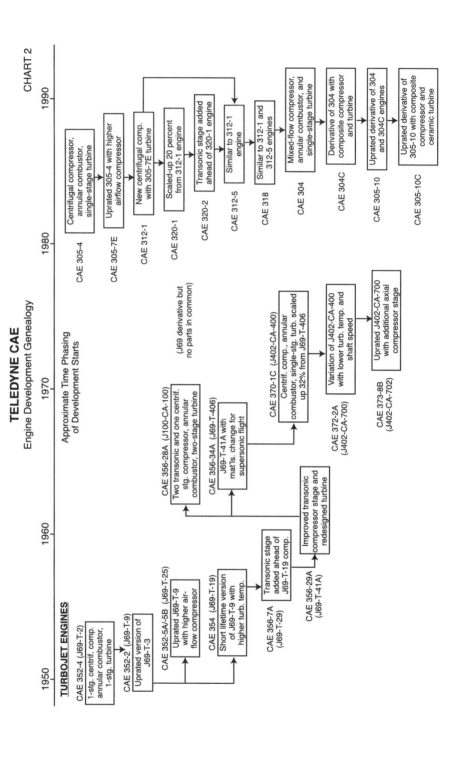

TELEDYNE CAE
Engine Development Genealogy

CHART 2

Approximate Time Phasing
of Development Starts

TURBOJET ENGINES

CAE 352-4 (J69-T-2)
1-stg. centrif. comp.,
annular combustor,
1-stg. turbine

CAE 352-2 (J69-T-9)
Uprated version of
J69-T-3

CAE 352-5A/-5B (J69-T-25)
Uprated J69-T-9
with higher air-
flow compressor

CAE 354 (J69-T-19)
Short lifetime version
of J69-T-9 with
higher turb. temp.

CAE 356-7A
(J69-T-29)
Transonic stage
added ahead of
J69-T-19 comp.

CAE 356-29A
(J69-T-41A)
Improved transonic
compressor stage and
redesigned turbine

CAE 356-28A (J100-CA-100)
Two transonic and one centrif.
stg. compressor, annular
combustor, two-stage turbine

CAE 356-34A (J69-T-406)
J69-T-41A with
mat'ls. change for
supersonic flight

(J69 derivative but
no parts in common)

CAE 370-1C (J402-CA-400)
Centrif. comp., annular
combustor, single-stg. turb. scaled
up 32% from J69-T-406

CAE 372-2A
(J402-CA-700)
Variation of J402-CA-400
with lower turb. temp. and
shaft speed

CAE 373-8B
(J402-CA-702)
Uprated J402-CA-700
with additional axial
compressor stage

CAE 305-4
Centrifugal compressor,
annular combustor,
single-stage turbine

CAE 305-7E
Uprated 305-4 with higher
airflow compressor

CAE 312-1
New centrifugal comp.
with 305-7E turbine

CAE 320-1
Scaled-up 20 percent
from 312-1 engine

CAE 320-2
Transonic stage added
ahead of 320-1 engine

CAE 312-5
Similar to 312-1
engine

CAE 318
Similar to 312-1 and
312-5 engines

CAE 304
Mixed-flow compressor,
annular combustor, and
single-stage turbine

CAE 304C
Derivative of 304 with
composite compressor
and turbine

CAE 305-10
Uprated derivative of 304
and 304C engines

CAE 305-10C
Uprated derivative of
305-10 with composite
compressor and
ceramic turbine

1950   1960   1970   1980   1990

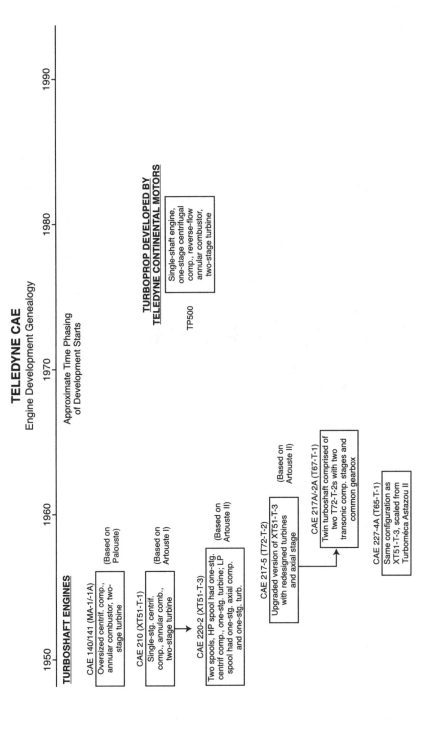

TELEDYNE CAE
Engine Development Genealogy

Approximate Time Phasing
of Development Starts

1950   1960   1970   1980   1990

<u>TURBOSHAFT ENGINES</u>

CAE 140/141 (MA-1/-1A)
Oversized centrif. comp.,
annular combustor, two-
stage turbine

(Based on
Palouste)

CAE 210 (XT51-T-1)
Single-stg. centrif.
comp., annular comb.,
two-stage turbine

(Based on
Artouste I)

CAE 220-2 (XT51-T-3)
Two spools, HP spool had one-stg.
centrif comp., one-stg. turbine; LP
spool had one-stg. axial comp.
and one-stg. turb.

(Based on
Artouste II)

CAE 217-5 (T72-T-2)
Upgraded version of XT51-T-3
with redesigned turbines
and axial stage

(Based on
Artouste II)

CAE 217A/-2A (T67-T-1)
Twin turboshaft comprised of
two T72-T-2s with two
transonic comp. stages and
common gearbox

CAE 227-4A (T65-T-1)
Same configuration as
XT51-T-3, scaled from
Turboméca Astazou II

<u>TURBOPROP DEVELOPED BY
TELEDYNE CONTINENTAL MOTORS</u>

Single-shaft engine,
one-stage centrifugal
comp., reverse-flow
annular combustor,
two-stage turbine

TP500

827

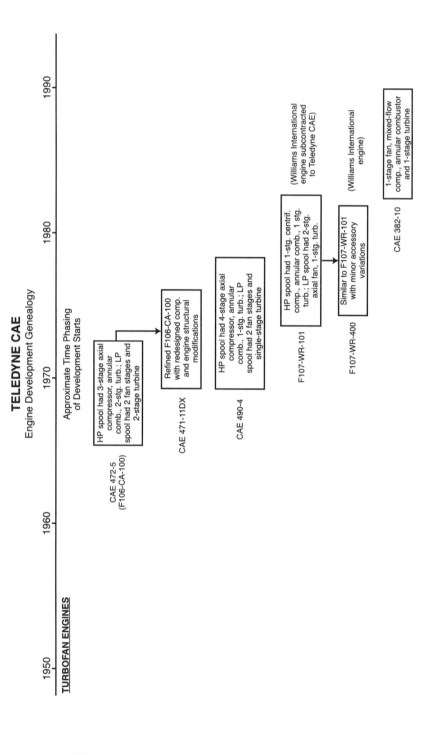

TELEDYNE CAE
Engine Development Genealogy

TURBOFAN ENGINES

Approximate Time Phasing
of Development Starts

1950   1960   1970   1980   1990

CAE 472-5
(F106-CA-100)

HP spool had 3-stage axial
compressor, annular
comb., 2-stg. turb.; LP
spool had 2 fan stages and
2-stage turbine

CAE 471-11DX

Refined F106-CA-100
with redesigned comp.
and engine structural
modifications

CAE 490-4

HP spool had 4-stage axial
compressor, annular
comb., 1-stg. turb.; LP
spool had 2 fan stages and
single-stage turbine

F107-WR-101

HP spool had 1-stg. centrif.
comp., annular comb., 1 stg.
turb.; LP spool had 2-stg.
axial fan, 1-stg. turb.

(Williams International
engine subcontracted
to Teledyne CAE)

F107-WR-400

Similar to F107-WR-101
with minor accessory
variations

(Williams International
engine)

CAE 382-10

1-stage fan, mixed-flow
comp., annular combustor
and 1-stage turbine

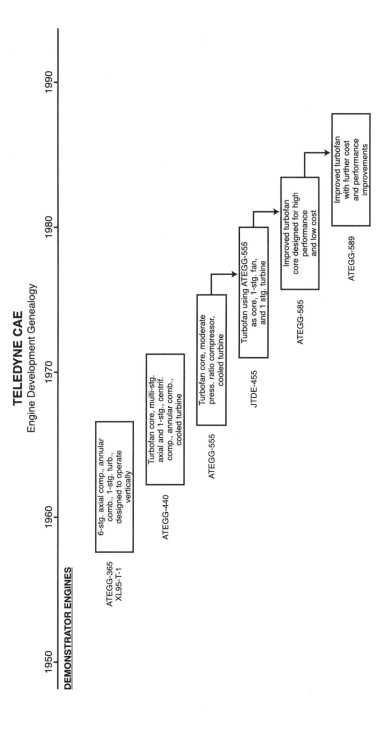

TELEDYNE CAE
Engine Development Genealogy

DEMONSTRATOR ENGINES

1950   1960   1970   1980   1990

ATEGG-365
XL95-T-1

6-stg. axial comp., annular
comb., 1-stg. turb.,
designed to operate
vertically

ATEGG-440

Turbofan core, multi-stg.
axial and 1-stg. centrif.
comp., annular comb.,
cooled turbine

ATEGG-555

Turbofan core, moderate
press. ratio compressor,
cooled turbine

JTDE-455

Turbofan using ATEGG-555
as core, 1-stg. fan,
and 1 stg. turbine

ATEGG-585

Improved turbofan
core designed for high
performance
and low cost

ATEGG-589

Improved turbofan
with further cost
and performance
improvements

Teledyne CAE (Turbojet Engines)

Company Model No.	Military Model No.	S.L. Rated Thrust, Lb	SFC at S.L. Rated Thrust	Engine Weight, Lb	Thrust-to-Weight Ratio	Configuration	Applications
352-4	J69-T-2	880	1.15	298	2.95	Turboméca Marboré II engine, single centrifugal compressor, slinger fuel injector, annular combustor, single-stage turbine.	Temco TT-1 trainer
352-2	J69-T-9	920	1.15	364	2.53	Uprated version of J69-T-3 engine.	Cessna T-37A trainer
352-5A/-5B	J69-T-25	1,025	1.14	358	2.86	Uprated J69-T-9 having higher airflow compressor.	Cessna T-37B/C trainers
354/-1	J69-T-19	1,000	1.29	317	3.15	Short lifetime version of J69-T-9 operated at increased turbine temperature.	Ryan Q-2A/B Firebee target drones
356-7A	J69-T-29	1,700	1.29	317	5.36	Uprated T69-T-19 with transonic-flow stage ahead of centrifugal compressor and redesigned turbine.	Ryan Q-2C (BQM-34A, MQM-34D) Firebee target drone Ryan 147 A/B Firebee reconnaissance drone
356-29/-29A	J69-T-41/-41A	1,920	1.10	350	5.49	Uprated J69-T-29 with improved transonic compressor stage and redesigned turbine.	Ryan BQM-34C/D Firebee target drones Ryan 147G (AQM-34) Firebee reconnaissance drone
356-34A	J69-T-406	1,920	1.11	360	5.33	Same as J69-T-41A with change in centrifugal stage material to permit supersonic flight.	Ryan BQM-34E/F Firebee supersonic target drones
356-28A	J100-CA-100	2,700	1.10	422	6.40	Derived from J69 family, but shared no parts in common. Two-stage transonic compressor, centrifugal compressor stage, annular combustor, two-stage turbine.	Ryan 147T (AQM-34P) Firebee high-alt. reconnaissance vehicle

Model	Engine					Construction	Application
370-1C	J402-CA-400	660	1.20	101.5	6.50	Single axial plus single centrifugal compressor, annular combustor, and single-stage turbine scaled from J69-T-406 to 32% of its airflow rate.	McDonnell Douglas AGM/RGM/UGM Harpoon air/sea/submarine-launched cruise missiles. SLAM.
372-2A	J402-CA-700	640	1.19	115	5.57	Variation of J402-CA-400, turbine temp. and shaft speed reduced to increase engine life.	Beech MQM-107 Streaker target drone
373-8B	J402-CA-702	960	1.03	138	6.96	Uprated version of J402-CA-400 with addition of axial-flow compressor stage and redesigned turbine.	Beech MGM-107D Streaker target drone
304		59	1.21	8.5	6.94	Single-stage mixed-flow compressor, annular combustor, single-stage turbine.	SENGAP program demonstration engine
304C		81	1.44	8	10.13	Single-stage mixed-flow composite compressor, annular combustor, single-stage composite turbine.	SENGAP program demonstration engine
305-4		40	1.60	12	3.33	Single-stage centrifugal compressor, annular combustor, single-stage turbine.	1/5 scale MiG27 target vehicle LOCAAPS program demonstration engine
305-7E		90	1.26	19	4.74	Uprated version of Model 305-4 with higher airflow compressor.	Designed for tank, RPV, helicopter, missile, and aerial targets NLOS program demonstration engine and NLOS/TRR
305-10		178	1.18	26	6.85	Single-stage mixed-flow compressor, annular combustor, single-stage turbine.	LONGFOG sustainer engine, NLOS, NLOS/TRR
305-10C		218	1.37	25	8.7	Derivative of Model 305-10 with composite inlet, composite single-stage mixed-flow compressor, annular combustor, and ceramic single-stage turbine.	LONGFOG sustainer engine

Teledyne CAE (Turbojet Engines) (*Continued*)

Company Model No.	Military Model No.	S.L. Rated Thrust, Lb	SFC at S.L. Rated Thrust	Engine Weight, Lb	Thrust-To-Weight Ratio	Configuration	Applications
312-1		177	1.16	34	5.2	New design with backward curved single-stage centrifugal compressor, annular combustor, and single-stage turbine.	Designed for small missiles, RPVs, and aerial targets
312-5	J700-CA-400	177	1.19	38	4.7	Derivative of Model 312-1 with minor modifications.	Improved Tactical Air-Launched Decoy (ITALD) drone
318		177	1.19	38	4.7	Derivative of Model 312-1 with minor modifications.	Ground-Launched Tacit Rainbow missile sustainer engine
320-1		240	1.13	50	4.8	Scaled up 20 percent from Model 312-1 engine.	Designed for small missiles, RPVs, and aerial targets
320-2		350	1.09	58	6.0	Transonic stage added ahead of Model 320-1 engine to increase airflow, and turbine temperature increased.	Designed for small missiles, RPVs, and aerial targets

Teledyne CAE (Turboprop/Turboshaft Engines)

Company Model No.	Military Model No.	S.L. Rated Thrust, Lb	SFC at S.L. Rated Thrust	Engine Weight, Lb	Thrust-to-Weight Ratio	Configuration	Applications
140/141	MA-1/-1A	191	1.50	198	0.96	Based on Turboméca Palouste with oversized centrifugal compressor to provide bleed air, annular combustor, and two-stage turbine.	Air compressor unit used to start jet engines
210	XT51-T-1	280	0.99	266	1.05	Based on Turboméca Artouste I. Single-shaft engine with centrifugal compressor, annular combustor, and two-stage turbine.	Cessna XL-19C experimental liaison aircraft; Bell XH-13F experimental helicopter
220-2	XT51-T-3	425	0.97	236	1.80	One axial-stage plus one centrifugal-stage compressor, annular combustor, two-stage core turbine core, single-stage free turbine driving axial-compressor stage and power takeoff.	Sikorsky XH-39 experimental helicopter
217-5	T72-T-2	525	0.69	230	2.28	Same configuration as XT51-T-3. New axial comp. stage and turbines, centrifugal stage and combustor derived from Turboméca Astazou II.	Republic Lark experimental assault support helicopter
227-4A	T65-T-1	310	0.66	136	2.28	Same configuration as XT51-T-3. Scaled from Astazou II and modified to free turbine configuration.	Developed for U.S. Army's Light Observation Helicopter
217-A/-2A	T67-T-1	1,540	0.56	519	2.97	Twin turboshaft with two scaled-up T72-T-2 engines coupled to common gear box, core axial stage replaced by two transonic stages.	Flown experimentally in Bell UH-1D helicopter

Turboprop Developed by **Teledyne Continental Motors**

Company Model No.	Military Model No.	S.L. Rated Thrust, Lb	SFC at S.L. Rated Thrust	Engine Weight, Lb	Thrust-to-Weight Ratio	Configuration	Applications
TP500		510	0.59	322	1.58	Modular single-shaft engine, single-stage centrifugal compressor, reverse-flow annular combustor, two-stage turbine, and 2-stage planetary gearbox.	Developed for general aviation aircraft; Test flown in Cessna Cheyenne II

Teledyne CAE (Turbofan Engines)

Company Model No.	Military Model No.	S.L. Rated Thrust, Lb	SFC at S.L. Rated Thrust	Engine Weight, Lb	Thrust-to-Weight Ratio	Configuration	Applications
472-5	F106-CA-100	615	0.61	139	4.42	Dual-shaft engine. HP spool had three-stage axial compressor, annular combustor, and two-stage turbine; LP spool had two-stage axial fan driven by two-stage turbine.	Entered in USAF engine competition for Subsonic Cruise Armed Decoy (SCAD)
471-11DX		620	0.646	175	3.55	Refinement of F106-CA-100 with redesigned compressor and structurally strengthened to withstand 100g loads.	Vought BGM-110 entry in U.S. Navy Sea-Launched Cruise Missile (SLCM) competition
490-4		2,966	0.71	640	4.63	Dual-shaft engine. HP spool had four-stage axial compressor, annular combustor, and single-stage turbine; LP spool had two-stage fan driven by single-stage turbine; bypass ratio 1.13.	Lockheed entry in U.S. Navy future undergraduate pilot trainer competition
	F107-WR-101	635	0.683	146	4.35	Williams International dual-shaft engine produced under license by CAE. HP spool had single centrifugal compressor stage, annular combustor, and single-stage turbine; LP spool had two-stage axial fan driven by single-stage turbine.	Boeing AGM-86B Air-Launched Cruise Missile (ALCM)
	F107-WR-400/-402	635	0.685	144	4.41	Basically similar to F107-WR-101 with minor accessory variations.	General Dynamics BGM-109 Sea-Launched Cruise Missile
382-10	F408-CA-400	1,008	0.97	145	6.95	Single-shaft engine with one fan stage, mixed-flow compressor, annular slinger combustor, and single-stage axial turbine.	Teledyne Ryan BQM-145 Mid-Range Unmanned Aerial Vehicle

Teledyne CAE (Demonstrator Engines)

Company Model No.	Military Model No.	S.L. Rated Thrust, Lb	SFC at S.L. Rated Thrust	Engine Weight, Lb	Thrust-to-Weight Ratio	Configuration	Applications
ATEGG-365	XL095-T-1	Approx. 4,000		Approx. 200	Approx. 22	1st generation demo. engine with six-stage axial-flow compressor, annular combustor, single-stage turbine, cusped exhaust nozzle; engine designed to operate vertically.	Demonstrated technical feasibility of high thrust-to-weight ratio turbojet lift engine for VTOL fighter aircraft.
ATEGG-440		3,000 to 5,000 class				2nd generation demo. engine with turbofan core having variable geometry multi-stage axial plus single-stage centrifugal compressor on one shaft, annular combustor, and cooled turbine.	Demonstrated high pressure ratio and high turbine temp. technology in core that had potential application in future high perf. trainer and RPV-type aircraft.
ATEGG-555		2,000 class				3rd generation demo. engine with turbofan core having moderate pressure ratio compressor and cooled turbine.	Demonstrated moderate press. ratio and high turbine temp. technology in core, having potential application in future trainer, business, and commuter type aircraft.
JTDE-455		2,000 class				4th generation demo. engine with dual shaft turbofan core using ATEGG-555 as HP spool; LP spool used Navy-sponsored single-stage fan and USAF-sponsored turbine.	Demonstrated low-cost, high performance technology in turbofan having potential applications similar to those of ATEGG-555.
ATEGG-585						Second 4th generation demo. engine with improved turbofan core.	Demonstrated further advances in low-cost, high-performance technology.
ATEGG-589						5th generation demonstrator engine with improved turbofan core.	Demonstrated further advances in low-cost, high-performance technology.

Lycoming Medium Power Engine Family (Significant Dates)

Company Model No.	Military Model No.	Design/ Devel. Start	Engine First Run	Engine First Flight	Engine Cert./Qual.	First Prod. Engine Delivered	Last Prod. Engine Delivered	Total Engines Produced
								Through 1989
Turboshaft engines								
LTC1B	T53-L-1	1952	Jan 1955	Sep 1956	April 1958	Mar 1959	1964	564
LTC1K-1	T53-L-5				Dec 1959	Feb 1960	1965	182
LTC1K-2	T53-L-9				Mar 1961	Jun 1961	1968	609
T5309						1961	1964	29
LTC1K-5	T53-L-11	1959				Aug 1963	1972	4,244
T5311						1963	1972	226
LTC1K-4	T53-L-13	1959	1960		Jun 1966	Aug 1966	1985	10,163
T5313						1968		503
LTC1K-4G	T53-L-703					1969/1978		191
Turboprop engines								
LTC1F-1	T53-L-3			Apr 1959	Jul 1959	Sep 1959	1967	420
LTC1F-2	T53-L-7				Jun 1962	Dec 1962	1968	446
LTC1F-4	T53-L-15				Mar 1967	Sep 1967	1969	245
LTC1F-5	T53-L-701				Feb 1969	Sep 1969	1980	175

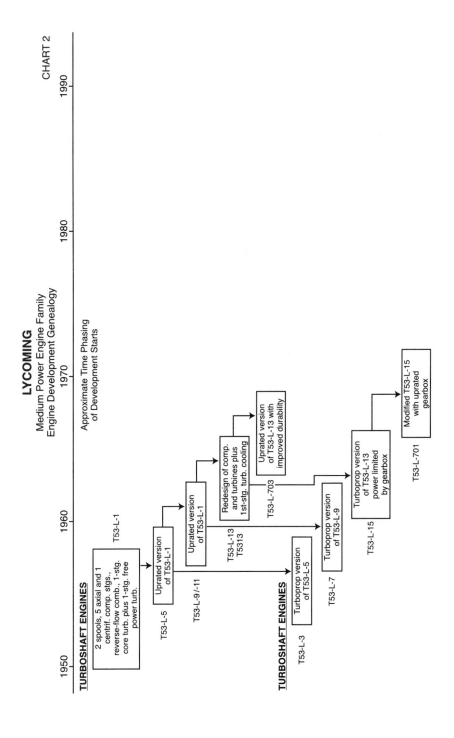

LYCOMING
Medium Power Engine Family
Engine Development Genealogy

CHART 2

Approximate Time Phasing
of Development Starts

1950   1960   1970   1980   1990

TURBOSHAFT ENGINES

2 spools, 5 axial and 1
centrif. comp. stgs.,
reverse-flow comb., 1-stg.
core turb. plus 1-stg. free
power turb.

T53-L-1

Uprated version
of T53-L-1

T53-L-5

Uprated version
of T53-L-1

T53-L-9/-11

Redesign of comp.
and turbines plus
1st-stg. turb. cooling

T53-L-13
T5313

Uprated version
of T53-L-13 with
improved durability

T53-L-703

Turboprop version
of T53-L-13
power limited
by gearbox

T53-L-15

Modified T53-L-15
with uprated
gearbox

T53-L-701

TURBOSHAFT ENGINES

Turboprop version
of T53-L-5

T53-L-3

Turboprop version
of T53-L-9

T53-L-7

837

Lycoming Medium Power Engine Family

Company Model No.	Military Model No.	S.L. Rated Power, SHP	SFC at S.L. Rated Power	Engine Weight, Lb	Power-to-Wt. Ratio	Configuration	Applications
Turboshaft Engines							
LTC1B	T53-L-1	860	0.770	484	1.78	2-spool free turbine engine with 5 axial plus 1 centrifugal compressor stages, reverse-flow annular combustor, 1-stage gas generator turbine and 1-stage free power turbine with geared front power takeoff.	Kaman HOK-1 Kaman H-43B Husky Bell HU-1A Iroquois Doak Model 16 Ryan Model 92 Vertol Models 76, 105, & 107
LTC1K-1	T53-L-5	960	0.694	487	1.97	Uprated version of T53-L-1 with higher turbine temp.	Bell HU-1B Iroquois
LTC1K-2	T53-L-9	1,100	0.682	485	2.27	Uprated versions of T53-L-5 with higher turbine temp.	Bell HU-1B, -1D, -1E Iroquois
LTC1K-5	T53-L-11	1,100	0.682	496	2.22	Commercial versions of T53-L-9 and T53-L-11.	Kaman HH-43F, Bell 204B
T5309		1,100	0.682	485	2.27		Bell Model 204B
T5311		1,100	0.682	496	2.22		Bell Model 204B
LTC1K-4	T53-L-13	1,400	0.580	530	2.64	Uprated version of T53-L-11 with variable compressor IGVs, transonic first 2 comp. stages, redesigned "hot end" with 1st-stage turbine-blade cooling, and addition of 2nd-stage gas generator turbine and 2nd-stage power turbine.	Bell HU-1C, -1H, -1D Iroquois Bell AH-1G HueyCobra

T5313		1,400	0.580	549	2.55	Commercial version of T53-L-13.	Bell Model 205A
LTC1K-4G	T53-L-703	1,485	0.596	545	2.72	Uprated version of T53-L-13 with higher turbine temp. and improved durability.	Bell AH-1Q, -1S HueyCobra
Turboprop Engines							
LTC1F-1	T53-L-3	960	0.686	524	1.83	Turboprop version of T53-L-3.	Grumman OV-1 Mohawk
LTC1F-2	T53-L-7	1,100	0.670	555	1.98	Turboprop version of T53-L-9 with uprated gearbox.	Grumman OV-1 Mohawk
	T53-L-15	1,160	0.620	605	1.92	Turboprop version of T53-L-13 with power rating limited by mechanical limits of gearbox.	Grumman OV-1 Mohawk
	T53-L-701	1,400	0.590	688	2.03	Modified version of T53-L-15 with uprated gearbox.	Grumman OV-1 Mohawk

Lycoming High Power Engine Family (Significant Dates)

Company Model No.	Military Model No.	Design/ Devel. Start	Engine First Run	Engine First Flight	Engine Cert./Qual.	First Prod. Engine Delivered	Last Prod. Engine Delivered	Total Engines Produced *Through 1989*
Turboprop Engine								
LTC4A-1	T55-L-1	Apr 1954	Feb 1955		Dec 1957			Prototype, test engines
Turboshaft Engines								
LTC4B-2		1956	Dec 1957		Mar 1959			Prototype, test engines
LTC4B-7	T55-L-3		Apr 1959		Feb 1960	Feb 1961	1963	146
LTC4B-8	T55-L-5			Oct 1961	Sep 1962	Jul 1963	1968	1,471
LTC4B-8D	T55-L-7					1973	1978	433
T5508D						1974	1982	88
LTC4B-11B	T55-L-11	1965			Jul 1968	Aug 1968	1984	1,009
	T55-L-712	1975				1978		849 thru 1989
	T55-L-714					1989		2
LTC4V-1		1965	Dec 1968					
Turbofan Engines								
PLF1A-2		1962	Feb 1964					
ALF502A	YF102-LD-100	1970	Jun 1971	May 1972	Mar 1972	1972	1972	6
ALF502D		1971		May 1973		1972	1978	11
ALF502L-2, L-2A, L-3		1977		Nov 1978	Feb 1980	1978	1985	189
ALF502R-3		1978		Sep 1981	Jan 1981	1980		819 ALF502Rs thru 1989
ALF502R-5, R-3A					1983			
LF507-1H				May 1991	Oct 1991	Oct 1991		48 thru Aug 1996
LF507-1F					Mar 1992	Mar 1992		270 thru Aug 1996

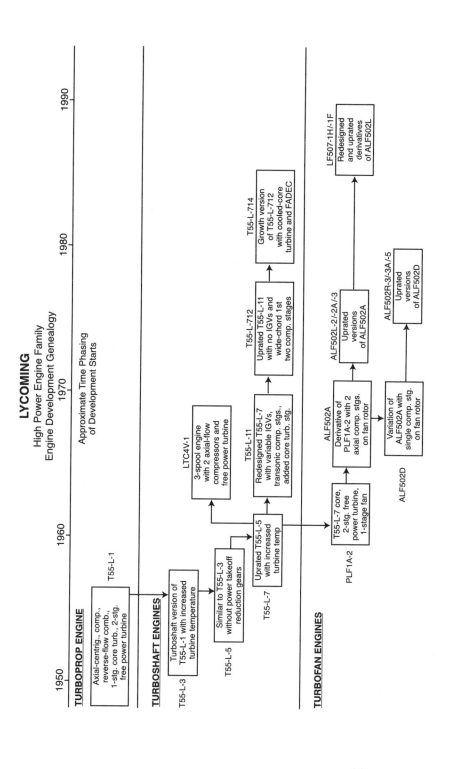

LYCOMING
High Power Engine Family
Engine Development Genealogy

Approximate Time Phasing
of Development Starts

1950   1960   1970   1980   1990

TURBOPROP ENGINE

T55-L-1
Axial-centrig., comp.,
reverse-flow comb.,
1-stg. core turb., 2-stg.
free power turbine

TURBOSHAFT ENGINES

T55-L-3
Turboshaft version of
T55-L-1 with increased
turbine temperature

T55-L-5
Similar to T55-L-3
without power takeoff
reduction gears

T55-L-7
Uprated T55-L-5
with increased
turbine temp

LTC4V-1
3-spool engine
with 2 axial-flow
compressors and
free power turbine

T55-L-11
Redesigned T55-L-7
with variable IGVs,
transonic comp. stgs.,
added core turb. stg.

T55-L-712
Uprated T55-L-11
with no IGVs and
wide-chord 1st
two comp. stages

T55-L-714
Growth version
of T55-L-712
with cooled-core
turbine and FADEC

TURBOFAN ENGINES

PLF1A-2
T55-L-7 core,
2-stg. free
power turbine,
1-stage fan

ALF502A
Derivative of
PLF1A-2 with 2
axial comp. stgs.
on fan rotor

ALF502D
Variation of
ALF502A with
single comp. stg.
on fan rotor

ALF502L-2/-2A/-3
Uprated
versions
of ALF502A

ALF502R-3/-3A/-5
Uprated
versions
of ALF502D

LF507-1H/-1F
Redesigned
and uprated
derivatives
of ALF502L

841

Lycoming High Power Engine Family

Company Model No.	Military Model No.	S.L. Rated Power, SHP	SFC at S.L. Rated Power	Engine Weight, Lb	Power-to-Weight Ratio	Configuration	Applications
Turboprop Engine							
LTC4A-1	T55-L-1	1,500	0.720	695	2.16	T53-type config. Two spools with 7 axial stages followed by one centrifugal-flow stage, annular reverse-flow combustor, 1-stage core turbine, 2-stage free power turbine with front power takeoff.	Developed for U.S. Army de Havilland Buffalo transport North American YAT-28 trainer prototype Experimental North American P-51 Mustang
Turboshaft Engines							
LTC4B-1	T55-L-3	1,900	0.670	600	3.17	Uprated turboshaft version of T55-L-1 with increased turbine temperature and power shaft reduction gear.	None
LTC4B-7	T55-L-5	2,200	0.622	570	3.86	Improved version of T55-L-3 with higher temperature turbine materials and elimination of power shaft reduction gearing.	Boeing Vertol CH-47A Chinook Curtiss-Wright X-19A VTOL aircraft
LTC4B-8	T55-L-7	2,650	0.615	580	4.57	Uprated version of T55-L-5 with increased turbine temperature.	Boeing Vertol CH-47B Chinook
LTC4B-8D		2,250 (flat rated)	0.630	605	3.72	Modified version of T55-L-7.	Bell 214A
T5508D		2,250 (flat rated)	0.630	605	3.72	Commercial version of T55-L-7.	Bell 214B

LTC4B-11B	T55-L-11	3,750	0.530	670	5.6	Uprated and redesigned derivative of T55-L-7 with second stage added to core turbine, variable IGVs, and 1st three axial comp. stages transonic.	Boeing Vertol CH-47C Chinook
	T55-L-712	3,750	0.530	750	5.33	Mechanically and aerodynamically redesigned version of T55-L-11 with wide-chord blades in 1st two axial comp. stages and no IGVs.	Boeing Vertol CH-47D Chinook Boeing Model 234
	T55-L-714	4,867 shp	0.503	800	6.08	Growth version of T55-L-712 with cooled-core turbine blades and FADEC.	Boeing MH-47E
LTC4V-1		5,000 shp	0.420	590	8.47	3-spool engine with 4 axial comp. stages on LP spool, 9 axial stages on HP spool, and free power turbine with front power takeoff. Titanium engine with high press. ratio (16:1) and high turb. temp. for its time (2,200 deg F).	Developed for Boeing Vertol heavy lift helicopter planned by U.S. Army
Turbofan engines							
PLF1A-2		4,320 lb	0.411	825	5.2	Used T55-L-7 gas generator with 2-stg. free power turbine geared to 1-stg. fan having 6.2:1 bypass ratio.	Experimental research and development engine
ALF502A		7,200 lb	0.420	1,175	6.1	Derivative of PLF1A using T55 core within 2-stg. cooled turbine, new fan, and 2 axial-flow comp stages mounted on fan shaft; 6.1:1 bypass ratio.	Northrop A-9A prototype NASA Quiet Short-Haul Research Aircraft
ALF502D		6,500 lb	0.417	1,245	5.2	Derated commercial version of ALF502D with single axial booster comp. stage on fan rotor; 6.1:1 bypass ratio.	Dassault Falcon 30 prototype Canadair Challenger 600 prototype

Lycoming High Power Engine Family (*Continued*)

Company Model No.	Military Model No.	S.L. Rated Power, Lb	SFC at S.L. Rated Thrust	Engine Weight, Lb	Power-to-Weight Ratio	Configuration	Applications
ALF502L/L-2/		7,500	0.428	1,311	5.78	Uprated versions of ALF502D with variable IGVs removed, wide-chord blades in 1st two HP compressor stages, and modified 3rd stage power turbine; 5.0:1 bypass ratio.	Canadair Challenger 600
L-2A		7,500	0.414	1,311			
L-3		7,500	0.415	1,311			
ALF502R-3/		6,700	0.411	1,336	5.01	Reduced rating versions of ALF502L-2 with 1st stage of LP compressor removed, and modified accessory gearbox; 5.7:1 bypass ratio.	BAe 146-100, -200, -300
R-3A		6,970	0.408	1,336	5.22		
R-5		6,970	0.408	1,336	5.22		
LF507-1H		7,000 flat rated	0.406	1,375	5.09	Redesigned derivative of ALF502 Series engines with 1-stage fan, 2-stage axial compressor on fan rotor, 2-stage LP turbine, 500 Series common core having 7-stage axial compressor and 1 centrifugal stage driven by 2-stage turbine, and hydromechanical control; 5.0:1 bypass ratio.	BAe 146-300
LF507-1F		7,000 flat rated	0.406	1,385	5.05	Similar to LF507-1H with Full Authority Digital Electronic Control (FADEC).	BAe 146-100, -200, -300 BAe RJ70

Lycoming Low Power Engine Family (Significant Dates)

Company Model No.	Military Model No.	Design/ Devel. Start	Engine First Run	Engine First Flight	Engine Cert./Qual.	First Prod. Engine Delivered	Last Prod. Engine Delivered	Total Engines Produced
Demonstrator Engines								
PLT32		1968						
PLT34		Dec 1971	Dec 1973					
PLT34A		1977	Nov 1979					
PLT210		1987	Jul 1991					
Turboshaft Engines								
APW34	T800-APW-800	1985	July 1986		Jul 1988 60-hr PFRT			11 Devel. engines
600 Series								
LTS101-600A,B	YT702-LD-700 (Model No. for all LT101s)	1972	Jun 1972	May 1973	1975	1973		2,000 LTS101s thru 1995
LTS101-600A -2, -3								
650 Series								
LTS101-65-B-1								
LTS101-650C-2, C-3, C-3A								

Lycoming Low Power Engine Family (Significant Dates) *(Continued)*

Company Model No.	Military Model No.	Design/Devel. Start	Engine First Run	Engine First Flight	Engine Cert./Qual.	First Prod. Engine Delivered	Last Prod. Engine Delivered	Total Engines Produced
750 Series								
LTS101-750A-1		1974						
LTS101-750B-1								
LTS101-750B-2								
LTS101-750C-1								
Turboprop Engines								
600 Series								
LTP101-600A-A1	Y702-LD-700 (Model No. for all LT101s)	1974			1976	1976		154 LTP101s thru 1989
700 Series								
LTP101-700A-1A					1980			
Turbofan Engine								
ALF101		1976						

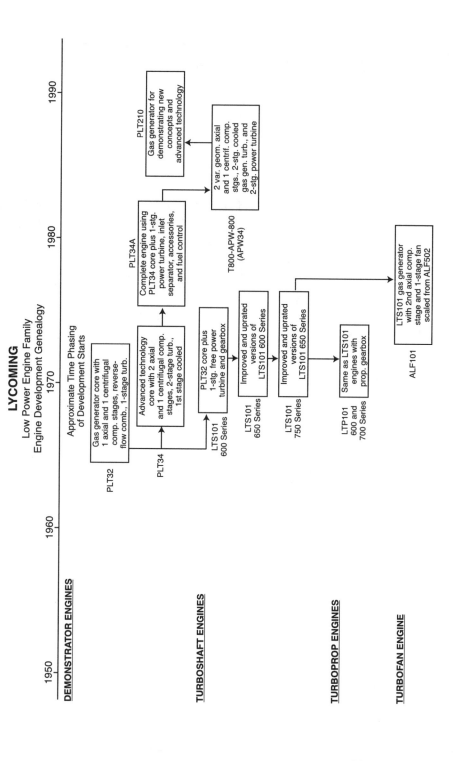

LYCOMING
Low Power Engine Family
Engine Development Genealogy

Approximate Time Phasing
of Development Starts

1950   1960   1970   1980   1990

DEMONSTRATOR ENGINES

PLT32
Gas generator core with 1 axial and 1 centrifugal comp. stages, reverse-flow comb., 1-stage turb.

PLT34
Advanced technology core with 2 axial and 1 centrifugal comp. stages, 2-stage turb., 1st stage cooled

PLT34A
Complete engine using PLT34 core plus 1-stg. power turbine, inlet separator, accessories, and fuel control

PLT210
Gas generator for demonstrating new concepts and advanced technology

T800-APW-800
(APW34)
2 var. geom. axial and 1 centrif. comp. stgs., 2-stg. cooled gas gen. turb., and 2-stg. power turbine

TURBOSHAFT ENGINES

LTS101
600 Series
PLT32 core plus 1-stg. free power turbine and gearbox

LTS101
650 Series
Improved and uprated versions of LTS101 600 Series

LTS101
750 Series
Improved and uprated versions of LTS101 650 Series

TURBOPROP ENGINES

LTP101
600 and 700 Series
Same as LTS101 engines with prop. gearbox

TURBOFAN ENGINE

ALF101
LTS101 gas generator with 2nd axial comp. stage and 1-stage fan scaled from ALF502

847

Lycoming Low Power Engine Family

Company Model No.	Military Model No.	S.L. Rated Power, SHP	SFC at S.L. Rated Power	Engine Weight, Lb	Power-to-Weight Ratio	Configuration	Applications
Demonstrator Engines							
PLT32		600 class				Advanced technology gas generator core having 1 axial and 1 centrifugal compressor stages, reverse-flow annular combustor, and single-stage turbine.	Led to development of LTS101, LTP101, and ALF101 engines
PLT34	STAGG	825				Gas generator core with 2 axial and 1 centrifugal compressor stages, reverse-flow annular combustor and 2-stage turbine with cooled first stage.	Demonstrated advanced gas generator technology for small turbine engines
PLT34A	ATDE	825	0.476	225	3.67	Complete engine using PLT34 core with 1-stage power turbine, inlet particle separator, accessories, and fuel control.	Demonstrated advanced technology in complete engine
PLT210		2,500 to 3,000				Gas generator for demonstrating new technology and concepts through the 1990s; initial configuration had axial-centrifugal compressor, single-stage high-work turbine, and free power turbine with front power drive.	Demonstrated technology applicable to all future Lycoming engines

Turboshaft Engines

APW34	T800-APW-800	1,200	0.465	298	4.03	Derived from PLT34A demo engine design; 2 variable geometry axial and 1 centrifugal compressor stages, reverse-flow annular combustor, 2-stage cooled gas generator turbine, and 2-stage free power turbine geared to front power takeoff.	Developed thru PFRT in U.S. Army engine competition for the planned LHX helicopter fleet
600 Series							
LTS101-600A/B	YT702-LD-700 (Model Number for All LT101s)	592	0.570	241	2.46	LTS series engines had only 4 rotating stages, 1 axial followed by 1 centrifugal compressor stage, 1 gas generator turbine stage, and 1 free power turbine stage; reverse-flow annular combustor.	Bell JetRanger testbed Aerospatiale AS350 AStar, SA366 (HH-65A) Dolphin Bell 222 MBB/Kawasaki BK117
600A-2		615	0.571	253	2.43		
650 Series							
LTS101-650B-1		550 flat rated	0.577	273	2.01	Improved and uprated versions of LTS101-600 Series engines.	MBB/Kawasaki BK117A, BK117B Bell 222
LTS101-650C-2/ C-3/C-3A		630	0.572	241	2.61		
750 Series							
LTS101-750A-1		618	0.570	268	2.31	Improved and uprated versions of LTS101-650 Series engines.	Aerospatiale SA366 (HH-65A) Bell 222B and 222UT MBB/Kawasaki BK117B
LTS101-750B-1		550 flat rated	0.577	297	1.85		
LTS101-750B-2		690	0.570	271	2.55		Aerospatiale SA366 (HH-65A) Bell 222, 222UT
LTS101-750C-1		684	0.577	276	2.49		

Lycoming Low Power Engine Family (Continued)

Company Model No.	Military Model No.	S.L. Rated Power, SHP	SFC at S.L. Rated Power	Engine Weight, Lb	Power-to-Weight Ratio	Configuration	Applications
Turboprop Engines							
LTP101-600A-1A	YT702-LD-700 (Model number for all LT101s)	620 eshp	0.544	325	1.85	Same gas generator and power turbine configuration as LTS101 series with addition of propeller gearbox with provisions for pusher or tractor installation, radial screened inlet, and anti-icing protection.	Piaggio P. 166-DL-3 Britten-Norman Islander Snow Air Tractor Riley Cessna 421 Dornier Turbo-Skyservant New Zealand Aerospace Cresco Fletcher Page Ag-Cat and Turbo Thrush Snow Thrush
LTP101-700A-1A		700 eshp	0.544	325	2.15	Uprated version of LTP101-600 series with redesigned gearbox.	Same applications as LTP101-600A-1A
Turbofan engine							
ALF101		1,620 lb	0.360	343	4.7	LT101 gas generator with additional axial compressor stage and free power turbine driving single-stage fan scaled from ALF502 fan; bypass ratio 8.5:1.	Used to obtain noise and exhaust emission data for NASA's Quiet Clean General Aviation Turbofan Program

Lycoming Vehicular Engine Family (Significant Dates)

Company Model No.	Military Model No.	Design/Devel. Start	Engine First Run	Engine First Flight	Engine Cert./Qual.	First Prod. Engine Delivered	Last Prod. Engine Delivered	Total Engine Produced
AGT1500		1963	Nov 1966			Nov 1979		7,677 thru 1989
TF15		1965						
PLT27		1965	Dec 1969					

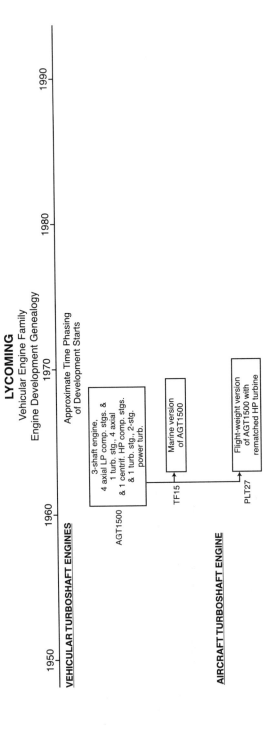

LYCOMING

Vehicular Engine Family
Engine Development Genealogy

Approximate Time Phasing
of Development Starts

VEHICULAR TURBOSHAFT ENGINES

AGT1500 — 3-shaft engine, 4 axial LP comp. stgs. & 1 turb. stg., 4 axial & 1 centrif. HP comp. stgs. & 1 turb. stg., 2-stg. power turb.

TF15 — Marine version of AGT1500

AIRCRAFT TURBOSHAFT ENGINE

PLT27 — Flight-weight version of AGT1500 with rematched HP turbine

1950   1960   1970   1980   1990

Lycoming Vehicular Engine Family

Company Model No.	Military Model No.	S.L. Rated Power, SHP	SFC at S.L. Rated Power	Engine Weight, Lb	Power-to-Weight Ratio	Configuration	Applications
Vehicular Turboshaft Engines							
AGT1500		1,500				3-shaft engine with contrarotating LP and HP shafts and free power turbine shaft with rear power takeoff. LP 2-stage turbine and rear power takeoff. LP rotor had 4-stage axial compressor and 1-stage turbine. HP rotor had 4-stage axial compressor plus one centrifugal stage and 1-stage turbine. Engine had single can-type combustor and a recuperator.	U.S. Army Abrams M1 Main Battle Tank
TF15		1,500				Marine version of AGT1500.	
Aircraft Turboshaft Engines							
PLT27-A2		2,050	0.450	366	5.6	Flight version of AGT1500 using AGT1500 gas turbine core and power turbine with wrap around reverse-flow annular combustor. Both front and rear power takeoff versions developed.	Proposals submitted for use of engine in: —Utility Tactical Transport Aircraft System (UTTAS) helicopter —Apache advanced attack helicopter —LAMPS III anti-submarine warfare helicopter

General Electric (B1 Significant Dates)

Company Model No.	Military Model No.	Design/ Devel. Start	Engine First Run	Engine First Flight	Engine Cert./Qual.	First Prod. Engine Delivered	Last Prod. Engine Delivered	Total Engines Produced
B-1		Spring 1944		Dec. 7, 1944				Prototypes only

General Electric (B1 turbojet engine)

Company Model No.	Military Model No.	S.L. Rated Thrust, Lb	SFC at S.L. Rated Thrust	Engine Wt., Lb	Thrust-to- Wt. Ratio	Configuration	Applications
B-1		400				Turbojet developed from converted GE Type B-31 turbosupercharger with redesigned turbine and the addition of a combustor.	Northrop JB -1A Power Bomb ground-to-ground missile

General Electric (T58 turboshaft Significant Dates)

Company Model No.	Military Model No.	Design/ Devel. Start	Engine First Run	Engine First Flight	Engine Cert./Qual.	First Prod. Engine Delivered	Last Prod. Engine Delivered	Total Engines Produced
	T58-GE-1			Jun 1963	Sep 1959 PFRT Jul 1960 MQT	Jan 1963	Dec 1965	134
	T58-GE-2, –2A	Jul 1953	Apr 1955	Jan 30, 1957	Aug 1956 PFRT Sep 1957 MQT	Oct 1956	Aug 1958	28
	T58-GE-3			Feb 20, 1964	Nov 1963 MQT	Sep 1963	Jul 1964	273
	T58-GE-5			Jun 1963	Mar 1965 MQT	Jan 1966	Mar 1980	607
	T58-GE-6	Aug 1957	Mar 1958	May 14, 1958	Aug 1958 MQT	Jul 1958	Sep 1960	161
	T58-GE-8, –8A, –8B, –8D	Aug 1958 (–8 model)	Nov 1958 (–8 model)	Mar 11, 1959 (–8 model)	Aug 1959 PFRT Jul 1960 MQT (–8 model)	Nov 1960 (–8 model)	Jun 1968	1,803 Total for –8 thru –8C
	T58-GE-8E, –8F							
	T58-GE-10	Mar 1961	May 1962	Nov 1963	Nov 1965 MQT	1966	1984	1,507
	T58-GE-14	ca. 1967						
	T58-GE-16				Oct 1968 PFRT Mar 1970 MQT	Jun 1974	Mar 1984	742
	T58-GE-100					Jul 1976	Dec 1983	123
	T58-GE-400					Jul 1974	Aug 1975	44

General Electric (CT58 turboshaft Significant Dates)

Company Model No.	Military Model No.	Design/ Devel. Start	Engine First Run	Engine First Flight	Engine FAA Certificated	First Prod. Engine Delivered	Last Prod. Engine Delivered	Total Engines Produced
CT58-GE-100, -100-1, -100-2			Apr 1955	May 14, 1958	Jul 1959 (-100) Dec 1960 (-100-2)	1959	1961	71 Total for -100 series
CT58-GE-110, -110-1, -110-2				Mar 11, 1959	Jul 1961 (-110) Mar 1968 (-110-2)	1961	1967	163 Total for -110 series
CT58-GE-140, -140-1, -140-2					Jun 1965 (-140) Oct 1965 (-140-1) Apr 1968 (-140-2)	1965	1984	352 Total for -140 series

GENERAL ELECTRIC
T58/CT58 Engine Development Genealogy

Approximate Time Phasing
of Development Starts

MILITARY TURBOSHAFT ENGINES

T58-GE-2, -6
Free turbine engine
with axial-flow comp.
2-stg. core turbine,
1-stg. power turbine

T58-GE-8 series
T58-GE-6 with
increased turbine
temperature

T58-GE-10
Increased air-
flow and higher
turbine temp.

T58-GE-14
2nd power turbine
stage added, higher
turbine temperature

T58-GE-16
2-stage power turbine,
cooled HP turbine,
higher turbine temp.

T58-GE-1
Derivative of
T-58-GE-8 series

T58-GE-3
Derivative of
T-58-GE-8 series

T58-GE-5
Derivative of
T58-GE-10

T58-GE-400
Improved hot section
materials, higher HP
turbine temperature

T58-GE-100
Similar to
T58-GE-400

T58-GE-402
Uprated
conversions of
T58-GE-10 engines

COMMERCIAL TURBOSHAFT ENGINES

CT58-GE-100 series
Commercial version
of T58-GE-2, -6

CT58-GE-110 series
Commercial version
of T58-GE-8 series

CT58-GE-140 series
Commercial version
of T58-GE-10

1950   1960   1970   1980   1990

General Electric (T58/CT58 turboshaft engines)

Company Model No.	Military Model No.	S.L. Rated Power, SHP	SFC at S.L. Rated Power	Engine Weight, Lb	Power-to-Weight Ratio	Configuration	Applications
T58 Engines							
	T58-GE-1	1,300	0.61	315	4.12	Derivative of T58-GE-8 engine series.	Sikorsky CH-3B (S-61-A) CH-3C (S-61R), and HH-3C
	T58-GE-2	1,024	0.66	325	3.15	Parent T58 engine model. Free turbine turboshaft with 10-stage axial-flow compressor with IGVs and first three stator stages variable, annular combustor, two-stage core turbine, single-stage power turbine.	Sikorsky HSS-1F Vertol H-21D
	T58-GE-3	1,325	0.60	315	4.2	Similar to T58-GE-8 engine series. Derived from T58-GE-1.	Agusta-Bell AB-204B Bell HX-1 (later UH-1F)
	T58-GE-5	1,400	0.61	335	4.18	Derivative of T58-GE-10 engine.	SikorskyHH-3E, CH-3E, HH-3F, and S-67
	T58-GE-6	1,050	0.64	267	3.93	Similar to T58-GE-2 with added installation flexibility, optional main reduction gear, 3-position exhaust nozzle, and torque sensor.	Sikorsky S-62 Kaman HU2K-1 and UH-2A
	T58-GE-8, -8B, -8C, -8D	1,250	0.61	305	4.10	First growth step in T58 series; derived from T58-GE-2 and -6 with moderate increase in turbine temperature.	Piasecki IGH-1A, Kaman HU2K-1 Sikorsky HSS-2 (SH-3A and CHSS-2), S-62 (HH-52A), HSS-2Z (later VH-3A) Boeing Vertol HRB-1 (CH-46A), V107 (CH-113/A) Bell Aerospace X-22A VTOL

Model				Description	Applications
T58-GE-8E, -F	1,350	0.60	315/305	4.3/4.4	Uprated versions of T58-GE-8D	Agusta/Sikorsky ASH-3; Sikorsky VH/HH-3A, VH/SH-3H; Kaman UH-2C, SH-2F, UH-2A
T58-GE-10	1,400	0.61	350	4.0	Second growth step; derivative of T58-8 series with airflow increased 8 percent, power turbine temperature raised 20 degrees F., new and improved fuel control.	Sikorsky SH-3D-H; Boeing Vertol UH/CH/HH-46D; NASA-Sikorsky S-72 (RSRA)
T58-GE-14					Third growth step; non-production engine introducing two-stage power turbine and higher turbine temperature.	
T58-GE-16	1,870	0.53	443	4.22	Fourth growth step; derivative of T58-GE-10 with cooled gas generator turbine blades, second power turbine stage added, and higher turbine temperature.	Boeing Vertol CH-46E
T58-GE-100	1,500	0.606	335	4.48	Uprated T58-GE-5 similar in design to T58-GE-400.	Sikorsky CH/HH-3E and HH-3F
T58-GE-400	1,500	0.606	335	4.48	Uprated T58-GE-5 with improved hot section materials and higher turbine temperature.	Sikorsky VH-3D and S-67
T58-GE-402	1,500	0.606	335	4.48	Uprated version of T58-GE-10 engine.	Sikorsky SH-3D/G/H
CT58 Engines						
CT58-GE-10, -100-1, -100-2	1,050	0.64	280	3.75	Commercial version of T58-GE-2 and -6.	Sikorsky S-62
CT58-GE-110, -110-1, -110-2	1,250	0.64	315	3.97	Commercial versions of T58-GE-8 engine models.	Sikorsky S-61, S-61L, S-61N, S-62; Boeing Vertol 107-II, V107, KV107
CT58-GE-140, -140-1, -140-2	1,350	0.61	315	4.28	Commercial versions of T58-GE-10 engine model.	Sikorsky S-61; Kawasaki KV-107; Boeing Vertol 107-II
	1,500	0.61	340	4.41		

General Electric (J85 turbojet engines Significant Dates)

Company Model No.	Military Model No.	Design/ Devel. Start	Engine First Run	Engine First Flight	Engine Cert./Qual.	First Prod. Engine Delivered	Last Prod. Engine Delivered	Total Engines Produced
	J85-GE-1	Dec 1954	Jan 1956	Apr 10, 1959	Aug 1958 PFRT	Aug 1958	Jul 1959	12
	J85-GE-2					1966	1966	21
CJ610-6	J85-GE-3	Apr 1956	Dec 1956	Aug 1958	Nov 1958 15-HRQT	Jan 1959	Apr 1960	105
	J85-GE-4, -4A, -4B	Jul 1956	Jan 1959	Sep 1959	Jun 1966 CERT	1968	1976	740
	J85-GE-5, -5A, -5B				Oct 1959 PFRT / Jul 1960 MQT	Dec 1960	1971	2,825
	J85-GE-7	Jun 1957	Jun 1958	Mar 1959	Jan 1959	Nov 1959	1962	577
	J85-GE-13, -13A	Mar 1961	Jun 1962	Jul 1963	Aug 1963 MQT	Jun 1962	1976	2,156
	J85-GE-15	Feb 1964	Jan 1965		Apr 1965 PFRT / May 1966 MQT	1965	1965	4
	J85-GE-17, -17A, -17B	Sep 1966		Oct 22, 1963	Mar 1966 MQT	1966	1978	2,058
	J85-GE-19		Oct 1966	Sep 28, 1968	Dec 1966 PFRT	1967	1967	12
	J85-GE-21, -21A	Feb 1962	Dec 1965	Mar 28, 1965	Nov 1967 PFRT / Feb 1972 MQT	May 1972	1988	3,483
	J85/J2			Sep 1963		1963	1964	8
	J85/J4			May 1964		1966	1967	30
	J85/LF1 (X353-5)	May 1957						

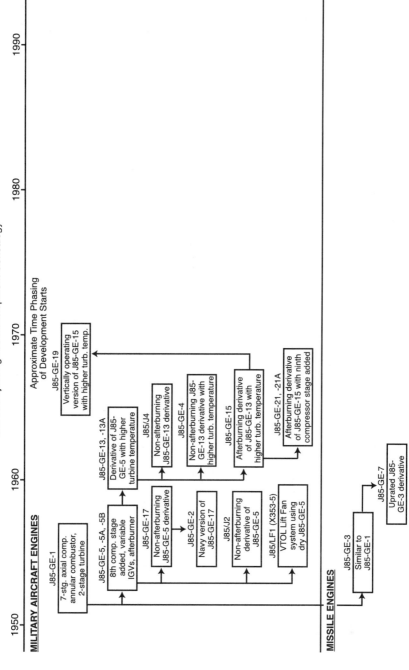

GENERAL ELECTRIC

J85 Turbojet Engines Development Genealogy

Approximate Time Phasing
of Development Starts

MILITARY AIRCRAFT ENGINES

J85-GE-1
7-stg. axial comp.
annular combustor,
2-stage turbine

J85-GE-5, -5A, -5B
8th comp. stage
added, variable
IGVs, afterburner

J85-GE-13, -13A
Derivative of J85-
GE-5 with higher
turbine temperature

J85-GE-19
Vertically operating
version of J85-GE-15
with higher turb. temp.

J85-GE-17
Non-afterburning
J85-GE-5 derivative

J85/J4
Non-afterburning
J85-GE-13 derivative

J85-GE-2
Navy version of
J85-GE-17

J85-GE-4
Non-afterburning J85-
GE-13 derivative with
higher turb. temperature

J85/J2
Non-afterburning
derivative of
J85-GE-5

J85-GE-15
Afterburning derivative
of J85-GE-13 with
higher turb. temperature

J85/LF1 (X353-5)
VTOL Lift Fan
system using
dry J85-GE-5

J85-GE-21, -21A
Afterburning derivative
of J85-GE-15 with ninth
compressor stage added

MISSILE ENGINES

J85-GE-3
Similar to
J85-GE-1

J85-GE-7
Uprated J85-
GE-3 derivative

General Electric (J85 turbojet engines)

Company Model No.	Military Model No.	S.L. Rated Thrust, Lb	SFC at S.L. Rated Thrust	Engine Weight, Lb	Thrust-to-Weight Ratio	Configuration	Applications
	J85-GE-1	2,180	1.09	285	7.65	Single-spool turbojet engine with 7-stage axial-flow compressor, annular combustor, and 2-stg. turb.	Northrop YT-38 and N-156F prototypes
	J85-GE-2	2,850		396	7.20	Navy engine similar to J85-GE-17.	Martin SP-5B (boost)
	J85-GE-3	2,250	1.12	259	8.69	Interim missile engine similar to J85-GE-1.	McDonnell GAM-72 / North American T-39 prototype
CJ610-6	J85-GE-4, -4A, -4B	2,950	0.98	404	7.30	Non-afterburning derivative of J85-GE-13 with increased turbine temperature.	Rockwell T-2C and OV-10 BZ (boost) / Teledyne Ryan MQM-34D
	J85-GE-5	2,450 dry / 3,600 A/B	1.05 / 2.25	525	4.67 / 6.68	Derivative of J85-GE-1 with 8th compressor stage added, variable inlet guide vanes, and afterburner.	Northrop YT-38, T-38A, N-156F, and Q-4B
	-5A, 5-B	2,500 dry / 3,850 A/B	1.03 / 2.2	570	4.39 / 6.75		Lockheed VTOL simulator / Bell/NASA X-14A / deHavilland Otter (boost) / Kaman UH-2A/C (boost) / GE/Ryan XV-5A VTOL
	J85-GE-7	2,450	0.975	326	7.52	Production missile engine, similar to J85-GE-5 without afterburner.	McDonnell GAM-72

Model	Thrust (lb)				Description	Applications
J85-GE-13, -13A	2,720 dry / 4,080 A/B	1.03 / 2.2	587	4.63 / 6.95	Afterburning engine derived from J85-GE-5 with higher turbine temperature.	Northrop YF-5A, F-5A, F-5B, SF-5, and N-156 / Fiat G.91Y
CJ610-8/9 J85-GE-15	2,925 dry / 4,300 A/B	1.03 / 2.18	615	4.76 / 6.99	Afterburning derivative of J85-GE-13 with improved turbine.	Northrop F-5, CF-5C/D, NF-5C/D
J85-GE-17, -17A, -17B	2,850	0.99	400	7.12	Non-afterburning derivative of J85-GE-5.	Fairchild AC-119K(boost) / C-123H/K(boost) / Cessna A-37A/B / Saab 105
J85-GE-19	3,015	1.00	387	7.79	Version of J85 modified to operate in a vertical attitude.	Lockheed XV-4B / Hummingbird II / Bell/NASA X-14B
J85-GE-21, -21A	3,500 dry / 5,000 A/B	1.00 / 2.13	684	5.12 / 7.31	Afterburning engine derived from J85-GE-15, airflow increased by adding zero compressor stage with variable stators in first three stages.	Northrop YF-5B-21, F-5E, RF-5E, F-5F / Rockwell International HIMAT
J85/J2	2,850	0.99	385	7.40	Non-afterburning version of J85-GE-5.	Cessna YAT-37D, T-37
J85/J4	2,950	1.01	385	7.66	Non-afterburning version of J85-GE-13.	Canadair CL-41G
J85/LF1 (X353-5)	7,430 lift thrust / 2,580 horiz. thrust	0.34 / 0.98	1,145	6.49 / 2.25	Convertible lift fan system comprised of lift fan, diverter valve, and J85-GE-5 engine.	Ryan XV-5A

General Electric (CJ610 turbojet and CF700 turbofan Significant Dates)

Company Model No.	Military Model No.	Design/ Devel. Start	Engine First Run	Engine First Flight	Engine Cert./Qual.	First Prod. Engine Delivered	Last Prod. Engine Delivered	Total Engines Produced
CJ610								
CJ610-1		1959	Apr 1960	Jan 27, 1963	Dec 1961 FAA	May 1962	1966	391
CJ610-4				Jul 31, 1964	May 1964 FAA	1965	1966	316
CJ610-5					Jun 1966 FAA	1966	1970	112
CJ610-6	J85-GE-4				Jun 1966 FAA	1966	1976	808
CJ610-8	J85-GE-15				Jan 1968 FAA			
CJ610-8A					Apr 1977 FAA	1975	1982	374
CJ610-9	J85-GE-15			Sep 28, 1970	Jan 1968 FAA	1968	1979	58
CF700								
CF700, 700-1		Nov 1959	May 1960					
CF700-2B				Jul 10, 1964	May 1964 MQT Jul 1964 FAA	Mar 1964	1965	28
CF700-2C					Jul 1965 FAA	1965	1977	406
CF700-2D					Jan 1968 FAA	1968	1974	144
CF700-2D2				Oct 1973	Oct 1969 FAA	1969	1981	583
CF700-2D2 (improved)					May 1975 FAA		1981	
CF700-2C(V)	TF37-GE-1	1963		Oct 30, 1964	May 1965 MQT	1967	1967	4

GENERAL ELECTRIC

CJ610 and CF700 Engine Development Genealogy

Approximate Time Phasing
of Development Starts

TURBOJET ENGINES

CJ610-1
Commercial derivative of
J85-GE-5 turbojet

CJ610-4
Derivative of J85-GE-5
with reduced engine weight

CJ610-8, -8A, -9
Non-afterburning
derivatives of
J85-GE-15

CJ610-6
Derivative of CJ610-4
with increased thrust

CJ610-5
Derivative of CJ610-1
with increased thrust

TURBOFAN ENGINES

CF700, 700-1
Aft fan engine with
CJ610 gas generator

CF700-2C (V)
Derived from J85/J2,
modified to operate
in vertical attitude

CF700-2B
Slightly modified and
improved version of CF700

CF700-2C
CF700-2B flat rated for
additional hot-day performance

CF700-2D
CF700-2C with redesigned
core turbine

CF700-2D2
Uprated C700-2D
with new tailpipe

CF700-2D2 Improved
Uprated version
of CF700-2D2

1950   1960   1970   1980   1990

General Electric (CJ610 turbojet engines)

Company Model No.	Military Model No.	S.L. Rated Thrust, Lb	SFC at S.L. Rated Thrust	Engine Weight, Lb	Thrust-to-Weight Ratio	Configuration	Applications
CJ610-1		2,850	0.99	399	7.14	Single-spool turbojet with 8-stage axial-flow compressor, annular combustor, and 2-stage turbine; derivative of J85-GE-5 turbojet.	Fairchild C-123H (boost); Jet Commander 1121; Learjet Model 23; Hansa 320; Canadair CT-114 and CL-41A, R
CJ610-4		2,850	0.99	389	7.33	Similar to CJ610-1 with different gearbox location and reduced engine weight.	Learjet Model 24; Piaggio-Douglas PD-808 Vespa Jet
CJ610-5		2,950	0.98	402	7.34	Uprated derivative of CJ610-1 with higher turbine temperature.	Hansa 320; Commodore Jet
CJ610-6	J85-GE-4	2,950	0.98	396	7.45	Uprated derivative of CJ610-4.	Learjet 24B, C, D and 25, 25B, 25C
CJ610-8	J85-GE-15	3,100	0.98	407	7.62	Non-afterburning version of J85-GE-15 turbojet.	Learjet Century III Models 24E, F and 25D, F, G
CJ610-8A		2,950	0.97	411	7.18	Designed for operation up to 51,000 feet altitude; longer life turbine and turbine nozzle area change.	Learjet Longhorn Models 28 and 29
CJ610-9	J85-GE-15	3,100	0.98	421	7.36	Non-afterburning version of J85-GE-15 turbojet.	Commodore Jet 1123; Hansa 320

General Electric (CF700 turbofan engines)

Company Model No.	Military Model No.	S.L. Rated Thrust, Lb	SFC at S.L. Rated Thrust	Engine Weight, Lb	Thrust-to-Weight Ratio	Configuration	Applications
CF700,700-1		4,000	0.69	710	5.92	Aft fan engine derived from TF37 with CJ610 core; fan rotor was aerodynamically coupled to gas generator rotor.	
CF700-2B		4,200	0.69	695	6.04	Slightly modified and improved version of CF700; first small turbofan to be certificated in U.S.	Dassault Falcon 20
CF700-2C		4,125 flat rated	0.653	725	5.69	Variation of CF700-2B flat rated to 86 deg. F for additional hot-day performance.	Dassault Falcon 20 and 20 Series C
CF700-2D		4,250 flat rated	0.652	737	5.77	Modified CF700-2C with redesigned higher efficiency HP turbine and improved materials in other engine components; flat rated to 86 deg. F.	Dassault Falcon 20 Series D
CF700-2D2		4,315 flat rated	0.64	737	5.85	Modified CF700-2D with new tailpipe and higher take-off thrust; flat rated to 86 deg. F.	Dassault Falcon 20 Series E and F Rockwell Sabre 75A
CF700-2D2 (improved)		4,500 flat rated	0.65	767	5.86	Uprated version of CF700-2D-2 flat rated to 59 degrees F.	Rockwell Sabre 75A
CF700-2C(V)	TF37-GE-1	4,200	0.69	675	6.22	Derived from J85/J2 turbojet with CF700 aft fan and engine modifications to permit operation in vertical attitude.	Bell Lunar Landing Training and Research Vehicles

General Electric (T64/CT64 turboprop Significant Dates)

Company Model No.	Military Model No.	Design/ Devel. Start	Engine First Run	Engine First Flight	Engine Cert./Qual.	First Prod. Engine Delivered	Last Prod. Engine Delivered	Total Engines Produced
	T64-GE-4	1957	Jan 1959	Sep 22, 1961	Jun 1961 PFRT; Dec 1964 MQT			
	T64-GE-8	1957	Jan 1959		Jun 1961 PFRT; Dec 1964 MQT			
	T64-GE-10			Apr 9, 1964	Nov 1974 MQT; Mar 1965 FAA			
	T64/P4D	1972		1975				
CT64-410-1						Dec 1963	Oct 1966	25
CT64-820-1				Apr 9, 1964	Jan 1967 FAA	1974	1986	168
CT64-820-3					Mar 1965 FAA	Jan 1967	Apr 1972	153
CT64-820-4					Nov 1974 FAA	Mar 1975	1980	196

General Electric (T64 turboshaft Significant Dates)

Company Model No.	Military Model No.	Design/ Devel. Start	Engine First Run	Engine First Flight	Engine Cert./Qual.	First Prod. Engine Delivered	Last Prod. Engine Delivered	Total Engines Produced
	T64-GE-1			Sep 29, 1964				
	T64-GE-2	1957	Mar 1959		Aug 1960 PFRT Dec 1964 MQT			
	T64-GE-3					Sep 1966	1969	30
	T64-GE-6	1957	Mar 1959	Oct 14, 1964	Aug 1960 PFRT Dec 1964 MQT	Oct 1963	Dec 1967	392
	T64-GE-1 (uprated)					Jun 1963	Feb 1967	38
	T64-GE-7				Dec 1966 PFRT Dec 1967 MQT	Dec 1967	Mar 1979	322 engines, 110 kits (-7 and -7A)
	T64-GE-7A				May 1971 MQT			
	T64-GE-12		Nov 5, 1965		Dec 1966 PFRT Dec 1967 MQT	Nov 1968	Jan 1969	13
	T64-GE-16			Sep 21, 1967	May 1967 PFRT Feb 1968 MQT	Apr 1967	Jun 1969	34
	T64-GE-413				Mar 1968 MQT	Feb 1969	Jul 1973	517 (-413,-413A)
	T64-GE-413A			Jan 12, 1969	May 1971 MQT			
	T64-GE-415			Mar 1, 1974	Sep 1973 MQT	Sep 1973	Nov 1975	24
	T64-GE-416				Oct 1978 MQT	Jan 1980	1989	527
	T64-GE-416A				Dec 1988 MQT	Dec 1988		121 through Dec 1991
	T64-GE-419	Jan 1988	Jun 1989		Sep 1991			
	T64-GE-100							
	T64/T4C2				Jul 1982			

869

GENERAL ELECTRIC

T64/CT64 Turboprop Engine Genealogy

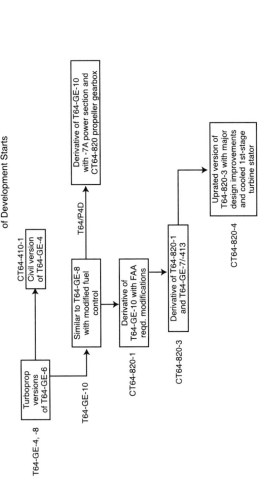

Approximate Time Phasing
of Development Starts

GENERAL ELECTRIC

T64 Turboshaft Engine Development Genealogy

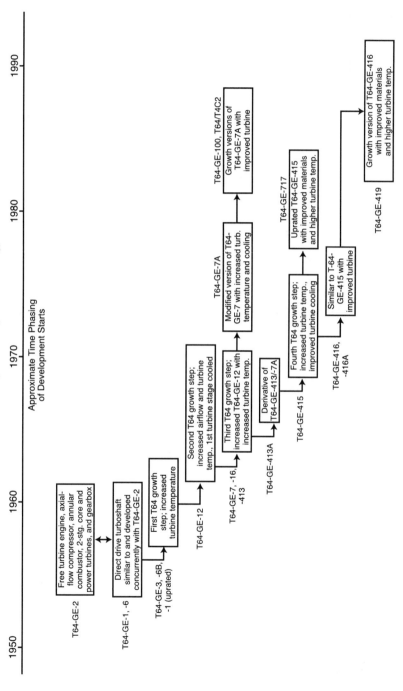

General Electric (T64 turboshaft engines)

Company Model No.	Military Model No.	S.L. Rated Power, SHP	SFC at S.L. Rated Power	Engine Weight, Lb	Power-to-Weight Ratio	Configuration	Applications
	T64-GE-2	2,650	0.506	864	3.07	Free turbine turboshaft with 14-stage axial-flow comp. having variable IGVs and 1st four stator stages, annular combustor, two-stage core turbine, two-stage power turbine, and gearbox; engine progressively uprated.	
	T64-GE-1, -6	2,850	0.495	723	3.94	The T64-GE-1 and -6 were direct drive turboshafts similar to the T64-GE-2 and developed concurrently with it.	Vought-Hiller-Ryan XC-142A tilt wing VTOL (-1) Sikorsky CH-53A (-6A) Hughes XV-9A (-6)
	T64-GE-1 (uprated) T64-GE-3, -6B	3,080 3,080	0.485 0.488	725 723	4.25 4.26	First T64 growth step with turbine temperature increased.	Vought-Hiller-Ryan XC-142A (-1 uprated) Sikorsky HH-53B (-3), CH-53A/B (-6B)
	T64-GE-12	3,400 flat rated	0.480	700	4.91	Second T64 growth step; higher compressor airflow and pressure ratio, higher turbine temperature, cooled turbine first-stage, new combustor design, and titanium compressor.	Flight tested in Sikorsky CH-53A

Engine	Power	SFC	Weight		Description	Applications
T64-GE-7, -16, -413	3,925	0.476	712	5.51	Third T64 growth step; uprated versions of T64-GE-12 with increased turbine temperature.	Sikorsky HH-53C, CH-53B/D/G (-7), CH-53D/G (-413) VFW Fokker VC-400 (-7) Lockheed AH-56A (-16)
	3,925	0.480	700	5.61		
T64-GE-7A	3,925 3,435 flat rated	0.476	716	5.48	Modified version of T64-GE-7 with increased turbine temperature and improved turbine cooling.	Sikorsky CH-53C/G, HH-53B/C/H, S-65C-3
T64-GE-413A	3,925	0.476	716	5.48	Derivative of T64-GE-413 and -7A.	Sikorsky CH-53D/G
T64-GE-415	4,380	0.466	720	6.08	Fourth T64 growth step with improved turbine cooling and combustor liner, and higher turbine temperature.	Sikorsky YCH-53E, CH-53E, RH-53D
T64-GE-416	4,380	0.466	720	6.08	Basically similar to T64-GE-415.	Sikorsky CH/MH-53E, S-80M-1
T64-GE-416A	4,380	0.471	720	6.08	Similar to T64-GE-416 with improved turbine.	Sikorsky CH/MH-53E, S-80M-1
T64-GE-419	4,750	0.474	755	6.29	Growth version of T64-GE-416 with improved turbine, improved hot-end materials, and increased turbine temperature.	Sikorsky CH/MH-53E
T64-GE-717	4,855	0.463	698	6.96	Growth version of T64-GE-415 with improved hot-end materials and higher turbine temperature.	Bell Boeing JVX
T64-GE-100	4,330 flat rated	0.478	720	6.01	Growth version of T64-GE-7A with improved turbine.	Sikorsky MH-53J, S-65C
T64/T4C2	4,330 flat rated	0.478	720	6.01	Basically similar to T64-GE-100; retrofitted all T64-GE-7 models.	Sikorsky S-65C-3

General Electric (T64/CT64 turboprop engines)

Company Model No.	Military Model No.	S.L. Rated Power, SHP	SFC at S.L. Rated Power	Engine Weight, Lb	Power-to-Weight Ratio	Configuration	Applications
	T64-GE-4, -8	2,850	0.490	1,171	2.43	Turboprop derivatives of T64-GE-6 with -4 propeller gearbox below engine centerline and -8 gearbox above centerline.	de Havilland DHC-4 testbed (CV-2A) (T64-GE-4)
	T64-GE-10 T64-IHI-10 T64-IHI-10J	2,850 2,765 2,605	0.490	1,167	2.44	Similar to T64-GE-8 with fuel control modified to provide for propeller reversing.	de Havilland DHC-5 (-10) Kawasaki P2J (IHI-10) Shin Meiwa PS-1 (IHI-10) Nihon Kokuki YS-11E (IHI-10J)
	T64/P4D	3,400 (flat rated)	0.484	1,188	2.86	Derivative of T64-GE-10 using -7A power section with stiffened compressor casing and CT64-820 uprated propeller gearbox.	Fiat G.222 U.S. Air Force C-27A
CT64-410-1		2,850	0.490	1,167	2.44	Civil version of T64-GE-4.	de Havilland CV-7A
CT64-820-1		2,970	0.503	1,130	2.63	Derivative of T64-GE-10 with FAA required modifications and higher turbine temperature.	de Havilland DHC-5A, CC-115
CT64-820-3		2,970	0.503	1,130	2.63	Derivative of T64-820-1 and T64-GE-7/-413	de Havilland DHC-5A
CT64-820-4		3,133	0.486	1,145	2.74	Improved version of T64-810-3 with modified gas generator turbine, cooled first-stage stator vanes and increased turbine temperature.	de Havilland DHC-5D, E

General Electric (TF34 and CF34 turbofan Significant Dates)

Company Model No.	Military Model No.	Design/Devel. Start	Engine First Run	Engine First Flight	Engine Cert./Qual.	First Prod. Engine Delivered	Last Prod. Engine Delivered	Total Engines Produced
TF34 Engines								
	TF34-GE-2	Apr 1968	Apr 1969	Jan 22, 1971	Aug 1972 MQT / Mar 1971 PFRT	Aug 1972		2,126 Total TF34s to Nov 1991
	TF34-GE-400A/B					Dec 1974	Sep 1993	
	TF34-GE-100,-100A	1970	Jul 1973		Oct 1974 MQT	Jun 1975	Mar 1983	
CF34 Engines								
CF34		Apr 1976 announced						
CF34-1A			1981	Apr 10, 1982	Aug 1982 FAA	1982	1985	414 Total CF34s to Nov 1991 (-1A, -3A, -3A1)
CF34-3A			Sep 1986	Sep 28, 1986	Sep 1986 FAA	1986		
CF34-3A1		1987	1988	May 10, 1991	Jul 1991 FAA	1990		
CF34-3B				Mar 17, 1995	May 1995 FAA			
CF34-3B1					May 1995 FAA			
CF34-8C		1995						

GENERAL ELECTRIC

TF34 and CF34 Engine Development Genealogy

Approximate Time Phasing
of Development Starts

TF34 MILITARY TURBOFAN ENGINES

TF34-GE-2
14-stg. axial-flow
compressor, 2-stg. HP
turb. 4-stg. LP turb.

TF34-GE-400A/B
Improved version
of TF34-GE-2

TF34-GE-100, 100A
Reengineered
version
of TF34-GE-2

CF34 COMMERCIAL TURBOFAN ENGINES

CF34
Proposed commercial
version of TF34-GE-100

CF34-1A
Improved version
of CF34

CF34-3A
Modified to improve
climb and hot-day
take-off performance

CF34-3A1
Durability and
maintainability
improvements made

CF34-3B, -3B1
Compressor 1st stg.
modified to increase
airflow and efficiency

CF34-8C
Major redesign

1960 1970 1980 1990 2000

General Electric (TF34 and CF34 turbofan engines)

Company Model No.	Military Model No.	S.L. Rated Thrust, SHP	SFC at S.L. Rated Thrust	Engine Weight, Lb	Power-to-Weight Ratio	Configuration	Applications
TF34 Engines							
	TF34-GE-2	9,275	0.363	1,478	6.28	High-bypass turbofan engine with 14-stage axial-flow compressor, annular combustor, 2-stage cooled HP turbine, and front fan; bypass ratio 6.2.	Lockheed YS-3A and S-3A Viking
	TF34-GE-400A/B	9,275	0.363	1,478	6.28	Improved version of TF34-GE-2 with configuration design changes and modified fuel control.	Lockheed S-3A and S-3B Viking Sikorsky S-72 RSRA (-400A)
	TF34-GE-100, -100A	9,065	0.371	1,437	6.31	Reengineered version of TF34-GE-2 to reduce cost and accomodate USAF specifications and airframe requirements.	Fairchild Republic YA-10A and A-10A
							Boeing YQM-94A
CF34 Engines							
CF34		7,990	0.358	1,525	5.24	Similar to TF34-GE-100 except for external configuration changes necessary for FAA certification.	
CF34-1A		9,140	0.360	1,625	5.62	Improved version of CF34.	Canadair Challenger 601
CF34-3A		9,220 T.O. w/APR flat rated	0.357	1,625	5.67	Flat rated to 70 degrees F for better climb and hot-day takeoff performance.	Canadair Challenger 601-3A

General Electric (TF34 and CF34 turbofan engines) *(Continued)*

Company Model No.	Military Model No.	S.L. Rated Thrust, SHP	SFC at S.L. Rated Thrust	Engine Weight, Lb	Power-to-Weight Ratio	Configuration	Applications
CF34–3A1		9,220 T.O. w/APR	0.357	1,625	5.67	Variation of CF34-3A with durability and maintainability improvements made for airline use.	Canadair Challenger CL-601 RJ
CF34–3B		9,220 T.O. w/APR flat rated	0.346	1,670	5.52	New first compressor stage to increase efficiency and airflow; flat rated to 86 degrees F.	Canadair Challenger 604
CF34–3B1		9,220 T.O. w/APR flat rated	0.346	1,670	5.52	CF34-3B with airline ratings.	
CF34–8C		13,000				Major upgrade with new fan, new 11-stage compressor, multi-hole combustor, improved HP turb., new 4-stage LP turb., and dual FADEC.	

General Electric (T700 turboshaft Significant Dates)

Company Model No.	Military Model No.	Design/ Devel. Start	Engine First Run	Engine First Flight	Engine Cert./Qual.	First Prod. Engine Delivered	Last Prod. Engine Delivered	Total Engines Produced
GE12		Aug 1967	1969				Sept 30, 1971 prog. ended	2
	T700-GE-700	Mar 1972	Feb 27, 1973	Oct 17, 1974	Jul 1974 PFRT Mar 1978 MQT	Mar 1978	1989	2,661
	T700-GE-401, 401A, B	Feb 1978	May 1978	Dec 12, 1979	May 1979 PFRT May 1981 MQT	1982	1989	449
	T700-GE-701, -701A	Aug 1980 (contract award)	Jun 1980		Feb 1983 MQT	1985	1989	1,618
	T700-GE-401C	1985/86	Mar 1987	Sep 1, 1987	Jul 1987 PFRT May 1988 MQT	1988		463 thru 1990
	T700-GE-701C	1985/86	Mar 1987	Sep 1, 1987	Jul 1987 PFRT Sep 1988 MQT	1987		613 thru 1990

General Electric (CT7 commercial turboshaft and turboprop engines Significant Dates)

Company Model No.	Military Model No.	Design/Devel. Start	Engine First Run	Engine First Flight	Engine Cert./Qual.	First Prod. Engine Delivered	Last Prod. Engine Delivered	Total Engines Produced
Turboshaft Engines								
CT7-1		ca. Jul 1974			Jun 1977 FAA			
T700/T1C		Mid-1970s		Feb 1977	Sep 1978 FAA			
CT7-2		Mid 1970s			Sep 1978 FAA	1981	1981	14
CT7-2A					May 1981 FAA	1981	1989	213
CT7-2D	T700-GE-401/-701				1985 FAA	1985	1990	42 (All CT7-2D models)
CT7-2D1	T700-GE-401C/-701C				Jun 1989 FAA			
CT7-6		1985	May 1987	Sep 30, 1988	Jun 1988 FAA	1988	1989	13
Turboprop Engines								
CT7-5		Jul 1979	Mar 1982		Aug 1983 FAA	1983	1989	
CT7-5A		Late 1980	Mar 1982	1982	Mar 1985 FAA	May 1985	1989	
CT7-5A2					Aug 1983 FAA	1984		
CT7-7			May 1987				1989	
CT7-9, -9B, -9C, -9D		1986			Jun 1988 FAA (-9/-9B/-9C) June 1992 (-9D)	Jun 1988		

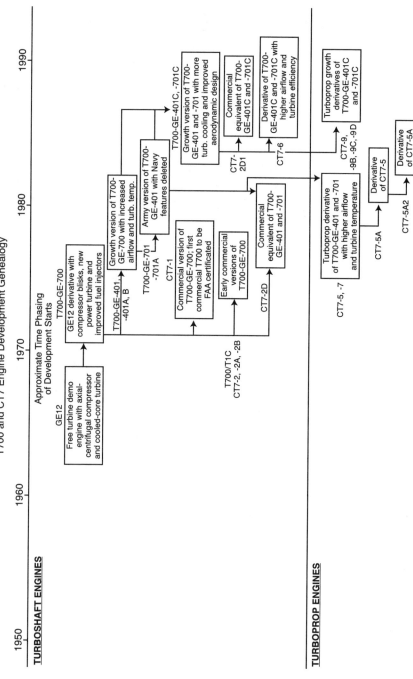

GENERAL ELECTRIC

T700 and CT7 Engine Development Genealogy

881

General Electric (GE12 and T700 turboshaft engines)

Company Model No.	Military Model No.	S.L. Rated Power, SHP	SFC at S.L. Rated Power	Engine Weight, Lb	Power-to-Weight Ratio	Configuration	Applications
GE12		1,500	0.425	282	5.32	Front-drive free turbine turboshaft with 5-stage axial and 1-stage centrifugal compressor, annular combustor, 2-stage core turbine, and 2-stage power turbine.	Prototype demonstrator engine
	T700-GE-700	1,536 (intermediate rating)	0.469	415	3.70	Derivative of GE12 with modular construction, compressor blisks, redesigned power and improved fuel injectors.	Sikorsky YUH-60A and UH-60A Boeing Vertol YUH-61A Hughes YAH-64A Bell YAH-63A, Westland WS-70
	T700-GE-401, -401A, B	1,690 (intermediate rating)	0.464	434	3.89	First T700 growth step; growth version of T700-GE-700 with increased airflow and turbine temperature.	Sikorsky SH-60B, S-70B-2, and VH-60 Bell AH-1W Kaman SH-2G Prototype EHI EH 101
	T700-GE-701, -701A	1,690 (intermediate rating)	0.464	427	3.96	Army version of T700-GE-401 with Navy anti-corrosion and fire resistant features deleted.	McDonnell Douglas AH-64A Sikorsky S-70
	T700-GE-401C	1,800 (intermediate rating)	0.460	458	3.93	Second T700 growth step. Growth version of T700-GE-401 and -701 with additional turbine cooling for higher turbine temperature, and improved aerodynamic design.	Sikorsky SH-60B, SH-60F, HH-60H, and HH-60J
	T700-GE-701C	1,800 (intermediate rating)	0.460	456	3.95	Basically similar to T700-GE-401C.	McDonnell Douglas AH-64A/C/D Sikorsky UH-60L, MH-60G/K

General Electric (CT7 commercial turboshaft and turboprop engines)

Company Model No.	Military Model No.	S.L. Rated Power, SHP	SFC at S.L. Rated Power	Engine Weight, Lb	Power-to-Weight Ratio	Configuration	Applications
Turboshaft Engines							
CT7-1		1,560	0.473	430	3.63	Commercial version of T700-GE-700 with changes made in accordance with FAA requirements and standards.	
T700/T1C, CT7-2, -2A, -2B		1,625	0.474	426	3.81	Early commercial version of T700-GE-700.	Bell Model 214ST (T1C, -2A) Westland Series 200 and 300 (-2B)
CT7-2D	T700-GE-401/ -701	1,623		442	3.67	Commercial equivalent of T700-GE-401/-701.	Sikorsky S-70C
CT7-2D1	T700-GE-401C/ -701C	1,625				Commercial equivalent of T700-GE-401C/-701C.	Sikorsky S-70
CT7-6		2,000	0.457	485	4.12	Derivative of T700-GE-401C/ -701C with higher airflow and improved LP turbine efficiency.	Italian EHI EH 101 Westland EH 101 prototype Sikorsky S-92 Helibus
Turboprop Engines							
CT7-5		1,654 eshp 1,700 flat rated	0.470	770	2.21	Turboprop derivatives of T700-GE-401/-701 with higher airflow and turbine temperature.	Saab-Fairchild SF340 (-5) CASA-IPTN CN-235 (-7)
CT7-5A		1,700 eshp flat rated	0.476	783	2.22	Derivative of CT7-5.	Saab 340
CT7-5A2		1,735 flat rated	0.478	773	2.24	Derivative of CT7-5A.	Saab 340
CT7-9, -9B, -9C, -9D		1,750	0.461	805	2.17	Growth derivatives of T700-GE-401C/-701C.	Saab 340B CASA-IPTN CN235 Series 100 (-9C) LET L610G (-9D)

General Electric (GE27 and derivative engines Significant Dates)

Company Model No.	Military Model No.	Design/ Devel. Start	Engine First Run	Engine First Flight	Engine Cert./Qual.	First Prod. Engine Delivered	Last Prod. Engine Delivered	Total Engines Produced
Turboprop/Turboshaft Engines								
GE27		Late 1970s	1984					
GE38	T407-GE-400	1987	1989/1990					
GLC38		1988	Dec 1989					
Turbofan Engines								
CFE738		May 1988 Studies began in Nov 1984	May 1990	Aug 30, 1992	Dec 17, 1993 FAA	Oct 1993		
CFE738-1B								

GENERAL ELECTRIC

GE27 and Derivative Engines Development Genealogy

Approximate Time Phasing
of Development Starts

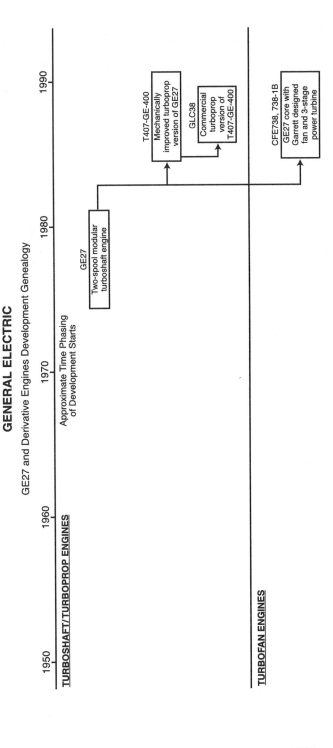

TURBOSHAFT/TURBOPROP ENGINES

| 1950 | 1960 | 1970 | 1980 | 1990 |

GE27
Two-spool modular
turboshaft engine

T407-GE-400
Mechanically
improved turboprop
version of GE27

GLC38
Commercial
turboprop
version of
T407-GE-400

TURBOFAN ENGINES

CFE738, 738-1B
GE27 core with
Garrett designed
fan and 3-stage
power turbine

General Electric (GE27 and derivative engines)

Company Model No.	Military Model No.	S.L. Rated Power, SHP/Thrust, Lb	SFC at S.L. Rated Power/Thrust	Engine Weight, Lb	Power/Thrust-to-Weight Ratio	Configuration	Applications
Turboshaft/Turboprop Engines							
GE27		5,000 to 6,000 shp class	0.380			Free turbine modular turboshaft engine with 5 axial and 1 centrifugal comp. stages, annular combustor, 2-stage cooled HP turbine, 3-stage LP turbine.	Unsuccessful entry in 1985 JVX competition
GE38	T407-GE-400	5,660 shp	0.380			Military turboprop derived from GE27 with gearbox and design mods for longer life.	Selected for: Lockheed P-7A (Program terminated in 1990)
GLC38		3,500 to 5,000 shp depending on application				Derated commercial version of T407 turboprop.	Unsuccessfully competed for: Saab 2000 IPTN N-250
Turbofan Engines							
CFE738		5,900 lb 5,725 lb flat rated	0.389	1,214	4.86	Free turbine high-bypass turbofan with GE27 core and Garrett-designed single-stage fan and 3-stage LP turbine.	Dassault Falcon 2000
CFE738-1B		5,725 lb	0.372	1,325	4.32	Production version of CFE738. Bypass ratio 5.3:1.	Dassault Falcon 2000

Williams International (Significant Dates)

Company Model No.	Military Model No.	Design/ Devel. Start	Engine First Run	Engine First Flight	Engine Cert./Qual.	First Prod. Engine Delivered	Last Prod. Engine Delivered	Total Engines Produced
Turbojet Engines								
Jet No. 1		1957	1957					
WR2-1		1960	1962					
WR2-2A		1963	1963					35
WR2-3		1963	1963					
WR2-4		1964	1964	1964				
WR2-5		1965	1965	1965				
WR2-6		1965	1965	1967		1968	1973	Approx. 300
WR2-6	YJ400-WR-400	1967	1967–68	1967–68	1968 Qual. Test	1968	1973	Approx. 2,200
WR24-7	YJ400-WR-401	1967	1968	1973	Apr 1973 Qual. Test	1974	1990	2,449
WR24-7A	YJ400-WR-402	1977	1977	1979		1978	1984	615
WR24-7B	YJ400-WR-403	1980	1980	1983		1983	1990	1,030
WR24-8	J400-WR-404	1982	1983	1985		1986		318 by 1991
WR36-1	F121-WR-100	Circa 1984			1985 Qual. Test			
P8300		Late 1980s						
WJ119-2		1988	1990					
P8910		1989	1991					
P9005		1990	1992					

Williams International (Significant Dates) (*Continued*)

Company/Model No.	Military Model No.	Design/Devel. Start	Engine First Run	Engine First Flight	Engine Cert./Qual.	First Prod. Engine Delivered	Last Prod. Engine Delivered	Total Engines Produced
Turbofan Engines								
WR19		Circa 1965	Aug 1967	Apr 1969				3
WR19-A2		1969	1970	1971				
WR19-7				Apr 1980				
WR19-9 BPR5				Dec 1973				
WR19-A7D	F107-WR-100	1972		Mar 1976				
WR19-A7-1	F107-WR-100 Variant	1972						
	F107-WR-400	1972	1975	1976		1982	1990	2,614
	F107-WR-101	1977	1979	1979	Mar 1981 Qual. Test	Mar 1981	Mar 1986	1,926
	F107-WR-102	1977		1979				10
	F112-WR-100	1982	1984	1985		1986		575 by 1991
	F107-WR-402	1987	1986	1988		1990		215
	F107-WR-104							
Turbofan Engines								
FJ44		1980	1981					
FJ44 uprated		1983	1985					
FJ44-1	F129			Jul 1988	Mar 1992	Jun 1992		
FJ44-2		1995	1995			1995		
Turboshaft Engines								
WR34-15		1979	1981	1981				6–10
WTS34-16		1984	1985	1987				14

WILLIAMS INTERNATIONAL

Engine Development Genealogy

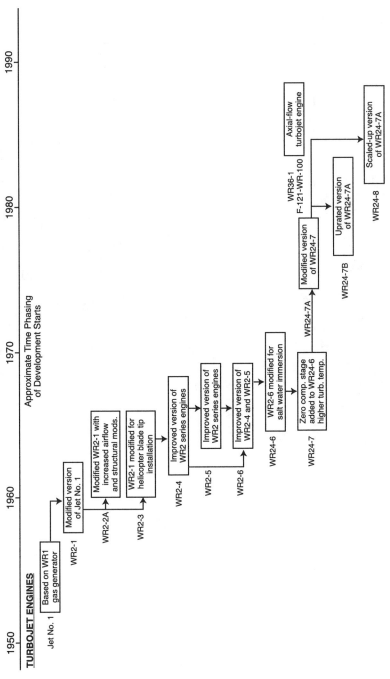

Approximate Time Phasing
of Development Starts

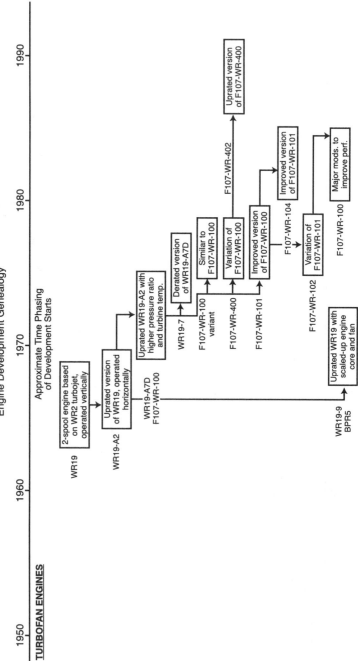

WILLIAMS INTERNATIONAL

Engine Development Genealogy

Approximate Time Phasing
of Development Starts

TURBOFAN ENGINES

1950   1960   1970   1980   1990

WR19
2-spool engine based on WR2 turbojet, operated vertically

WR19-A2
Uprated version of WR19, operated horizontally

WR19-A7D
F107-WR-100
Uprated WR19-A2 with higher pressure ratio and turbine temp.

WR19-7
Derated version of WR19-A7D

F107-WR-100 variant
Similar to F107-WR-100

F107-WR-400
Variation of F107-WR-100

F107-WR-101
Improved version of F107-WR-100

F107-WR-104
Improved version of F107-WR-101

F107-WR-402
Uprated version of F107-WR-400

F107-WR-102
Variation of F107-WR-101

F107-WR-100
Major mods. to improve perf.

WR19-9
BPR5
Uprated WR19 with scaled-up engine core and fan

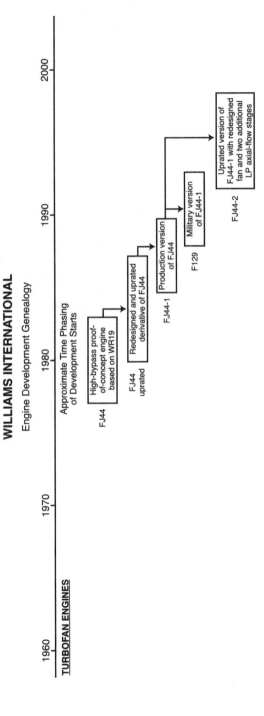

WILLIAMS INTERNATIONAL
Engine Development Genealogy

TURBOFAN ENGINES

Approximate Time Phasing
of Development Starts

1960   1970   1980   1990   2000

FJ44 — High-bypass proof-of-concept engine based on WR19

FJ44 uprated

Redesigned and uprated derivative of FJ44

FJ44-1 — Production version of FJ44

F129 — Military version of FJ44-1

FJ44-2 — Uprated version of FJ44-1 with redesigned fan and two additional LP axial-flow stages

Williams International (turbojet engines)

Company Model No.	Military Model No.	S.L. Rated Thrust, Lb	SFC at S.L. Rated Thrust	Engine Weight, Lb	Thrust-to-Weight Ratio	Configuration	Applications
Jet No. 1		60		23	2.6	Used gas generator section of WR-1 free turbine turboshaft engine. Single-stg. centrifugal compressor, annular combustor, and single-stg. axial turbine.	First Williams aircraft-type engine. Test engine developed to demonstrate feasibility.
WR2-1		70				Modified version of Jet No. 1 with redesigned bearings and lube system.	Test version modified for potential use in Canadair CL-89 drone.
WR2-2A		95	1.25	29	3.3	WR2-1 compressor design modified to increase airflow and efficiency; structural design modified to lower cost; two-position jet nozzle for starting.	Used for ground development and cold starting tests simulating CL-89 requirements.
WR2-3		95	1.25	29	3.3	WR2-1 modified for installation on helicopter rotor tip.	Intended for use on Hiller helicopter that was canceled.
WR2-4		105	1.25	29	3.6	WR2-4 and -5 were progressively improved versions of WR2 engine.	Flight test versions of Canadair CL-89
WR2-5		115	1.25	29	3.96		
WR2-6		125	1.25	30	4.3	Upgraded version of WR2-4 and -5. Incorporated new starting system and higher output alternator.	Canadair CL-89 (AN-USD-501) reconnaissance vehicle
WR24-6	YJ-400-WR-403	121	1.25	30	4.0	Same aero-thermal design as WR2-6. Design changes included improved lube system and materials mods to meet salt water immersion requirements.	Northrop Chukar I (MQM-74A) target drone

Williams International (turbojet engines) (*Continued*)

Company Model No.	Military Model No.	S.L. Rated Thrust, Lb	SFC at S.L. Rated Thrust	Engine Weight, Lb	Thrust-to-Weight Ratio	Configuration	Applications
WR24-7	YJ400-WR-401	176	1.2	44	4.0	Variation of WR24-6 with addition of zero axial-flow compressor stage and higher turbine temp.	Northrop Chukar II (MQM-74C) target drone
WR24-7A	YJ400-WR-402	180	1.2	44	4.1	Slightly modified version of WR24-7.	Early versions of Northrop Chukar III (BQM-74C) target drone
WR24-7B	YJ400-WR-403	190	1.2	44	4.3	Uprated version of WR24-7A.	Northrop BQM-74C
WR24-8	J400-WR-404	240	1.2	50	4.8	Scaled-up version of WR24-7A with compressor modified to increase airflow.	In 1986 replaced WR24-7A in Northrop Chukar III and BQM-74E
WR36-1	F121-WR-100	150		42	3.57	Axial-flow turbojet engine.	Northrop AGM-136A Tacit Rainbow anti-radiation missile
P8300		1,000 class		150	6.7	Data not available due to security classification.	Intended for Rockwell AGM-130 glide bomb
WJ119-2		92	1.39	27.5	3.3	Six-stage axial-flow compressor, reverse-flow annular combustor, single-stage axial turbine.	Demonstration sustainer engine for NLOS missile and NLOS/TRR.
P8910		144	1.35	28	5.1	Growth version of WJ119-2 within same engine envelope; increased airflow and turbine temp.	NLOS, NLOS/TRR, and LONGFOG program demonstration engine
P9005		51	1.43	10	5.1	Single-stage mixed-flow compressor, reverse-flow annular combustor, single-stage axial turbine.	SENGAP program demonstration engine

Williams International (turbofan engines)

Company Model No.	Military Model No.	S.L. Rated Thrust, Lb	SFC at S.L. Rated Thrust	Engine Weight, Lb	Thrust-to-Weight Ratio	Configuration	Applications
WR-19		430	0.66	67	6.4	2 contrarotating spools. HP spool based on WR-2 turbojet had 1-stage centrifugal comp., annular comb. with slinger-type fuel injection, and 1-stg. turb. LP spool had 2-stage axial-flow comp. 2 fan stgs. providing both bypass air and air into the core flow path, and 2-stg. turb. Operated in vertical attitude. Bypass ratio 1.1.	Bell Jet Flying Belt
WR19-A2		476	0.66			Variation and uprated version of WR19 engine that operated in horizontal attitude. Accessory gearbox added. Bypass ratio 1.1.	Kaman Saver experimental autogiro fighter ejection seat. Also part of cruise missile technology feasibility program
WR19-7		570	0.65	125	4.56	Derated version of WR19-A7D. Bypass ratio 1.1.	WASP II prototype under U.S. Army Individual Lift Device (ILD) program
WR19-9 BPR5		670	0.47			Uprated WR19 with scaled-up engine core and larger single-stage fan. High-bypass ratio.	Wasp I prototype vehicle under U.S. Marine Corps Small Tactical Air Mobility Platform (STAMP)
WR19-A7D	F107-WR-100	600 class		130	Approx 4.6	Uprated version of WR19 engine with increased pressure ratio and turbine temperature. Bypass ratio 1.03.	Prototype version of Boeing AGM-86A Air-Launched Cruise Missile (ALCM)

Williams International (turbofan engines) (*Continued*)

Company Model No.	Military Model No.	S.L. Rated Thrust, Lb	SFC at S.L. Rated Thrust	Engine Weight, Lb	Thrust-to-Weight Ratio	Configuration	Applications
WR19-A7-1	F107-WR-100 variant	600 class		130	4.6	Basically similar to F107-WR-100 having 80 to 90% parts commonality. Bypass ratio approximately 1.0.	Prototype versions of General Dynamics BGM-109 Navy sea-launched and Air Force ground- and air-launched cruise missiles
	F107-WR-400	600 class		144	4.2	Variation of F107-WR-100.	General Dynamics BGM-109 Tomahawk Ground-Launched (GLCM) and Sea-Launched (SLCM) Cruise Missiles
	F107-WR-101	600 class		145	4.2	Improved version of F107-WR-100.	Prototype and operational Boeing AGM-86B cruise missiles
	F107-WR-102	600 class		144	4.2	Variation of F107-WR-101.	Prototype General Dynamics AGM-109 cruise missile
	F112-WR-100					Major modifications to improve performance.	Developed for General Dynamics AGM-129A Advanced Cruise Missile
	F107-WR-402	600 class		142	4.2	Uprated version of F107-WR-400.	Developed for upgraded AGM-109 Tomahawk missile
	F107-WR-104	600 class				Improved version of F107-WR-101.	Retrofitted in AGM-86B cruise missile

Williams International (turbofan and turboshaft engines)

Company Model No.	Military Model No.	S.L. Rated Thrust/Power, Lb/SHP	SFC at S.L. Rated Thrust/Power	Engine Weight, Lb	Thrust/Power-to-Weight Ratio	Configuration	Applications
Turbofan Engines							
FJ44		1,500 lb				Turbofan engine with design based on WR19 configuration.	Proof-of-concept engine developed for 4 to 6 passenger business jets
FJ44 uprated		1,800 lb	0.47	445	4.27	Redesigned FJ44 with titanium blisk-type fan and single axial-flow compressor stage on LP shaft driven by two-stage axial turbine; HP core comprised of single-stage titanium centrifugal compressor, annular combustor, and single-stage turbine. Bypass ratio 3.28.	Beech Triumph Swearingen SJ30 Cessna CitationJet
FJ44-1		1,900 lb				Production version of FJ44 series.	Cessna CitationJet
	F129	1,900 lb				Military version of FJ44-1.	Cessna 526 (entered in JPATS competition) Saab SK60 jet trainer Lockheed Martin Tier III Minus Darkstar UAV
FJ44-2		2,300 lb		520	4.42	Uprated by redesigning fan and adding two more axial-flow comp. stages on LP shaft. Bypass ratio 2.3:1.	Raytheon PD374 Premier I Swearingen SJ30-2
Turboshaft Engines							
WR-34-15		32 shp		38	0.84	WR-34-15 and -16 were of simple design with single-stage centrifugal compressor, annular combustor, and single-stg. radial-flow turb.	Canadair CL-227 Sentinel rotary-wing RPV
WTS-34-16		51.5 shp	0.98	53	0.97		

Pratt & Whitney Canada (**Significant Dates**)

Company Model No.	Military Model No.	Design/ Devel. Start	Engine First Run	Engine First Flight	Engine Cert./Qual.	First Prod. Engine Delivered	Last Prod. Engine Delivered	Total Engines Produced
Turbojet Engines								
JT12 Series	J60	Jul 1957	May 1958	Jan 1960		Oct 1960	1977	2,269
Turbofan Engines								
JT15D-1, -1A, -1B		Jun 1966	Sep 1967	Aug 1968	May 1971	1971		5,161 of all JT15D models thru 1995
JT15D-4, -4B, -4C, -4D			Jan 1972		Sep 1973	Sep 1973		
JT15D-5, -5A, -5B		1976	Late 1977	Apr 1978	Dec 1983	Jun 1984		
PW300Series								
PW300		Sep 1985 (Full design)	Mar 1988	May 1989	Aug 1990	1990		288 of all PW300 models thru 1995
PW305					Aug 1990			
PW305A				Jun 1990	Dec 1992			
PW305B				Jun 1991	Jan 1993			
PW306A		Oct 1993	Mar 1994	Aug 1994	Nov 1995			
PW500Series								
PW530A		Aug 1992	Oct 1993	May 1994		1996		
PW545		Apr 1994	Dec 1994	Feb 1996		1996		

Pratt & Whitney Canada (Significant Dates) (*Continued*)

Company Model No.	Military Model No.	Design/ Devel. Start	Engine First Run	Engine First Flight	Engine Cert./Qual.	First Prod. Engine Delivered	Last Prod. Engine Delivered	Total Engines Produced
Turboprop Engines								
PT6A Small Engine Series								
PT6A-6		1958	Feb 1960	May 1961	Dec 1963	Dec 1963		25,234 of all PT6 turboprop models thru 1995
PT6A-20					Oct 1965			
PT6A-25					May 1976			
PT6A-25A					Nov 1976			
PT6A-27	T74-CP-702				Dec 1967			
PT6A-34					Nov 1971	Nov 1971		
PT6A-114					Jan 1984	Apr 1984		
PT6A-135					Jan 1978			
PT6A-11AG					Mar 1979			
PT6A-15AG					Sep 1978			
PT6A-34AG					Feb 1977			
PT6A Large Engine Series								
PT6A-41					Oct 1973			
PT6A-45A					Apr 1976			
PT6A-50					Sep 1976			
PT6A-60					Nov 1982			
PT6A-61					Nov 1982			

Turboprop Engines						
PT6A-62				Feb 1985		
PT6A-65R	1977			Aug 1982	Aug 1982	
PT6A-65B				Aug 1982		
PT6A-67/R				Jan 1987		
PT6A-67B/D				Oct 1990		
PT6A-68						
PW100 Series						
PT7A-1 (PW100)	Jun 1979	Mar 1981	Feb 1982			3,791 of all PW100 models thru 1995
PW115				Dec 1983	Oct 1984	
PW118				Mar 1986		
PW118A				Jun 1987		
PW119						
PW119B				Jun 1993		
PW120				Dec 1983	Jan 1984	
PW120A				Sep 1984		
PW121				Feb 1987		
PW124	Jul 1982			Nov 1985		
PW124B				May 1988		
PW123				Jun 1987		
PW123AF				Feb 1990		
PW123E						
PW123B				Nov 1991		
PW123, 123C/D						
PW125B				May 1987		
PW126				May 1987		
PW126A				Jun 1989		
PW127				Feb 1992		
127B/C/E						
PW127D				Jan 1994		
PW150						

Pratt & Whitney Canada (Significant Dates) (*Continued*)

Company Model No.	Military Model No.	Design/ Devel. Start	Engine First Run	Engine First Flight	Engine Cert./Qual.	First Prod. Engine Delivered	Last Prod. Engine Delivered	Total Engines Produced
Turboshaft Engines								
PT6B Single Power Section Turbine								
PT6B Mk.2			Feb 1961	May 1961				7,238 of all PT6 turboshaft models thru 1995
PT6B-9					May 1965	1967		
PT6B-36					Sep 1984	Jun 1984		
PT6T Twinned Engine Series								
PT6T-3		Apr 1967	Jun 1968	May 1969	Jul 1970	1970		
PT6T-6				Nov 1969	Dec 1974			
	T400-CP-400	1967			Mar 1970			
	T400-CP-401				Jun 1972			
	T400-WV-402				Feb 1975 (Fully rated)			
PW200 Series								
PW209T		Oct 1983	Nov 1985					86 of all PW200 series models thru 1995
PW205B		Jul 1985	Feb 1987					
PW206A		1990		Oct 1988	Dec 1991 Orig. Cert. Nov 1993 Recert.	Jan 1992		
PW206B			Feb 1992	Apr 1994				
PW206C								

PRATT & WHITNEY CANADA

Engine Development Genealogy

Approximate Time Phasing
of Development Starts

TURBOFAN ENGINES

1950　　　1960　　　1970　　　1980　　　1990

JT15D SERIES

Fan scaled from JT9D, combustor and compressor scaled from PT6	JT15D-1, -1A, -1B

Single axial stage added to core ahead of centrifugal compressor	JT15D-4, -4B, -4C, -4D

Redesigned fan and axial stage, larger diameter centrifugal compressor	JT15D-5, -5A, -5B

901

PRATT & WHITNEY CANADA

Engine Development Genealogy

Approximate Time Phasing
of Development Starts

TURBOFAN ENGINES

| 1960 | 1970 | 1980 | 1990 | 2000 |

PW300 SERIES

PW300 — Two-spool engine, four-stg. axial and single-stg. centrif. HP comp., single-stg. front fan

PW305 — Uprated version of PW300 with minor modifications

PW305A — Minor variations from PW305

PW305B — Uprated production version of PW305

PW306A — PW305 derivative with redesigned fan, 1st axial comp. stg., and LP and HP turbines

PW500 SERIES

PW530, A — Two-spool engine, 1-stg. fan, 2-stg. axial/1-stg. centrif. comp., rev-flow comb., 1-stg. HP turb., 2-stg. LP turb.

PW545 — Increased fan diameter, added comp. boost stg. and LP turb. stg.

PRATT & WHITNEY CANADA

Engine Development Genealogy

Approximate Time Phasing
of Development Starts

TURBOPROP ENGINES

PT6A SMALL ENGINE SERIES

P6A-6 — Opposed-type free turbine engine with front gearbox; first certificated PT6 engine

PT6A-20 — Derivative of PT6A-6 with titanium compressor impeller

PT6A-27
TP74-CP-702 — Uprated model with increased airflow; new centrifugal comp. and gearbox designs

PT6A-25, -25A — Aerobatic derivative of PT6A-27

PT6A-34 — Derivative of PT6A-27 with PT6T-3 HP turb. assembly and cooled HP turbine stator blades

PT6A-114 — Derivative of PT6A-34/-36 with low-speed gearbox and single exhaust port

PT6A-135 — Derivative of PT6A-34 with improved hot end and low-speed gearbox

PT6A-11AG, -15AG, -34AG — Derivatives of PT6A-11, A-27 and A-34 respectively; modified for more rugged agricultural aircraft use

1950   1960   1970   1980   1990

PRATT & WHITNEY CANADA

Engine Development Genealogy

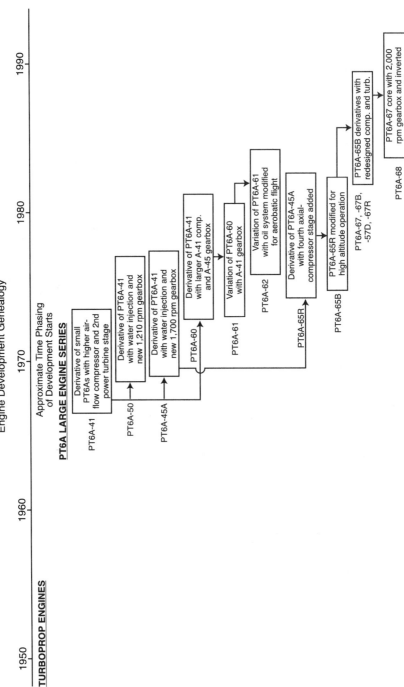

PRATT & WHITNEY CANADA

Engine Development Genealogy

TURBOPROP ENGINES

Approximate Time Phasing
of Development Starts

PW100 SERIES

1960   1970   1980   1990

PT7A-1
(PW100)

New three-shaft
free turbine
development engine

PW115

Production derivative
of PT7A-1

PW118/
118A

PW115 derivatives with
increased power rating

PW119/
119B

PW118 derivatives with
increased power rating

PW120/
120A

Derivative of PW115
with new 1,200 rpm
gearbox and reserve
power setting

PW121

Variant of PW120 with
increased power rating

PW124/
124A/B

Growth verion of PW120
with higher airflow,
cooled LP turb. stators,
and new gearboxes

PW123/123AF
PW123B/C/D/E
PW125B

PW124 derivatives with
PW124 turbomachinery and
PW120A gearbox (PW123)

PW126/126A
PW127/B/C/D

Growth
engines

PW150

New design with three-stage
axial LP comp., single-stage
centrif. HP comp., and increased
turbine cooling

905

PRATT & WHITNEY CANADA

Engine Development Genealogy

TURBOSHAFT ENGINES

Approximate Time Phasing
of Development Starts

PT6B SINGLE POWER SECTION SERIES

First PT6
turboshaft
engine

PT6 Mk. 2

Turboshaft
version of
PT6A-6

PT6B-9

PT6T-3B power section
with modified exhaust
and redesigned gearbox

PT6B-36

PT6T TWINNED POWER SECTION SERIES

Twin power sections
similar to PT6A-27
with single gearbox

PT6T-3

PT6T-6

Uprated version of
PT6-3 with hot-end
improvements

Military versions
of PT6T-3

T400-CP-400,
-401

Uprated version of
T400-CP-400 with hot-
end improvements

T400-WV-402

1950   1960   1970   1980   1990

906

PRATT & WHITNEY CANADA

Engine Development Genealogy

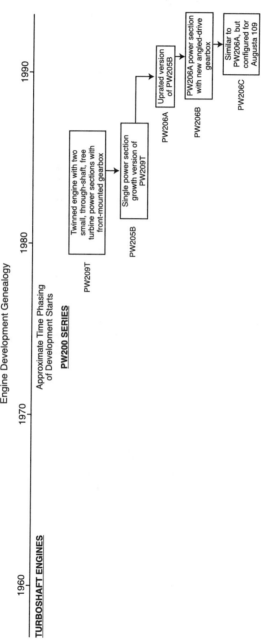

Pratt & Whitney Canada (turbojet/turbofan engines)

Company Model No.	Military Model No.	S.L. Rated Power, SHP	SFC at S.L. Rated Power	Engine Weight, Lb	Power-to-Weight Ratio	Configuration	Applications
Turbojet Engines							
JT12A-5	J60-P-3 J60-P-3A J60-P-5 J60-P-6 J60-P-9	3,000	0.96	448	6.7	Single-shaft engine with 9-stage axial compressor, cannular combustor, and two-stage turbine.	Lockheed JetStar C-140 North American Sabreliner Fairchild SD-5 drone North American T-2B Buckeye trainer 2 boosters for Martin RB-57F
JT12A-6A		3,000	0.96	453	6.6	Commercial version of JT12A-5.	Lockheed JetStar North American Sabreliner
JT12A-8A		3,300	0.89	468	7.05	Uprated version of JT12A-6A.	Lockheed JetStar North American Sabreliner

Turbofan Engines

JT15D-1, -1A, -1B	2,200	0.54	514	4.28	Two-spool engine with single-stage centrifugal comp. driven by single-stage turbine, annular reverse-flow combustor, single-stage front fan (scaled from P&WA JT9D fan) driven by two-stage turbine. Bypass ratio 3.3.	Cessna Citation Model 500 and Citation I
JT15D-4, -4B, -4C, -4D	2,500	0.562	557	4.49	Single axial stage added to core ahead of centrifugal compressor to raise airflow and pressure ratio. Bypass ratio 2.6.	Cessna Citation II (D-4) Aerospatiale Corvette (D-4) Mitsubishi Diamond IA (D-4D) Cessna Citation S/II (D-4B), others
JT15D-5, -5A, -5B	3,190 2,900 flat rated	0.554 0.551	632	5.05 4.59	Shroudless fan redesigned to increase airflow, axial stage modified to raise efficiency, centrifugal compressor diameter increased. Bypass ratio 2.0.	Beech Beechjet 400A (D-5) Cessna T47-A (D-5) Cessna Citation V (D-5A) Northrop Grumman Agusta 211.A (D-5C) Beech T-1A Jay Hawk (D-5B) Others

Pratt & Whitney Canada (turbofan engines)

Company Model No.	Military Model No.	S.L. Rated Thrust, Lb	SFC at S.L. Rated Thrust	Engine Weight, Lb	Thrust-to-Weight Ratio	Configuration	Applications
PW300 Series							
PW300		4,750	0.406	950	5.0	Two-spool engine with four-stage axial and single-stage centrifugal HP compressor (first two axial stages had variable IGVs), through-flow combustor, two-stage HP turbine (both turbine rotor and stator stages cooled), single-stage front fan driven by three-stg. LP turbine. Bypass ratio 4.5.	
PW305		5,200 4,750 flat rated	0.402	950	5.47 5.0	Uprated version of PW300 with minor modifications. Bypass ratio 4.5.	Retrofitted Dassault Falcon 20 British Aerospace BAe 1000 prototype
PW305A		4,679 flat rated	0.388	993	4.71	Minor variations from PW305.	Learjet 60

PW305B	5,266 flat rated	0.391	993	5.30	Uprated production version of PW305.	British Aerospace Hawker 1000
PW306A	5,700 flat rated	0.394	1,043	5.47	Derivative of PW305 with redesigned fan and first axial compressor stage, modified LP turbine, redesigned HP turbine, and exhaust mixer.	Israel Aircraft Industries Galaxy
PW500 Series						
PW530	2,600	0.468	605	4.30	Two-spool, high-bypass turbofan engine.	Cessna Citation Bravo
PW530A	2,750		605	4.55	Single-stage fan, 2-stage axial and 1-stage centrifugal compressor, reverse-flow combustor, 1-stage HP turbine, 2-stage LP turbine.	
PW545	3,876 flat rated	0.436	765	5.07	Increased fan diameter, added compressor boost stage, added third LP turbine stage.	Cessna Citation Excel

Pratt & Whitney Canada (turboprop engines)

Company Model No.	Military Model No.	S.L. Rated Power, SHP	SFC at S.L. Rated Power	Engine Weight, Lb	Power-to-Weight Ratio	Configuration	Applications
PT6A Small Engine Series							
PT6A-6		550	0.65	270	2.04	Opposed free turbine turboprop with three-stage axial and single-stage centrifugal HP compressor, reverse-flow annular combustor, single-stage HP turbine, front-mounted propeller gearbox driven by single-stage power turbine.	Beech King Air 90 de Havilland Canada Turbo-Beaver Pilatus Turbo Porter Beech 18 Various aircraft conversions
PT6A-20		550	0.649	286	1.92	Derivative of PT6A-6 with maximum continuous power rating equal to take-off rating; first P&WC engine to use titanium compressor impeller.	Beech King Air 90 Series Beech 99 de Havilland Canada Twin Otter & Turbo-Beaver Pilatus Turbo-Porter Embraer Bandeirante Swearingen Merlin II Various aircraft conversions
PT6A-25, -25A		550	0.630	341	1.61	Aerobatic derivative of PT6A-27 with oil system modified for aerobatic maneuvers and PT6A-20 reduction gear.	Beech T-34C (A-25) Pilatus PC-7 (A-25A)

Model	Military designation	Power				Description	Applications
PT6A-27	T74-CP-702	680 flat rated	0.602	314	2.17	Uprated model with axial compressor diameter increased to raise airflow, centrifugal compressor aerodynamically redesigned, new reduction gearbox, and pipe diffuser. Initial engines had narrow-chord first-stage compressor blades, later engines had wide-chord blades.	Beech 99/99A, U-21A/D (-700/-702); de Havilland Canada Twin-Otter; Embraer EMB-110; Frakes Turbo Mallard; Helio Stallion; LET 410; Pilatus Turbo-Porter; CATIC Y-12
PT6A-34		750 flat rated	0.595	320	2.34	Derivative of PT6A-27 using PT6T-3 HP turbine assembly with cooled first-stage turbine stator blades, and single exhaust port.	Embraer EMB-110/-111; Frakes Turbo Mallard; IAI Arava; Saunders ST-27/ST-28; Omni Turbo Titan; Airmaster Avalon 680; Spectrum-One
PT6A-114		600 flat rated	0.64	345	1.74	Derivative of PT6A-34/-36 with single-port exhaust and low-speed gearbox.	Cessna Caravan I
PT6A-135		750 flat rated	0.585	330	2.27	Derivative of PT6A-34 with improved turbine assembly from T400-WV-402 and low-speed gearbox.	Beech King Air F90; Embraer EMB-121A1, Xingu II; Piper Cheyenne II XL; Schafer Comanchero, 750
PT6A-11AG		500	0.647	321	1.56	Derivatives of PT6A-11, PT6A-27, and PT6A-34 respectively with modifications to accommodate the rugged operating conditions of agricultural aircraft and the ability to use diesel fuel.	Ayres Turbo Thrush; Frakes Turbo-Cat; Air Tractor AT-502; Grumman Ag-Cat; Other agricultural aircraft
PT6A-15AG		680	0.602	328	2.07		
PT6A-34AG		750 flat rated	0.595	331	2.27		

Pratt & Whitney Canada (turboprop engines) (Continued)

Company Model No.	Military Model No.	S.L. Rated Power, SHP	SFC at S.L. Rated Power	Engine Weight, Lb	Power-to-Weight Ratio	Configuration	Applications
PT6A Large Engine Series							
PT6A-41		850 flat rated	0.591	391	2.17	New model with higher airflow compressor, two-stage power turbine, and gearbox designed for higher power level.	Beech Super King Air 200 and C-12F; Piper Cheyenne III
PT6A-45A		1,173	0.554	434	2.70	Derivative of PT6A-41 with water injection and new 1,700 rpm low-speed gearbox.	Frakes Mohawk 298; Shorts 330
PT6A-50		1,120	0.560	607	1.85	Derivative of PT6A-41 with water injection and new 1,210 rpm gearbox to reduce propeller speed.	de Havilland DCH Dash-7
PT6A-60		1,050	0.548	465	2.26	Derivative of PT6A-41 with larger A-41 compressor, new flame tube, and A-45 gearbox.	Beech King Air 300, 350
PT6A-61		850	0.591	426	2.0	Derivative of PT6A-60 with A-41 gearbox.	Piper Cheyenne IIIA, TP-600 Malibu
PT6A-62		950 flat rated	0.567	454	2.09	Modified version of PT6A-61 with oil system designed for aerobatic flight.	Pilatus PC.9; HAS HTT-35
PT6A-65R		1,230 / 1,376 reserve	0.512 / 0.512	481	2.56 / 2.86	Derivative of PT6A-45R without water injection, with addition of a fourth compressor stage to increase airflow and pressure ratio, and an improved hot end.	Shorts 360

Engine	Power rating				Description	Application
PT6A-65B	1,100 flat rated	0.536	481	2.29	Derivative of PT6A-65R designed for high altitude operation.	Beech 1900
PT6A-67	1,100	0.547	506	2.17	Derivatives of PT6A-65B with compressor and turbine redesigned to increase airflow by 10 percent.	Beech RC-12 (A-67)
PT6A-67B	1,200	0.552	515	2.33		Pilatus PC-12 (A-67B)
PT6A-67D	1,279	0.530	515	2.48		Beech 1900D (A-67D)
PT6A-67R	1,424 flat rated	0.520	515	2.77		Shorts 360-300 (A-67R)
PT6A-68	1,250	0.541	572	2.19	PT6A-67 core with 2,000 rpm gearbox, inverted flight capability, and full-authority digital electronic control.	Beech/Pilatus PC.9 Mk.II Embraer EMB-312-312H

PW100 Series

Engine	Power rating				Description	Application
PT7A-1 (PW100)	2,000 class				New three-shaft free turbine turboprop with single-stage LP centrifugal compressor driven by single-stage turbine, single-stage HP centrifugal compressor driven by single-stage cooled turbine, annular reverse-flow combustor, and two-stage power turbine driving propeller through an offset gearbox.	Development engine
PW115	1,600	0.516	861	1.86	Production derivative of PW100.	Embraer EMB-120 Brasilia
PW118/118A	1,800	0.498	861	2.09	Mechanically similar to PW115 with increased power rating.	Embraer EMB-120 Brasilia
PW119/119B	2,180	0.490	916	2.38	Derivative of PW118 with increased power rating.	Dornier 328
PW120/120A	2,000	0.485	921	2.17	Derivatives of PW115 with new 1,200 rpm gearbox and reserve power rating.	Aerospatiale/Aeritalia ATR 42 de Havilland Dash 8-100
PW121	2,150	0.476	936	2.30	Variant of PW120 with increased power rating.	Aerospatiale/Aeritalia ATR 42 de Havilland Dash 8-100

Pratt & Whitney Canada (turboprop engines) (*Continued*)

Company Model No.	Military Model No.	S.L. Rated Power, SHP	SFC at S.L. Rated Power	Engine Weight, Lb	Power-to-Weight Ratio	Configuration	Applications
PW124/124A/B		2,400	0.468	1,060	2.26	Growth derivative of PW120 with larger diameter LP compressor providing increased airflow, intercompressor bleed valve, cooled LP turbine stators, and new gearbox.	Aerospatiale/Aeritalia ATR72–200 British Aerospace ATP Fokker 50
PW123/123AF/E PW123B PW123C/D		2,380 2,500 2,150	0.470 0.463 0.483	992 992 992	2.40 2.52 2.17	Derivatives of PW124 with 120A gearbox and PW124 turbomachinery module.	de Havilland Dash 8–300 (PW123/123B) Canadair CL–215T/415 (PW123AF) Dash 8–200 (PW123C/D) Dash 8–315 (PW123E)
PW125B PW126 PW126A PW127/B/C/D		2,500 2,653 2,662 2,750	0.463 0.463 0.461 0.459	1,060 1,060 1,060 1,060	2.36 2.50 2.51 2.59	Uprated version of PW124. Growth engines.	Fokker 50 Jetstream ATP (PW126/126A) Aerospatiale/Aeritalia ATR72/ATR42–500 (PW127) Antonov An–140 (PW127A) Fokker 50 (PW127B) Xian XAC Y7–200A (PW127C) Jetstream 61 (PW127D) de Havilland Dash 8–400
PW150		6,500 to 7,500 eshp				Major redesign with three-stage axial-flow LP compressor, single-stage centrifugal HP compressor, increased turbine cooling, and new 5,000 shp-class reduction gearbox.	

Pratt & Whitney Canada (turboshaft engines)

Company Model No.	Military Model No.	S.L. Rated Power, SHP	SFC at S.L. Rated Power	Engine Weight, Lb	Power-to-Weight Ratio	Configuration	Applications
PT6B Single Power Section Series							
PT6 Mk.2						Opposed free turbine turboshaft with three-stage axial-flow and single-stage centrifugal flow compressor powered by a single-stage turbine, reverse-flow annular combustor, and single-stage power turbine driving front-mounted power takeoff gearbox.	Flown in Hiller Ten99
PT6B-9		550	0.665	245	2.24	Power section same as that in PT6A-6.	Lockheed 286
PT6B-36		960	0.594	372	2.58	Basically a PT6T-3B power section with modified exhaust section, and redesigned gearbox.	Sikorsky S-76B
PT6T Twinned Engine Series							
PT6T-3		1,800	0.595	645	2.79	Two free turbine power sections coupled to a common gearbox having a single power output shaft. Each power section was basically a PT6A-27 with the addition of cooling in the gas generator turbine stators and a single exhaust port. First PT6 with cooled stators.	Bell 212 / Augusta-Bell 212 / Sikorsky S-58T
PT6T-6		1,875	0.592	657	2.85	Uprated version of PT6T-3 with hot-end improvements to provide higher power.	Augusta-Bell 212 / Sikorsky S-58T
	T400-CP-400	1,800	0.594	714	2.52	U.S. Navy, U.S. Air Force, and Canadian Armed Forces version of PT6T-3.	Bell AH-1J Seacobra, UH-1N, CUH-1N Iroquois
	T400-CP-401	1,800	0.594	714	2.52	U.S. Army version of T400-CP-400.	Bell VH-1N and UH-1N Iroquois

Pratt & Whitney Canada (turboshaft engines) (*Continued*)

Company Model No.	Military Model No.	S.L. Rated Power, SHP	SFC at S.L. Rated Power	Engine Weight, Lb	Power-to-Weight Ratio	Configuration	Applications
	T400-WV-402	1,970	0.591	745	2.64	Uprated version of T400-CP-400 having hot-end improvements to accommodate higher power requirements for use by U.S. Navy.	Bell AH-1J and AH-1T Seacobra
PW200 Series							
PW209T		937	0.580	468	2.0	Twin-shaft engine with two through-shaft power sections driving a front-mounted gearbox having a single power output shaft. Power sections had a single-stage centrifugal compressor driven by a single-stage turbine, annular reverse-flow combustor, and single-stage free power turbine.	Intended for Bell TwinRanger
PW205B		590	0.556	220	2.68	Single power section growth version of PW209T with increased airflow and turbine temperature.	First flown on MBB BO 105LS-B1
PW206A		621 640 Recert.	0.543			Uprated version of PW205B.	McDonnell Douglas MD Explorer
PW206B		635	0.548			Variation of PW206A using 206A power section with a new angled drive gearbox to fit the EC135 helicopter.	Aerospatiale/Deutsche Eurocopter EC135 MBB BO 108
PW206C		640				Similar to PW206A, but configured for Augusta A109 Power.	Agusta A109 Power

Allison (Significant Dates)

Company Model No.	Military Model No.	Design/ Devel. Start	Engine First Run	Engine First Flight	Engine Cert./Qual.	First Prod. Engine Delivered	Last Prod. Engine Delivered	Total Engines Produced (Production thru 1995)
Turboshaft Engines								
Type I		1957	Apr 1959					
Series I								
Type II	T63-A-3	1958	Feb 1960	Feb 1962	Nov 1961	Dec 1962	1962	6
250-C10	T63-A-5	1961	Feb 1962	Dec 1962	Dec 1962	1962	1962	54
	T63-A-5A	1961			Sep 1965	Dec 1965	Aug 1969	2,455
250-C18/18A	T63-A-700	1965				1965	1975	3,895
Series II								
250-C20		1966			May 1970	1971		3,527
250-C20B	T63-A-720	1968			Feb 1974	1973	Aug 1978	11,774 –C20/B/R/ All and A-720 total
250-C20R		1984	1985	1986	Sep 1986	Sep 1986		
Series III								
250-C28 B/C			Oct 1974	1975	May 1976	1976		869
Series IV								
250-C30/All			May 1976	1977	Mar 1978	1978		1,855 –C30/C30R
250-C30R	T703-AD-700				Jul 1981	1985		514 –A-703
Turboprop Engines								
Series I								
250-B7	T63-A-1	1958	Jun 1960		Dec 1961	1962	1962	3
250-B15G					Mar 1969	Jul 1969	Jul 1974	92

Allison (Significant Dates) (*Continued*)

Company Model No.	Military Model No.	Design/ Devel. Start	Engine First Run	Engine First Flight	Engine Cert./Qual.	First Prod. Engine Delivered	Last Prod. Engine Delivered	Total Engines Produced
								Production thru 1995
Series II								
250-B17B/C/D		1966			Apr 1971	Dec 1970		958 all -B17s
250-B17F		1985	1985	1986	May 1988	Feb 1988		
Demonstrator Engine								
GMA 500		1977	1979					
Turboshaft Engines								
ATE109	T800 prototype	1984	Oct 1984	Mar 1985				
	T800-LHT-800	Jul 1984	Aug 1986	Oct 1988	Jul 1993 (Army) Sep 1993 (FAA)			
	T406-AD-400	1986	Dec 1986	Mar 1989	Jul 1988 (FAA)			
Turboprop Engine								
GMA 2100					Apr 1993			
Turbofan Engine								
GMA 3007		1989	Jul 1991	Aug 1992				
Turbojet Engines								
Model 150		1989	Aug 1990					
Model 120		1990	Jul 1993					
GMA 900	J102-AD-100	1988	1991					

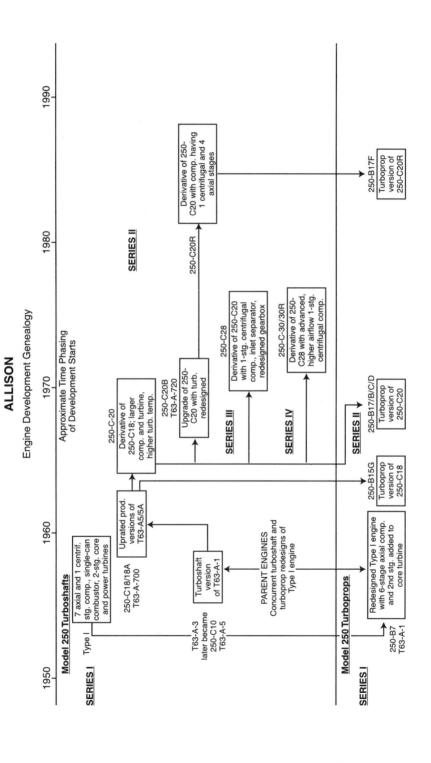

ALLISON

Engine Development Genealogy

Approximate Time Phasing
of Development Starts

Model 250 Turboshafts

SERIES I — Type I — 7 axial and 1 centrif. stg. comp., single-can combustor, 2-stg. core and power turbines

T63-A-3 later became 250-C10 T63-A-5

250-C18/18A T63-A-700

Turboshaft version of T63-A-1

Uprated prod. versions of T63-A5/5A

250-C-20 — Derivative of 250-C18; larger comp. and turbine, higher turb. temp.

250-C20B T63-A-720 — Upgrade of 250-C20 with turb. redesigned

SERIES III — 250-C28 — Derivative of 250-C20 with 1-stg. centrifugal comp., inlet separator, redesigned gearbox

SERIES IV — 250-C-30/30R — Derivative of 250-C28 with advanced, higher airflow 1-stg. centrifugal comp.

SERIES II — 250-C20R — Derivative of 250-C20 with comp. having 1 centrifugal and 4 axial stages

250-B17F — Turboprop version of 250-C20R

PARENT ENGINES — Concurrent turboshaft and turboprop redesigns of Type I engine

Model 250 Turboprops

SERIES I — 250-B7 T63-A-1

Redesigned Type I engine with 6-stage axial comp. and 2nd stg. added to core turbine

250-B15G — Turboprop version of 250-C18

SERIES II — 250-B17/B/C/D — Turboprop version of 250-C20

1950   1960   1970   1980   1990

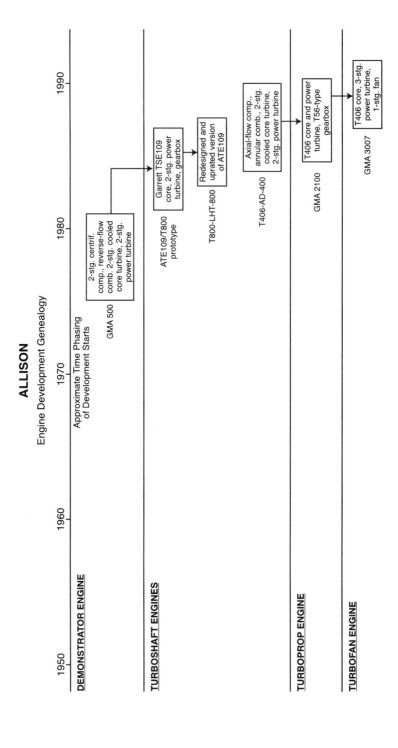

ALLISON

Engine Development Genealogy

Approximate Time Phasing
of Development Starts

1950   1960   1970   1980   1990

DEMONSTRATOR ENGINE

GMA 500

2-stg. centrif.
comp., reverse-flow
comb. 2-stg. cooled
core turbine, 2-stg.
power turbine

TURBOSHAFT ENGINES

ATE109/T800
prototype

Garrett TSE109
core, 2-stg. power
turbine, gearbox

T800-LHT-800

Redesigned and
uprated version
of ATE109

T406-AD-400

Axial-flow comp.,
annular comb., 2-stg.
cooled core turbine,
2-stg. power turbine

TURBOPROP ENGINE

GMA 2100

T406 core and power
turbine, T56-type
gearbox

TURBOFAN ENGINE

GMA 3007

T406 core, 3-stg.
power turbine,
1-stg. fan

Allison (turboshaft engines)

Company Model No.	Military Model No.	S.L. Rated Power, SHP	SFC at S.L. Rated Power	Engine Weight, Lb	Power-to-Weight Ratio	Configuration	Applications
Type I		250 class				Free turbine engine with 7 axial and 1 centrifugal compressor stgs., single can-type comb., single-stg. core turb., 2-stg. free power turb., gearbox off middle of engine with power takeoff axis parallel to rotor shafts.	Original Model 250 running engine
Series I							
250-C-10	T63-A-3	250	0.71	136	1.84	Redesigned Type I engine with 6-stg. axial compressor using new aerodynamics, single-stg. centrifugal compressor, mechanically and aerodynamically redesigned turbine section with 2nd stg. added to core turbine, and dual outlet exhaust duct.	Flight tested in Bell HUL-1M
	T63-A-5	250	0.71	136	1.84	Slightly modified model of T63-A-3.	Bell OH-4A prototype Hughes OH-6A prototype Hiller OH-5A prototype
250-C18/18A	T63-A-5A	317	0.70	141	2.25	Uprated production derivative of T63-A-5.	Hughes OH-6A (A-5A) Bell 206A JetRanger
	T63-A-700	317	0.70	141	2.25	Upgraded production derivative of T63-A-5A.	Fairchild-Hiller FH-1100 Hughes 500 Augusta AB206 JetRanger MBB BO 105 (twin) Bell OH-58A (A-700)

Allison (turboshaft engines) (*Continued*)

Company Model No.	Military Model No.	S.L. Rated Power, SHP	SFC at S.L. Rated Power	Engine Weight, Lb	Power-to-Weight Ratio	Configuration	Applications
Series II							
250-C20		400	0.63	158	2.53	Derivative of 250-C18 with larger diameter (13% higher airflow) axial and centrif. comp. stgs. and power turbine plus higher turbine temp.	Bell 206B JetRanger II MBB BO 105A Augusta-Bell 206B/206B-1 Augusta A109 Soloy UH-12E conversions
250-C20B	T63-A-720	420	0.65	158	2.66	Upgraded version of 250-C20 with redesigned turbine.	Bell/Augusta-Bell 206B JetRanger III and 206 LongRanger MD 500D BO 105CBS/E
250-C20R		450	0.60	173	2.60	Derivative of 250-C20 with redesigned compressor having 4 axial and 1 centrifugal compressor stages.	Agusta A 109A Mk.II Gemini ST

Allison

Company Model No.	Military Model No.	S.L. Rated Power, SHP	SFC at S.L. Rated Power	Engine Weight, Lb	Power-to-Weight Ratio	Configuration	Applications
Series III							
250-C28		500	0.61	219	2.28	Derivative of 250-C20 with advanced higher airflow single-stage centrifugal comp., redesigned gearbox, inlet particle separator.	Bell 206L LongRanger, MBB Bo105 (twin)
Series IV							
250-C30		650	0.59	240	2.71	Derivative of 250-C20 with a more advanced and higher airflow single centrifugal compressor stage.	Bell 206/OH-58D (AHIP), 206L LongRanger, Aerospatiale AD350G, Sikorsky S-76 (twin), Hughes 530
250-C30R	T703-AD-700	650	0.59	252	2.57	250-C30 with electronic fuel control.	
Turboprop Engines							
Series I							
250-B7	T63-A-1	250	0.71	136	1.84	Turboprop version of Model 250 engine developed concurrently with T63-A-3.	Experimental and test engine
250-B15G		317	0.70	182	1.74	Turboprop version of 250-C18.	Cessna O-2T, Helio Twin Stallion, H-S70 Courier, Robertson STOL Series, SIAI-Marchetti L-1019, Beech Baron and Bonanza conversions

Company Model No.	Military Model No.	S.L. Rated Power, SHP	SFC at S.L. Rated Power	Engine Weight, Lb	Power-to-Weight Ratio	Configuration	Applications
Series II							
250-B17B/C/D		400 to 420	0.66	182	2.20	Turboprop version of 250-C20.	SIAI-Marchetti SF.260TP, SF.600TP; FUJI KM-2Kai, T-5; HAL HTT-34; Britten-Norman Turbo Islander; Nomad N-22B; Enaer T35DT; Valmet L-90TP
250-B17F		420	0.65	195	2.15	Turboprop version of 250-C20R.	
Demonstrator Engine							
GMA 500		850 shp				Advanced technology turboshaft. 2-stage centrifugal comp., annular reverse-flow transpiration–cooled combustor, 2-stage cooled core turb., 1-stage free power turb., inlet particle separator.	ATDE demonstrator engine
Turboshaft Engines							
ATE109	T800 Prototype	1,200 shp				2-stg. centrifugal comp., reverse-flow annular comb., 2-stg. cooled core turb., 2-stg. free power turb., inlet particle separator.	
T800-LHT-800		1,382 shp	0.46	310	4.46	Garrett TSE109 core with increased turb. temp. Allison designed 2-stg. power turb., inlet particle separator, electronic control, gearbox, and accessories.	Boeing Sikorsky RAH-66 Comanche; Agusta A-129; Bell UH-1H Huey

Designation	Model	Rating		Weight (lb)		Description	Applications
T406-AD-400		6,150 shp flat rated	0.42	971	6.33	T56-A-427 comp. and turb. aerodynamics with T701-AD-700 comp. design. 14-stg. axial-flow comp. with IGVs and 1st 5 stators variable, annular comb., 2-stg. cooled core turb., 2-stg. free power turbine, and FADEC system.	Bell-Boeing V-22 Osprey
Turboprop Engine	GMA 2100	5,700 shp flat rated	0.41	1,548	3.68	T406-AD-400 core and power turbine with T56-type gearbox and FADEC system.	Saab 2000 IPTN N-250 Lockheed c-130J/L100F
Turbofan Engine	GMA 3007	7,150 lb	0.39	1,581	4.52	T406-AD-400 core, 3-stg. power turb., single-stg. fan, and FADEC system. Bypass ratio 5:1.	Embraer EMB-145 Cessna Citation X Teledyne Ryan Tier 2+
Turbojet Engines	Model 150	500				Low-cost, simple turbojet flow path; engine designed as a complete installed module with bifurcated inlets and exhausts. (Engine details and performance classified).	Sea Bear standoff missile Maverick Longhorn missile
	Model 120	274	1.19	16.1	17	Low-cost simple turbojet with single-stage centrifugal compressor, annular reverse-flow combustor, single-stage turbine.	
J102-AD-100	GMA 900	274				Mach 3.5 engine using Lamilloy® in combustor, turbine guide vanes and single-stage turbine rotor. Fuel-to-air heat exchanger lowered temperature of compressor outlet air used to cool Lamilloy® structures. (Engine details and performance classified).	Army LONGFOG boost/sustainer missile system

Garrett (AlliedSignal Engines) (Significant Dates)

Turboprop Engines

Company Model No.	Military Model No.	Design/Devel. Start	Engine First Run	Engine First Flight	Engine Cert./Qual.	First Prod. Engine Delivered	Last Prod. Engine Delivered	Total Engines Produced
TPE331-25, -43, Series I and II	YT76-G-2, -4	1962	Jul 1963	Apr 1964	Feb 1965	1965		12,774 of all TPE331 models thru 1996
	YT76-G-6, -8			Jul 1965		1965		
	T76-G-10, -12	1966	1966	Aug 1967	1967 (MQT)	1967		
TPE331-1,-2		1965			Dec 1967			
TPE331-3		1967			Mar 1969			
TPE331-5		1969	1970		May 1970			
TPE331-6		1969	1970		May 1970			
TPE331-8, -9	T76-G-420, -421	1974	1975		1977			
TPE331-10, -11		1974	1975		Nov 1976			
		1975	1976		Jan 1978,-10			
					1979,-11			
TPE331-12		1983	1984		Dec 1984			
TPE331-14		1979	1981	1982	Apr 1984			
TPE331-15		1979	1981		Nov 1988			
TPE331-14GR, -14HR					1992	1993		
TPF351-20		1987	May 1989	Jul 1990				

Turboshaft Engines							
GTP331	1959	Dec 1960	Oct 1961		1967		
TSE331-7	1967	1968	May 1968			1975	48 of all TSE331 models
TSE331-3U	1968			Apr 1970		1975	
TSE36-1	1969	Sep 1970	Dec 1970				
TSE231-1	1983	Oct 1984	Mar 1985				
TSE109/ATE109	1985	Aug 1986	Oct 1988				
T800-LHT-800	1992	Mar 1994			1993		
T800-LHT-801							
CTS800-0A			1988	Jul 1993			
CTS800-4N				Sep 1993			
Turbofan Engines							
ATF3-1	1966	May 1968	May 1968	1972 (PFRT)	1980		204 of all ATF3 models
YF104-GA-100				May 1981			
ATF3-3	1976	Nov 1977	May 1978	Dec 1981			
ATF3-6	1979		Feb 1982		Dec 1983		
ATF3-6A						1985	
TFE731-2	May 1969	Sep 1970	May 1971		Aug 1972		2,597 of all TFE731s thru 1996
TFE731-3	Oct 1972	Sep 1973	Jul 1974	Sep 1974	Apr 1974		3,455
TFE731-3A, -3B			Aug 1990	Mar 1981			
TFE731-4	Oct 1979	Aug 1981	Oct 1982	Nov 1991	1982		156
TFE731-5	Mar 1979		May 1984	Nov 1983			1,418
TFE731-5A			Mar 1990	Mar 1985			
TFE731-5B				Feb 1991			
TFE731-20	Mar 1992	Jun 1993	Dec 1993	Dec 1996	1994		31
TFE731-40	Mar 1992	Sep 1993	Apr 1994	Jul 1995	1994		73
TFE731-60	Dec 1992	Dec 1993	Jun 1994	May 1995	1994		66

Garrett (AlliedSignal Engines) (Significant Dates) (*Continued*)

Company Model No.	Military Model No.	Design/ Devel. Start	Engine First Run	Engine First Flight	Engine Cert./Qual.	First Prod. Engine Delivered	Last Prod. Engine Delivered	Total Engines Produced
Turbofan Engines								
TFE1042-5, -6, -7		1978	Aug 1979					
TFE1042-70		1983	Apr 1985	1987		Oct 1988		
	F124-GA-100							
	F125-GA-100							
TFE76		1979	1981					
TFE109	F109-GA-100	1982	Dec 1983	Oct 1985	1987	1987		25
TFE109-1								
TFE109-2								
TFE109-3								
CFE738		1988	May 1990	Mar 1993	Dec 1993	Sep 1993		104 thru 1996
Turbojet Engines								
ETJ331	YJ401-GA-400	1971						
ETJ341		1974						
ETJ131		1976						
ETJ1081		1984	1985					
ETJ1091		1985	1990					

GARRETT (ALLIEDSIGNAL ENGINES)

Engine Development Genealogy

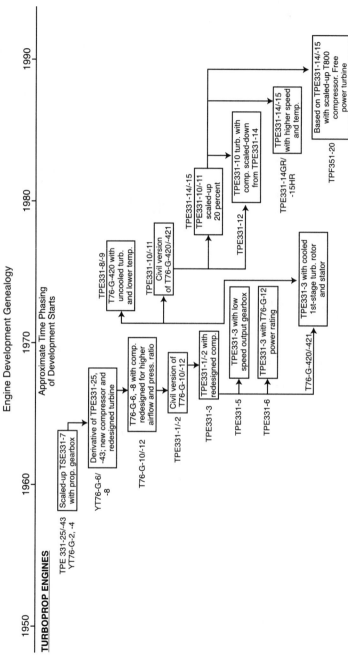

GARRETT (ALLIEDSIGNAL ENGINES)

Engine Development Genealogy

Approximate Time Phasing
of Development Starts

TURBOSHAFT ENGINES

Timeline: 1950, 1960, 1970, 1980, 1990, 2000

GTP331 — 2 centrifugal comp. and 3 turbine stages on one shaft

TSE331-7 — GTP331 with new gearbox

TSE331-3U — Turboshaft version of TPE331-3

TSE36-1 — Derived from GTCP36-4 APU

TSE231-1 — New free turbine engine with front-end drive

TSE109/ATE109 — Turboshaft version of TFE109 with redesigned cooled turbine

T800-LHT-800 — Garrett TSE109 with Allison ATDE technology

CTS800-0A — Civil version of T800-LHT-800

CTS800-4N — Variation of CTS800-0A with speed reduction gearbox

T800-LHT-801 — Growth version of T800-LHT-800

932

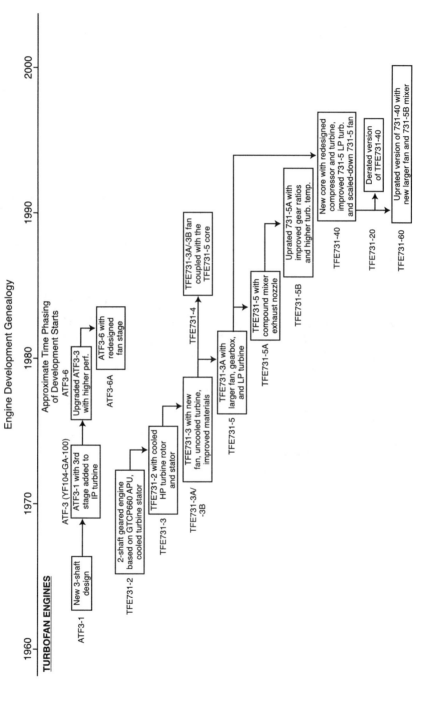

GARRETT (ALLIEDSIGNAL ENGINES)

Engine Development Genealogy

GARRETT (ALLIEDSIGNAL ENGINES)

Engine Development Genealogy

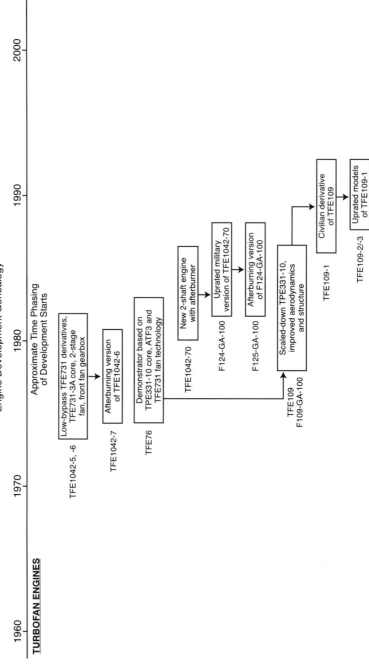

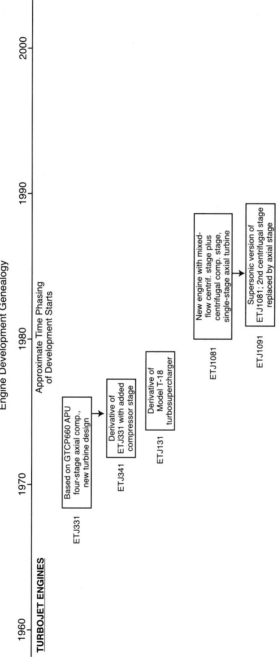

GARRETT (ALLIEDSIGNAL ENGINES)

Engine Development Genealogy

Approximate Time Phasing
of Development Starts

TURBOJET ENGINES

1960   1970   1980   1990   2000

ETJ331 — Based on GTCP660 APU four-stage axial comp., new turbine design

ETJ341 — Derivative of ETJ331 with added compressor stage

ETJ131 — Derivative of Model T-18 turbosupercharger

ETJ1081 — New engine with mixed-flow centrif. stage plus centrifugal comp. stage, single-stage axial turbine

ETJ1091 — Supersonic version of ETJ1081; 2nd centrifugal stage replaced by axial stage

Garrett (AlliedSignal Engines) (turboprop engines)

Company Model No.	Military Model No.	S.L. Rated Power, SHP	SFC at S.L. Rated Power	Engine Weight, Lb	Power-to-Weight Ratio	Configuration	Applications
TPE331-25/-43 Series I&II	YT76-G-2/-4	575	0.665	330	1.74	Derivative of TSE331-7 with redesigned compressor, higher turb. temp, and two-stage reduction gear.	Gulfstream Commander 680, Air Asia Porter, Volpar Super Turbo 18, Mitsubishi MU-2B, D, Carstedt Jetliner 600A, Fairchild Helicopter
	YT76-G-6/-8	660	0.63	328	2.01	Derivative of TPE331-25, -43 with compressor splitter vanes added.	North American OV-10A prototype
	YT76-G-10/-12	715	0.60	324	2.21	Derivative of YT76-G-6, -8 with redesigned compressor and turbine providing higher airflow, press. ratio, and efficiency	North American OV-10, -10A counter insurgency (COIN) aircraft
TPE331-1		665 flat rated	0.59	336	1.98	Civil versions of T76-G-10, -12; TPE331-1 was flat-rated version of TPE331-2.	Beech C-90, Carstedt Jetliner, Fairchild 600B, Merlin IIB, Volpar Turbo Liner, CASA 212-100, Shorts Skyvan 3, Marsh Turbo Ag Cat, Mitsubishi MU-2DP, F, G, Fairchild Peacemaker
TPE331-2		715	0.59	336	2.13		

Engine	Power (shp)	SFC			Description	Applications
TPE331-3	840	0.59	355	2.37	Derivative of TPE331-1/-2 with compressor redesigned using transonic aerodynamics to raise airflow and pressure ratio.	Fairchild Merlin III, IIIA; Swearingen Metro I, II; BAe Jetstream 31
TPE331-5	776 flat rated	0.59	370	2.09	TPE331-3 power section with new low-speed gearbox.	Gulfstream Commander 690/840; CASA 212-100; Mitsubishi MU-2N; Dornier DO228
TPE331-6	715 flat rated	0.59	360	1.99	TPE331-3 power section matched with 715-hp gearbox for better high-altitude, hot-day perf.	Mitsubishi MU-2S,K,L,M; Beech B100
T76-G-420/-421	1,040	0.56	389	2.67	TPE331-3 compressor with redesigned turbine, increased turbine temperature, cooled turbine first-stage rotor and stator, and shortened combustor.	North American Rockwell OV-1OD COIN aircraft
TPE331-8/-9	865 flat rated	0.566	370	2.34	Derivative of T76-G-420 with uncooled turbine and reduced turbine temperature.	Cessna Conquest 441
TPE331-10	1,000	0.56	380	2.63	Civilian version of T76-420.	Gulfstream Commander 980, 1000; CASA 212-200
TPE331-11	1,100 (wet rating)	0.558	400	2.75	Redesigned gearbox with higher power rating.	BAe Jetstream 31; Fairchild Merlin IIIB, IIIC; Metro II, III, IV; Mitsubishi Solitaire, Marquise

Garrett (AlliedSignal Engines) (turboprop engines) (*Continued*)

Company Model No.	Military Model No.	S.L. Rated Power, SHP	SFC at S.L. Rated Power	Engine Weight, Lb	Power-to-Weight Ratio	Configuration	Applications
TPE331-12/-12B		1,070 flat rated	0.534	385	2.78	Turbine based on TPE331-10/-11 with compressors scaled down from TPE331-14.	Short Bros./Embraer Tucano Trainer BAe Jetstream Super 31 Metro V
TPE331-14		1,250				Redesigned version of TPE331-10/-11 having a 20% increase in size.	Pipe Cheyenne 400LS
TPE331-15		1,645	0.50	554	2.97	Redesigned gearbox with higher power rating.	Grumman S-2
TPE 331-14GR/-14HR		1,759 flat rated 1,960 APR	0.52	620	2.84	Higher rotating speed, higher turbine temperature, improved gas generator turb. materials.	BAe Jetstream 41
TPF351-20		2,100	0.496	763	2.75	New rear drive three-stage power turbine, two-stage TPE331-14 and -15 gas generator turbine, and scaled-up T800 compressor.	Embraer/FAMA CBA123

938

Garrett (AlliedSignal Engines) (turboshaft engines)

Company Model No.	Military Model No.	S.L. Rated Power, SHP	SFC at S.L. Rated Power	Engine Weight, Lb	Power-to-Weight Ratio	Configuration	Applications
GTP331		400	0.66	200	2.0	Begun as advanced 330 shp APU. Two centrifugal compressors and three axial turbines on a single shaft; reverse-flow annular combustor.	
TSE331-7		440	0.64	285	1.54	Derivative of GTP331 with new gearbox and inlet.	Republic Lark (Sud Alouette)
TSE331-3U		800	0.59	355	2.25	Turboshaft version of TPE331-3 turboprop.	Sikorsky S-55 (H-19)
TSE36-1		240	0.83	178	1.35	Derived from GTCP36-4 APU; single-stage centrifugal compressor, single-can reverse-flow comb., single-stage radial-inflow turbine.	Enstrom T-28 prototype
TSE231-1		474	0.605	171	2.77	New free turbine engine with front end drive; two-stage centrifugal compressor, reverse-flow annular combustor, single-stage gas generator turbine and single-stage power turbine.	Flown in U.S. Marine Small Tactical Aerial Mobility Platform (STAMP)
TSE109/ATE109		1,000 to 1,200				Free turbine demonstrator engine based on F109 turbofan; two-stage centrifugal compressor, reverse-flow annular combustor, two-stage cooled gas generator turbine, two-stage power turbine.	Served as prototype for T800 engine
	T800-LHT-800	1,334	0.45	310	4.3	Developed as joint venture with Allison; combined Garrett's TSE109 design having cooled gas generator turbine with Allison two-stage power turbine, gearbox, inlet particle separator, and accessories.	Boeing Sikorsky RAH-66 Comanche Bell LongRanger IV 442

Garrett (AlliedSignal Engines) (turboshaft and turbofan engines)

Company Model No.	Military Model No.	S.L. Rated Power, SHP	SFC at S.L. Rated Power	Engine Weight, Lb	Power-to-Weight Ratio	Configuration	Applications
Turboshaft Engines							
	T800-LHT-801	1,550 shp	0.46	330	4.7	Growth version of T800-LHT-800 with new compressor technology providing 17 percent increase in airflow, also turbine temperature increased 30 deg. F.	
CTS800-OA		1,360 shp	0.45	310	4.39	Civil derivative of T800-LHT-800.	
CTS800-4N		1,290 shp	0.467	383	3.37	Variation of CTS800-OA with speed reduction gearbox for commercial helicopter applications.	Option for Bell 442
Turbofan Engines							
ATF3-1		4,050 lb	0.46	817	4.96	Three-spool engine; single-stage fan driven by two-stage IP turbine, five-stage LP axial compressor driven by two-stage LP turbine, single-stage centrifugal HP compressor driven by single-stage HP turbine, reverse-flow annular comb.; bypass ratio 3.19.	
ATF3	YF104-GA-100	4,050 lb	0.47	874	4.63	Same as ATF3-1 except for three-stage IP turbine; bypass ratio 3.16.	Teledyne Ryan YQM-98A Compass Cope R prototype RPV
ATF3-6		5,050 lb	0.48	1,100	4.59	Components rematched and optimized to operate at higher turbine temperature; bypass ratio 2.83.	Dassault HU-25A Guardian

Garrett (AlliedSignal Engines) (turbofan engines)

Company Model No.	Military Model No.	S.L. Rated Thrust, Lb	SFC at S.L. Rated Thrust	Engine Weight, Lb	Thrust-to-Weight Ratio	Configuration	Applications
ATF3-6A		5,440	0.506	1,125	4.84	Same as ATF3-6 except fan redesigned with higher efficiency and enough ruggedness to withstand bird ingestion; bypass ratio 2.88.	Dassault-Breguet Falcon 200; Dassault HU-25A Guardian
TFE731-2		3,500	0.504	743	4.71	Two-spool engine with geared front fan; fan and four-stage LP axial compressor driven by three-stage LP turbine; core had single-stage centrifugal compressor, reverse-flow annular combustor, and single-stage cooled turbine; bypass ratio 2.67.	Gates Learjet Models 35, 36; Dassault-Breguet Falcon 10; Republic of China AT-TC-3; CASA C-101 Aviojet; Argentina FMA IA.63 Pampa; Nanchang K-8; Aero Industry Development Center AT-3
TFE731-3		3,700	0.515	754	4.91	Same basic configuration as TFE731-2 with addition of new cooled-core turbine rotor, higher turbine temperature, and mechanical design changes for easier maintenance; bypass ratio 2.76.	Lockheed JetStar I and II conversions; Hawker-Siddeley HS 125-700; Rockwell Sabreliner 65; Dassault-Breguet Falcon 50; CASA C-101BB trainer; IAI 1124 Westwind
TFE731-3A		3,700	0.515	775	4.77	Modified fan design giving improved altitude performance, TFE731-3B core had uncooled turbine with improved materials; bypass ratio 2.93 for -3A and 3.01 for -3B.	—3A engine used in: Gates Learjet 55; —3B engine used in: Cessna Citation III; IAI 1125 Westwind
TFE731-3B		3,650	0.507	769	4.75		
TFE731-4		4,080	0.517	822	4.96	Used TFE731-3A/-3B fan coupled with TFE731-5 core; bypass ratio 2.74.	Cessna Citation VII

Garrett (AlliedSignal Engines) (turbofan engines) (*Continued*)

Company Model No.	Military Model No.	S.L. Rated Thrust, Lb	SFC at S.L. Rated Thrust	Engine Weight, Lb	Thrust-to-Weight Ratio	Configuration	Applications
TFE731-5		4,304	0.484	852	5.05	Used TFE731-3 compressor, modified HP turbine, ATF3-6A fan scaled to 0.97, new gearing and LP turbine to match fan; bypass ratio 3.33.	British Aerospace BAe 125-800 CASA C-101CC trainer Dassault-Breguet Falcon 20 conversions
TFE731-5A		4,500	0.469	884	5.09	Modified version of TFE731-5 using compound mixer exhaust nozzle technology; bypass ratio 3.61.	Dassault-Breguet Falcon 900
TFE731-5B		4,750	0.47	899	5.28	Uprated version of TFE731-5A with improved gear ratio for higher fan speed, and increased turbine temperature; bypass ratio 3.62.	Dassault Falcon 900B and 20B Hawker 800XP
TFE731-20		3,500	0.458	895	3.91	Derated version of TFE731-40; bypass ratio 3.66.	Learjet 45
TFE731-40		4,250	0.472	885	4.80	New core with redesigned higher pressure ratio HP centrifugal compressor, improved turbine mechanical and aerodynamic design, and scaled down TFE731-5/5B fan; bypass ratio 3.5.	IAI Astra SPX Dassault Falcon 50EX
TFE731-60		5,000	0.424	988	5.06	Uprated version of TFE731-40 with new larger diameter fan and TFE731-5A mixer; bypass ratio 4.4.	Dassault Falcon 900EX
TFE1042-5		4,100	0.66	870	4.7	Low-bypass derivatives of TFE731 with TFE731-3A core, two-stage fan, front fan gearbox; bypass ratio 0.8.	
TFE1042-6		4,100	0.685	870	5.6		
TFE1042-7		8,340 (A/B)	2.21	1,235	6.75	Afterburning version of 1042-6.	
		4,820 (dry)	0.696		3.9		

TFE1042-70		8,350 (A/B) 5,000 (dry)	2.24	1,350	6.2 3.7	New two-spool engine with afterburner and variable exhaust nozzle, three-stage LP fan driven by single-stage cooled turbine, four-stage axial plus single-stage centrifugal HP compressor driven by cooled single-stage turb., annular combustor, bypass ratio 0.36.	
	F124-GA-100	6,300	0.81	1,100	5.73	Uprated military versions of TFE1042-70 with higher airflow fan, redesigned LP turbine and increased turbine temperature.	Aero Vodochody L-159 (F124 engine)
	F125-GA-100	9,250 (A/B) 6,025 (dry)					Republic of China Indigenous Defense Fighter (F125 engine)
TFE76		1,500 class				Demonstrator engine based on TPE331-10 core and utilizing ATF3 and TFE731 fan technology.	
TFE109	F109-GA-100	1,330	0.39	430	3.09	Two-spool contrarotating high-bypass engine based on core of TPE331 turboprop family with significant perf. improvements; single-stage fan directly coupled to two-stage LP turbine, core had two-stage centrifugal HP compressor driven by two-stage turbine and reverse-flow annular combustor; bypass ratio 5.0.	Designed to power Fairchild T-46A trainer (canceled in 1986)
TFE109-1		1,330	0.39	439	3.03	Civilian derivative of the TFE109.	Promavia Jet Squalus trainer
TFE109-2		1,500				Uprated models of TFE109-1.	
TFE109-3		1,600					

Garrett (AlliedSignal Engines) (turbofan/turbojet engines)

Company Model No.	Military Model No.	S.L. Rated Thrust, SHP	SFC at S.L. Rated Thrust	Engine Weight, Lb	Power-to-Weight Ratio	Configuration	Applications
Turbofan Engine							
CFE738		5,725	0.371	1,325	4.32	Developed as joint venture with GE; two-spool engine with GE HP spool having 5-stage axial and single-stage centrifugal compressor, annular combustor, cooled two-stage turbine; LP spool had single-stage fan and three-stage turbine based on Garrett TFE731-5; bypass ratio 5.3.	Dassault Falcon 2000
Turbojet Engines							
ETJ331	YJ401-GA-100	500 to 1,000				Based on GTCP660 APU; four-stage axial compressor, annular combustor, newly designed turbine.	Developed for target drones and cruise missiles including Harpoon
ETJ341		1,000 class				Same as ETJ331 except compressor stage added.	Flight tested in Boeing Turbo Condor drone
ETJ131		95	1.7			Evolved from Model T-18 turbosupercharger; single-stage centrifugal compressor, single-stage radial turbine, single can-type combustor.	Developed for low-cost military decoys and target vehicles
ETJ1081		1,000 to 1,300	0.95	131	8 to 10	Mixed-flow plus centrifugal compressor stage, air blast atomizing combustor, single-stage axial turbine, fuel-lubricated ceramic bearings.	Developed for advanced RPVs and cruise missiles
ETJ1091						Supersonic version of ETJ1081 with 2nd centrifugal compressor stage replaced by axial stage, improved high temperature materials and higher turbine temperature.	Developed for supersonic missiles

Miniature Low-Cost Expendable Turbojet Engines

Company Model No.	Military Model No.	Design/Devel. Start	Engine First Run	Engine First Flight	Engine Cert./Qual.	First Prod. Engine Delivered	Last Prod. Engine Delivered	Total Engines Produced
Allison								
120		1990	1992					
Sunstrand								
TJ-20		1986	1987					
TJ-50		1990	1992					
TJ50C		1990	1992					
TJ70		1989	1988					
TJ90		1989	1989					
Technical Directions								
TD4		1990	1992					
TD7		1988	1992					
Teledyne CAE								
304		1990	1992					
304C		1990	1992					
305-4		Jan 1986	Jul 1986					
305-7E		Mid-1986	Late 1986	May 1987				
305-10		1990	1991					
305-10C		1991	1992					
318		1989	1991					
Turbine Technologies								
SR30		1989	1990					
Williams International								
WJ119, -2		1988	1990 (WJ119)					
P8910		1989	1991					
P9005		1990	1992					

Miniature Low-Cost Expendable Turbojet Engines

Company Model No.	Military Model No.	S.L. Rated Thrust, SHP	SFC at S.L. Rated Thrust	Engine Weight, Lb	Power-to-Weight Ratio	Configuration	Applications
Allison							
120		274	1.119	16	17.1	Single-stage centrifugal compressor, reverse-flow annular combustor, single-stage turbine.	LONGFOG boost/sustain propulsion system
Sunstrand							
TJ-20		43	1.38	12	3.6	All Sunstrand engines used monorotor back-to-back compressors and radial turbines with reverse-flow annular combustors.	LOCAAPS program demonstration engine
TJ-50		50	1.52	6	8.3		SENGAP program demonstration engine
TJ-50C		66	1.63	6	11.0	TJ-20, -70, -90 had single-stage centrifugal compressors and radial turbines.	SENGAP program demonstration engine
TJ-70		72	1.38	12	6.0	TJ-50 and -50C had mixed-flow compressors and ceramic monorotor, compressor diffuser, turbine nozzle, and exhaust duct.	NLOS and NLOS/TRR program demonstration engine
TJ-90		97	1.44	10.5	9.2		NLOS and NLOS TRR programs demonstration engine
Technical Directions							
TD4		51	1.72	7	7.3	Both Technical Directions engines had a single-stage centrifugal compressor, annular combustor, and single-stage turbine.	SENGAP program demonstration engine
TD7		101	1.6	13.5	7.5		NLOS program demonstration engine
						TD4 had a mixed-flow turbine and TD7 had a radial-flow turbine.	

Model					Description	Program
Teledyne CAE						
304	59	1.21	8.5	6.94	All Teledyne CAE engines had single-stage compressors, annular combustors, and single-stage turbines.	SENGAP program demonstration engine
304C	81	1.44	8	10.13		SENGAP program demonstration engine
305-4	40	1.60	12	3.33	Models 305-4, 305-7, and 318 had centrifugal compressors.	LOCAAPS program demonstration engine
305-7/-7E	90	1.26	19	4.74		NLOS program and NLOS/TRR demonstration engine
305-10	178	1.18	26	6.85	Models 304, 304C, 305-10, and 305-10C had mixed-flow compressors.	LONGFOG sustainer engine, NLOS, and NLOS/TRR
305-10C	218	1.37	25	8.72	Model 305-10C had composite inlet and compressor and a ceramic turbine.	LONGFOG sustainer engine
318	177	1.19	38	4.66		GLTR missile sustainer engine
Turbine Technologies						
SR30	33	1.59	12	2.75	Single-stage centrifugal compressor, reverse-flow annular combustor, and single-stage axial turbine	LOCAAPS program and FOG-M demonstration engine
Williams International						
WJ119, -2	92	1.39	27.5	3.34	All Williams engines had reverse-flow annular combustors and single-stage axial turbines. WJ119-2 and P8910 had 6-stage axial compressors and P9005 had single-stage mixed-flow compressor.	NLOS program and NLOS/TRR demonstration engine
P8910	144	1.35	28	5.14		NLOS, NLOS/TRR, and LONGFOG program demonstration engine
P9005	51	1.43	10	5.14		SENGAP program demonstration engine

APPENDIX C

Acronyms and Abbreviations

AAF	Army Air Forces
AAFSS	Advanced Aerial Fire Support System
AAH	Advanced Attack Helicopter
ACM	Advanced Cruise Missile
AEBG	Aircraft Engine Business Group (GE)
AEDC	Arnold Engineering Development Center
AFAPL	Air Force Aero Propulsion Laboratory
AFB	Air Force Base
AGARD	Advisory Group for Aerospace Research and Development
AHIP	Army Helicopter Improvement Program
AHS	American Helicopter Society
AIDC	Aero Industry Development Center
AIAA	American Institute of Aeronautics And Astronautics
AIPS	Advanced Integrated Propulsion System
AIR	Aero International Regional
ALCM	Air-Launched Cruise Missile
ALH	Advanced Light Helicopter
AMOC	Authorized Maintenance and Overhaul Centers
AN	Army-Navy
AOPA	Aircraft Owners and Pilots Association
APR	Automatic Power Reserve
APU	Auxiliary Power Unit
APW	Avco-Lycoming/Pratt & Whitney
ARPA	Advanced Research Projects Agency
ASEE	American Society of Electrical Engineers
ASH	Assault Support Helicopter and Advanced Scout Helicopter
ASME	American Society of Mechanical Engineers
ASMET	Accelerated Simulated Mission Endurance Test
ASW	Anti-Submarine Warfare
ATDE	Advanced Technology Demonstrator Engines
ATEGG	Advanced Turbine Engine Gas Generator
Avco or AVCO	Aviation Corporation (Lycoming)
AWACS	Airborne Warning and Control System
AX	Experimental Attack

BED	Basic Engineering Development
BITE	Built-In Test Equipment
Btu	British thermal unit
BuAer	Bureau of Aeronautics (Navy)
BuShips	Bureau of Ships (Navy)
CAA	Civil Aeronautics Administration
CAB	Civil Aeronautics Board
CAD/CAM	Computer-Aided Design/Computer-Aided Manufacturing
CAE	Continental Aviation and Engineering
CECO	Chandler Evans Company
CCIP	Critical Component Improvement Program
CEO	Chief Executive Oiffcer
CFE	Commercial Fan Engines (Company)
C-FIN	Combat Flight Inspection and Navigation
CIP	Component Improvement Program
CIS	Commonwealth of Independent States
COIN	Counterinsurgency
CSC	Customer Support Center
DARPA	Defense Advanced Research Projects Agency
DASH	Drone Anti-Submarine Helicopter
DEEC	Digital Electronic Engine Controls
DOD or DoD	Department of Defense
DOT	Department of Transportation
DSARC	Defense Systems Acquisition Review Council
ECD	Eurocopter Deutschland
ECU	Electronic Control Unit
EDSA	European Distribution System Aircraft
EHI	European Helicopter Industries
EMDP	Engine Model Derivative Program
EMI	Electromagnetic Interference
EMS	Engine Monitoring System
ENSIP	Engine Structural Integrity Program
EPNdB	Effective Perceived Noise decibel
EPR	Engine Pressure Ratio
ERDL	Engine Research and Development Laboratory
ESFC	Equivalent Specific Fuel Consumption
ESHP	Equivalent Shaft Horsepower
ETEC	Expendable Turbine Engine Concepts

ETJ	Expendable Turbojet
F	Fahrenheit
FAA	Federal Aviation Administration
FADEC	Full Authority Digital Electronic Control
FCS	Flight Control System
FETT	First Engine To Test
FFR	Full Flight Release
FOD	Foreign Object Damage
FOG	Fiber Optic Guided (Missile)
FOG-M	Fiber Optic Guided Missile
FSD	Full-Scale Development
GAP	General Aviation Program
GE	General Electric
GEAC	General Electric Aircraft Engines
GEFS	General Electric Financial Services
GFE	Government Furnished Equipment
GLC	General Electric/Lycoming Commercial
GLCM	Ground-Launched Cruise Missile
GLTR	Ground-Launched (or Launch) Tacit Rainbow
GM	General Motors
gph	gallons per hour
GTV	Ground Test Vehicle
HALE	High Altitude Long Endurance
HFB	Hamburger Flugzeugbau
HLH	Heavy Lift Helicopter
HP	High Pressure (Rotor)
HPTET	High Performance Turbine Engine Technologies
IAI	Israel Aircraft Industries
IAS	Institute of Aerospace (formerly, Aeronautical) Sciences (later AIAA)
IBR	Integrally Bladed Rotor
IDF	Indigenous Defensive Fighter
IFR	Initial Flight Release
IFSD	In-Flight Shut Down
IGTI	International Gas Turbine Institute
IGV	Inlet Guide Vane(s)
IHI	Ishikawajima-Harima Heavy Industries
IHPTET	Integrated High Performance Turbine Engine Technology
ILD	Individual Lift Device

INF	Intermediate-Range Nuclear Forces (Treaty)
IOE	Initial Operational Evaluation
IP	Intermediate Pressure (Rotor)
IPM	Integrated Propulsion Module
IPS	Inlet Particle Separator
IPTN	Industri Pesawat Terbang Nusantara
IR	Infra-Red
IRAD also IR&D	Independent (or Internal) Research and Development
ISR	Initial Service Release
ITALD	Improved Tactical Air-Launched Decoy
ITEC	International Turbine Engine Corporation
ITT	Inter-Turbine Temperature
JAA	Joint Aviation Authority
JAS	Jakt/Attack/Spanning
JETEC	Joint Expendable Turbine Engine Concepts
JPATS	Joint Primary Aircraft Training System
JSOW	Joint Standoff Weapon
JTAGG	Joint Turbine Advanced Gas Generator
JTDE	Joint Technology Demonstrator Engine
JVX	Joint-Services VTOL Experimental
KHD	Klöckner-Humboldt-Deutz
KTAS	Knots True Airspeed
LAMPS	Light Airborne Multi-Purpose System
LART	Low Aspect Ratio Turbine
LHTEC	Light Helicopter Turbine Engine Company
LHX	Light Helicopter Experimental
LOCAAPS	Low Cost Air Augmented Propulsion System
LOH	Light Observation Helicopter
LONGFOG	Long-Range Fiber Optic Guided (Missile)
LP	Low Pressure (Rotor)
LRAACA	Long-Range Air Anti-Submarine Warfare Capable Aircraft
LRCSW	Long-Range Conventional Standoff Weapon
LWGG	Light Weight Gas Generator
MANPRINT	Manpower Personnel Integration
MAP	Military Assistance Program
MATE	Materials for Advanced Turbine Engines
MBB	Messerschmitt-Boelkow-Blohm
MEP	Mean Effective Pressure

MICOM	U.S. Army Missile Command
MQT	Model Qualification Test
MRS	Medium Range Surveillance
MRUAV	Mid-Range Unmanned Aerial Vehicle
MSOW	Modular Standoff Weapon
MTBF	Mean Time Between Failures
MTDE	Modern Technology Demonstrator Engines
MTU	Motoren- und Turbinen-Union GmbH
NACA	National Advisory Committee for Aeronautics
NASA	National Aeronautics and Space Administration
NASM	National Air and Space Museum
NATO	North Atlantic Treaty Organization
NAVAIR	Naval Air Systems Command
NAWC	Naval Air Warfare Center
NDRC	National Defense Research Committee
NGT	Next Generation Trainer
NIH	Not Invented Here
NLOS	Non-Line-Of-Sight
NOTAR	No Tail Rotor
NRC	National Research Council (Canadian)
NSA	New Shipborne Aircraft
NTH	New Training Helicopter
NTS	Negative Torque Sensing
OPEC	Organization of Petroleum Exporting Countries
PATS	Primary Aircraft Training System
PERT	Program Evaluation and Review Technique
P&W	Pratt & Whitney
P&WA	Pratt & Whitney Aircraft
P&WC	Pratt & Whitney Canada
P&WWV	Pratt & Whitney West Virginia
PFR	Preliminary Flight Rating
PFRT	Preliminary Flight Rating Test
PTA	Propfan Test Assessment (Program)
QCGAT	Quiet Clean General Aviation Turbofan
QSRA	Quiet Short-haul Research Aircraft
QT	Qualification Test
R	Rankine
R&D	Research and Development
RAAF	Royal Australian Air Force

RAM/ILS	Reliability, Availability, and Maintainability/Integrated Logistic Support
RFP	Request For Proposal(s)
RPM	Revolutions Per Minute
RPV	Remotely Piloted Vehicle
RSRA	Rotor System Research Aircraft
SAED	Small Aircraft Engine Division (GE)
SAE	Society of Automotive Engineers
SALT	Strategic Arms Limitation Treaty
SAM	Surface-to-Air Missile
SCAD	Subsonic Cruise Armed Decoy
SCLM	Sea-Launched Cruise Missile
SCM	Strategic Cruise Missile
SENGAP	Small Engine Applications Program
SFC	Specific Fuel Consumption
SHP	Shaft Horsepower
SLAM	Standoff Land Attack Missile
SLCM	Sea-Launched Cruise Missile
SLEP	Structural Life Extension Program
SOAP	Spectrometric Oil Analysis Program
SPA	Special Purpose Aircraft
SPECO	Steel Products Engineering Company
SRR	Short Range Recovery
SSST	Supersonic Sea Skimming Target
SST	Supersonic Transport
STAGG	Small Turbine Advanced Gas Generator
STAMP	Small Tactical Air Mobility Platform
START	Strategic Arms Treaty
TBO	Time Between Overhaul
TCM	Teledyne Continental Motors
TDE	Technology Demonstrator Engine
TEC	Turbo Engineering Corporation
TERCOM	Terrain Contour Matching
TIT	Turbine Inlet Temperature
TRECOM	Transportation Research Command
TRR	Technical Risk Reduction
TSFC	Thrust Specific Fuel Consumption
TSSAM	Tri-Service Standoff Attack Missile
TTID	Turbofan Technology Integrator/Demonstrator

TTTS	Tanker Transport Training System
UAV	Unmanned Aerial Vehicle
UCLA	University of California, Los Angeles
UCX	Utility Cargo-Transport Experimental
USAAVLABS	U.S. Army Aviation Materiel Laboratories
UTC	United Technologies Company
UTTAS	Utility Tactical Transport Aircraft System
UTX	Utility Trainer Experimental
VEN	Variable Exhaust Nozzle
VOD	Vertical On-board Delivery
VSTT	Variable-Speed Training Target
VSX	Aircraft (Navy designation) Surveillance Experimental
V/STOL	Vertical/Short Takeoff and Landing
VTOL	Vertical Takeoff and Landing
WAPEC	Williams-Allison Propfan Engine Company
WASP	Williams Aerial Systems Platform
WPAFB	Wright-Patterson Air Force Base

ACKNOWLEDGMENTS

We would like to express our sincere appreciation to the following individuals who contributed to making this book possible. For each organization, special recognition is given to those individuals who were especially generous with their time and effort.

(★) Indicates an interviewed person

ALLISON

Special recognition:

William S. Castle★
Richard B. Fisher★
S. Michael Hudson★
David H. Quick
F. Blake Wallace★
John M. Wetzler★

Contributors:

Cindy L. Benningfield
George V. Bianchini★
Nicholas J. Blaskoski★
Eric Q. Dickerson
Norman F. Egbert★
Robert C. Ehrenstrom

Jerry D. Flanders★
Ray H. Funkhouser★
George H. Mayo★
Tera L. Miles
Barry J. New (Rolls-Royce)
Joan Pauls
Anthony A. Perona
Ron Riffel
F. Jack Schweibold★
Robert Stangarone (Rolls-Royce)
Eloy C. Stevens★
Harold L. Stocker★
Tommy H. Thomason
Gary A. Williams★
Jerry L. Wouters★

GARRETT (ALLIEDSIGNAL ENGINES)

Special recognition:

Neil F. Cleere for coordinating the Garrett interviews and for providing considerable documentation and photographs. After retirement, he became a National Air and Space Museum volunteer and further contributed importantly to the development of the Garrett chapter.

Robert O. Bullock★
Don H. Comey★
J.-P. Frignac★
Kyle Hultquist
Louis P. Savenelli
Roy Schinnerer★
Montgomerie C. Steele★
Homer J. Wood★

Contributors:

John V. Alexander
Carl F. Baerst★
Ken Blakely
Curtis E. Bradley
Thomas W. Bruce★
Steve Burns

Don G. Caldwell★
Robert A. Choulet★
Chuck Corrigan★
Malcom E. Craig★
Douglas G. Culy
W. E. Cummings★
Edmund R. Dobak★
Roy H. Ekrom★
Al Finkelstein★
Karl R. Fledderjohn★
Alice Fordy
Dave Franson
Walt Gipson
Tom Howell
Lou Kell★
Howard Krasnow★
Steve Lowe
Art Manni
Jerry McCann★
R. D. Miller★
William M. Norgren★
B. Edward Nuckols
W. T von der Nuell★
Wilton Parker★
William J. Pattison★
Dave Pishko
Helmut Schelp★
Ivan E. Speer★
Al Stimac★
Bill Waterman★
Harry Wetzel★
Elmer L. Wheeler★

GENERAL ELECTRIC SMALL AIRCRAFT ENGINES

Special recognition:

James H. Arlinghaus
Marsha Carter
Fredric Ehrich★

R. Eric Falk
Rick Kennedy
Sandy S. Moltz
William R. Travers★
Ed Woll★

Contributors:

Art Adinolfi★
John A. Benson★
Wayman E. Brown
Dave Carpenter
Nick Constantine
William J. Crawford III★
George W. Eddy
Bernard F. Gregoire
Floyd Heglund★
Richard Hickok★
Norris E. Kenyon
Jim Krebs★
Hank J. McGonagle
William M. Meyer
John Morris
Gerhard Neumann★
Jack Parker★
Frank Pickering
William L. Rodenbaugh★
Brian H. Rowe★
Dale D. Streid★
Robert C. Turnbull
Dwight E. Weber
Ralph Wheeler
George Wise

TEXTRON LYCOMING

Special recognition:

Rich Berman
Anselm Franz★
Louis P. Savenelli

Contributors:

Aavo Anto★
James B. Catlin
Thomas A. Dickey★
Harold F. Grady★
Ward Hemmingway
David W. Knobloch★
Charles Kuintzle★
Paul L. Lovington
James E. Lunny★
William B. McDaniel
John R. Myers★
Heinz Moellmann
Dexter A. Picozzi
Ceile Plonski
George J. Pond
J. W. Schrader★
John Sherman
Richard A. Streib
Susan Wilber

PRATT & WHITNEY CANADA

Special recognition:

Louise Boutin
Gordon Hardy★
Francine Osborne
Colin B. Wrong★

Contributors:

Roy C. Abraham★
Jean Demers
Derek C. Emmerson★
Pierre Herron
José-L. Jacome
David P. Kenny
Richard H. McLachlan★
S. Monaghan
G. P. Peterson★

J. Wayne Petitpas
E. L. Smith★
Ken Sullivan★
Don Tedstone

TELEDYNE CAE

Special recognition:

James R. Apel
Eli H. Benstein★
Susan W. Brane (Teledyne Continental Motors)
Frank X. Marsh★
Robert H. Snook★

Contributors:

Robert J. Anderson★
Carl Bachle★
Marie Banks
Kevin L. Brane★ (Teledyne Continental Motors)
J. William Brogdon★ (Teledyne Continental Motors)
Walt Dindoffer
Hans Due★
Harley Greenburg★
Eli Razinsky
Robert R. Schwanhausser★
Raymond Smith★
Robert S. van Huysen

WILLIAMS INTERNATIONAL

Special recognition:

Sam B. Williams★

Contributors:

David C. Jolivette
Patricia Lee
Richard J. Mandle
Christina J. Pearce
T. K. Wills
Linda Wright

ORGANIZATIONS, COMPANIES, AND INDIVIDUALS

Special recognition:

Diana Cornelisse, U.S. Air Force History Office

Alexis Doster III, Smithsonian Institution

William Downs, U.S. Air Force Power Plant Laboratory

Jodi L. Glasscock, AIAA

G. V. Henderson, Jet Pioneers' Association

George C. Larson, Air & Space Magazine

Richard and Elnora Leyes

Harvey H. Lippincott, New England Air Museum

Marilyn A. Phipps, Boeing Archives

Paul A. Pitt, Solar Turbines Inc.

Deborah Reid, TechniType Transcripts

Roberta W. Rubinoff, Smithsonian Institution

Fred T. Sarginson, Fredric Flader★

Cheryl Shanks, formerly of AIAA

John Sloop, NASA, author *Liquid Hydrogen as a Propulsion Fuel*

John D. Weber, U.S. Air Force History Office

Donald D. Weidhuner, head of U.S. Army Engines Research and
Development

Sir Frank Whittle, inventor of the first British turbojet engine

Rodger S. Williams, AIAA

Contributors:

Mr. Eric Clay Ames, U.S. Army Aviation Applied Technology Directorate

Alan Antin, National Academy of Sciences

Gayle I. Armstrong, formerly of AIAA

William J. Armstrong, Naval Air Systems Command

Jean August, U.S. Air Force History Office

Janet R. Bednarek, U.S. Air Force History Ofice

Mike Bednarek, U.S. Air Force

John Blanton, Fredric Flader

Waldemar O. Breuhaus

Billy G. Broach, Redstone Arsenal

Vern E. Brooks, Technical Directions Inc.

Jeffrey P. Buchheit, Historical Electronics Museum, Inc.

Regina G. Burns, U.S. Army Aviation Museum

Harry Chun, U.S. Navy, Naval Air Systems Command, Propulsion and
Power Division

M. B. Comberiate, U.S. Navy, Naval Air Systems Command, Power Plant
  Division
Brad Considine, Sunstrand Corporation
William D. Cowie, U.S. Air Force Aeronautical Systems Division
Virginia P. Dawson, author *Engines and Innovation*
Larry D. Dickerson, Forecast International/DMS
R. R. Dickson, Boeing
Robert Dreshfield, NASA Lewis Research Center
Penny Dudley, American Helicopter Society
Marcio C. Duffles, Naval Air Systems Command
Andy Farrell, Wright Aeronautical Corporation
Bernard Finn, National Museum of American History
Tracie L. Gates, National Aeronautic Association
Mike Gorn, U.S. Air Force History Office
George D. Hall, Solar Turbines Inc.
Richard Hallion, U.S. Air Force History Office
Jeanette Hammond, Cessna Aircraft Company
Mack High
Regina M. Horan
Roxie Laybourne, Smithsonian Institution
Tony LeVier, Lockheed
Lexicon Aviation
Jay S. Lilley, U.S. Army Missile Command
Bruce T. Lundin, NASA
Reinout Kroon, Westinghouse
Michelle Maynard, U.S. Department of State
Charles McClellan, Naval Air Propulsion Association
Don McQueeney, Indianapolis Motor Speedway Corporation
Walter C. Merrill, NASA Lewis Research Center
William G. Milner, U.S. Navy, Naval Air Development Center
Model Engine Company of America
Keith Moore, U.S. Army
Henry Morrow, U.S. Army Aviation Applied Technology Directorate
Robert Muirson, Sunstrand Corporation
Marge Nebinger, Forecast International/DMS
Jacob Neufeld, U.S. Air Force History Office
Henry M. Ogrodzinski, General Aviation Manufacturers Association
William M. Owen, Sunstrand Corporation
Tom Pendergast, Solar Turbines Inc.

Roy R. Peterson, U.S. Navy, Bureau of Ships, Gas Turbine Section

M. Potzmann, Westinghouse

James Price, Sunstrand Corporation

Robert H. Prine, U.S. Navy, Naval Air Systems Command, Propulsion and
  Power Division

Erik Prisell, Turbomin AB

Rita Quinn, Keiler & Company

Ed J. Reis, Westinghouse

Oliver E. Rodgers, Westinghouse

Alex Roland, Duke University

Laura N. Romesburg, U.S. Air Force History Office

Charles A. Ruch, Westinghouse

John C. Schettino, Wright Aeronautical Corporation

Kent Schubert, Schubert & Company

Lewis F. Smith, U.S. Navy, Bureau of Aeronautics, Power Plant Design
  Branch

Raymond M. Standahar, Department of Defense, DDR&E, Propulsion

Marvin A. Stibich, U.S. Air Force Aero Propulsion & Power Directorate

Robert Tatge, CV International, Inc.

Bruce R. Thomas, Chrysler Corporation

Robert G. Thompson, Sunstrand Corporation

Clifton von Kann, National Aeronautic Association

James J. Kramer

Deborah M. Vito, Schubert & Company

Stewart Way, Westinghouse

Robert L. Wells, Westinghouse

Edward West, Jr., West Engineering Co., Inc.

Dan Whitney, author *Vee's for Victory*

Jim Young, U.S. Air Force History Office, Edwards AFB

August H. Zoll, Aviation Hall of Fame of New Jersey

NATIONAL AIR AND SPACE MUSEUM

Special recognition:

John D. Anderson

Mark Avino

Dorothy Cochrane

Deborah G. Douglas

Phil Edwards

Don Engen, Director
Paul Garber
Patricia J. Graboske
Dan Hagedorn
Von Hardesty
Paul Lagasse, NASM Intern
Don Lopez, Deputy Director
Bernard Maggin, NASM Volunteer
Ted A. Maxwell
Robert B. Meyer Jr., NASM Volunteer, former Curator for Aero Propulsion
Brian Nicklas
Claudia M. Oakes
Dom Pisano
Herb Rochen, NASM Volunteer
Dave Spencer
Hans von Ohain, Charles A. Lindbergh Fellow, inventor of the first German
    turbojet engine
Howard S. Wolko

Contributors:

Alice Adams
James Carter
Elaine Cline
Tim Cronen
Tom D. Crouch
Bob Curran
Ron Davies
Sybil Descheemaeker
Thang Duong
Ellen Folkama
Marilyn Graskowiak
Martin Harwit, Director
Dave Heck
Robert S. Hoffmann, Director
Dale Hrabak
Melissa Keiser
Russ Lee
Amy Levin
Eric Long

Anita Mason
Paul McCutcheon
Robert C. Mikesh
Kathy Murphy
Mary Pavlovich
Natalie Rjedkin-Lee
Ted Robinson, NASM Fellow
Carolyn Russo
Liz Scheffler
Phouy Sengsourinh
Tom Soapes
Alex Spencer
Wendy Stephens
Deborah Swinson
Jim Tyler, Director
F. Robert van der Linden
Barbara Weitbrecht
Collette Williams
Patti Williams
Larry Wilson
Frank Winter
E. T. Wooldridge
Jim Zimbelman

XP-54 fighter, 150
XP-59A (Bell)
  I-A engine, 237
XP-79B (Northrop)
  19B engine, 39
XP-80A (Lockheed), 616
XP-81 (Consolidated Vultee)
  TG-100 engine, 238
XP-84 (Republic)
  TG-180 engine, 238
XQ-2 (Ryan), 46, 50
XTB3F-1 (Grumman)
  19XB engine, 39
XV-4A (Lockheed), 277
XV-5A (Ryan)
  J85 engine, 277
XV-6A (Hawker-Siddeley), 277

Y

YA-9 (Northrop), 318
YA-10 (Fairchild), 218
YAH-64A helicopter (Hughes), 221,
  775
YAT-28E prototype (North American)
  T55-L-9 engine, 178
YF-5B-21 (Northrop)
  J85 engine, 275
YQM-98A Compass Cope R (Ryan)
  YF104-GA-100 (ATF3-3) engine,
  670
YT-38 (Northrop)
  J85 engine, 270, 271
YUH-60A helicopter (Sikorsky), 221,
  775
  YT700 engine, 343

ENGINE INDEX

GENERAL INDEX